Handbook of Graphene

Scrivener Publishing
100 Cummings Center, Suite 541J
Beverly, MA 01915-6106

Publishers at Scrivener
Martin Scrivener (martin@scrivenerpublishing.com)
Phillip Carmical (pcarmical@scrivenerpublishing.com)

Handbook of Graphene comprises 8 volumes:

Volume 1: Growth, Synthesis, and Functionalization
Edited by Edvige Celasco and Alexander Chaika
ISBN 978-1-119-46855-4

Volume 2: Physics, Chemistry, and Biology
Edited by Tobias Stauber
ISBN 978-1-119-46959-9

Volume 3: Graphene-Like 2D Materials
Edited by Mei Zhang
ISBN 978-1-119-46965-0

Volume 4: Composites
Edited by Cengiz Ozkan
ISBN 978-1-119-46968-1

Volume 5: Energy, Healthcare, and Environmental Applications
Edited by Cengiz Ozkan and Umit Ozkan
ISBN 978-1-119-46971-1

Volume 6: Biosensors and Advanced Sensors
Edited by Barbara Palys
ISBN 978-1-119-46974-2

Volume 7: Biomaterials
Edited by Sulaiman Wadi Harun
ISBN 978-1-119-46977-3

Volume 8: Technology and Innovation
Edited by Sulaiman Wadi Harun
ISBN 978-1-119-46980-3

Handbook of Graphene

Volume 2: Physics, Chemistry, and Biology

Edited by
Tobias Stauber
Institute of Materials Science,
Spanish National Research Council,
Madrid, Spain

Scrivener Publishing

WILEY

This edition first published 2019 by John Wiley & Sons, Inc., 111 River Street, Hoboken, NJ 07030, USA and Scrivener Publishing LLC, 100 Cummings Center, Suite 541J, Beverly, MA 01915, USA

For more information about Scrivener publications please visit www.scrivenerpublishing.com.

Wiley Global Headquarters
111 River Street, Hoboken, NJ 07030, USA

For details of our global editorial offices, customer services, and more information about Wiley products visit us at www.wiley.com.

Library of Congress Cataloging-in-Publication Data

ISBN 978-1-119-46959-9

Cover image: Pixabay.Com
Cover design by Russell Richardson

Set in size of 11pt and Minion Pro by Manila Typesetting Company, Makati, Philippines

10 9 8 7 6 5 4 3 2 1

Contents

Preface

When Andre Geim first presented the "Electric Field Effect in Planar Single-Layer Graphene" at the March meeting of the American Physical Society in 2005, the room was barely filled and his contributed talk of 10 minutes did not receive much attention. This rapidly changed with the observation of the half-integer quantum Hall effect by his and Philip Kim's group in the same year, promoting graphene and its family of other two-dimensional van der Waals materials to one of the most active current research areas in physics, chemistry, and also biology/medicine. After more than 10 years of worldwide research activity, the *Handbook of Graphene, Volume 2*, which is dedicated to selected topics in physics, chemistry, and biology, attempts to give an overview of the multitude of different research directions that are currently being taken at the international level.

Pristine graphene is nominally a semimetal, but in practice its electronic properties and structure are often modified, as examined in Chapters 2, 3, and 11. These changes can be due to topological defects (see Chapter 1), chemical adsorption (see Chapter 7), isolated vacancies (see Chapter 12), strain (see Chapter 8), or by confined geometries/nanoribbons (see Chapter 5). Electron–electron interaction can also modify graphene's properties, as outlined in Chapters 4 and 14, which focus on the Fermi velocity renormalization and optical response as well as the magnetotransport in the extreme quantum limit, respectively. Furthermore, graphene or other two-dimensional structures often need to be described as membrane, as described in Chapters 6, 9, 17, and 18.

Among the possible applications, optoelectronic devices are arguably among the most likely ones, as reviewed and analyzed in Chapters 19 and 13 respectively. Graphene can also host highly confined surface plasmon polaritons with low losses, whose properties are discussed in Chapters 15 and 16. Finally, the use of graphene for the detection of biomolecules as well as tissue engineering and regenerative medicine are described in Chapters 10 and 20 respectively.

Science is an international endeavor that needs the interconnectedness and stimuli of a large scientific community. Graphene has managed to attract the interest of numerous researchers from all over the world, which can nicely be exemplified by looking at the various contributors to this book, from countries ranging from Taiwan to India, from Nigeria to Ukraine, from Egypt to the United States, from Iran to Canada, from Korea to Poland, from Russia to France, and from Greece to Spain. May this *Handbook of Graphene, Volume 2*, help to further increase the connection between scientists from different countries and engage them in the common goal to better understand and exploit the fascinating properties of the ever-growing graphene family.

In conclusion, I would like to thank all the authors whose expertise in their respective fields has contributed to this book and express my sincere appreciation to the International Association of Advanced Materials.

Tobias Stauber
Madrid, Spain
February 1, 2019

1

Topological Design of Graphene

Bo Ni[1†], Teng Zhang[2†], Jiaoyan Li[1], Xiaoyan Li[3] and Huajian Gao[1*]

[1]*School of Engineering, Brown University, Providence, RI, USA*
[2]*Department of Mechanical and Aerospace Engineering, Syracuse University, Syracuse, NY, USA*
[3]*Applied Mechanics Laboratory, Department of Engineering Mechanics, Centre for Advanced Mechanics and Materials, Tsinghua University, Beijing, China*

Abstract
Topological defects (e.g., pentagons, heptagons, and pentagon–heptagon pairs) have been widely observed in large-scale graphene and have been recognized to play important roles in tailoring the mechanical and physical properties of two-dimensional (2D) materials in general. Thanks to intensive studies over the past few years, optimizing properties of graphene through topological design has become a new and promising direction of research. In this chapter, we review some of the recent advances in experimental, computational, and theoretical studies on the effects of topological defects on mechanical and physical properties of graphene and applications of topologically designed graphene. The discussions cover out-of-plane effects, inverse problems of designing distributions of topological defects that make a graphene sheet conform to a targeted three-dimensional surface, grain boundary engineering for graphene strength, curved graphene for toughness enhancement, and applications in engineering energy materials, multifunctional materials, and interactions with biological systems. Despite the rapid developments in experiments and simulations, our understanding on the relations between topological defects and mechanical and physical properties of graphene and other 2D materials is still in its infancy. The intention here is to draw the attention of the research community to some of the open questions in this field.

Keywords: Topological design, defects, morphology and curvature, strength and toughness, multifunction, nonlinear multiphysics coupling, multiscale fabrication, interconnected and multilayer graphene

1.1 Introduction

As the very first and most prominent example of two-dimensional (2D) materials, pristine graphene [1] consists of an atomic layer of carbon atoms densely packed in the hexagonal crystal lattice via sp^2 covalent bonding. Topological defects in graphene are those induced by rearrangements of atomic bonds that break the hexagonal symmetry of the 2D lattice. Fundamental units of topological defects in graphene include disclinations [2] (pentagons

**Corresponding author*: Huajian_Gao@brown.edu
[†]*These authors contributed equally*

Tobias Stauber (ed.) Handbook of Graphene: Volume 2, (1–44) © 2019 Scrivener Publishing LLC

and heptagons, Figure 1.1a–b) and dislocations [3] (pairs of pentagon and heptagon, Figure 1.1c), which are disruptions of the rotational and translational symmetry of the lattice, respectively. Grain boundaries (GBs) [3, 4] are topological line defects formed between grains with different crystal orientations (Figure 1.1d). Indeed, various forms of topological defects are widely observed in large-scale graphene samples fabricated by chemical vapor deposition (CVD) [4–7] (Figure 1.1e–f). Understanding how they alter the mechanical and physical properties of graphene, including strength [8–11], morphology [2, 12, 13], toughness [11, 14, 15], heat conductivity [16], chemical reactivity [17], and electrical properties [18–21], is of great importance in advancing fundamental sciences and applications of 2D materials.

Over the past few years, more and more theoretical studies and experimental observations have shown that mechanical and physical properties of graphene can be tailored by topological defects. For example, molecular dynamics (MD) simulations reveal toughness enhancements in sinusoidal graphene containing periodically distributed disclination quadrupoles [15] and in polycrystalline graphene with well-stitched GBs [14]. Experimental measurements show that the thermal conductivity of polycrystalline graphene decreases dramatically with grain size due to the influence of GBs [22]. MD simulations predict that graphene samples forming a gyroid surface have 300-fold reduction in thermal conductivity due to the presence of topological defects and curvature [23]. Topological defects have also been shown to alter electronic transport behaviors from high transparency to perfect reflection of charge carriers [19]. Recent experimental advances [24–27] have made it increasingly possible to control atomic structure and distribution of topological defects, paving the way for large-scale fabrication of "topologically designed" graphene structures and devices.

Here, the concept of topological design may be defined as "taking advantage of the cooperative interactions of topological defects, e.g., disclinations, dislocations, and GBs, to achieve novel mechanical and physical properties of graphene through design and fabrication of 2D lattices with controlled distribution of topological defects." We will focus our discussion on graphene, yet many of the discussions and findings are also applicable to other 2D materials [30, 31], such as monolayer hexagonal boron nitride (h-BN) [32, 33] and semiconducting transition metal dichalcogenides (TMDCs) MX_2 (M = Mo, W; X = S, Se) [34, 35]. To avoid the complexity of dangling bonds, we will restrict our discussion to topological defects that do not involve vacancies and free edges, for which relevant papers and reviews including porous graphene [36, 37] and kirigami/origami graphene [38–43] can already be found in the literature.

Manipulating nanoscale topological defects to improve the mechanical properties at macroscale is not new and has been widely employed in bulk materials including metals [44, 45], ceramics [46], and diamond [47]. For example, it has been well recognized that GBs and twin boundaries play crucial roles in developing novel metallic materials with ultrahigh strength, good ductility, and superior fatigue resistance [44, 45, 48]. Nano-twinned cubic boron nitride [46] and diamond [47] exhibit higher hardness and toughness than their defect-free counterparts. The past successes in topological design of bulk materials provide a solid foundation for the extension of similar concept to 2D materials.

While sharing a number of common features, topological design of graphene exhibits several important distinctions compared to that of bulk materials. First, unlike the bulk materials with abundant slip systems in three dimensions, the migration paths of topological defects in 2D materials are confined within a basal plane [29, 49], as shown in Figure 1.2a. This dimensionality restriction largely reduces the accessibility and variability of mechanical behaviors in 2D materials. Second, due to the large differences in the

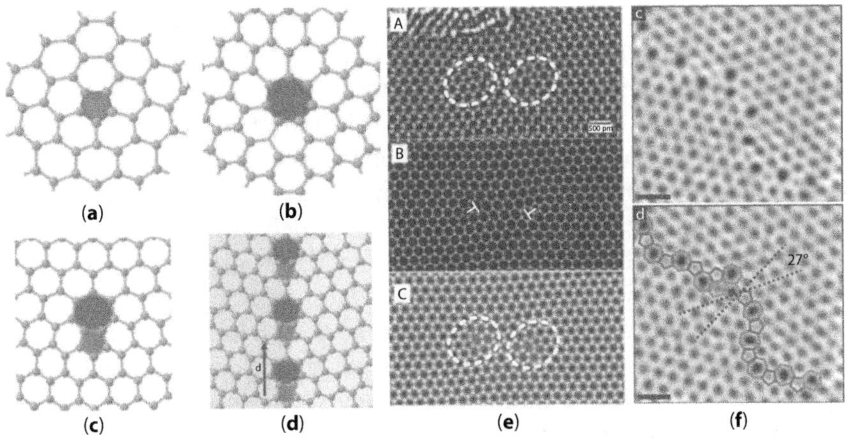

Figure 1.1 Topological defects in single-layer graphene. (a–d) Schematics of topological defects in graphene, including positive disclination (a), negative disclination (b), dislocation (c), and GB (d) [28]. (e–f) Experimentally observed atomic structure of dislocations (e) [29] and GBs (f) [4] in graphene.

rigidity of in-plane [50] and out-of-plane [51, 52] deformations in graphene, the presence of topological defects can trigger substantial out-of-plane deformation (Figure 1.2b), especially in free-standing graphene, to minimize its strain energy [3, 13]. The resulting three-dimensional (3D) geometry will in turn alter mechanical and physical properties like the elastic modulus, strength [9, 53], fracture toughness [14, 15], adhesion and friction [54], chemical reactivity [17], local density of states [55], and flexoelectricity [56, 57]. Third, the flexibility of graphene with respect to the out-of-plane deformation [58]

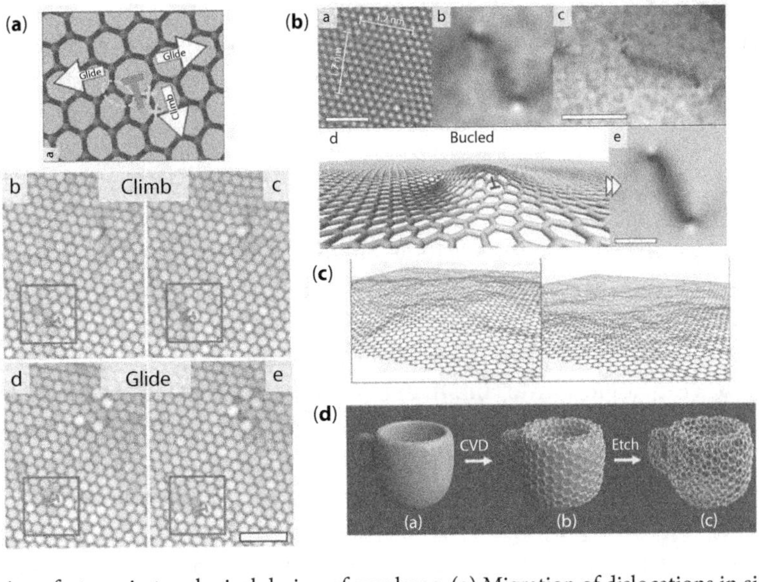

Figure 1.2 Unique features in topological design of graphene. (a) Migration of dislocations in single-layer graphene is limited within the lattice plane [49]. (b) The long range out-of-plane deformation triggered by topological defects in graphene in experimental observations and simulations [49]. (c) Atomistic simulations of thermal fluctuations in a free-standing graphene [68]. (d) Schematic of growing graphene on designed curved surfaces via CVD methods [69].

makes its mechanical and physical properties highly sensitive to thermal fluctuations at room temperature [59, 60]. As a result, the effective properties of graphene often need to be considered as a consequence of interactions between the intrinsic properties and thermal fluctuations (Figure 1.2c) [38, 61–64]. Fourth, the atomically thin structure of graphene also poses great challenges in fabrication and postprocessing of the material (Figure 1.2d) [65–67]. Thus, it should be empathized that the topological design of graphene involves intrinsically nonlinear coupling between many properties including stress, deformation, electricity, and chemical reactivity. Addressing these challenges requires a highly interdisciplinary collaboration from multiple communities of researchers such as mechanics, physics, chemistry, material science, and nanoengineering. Here, we review some of the recent advances in engineering mechanical and physical properties of graphene through topological design, hoping to draw the attention of various research communities to some of the open questions in this field.

This chapter is organized as follows. In Section 1.2, we summarize some studies on how to optimize the mechanical and physical properties of graphene through topological design. Section 1.3 reviews selected applications of topologically designed graphene, examples including applications as novel materials in energy and multifunctional devices. In Section 1.4, we discuss a few promising techniques to fabricate graphene with deliberately designed topological defects. Finally, some conclusions and outlook remarks will be made in Section 1.5.

1.2 Topological Design for Engineering Strength, Morphology, and Toughness of Graphene

In this section, we will review some of the recent progress on engineering graphene with targeted properties through topological design, with a focus on the mechanical properties (i.e., strength and toughness) and 3D morphology. Examples include how to tailor the strength of graphene by designing GBs in it, how to inversely design the distribution of topological defects to create a 3D single-layer graphene with targeted shape, and how to enhance the fracture toughness of graphene via intentionally introduced topological defects.

1.2.1 Tuning Strength of Graphene via GBs

For 2D materials, a GB is a one-dimensional chain of edge dislocations. Thus, a GB can be regarded as a simple linear array of topological defects. For single-layered 2D materials, there is only edge dislocation and no screw dislocation, because all dislocations and their projections are located in the basal plane. The edge dislocation in graphene is described by the Burger vector **b**, a topological invariant. In 2D materials, an array of edge dislocations constitutes a tilting GB, usually described by a misorientation angle or tilting angle θ, that separates two grains with different crystal orientations. Recent high-resolution transmission electron microscopy (HRTEM) observations showed the detailed atomic structures of some GBs in graphene [4, 70, 71], h-BN [72], and TMDCs [73]. GBs in 2D materials usually form during growth. For example, single-layer graphene can be synthesized by CVD [4, 70, 71] on metal substrates with a certain crystalline orientation. During synthesis, independent grains simultaneously nucleate at different points on the metal surface, and the misfit between graphene and metal leads to different lattice orientations in different grains. When two grains with different orientations meet, a line defect, i.e., a GB, forms along the interface.

Before we discuss how we can tune the strength of graphene with GBs, let us quickly review the atomic structures and energies of GBs in graphene. In the single-layer graphene, GBs are generally made of two types of edge dislocations, one type with Burgers vector **b** = (1,0) or (0,1) consisting of a pair of neighboring pentagon and heptagon (Figure 1.3a) and another with the distance between pentagon and heptagon increasing by one lattice length, resulting in Burgers vector of **b** = (1,1), as shown in Figure 1.3b. Periodically aligning edge dislocations along a certain direction leads to a GB. Recent first-principles calculations revealed the atomic structures of some energetically favorable GBs in graphene consistent with HRTEM observations and also provided a diagram of GB energies per unit length γ as a function of the tilting angle θ for various GB structures (Figure 1.3c) [3]. When the GBs are confined in a 2D plane, their energies in the small-angle regime ($\theta < 10°$) can be described by the Read–Shockley equation [3]:

$$\gamma = \frac{\mu b \theta'}{4\pi(1-\nu)}\left(1 - \ln\theta' + \ln\frac{b}{2\pi r_0}\right) \tag{1.1}$$

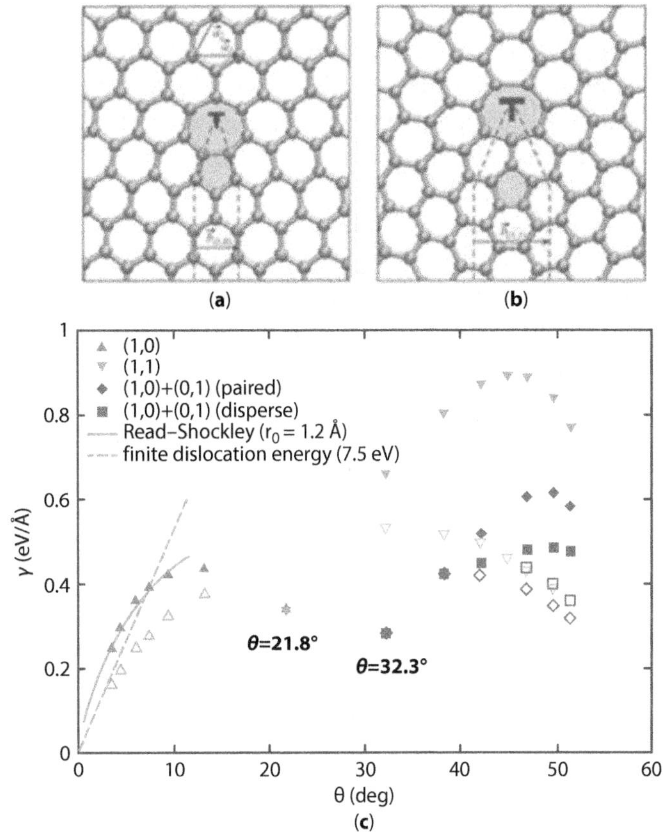

Figure 1.3 Atomic structures of GB dislocations and GB energies in graphene. (a–b) Atomic structures of dislocations with b = (1,0) and b = (1,1). (c) GB energies per unit length as a function of the tilting angle. Solid and hollow data points are for flat and buckled graphene, respectively. Solid curve is a fitting curve with dislocation core radius of r_0 = 0.12 nm based on the Read–Shockley equation, while the dashed curve reflects the asymptotic linear expression with E_f=7.5 eV based on Equation 1.2 for buckled GBs. (Reprinted with permission from [3].)

where μ is the shear modulus, ν is the Poisson's ratio, b is the magnitude of Burgers vector, and r_0 is the radius of dislocation core. In Equation 1.1, $\theta' = \theta$ or $\theta' = \pi/3 - \theta$ for armchair and zigzag GBs, respectively. If there are no planar constraints, the GB will exhibit a buckled shape due to the out-of-plane deformation associated with dislocations. Such buckling reduces the energy of GBs (open data points in Figure 1.3c), making them more stable. There exist two particularly stable large-angle GBs with $\theta = 21.8°$ and 32.3°, which hold the lowest energies in the $\theta < 21.8°$ and $\theta > 32.3°$ regimes, respectively. For buckled GBs with $\theta < 3.5°$, the GB energies exhibit a linear dependence on the tilting angle, yielding the following expression [3]:

$$\gamma = \frac{E_f \theta'}{b} \tag{1.2}$$

where E_f represents the formation energy of a dislocation. The fitting in Figure 1.3c gave E_f as 7.5 eV [3].

The effects of GBs on mechanical strength of polycrystalline graphene have been widely explored by both experiments and simulations. Lee *et al.* [50] have experimentally studied the mechanical properties of CVD-graphene films with different grain sizes by combining TEM structural characterization with nanoindentation. They found that the elastic stiffness of CVD-graphene is close to that of pristine graphene while the mechanical strength is slightly reduced. Rassol *et al.* [10] performed nanoindentation on a bi-crystalline graphene and found that the mechanical strength of GBs with large mismatch angles are larger than that of GBs with low mismatch angles. This observed misorientation dependence of GB strength corroborates predictions from atomistic simulations [8, 9, 74]. Using atomistic calculations, Grantab *et al.* [8] showed that the large-angle GBs are stronger. Wei *et al.* [9] combined continuum modeling and atomistic simulations to study how defects in GB interact; their results emphasized that it is not only the density of defects that affects the mechanical properties, but the detailed arrangements of the defects are also important to GB strength. Later on, further efforts [75] have extended the studies from symmetric to asymmetric GBs. Besides straight GBs, Zhang *et al.* [74] investigated sinuous GBs, which are frequently observed in experiments; they concluded that the sinuous GBs can be more energetically favorable than straight ones and have improved mechanical properties.

Beyond the above studies on the properties of GBs as line defects, designing networks of GBs [76–78] has also attracted increasing attention as a promising way to control and/or tune physical properties of 2D materials. Recent advances of fabrication techniques showed a great potential to control GBs in polycrystalline graphene during the growth process. For example, seed-assisted growth has been successfully carried out in experiments by suppressing random nucleated islands via an array of pre-patterned seeds [79–81]. Besides the locations of nucleated islands, the shape, orientation, and edge geometry of CVD graphene domains are also controllable by the crystallographic orientations of copper substrates [79, 82]. Based on a phase field crystal (PFC) model [83], Li *et al.* [84] numerically simulated the dynamic formation of GBs in CVD graphene and demonstrated possible routes of engineering GBs by controlling grain orientations in pre-patterned growth seeds. This study provided a theoretical platform to explore the potential rational design of GBs using pre-patterned growth seeds. In a simple geometrical model to understand the dynamic coalescence of growing seeds, the direction and misorientation angles of a GB are determined from the geometries of polygonal graphene flakes [85]. Integrating this geometrical rule and the PFC model, Li *et al.* [84] demonstrated that starting with random seeds, serpentine GBs and triple junctions (TJs) are likely to appear, which is consistent with experimental observations in CVD-grown polycrystalline graphene. Extensive studies have shown that the strength of polycrystalline graphene is not only dependent on the grain size but also highly sensitive to the detailed distribution of topological defects in the GB network, such as TJs and vacancies [78, 86–88]. As a prominent example of engineering GBs with seed-assisted growth, a design of TJ-free polycrystalline graphene [84] has been proposed (illustrated in Figure 1.4) where GBs are of 30° misorientation angles, leading to enhanced, grain-size-insensitive mechanical strength that defies the reported Hall–Petch-type relation for polycrystalline graphene [89].

Besides graphene, it may be possible to generalize the concept of GB engineering for strength to other 2D materials. For this purpose, it might be interesting to compare similarities and differences in the fundamental structures and energies of GBs in other two typical 2D materials, h-BN and TMDC MS_2 (M = Mo or W), with those in graphene. In the single-layer h-BN, apart from the dislocation with pentagon–heptagon pair and $\mathbf{b}$ = (1,0), a new dislocation structure with square–octagon pair and $\mathbf{b}$ = (1,1) has been predicted by first-principle calculation [90]. In h-BN, a pentagon–heptagon pair contains energetically unfavorable homo-elemental bonds B-B or N-N, while a square–octagon pair involves the hetero-elemental bond B-N and is free of homo-elemental bonds. As a result,

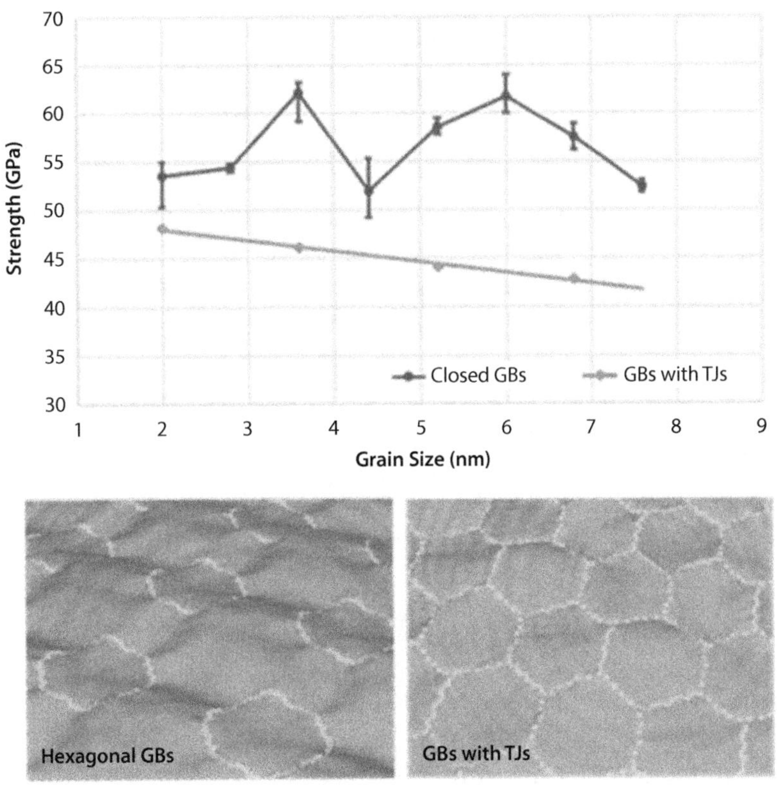

Figure 1.4 Mechanical strength of TJ-free graphene with hexagonal GB loops and polycrystalline graphene with TJs under biaxial tension for grain sizes from 2nm to 10nm. (Reprinted with permission from [84].)

the dislocation of square–octagon pair has lower energy than that of pentagon–heptagon pair [90]. The stabilization of dislocation with square–octagon pair is to some extent associated with its out-of-plane buckling [90]. According to the mirror symmetry and hetero-elemental composition, GBs in h-BN can be classified into two types, i.e., symmetric armchair GBs (A-GBs) and asymmetric zigzag GBs (Z-GBs), as shown in Figure 1.5a. The symmetric A-GB consists of dislocations with pentagon–heptagon pairs, while the asymmetric Z-GB is composed of dislocations with square–octagon pairs [90] (Figure 1.5a). The symmetric A-GB has been observed by HRTEM [72]. Due to the elemental polarity (either B- or N-rich) along a GB, the symmetric A-GB is capable of carrying net charges [90], which suggests possible applications in electronic and optical devices. Figure 1.5b presents the GB energy per unit length as a function of the tilting angle [90]. It is seen that the energies of GBs composed of square–octagon pairs are always lower than those of pentagon–heptagon pairs. The dislocations aligned along a GB induce out-of-plane buckling. The asymmetric Z-GBs exhibit more buckling compared to the symmetric A-GBs, which helps to reduce the energies of the Z-GBs.

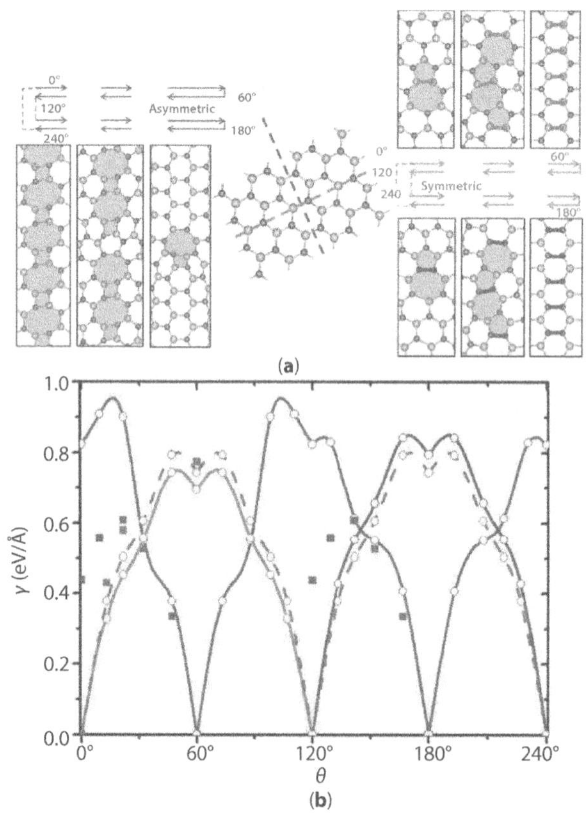

Figure 1.5 Atomic structures and energies of GBs in h-BN. (a) Atomic structures of GBs. The middle picture shows a perfect lattice. The symmetric A-GBs (right figures) and asymmetric Z-GBs (left figures) are generated by rotation of two grains with respect to green and purple lines, respectively. (b) GB energies per unit length as a function of tilting angle. Scatted data points are from first-principle calculations and are connected by guide lines. Hollow circles are for GBs constituted by dislocations with pentagon–heptagon pairs, while solid squares are for GBs constituted of dislocations with square–octagon pairs. Purple for Z-GBs, red for B-rich A-GBs, blue for N-rich A-GBs, and green for the average energy of B-rich GBs and their N-rich analogs. (Reprinted with permission from [90].)

A sheet of TMDC MS_2 (M = Mo or W) is a sandwiched structure containing a mid-plane of metal atoms and two layers of sulfur atoms triangularly packed in their respective planes. Such three-atomic-layer structure makes dislocation structure more complicated compared with mono-atomic-layer graphene and h-BN. Recent first-principle calculations [91] predicted that there exist three types of edge dislocations in TMDC MS_2 that extend through the triatom layers and form concave dreidel-shaped polyhedra [91]. In the planar view, the three types of dislocations are constituted by pentagon–heptagon pairs with M-M bonds, pentagon–heptagon pairs with S-S bonds, and square–octagon pairs with M-S bonds [91], with Burgers vectors of (1,0), (0,1), and (1,1), respectively. Due to the local-chemical energy, the dislocation cores in the TMDC MS_2 can reconstruct or react with the point defects [91]. For example, an isolated dislocation with square–octagon pair is unstable and can split into two dislocations (one with **b** = (1,0) and the other with **b** = (0,1)) via exothermic reconstruction [91]. Similar to h-BN, GBs in TMDC MS_2 can be classified into two types: A-GBs and Z-GBs (Figure 1.6). But these GB structures are more complex than those of h-BN. Recent HRTEM observations [73] showed that the Z-GBs with square–octagon pairs are dominant in the samples fabricated by CVD. Figure 1.6 shows the GB energies per unit of length as function of tilting angles [91]. In the large-angle regime, the GB energies have a large variation due to the reconstructions of dislocation cores or their reaction with point defects.

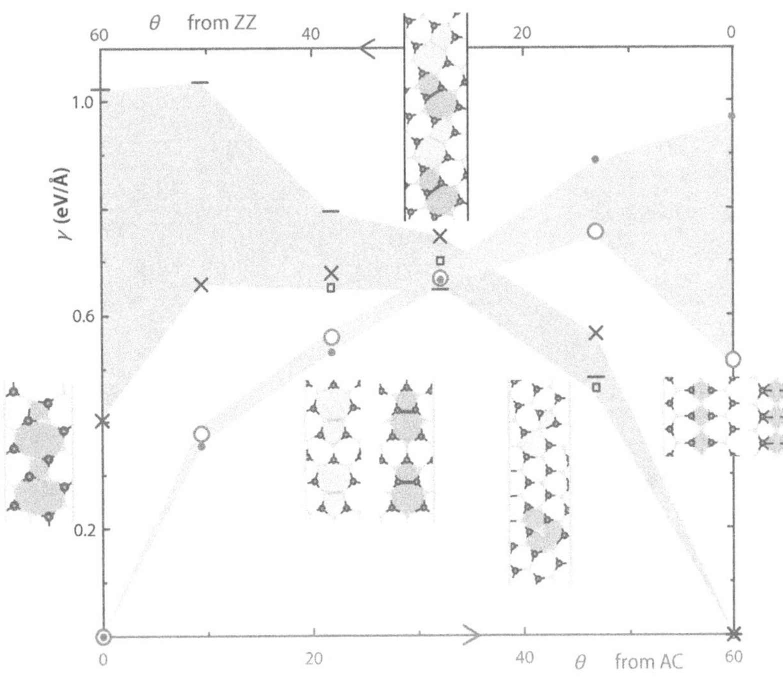

Figure 1.6 Atomic structures and energies per unit length of GBs as function of tilting angle in TDMC MS_2. Atomic structures of some specific GBs are shown in the insets. GB energies per unit length change with the tilting angles along armchair (AC) and zigzag (ZZ) directions. Red solid and open circles correspond to A-GBs composed of pentagon–heptagon pairs and rhomb–hexagon plus hexagon–octagon pairs, respectively. Blue dashes, crosses, and open squares correspond to Z-GBs composed of pentagon–heptagon pairs, rhomb–hexagon plus hexagon–octagon pairs, and square–octagon pairs, respectively. Shaded areas show the energy range due to reconstruction of dislocation cores. (Reprinted with permission from [91].)

1.2.2 Topological Design for 3D Shapes of Graphene

Since graphene is a highly flexible atomic thin crystal membrane, it will adopt a 3D configuration to release strain energy induced by topological defects. As shown in Figure 1.7, global 3D configurations can be formed by introducing a single disclination, such as a cone shape for a pentagon and a saddle surface for a heptagon. Even an isolated dislocation can cause substantial out-of-plane deformation [13]. It has been widely recognized that the shape of graphene plays a crucial role in determining its mechanical [15, 92], thermal [23], chemical [93], and physical properties [94–96]. If we can design graphene with arbitrary shapes by deliberately controlling the topological defects in it, there will be tremendous opportunities to tailor its properties for specific applications. For example, the super strong yet brittle pristine graphene [97] may not be the best candidate for reinforcing lightweight, strong, and tough composites, where an alternative structure with designed topological defects to achieve balanced properties among strength, toughness, and interfacial adhesion may be more desirable.

Achieving an inverse design of 3D curved graphene with a targeted shape is a very challenging task, as we need to search for the number, type, and location of corresponding topological defects. The first challenge comes from highly nonlinear interactions between the topological defects and 3D shapes of graphene in the forward analysis for a given defect distribution [13, 98]. The second challenge is due to the multiple time scales involved in directly optimizing the carbon atom positions, which is generally at the level of seconds, which far exceeds the current capability of MD (nanoseconds) [99, 100]. Other techniques,

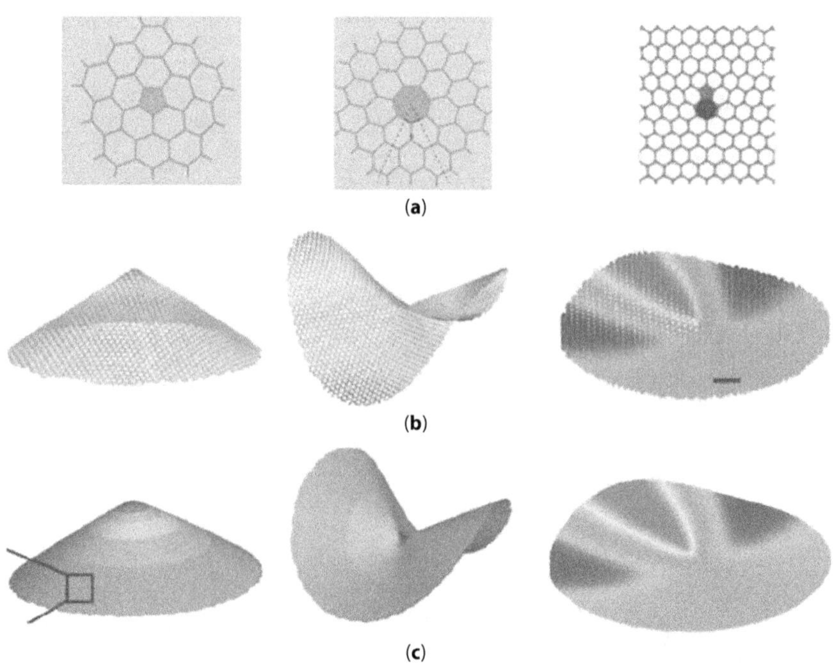

Figure 1.7 3D curved shapes induced by elementary topological defects in graphene [13]. (a) Atomic structures of positive and negative disclination and edge dislocation in graphene. (b) 3D configurations from MD simulations. (c) 3D configurations from continuum model.

like geometrical methods [101–103] and Monte Carlo simulations [104, 105], may provide a path to bridging the time scales, but still require a large amount of computation efforts, especially for large graphene structures. The third challenge stems from the fabrication techniques for realizing topologically designed graphene. In this section, we will review some of the recent developments in predicting 3D curved graphene with topological defects via continuum models and inverse design of 3D curved graphene through PFC methods. It remains an open question how to fabricate 3D curved graphene, and we will review some promising techniques in Section 1.3.

Studies of buckling of plates with defects can be traced back to the 1960s, when Mitchell and Head [106] investigated the critical buckling condition of a plate with a central dislocation based on an energy method. In 1988, Seung and Nelson [107] derived a generalized von Karman equation for thin elastic sheets with various topological defects and validated the theoretical predictions of shape and energy via a triangular lattice model. Zubov [108–110] conducted a series of studies on thin shells and plates with topological defects and showed that the problem of a thin shell with defects can be linked to its dual problem of a thin shell with external loading [110]. Chen and Chrzan [98] formulated a continuum theory for dislocations in graphene by modeling dislocations as topological constraints and minimizing the total strain energy in the Fourier space, which was shown to accurately capture the self-energy of periodical dislocation dipoles in a graphene sheet with out-of-plane deformation compared with MD simulations. Zhang *et al.* [13] developed a continuum model of topological defects in graphene in terms of a classical von Karman equation with eigenstrain field based on a mathematical analogy between topological defects and incompatible growth metric field. The model proposed by Zhang *et al.* [13] successfully captured the global wrinkling profiles and atomic scale wrinkles near disclination/dislocation cores, with much higher efficiency compared to full atom MD simulations [13].

In the generalized von Karman model of a 2D lattice with topological defects [107], the out-of-plane deformation w and Airy stress function Φ are expressed as

$$B\nabla^4 w=\left[w,\Phi\right]$$
$$\nabla^4\Phi=-S\left[\kappa_G-\sum_{i=1}^{N}s_i\delta\left(\mathbf{r}-\mathbf{r}_i\right)\right] \tag{1.3}$$

where B is the bending stiffness, $S = Eh$ denotes the in-plane stretching stiffness, K_G is the Gaussian curvature, $s_i\delta(\mathbf{r}-\mathbf{r}_i)$ represents the i_{th} disclination at position r_i with strength s_i, ∇^4 is the bi-harmonic operator, and $[f, g]=f_{,11}g_{,22}+f_{,22}g_{,11}-2f_{,12}g_{,12}$. Interestingly, a similar governing equation has also been derived for inhomogeneous growth of a thin film [111, 112]:

$$B\nabla^4 w=\left[w,\Phi\right]$$
$$\nabla^4\Phi=-S\left[\kappa_G+\lambda_g\right] \tag{1.4}$$

where $\lambda_g=\varepsilon^g_{11,22}+\varepsilon^g_{22,11}-2\varepsilon^g_{12,12}$ is the incompatibility metric due to in-plane growth or swelling. It is noted that the two sets of equations (i.e., Equations 1.3 and 1.4) are identical if one sets $\lambda_g = -\sum_{i=1}^{N} s_i \delta(\mathbf{r} - \mathbf{r}_i)$. In principle, $\varepsilon_{11}^g, \varepsilon_{22}^g$, and ε_{12}^g can be chosen independently, as they do not necessarily satisfy the incompatible conditions. One possible choice is $\varepsilon_{12}^g = 0, \varepsilon_{11}^g = \varepsilon_{22}^g = \varepsilon^g$, which will lead to a Poisson's equation for ε^g,

$$\nabla^2 \varepsilon^g = -\sum_{i=1}^{N} s_i \delta(\mathbf{r} - \mathbf{r}_i) \tag{1.5}$$

The fundamental solution of Equation 1.5 in an infinite domain can be written as,

$$\varepsilon^g = -\sum_{i=1}^{N} \frac{s_i}{2\pi} \log|\mathbf{r} - \mathbf{r}_i| + \text{constant} \tag{1.6}$$

where the constant value should be determined from boundary conditions.

Topological defects in graphene can be represented by the following growth strain field in a perfect continuum film [13]:

$$\varepsilon_{12}^g = 0,\ \varepsilon_{11}^g = \varepsilon_{22}^g = -\sum_{i=1}^{N} \frac{s_i}{2\pi} \log|\mathbf{r} - \mathbf{r}_i| + \text{constant} \tag{1.7}$$

A Gaussian function $\frac{1}{\pi r_c^2} \exp\left[-\frac{(\mathbf{r} - \mathbf{r}_i)^2}{r_c^2}\right]$ with an intrinsic length scale r_c can be used to replace $\delta(\mathbf{r} - \mathbf{r}_i)$ to eliminate the singularity at the center of defects, which then modify the solution to Equation 1.5 as,

$$\varepsilon^g = -\frac{s_i}{2\pi}\left[\log\left(|\mathbf{r} - \mathbf{r}_i|\right) - \frac{1}{2}\mathrm{Ei}\left(-\frac{(\mathbf{r} - \mathbf{r}_i)^2}{r_c^2}\right)\right] + \text{constant} \tag{1.8}$$

where $Ei(x)$ is the exponential integral.

Zhang *et al.* [13] implemented the above continuum model into a triangle lattice model and simulated the wrinkle patterns of graphene with isolated pentagons (negative disclination), heptagons (positive disclination), and pentagon–heptagon pairs (dislocation). The 3D configurations predicted by the continuum model are in good agreement with full atom simulations based on the adaptive intermolecular reactive empirical bond order (AIREBO) potential [113] (Figure 1.8). Inspired by the significant 3D deformation of graphene with simple topological defects, it was further shown from continuum and atomistic simulations that periodically distributed disclination quadrupoles lead to a sinusoidal graphene ruga[1] (Figure 1.8a), and an array of dislocations on a cylindrical graphene can deform the carbon tube into a catenoid graphene funnel (Figure 1.8b).

---

[1] The Latin word *ruga* is used to refer to a large-amplitude state of wrinkles, creases, ridges, or folds [114].

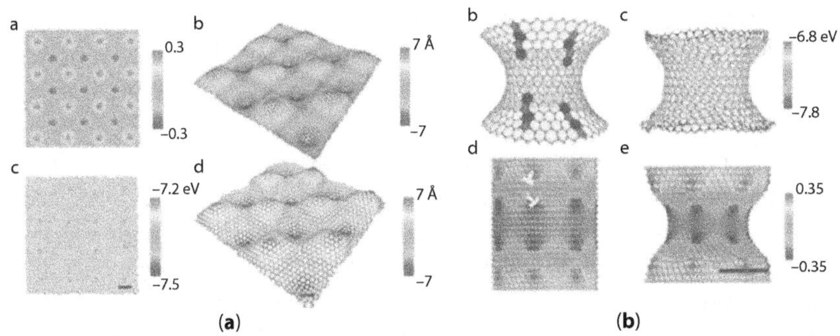

Figure 1.8 Sinusoidal graphene and catenoid graphene funnel achieved via topological design [13]. (a) A sinusoidal graphene induced by a periodic array of disclinations from continuum and atomistic simulations. (b) A catenoid graphene funnel from atomic simulations and continuum modeling.

These successes suggested the possibility to design arbitrarily curved graphene with topological defects. However, a direct search of atomic positions for curved graphene from MD is prohibited by the huge time scale gap between atom diffusion in graphene (seconds to hours) and the typical time scale associated with MD simulations (~ nanoseconds) [99, 100]. Other studies have attempted to employ geometrical methods [101–103] and Monte Carlo simulations [104, 105] to search for equilibrium positions of carbon atoms on a curved surface. Zhang *et al.* [15] developed a general design methodology by combining a PFC method [83] and MD simulations. The PFC method [83] can describe the defect motions in crystalline structures on both flat and curved configurations through over-damped conservative (diffusive) dynamics [115], which is a key to bringing realistic time scale in simulations with atomistic spatial resolutions.

The PFC model can be defined through the following free energy functional [83]:

$$F=\int\left[\frac{\phi}{2}\left(-\varepsilon+\left(1+\nabla^2\right)^2\right)\phi+\frac{1}{4}\phi^4\right]d\mathbf{x}, \tag{1.9}$$

where $\nabla=\partial/\partial x\,\boldsymbol{e}_i+\partial/\partial y\,\boldsymbol{e}_j$ is the gradient vector operator in 2D, ϕ the reduced density, and ε the reduced temperature. The governing equation for the dynamics of density evolution can be defined as,

$$\partial\phi/\partial t=\nabla^2\{[-\varepsilon+(1+\nabla^2)^2]\phi+\phi^3\}. \tag{1.10}$$

To handle complex geometry, Equation 1.10 can be solved using finite element method (FEM) by re-writing it in the following form:

$$\begin{aligned}&\frac{\partial\phi}{\partial t}=\nabla^2\mu\\&\mu=(-\varepsilon+1)\phi+2u+\nabla^2u+\phi^3,\\&u=\nabla^2\phi\end{aligned} \tag{1.11}$$

where two new variables (μ,u) are introduced to convert the order of the sixth-order partial differential equation (PDE) to a set of second order PDEs [116]. Equation 1.11 can be implemented in the standard FEM framework and solved efficiently by leveraging open-source software packages like FEniCS [116].

Taking the sinusoidal graphene as an example (Figure 1.9), the design methodology can be summarized as follows. First, PFC simulation is applied on the targeted curved manifold (Figure 1.9a), whose solution leads to an equilibrium triangular pattern of continuum density waves corresponding to a minimum energy state (Figure 1.9b). Second, a discrete triangular lattice network is obtained by identifying wave crests of the continuum triangular pattern of density as particles (Figure 1.9c). Third, a full-atom graphene structure is generated via Voronoi construction on the triangular lattice. Finally, a thermodynamically stable structure (Figure 1.9d) is achieved through MD equilibration at a finite temperature. Zhang *et al.* [15] showed that the predicted atomic structure of sinusoidal graphene agrees well with that from previous Monte Carlo simulations [105] of particle patterns confined on a sinusoidal surface. The new method is efficient and flexible enough to handle complex geometries and can be used to design graphene with targeted shapes through controlled distributions of topological defects. The designed sinusoidal graphene exhibits interesting properties such as enhanced toughness [15], tunable friction [54], and even negative Poisson's ratio [117].

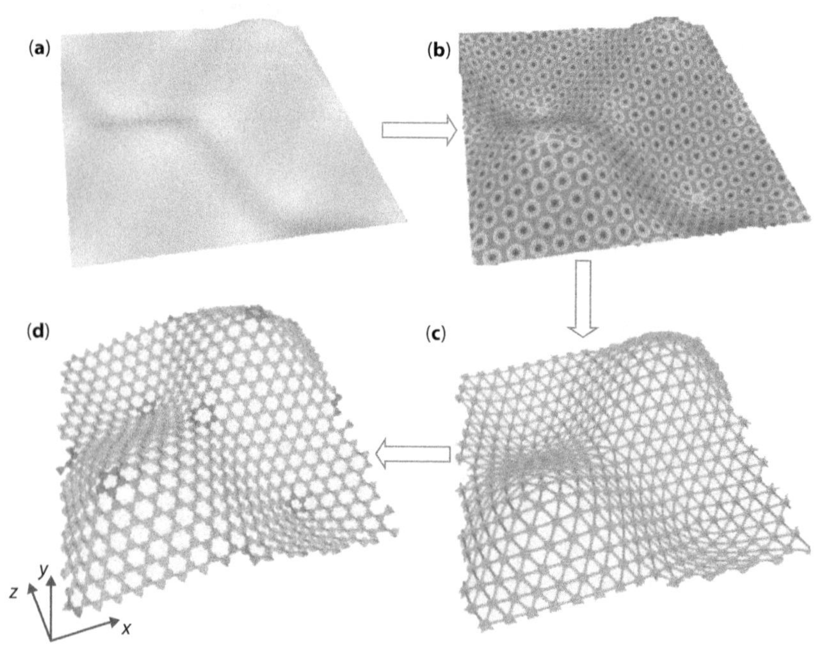

Figure 1.9 A general methodology to design an arbitrary 3D curved graphene structure through controlled distributions of topological defects via a combination of PFC and atomistic methods [15]. (a) The targeted curved surface. (b) A continuum triangular pattern of density waves on the targeted curved surface generated by PFC. (c) A discrete triangular lattice network from the continuum density waves. (d) The full-atom structure generated by Voronoi construction from the triangular network, followed by equilibration through MD simulations.

1.2.3 Topological Design for Toughening Graphene

Toughness is defined as the elastic energy released per unit area associated with advancement of a crack [118], thereby characterizing the resistance of a material to fracture in the presence of crack-like flaws. Sufficiently high toughness is important to ensure mechanical reliability of graphene for applications in practical devices/systems. However, experimental studies have shown that pristine graphene has a fracture energy as low as 16 J/m^2 [87], close to that of an ideally brittle solid, even though it is the strongest materials with a Young's modulus of 1 *TPa* and a strength of 130 *GPa* [50]. During the design, large-scale fabrication, and postprocessing operations of real devices and systems with graphene (e.g., CVD growth [5, 119, 120], transfer between different substrates [121–123], patterning and etching [124–126]), various forms of geometrical flaws (e.g., holes, notches, and cracks) may be introduced. This makes the actual failure strength of graphene determined by its ability to resist crack growth. Moreover, when corrosive species, like water vapor, are present in the working environment, stress corrosion cracking could further reduce the fracture resistance of material [127]. Thus, the inevitable flaws and corrosive environments make fracture one of the most prominent concerns in large-scale applications of graphene and are calling for efforts to explore effective methods to toughen graphene and other 2D materials [128–135]. In this section, we will review some of the recent progresses on toughening graphene through topological design.

In bulk materials, it is well known that topological defects like dislocations and GBs play important roles in tuning deformation mechanisms and fracture behaviors of many types of materials, such as metal [44, 45, 136, 137], ceramics [46], and diamond [47] (Figure 1.10). For instance, nanoscale engineering of GBs and twin boundaries [44, 45] have been widely employed to design superior metals with high strength and toughness (Figure 1.10a). The fracture toughness of ceramics (like Al_2O_3, Si) [138, 139] shows a more than two-fold enhancement by creating a high density of tangled dislocations in the sub-surface region (Figure 1.10b). Nanotwinned cubic boron nitride [46] and diamond [47] exhibit higher hardness and toughness than their defect-free counterparts (Figure 1.10c). The success of topological design in toughening bulk materials poses the question whether graphene and 2D materials in general can be toughened by designed topological defects. Recent studies in this direction have unveiled a number of toughening mechanisms induced by topological defects, including stress shielding, crack branching, atomic chain bridging, and stress reduction due to 3D geometry and nanocrack shielding.

The interaction between crack tip and stress induced by topological defect results in stress shielding, which is an important toughening mechanism in topologically designed graphene. Recent theoretical and numerical studies have shown that topological defects can alter the crack tip stress field and induce effective toughness enhancement [140–146]. For example, via MD simulations, it has been demonstrated that the stress resulting from individual topological defects like dislocations (a pentagon–heptagon pair) (Figure 1.11a) [140], Stone–Thrower–Wales (STW) defects (5-7-5-7 rings) (Figure 1.11b) [141, 142, 144], and 5-8-5 defects (Figure 1.11c) [143] can alter the crack tip stress intensity factor. Combining MD simulations and continuum modeling, Meng *et al.* [140] showed that dislocation shielding in graphene agrees well with the prediction of continuum linear elastic fracture mechanics (LEFM) (Figure 1.11a). By arranging topological defects into regular or irregular GBs, researchers [14, 145, 146] have studied more complicated interactions between

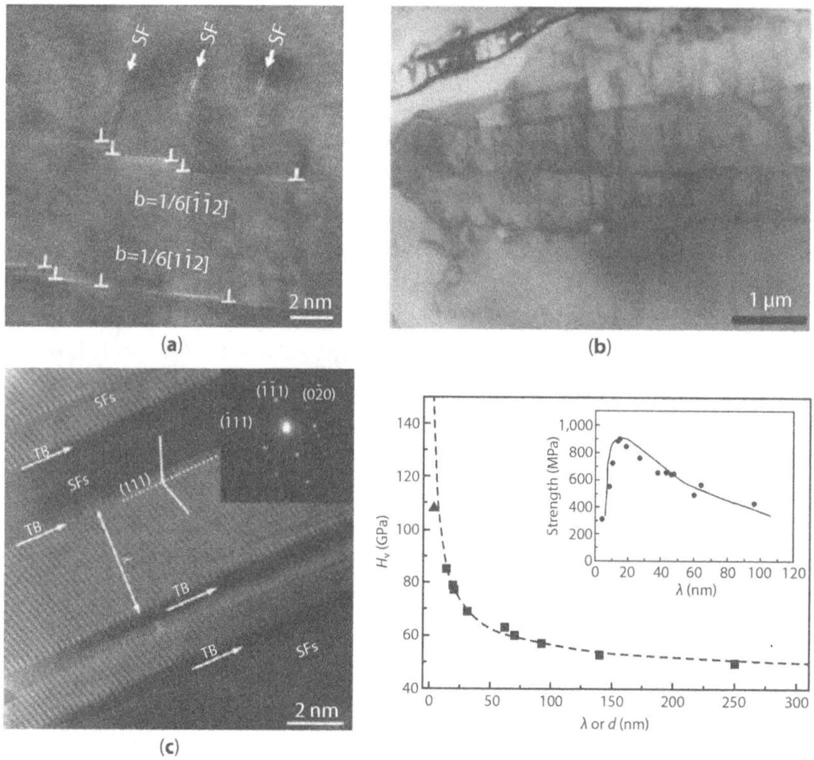

Figure 1.10 Topological defect design in bulk materials with enhanced mechanical properties. (a) The interaction of dislocations with nanoscale twin boundaries in pure Cu deformed in tension [45]. (b) Tangled dislocations form well-defined sub-boundaries beneath the surface of bulk Al_2O_3 after shot blasting and annealing [138]. (c) HRTEM image of twin boundaries in nanotwinned cubic boron nitride and the hardness of nanotwinned cubic boron nitride as a function of averaged grain size, *d*, or twin thickness λ [46].

topological defects and crack tip during propagation (Figure 1.11d–f). In addition to dislocation shielding, other toughening mechanisms including crack branching and atomic chain bridging can be activated during the crack propagation process. For instance, through MD simulations, Jung *et al.* [14] showed that the weak points within pentagon–heptagon defects can break near a crack tip resulting in crack branching and atomic chain bridging in polycrystalline graphene samples (Figure 1.11e–f). Combining these toughening mechanisms together, a 50% enhancement in fracture toughness was reported in a polycrystalline graphene sample with well-stitched randomly distributed GBs [14].

Compared with bulk materials, the interaction between crack tip and topological defects in graphene has several unique features. For atomic thin membranes like graphene, it is essential to consider non-local coupling between out-of-plane deformation and topological defects. On one hand, it has been demonstrated that the residual stress resulting from the in-plane lattice distortion of a topological defect can be partially released through out-of-plane deformation [3, 13]. This 3D relaxation tends to weaken the effect of the residual stress of topological defects on a crack tip. On the other hand, the out-of-plane deformation induced by topological defects also changes the sample shape in the global level [13], including the region where a crack tip resides, which may reduce the effective stress intensity near the crack tip and toughen the 2D material. This non-local interaction has been demonstrated by the MD

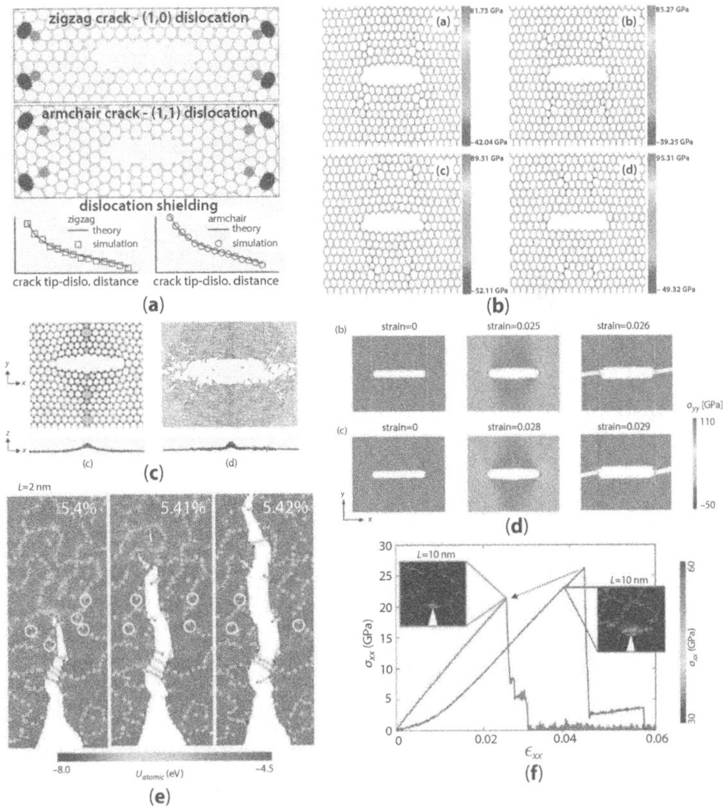

Figure 1.11 Interactions between cracks and topological defects in graphene. (a) Dislocation shielding of crack tips [140]. (b) Stress distribution in a graphene sheet with an embedded crack and STW defects placed at various positions [141]. (c) Stress distribution and failure process in a graphene sheet with a finite crack and 5-8-5 defects [143]. (d) Fracture of armchair-oriented bi-crystalline graphene with a finite crack and symmetric tilt GBs of different misorientation angles [145]. (e) Atomic energy distribution during crack propagation in a polycrystalline graphene with irregular grain shapes [14]. (f) Stress–strain curves of a cracked polycrystalline graphene of irregular grain shapes with (blue curve) and without (red curve) out-of-plane relaxation [14].

simulations of Jung *et al.* [14], where the toughening effect of GBs diminishes when the out-of-plane deformation is constrained even though the distribution of topological defects remains the same (Figure 1.11f). It is the competition of these effects that determines the overall toughness enchantment.

The interaction between cracks and 3D curved geometry of graphene results in another group of toughening mechanisms. As discussed in Section 1.2.2, a 3D curved sinusoidal graphene with a wavelength of 4 nm and an out-of-plane amplitude of 0.75 nm can be designed by periodically arranging disclination quadrupoles in graphene [15]. The mode I fracture toughness of this sinusoidal graphene is about 25.0 J/m^2 based on MD simulations, nearly twice that of pristine graphene (Figure 1.12a–b). It was found that the sinusoidal geometry and distributed defects give rise to stress reduction near the crack tip, nanocrack initiation at the defected sites, and atomic scale crack bridging. As depicted in Figure 1.12c, the non-planar sinusoidal geometry leads to a non-uniform stress field in the sample and a moving crack tip can be trapped in less stressed regions. In addition, the topological defects

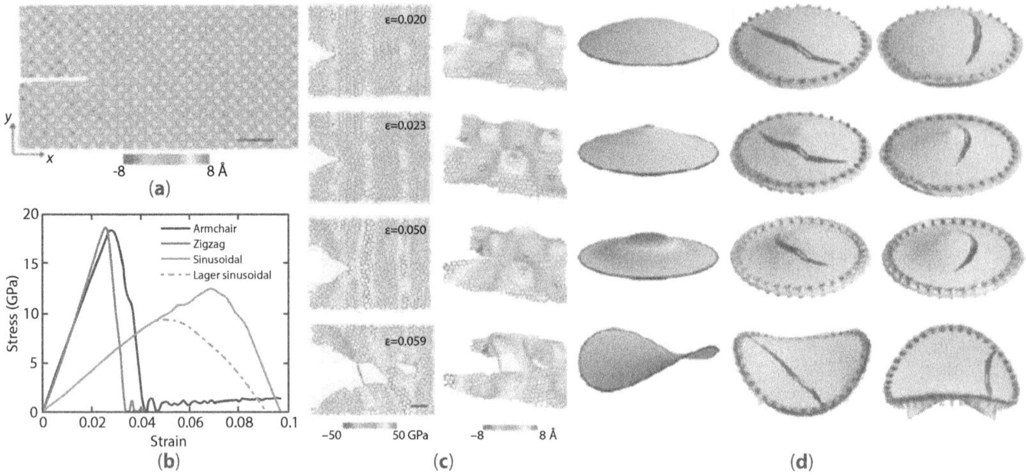

Figure 1.12 Crack propagation behaviors in sinusoidal graphene and thin rubber sheets. (a) A nanostrip of sinusoidal graphene with an edge crack [15]. (b) Stress-strain curves of pristine and sinusoidal graphene samples with an edge crack [15]. (c) Sequential snapshots of crack propagation in the sinusoidal graphene [15]. (d) Crack paths in rubber sheets draped on curved substrates [147].

within this non-uniform deformation field tend to fail at bonds with high pre-stress (e.g., bonds shared by heptagon and hexagon rings) [9, 75], leading to discrete rupture events ahead of a trapped crack tip, forming a nanocrack that shields the main crack. The deformation of the material connection between the nanocrack and main crack results in an atomic-scale chain bridging mechanism on crack advance.

The above example of interaction between crack and topologically designed 3D shape of graphene illustrates that fracture in topologically designed 2D materials needs to be treated as a fully 3D problem with strong nonlinearity. The sinusoidal graphene is a consequence of the out-of-plane relaxation of lattice frustration due to the periodically distributed disclination quadrupoles, which can only be understood by modeling fracture along a 2D topological manifold in a 3D space. In such a 3D model of fracture in 2D materials, there exists strong nonlinear coupling between topological defects, out-of-plane geometry and deformation, and crack propagation behaviors. By studying the fracture behavior of rubber sheets draped on curved surfaces, Mitchell *et al.* [147] demonstrated that surface curvature alone could stimulate or suppress crack propagation via the curvature-induced stress at the continuum level (Figure 1.12d). At the atomic level, the case of sinusoidal graphene shows that, apart from the curvature-induced stress, topological defects provide sites of nanocrack nucleation ahead of the main crack, resulting in additional toughness enhancement.

The topology-induced toughening, or simply topological toughening, discussed in this part demonstrates that the essence of topological design of 2D materials lies with the intrinsic connection between out-of-plane deformation and topological defect distribution. This connection is highly nonlinear and often involves strong multiphysics coupling. While topological toughening could be a promising way to introduce controllable/designable toughening mechanisms into 2D materials to mitigate or overcome their intrinsic brittleness [148], much effort is needed in the future to explore the full potential of enhancing the mechanical and physical properties of graphene and other 2D materials through topological design.

1.3 Applications of Topologically Designed Graphene

Beyond theoretical predictions of enhanced mechanical properties of graphene through topological design, topological defects have already been shown to play critical roles in a number of novel applications, such as chirality-specific single-walled carbon nanotube growth, energy material engineering, multifunctional materials, and interaction with biological systems. In this section, we will briefly review some of the advancements in this field. We emphasize that a rational design and fabrication of topological defects in graphene for specific applications have not yet been fully realized, and there will be tremendous opportunities to optimize the performance of novel devices and techniques based on topologically designed graphene once their fabrication techniques are matured.

1.3.1 Topologically Designed Graphene Flake[1] to Guide the Growth of Single-Walled Carbon Nanotube (SWCNT)

As an important one-dimensional nanomaterial, SWCNT has attracted a great deal of research interests due to its extraordinary physical properties and promising potential in various applications [149–151]. Many of these properties and applications are strongly dependent on the structure of the SWCNT, such as the diameter and the orientation angle of the hexagonal lattice relative to the tube axis, also known as chirality (n, m). For example, SWCNTs could be either metallic or semiconducting depending on the chirality [152]. In semiconducting SWCNTs, band gaps are inversely proportional to the diameter. Thus, fabrication of chirality-specific SWCNTs is of great importance in achieving the full technical potential of CNTs. Recently, it has been shown that a $C_{96}H_{54}$ precursor [153] could be transformed into a 3D curved nano-graphene flake with identical atomic structure as the end-cap of a (6, 6) nanotube through an intramolecular cyclodehydrogenation process on a Pt (111) surface. The (6, 6) caps could be used as seeds to grow defect-free SWCNTs with lengths up to a few hundred nanometers through a surface-catalyzed growth process (Figure 1.13a). The well-designed atomic structure of the 3D end-cap seeds enables the synthesis of SWCNTs with a specific chirality (6, 6) (more than 90%), instead of a mixture of uncontrolled structures. Through different synthesis routes, 3D topologically designed nano-graphene flakes with end-cap structures for (5,5), (9,0), (8,8), (10,10), and (12,12) SWCNTs have also been reported [154–156], further demonstrating the potential of topological design (Figure 1.13b). It remains a challenge to systematically design topological structures of the SWCNT end-cap with different chiralities.

1.3.2 Topologically Designed Graphene for Novel Energy-Related Applications

For energy-related applications, topologically designed graphene has been used to enhance performance of rechargeable lithium ion battery (LIB) and supercapacitor systems. One criterion for choosing anode materials for LIB is to achieve high specific charge capacity [157].

[1] Graphene flake here refers to nanoscale carbon molecules made of single-layer sp^2-hybridized carbon atoms.

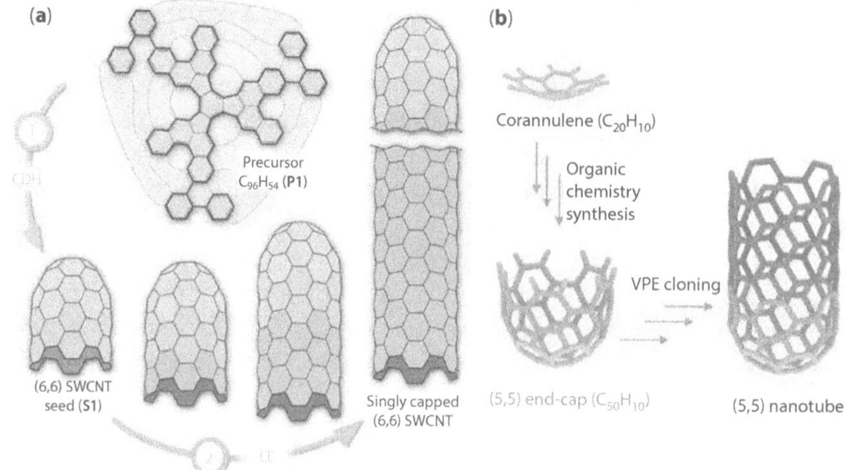

Figure 1.13 Topologically designed graphene flake to guide the growth of chirality-specific SWCNTs. (a) Schematic illustration of a bottom-up synthesis of (6,6) SWCNTs from a designed end-cap [153]. (b) Atomic structure of the end-cap for (5,5) SWCNTs [155].

Topological defects have been predicted to be useful in improving the capacity in graphene-made electrode. First-principle calculations [158] based on density functional theory (DFT) revealed that the lithium adsorption on graphene could be enhanced by topological defects including divacancy (5-8-5 rings) and STW defect (5-7-5-7 rings) due to the increased charge transfer between the adatom and defected sites in graphene. Later theoretical studies found that not only topological defects like 5-, 7-, and 8- rings (Figure 1.14a–c), but curvatures (Figure 1.14d) of the graphene sheet could also enhance lithium adsorption resulting in better lithium storage property [159].

Besides potential applications as an anode material, topologically designed graphene could also be integrated with other anode materials like silicon to optimize battery performance. Although silicon is known to possess the highest theoretical charge capacity [160], it suffers from chemomechanical degradations due to a large volume change (300%) during battery operation [161–163]. Fracture, loss of electrical contact, and repeated chemical side reaction with the electrolyte can occur during lithiation and delithiation processes and a tremendous amount of research effort has been devoted to resolving these issues [164–170]. In a recent experimental study, Li *et al.* [171] designed 3D curved graphene cages to encapsulate micro-Si particles and achieved outstanding long time stability of the resulting anode. The 3D graphene cages possessing wavy profiles are mechanically strong and flexible; During battery operations, the encapsulated Si particles could undergo large deformation and even fracture without losing electrical contact because of the constraint of the graphene cages (Figure 1.15a–c). Additionally, it was shown that the solid electrolyte interphase (SEI) layer formed on a graphene cage remains intact during repeated lithiation/delithiation, resulting in stable cycling with 90% capacity retention after 100 cycles.

Topological defects and the resulting 3D curvatures have also been shown to contribute to the performance of electrochemical supercapacitors [172] made of graphene. DFT

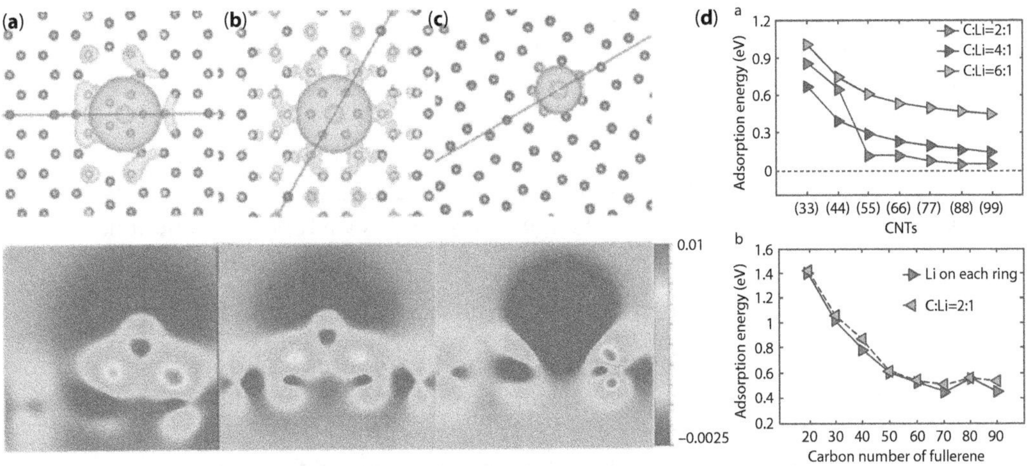

Figure 1.14 Lithium adsorption on graphene enhanced by topological defects and curvature [159]. (a–c) Top and side views of charge density of Li adsorbed in a pentagon ring (a), a hexagon ring (b), and a heptagon ring (c). Li on the defective rings (a) and (c) transfer more charge to C than on the pristine hexagon ring (b). (d) Adsorption energy of Li atoms adsorbed on CNTs and fullerene molecules is observed to increase with surface curvature.

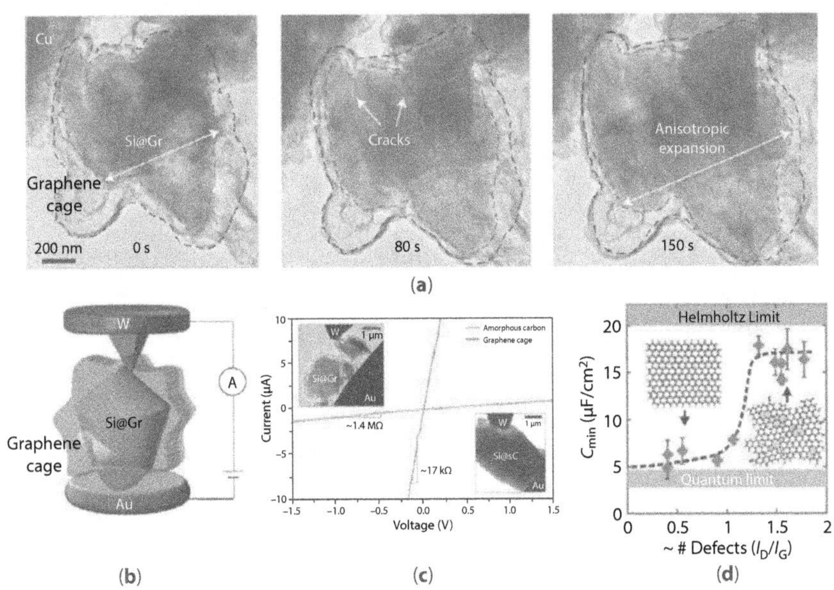

Figure 1.15 Topologically designed graphene enhances the performance of Si anode and supercapacitors. (a) *In situ* TEM observation of deformation and fracture of a graphene-caged Si particle during lithiation. The Si particle (outline in red) fractures abruptly and violently within the mechanically strong graphene cage (outlined in black), which remains intact throughout the process [171]. (b) Schematic diagram of a nanoscale electrochemical cell with a graphene-caged Si particle [171]. (c) Current-voltage curve of graphene-encapsulated Si microparticle (SiMP) and amorphous-carbon-coated SiMP [171]. (d) Measured capacitance of graphene supercapacitor as a function of defect density [174].

calculations [173] have predicted that topological defects (such as 5-7-5-7 rings, and 5-8-5 rings) could substantially enhance the quantum capacitance of graphene by inducing quasi-localized states near the Fermi level, achieving a nearly four-fold increase [174] in double-layer capacitance after combining with functionalization (Figure 1.15d).

Note that in the studies discussed above, topological defects and 3D curvature serve as a new platform to couple mechanical deformation, chemical reaction, and electronic structures of graphene in enhancing specific targeted properties via topological design.

1.3.3 Topologically Designed Graphene for Multifunctional Materials

Topological defects and 3D curvature have also shown promises in altering electrical transport behaviors [4, 18, 19, 21], tuning thermal conductivity [16, 22, 175, 176], generating mechanical–electrical coupling [56, 57, 177], and modifying chemical reactivity [17] in graphene systems. Yazyev and Louie [19] theoretically explored the potential of controlling electronic transport in graphene with GBs, finding two distinct transport behaviors, either high transparency or perfect reflection of charge carriers over remarkably large energy ranges, as shown in Figure 1.16a. In experiments, while Huang *et al.* [4] detected no measurable electrical resistance from GBs within instrument limits, Jauregui *et al.* [18] showed that GBs impede electrical transport and induce prominent weak localization, indicative of intervalley scattering in graphene. Tsen *et al.* [21] found that GBs with better inter-domain stitching lead to more uniform transport. These studies show that the effects of GBs

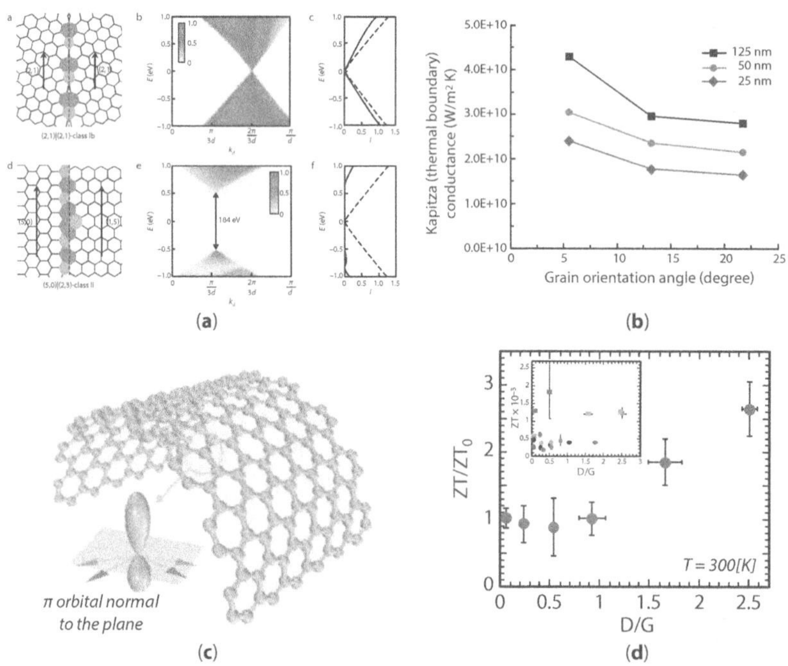

Figure 1.16 Electrical, thermal, thermoelectrical, and flexoelectrical properties of graphene with topological defects. (a) Two distinct electronic transport behaviors through GBs in graphene from first principles [19]. (b) Boundary conductance of GBs as a function of grain orientation angle [175]. (c) Curvature-induced polarization upon bending [180]. (d) Normalized ZT values with respect to defect density [179].

on electrical properties of graphene are highly dependent on their atomic structures, suggesting possibilities to control electrical properties of graphene through GB engineering.

On thermal properties of graphene, both theoretical studies and experiments have confirmed that the presence of topological defects could result in substantial reduction in thermal conductance [16, 176, 178]. This is mainly because phonons, as the dominant carriers of thermal energy in 2D materials, are scattered when they encounter topological defects, thereby limiting the phonon mean free path. This reduction in thermal conductance is closely related to the distribution of topological defects and can be characterized by the Kapitza resistance. Bagri *et al.* [175] studied thermal transport across several GBs with different grain orientations using non-equilibrium MD and found that the thermal conductance could be tuned in a certain range by the atomic structures of GBs, as shown in Figure 1.16b. Serov *et al.* [16] found that the type and size of GBs play important roles in determining the thermal conductance when grain sizes are smaller than a few hundred nanometers. Fthenakis *et al.* [176] showed that the thermal conductivity of graphene depends sensitively on whether the defects are isolated, form lines, or form extended arrangements in haeckelites. These studies point to the potential of tuning thermal properties of 2D materials through GB engineering. Controlling heat conduction in materials is of great importance for thermal management of electronic devices as well as thermal energy recycling, which can be achieved through thermoelectric conversion that depends inversely on the thermal conductivity. Graphene has displayed larger Seebeck coefficient and higher overall power factor than other semiconductors and metals. By reducing the thermal conductivity of graphene while maintaining its electrical conductivity or increasing the ratio between electrical conductivity and thermal conductivity through defect engineering, the standard figure of merit for thermoelectricity (ZT) can be enhanced up to three times that of pristine graphene (Figure 1.16d) [179]. Ma *et al.* [22] experimentally measured the influence of grain size on the thermal and electrical transport behaviors in polycrystalline graphene, and further demonstrated the possibility of improving the thermoelectricity of 2D materials through grain size engineering.

Topological design can also play an important role in tailoring the piezoelectricity and flexoelectricity of graphene. Pristine graphene possesses no piezoelectric property due to its intrinsically centrosymmetric hexagonal structure. Piezoelectricity can be induced by breaking the structural symmetry, and doping is an effective way to create internal polarization [177]. Compared with piezoelectricity, flexoelectricity is a more universal phenomenon of dielectrics in which strain gradient can polarize the material and conversely, non-uniform electric fields can cause mechanical deformation. Based on DFT calculations of curved graphene surface, Kalinin and Meunier [57] proposed the concept of electronic flexoelectricity, in which the bending of single graphene layer results in a transfer of electron gas density across the basal plane and yields a curvature-dependent electrical dipole response. Later on, Kvashnin *et al.* [56] established the universality of the linear dependence of flexoelectric atomic dipole moments on local curvature in various carbon networks such as nanotubes, fullerenes, and nanocones. This field is still in its infancy with a lot of open questions. For example, since topological defects are intimately connected to the out-of-plane deformation and curvature of graphene, it will be interesting to investigate whether and how they induce flexoelectricity, and if they do, how to optimize such effect through topological design.

The chemical reactivity of graphene has also been shown to be sensitive to the presence of topological defects and curvature. Through experimental study of curved nanoporous graphene, Ito *et al.* [17] have shown that highly curved graphene with a high density of topological defects can promote chemical doping contents, either electron donors or acceptors. Wu *et al.* [93] have shown that *in situ* generated aryl radicals were more likely to be found in regions with higher local curvature, demonstrating the selective effect of curvature on surface functionalization of graphene. Based on a molecular mechanics model, Pacheco Sanjuan *et al.* [181] found that the pyramidalization angle, which is directly proportional to the mean curvature of a curved 3D graphene, is important to determine the chemical reactivity (such as chemisorption of H_2) of the material. This theoretical study sheds light on the experimental observations and suggests a route to rationally design the chemical reactivity of graphene by controlling its mean curvature.

Potential use of topological design in tuning multifunctionalities of graphene has also been explored in bulk materials made of graphene. For example, Qin *et al.* [182] and Jung *et al.* [23] have studied the properties of 3D graphene with gyroid-shaped unit cells using MD simulations and 3D-printed models. It was shown that the graphene gyroids could achieve high specific strength and low thermal conductivity due to the curvature and topological defects. Indeed, topological defects are geometrically necessary in 3D systems made of 2D materials, where topological design can be expected to play essential roles.

1.3.4 Topologically Designed Graphene for Biological Applications

Due to their remarkable multi-physical properties and large specific surface areas, graphene and its derivatives, such as CNTs, fullerene molecules, and graphene oxide (GO), have been applied to various biological applications as diverse as biosensors [183, 184], drug delivery [185, 186], and biological imaging [187, 188]. Experimental and theoretical studies have demonstrated that topological features, especially curvature, play important roles in determining the interaction of graphene with proteins [189–191], nucleic acids (such as DNA), and cell membranes, where weak non-covalent bonding is the dominant force of interaction.

Previous studies have found that proteins and DNA molecules interact with graphene mainly through π-π stacking and dispersion interactions. The surface curvature of graphene has been proven to play important roles in modifying these non-covalently bonding interactions and tuning its adsorption capacity. For proteins interacting with graphene, Zuo *et al.* [189] demonstrated from MD simulations that the flat and flexible surface of graphene has better chance to form flat π-π stacking with aromatic residues in protein villin headpiece (HP35) while the convex surfaces of SWCNT and C_{60} interact with HP35 mainly through the dispersion interaction (Figure 1.17a–c). Beyond the flat and convex surfaces, Jana *et al.* [190] studied curvature dependence of polypeptide adsorption on flat, convex, and concave graphene surfaces, and found that the concave surface shows the strongest absorptivity (Figure 1.17d). Gu *et al.* [191] studied the adsorption of bovine serum albumin (BSA), a model protein, on SWCNTs and graphene using MD simulations and fluorescence spectroscopy experiments, with results demonstrating that the adsorption capacity of the protein depends on surface curvature. Similar phenomena have also been reported for

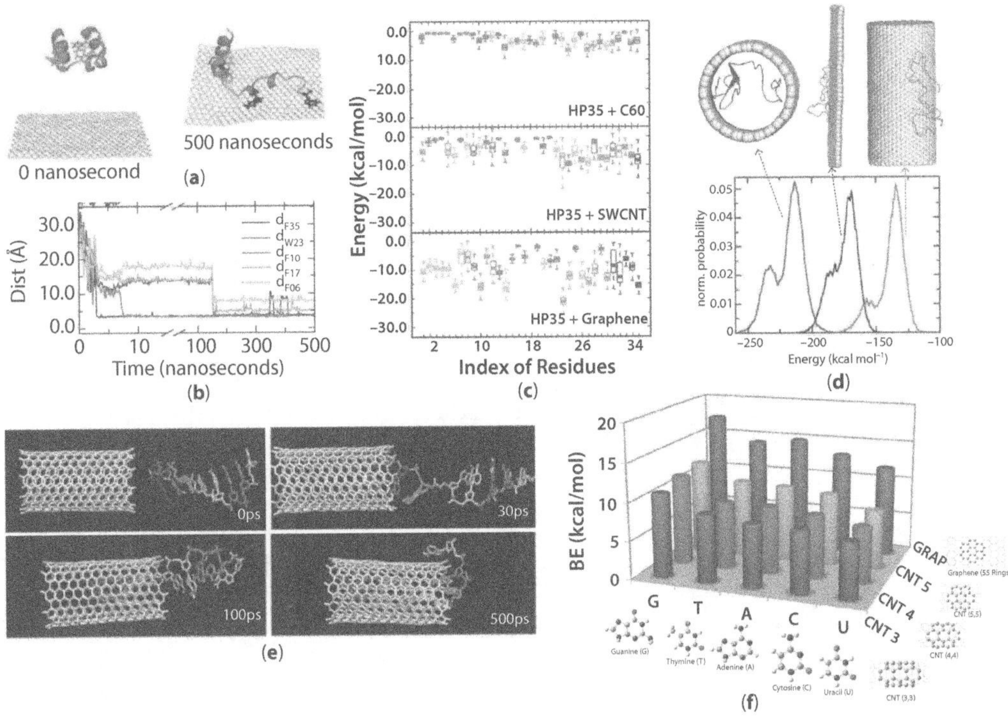

Figure 1.17 Curvature-dependent interactions between proteins, DNA molecules, and graphene surfaces. (a) Representative snapshots of HP35 adsorbed on graphene surface [189]. The protein is shown in cartoons with red helix and green loop, and graphene is shown in orange. Aromatic residues that form π–π stacking are shown as blue stick, while the rest shown in green. (b) Distances between graphene and aromatic residues, including F35, W23, F10, F17, and F06 [189]. (c) Interaction energies between different residues of HP35 and graphene, (5,5)-SWCNT, and C60. Color of points indicates probability of a residue in contact with graphene: 0–20% (red), 20–40% (orange), 40–60% (green), 60–80% (cyan), and 80–100% (blue) [189]. (d) Normalized distribution of interaction energy of amphiphilic full-length amyloid beta peptide with concave (blue), flat (black), and convex (red) graphene surfaces [190]. (e) Simulation snapshots of a DNA oligonucleotide interacting with a (10, 10) CNT in water solute environment [192]. (f) Binding energy of DNA/RNA nucleobase G, T, A, C, U with curved outer surface of CNTs and flat graphene [193].

the interaction of graphene with DNA/RNA nucleobases and molecules. Gao *et al.* [192] showed that a DNA molecule could be spontaneously inserted into a SWCNT in aqueous solution using MD simulations (Figure 1.17e). Umadevi and Sastry [193] demonstrated that the binding energy of DNA/RNA nucleobases on the outer surface of a SWCNT increases with the radius of the SWCNT through quantum chemical calculations (Figure 1.17f). The curvature dependence of biomolecular adsorption could lead to novel devices to detect different molecules and even to remove harmful molecules in disease treatment.

Recent studies have demonstrated that graphene nanosheets could damage bacterial cell membrane by insertion/cutting as well as destructive extraction of lipid molecules [194–196]. This reveals a new toxic mechanism of graphene and opens the possibility of exploiting graphene as a novel antibacterial material [197, 198]. The surface curvature of graphene is also found to affect lipid extraction via modifying the dispersive adhesion between graphene and lipid molecules. For example, using MD simulations and theoretical analysis, Luan *et al.* [199]

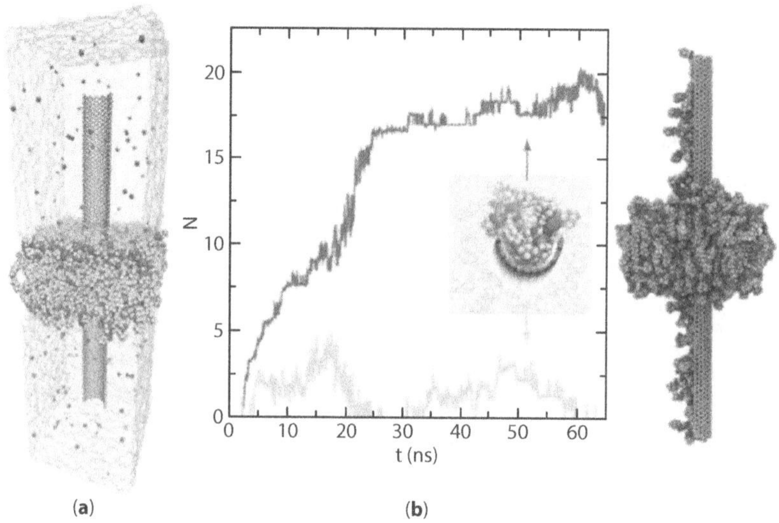

Figure 1.18 Curvature-dependent lipid extraction on graphene surfaces [199]. (a) MD simulation of curved graphene inserted into the bi-layer lipid membrane. (b) Time-dependent numbers of extracted lipids on a concave surface (blue) and a convex surface (orange) of graphene.

have demonstrated that the lipid extraction could be understood as a wetting process and concave graphene surfaces possess the strongest extraction effect (Figure 1.18).

These studies clearly show a universal effect of surface curvature on the interactions of graphene with biomolecules, which is critical not only for applications of graphene in biotechnology but also in understanding the biosafety of nanomaterials. Topological design may provide an effective means to control the surface curvature and morphology of graphene and further tune these interactions. Novel techniques and devices made of topological designed graphene could be expected to further unleash the potential in this direction.

1.4 Fabrication Techniques of Topologically Designed Graphene

While topologically designed graphene has been found promising in various applications, as discussed in Section 1.3, it is still very challenging and interesting to develop effective fabrication techniques to deliberately control the distributions of topological defect in graphene and other 2D materials. With the rapid development of capabilities in 2D materials synthesis, characterization, and modification, several manipulation techniques and fabrication pathways have emerged as promising candidates for large-scale applications of topologically designed graphene, such as CVD growth on curved templates, controlled irradiation, and organic chemical synthesis.

First, controlled CVD growth continues to show promises for engineering topologically designed graphene at large scales. As a popular method to grow large-scale graphene, currently CVD methods [4, 5] can only produce samples with randomly distributed topological defects, including GBs. However, with more controlled growth conditions, CVD methods could be further developed to generate desired patterns of topological defects. For example, pre-patterned growth seeds [81, 200] may be used to tune the GB density and patterns in polycrystalline graphene grown by CVD (Figure 1.19a). With appropriate substrate materials and growth conditions, graphene could grow on substrate of various geometrical and topological configurations, including porous metal foams [24, 201], network of nanowires [202] and microparticles [191], 3D-printed scaffolds [203], and even zeolite crystal with nanoscale pores [204] (Figure 1.19a–f). To conform to the curvature of these surfaces, topological defects [24] are generated naturally and observed experimentally (Figure 1.19g). CVD growth of graphene on a curved substrate could be a promising way to fabricate topologically designed graphene at large scale. Similar attempts in CVD growth of TMDCs [205] have proven to be successful. For example, WS_2 growing on a cone-shaped surface has been observed to yield GBs [205].

Second, controlled irradiation or thermal excitation has been found to be able to create topological defects with desired type and locations at the atomic level. Early theoretical study [206] indicated that in the vicinity of vacancies, the energy barriers of the formation of various topological defects are sufficiently low in graphene that topological defects could be introduced via thermally activated reconstruction. Recent experiments have confirmed this prediction. For example, by adjusting the irradiation dosage and focus region

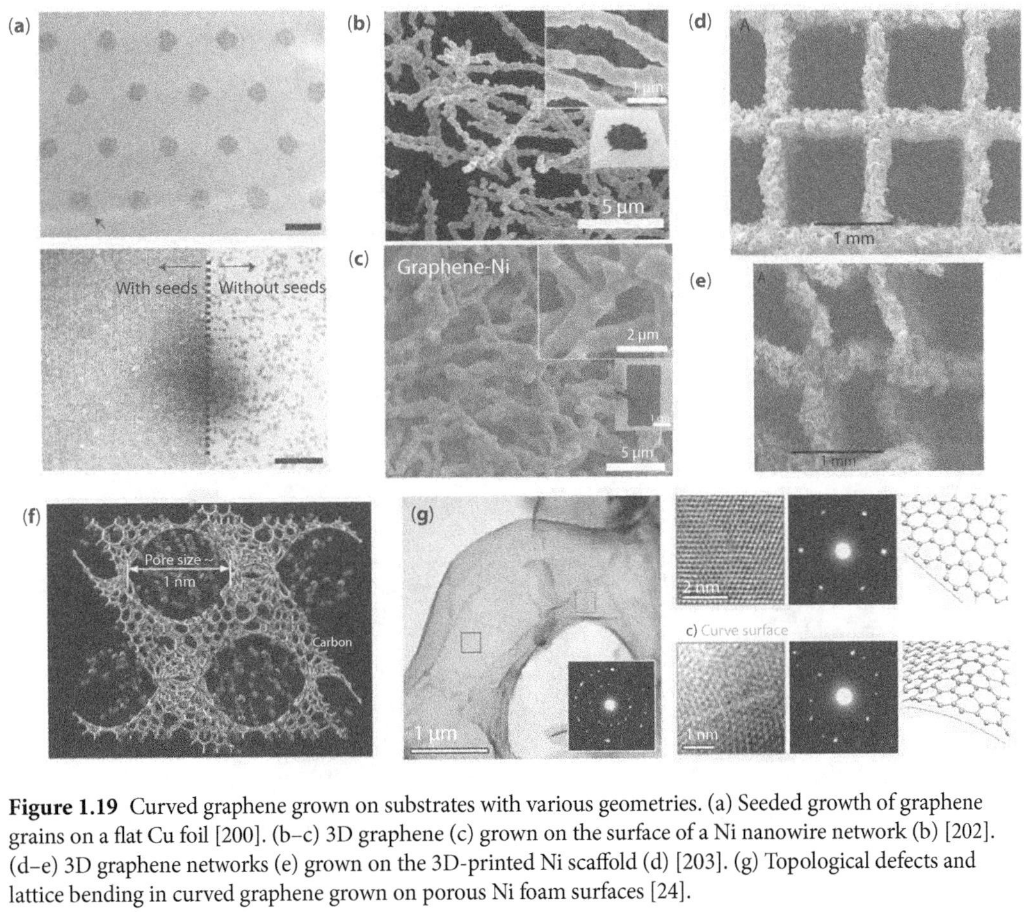

Figure 1.19 Curved graphene grown on substrates with various geometries. (a) Seeded growth of graphene grains on a flat Cu foil [200]. (b–c) 3D graphene (c) grown on the surface of a Ni nanowire network (b) [202]. (d–e) 3D graphene networks (e) grown on the 3D-printed Ni scaffold (d) [203]. (g) Topological defects and lattice bending in curved graphene grown on porous Ni foam surfaces [24].

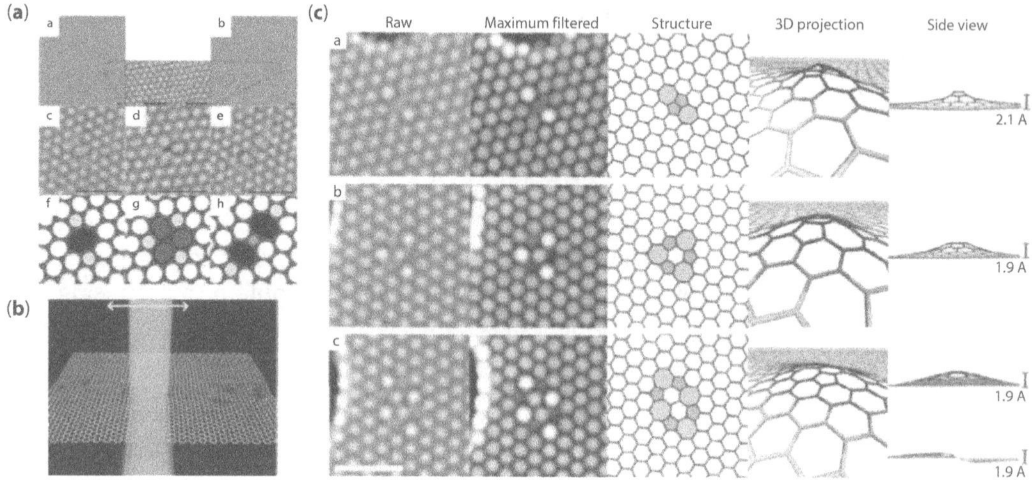

Figure 1.20 Creating topological defects in graphene via controlled irradiation or thermal excitation. (a) Formation of topological defects in graphene after a 30s electron beam exposure [207]. (b) Schematic illustration of controlled creation of large dislocation numbers in graphene by scanning electron beam irradiation [12]. (c) Implantation of atomic-scale blisters in graphene by depositing extra carbon atoms into single-layer graphene using a standard carbon coater [27].

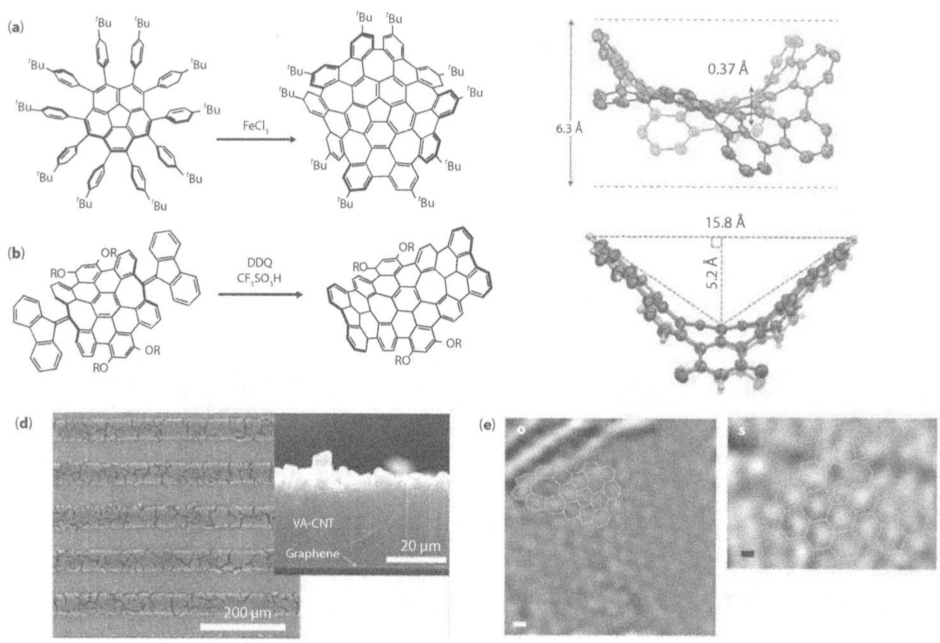

Figure 1.21 Molecular-level fabrications of nanographene flakes and graphene–CNT hybrids. (a–c) Examples of the syntheses (a–b) of curved molecular nanographene flakes with 5-, 7-, and 8-member rings (c) [208, 209]. (d) Selective growth of vertically aligned CNTs on patterned graphene [204]. (e) Topological defects at connections of seamless CNT–graphene hybrids [25].

of the electron beam, Robertson *et al.* [207] demonstrated that stable topological defects like dislocation pairs could be created within a controlled area with high spatial precision (~10*10 nm^2) (Figure 1.20a). Warner *et al.* [12] showed that with a slowly moving electron beam, a large number of dislocations could be created within a well-defined nanoscale area without creating holes in graphene (Figure 1.20b). Besides reorganizing the original carbon atoms via high-energy beams, it is also possible to add extra carbon atoms into graphene to create topological defects. Using a standard carbon coater, Lehtinen *et al.* [27] implanted extra carbon atoms into graphene samples to create dislocation dipoles, which can form atomic-scale blisters with strong out-of-plane buckling (Figure 1.20c). This seems to be a promising method to manipulate local configurations of topological defects, if higher precisions can be achieved in controlling detailed defect types.

Third, at the molecular level, effective organic chemical synthesis pathways have been developed for different sp^2 carbons including curved graphene-like carbon with non-hexagon rings. For instance, chemical synthesis routes have been proposed and tested to create nanographene molecules with heptagon, pentagon, and even octagon rings [26, 208, 209], whose equilibrium configurations are warped in 3D space (Figure 1.21a–c). Another study [153] has demonstrated that designed carbon molecules with both hexagons and non-hexagon rings are capable of folding into curved 3D configurations, like a spherical cap, which could serve as a basis to grow carbon nanotubes with controlled chirality. At the same time, there has been a long-term trend of research on fabricating sp^2 carbon hybrids of fullerene molecules, carbon nanotubes, and graphene. For example, seamless covalent bonding connection has been achieved during sequential growth of graphene layers and CNTs [25, 204] (Figure 1.21d). Since the components in such hybrids show different curvatures, topological defects are naturally observed in the connection regions within these hybrids to accommodate the transition in curvature [25] (Figure 1.21e). The chemical synthesis routes and fabrication procedures could provide bottom–up pathways to engineer topological defects by combining growth and merging of small molecules.

1.5 Outlook

In this chapter, we have reviewed some of the recent progresses in experimental, computational and theoretical studies on graphene with topological defects and structures, summarizing some of the important and unique roles played by topological defects in atomically thin graphene structures. Based on these studies, one might propose topological design of 2D materials as an important research field to explore and take advantage of the full potential of graphene and other 2D materials. We have emphasized that topological defects play very important roles in determining the 3D curved geometry, residual stress field, as well as phonon and electron transport properties of graphene, which in principle could be designed for optimized mechanical, physical, and chemical properties. The potential of topological design has been demonstrated through some of the preliminary successes in this field, such as tuning the strength of polycrystalline graphene, designing the shape of 3D curved graphene, enhancing fracture toughness of graphene, growing chirality-specific SWCNTs, engineering materials for energy-related applications, designing multifunctional materials, and tuning interactions with biological systems. In spite of the rapid developments in experiments and simulations, our understanding of the fundamental relations between topological defects and the mechanical, physical, and chemical properties of 3D curved graphene and other 2D materials is still in its infancy. There remain many important open questions related to topological design of graphene that deserve further research efforts. Some of the open questions, opportunities, and potential research topics/directions in this promising field are summarized below.

1. How can we systematically enhance the fracture toughness of graphene via topological design? In Section 1.2.3, we have reviewed some preliminary studies showing that various toughening mechanisms can be introduced in graphene via topological defects, including crack tip shielding by defects [140], stress reduction by defects-induced 3D geometry [15], crack branching, atomic-chain bridging [144], and nanocrack shielding [14]. These mechanisms lead to substantial enhancement in fracture toughness. Further improvement in toughness may be achieved by activating more effective toughening mechanisms or maximizing the synergistic effect of existing toughening mechanisms. Inspirations from biological materials and bulk engineering materials include crack trapping, shielding, deflection, crack bridging in nacre [210–212], and plasticity-induced energy dissipation in metals [213–215]. It will be interesting and challenging to see if it is possible to introduce more toughening mechanisms (and even ductile deformation modes) into graphene via topological design. At the same time, theoretical models of topological toughening may need to touch on fracture mechanics of curved membrane and shell structures [216–218] to account for the intrinsic coupling between topological defects and out-of-plane geometry.
2. How does thermal fluctuation affect the topological design of graphene? Thermal fluctuation can significantly influence the mechanical and physical properties of graphene because of its very small bending stiffness (~1 eV). At room temperature, it has been shown from self-consistent theory and MD simulations [61] that the stiffness of a micron-scale graphene sheet can be enhanced by several orders of magnitude through thermal vibration, which seems to be supported by experiments on graphene kirigami [38]. An important general question is whether and how the 3D geometry of topologically designed graphene is influenced by thermal fluctuations. Besides directly modifying the effective properties of graphene, thermal fluctuations can also add time-varying perturbations to the morphology and mechanical properties (i.e., modulus and bending stiffness) of topologically designed graphene as they undergo random vibrations at finite temperature. These stochastic effects may be posed as an optimization problem with uncertainty.
3. How can one best tailor the multiphysical properties of graphene through topological design? By deliberately introducing topological defects, topological design can activate the effects of nanoscale straining and buckling on electronic band structure and chemical bonding in graphene. The topological defect-mediated coupling between electronic band structure, chemical reactivity, thermal conductivity, lattice distortion, and 3D geometry raises interesting questions and challenges for understanding and manipulating

multiphysical properties of graphene: (1) How does the topological design alter the electronic band structure [19], pseudomagnetic field [94], and surface plasmons [219, 220] in graphene? (2) How do the curvature and residual stress introduced by topological defects interact with functional groups or corrosive agents? And what are the chemically stable topological designs in an ambient environment? (3) Can one further tune these properties of topologically designed graphene through controlled deformation or strain engineering [221, 222]?

4. How can one design/achieve an optimal balance between various mechanical, physical, and chemical properties in graphene/graphene devices through topological design? Numerous examples have shown that topological design of graphene could modify multiple properties simultaneously via the fully coupled mechanical, chemical, and electrical interactions. Such coupling suggests that topological design of graphene can be a multi-objective optimization problem. For example, the out-of-plane geometry and curvature could generate electrical dipole moments through flexoelectricity [56, 57], and the curvature-dependent polarization could be used in DNA sequencing [223, 224]. At the same time, for large-scale applications of such devices, it will be important to enhance the fracture toughness of graphene, which also depends on the out-of-plane geometry [15]. A design question might involve how to take the most advantage of flexoelectricity effect in DNA sequencing application while maintaining sufficient fracture toughness. It should be noticed that topological defects may diminish some properties while optimizing targeted ones. For example, defects can be used to enhance toughness of graphene [14, 15], while they can also result in reduction of strength and/or modulus of graphene [8–11, 54, 131]. Therefore, another important question is how one can achieve optimally balanced properties with topological design.
5. Can topological design go beyond a single layer of graphene to an assembly of a large number of graphene structures, such as multilayer graphene and 3D assemblies of interconnected graphene? For multilayer graphene, strong differences exist between the intra- and interlayer deformation, as the former is determined by the strong carbon–carbon covalent bonds, whereas the latter are governed by relatively weak forces (i.e., van der Waals interaction). In addition, the interlayer interaction is sensitive to atomic registry [225], surface geometry [54], and chemical bonding [226] between the layers. It could be expected that topological design could also affect the adhesion and friction between different layers. Fundamental and application questions could be raised in designing multilayer graphene, including: How do the topological defects and curvature affect the interlayer adhesion in multilayer graphene systems? What are the friction mechanisms in multilayer graphene with distributed topological defects? Can one tune or optimize the interlayer adhesion and friction properties of multilayer graphene via topological design?
   Graphene foam [24, 203], CNT–graphene hybrids [25, 204], and graphene gyroids [23, 182] present examples of 3D interconnected graphene assemblies. Different from multilayer graphene, 3D interconnected graphene consists of strong sp^2 carbon lattices spreading on 3D surfaces with complicated

geometrical and topological configurations (e.g., gyroid surfaces), which occupies the whole 3D space. Curvature and topological defects play critical roles in maintaining the final self-equilibrium shapes and thus determine the effective bulk properties of the assembly. For example, by achieving a cork-like hierarchical structured 3D graphene monolith via freeze casting technique, Qiu *et al.* [226] demonstrated that the resulting 3D graphene foam with well-organized cellular structures could recover its shape and dimension after 90% compression. After 1,000 cycles of compression test, it only showed 7% reduction in dimension. This super-elastic behavior results from the well-organized structure of 3D graphene and is rarely achieved in poorly organized structures. Via MD simulations, Qin *et al.* [182] and Jung *et al.* [23] showed that graphene gyroids can achieve excellent specific strength and unusually low thermal conductivity. These examples demonstrate the potential in exploiting geometrical/topological design of 3D interconnected graphene assemblies to further tune/optimize their properties and achieve novel applications. From the topological design point of view, there are plenty of challenges and opportunities beyond the monolayer topological design. For example, what is the best 3D surface of the interconnected graphene to optimize given properties, such as modulus–density ratio and thermal transfer? The necessary topological defects for forming these continuum surfaces will in turn alter the effective mechanical and physical properties and call for reevaluation of the targeted functions. Therefore, another important question is how we can develop a fully coupled design methodology to integrate the global surface topology optimization and local topology dependent properties. Finally, the design schemes could also include the consideration of how to effectively synthesize such graphene structures with desired distribution of topological defects/3D geometry.

6. How can one efficiently and economically fabricate topologically designed graphene at different scales? In spite of the rapid developments in synthesizing various graphene structures with topological defects, as reviewed in Section 1.4, there are still a few key challenges in the application of the existing methods to achieve topological design at different length scales. For large-scale CVD methods, an important question is how to keep the precise substrate confinement of graphene during growth, considering that the high temperature in CVD growth [5] could decrease the adhesion between graphene and substrate and even destroy the designed geometrical features of the substrate [69]. To resolve problems like these, one needs a fundamental understanding of how curvature and topological defects affect the energetics of the nucleation, growth, and coarsening of 2D materials. For methods focusing on the atomic or molecular levels, efficiency remains a trade-off against precision in large-scale applications. Additionally, the scalability of the methods needs to be improved. These challenges themselves in turn present important research topics and could inspire further studies both in theoretical and experimental aspects.
7. Can one generalize the topological design concept to other 2D materials? Beyond graphene, topological defects have also been widely observed in other 2D materials, ranging from h-BN and black phosphorus to TMDC monolayers like

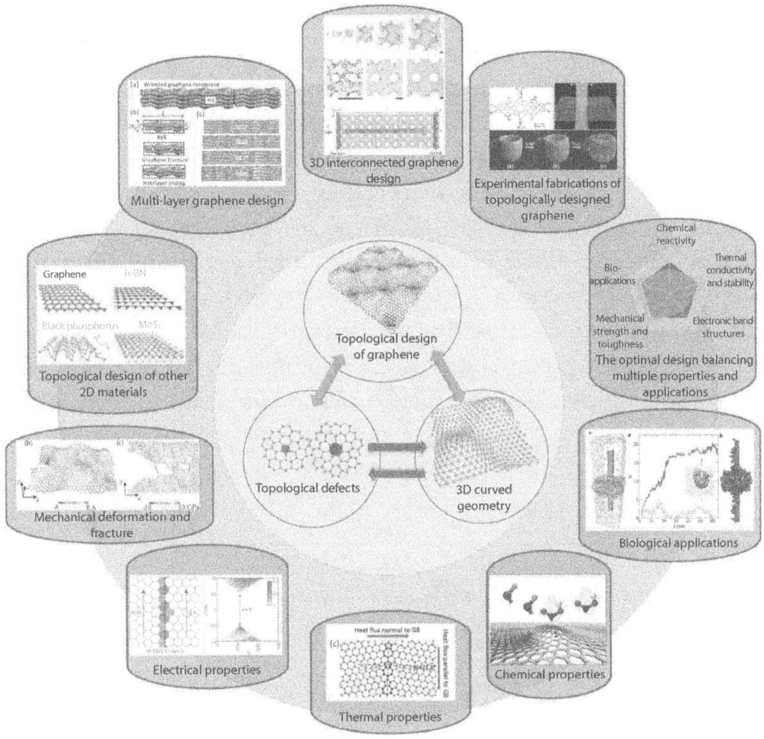

Figure 1.22 Outlooks for topological design of graphene: morphology and curvature; strength and toughness; multifunction; nonlinear multiphysical coupling; multiscale fabrication; interconnected and multilayer graphene; extension to other 2D materials [12, 15, 23, 26, 28, 54, 69, 199, 232, 233].

MoS_2, WS_2, and so on. Similar to graphene, the presence of topological defects in these 2D materials [72, 227–230] not only induces out-of-plane deformation but also modifies physical properties compared with pristine samples. Moreover, the chemical bonding, dislocation core structures, mobility, and interaction of topological defects can vary greatly for different 2D materials. For example, due to polarized bonding, GBs in h-BN could carry net charges [90]. Because of the puckered lattice structure, dislocation cores in black phosphorus are predicted to have not only 5-7 but also 4-8, 5-8-7, and 5-8-8-7 rings [229]. Recently, an experimental study observed high-frequency dislocation emission in MoS_2 during crack propagation, suggesting possible plastic deformation activities in TMDCs [89]. These predictions and observations are broadening the horizon of topological design of 2D materials beyond graphene and suggest new opportunities and challenges for further study. For instance, (1) how could one modify the methods like PFC to solve the inverse design problem in topological design for 2D materials with multiple elements (e.g., h-BN) [231] or non-flat lattice (black phosphorus and TMDCs)? (2) What are the similarities and differences in the topological defect-mediated coupling between the 3D curved geometry and multiphysical properties in different 2D materials? (3) What novel functionalities could be achieved through the topological design of a heterogeneous system combining different 2D materials?

In summary, the concept of topological design of graphene discussed in this chapter focuses on the topological defect-mediated coupling between 3D geometry, residual stress distribution, deformation, strength, fracture, and phonon and electron behaviors in graphene, as indicated by Figure 1.22. In principle, a virtual simulation platform could be developed to optimize targeted properties of graphene by directly manipulating its topological structure. The rapid developments of experimental and fabrication techniques are paving ways to achieve topologically design of graphene from atom to device levels. This is inherently a highly interdisciplinary effort across multiple research communities including mechanics, nanotechnology, material science, physics, chemistry, and even biology. There exist plenty of research opportunities in this emerging field, where fundamental studies could be expected to greatly expand our knowledge about 2D materials as well as open new avenues of application in novel devices based on 2D materials in the foreseeable future.

References

1. Geim, A.K. and Novoselov, K.S., The rise of graphene. *Nat. Mater.*, 6, 183, 2007.
2. Liu, Y. and Yakobson, B.I., Cones, pringles, and grain boundary landscapes in graphene topology. *Nano Lett.*, 10, 2178, 2010.
3. Yazyev, O.V. and Louie, S.G., Topological defects in graphene: Dislocations and grain boundaries. *Phys. Rev. B*, 81, 195420, 2010.
4. Huang, P.Y. *et al.*, Grains and grain boundaries in single-layer graphene atomic patchwork quilts. *Nature,* 469, 389, 2011.
5. Li, X. *et al.*, Large-area synthesis of high-quality and uniform graphene films on copper foils. *Science*, 324, 1312, 2009.
6. Kim, K.S. *et al.*, Large-scale pattern growth of graphene films for stretchable transparent electrodes. *Nature*, 457, 706, 2009.
7. Kim, K., Lee, Z., Regan, W., Kisielowski, C., Crommie, M., Zettl, A., Grain boundary mapping in polycrystalline graphene. *ACS Nano*, 5, 2142, 2011.
8. Grantab, R., Shenoy, V.B., Ruoff, R.S., Anomalous strength characteristics of tilt grain boundaries in graphene. *Science*, 330, 946, 2010.
9. Wei, Y., Wu, J., Yin, H., Shi, X., Yang, R., Dresselhaus, M., The nature of strength enhancement and weakening by pentagon–heptagon defects in graphene. *Nat. Mater.*, 11, 759, 2012.
10. Rasool, H.I., Ophus, C., Klug, W.S., Zettl, A., Gimzewski, J.K., Measurement of the intrinsic strength of crystalline and polycrystalline graphene. *Nat. Commun.*, 4, 2811, 2013.
11. Shekhawat, A. and Ritchie, R.O., Toughness and strength of nanocrystalline graphene. *Nat. Commun.*, 7, 10546, 2016.
12. Warner, J.H., Fan, Y., Robertson, A.W., He, K., Yoon, E., Lee, G.D., Rippling graphene at the nanoscale through dislocation addition. *Nano Lett.*, 13, 4937, 2013.
13. Zhang, T., Li, X., Gao, H., Defects controlled wrinkling and topological design in graphene. *J. Mech. Phys. Solids*, 67, 2, 2014.
14. Jung, G., Qin, Z., Buehler, M.J., Molecular mechanics of polycrystalline graphene with enhanced fracture toughness. *Extreme Mech. Lett.*, 2, 52, 2015.
15. Zhang, T., Li, X., Gao, H., Designing graphene structures with controlled distributions of topological defects: A case study of toughness enhancement in graphene ruga. *Extreme Mech. Lett.*, 1, 3, 2014.
16. Serov, A.Y., Ong, Z.-Y., Pop, E., Effect of grain boundaries on thermal transport in graphene. *Appl. Phy. Lett.*, 102, 033104, 2013.

17. Ito, Y. *et al.*, Correlation between chemical dopants and topological defects in catalytically active nanoporous graphene. *Adv. Mater.*, 28, 10644, 2016.
18. Jauregui, L.A., Cao, H., Wu, W., Yu, Q., Chen, Y.P., Electronic properties of grains and grain boundaries in graphene grown by chemical vapor deposition. *Solid State Commun.*, 151, 1100, 2011.
19. Yazyev, O.V. and Louie, S.G., Electronic transport in polycrystalline graphene. *Nat. Mater.*, 9, 806, 2010.
20. Zhang, H., Lee, G., Gong, C., Colombo, L., Cho, K., Grain boundary effect on electrical transport properties of graphene. *J. Phys. Chem. C*, 118, 2338, 2014.
21. Tsen, A.W. *et al.*, Tailoring electrical transport across grain boundaries in polycrystalline graphene. *Science*, 336, 1143, 2012.
22. Ma, T. *et al.*, Tailoring the thermal and electrical transport properties of graphene films by grain size engineering. *Nat. Commun.*, 8, 14486, 2017.
23. Jung, G.S., Yeo, J., Tian, Z., Qin, Z., Buehler, M.J., Unusually low and density-insensitive thermal conductivity of three-dimensional gyroid graphene. *Nanoscale*, 9, 13477, 2017.
24. Ito, Y. *et al.*, High-quality three-dimensional nanoporous graphene. *Angew. Chem. Int. Ed.*, 53, 4822, 2014.
25. Zhu, Y. *et al.*, A seamless three-dimensional carbon nanotube graphene hybrid material. *Nat. Commun.*, 3, 1225, 2012.
26. Cheung, K.Y., Chan, C.K., Liu, Z., Miao, Q., A twisted nanographene consisting of 96 carbon atoms. *Angew. Chem.*, 129, 9131, 2017.
27. Lehtinen, O., Vats, N., Algara-Siller, G., Knyrim, P., Kaiser, U., Implantation and atomic-scale investigation of self-interstitials in graphene. *Nano Lett.*, 15, 235, 2014.
28. Kim, P., Graphene: Across the border. *Nat. Mater.*, 9, 792, 2010.
29. Warner, J.H., Margine, E.R., Mukai, M., Robertson, A.W., Giustino, F., Kirkland, A.I., Dislocation-driven deformations in graphene. *Science*, 337, 209, 2012.
30. Bhimanapati, G.R. *et al.*, Recent advances in two-dimensional materials beyond graphene. *ACS Nano*, 9, 11509, 2015.
31. Akinwande, D. *et al.*, A review on mechanics and mechanical properties of 2D materials—Graphene and beyond. *Extreme Mech. Lett.*, 13, 42, 2017.
32. Song, L. *et al.*, Large scale growth and characterization of atomic hexagonal boron nitride layers. *Nano Lett.*, 10, 3209, 2010.
33. Watanabe, K., Taniguchi, T., Kanda, H., Direct-bandgap properties and evidence for ultraviolet lasing of hexagonal boron nitride single crystal. *Nat. Mater.*, 3, 404, 2004.
34. Wang, Q.H., Kalantar-Zadeh, K., Kis, A., Coleman, J.N., Strano, M.S., Electronics and optoelectronics of two-dimensional transition metal dichalcogenides. *Nat. Nanotechnol.*, 7, 699, 2012.
35. Chhowalla, M., Shin, H.S., Eda, G., Li, L.-J., Loh, K.P., Zhang, H., The chemistry of two-dimensional layered transition metal dichalcogenide nanosheets. *Nat. Chem.*, 5, 263, 2013.
36. Russo, P., Hu, A., Compagnini, G., Synthesis, properties and potential applications of porous graphene: A review. *Nano-Micro Lett.*, 5, 260, 2013.
37. Jiang, L. and Fan, Z., Design of advanced porous graphene materials: From graphene nanomesh to 3D architectures. *Nanoscale*, 6, 1922, 2014.
38. Blees, M.K. *et al.*, Graphene kirigami. *Nature*, 524, 204, 2015.
39. Qi, Z., Campbell, D.K., Park, H.S., Atomistic simulations of tension-induced large deformation and stretchability in graphene kirigami. *Phys. Rev. B*, 90, 245437, 2014.
40. Xu, L., Shyu, T.C., Kotov, N.A., Origami and kirigami nanocomposites. *ACS Nano*, 11, 7587, 2017.
41. Shyu, T.C., Damasceno, P.F., Dodd, P.M., Lamoureux, A., Xu, L., Shlian, M., Shtein, M., Glotzer, S.C., Kotov, N.A., A kirigami approach to engineering elasticity in nanocomposites through patterned defects. *Nat. Mater.*, 14, 785, 2015.

42. Zhu, S. and Li, T., Hydrogenation-assisted graphene origami and its application in programmable molecular mass uptake, storage, and release. *ACS Nano*, 8, 2864, 2014.
43. Shenoy, V.B. and Gracias, D.H., Self-folding thin-film materials: From nanopolyhedra to graphene origami. *Mrs Bull.*, 37, 847, 2012.
44. Kumar, K., Van Swygenhoven, H., Suresh, S., Mechanical behavior of nanocrystalline metals and alloys1. *Acta Mater.*, 51, 5743, 2003.
45. Lu, K., Lu, L., Suresh, S., Strengthening materials by engineering coherent internal boundaries at the nanoscale. *Science*, 324, 349, 2009.
46. Tian, Y. *et al.*, Ultrahard nanotwinned cubic boron nitride. *Nature*, 493, 385, 2013.
47. Huang, Q. *et al.*, Nanotwinned diamond with unprecedented hardness and stability. *Nature*, 510, 250, 2014.
48. Pan, Q., Zhou, H., Lu, Q., Gao, H., Lu, L., History-independent cyclic response of nanotwinned metals. *Nature*, 551, 214, 2017.
49. Lehtinen, O., Kurasch, S., Krasheninnikov, A., Kaiser, U., Atomic scale study of the life cycle of a dislocation in graphene from birth to annihilation. *Nat. Commun.*, 4, 2098, 2013.
50. Lee, C., Wei, X., Kysar, J.W., Hone, J., Measurement of the elastic properties and intrinsic strength of monolayer graphene. *Science*, 321, 385, 2008.
51. Lu, Q., Arroyo, M., Huang, R., Elastic bending modulus of monolayer graphene. *J. Phys. D*, 42, 102002, 2009.
52. Wei, Y., Wang, B., Wu, J., Yang, R., Dunn, M.L., Bending rigidity and Gaussian bending stiffness of single-layered graphene. *Nano Lett.*, 13, 26, 2012.
53. Song, Z., Artyukhov, V.I., Wu, J., Yakobson, B.I., Xu, Z., Defect-detriment to graphene strength is concealed by local probe: The topological and geometrical effects. *ACS Nano*, 9, 401, 2014.
54. Qin, H., Sun, Y., Liu, J.Z., Liu, Y., Mechanical properties of wrinkled graphene generated by topological defects. *Carbon*, 108, 204, 2016.
55. Cortijo, A. and Vozmediano, M.A., Effects of topological defects and local curvature on the electronic properties of planar graphene. *Nucl. Phys. B*, 763, 293, 2007.
56. Kvashnin, A.G., Sorokin, P.B., Yakobson, B.I., Flexoelectricity in carbon nanostructures: Nanotubes, fullerenes, and nanocones. *J. Phys. Chem. Lett.*, 6, 2740, 2015.
57. Kalinin, S.V. and Meunier, V., Electronic flexoelectricity in low-dimensional systems. *Phys. Rev. B*, 77, 033403, 2008.
58. Ni, Z., Wang, H., Kasim, J., Fan, H., Yu, T., Wu, Y., Feng, Y., Shen, Z., Graphene thickness determination using reflection and contrast spectroscopy. *Nano Lett.*, 7, 2758, 2007.
59. Meyer, J.C., Geim, A.K., Katsnelson, M.I., Novoselov, K.S., Booth, T.J., Roth, S., The structure of suspended graphene sheets. *Nature*, 446, 60, 2007.
60. Fasolino, A., Los, J., Katsnelson, M.I., Intrinsic ripples in graphene. *Nat. Mater.*, 6, 858, 2007.
61. Wan, D., Nelson, D.R., Bowick, M.J., Thermal stiffening of clamped elastic ribbons. *Phys. Rev. B*, 96, 014106, 2017.
62. Yllanes, D., Bhabesh, S.S., Nelson, D.R., Bowick, M.J., Thermal crumpling of perforated two-dimensional sheets. *Nat. Commun.*, 8, 1381, 2017.
63. Ahmadpoor, F., Wang, P., Huang, R., Sharma, P., Thermal fluctuations and effective bending stiffness of elastic thin sheets and graphene: A nonlinear analysis. *J. Mech. Phys. Solids*, 107, 294, 2017.
64. Košmrlj, A. and Nelson, D.R., Thermal excitations of warped membranes. *Phys. Rev. E*, 89, 022126, 2014.
65. Zhong, Y.L., Tian, Z., Simon, G.P., Li, D., Scalable production of graphene via wet chemistry: Progress and challenges. *Mater. Today*, 18, 73, 2015.
66. Chen, Y., Gong, X.L., Gai, J.G., Progress and challenges in transfer of large-area graphene films. *Adv. Sci.*, 3, 2016.

67. Pham, V.P., Jang, H.-S., Whang, D., Choi, J.-Y., Direct growth of graphene on rigid and flexible substrates: Progress, applications, and challenges. *Chem. Soc. Rev.*, 46, 6276, 2017.
68. Ackerman, M., Kumar, P., Neek-Amal, M., Thibado, P., Peeters, F., Singh, S., Anomalous dynamical behavior of freestanding graphene membranes. *Phys. Rev. Lett.*, 117, 126801, 2016.
69. Wilson, P.M., Mbah, G.N., Smith, T.G., Schmidt, D., Lai, R.Y., Hofmann, T., Sinitskii, A., Three-dimensional periodic graphene nanostructures. *J. Mater. Chem. C*, 2, 1879, 2014.
70. Červenka, J., Katsnelson, M., Flipse, C., Room-temperature ferromagnetism in graphite driven by two-dimensional networks of point defects. *Nat. Phys.*, 5, 840, 2009.
71. Lahiri, J., Lin, Y., Bozkurt, P., Oleynik, I.I., Batzill, M., An extended defect in graphene as a metallic wire. *Nat. Nanotechnol.*, 5, 326, 2010.
72. Gibb, A.L., Alem, N., Chen, J.-H., Erickson, K.J., Ciston, J., Gautam, A., Linck, M., Zettl, A., Atomic resolution imaging of grain boundary defects in monolayer chemical vapor deposition-grown hexagonal boron nitride. *J. Am. Chem. Soc.*, 135, 6758, 2013.
73. Van Der Zande, A.M. *et al.*, Grains and grain boundaries in highly crystalline monolayer molybdenum disulphide. *Nat. Mater.*, 12, 554, 2013.
74. Zhang, Z., Yang, Y., Xu, F., Wang, L., Yakobson, B.I., Unraveling the sinuous grain boundaries in graphene. *Adv. Funct. Mat.*, 25, 367, 2015.
75. Wu, J. and Wei, Y., Grain misorientation and grain-boundary rotation dependent mechanical properties in polycrystalline graphene. *J. Mech. Phys. Solids*, 61, 1421, 2013.
76. Khosravian, N., Samani, M.K., Loh, G.C., Chen, G.C.K., Baillargeat, D., Tay, B.K., Effects of a grain boundary loop on the thermal conductivity of graphene: A molecular dynamics study. *Comput. Mater. Sci.*, 79, 132, 2013.
77. Sha, Z.D., Quek, S.S., Pei, Q.X., Liu, Z.S., Wang, T.J., Shenoy, V.B., Zhang, Y.W., Inverse pseudo hall-petch relation in polycrystalline graphene. *Sci. Rep.*, 4, 5991, 2014.
78. Song, Z., Artyukhov, V.I., Yakobson, B.I., Xu, Z., Pseudo hall–petch strength reduction in polycrystalline graphene. *Nano Lett.*, 13, 1829, 2013.
79. Murdock, A.T. *et al.*, Controlling the orientation, edge geometry, and thickness of chemical vapor deposition graphene. *ACS Nano*, 7, 1351, 2013.
80. Song, X. *et al.*, Seed-assisted growth of single-crystalline patterned graphene domains on hexagonal boron nitride by chemical vapor deposition. *Nano Lett.*, 16, 6109, 2016.
81. Geng, D. *et al.*, Uniform hexagonal graphene flakes and films grown on liquid copper surface. *Proc. Natl. Acad. Sci.*, 109, 7992, 2012.
82. Shu, H., Chen, X., Tao, X., Ding, F., Edge structural stability and kinetics of graphene chemical vapor deposition growth. *ACS Nano*, 6, 3243, 2012.
83. Elder, K., Katakowski, M., Haataja, M., Grant, M., Modeling elasticity in crystal growth. *Phys. Rev. Lett.*, 88, 245701, 2002.
84. Li, J., Ni, B., Zhang, T., Gao, H., Phase field crystal modeling of grain boundary structures and growth in polycrystalline graphene. *J. Mech. Phys. Solids*, 2017.
85. Guo, W. *et al.*, Governing rule for dynamic formation of grain boundaries in grown graphene. *ACS Nano*, 9, 5792, 2015.
86. Kotakoski, J. and Meyer, J.C., Mechanical properties of polycrystalline graphene based on a realistic atomistic model. *Phys. Rev. B*, 85, 195447, 2012.
87. Sha, Z., Wan, Q., Pei, Q., Quek, S., Liu, Z., Zhang, Y., Shenoy, V., On the failure load and mechanism of polycrystalline graphene by nanoindentation. *Sci. Rep.*, 4, 7437, 2014.
88. Yang, Z., Huang, Y., Ma, F., Sun, Y., Xu, K., Chu, P.K., Size-dependent deformation behavior of nanocrystalline graphene sheets. *Mater. Sci. Eng. B*, 198, 95, 2015.
89. Ly, T.H., Zhao, J., Cichocka, M.O., Li, L.-J., Lee, Y.H., Dynamical observations on the crack tip zone and stress corrosion of two-dimensional MoS 2. *Nat. Commun.*, 8, 14116, 2017.

90. Liu, Y., Zou, X., Yakobson, B.I., Dislocations and grain boundaries in two-dimensional boron nitride. *ACS Nano*, 6, 7053, 2012.
91. Zou, X., Liu, Y., Yakobson, B.I., Predicting dislocations and grain boundaries in two-dimensional metal-disulfides from the first principles. *Nano Lett.*, 13, 253, 2012.
92. Choi, J.S. *et al.*, Friction anisotropy–driven domain imaging on exfoliated monolayer graphene. *Science*, 333, 607, 2011.
93. Wu, Q. *et al.*, Selective surface functionalization at regions of high local curvature in graphene. *Chem. Commun.*, 49, 677, 2012.
94. Levy, N., Burke, S., Meaker, K., Panlasigui, M., Zettl, A., Guinea, F., Neto, A.C., Crommie, M., Strain-induced pseudo–magnetic fields greater than 300 tesla in graphene nanobubbles. *Science*, 329, 544, 2010.
95. Pereira, V.M., Neto, A.C., Liang, H., Mahadevan, L., Geometry, mechanics, and electronics of singular structures and wrinkles in graphene. *Phys. Rev. Lett.*, 105, 156603, 2010.
96. Klimov, N.N., Jung, S., Zhu, S., Li, T., Wright, C.A., Solares, S.D., Newell, D.B., Zhitenev, N.B., Stroscio, J.A., Electromechanical properties of graphene drumheads. *Science*, 336, 1557, 2012.
97. Zhang, P. *et al.*, Fracture toughness of graphene. *Nat. Commun.*, 5, 3782, 2014.
98. Chen, S. and Chrzan, D., Continuum theory of dislocations and buckling in graphene. *Phys. Rev. B*, 84, 214103, 2011.
99. Shibuta, Y. and Maruyama, S., Molecular dynamics simulation of formation process of single-walled carbon nanotubes by CCVD method. *Chem. Phys. Lett.*, 382, 381, 2003.
100. Piper, N., Fu, Y., Tao, J., Yang, X., To, A., Vibration promotes heat welding of single-walled carbon nanotubes. *Chem. Phys. Lett.*, 502, 231, 2011.
101. Biyikli, E., Liu, J., Yang, X., To, A.C., A fast method for generating atomistic models of arbitrarily-shaped carbon graphitic nanostructures. *RSC Adv.*, 3, 1359, 2013.
102. Chuang, C., Fan, Y.-C., Jin, B.-Y., Generalized classification scheme of toroidal and helical carbon nanotubes. *J. Chem. Inf. Mod.*, 49, 361, 2009.
103. Varshney, V., Unnikrishnan, V., Lee, J., Roy, A.K., Developing nanotube junctions with arbitrary specifications. *Nanoscale*, 10, 403, 2018.
104. Petersen, T.C., Snook, I.K., Yarovsky, I., McCulloch, D.G., Monte Carlo based modeling of carbon nanostructured surfaces. *Phys. Rev. B*, 72, 125417, 2005.
105. Hexemer, A., Vitelli, V., Kramer, E.J., Fredrickson, G.H., Monte Carlo study of crystalline order and defects on weakly curved surfaces. *Phys. Rev. E*, 76, 051604, 2007.
106. Mitchell, L. and Head, A., The buckling of a dislocated plate. *J. Mech. Phys. Solids*, 9, 131, 1961.
107. Seung, H. and Nelson, D.R., Defects in flexible membranes with crystalline order. *Phys. Rev. A*, 38, 1005, 1988.
108. Zubov, L.M., *Nonlinear theory of dislocations and disclinations in elastic bodies*, vol. 47, Springer Science & Business Media, 1997.
109. Zubov, L., *Von Kármán equations for an elastic plate with dislocations and disclinations. Doklady Physics*, p. 67, Springer, 2007.
110. Zubov, L., The linear theory of dislocations and disclinations in elastic shells. *J. Appl. Math. Mech.*, 74, 663, 2010.
111. Dervaux, J. and Amar, M.B., Morphogenesis of growing soft tissues. *Phys. Rev. Lett.*, 101, 068101, 2008.
112. Liang, H. and Mahadevan, L., The shape of a long leaf. *Proc. Natl. Acad. Sci.*, 106, 22049, 2009.
113. Stuart, S.J., Tutein, A.B., Harrison, J.A., A reactive potential for hydrocarbons with intermolecular interactions. *J. Chem. Phys.*, 112, 6472, 2000.
114. Diab, M., Zhang, T., Zhao, R., Gao, H., Kim, K.-S., Ruga mechanics of creasing: From instantaneous to setback creases. *Proc. R. Soc. A*, p. 20120753, The Royal Society, 2013.

115. Emmerich, H., Löwen, H., Wittkowski, R., Gruhn, T., Tóth, G.I., Tegze, G., Gránásy, L., Phase-field-crystal models for condensed matter dynamics on atomic length and diffusive time scales: An overview. *Adv. Phys.*, 61, 665, 2012.
116. Galenko, P., Gomez, H., Kropotin, N., Elder, K., Unconditionally stable method and numerical solution of the hyperbolic phase-field crystal equation. *Phys. Rev. E*, 88, 013310, 2013.
117. Qin, H., Sun, Y., Liu, J.Z., Li, M., Liu, Y., Negative Poisson's ratio in rippled graphene. *Nanoscale*, 9, 4135, 2017.
118. Anderson, T.L., *Fracture mechanics: Fundamentals and applications*, CRC Press, 2017.
119. Eda, G., Fanchini, G., Chhowalla, M., Large-area ultrathin films of reduced graphene oxide as a transparent and flexible electronic material. *Nat. Nanotechnol.*, 3, 270, 2008.
120. Reina, A., Jia, X., Ho, J., Nezich, D., Son, H., Bulovic, V., Dresselhaus, M.S., Kong, J., Large area, few-layer graphene films on arbitrary substrates by chemical vapor deposition. *Nano Lett.*, 9, 30, 2008.
121. Suk, J.W., Kitt, A., Magnuson, C.W., Hao, Y., Ahmed, S., An, J., Swan, A.K., Goldberg, B.B., Ruoff, R.S., Transfer of CVD-grown monolayer graphene onto arbitrary substrates. *ACS Nano*, 5, 6916, 2011.
122. Kang, J., Shin, D., Bae, S., Hong, B.H., Graphene transfer: Key for applications. *Nanoscale*, 4, 5527, 2012.
123. Gao, L., Ni, G.-X., Liu, Y., Liu, B., Neto, A.H.C., Loh, K.P., Face-to-face transfer of wafer-scale graphene films. *Nature*, 505, 190, 2014.
124. Hofmann, M., Hsieh, Y.-P., Hsu, A.L., Kong, J., Scalable, flexible and high resolution patterning of CVD graphene. *Nanoscale*, 6, 289, 2014.
125. Celebi, K., Buchheim, J., Wyss, R.M., Droudian, A., Gasser, P., Shorubalko, I., Kye, J.-I., Lee, C., Park, H.G., Ultimate permeation across atomically thin porous graphene. *Science*, 344, 289, 2014.
126. Ramasse, Q.M., Zan, R., Bangert, U., Boukhvalov, D.W., Son, Y.-W., Novoselov, K.S., Direct experimental evidence of metal-mediated etching of suspended graphene. *ACS Nano*, 6, 4063, 2012.
127. Hwangbo, Y. *et al.*, Fracture characteristics of monolayer CVD-graphene. *Sci. Rep.*, 4, 4439, 2014.
128. Khare, R., Mielke, S.L., Paci, J.T., Zhang, S., Ballarini, R., Schatz, G.C., Belytschko, T., Coupled quantum mechanical/molecular mechanical modeling of the fracture of defective carbon nanotubes and graphene sheets. *Phys. Rev. B*, 75, 075412, 2007.
129. Terdalkar, S.S., Huang, S., Yuan, H., Rencis, J.J., Zhu, T., Zhang, S., Nanoscale fracture in graphene. *Chem. Phys. Lett.*, 494, 218, 2010.
130. Cohen-Tanugi, D. and Grossman, J.C., Mechanical strength of nanoporous graphene as a desalination membrane. *Nano Lett.*, 14, 6171, 2014.
131. Zhang, T., Li, X., Kadkhodaei, S., Gao, H., Flaw insensitive fracture in nanocrystalline graphene. *Nano Lett.*, 12, 4605, 2012.
132. Sen, D., Novoselov, K.S., Reis, P.M., Buehler, M.J., Tearing graphene sheets from adhesive substrates produces tapered nanoribbons. *Small*, 6, 1108, 2010.
133. Moura, M.J. and Marder, M., Tearing of free-standing graphene. *Phys. Rev. E*, 88, 032405, 2013.
134. Huang, X., Yang, H., van Duin, A.C., Hsia, K.J., Zhang, S., Chemomechanics control of tearing paths in graphene. *Phys. Rev. B*, 85, 195453, 2012.
135. Zhao, S. and Xue, J., Mechanical properties of hybrid graphene and hexagonal boron nitride sheets as revealed by molecular dynamic simulations. *J. Phys. D*, 46, 135303, 2013.
136. Lu, L., Chen, X., Huang, X., Lu, K., Revealing the maximum strength in nanotwinned copper. *Science*, 323, 607, 2009.
137. Li, X., Wei, Y., Lu, L., Lu, K., Gao, H., Dislocation nucleation governed softening and maximum strength in nano-twinned metals. *Nature*, 464, 877, 2010.
138. Moon, W.-J., Ito, T., Uchimura, S., Saka, H., Toughening of ceramics by dislocation sub-boundaries. *Mater. Sci. Eng. A*, 387, 837, 2004.

139. Saka, H., Toughening of a brittle material by means of dislocation subboundaries. *Philos. Mag. Lett.*, 80, 461, 2000.
140. Meng, F., Chen, C., Song, J., Dislocation shielding of a nanocrack in graphene: Atomistic simulations and continuum modeling. *J. Phys. Chem. Lett.*, 6, 4038, 2015.
141. Verma, A. and Parashar, A., The effect of STW defects on the mechanical properties and fracture toughness of pristine and hydrogenated graphene. *Phys. Chem. Chem. Phys.*, 19, 16023, 2017.
142. Rajasekaran, G. and Parashar, A., Molecular dynamics study on the mechanical response and failure behaviour of graphene: Performance enhancement via 5–7–7–5 defects. *RSC Adv.*, 6, 26361, 2016.
143. Wang, S., Yang, B., Yuan, J., Si, Y., Chen, H., Large-scale molecular simulations on the mechanical response and failure behavior of a defective graphene: Cases of 5 -8 -5 defects. *Sci. Rep.*, 5, 14957, 2015.
144. Rajasekaran, G. and Parashar, A., Enhancement of fracture toughness of graphene via crack bridging with stone-thrower-wales defects. *Diamond Relat. Mater.*, 74, 90, 2017.
145. Han, J., Sohn, D., Woo, W., Kim, D.-K., Molecular dynamics study of fracture toughness and trans-intergranular transition in bi-crystalline graphene. *Comput. Mater. Sci.*, 129, 323, 2017.
146. Wang, Y. and Liu, Z., The fracture toughness of graphene during the tearing process. *Model. Simul. Mater. Sci. Eng.*, 24, 085002, 2016.
147. Mitchell, N.P., Koning, V., Vitelli, V., Irvine, W.T., Fracture in sheets draped on curved surfaces. *Nat. Mater.*, 16, 89, 2017.
148. Zhang, T. and Gao, H., Toughening graphene with topological defects: A perspective. *J. Appl. Mech.*, 82, 051001, 2015.
149. Jorio, A., Dresselhaus, G., Dresselhaus, M.S., *Carbon Nanotubes: Advanced Topics in the Synthesis, Structure, Properties and Applications*, Springer Berlin Heidelberg, 2007.
150. Jariwala, D., Sangwan, V.K., Lauhon, L.J., Marks, T.J., Hersam, M.C., Carbon nanomaterials for electronics, optoelectronics, photovoltaics, and sensing. *Chem. Soc. Rev.*, 42, 2824, 2013.
151. Wang, J., Carbon-nanotube based electrochemical biosensors: A review. *Electroanalysis*, 17, 7, 2005.
152. Odom, T.W., Huang, J.-L., Kim, P., Lieber, C.M., Structure and electronic properties of carbon nanotubes. *J. Phys. Chem. B*, 104, 2794, 2000.
153. Sanchez-Valencia, J.R., Dienel, T., Gröning, O., Shorubalko, I., Mueller, A., Jansen, M., Amsharov, K., Ruffieux, P., Fasel, R., Controlled synthesis of single-chirality carbon nanotubes. *Nature*, 512, 61, 2014.
154. Mueller, A. and Amsharov, K.Y., Synthesis of robust precursors for the controlled fabrication of (6, 6),(8, 8),(10, 10), and (12, 12) armchair single-walled carbon nanotubes. *Eur. J. Organic Chem.*, 2015, 3053, 2015.
155. Liu, B. *et al.*, Nearly exclusive growth of small diameter semiconducting single-wall carbon nanotubes from organic chemistry synthetic end-cap molecules. *Nano Lett.*, 15, 586, 2014.
156. Abdurakhmanova, N., Mueller, A., Stepanow, S., Rauschenbach, S., Jansen, M., Kern, K., Amsharov, K.Y., Bottom up fabrication of (9, 0) zigzag and (6, 6) armchair carbon nanotube end-caps on the Rh (1 1 1) surface. *Carbon*, 84, 444, 2015.
157. Tarascon, J.-M. and Armand, M., *Materials For Sustainable Energy: A Collection of Peer-Reviewed Research and Review Articles from Nature Publishing Group*, p. 171, World Scientific, 2011.
158. Datta, D., Li, J., Koratkar, N., Shenoy, V.B., Enhanced lithiation in defective graphene. *Carbon*, 80, 305, 2014.
159. Pang, Z., Shi, X., Wei, Y., Fang, D., Grain boundary and curvature enhanced lithium adsorption on carbon. *Carbon*, 107, 557, 2016.
160. Lin, D., Liu, Y., Cui, Y., Reviving the lithium metal anode for high-energy batteries. *Nat. Nanotechnol.*, 12, 194, 2017.
161. Beaulieu, L., Eberman, K., Turner, R., Krause, L., Dahn, J., Colossal reversible volume changes in lithium alloys. *Electrochem. Solid-State Lett.*, 4, A137, 2001.

162. Obrovac, M. and Christensen, L., Structural changes in silicon anodes during lithium insertion/extraction. *Electrochem. Solid-State Lett.*, 7, A93, 2004.
163. Obrovac, M., Christensen, L., Le, D.B., Dahn, J.R., Alloy design for lithium-ion battery anodes. *J. Electrochem. Soc.*, 154, A849, 2007.
164. Chan, C.K., Peng, H., Liu, G., McIlwrath, K., Zhang, X.F., Huggins, R.A., Cui, Y., High-performance lithium battery anodes using silicon nanowires. *Nat. Nanotechnol.*, 3, 31, 2008.
165. Cui, L.-F., Ruffo, R., Chan, C.K., Peng, H., Cui, Y., Crystalline-amorphous core–shell silicon nanowires for high capacity and high current battery electrodes. *Nano Lett.*, 9, 491, 2008.
166. Zhou, S., Liu, X., Wang, D., Si/TiSi2 heteronanostructures as high-capacity anode material for Li ion batteries. *Nano Lett.*, 10, 860, 2010.
167. Park, M.-H., Kim, M.G., Joo, J., Kim, K., Kim, J., Ahn, S., Cui, Y., Cho, J., Silicon nanotube battery anodes. *Nano Lett.*, 9, 3844, 2009.
168. Xiao, J., Xu, W., Wang, D., Choi, D., Wang, W., Li, X., Graff, G.L., Liu, J., Zhang, J.-G., Stabilization of silicon anode for Li-ion batteries. *J. Electrochem. Soc.*, 157, A1047, 2010.
169. Magasinski, A., Dixon, P., Hertzberg, B., Kvit, A., Ayala, J., Yushin, G., High-performance lithium-ion anodes using a hierarchical bottom-up approach. *Nat. Mater.*, 9, 353, 2010.
170. Liu, N., Wu, H., McDowell, M.T., Yao, Y., Wang, C., Cui, Y., A yolk-shell design for stabilized and scalable Li-ion battery alloy anodes. *Nano Lett.*, 12, 3315, 2012.
171. Li, Y., Yan, K., Lee, H.-W., Lu, Z., Liu, N., Cui, Y., Growth of conformal graphene cages on micrometre-sized silicon particles as stable battery anodes. *Nat. Energy*, 1, 15029, 2016.
172. Liu, C., Yu, Z., Neff, D., Zhamu, A., Jang, B.Z., Graphene-based supercapacitor with an ultra-high energy density. *Nano Lett.*, 10, 4863, 2010.
173. Zhou, L.-J., Hou, Z., Wu, L.-M., First-principles study of lithium adsorption and diffusion on graphene with point defects. *J. Phys. Chem. C*, 116, 21780, 2012.
174. Pope, M.A. and Aksay, I.A., Four-fold increase in the intrinsic capacitance of graphene through functionalization and lattice disorder. *J. Phys. Chem. C*, 119, 20369, 2015.
175. Bagri, A., Kim, S.-P., Ruoff, R.S., Shenoy, V.B., Thermal transport across twin grain boundaries in polycrystalline graphene from nonequilibrium molecular dynamics simulations. *Nano Lett.*, 11, 3917, 2011.
176. Fthenakis, Z.G., Zhu, Z., Tománek, D., Effect of structural defects on the thermal conductivity of graphene: From point to line defects to haeckelites. *Phys. Rev. B*, 89, 125421, 2014.
177. Ong, M.T. and Reed, E.J., Engineered piezoelectricity in graphene. *ACS Nano*, 6, 1387, 2012.
178. Yasaei, P. *et al.*, Bimodal phonon scattering in graphene grain boundaries. *Nano Lett.*, 15, 4532, 2015.
179. Yuki, A., Yuki, I., Kuniharu, T., Seiji, A., Takayuki, A., Enhancement of graphene thermoelectric performance through defect engineering. *2D Mater.*, 4, 025019, 2017.
180. Ahmadpoor, F. and Sharma, P., Flexoelectricity in two-dimensional crystalline and biological membranes. *Nanoscale*, 7, 16555, 2015.
181. Pacheco Sanjuan, A.A., Mehboudi, M., Harriss, E.O., Terrones, H., Barraza-Lopez, S., Quantitative chemistry and the discrete geometry of conformal atom-thin crystals. *ACS Nano*, 8, 1136, 2014.
182. Qin, Z., Jung, G.S., Kang, M.J., Buehler, M.J., The mechanics and design of a lightweight three-dimensional graphene assembly. *Sci. Adv.*, 3, e1601536, 2017.
183. Shao, Y., Wang, J., Wu, H., Liu, J., Aksay, I.A., Lin, Y., Graphene based electrochemical sensors and biosensors: A review. *Electroanalysis*, 22, 1027, 2010.
184. Liu, J., Liu, Z., Barrow, C.J., Yang, W., Molecularly engineered graphene surfaces for sensing applications: A review. *Anal. Chim. Acta*, 859, 1, 2015.
185. Goenka, S., Sant, V., Sant, S., Graphene-based nanomaterials for drug delivery and tissue engineering. *J. Controll. Rel.*, 173, 75, 2014.

186. Yang, K., Feng, L., Liu, Z., The advancing uses of nano-graphene in drug delivery. *Expert Opin. Drug Delivery*, 12, 601, 2015.
187. Hong, G., Diao, S., Antaris, A.L., Dai, H., Carbon nanomaterials for biological imaging and nanomedicinal therapy. *Chem. Rev.*, 115, 10816, 2015.
188. Yoo, J.M., Kang, J.H., Hong, B.H., Graphene-based nanomaterials for versatile imaging studies. *Chem. Soc. Rev.*, 44, 4835, 2015.
189. Zuo, G., Zhou, X., Huang, Q., Fang, H., Zhou, R., Adsorption of villin headpiece onto graphene, carbon nanotube, and C60: Effect of contacting surface curvatures on binding affinity. *J. Phys. Chem. C*, 115, 23323, 2011.
190. Jana, A.K., Tiwari, M.K., Vanka, K., Sengupta, N., Unraveling origins of the heterogeneous curvature dependence of polypeptide interactions with carbon nanostructures. *Phys. Chem. Chem. Phys.*, 18, 5910, 2016.
191. Gu, Z., Yang, Z., Chong, Y., Ge, C., Weber, J.K., Bell, D.R., Zhou, R., Surface curvature relation to protein adsorption for carbon-based nanomaterials. *Sci. Rep.*, 5, 10886, 2015.
192. Gao, H., Kong, Y., Cui, D., Ozkan, C.S., Spontaneous insertion of DNA oligonucleotides into carbon nanotubes. *Nano Lett.*, 3, 471, 2003.
193. Umadevi, D. and Sastry, G.N., Quantum mechanical study of physisorption of nucleobases on carbon materials: Graphene versus carbon nanotubes. *J. Phys. Chem. Lett.*, 2, 1572, 2011.
194. Tu, Y. *et al.*, Destructive extraction of phospholipids from Escherichia coli membranes by graphene nanosheets. *Nat. Nanotechnol.*, 8, nnano, 2013.125, 2013.
195. Liu, S., Hu, M., Zeng, T.H., Wu, R., Jiang, R., Wei, J., Wang, L., Kong, J., Chen, Y., Lateral dimension-dependent antibacterial activity of graphene oxide sheets. *Langmuir*, 28, 12364, 2012.
196. Li, Y., Yuan, H., von dem Bussche, A., Creighton, M., Hurt, R.H., Kane, A.B., Gao, H., Graphene microsheets enter cells through spontaneous membrane penetration at edge asperities and corner sites. *Proc. Natl. Acad. Sci.*, 110, 12295, 2013.
197. Hu, W., Peng, C., Luo, W., Lv, M., Li, X., Li, D., Huang, Q., Fan, C., Graphene-based antibacterial paper. *ACS Nano*, 4, 4317, 2010.
198. Liu, S., Zeng, T.H., Hofmann, M., Burcombe, E., Wei, J., Jiang, R., Kong, J., Chen, Y., Antibacterial activity of graphite, graphite oxide, graphene oxide, and reduced graphene oxide: Membrane and oxidative stress. *ACS Nano*, 5, 6971, 2011.
199. Luan, B., Huynh, T., Zhou, R., Complete wetting of graphene by biological lipids. *Nanoscale*, 8, 5750, 2016.
200. Yu, Q. *et al.*, Control and characterization of individual grains and grain boundaries in graphene grown by chemical vapour deposition. *Nat. Mater.*, 10, 443, 2011.
201. Chen, Z., Ren, W., Gao, L., Liu, B., Pei, S., Cheng, H.-M., Three-dimensional flexible and conductive interconnected graphene networks grown by chemical vapour deposition. *Nat. Mater.*, 10, 424, 2011.
202. Min, B.H., Kim, D.W., Kim, K.H., Choi, H.O., Jang, S.W., Jung, H.-T., Bulk scale growth of CVD graphene on Ni nanowire foams for a highly dense and elastic 3D conducting electrode. *Carbon*, 80, 446, 2014.
203. Yang, Z., Yan, C., Liu, J., Chabi, S., Xia, Y., Zhu, Y., Designing 3D graphene networks via a 3D-printed Ni template. *RSC Adv.*, 5, 29397, 2015.
204. Kim, K. *et al.*, Lanthanum-catalysed synthesis of microporous 3D graphene-like carbons in a zeolite template. *Nature*, 535, 131, 2016.
205. Yu, H., Gupta, N., Hu, Z., Wang, K., Srijanto, B.R., Xiao, K., Geohegan, D.B., Yakobson, B.I., Tilt grain boundary topology induced by substrate topography. *ACS Nano*, 11, 8612, 2017.
206. Lusk, M.T. and Carr, L.D., Nanoengineering defect structures on graphene. *Phys. Rev. Lett.*, 100, 175503, 2008.

207. Robertson, A.W., Allen, C.S., Wu, Y.A., He, K., Olivier, J., Neethling, J., Kirkland, A.I., Warner, J.H., Spatial control of defect creation in graphene at the nanoscale. *Nat. Commun.*, 3, 1144, 2012.
208. Kawasumi, K., Zhang, Q., Segawa, Y., Scott, L.T., Itami, K., A grossly warped nanographene and the consequences of multiple odd-membered-ring defects. *Nat. Chem.*, 5, 739, 2013.
209. Cheung, K.Y., Xu, X., Miao, Q., Aromatic saddles containing two heptagons. *J. Am. Chem. Soc.*, 137, 3910, 2015.
210. Barthelat, F. and Espinosa, H., An experimental investigation of deformation and fracture of nacre-mother of pearl. *Exp. Mech.*, 47, 311, 2007.
211. Gao, H., Ji, B., Jäger, I.L., Arzt, E., Fratzl, P., Materials become insensitive to flaws at nanoscale: Lessons from nature. *Proc. Natl. Acad. Sci.*, 100, 5597, 2003.
212. Ji, B. and Gao, H., Mechanical properties of nanostructure of biological materials. *J. Mech. Phys. Solids*, 52, 1963, 2004.
213. Wang, Y., Chen, M., Zhou, F., Ma, E., High tensile ductility in a nanostructured metal. *Nature*, 419, 912, 2002.
214. Ma, E., Wang, Y., Lu, Q., Sui, M., Lu, L., Lu, K., Strain hardening and large tensile elongation in ultrahigh-strength nano-twinned copper. *Appl. Phy. Lett.*, 85, 4932, 2004.
215. Kim, N.D. *et al.*, Growth and transfer of seamless 3D graphene-nanotube hybrids. *Nano Lett.*, 16, 1287, 2016.
216. Li, B. and Arroyo, M., Towards understanding the geometry effects on fracture in thin elastic shells. arXiv preprint arXiv:1703.09371, 2017.
217. Folias, E., On the theory of fracture of curved sheets. *Eng. Fract. Mech.*, 2, 151, 1970.
218. Folias, E., On the effect of initial curvature on cracked flat sheets. *Int. J. Fract. Mech.*, 5, 327, 1969.
219. Langer, T., Baringhaus, J., Pfnür, H., Schumacher, H., Tegenkamp, C., Plasmon damping below the Landau regime: The role of defects in epitaxial graphene. *New J. Phy.*, 12, 033017, 2010.
220. Smirnova, D., Mousavi, S.H., Wang, Z., Kivshar, Y.S., Khanikaev, A.B., Trapping and guiding surface plasmons in curved graphene landscapes. *ACS Photonics*, 3, 875, 2016.
221. Bissett, M.A., Tsuji, M., Ago, H., Strain engineering the properties of graphene and other two-dimensional crystals. *Phys. Chem. Chem. Phys.*, 16, 11124, 2014.
222. Guinea, F., Strain engineering in graphene. *Solid State Commun.*, 152, 1437, 2012.
223. Schneider, G.F., Kowalczyk, S.W., Calado, V.E., Pandraud, G., Zandbergen, H.W., Vandersypen, L.M., Dekker, C., DNA translocation through graphene nanopores. *Nano Lett.*, 10, 3163, 2010.
224. Kothari, M., Cha, M.-H., Kim, K.-S., Critical behavior of curvature localization in graphene. *APS Meeting Abstracts*, 2017.
225. Kolmogorov, A.N. and Crespi, V.H., Registry-dependent interlayer potential for graphitic systems. *Phys. Rev. B*, 71, 235415, 2005.
226. Qiu, L., Liu, J.Z., Chang, S.L., Wu, Y., Li, D., Biomimetic superelastic graphene-based cellular monoliths. *Nat. Commun.*, 3, 1241, 2012.
227. Liu, Y., Xu, F., Zhang, Z., Penev, E.S., Yakobson, B.I., Two-dimensional mono-elemental semiconductor with electronically inactive defects: The case of phosphorus. *Nano Lett.*, 14, 6782, 2014.
228. Azizi, A. *et al.*, Dislocation motion and grain boundary migration in two-dimensional tungsten disulphide. *Nat. Commun.*, 5, 4867, 2014.
229. Guo, Y., Zhou, S., Zhang, J., Bai, Y., Zhao, J., Atomic structures and electronic properties of phosphorene grain boundaries. *2D Mater.*, 3, 025008, 2016.
230. Lin, Y.-C. *et al.*, Three-fold rotational defects in two-dimensional transition metal dichalcogenides. *Nat. Commun.*, 6, 6736, 2015.

231. Taha, D., Mkhonta, S., Elder, K., Huang, Z.-F., Grain boundary structures and collective dynamics of inversion domains in binary two-dimensional materials. *Phys. Rev. Lett.*, 118, 255501, 2017.
232. Mu, X., Song, Z., Wang, Y., Xu, Z., Go, D.B., Luo, T., Thermal transport in oxidized polycrystalline graphene. *Carbon*, 108, 318, 2016.
233. Lee, J.Y., Shin, J.-H., Lee, G.-H., Lee, C.-H., Two-dimensional semiconductor optoelectronics based on van der Waals heterostructures. *Nanomaterials*, 6, 193, 2016.

2

Graphene at the Metal–Oxide Interface: A New Approach to Modify the Chemistry of Supported Metals

Wen Luo[1] and Spyridon Zafeiratos[2]*

[1]*Institute of Chemical Sciences and Engineering (ISIC), École Polytechnique Fédérale de Lausanne (EPFL) Valais/Wallis, Sion, Switzerland*
[2]*Institut de Chimie et Procédés pour l'Energie, l'Environnement et la Santé (ICPEES), ECPM, UMR, CNRS-Université de Strasbourg, Strasbourg, France*

Abstract

As a single layer of sp^2 carbon atoms, graphene displays a unique and visualized property—the two-dimensional (2D) structure. Owning to this, graphene plays, or will play, a distinct role in some applications that is difficult to achieve by other carbon allotropes (graphite, carbon nanotube, diamond and fullerene, etc.). In this chapter, we will discuss a novel application of graphene when it is inserted as an interlayer between a metal and its oxide support, in order to tune the metal-support interaction (MSI). In the past, hybrid materials consisting of metal and/or oxide–graphene composites have been tested for a wide range of catalytic reactions, including thermocatalytic, photocatalytic, and electrocatalytic reactions. However, in most of the cases, graphene sheets were simply decorated with oxide nanoparticles. In this chapter, we will focus on nanocomposite materials made of graphene-coated oxides, which in a later step can act as a support for metal particles. The dramatic influence of graphene interlayer on the MSI effect will be initially demonstrated on a planar model system consisting of cobalt/graphene/oxide (ZnO or SiO_2). The fabrication of graphene-covered oxide, and the consequent deposition of cobalt, will be discussed first. Consequently, we will present a comparative study of the MSI effect between Co-ZnO and Co/graphene/ZnO under ultrahigh vacuum conditions. This will reveal that graphene plays a significant role on the Co morphology and Co-ZnO interactions. Next, we will show studies of the redox properties of Co under O_2/H_2 gas atmospheres and the effect of graphene substrate on these properties. In a further step, we will expose recent findings on a more complex bimetallic PtCo/graphene/ZnO system, including both planar and powder sample configurations. The purpose is to discuss the methods to synthesize metal/graphene/ZnO in powder morphology and demonstrate the potential applications of such kind of materials in catalytic reactions. Finally, the chapter will conclude with a brief summarization and perspectives.

Keywords: Metal-support interaction, model catalyst, x-ray photoelectron spectroscopy, oxidation, reduction, cobalt, oxide, ultrahigh vacuum

**Corresponding author*: spiros.zafeiratos@unistra.fr

Tobias Stauber (ed.) Handbook of Graphene: Volume 2, (45–72) © 2019 Scrivener Publishing LLC

2.1 Introduction

Metal-support interaction (MSI), and particularly the metal–oxide interaction, has attracted great and constant diligence and attention in the past 40 years since it plays a key role in many technological fields including energy transformation, solid-state chemistry, and heterogeneous catalysis [1, 2]. The MSI is primarily influenced by the nature of the particular metals and oxides. For example, low work function metals (i.e., Na, K, and Al) may be oxidized by reducible oxide supports (i.e., TiO_2 and ZnO), while metals with high work functions (i.e., Pt, Pd, and Au) may be encapsulated underneath these oxides [1]. Therefore, the influence of MSI on the material performance can be positive or negative depending on the particular material combination. In the case of heterogeneous catalysis, a moderate interaction between a metal and its oxide support can help to disperse and stabilize the metal active centers and enhance the activity and stability of the catalyst. Conversely, very strong MSI may lead to the deactivation of the catalyst due to the oxidation or encapsulation of the metal active centers by the support [3, 4]. Consequently, understanding and then tailoring the MSI phenomena is of essential importance for the production of advanced catalytic materials and processes.

One of the most-studied metal–oxide heterogeneous catalytic systems is the cobalt–oxides catalyst, which has been employed in several industrially important reactions, such as Fischer–Tropsch synthesis (FTS) and hydrogenolysis reactions [5, 6]. The interaction between Co and oxides has been intensively investigated and it is widely acknowledged that the support affects the Co particle size, reducibility, and stability. It seems that there is an optimal interaction strength between the two that can benefit the application. For instance, Co supported on SiO_2 is a catalyst widely used in FTS, the relatively weak Co-SiO_2 interaction helps the reduction of Co oxides precursors but leads to the formation of large Co particles [5]. In contrast, the stronger interaction between Co and Al_2O_3 improves the dispersion of Co nanoparticles, while on the downside decreases the reducibility [5]. With reducible oxide supports, such as ZnO, the oxidation of Co by ZnO and the formation of $Co_xZn_{1-x}O$ have been reported to cause the deactivation of the catalyst [7, 8].

Some strategies have been reported to modify the Co–support interactions and therefore tune the catalytic performance. One is the addition of noble metal promoters (Ru, Rh, Pt, etc.), which enhances not only the reducibility of Co but also the catalytic selectivity, activity, and stability due to the synergistic effects [5, 9]. A disadvantage of this strategy is the high price of the noble metals and the complexity in the preparation of the catalyst. Another strategy is to use the non-oxide supports, in particular, carbon-based materials. The use of carbon materials as catalytic supports provides high specific surface area for the homogenous dispersion of Co nanoparticles while the weak interaction between Co and carbon prevents phenomena such as encapsulation and oxidation of Co [5, 10, 11]. However, the poor mechanical stability of all carbon-supported catalysts limits their shape formulation for industrial applications. Therefore, an efficient strategy would be the utilization of carbon-coated oxides, which takes the advantage of the surface properties of the carbon materials (i.e., high surface area and high chemical stability) and the advantages of the bulk properties of oxides (good mechanical stability, adjustable acidity, etc.) [10].

Among all the carbon allotropes, graphene has the unique and visualized property—the two-dimensional (2D) structure, which makes it a perfect candidate for coating since the atomic thickness of the graphene layer can minimize the impact on the macroscopic properties of the catalyst. Furthermore, as a single layer of pure sp^2 carbon, graphene can be a good platform for establishing highly controlled model system to carry out the fundamental investigations. Although composited materials consisting of graphene and metal/oxide have also been applied in a wide range of applications, such as catalysis, batteries, supercapacitors, etc. [12], little is known about the graphene's effects on the MSIs. This is partly due to the high complexity of the composite materials, and therefore it is difficult to separate and identify with precision the role of graphene. Development of model graphene/metal/oxide systems might be a useful approach in order to gain detailed insights about these interactions. The authors have been systematically investigating this issue with spectroscopic and microscopic techniques based on rational designed Co/graphene/oxide model sample systems [13–18] and this chapter will mainly review the work on this topic. The experimental procedure of creating the model Co/graphene/oxide samples will be briefly presented in Section 2.2. The results obtained at ultra-high vacuum conditions (UHV) (Section 2.3.1) and at various temperatures provide direct indications that graphene significantly modifies the Co-ZnO interactions as well as the Co morphology. Section 2.3.2 deals with the important role of graphene on the Co redox properties under O_2/H_2 gas atmospheres. It is shown that graphene modifies the interactions between Co with both a reducible oxide (ZnO) and a relatively inert oxide (SiO_2). A further step is to adapt these findings on a more complex bimetallic PtCo/graphene/ZnO system, including both planar (Section 2.4.1) and powder sample (Section 2.4.2) configurations, which enables the transfer of fundamental understandings established from well-defined model systems to a complex working catalytic configuration. These model samples also allow to systematically investigate the evolution of the graphene properties under various conditions, which will be discussed in Section 2.5. Finally, the chapter concludes with a brief summarization and future perspectives.

2.2 Fabrication of Model Metal/Graphene/Oxide Samples

The typical procedure to prepare planar model metal/graphene/oxide samples is shown in Figure 2.1. This widely used method is usually referred to as wet transfer process (Figure 2.1a) [19]. Initially, graphene (single layer or bilayer) with high quality and large area is prepared by chemical vapor deposition (CVD) method on a metal support in most of the cases on a copper foil [20]. Then, graphene is coated with a thin poly-methyl methacrylate (PMMA) film (~0.5 μm), and the Cu foil is etched away in an appropriate solution (such as $FeCl_3$ solution). The remaining PMMA/graphene film is rinsed with clean water and can be transferred in a water medium onto an arbitrary oxide substrate, in this case onto the clean ZnO(0001) surface. Consequently, the majority of the PMMA layer is removed by dissolution in acetone (in some cases also, other alcohols have been used). The remaining trace amount of PMMA residues can be evaporated by several hours of annealing treatment at around 300 °C, preferentially under vacuum to avoid oxidation of graphene [13]. With this method, the continuity and the high quality of the CVD-grown graphene can be preserved after transfer. The resulting graphene/ZnO(0001) sample (abbreviated as G/ZnO) serves here as a model graphene-coated oxide substrate on which metals can be deposited in order to study the influence of graphene in the MSI. Therefore, through a final physical

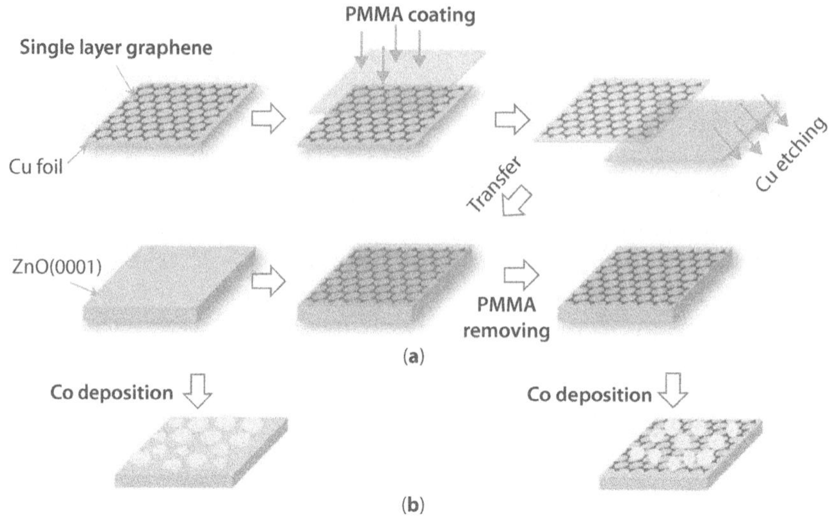

Figure 2.1 Schematic illustration of (a) transfer of graphene onto ZnO(0001) by wet transfer process and (b) deposition of Co by PVD method.

vapor deposition (PVD) method (Figure 2.1b), a thin layer of cobalt (~0.5 nm) is deposited at room temperature onto the G/ZnO surface under UHV conditions to get the Co/G/ZnO model sample. For comparison, it is useful that an equivalent Co/ZnO sample (without graphene interlayer) is prepared using the same cobalt deposition method.

The successful transfer of the single graphene layer can be confirmed by various techniques (Figure 2.2). The long-range homogeneity of the graphene layer is verified by scanning electron microscopy (SEM). As shown in Figure 2.2a, the graphene layer is flat and uniform with some wrinkles and bilayer islands. The detailed morphology and the thickness of the graphene layer can be analyzed by atomic force microscopy (AFM). The image of Figure 2.2b confirms the flat and continuous morphology (root-mean-squared (RMS) roughness ~0.5 nm) as well as the existence of the wrinkles in graphene. The height of the single-layer graphene obtained from the line profile in Figure 2.2b is ~1 nm, which is higher than the theoretical value of single-layer graphene (0.35 nm) due to the ambient measurement environment and the settings of the AFM instruments [13]. The purity of the graphene layer can be verified by X-ray photoelectron spectroscopy (XPS). A single and asymmetric C 1s peak (Figure 2.2c) confirms the graphitic properties of the graphene layer and the absence of a significant population of oxygenated carbon species [21].

Raman spectroscopy has been considered as the most efficient technique to study the quality and the layer numbers of the graphene [22]. In Figure 2.2d, two high intensity peaks can be observed, which are the so-called G (1580 cm^{-1}) and 2D (cm^{-1}) band, rising from the in-plane vibrational (E_{2g}) mode and the two phonon intervalley double resonance scattering of graphene, respectively. The narrow symmetric 2D band and the high 2D/G band intensity ratio indicate that the transferred graphene is in a single-layer form [23]. In addition, no obvious peak can be found at 1350 cm^{-1} where a D band is normally observed for the defective graphene. This confirms that the graphene layer transferred on ZnO is of high quality.

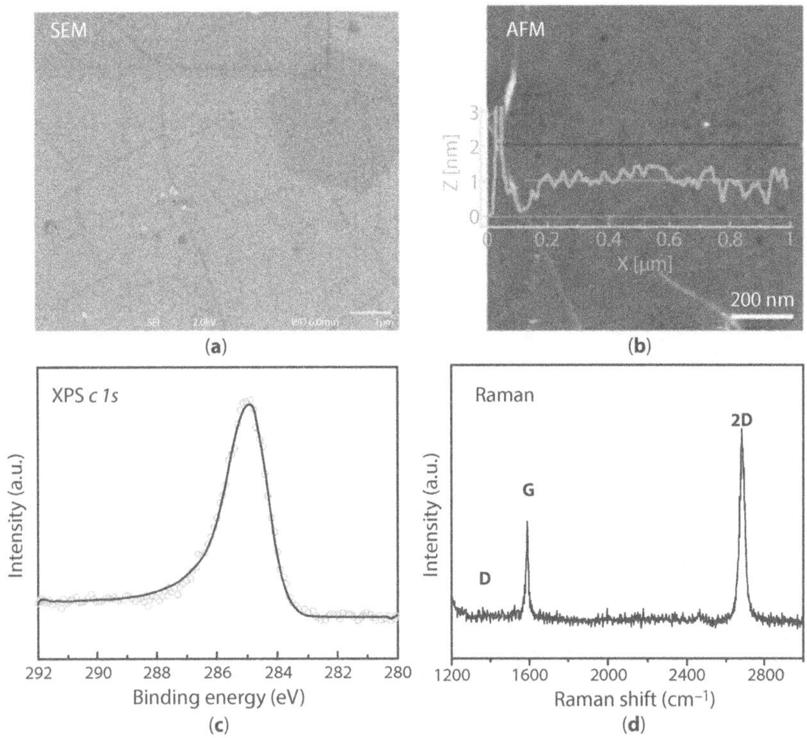

Figure 2.2 Characterization of G/ZnO sample prepared by the wet transfer process. (a) SEM image. (b) Top-view tapping-mode AFM topographic image. (c) C 1s XP spectrum. (d) Raman spectrum.

2.3 Effect of Graphene on the Cobalt–Oxide Support Interaction

2.3.1 Studies under UHV Conditions

Under atmospheric conditions, the effect of graphene on the metal-support interaction, in particular to the metal oxidation state, is difficult to differentiate from the effect of the gas phase environment. To eliminate the influence of the surrounding gas environment, the well-defined Co/ZnO and Co/G/ZnO model samples are investigated under UHV conditions. Moreover, under such conditions, surface-sensitive techniques, such as XPS, can be applied to study the surface properties *in situ*. As evident by the XPS spectra in Figure 2.3a, the Co oxidation state after annealing at various temperatures changes, inducing some distinct evolution trends in the Co $2p_{3/2}$ peak. For Co supported on G/ZnO, the metallic state can maintain till 350 °C shown as a sharp Co $2p_{3/2}$ peak at 778.3 eV. Contrarily, on a bare ZnO surface, Co started to be oxidized from 200 °C indicated by the additional shoulder at 780.6 eV, which is typical for CoO. With the temperature increasing to 350 °C, the oxide peak fully replaced the metallic peak demonstrating the complete oxidation of Co to CoO. Therefore, it is obvious that the single graphene layer plays an important role on the Co and ZnO interactions. Under UHV conditions, the oxidation of Co by ZnO has been attributed to the solid reaction between Co and ZnO: Co + ZnO → CoO + Zn, where the generated Zn would evaporate under the vacuum condition [24, 25]. While with a graphene layer in-between of Co and ZnO, the solid reaction is blocked; thus, the oxidation of Co is avoided.

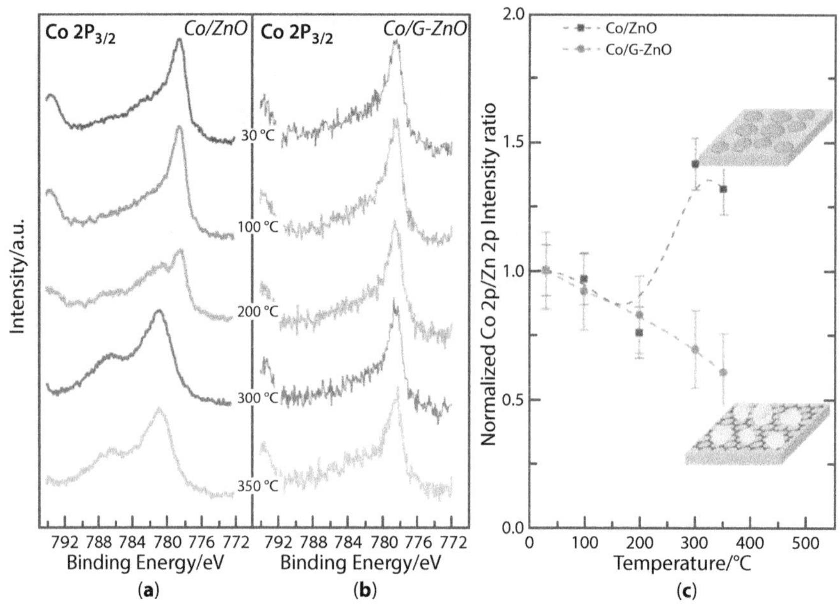

Figure 2.3 XPS Co $2p_{3/2}$ core level spectra of (a) Co/ZnO and (b) Co/G/ZnO upon annealing at different temperatures under UHV conditions. (c) The corresponding XPS Co 2p/Zn 2p intensity ratios at different temperature for Co/ZnO and Co/G/ZnO samples (the intensity ratios are normalized to the initial ratio at 30 °C), the graphical insets represent the proposed evolution of the Co morphology during the temperature rising in the two cases. (Reproduced from [13].)

In addition to the effects on the Co oxidation state, graphene also modifies the Co morphology, which can be deduced from the XPS intensity ratio between Co 2p and Zn 2p (I_{Co}/I_{Zn}) signals [13]. As shown in Figure 2.3c, a decreasing trend of I_{Co}/I_{Zn} can be observed for Co/G/ZnO sample till annealing to 350 °C, indicating Co particles are agglomerated on graphene to decrease their surface energy. On the contrary for Co/ZnO, I_{Co}/I_{Zn} shows a decreasing trend before 200 °C while it increases to a higher value after annealing at 300 °C. In combination with the oxidation state of Co shown in Figure 2.3a, it is evident that the oxidation of Co to CoO at higher temperature induced the dispersion of CoO layer over the ZnO surface. A visualized evidence to confirm the morphology difference can be obtained from the AFM results. Figure 2.4 shows the tapping-mode AFM topographic images of fresh and annealed Co/ZnO and Co/G/ZnO samples. The surface of the as-deposited Co on ZnO is relatively flat and continuous, which becomes more flat (RMS roughness of 0.35 nm) after 350 °C annealing. For Co/G/ZnO, a significantly different morphology of fresh Co is observed (Figure 2.4c), where Co forms highly dispersed homogenous particles on G/ZnO surface. As anticipated, after annealing, the small Co nanoparticles are agglomerated to larger ones.

The results from UHV-based investigation on the model samples demonstrate that inserting a single layer graphene in between metal (Co) and reducible oxide (ZnO) can effectively prevent diffusion phenomena at the metal/oxide interface upon thermal treatment and eventually suppresses metal oxidation. The weaker interaction between Co and graphene leads the as-deposited Co to form 3D nanoparticles, which are also prone to agglomerate. These results can inspire the new strategies to control the MSI in applications where the strong MSI (SMSI) should be avoided.

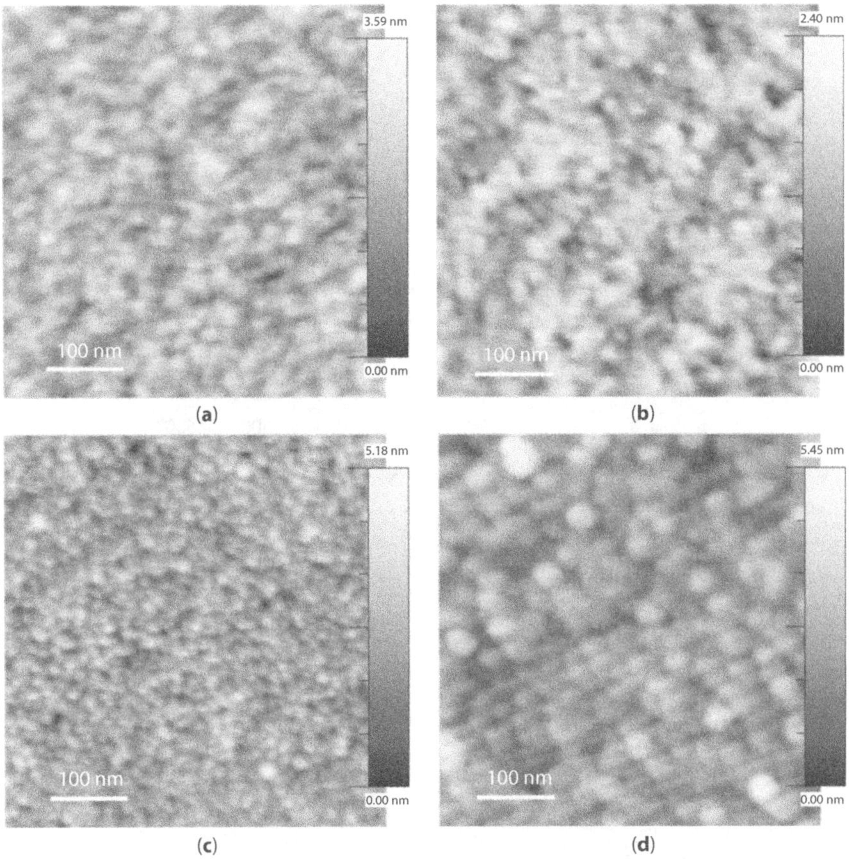

Figure 2.4 Top view of tapping-mode AFM images of (a) fresh Co/ZnO, (b) Co/ZnO after annealing at 350 °C, (c) fresh Co/G- ZnO, and (d) Co/G-ZnO after annealing at 350 °C. (Reproduced from [13].)

2.3.2 Physicochemical Studies under Gas Atmospheres

Although the UHV conditions serve as an ideal environment for fundamental studies using model samples, in real applications, generally reactive gases are involved, such as air, O_2, and H_2, etc. Therefore, to explore the influence of graphene on the MSI phenomena under more realistic conditions, the oxidation/reduction studies should carried out also in O_2/H_2 environments [14, 16]. Moreover, to extend the investigation also to inert oxide supports, SiO_2 films, representing as non-reducible oxide, were used to prepare Co/SiO_2 and $Co/G/SiO_2$ model samples with identical procedures as the one shown in Figure 2.1.

Initially, the oxidation/reduction experiments are conducted at low pressure gas exposure, namely, 5×10^{-7} mbar O_2/H_2. These conditions allow to probe in detail the initial stages of oxidation and reduction, while they can be also applied in conventional UHV systems allowing *in situ* spectroscopic techniques. The evolution of cobalt oxidation state was deduced from the XPS spectra recorded after 0.5 h annealing of the sample in O_2/H_2 atmospheres. The Co $2p_{3/2}$ XPS peaks were deconvoluted with standard peaks of Co (0 valence) and CoO (+2 valence) and the mean value was calculated as the Co average valence (Figure 2.5a). Comparison of the four curves in Figure 2.5a shows a clear difference in the Co

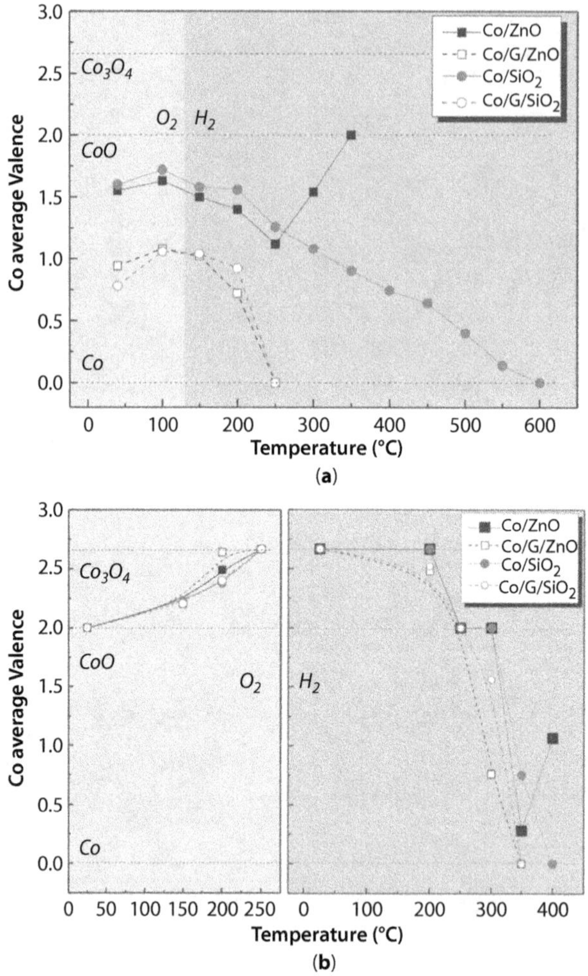

Figure 2.5 Evolution of the Co average valence state (Co^{x+}) as a function of the annealing temperature under (a) 5×10^{-7} mbar and (b) 7 mbar O_2/H_2 environment. (Reproduced from [14, 16].)

valence evolution between bare oxides (ZnO and SiO_2) and graphene-coated oxides (G/ZnO and G/SiO_2) under both O_2 and H_2 atmospheres. After 100 °C oxidation in O_2 environment, 85% of Co on bare oxides is oxidized to CoO while this value is only 50% in case of Co on graphene-coated oxides. In H_2, CoO reduction trends are the same for Co/G/ZnO and Co/G/SiO_2 where CoO is fully reduced to Co after 250 °C reduction. However, there are distinct differences between bare ZnO and SiO_2 supports. On SiO_2, CoO is gradually reduced at a much higher temperature (600 °C). While on ZnO, reduction of CoO starts at 150 °C and continues up to 250 °C, whereas above this temperature Co starts to be reoxidized until fully oxidized at 350 °C. As we have shown in Figure 2.3 that Co can be oxidized by ZnO through the solid reaction (Co + ZnO → CoO + Zn), the V-shape CoO reduction trend on ZnO could be explained by the low temperature reduction of CoO through CoO + H_2 → Co + H_2O and the high temperature oxidation of Co through the solid reaction between Co and ZnO. In the case of the "inert" SiO_2 support, the interfacial oxidation does not occur and CoO is reduced by gas-phase H_2 or by simple thermal decomposition.

To bridge the pressure gap between UHV and ambient pressure conditions, near-ambient pressure oxidation/reduction studies (7 mbar O_2/H_2 conditions) were also conducted on these four samples [14]. The experiments were carried out in a high-pressure chamber that is attached to the UHV in order to avoid the exposure of the sample in air during transfer to the spectrometer. As shown in Figure 2.5b, compared with 5×10^{-7} mbar O_2 condition, 7 mbar O_2 is much more efficient in Co oxidization, since all the samples are fully oxidized to CoO already at room temperature and then gradually oxidized to Co_3O_4 after the temperature rises to 250 °C. Moreover, no effect from the graphene layer on the oxidation behavior of Co is observed. The reduction of the obtained Co_3O_4 at various annealing temperatures in 7 mbar H_2 was studied next. As shown in Figure 2.5b, the evolution of Co average valence indicates that in all cases, Co_3O_4 follows a two-step reduction process: $Co_3O_4 \rightarrow CoO \rightarrow Co$, in agreement with previous reports [9, 26]. However, the reduction temperature of Co oxides is significantly influenced by the substrate. Similar as the low-pressure oxidation results, Co valence state on graphene-coated oxides are systematically lower (more reduced) as compared to the bare oxide substrates. For Co/SiO_2 sample, a higher reduction temperature (400 °C) is needed to get fully reduced Co. While for Co/ZnO, the reoxidation of Co at high annealing temperature is shown again as in the case of UHV and low-pressure condition. This is remarkable considering that 7 mbar H_2 is a highly reducing gas atmosphere. A plausible explanation is that at high temperature Co forms a mixed $Co_xZn_{1-x}O$ spinel phase with the ZnO support ($Co + ZnO \rightarrow Co_xZn_{1-x}O$), which according to reports is resistant to reduction under H_2 [7]. This was confirmed by near edge X-ray absorption fine structure (NEXAFS) spectra, which showed clear characteristics of tetrahedrally coordinated Co^{2+} ions as expected for $Co_xZn_{1-x}O$ instead of octahedrally (CoO) coordinated Co^{2+} ions formed in CoO oxide [14].

The redox studies shown that the single-layer graphene also modifies the oxidation and reduction properties of oxide-supported cobalt. Under a low pressure O_2 condition, the insertion of a graphene interlayer limits the oxidation of Co, indepentent if Co is supported on a reducible or inert oxide. Under H_2 condition, at both low and near ambient pressure, the reduction of Co oxides is facilitated by the graphene layer. In particular, for the ZnO support, the solid reaction between Co and ZnO is inhibited.

In order to understand the graphene's role on the redox behavior of Co, the morphology of the supported Co should be addressed since the particle size of Co also influences the redox properties. Figure 2.6 summarizes in a schematic illustration the morphology (from AFM results) and the oxidation state evolution (from XPS results) of Co during the redox treatment [14, 16]. For the fresh deposited samples, no obvious difference of Co morphology is observed between the two oxides (bare and graphene coated). However, as illustrated in Section 2.3.1, graphene could adjust the Co morphology due to the low Co-C interactions and therefore, Co forms nanoparticles on the graphene-covered oxides and relatively flat layer structure on bare oxides. Consequently, the morphology difference has an effect on the oxidation properties of Co. According to the mechanism of Co oxidation by gas-phase oxygen, the oxidation starts from the surface of Co through the dissociatively adsorbed oxygen to form CoO [27, 28]. Since Co is in the form of nanoparticles on G/ZnO, G/SiO_2, after a layer of CoO is formed, dissociation and deeper diffusion of oxygen into the core of the nanoparticles is restricted by the slow kinetics at low temperature. Thus, with low pressure O_2 (5×10^{-7} mbar), oxidation is limited to the outermost layers of Co for the Co/G/ZnO and Co/G/SiO_2. For the flat structured-Co on ZnO and SiO_2, gas phase oxidation is more

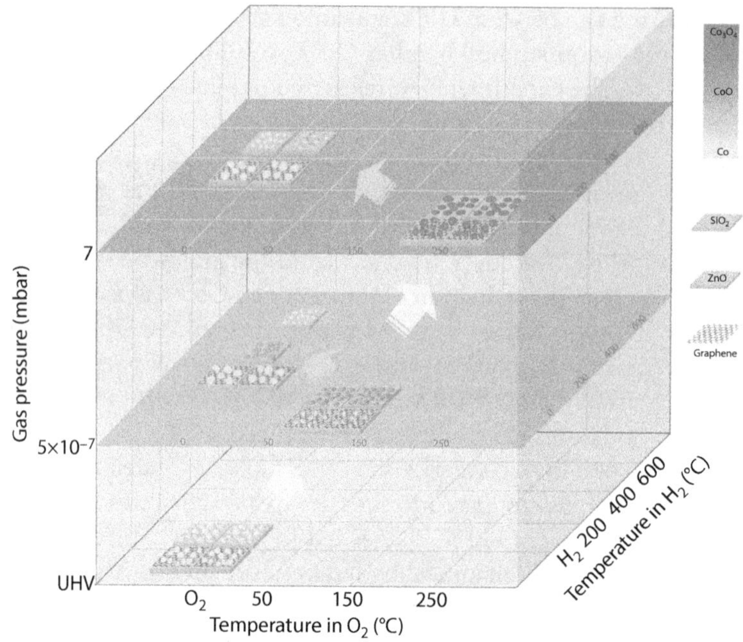

Figure 2.6 Schematic illustration of the evolution of Co oxidation state and morphology under redox treatments in various pressure regimes. The qualitative Co particle sizes were generated from the AFM images while the oxidation states were from the XPS data.

efficient due to the short diffusion pathway. Moreover, at the Co-oxides interface, oxidation of Co may also occur by direct transfer/spillover of oxygen species from the support to the nanoparticles and/or by the OH groups on the oxides surface (in particular, ZnO surface). These differences can only be observed at low-pressure O_2 and low oxidation temperature conditions (as shown in Figure 2.5), where oxygen diffusion kinetics are strongly limited.

As mentioned above, studies on model systems facilitate the interpretation of the experimental results and offer more possibilities for the applied experimental methods. Hence, to further confirm the different Co oxidation scenarios on the bare and graphene-coated oxide supports, the angle-resolved XPS (ARXPS) experiments were conducted at two different take-off angles (normal and grazing) to probe the distribution of Co and CoO in the Co particles. Please note that such studies can be only performed on planar samples and not on powders [29]. Since the effective sampling depth (d) of the XPS measurement is related to the take-off angle (θ) according to $d = 3\lambda \cdot \cos\theta$ (λ is the inelastic mean free path of the photoelectrons), more surface information can be obtained at the grazing angle (80°) than the normal angle (0°) [30]. Figure 2.7a compares the typical ARXPS spectra of Co on ZnO and highly oriented pyrolytic graphite (HOPG), which represents measurements on graphene-coated oxides [16]. As shown in Figure 2.7, different tendencies of ionic (Co^{2+}) and metallic (Co^0) cobalt distribution can be found for the two substrates. In particular, although at grazing angle the Co^{2+} component is enhanced in HOPG-supported cobalt, on ZnO it declines. More experimental data points collected at different Co oxidation states were plotted in Figure 2.7b. In this graph and in case of a homogenous oxidation of cobalt in all its volume, the two collection angles should not result in any difference and therefore

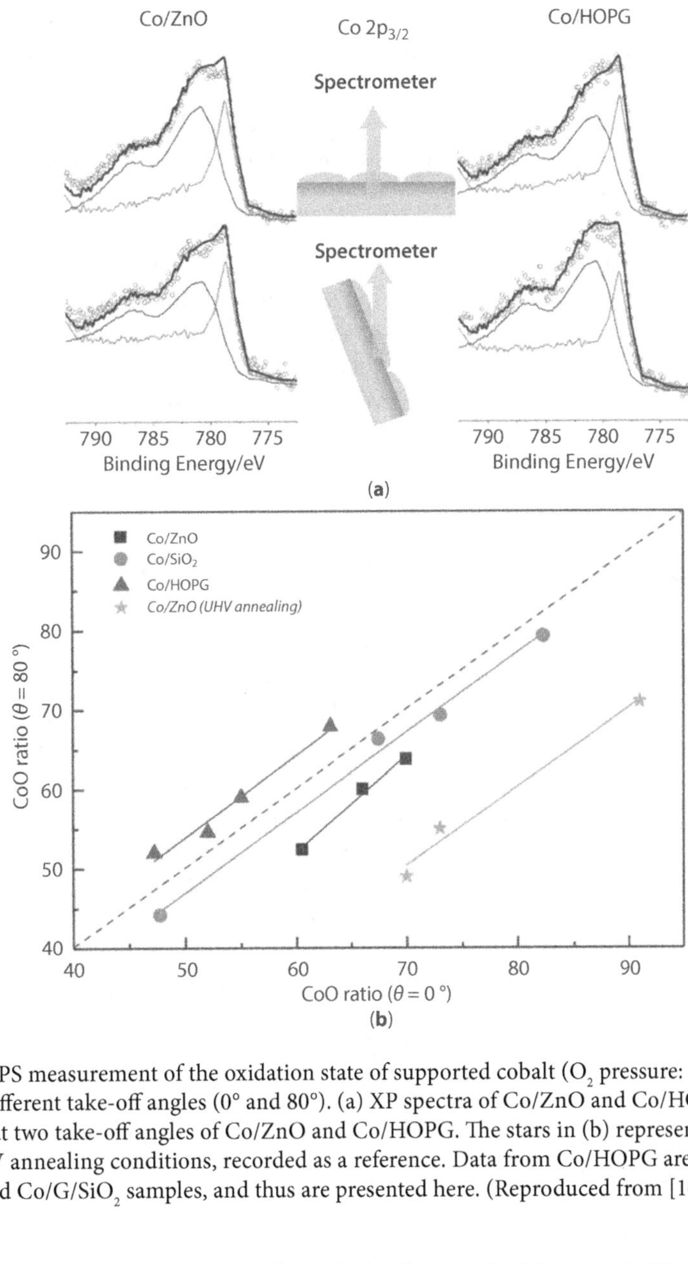

Figure 2.7 ARXPS measurement of the oxidation state of supported cobalt (O_2 pressure: 5×10^{-7} mbar, T $\leq$ 100 °C) at two different take-off angles (0° and 80°). (a) XP spectra of Co/ZnO and Co/HOPG. (b) CoO atomic fraction at two take-off angles of Co/ZnO and Co/HOPG. The stars in (b) represent Co oxidation by ZnO under UHV annealing conditions, recorded as a reference. Data from Co/HOPG are similar from that of Co/G/ZnO and Co/G/SiO_2 samples, and thus are presented here. (Reproduced from [16].)

the experimental points should coincide with the diagonal of the graph. However, as shown in the case of Co/HOPG, the CoO atomic fraction is systematically higher at 80° than 0° take-off angle, indicating that when oxidized cobalt is supported on carbon substrates, the oxide layer is preferentially located at the surface of the particles. On the other hand, in case of ZnO, the trend is reversed, demonstrating that a considerable amount of cobalt oxide is found at the subsurface region, likely from the interface oxidation. This trend is more pronounced on UHV-annealed Co/ZnO sample, where only interface oxidation is expected. For Co/SiO_2, the experimental points are very close to the diagonal, suggesting that CoO are rather uniformly distributed within the Co nanoparticles.

The reduction of Co oxides is mainly influenced by the interactions with the support. Cobalt experiences a strong chemical interaction with ZnO; therefore, increasing the annealing temperature leads to the formation of more thermodynamically stable compounds even under the reduction environments, e.g., CoO (5×10^{-7} mbar H_2 condition) and $Co_xZn_{1-x}O$ (7 mbar H_2 condition). In case of Co/SiO_2, since no solid reaction occurs at the interface, Co oxides can be completely reduced in both low- and high-pressure H_2 conditions. However, with a single layer of graphene as a buffer layer, the reduction of Co oxides is further facilitated, even at a lower temperature than that on SiO_2. This can be attributed to the weak Co-C interaction and the effectively blocked oxygen diffusion at the interface. Concerning the morphology of Co after reduction treatments, reduced Co always agglomerates to form bigger particles, which is more significant in the case of graphene-coated supports (Figure 2.6); however, reoxidized Co forms a relatively flat CoO layer (Co/ZnO reduced at low-pressure H_2). Interestingly, the oxide substrate beneath the graphene layer (ZnO or SiO_2) has no pronounced effect on the redox processes (for both of the reduction temperature and the final morphology).

2.4 Effect of Graphene on the PtCo-Oxide Support Interaction

As the effects of graphene on the MSI have been demonstrated with the monometallic Co-oxides model systems, the general validity of these findings should be confirmed on more complex systems, in particular, bimetallic systems. Bimetallic catalysts have found their applications in a number of important processes such as fuel cells and hydrocarbon-reforming reactions [31, 32]. Due to synergistic effects, bimetallic catalysts often display different electronic and chemical properties with respect to their monometallic counterpart. This provides an opportunity to obtain new catalytic materials with enhanced activity, selectivity, and stability. Similar to monometallic catalysts, the MSI also plays an important role on bimetallic catalysts, which not only influences the morphology and oxidation states of the metals but also changes the distribution of the two metals [32]. In this section, bimetallic PtCo was chosen for investigation due to its potential applications in many catalytic reactions, such as FTS [5], CO oxidation [33, 34], and electrochemical reactions [31]. The PtCo bimetallic samples were prepared with the same method shown in Figure 2.1, but Pt and Co were deposited simultaneously with an atomic ratio of 1:3 (0.1 nm:0.3 nm) [17].

2.4.1 Studies under UHV Conditions

Initially, the effect of graphene on the interactions between PtCo and supports is discussed based on model formulation and performing studies under UHV conditions. The XPS spectra recorded after annealing the sample at various temperatures for 5 min at a time were shown in Figure 2.8. Interestingly, the Co oxidation state evolution in bimetallic PtCo shows similar trends as in the case of monometallic Co shown in Figure 2.3. Figure 2.8a shows that Co on ZnO could be partially oxidized at 450 °C and fully oxidized to Co^{2+} at 550 °C, due to the interface solid reaction between Co and ZnO, as discussed above. In Figure 2.8b, as anticipated, Co on G/ZnO could mostly maintain the metallic state up to 450 °C, while partial oxidation of Co is observed after annealing at 550 °C. The latter might be attributed to areas where cobalt is in contact with ZnO through the defects or open areas of the graphene layer. It should be noted here that, as compared to the Co/ZnO, addition of Pt (PtCo/ZnO) significantly increases the

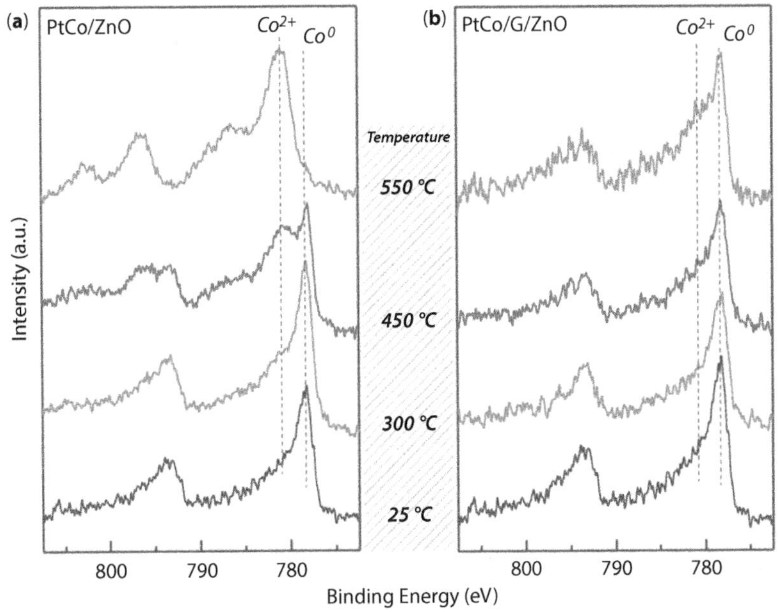

Figure 2.8 Intensity-normalized Co 2p XPS spectra of (a) PtCo/ZnO and (b) PtCo/G/ZnO recorded at room temperature after annealing in UHV at the indicated temperature. (Reproduced from [17].)

oxidation temperature of cobalt from 200 °C (for monometallic Co) to 450 °C (for bimetallic PtCo). This could be attributed to the Pt-Co synergetic effects that limit the Co migration and its interaction with ZnO, thus accounting for the higher oxidation temperature. At the same time, no oxidation of Pt is observed due to the low oxygen affinity toward Pt that hinders Pt oxidation and facilitates that of Co.

With a bimetallic model system, the effect of the graphene layer on the distribution of Pt and Co within the PtCo particles could also be investigated, which is important for understanding the performance of the material. Two surface-sensitive techniques, i.e., XPS and low energy ion scattering spectroscopy (LEIS), were utilized to probe the Pt and Co arrangement. The Co/Pt peak ratios obtained from both methods are shown in Figure 2.9. The Co 2p/Pt 4f XPS peak area ratios (R_{XPS}) of PtCo/ZnO and PtCo/G/ZnO display some distinctly different characteristics. Upon UHV annealing, the R_{XPS} of PtCo/G/ZnO gradually decreases till 400 °C, but then increase again to values higher than the initial. In case of PtCo/G/ZnO a monotonically decrease of the R_{XPS} is observed within all studied temperatures. The LEIS peak area ratios (R_{LEIS}) confirm the evolution trends shown in the XPS results. Similar to the explanation given for the monometallic Co results shown in Figure 2.3c, the different evolution of Co/Pt ratios between ZnO and G/ZnO substrates may also result from two main reasons: thermal induced agglomeration and/or oxidation. The decreasing trend of Co/Pt ratio for both substrates could mainly be due to the higher agglomeration level of Co compared to Pt, as the deposition amount of Co is three times that of Pt. The increase of the Co/Pt ratios on Pt/Co/ZnO after 400 °C observed in Figure 2.9a can be explained by the redispersion of oxidized CoO accompanied by the continuous agglomeration of Pt. These results demonstrate that through preventing the oxidation of Co, the graphene layer could also modify the arrangement of Pt and Co.

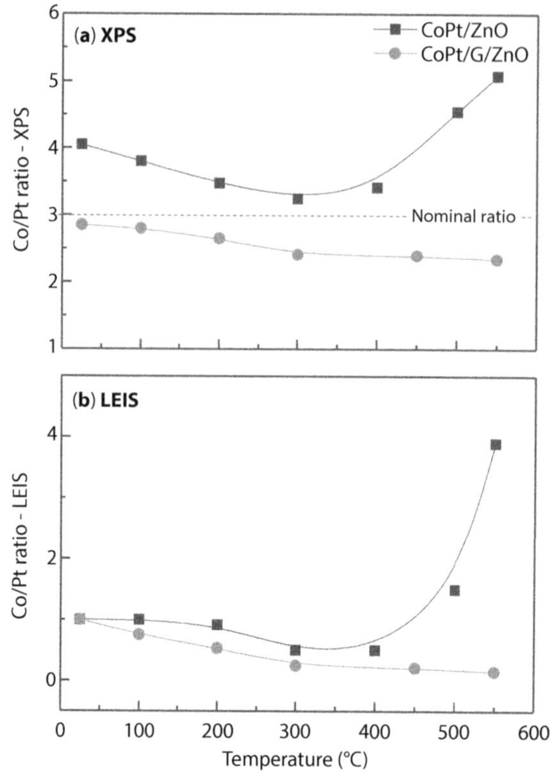

Figure 2.9 (a) Variation of the Co 2p/Pt 4f intensity ratios (R_{XPS}) for PtCo/ZnO and PtCo/G/ZnO samples as a function of the UHV annealing temperature. (b) The Co/Pt ratio obtained after deconvolution of the LEIS spectra (R_{LEIS}), as a function of the UHV annealing temperature. The R_{LEIS} for each sample is normalized to initial measured value just after metals deposition. (Reproduced from [17].)

2.4.2 Physicochemical Studies under O_2/H_2 Gas Atmospheres

Like the monometallic Co samples, O_2/H_2 treatments at different pressure regimes were also introduced to study the effects of graphene on the redox properties [15, 17]. Interestingly, as shown in Figure 2.10, there are some similar results on the redox properties obtained on bimetallic PtCo samples with those previously shown for supported monometallic Co samples. In particular,

1. The graphene layer effectively limits the oxidation of Co under low pressure (5×10^{-7} mbar) O_2 condition shown as systematically lower Co average valence (Figure 2.10a);
2. The graphene layer facilitates the reduction of Co at both low and near ambient pressure (7 mbar) H_2 conditions, shown as the lower reduction temperature presented in Figure 2.10b and d.
3. The ARXPS results (Figure 2.10a) show that under low-pressure O_2 environment, the oxidation of Co on G/ZnO is mainly from the surface, while that on ZnO is preferentially at the interface.

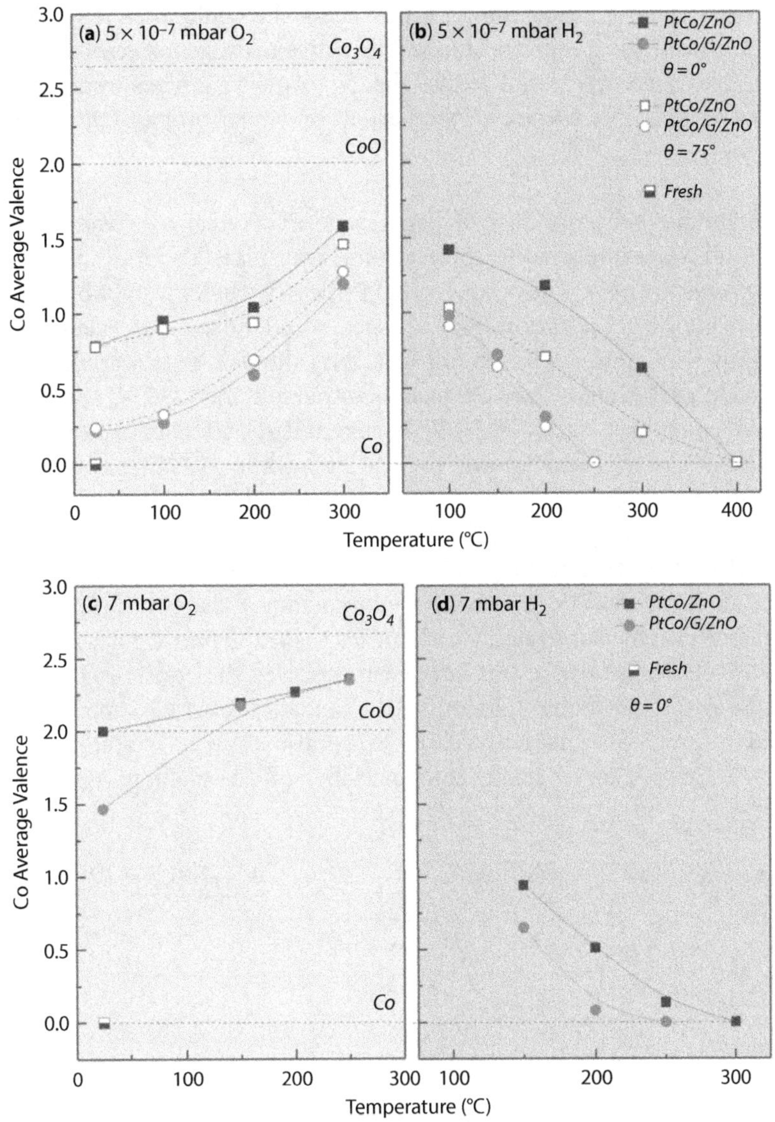

Figure 2.10 Co average valence state evolution of PtCo/ZnO and PtCo/G/ZnO samples during the 5×10^{-7} mbar (a) O_2 and (b) H_2 treatments, and 7 mbar (c) O_2 and (d) H_2 treatments. The initial oxidation states (just after UHV deposition) are represented by the two colored points. (Reproduced from [17].)

However, addition of Pt also introduces some different results:

1. Comparison of Figure 2.5 with Figure 2.10a and c reveals that the oxidation of cobalt is limited for both PtCo/ZnO and PtCo/G/ZnO at two O_2 pressure conditions as compared to the monometallic samples. This demonstrates that Pt co-deposition could partially prevent the Co oxidation.
2. Pt facilitates the reduction of Co oxides in all cases, and no reoxidation of Co is observed even on ZnO support. This can be attributed to the well-known

hydrogen spillover. According to this effect, Pt could preferentially adsorb and dissociate the gas phase H_2, creating active hydrogen species that can later migrate to cobalt areas and reduce the Co oxide. As a result, reduction of Co oxides starts at the surface of the particle and then propagated toward their interior.

Apart from the above-discussed Co oxidation state, Pt could also be oxidized under 7 mbar O_2 condition, and more importantly, the graphene layer plays a significant role to influence the oxidation properties of Pt. The deconvoluted Pt 4f XPS spectra were shown in Figure 2.11a and b and the evolution of the various Pt oxide species and the mean Pt valence for PtCo/ZnO sample were shown in Figure 2.11c. Up to 150 °C annealing, no Pt oxidation was observed on G/ZnO and slight oxidation of Pt to PtO can be observed after 250 °C annealing, indicated as an additional component at 72.6 eV [35]. For PtCo/ZnO, 50 °C oxidation condition could already oxidize ~9% of the total amount of Pt to PtO, while 150 °C condition increases this ratio to 19%. Increasing the annealing temperature to 250 °C could further oxidize Pt to a higher oxidation state, PtO_2, which is located at ~74.2 eV in XPS spectrum [35, 36]. These results clearly demonstrate that the graphene layer also limits the oxidation of Pt, similarly like Co. Since identical conditions were applied to oxidize these two samples, the difference should be attributed to the substrates, which means that ZnO participates in the oxidation of Pt as in the case of Co. Similar results have been observed previously in other noble metal/reducible oxide systems. For example, in O_2, the oxidation of ZnO-supported Au at 200°C was attributed to Au–O–Zn interactions [3]. Ceria is also known to stabilize Pt oxides by formation of Pt–O–Ce species, especially at the periphery of the metal particle [37]. Moreover,

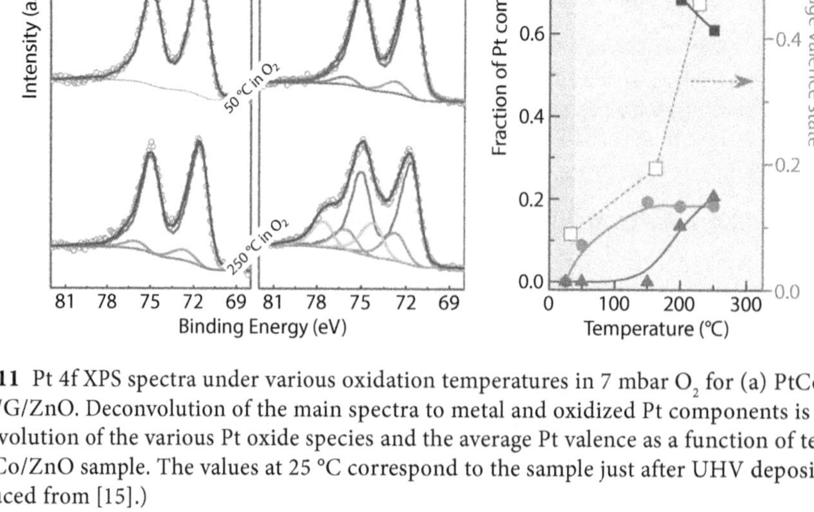

Figure 2.11 Pt 4f XPS spectra under various oxidation temperatures in 7 mbar O_2 for (a) PtCo/ZnO and (b) PtCo/G/ZnO. Deconvolution of the main spectra to metal and oxidized Pt components is included. (c) The evolution of the various Pt oxide species and the average Pt valence as a function of temperature at the PtCo/ZnO sample. The values at 25 °C correspond to the sample just after UHV deposition. (Reproduced from [15].)

preferential oxidation of Pd at the Pd/Fe_3O_4 interface was suggested to stabilize PdO by the interaction with the Fe_3O_4 support [38]. Under 7 mbar H_2 condition, oxidized Pt could be easily reduced to Pt (at 150 °C); therefore, no significant effect from graphene was observed.

2.4.3 Preparation and Testing of Powder PtCo/Graphene/ZnO

The systematic investigation of the planar model systems illustrates in detail the effects of the graphene layer on the morphology, the arrangement, and the redox properties of Pt and Co. However, a real catalyst is usually in powdered form with nanoparticles supported on high surface area supports. Therefore, it is a great challenge to verify if the fundamental findings found on the planar model systems prepared by well-defined physical methods are also accountable for a realistic three-dimensional (3D) material synthesized by a chemical approach. This is a long-debated problem in catalysis research usually referred to as the "material gap," which described the differences between model and real catalysts concerning the raw materials, the preparation methods, the final morphology, etc. [39].

Therefore, in order to verify if the results obtained above on the model systems can overcome the "material gap," the analogous supported PtCo samples in a 3D powder form were synthesized [15]. In brief, powder graphene-coated ZnO (abbreviated as 3D-G@ZnO) was prepared from graphene oxides (GO) and $Zn(Ac)_2 \cdot 2H_2O$ using a method previously proposed by Son *et al.* [40]. While the powder ZnO sample (3D-ZnO) was prepared with the same method without addition of the GO. The co-impregnation method was used to decorate the aforementioned supports with bimetallic PtCo particles (further abbreviated as PtCo/3D-ZnO and PtCo/3D-G@ZnO). The nominal loading of Pt and Co is 4.0 and 3.6 wt%, respectively, with the atomic ratio of Pt and Co to be about 1:3 as in the case of the planar model samples.

Figure 2.12 shows the morphology of the two supports and the final PtCo samples. The SEM images of the as-prepared 3D-ZnO (Figure 2.12a) show that ZnO forms well-crystallized flakes, which are stacking in large aggregates. In contrast, the 3D-G@ZnO consists of agglomerated nanoparticles with much smaller size as compared to the bare 3D-ZnO. This can be attributed to the wrapping of graphene layers, which limits the growth of ZnO crystals during the sample preparation. Moreover, few layers of graphene can be clearly observed coating on the surface of the ZnO crystals as shown in the transmission electron microscopy (TEM) images of PtCo/3D-G@ZnO (Figure 2.12e and f).

After the as-prepared 3D powder samples, the redox experiments were conducted in order to study the effects of graphene coating layer as compared to PtCo directly supported on ZnO. As shown in Figure 2.13, initially, the samples were annealed under 0.3 bar Ar condition to decompose the $H_2PtCl_6 \cdot 6H_2O$ and cobalt(II) acetate precursors. The Co^{2+} compounds are gradually decomposed and about 70% of metallic Co is formed at 350°C on both substrates. In 0.3 bar H_2 condition, the residual CoO can be completely reduced to Co at 250 °C on 3D-G@ZnO, while with 3D-ZnO as support, it cannot be fully reduced even after 450 °C. A second redox cycle (0.3 bar O_2 oxidation and then 0.3 bar H_2 reduction) confirms the reduction promotion effect of the graphene layers, which shows that the oxidized Co (a mixture of CoO and Co_3O_4) are reduced more efficiently on 3D-G@ZnO as compared to 3D-ZnO. The qualitative similarities of the results between the planar and the powder samples indicate that the studies described above on the model samples can be used as a proof of concept to design and synthesize realistic 3D powder catalysts with in principle similar characteristics.

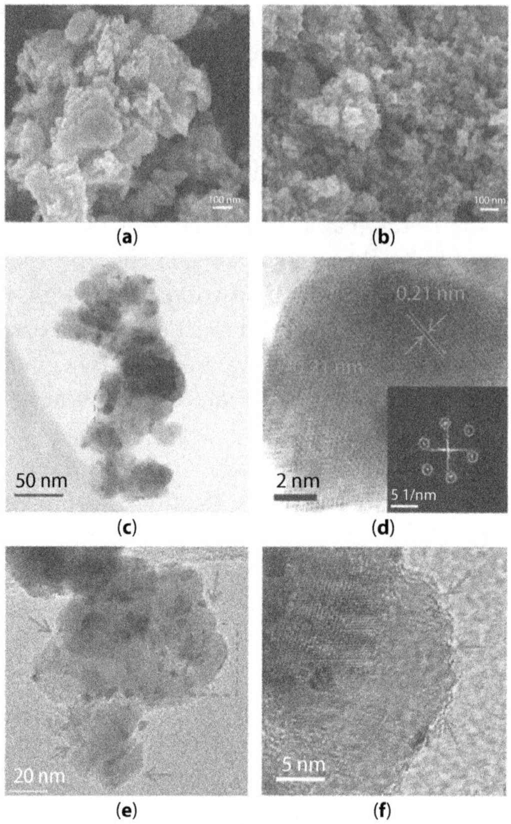

Figure 2.12 SEM images of (a) 3D-ZnO and (b) 3D-G@ZnO. (c) STEM bright field image of PtCo/3D-ZnO, (d) zoom on a PtCo particle with the corresponding FFT, and (e, f) TEM images of PtCo/3D-G@ZnO after decomposition and reduction treatments. The graphene layers are indicated by red arrows. (Reproduced from [15].)

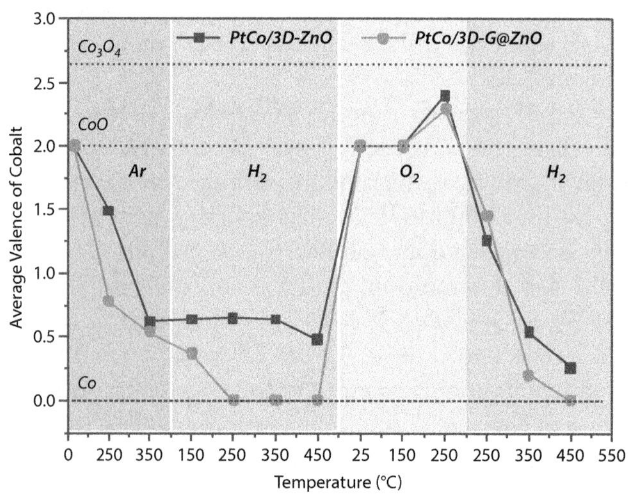

Figure 2.13 Co average valence state evolution of PtCo/3D-ZnO and PtCo/3D-G@ZnO during the annealing treatments under various gas environments. (Reproduced from [15].)

2.5 Stability of Graphene

Apart from the systematic investigations on the effect of graphene insertion on MSI effects, the well-defined model systems facilitate the study of the stability of graphene intergraded between metal and the oxide. Raman spectroscopy as a facile and powerful technique has been widely used to study the strain, the defect density, and the charge doping level of the graphene [22]. Figure 2.14 shows a schematic overview of the graphene defects density as a function of temperature measured for the aforementioned samples. These results qualitatively represent the Raman findings collected after UHV and gas treatments and for a short annealing period [13–16]. It is clear that for the planar model samples, the supported single-layer CVD graphene remains stable after UHV annealing and/or redox treatments, independent of the type of oxide support. However, the contact with cobalt introduces defects on the graphene layer even just after its deposition at room temperature. In addition, the defect density is further enhanced by the UHV annealing and the gas oxidation treatments. High temperature and high oxygen pressure is more critical in creating defects on graphene, but the reduction treatments in H_2 barely change the defect density. Moreover, under the identical sample preparation and redox treatment conditions, the graphene layer in between Co and ZnO is more defective than the one at Co-SiO_2 interface. Concerning the powder sample, since GO is used as the graphene precursor for preparation of PtCo/3D-G@ZnO, the GO-derived graphene is highly defective but relatively stable under the experimental conditions. In the following part (Figure 2.15), the typical Raman spectra will be presented to understand the origin of the defect formation.

The Raman spectra of the fresh Co/G/ZnO samples with various amounts of Co deposition are shown in Figure 2.15a. With the increasing Co deposition amount, two additional features at ~1350 and 1625 cm^{-1} appear and gradually increase in the Raman spectra. The peak at

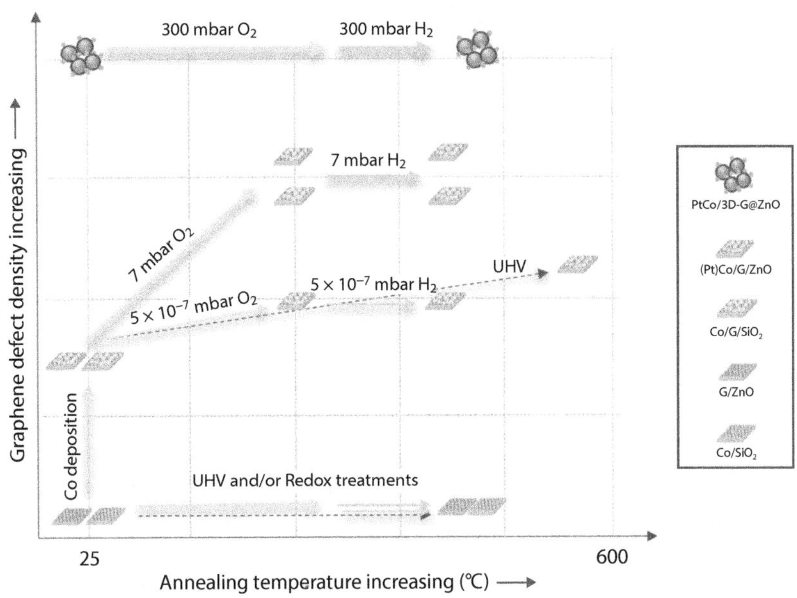

Figure 2.14 Schematic illustration of the graphene defects density on different samples after various treatments; the data points in this figure qualitatively represent the Raman results collected from the above-mentioned samples.

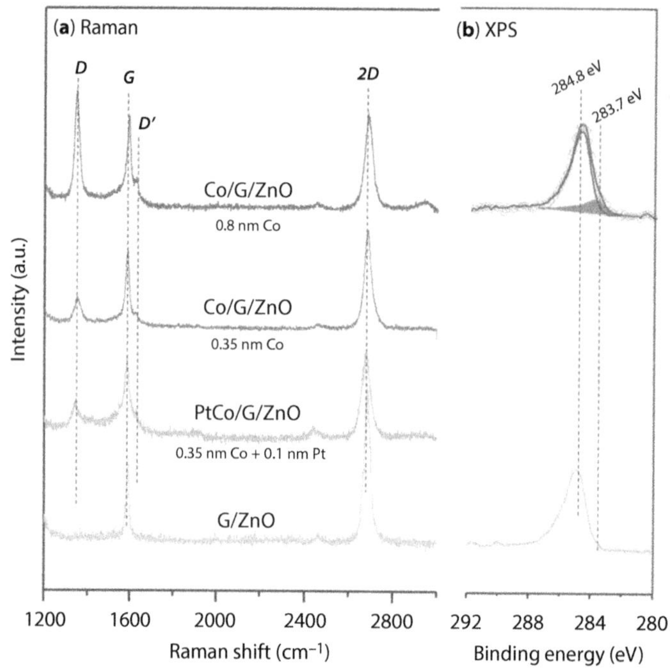

Figure 2.15 (a) Raman spectra of graphene transferred on ZnO substrates before and after vacuum deposition of different amounts of Pt(Co) at room temperature. (b) C 1s XPS spectra of G/ZnO and Co/G/ZnO (0.8 nm Co).

1350 cm^{-1} is assigned to the D band and is activated due to a single-phonon intervalley process caused by defects in the graphene lattice (edges, vacancies, etc.). The other one, known as D′ band, is activated by an intravalley scattering process that also requires defects. Therefore, it is evident that vacuum deposition of Co at room temperature introduces defects onto the graphene layer. Additional evidence can be found from the C 1s XPS spectra in Figure 2.15b; compared to the sample without Co and the one with 0.8 nm Co, a new peak feature appears at 283.7 eV, which is characteristic for carbon dissolved in metals (i.e., carbides formation) [41]. This indicates that the defects formation on graphene could be attributed to the chemical interaction between Co and graphene through a carbon dissolution–precipitation mechanism. It is noticeable that the defects should be mainly introduced by Co instead of Pt, as Pt is reported to bind weakly on graphene through physisorbed interactions [41, 42].

To get further insights about the defects formation during the redox treatments, the optical microscopy images of Co/G/ZnO and Co/G/SiO_2 samples (with 0.8 nm Co deposited) after 7 mbar O_2/H_2 redox treatments and the corresponding Raman spectra from different regions of the samples are shown in Figure 2.16. Different sample regions, including the bilayer-graphene island (abbreviated as BL-Co/G/ZnO and BL-Co/G/SiO_2) and corner region where no Co was deposited (abbreviated as G/ZnO and G/SiO_2), allow to study the role of metal, oxides, and number of graphene layers, on the graphene stability. As it is shown, the intensity of D and D′ bands significantly increased after the redox treatments and being accompanied by the appearance of a weak peak near 2920 cm^{-1} (D + D′ band). These are clear evidences that indicate that much more defects were generated on the graphene layer as compared to the fresh samples.

The overlapped G and D′ bands are then deconvoluted with Lorentzian line shapes, and the calculated I_D/I_G and $I_D/I_{D'}$ intensity ratios are shown in Table 2.1. The I_D/I_G ratio can be

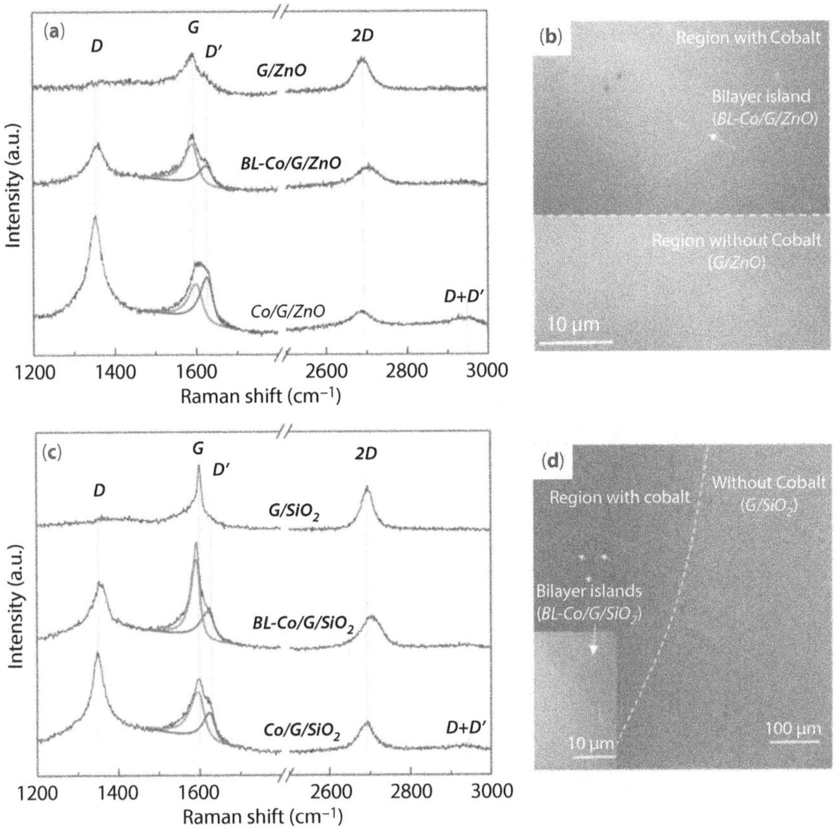

Figure 2.16 Raman spectra recorded at different sample regions for (a) Co/G/ZnO and (b) Co/G/SiO_2 after 7 mbar redox treatment. The spectra are fitted with Lorentzian line shapes. (c) Optical images of Co/G/ZnO and (d) Co/G/SiO_2 after the redox treatment. The different regions where Raman spectra were recorded are indicated in the images. (Reproduced from [14].)

used to estimate the average interdefect distance (L_D) and the defect density (n_D) according to empirical relations presented by Cançado *et al.* [43]:

$$L_D^2(nm^2) = (1.8 \pm 0.5) \times 10^{-9} \lambda_L^4 (\frac{I_D}{I_G})^{-1} \quad (2.1)$$

$$n_D(\mu m^{-2}) = \frac{(1.8 \pm 0.5)}{\lambda_L^4} \times 10^{14} (\frac{I_D}{I_G}) \quad (2.2)$$

where λ_L is the excitation wavelength in nm (here, it is 532 nm). Comparing the n_D value in Table 2.1, one can find that the defect density on single-layer graphene is higher than that of the bilayer graphene island, indicating the higher stability of the bilayer graphene, which is in agreement with previous results [44, 45]. In addition, the sample areas without

Table 2.1 Intensity ratio, average interdefect distance, and defect density of Co/G/ZnO and Co/G/SiO_2 after 7 mbar redox treatment. Reproduced from [14].

Samples	**Intensity ratio (I_D/I_G)**	**Intensity ratio ($I_D/I_{D'}$)**	**Average interdefect distance L_D (nm)**	**Defect density n_D (μm^{-2})**
Co/G/ZnO (single-layer part)	2.5	2.24	7.6	5617
Co/G/ZnO (bilayer island)	0.83	1.67	13.2	1865
Co/G/SiO_2 (single-layer part)	1.77	2.87	9.0	3977
Co/G/SiO_2 (bilayer island)	0.61	1.68	15.4	1370

Co deposition are even more resistant to redox treatment shown as a very low intensity of the D peak. These results confirm that Co plays a catalytic role on the introduction of defects in graphene. Moreover, the defect densities on different areas of Co/G/ZnO are systematically higher than that of Co/G/SiO_2, suggesting that ZnO also plays an important role in inducing defects on graphene [46].

As shown in Figure 2.15, XPS is sensitive to the binding state of the carbon, thus characteristic C 1s spectra of the Co/G/SiO_2 after different steps of the redox treatment were also recorded and displayed in Figure 2.17. All the spectra are dominated by the graphitic C 1s peak at 284.8 eV, but additional Co-diluted carbon peak (283.7 eV) and oxidized carbon species (288.4 eV) can also be observed. The ratio evolution of these species during the redox treatments (Figure 2.17) indicates that annealing in O_2 could gradually introduce oxidized carbon species while the diluted carbon species in the fresh sample disappeared. The following H_2 reduction treatment could remove the oxidized carbons leaving behind only the graphene-related carbon peak.

A proposed mechanism of the graphene deterioration procedure based on the above results is shown at the bottom of Figure 2.17. Initially, the deposited Co particles dilute the

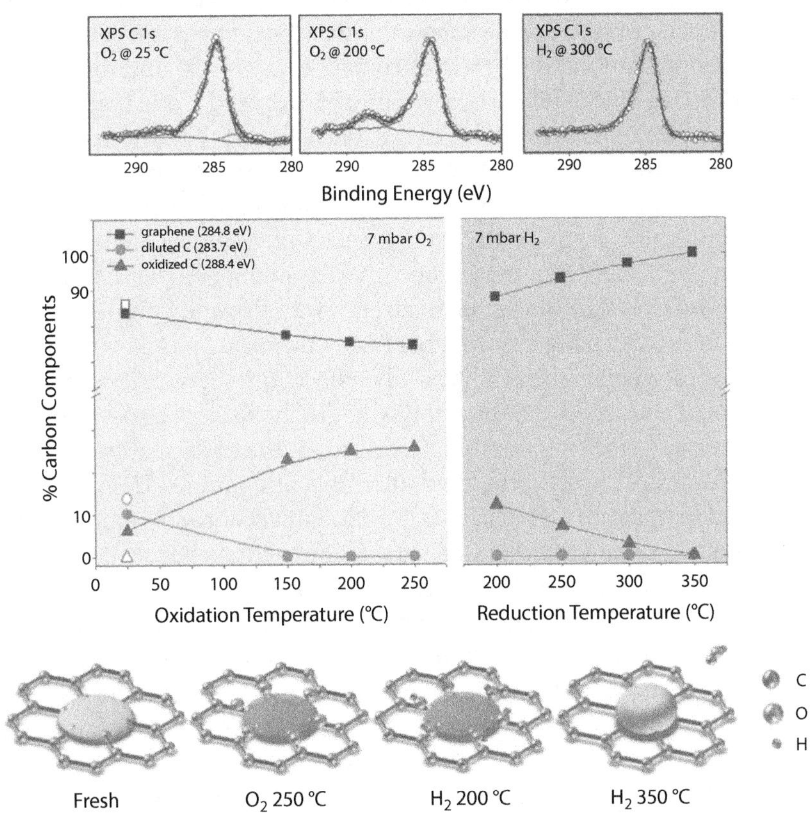

Figure 2.17 (Top) Characteristic C 1s XPS spectra recorded at various stages of the Co/G/SiO_2 sample treatment. (Middle) Evolution of the carbon components derived by deconvolution of the C 1s XPS spectra of Co/G/SiO_2 and Co/G/ZnO samples as a function of temperature during the oxidation and reduction treatments. Since the C 1s peak was very similar in both samples, an average of the deconvolution results is shown for clarity. (Bottom) Schematic representation of the suggested mechanism responsible for formation of graphene defects in contact with cobalt during the redox treatment. (Reproduced from [14].)

carbon atoms at the interface of Co and graphene, and then these carbon atoms become more vulnerable to O_2 oxidation. After the oxidation step, the oxygenated carbon species could be reduced by H_2 to volatile C-O and/or C-O-H compounds, leaving the defective graphene layer in the vicinity of Co particles. This mechanism explains the absence of diluted carbon species in the C 1s XPS spectra after the reduction step, since the carbon atoms around cobalt particles have been already consumed. It also explains why although the H_2 reduction step does not introduce more defects on graphene, it cannot recover the original graphene quality before cobalt introduction.

2.6 Conclusions and Perspectives

This chapter summarizes a recently proposed new concept to modify the metal oxide interface interaction using an in-between graphene layer. With the well-defined planar Co/graphene/ZnO model system, graphene was shown to effectively prevent the oxidation of Co by the ZnO support and decrease Co dispersion during UHV annealing. These effects can also be extended to the O_2/H_2 gas environments. As a buffering layer, graphene limits the oxidation of metal at the interface with the oxide support under low O_2 pressure condition. This effect is attributed to the interruption of the direct contact between Co and the oxide, as well as the enlarged Co particle size. Under H_2 environment, oxidized Co is readily reduced when supported on graphene-coated oxides due to the weaker interaction between Co and carbon as compared to that with oxides. These findings are not limited to the monometallic Co but can be well adapted to the bimetallic PtCo system. Although there is an evident higher degree of complexity in bimetallic particles, the general effects of graphene on modification of MSI effects found on monometallic system still count. A final attempt on the powder PtCo/3D-G@ZnO sample demonstrates that the realistic catalysts with designed properties can be developed on the basis of insights gained from model catalytic formulation.

Besides the role of graphene on the MSI, the effects of metal, oxides, as well as of the various experimental conditions on the stability of graphene were also investigated, thanks to a model sample platform. Without metal deposition, the high quality of CVD graphene supported on oxides can be well preserved after annealing under UHV and redox treatments at the tested temperature. However, deposition of Co at room temperature will readily introduce defects on graphene due to chemical interactions between Co and graphene. The defect density of graphene will be further enhanced by annealing under UHV and O_2 conditions, in particular the high temperature and high gas pressure environment. H_2 reduction can reduce the oxidized carbon species but cannot recover the high quality of graphene. For the powder samples, the number of defects on graphene was initially very high since a graphene oxide precursor was used in the synthesis. Therefore, no significant deterioration of graphene quality was observed after the treatments.

The major issue described in this chapter is the concept of using graphene to modify the MSI, which is based on the experimental results on Co-oxides model samples. As we have shown, this concept is not limited to monometallic Co or reducible oxide ZnO; it can be extended to bimetallic metals (e.g., PtCo) and non-reducible oxides, like SiO_2. Therefore, the method of building a metal/graphene/oxide model system as well as the fundamental insights gained from these samples can also be applied to other appropriated metal-oxide systems. Moreover, the coordinated approaches shown in this chapter also provide new perspectives on bridging the pressure and material gaps between model and realistic samples, which are of great importance for graphene-based materials. Finally, considering that the graphene-based composited materials have been intensively studied in various applications in the last few years, a fundamental understanding of the metal–graphene–oxide interaction will be beneficial for the rational design of functional graphene-based materials.

Beyond the fundamental findings shown in this chapter, the materials presented here can also find their potential applications. As one of the best FTS catalysts, Co is commonly supported on Al_2O_3, SiO_2, and TiO_2 and its performance is strongly influenced by the MSI. In particular, formation of Co-oxide mixed compounds is reported to decrease the Co activity and stability due to the irreducibility of the compounds. The Co/G/ZnO and Co/G/SiO_2 samples introduced in this chapter can be promising catalysts for this reaction as we have shown that the reducibility of Co is greatly enhanced. Moreover, in realistic applications, GO is normally used as the graphene precursor, which provides abundance of defects centers for Co nanoparticles anchoring and is relatively stable, as shown in Section 2.5. Such application is also consistent with the idea of using carbon nanotubes or amorphous carbon as metal/support interlayers for FTS catalysts [10, 47]. In addition, graphene-coated semiconductors have shown their advances in the photocatalytic reactions (i.e., water splitting and elimination of pollutants) due to the multifold role of graphene, including tuning light absorption range and intensity, enhancing adsorption capacity, and acting as photoelectron mediator and acceptor [48, 49]. Therefore, the planar samples presented in this chapter can be applied for fundamental investigation of photocatalytic reactions, while the powder samples can be used directly in heterogeneous catalytic reactions.

References

1. Fu, Q. and Wagner, T., Interaction of nanostructured metal overlayers with oxide surfaces. *Surf. Sci. Rep.*, 62, 11, 431–498, 2007.
2. Tauster, S.J., Fung, S.C., Garten, R.L., Strong metal-support interactions: Group 8 noble metals supported on titanium dioxide. *J. Am. Chem. Soc.*, 100, 1, 170–175, 1978.
3. Liu, X., Liu, M.-H., Luo, Y.-C., Mou, C.-Y., Lin, S.D., Cheng, H., Chen, J.-M., Lee, J.-F., Lin, T.-S., Strong metal-support interactions between gold nanoparticles and ZnO nanorods in CO oxidation. *J. Am. Chem. Soc.*, 134, 24, 10251–10258, 2012.
4. Wang, Y., Widmann, D., Behm, R.J., Influence of TiO_2 bulk defects on CO adsorption and CO oxidation on Au/TiO_2: Electronic metal-support interactions (EMSIs) in supported Au catalysts. *ACS Catal.*, 7, 4, 2339–2345, 2017.
5. Khodakov, A.Y., Chu, W., Fongarland, P., Advances in the development of novel cobalt Fischer–Tropsch catalysts for synthesis of long-chain hydrocarbons and clean fuels. *Chem. Rev.*, 107, 5, 1692–1744, 2007.
6. Chu, W., Xu, J., Hong, J., Lin, T., Khodakov, A., Design of efficient Fischer Tropsch cobalt catalysts via plasma enhancement: Reducibility and performance (Review). *Catal. Today*, 256, P1, 41–48, 2015.
7. Turczyniak, S., Luo, W., Papaefthimiou, V., Ramgir, N.S., Haevecker, M., MacHocki, A., Zafeiratos, S., A comparative ambient pressure x-ray photoelectron and absorption spectroscopy study of various cobalt-based catalysts in reactive atmospheres. *Top. Catal.*, 59, 5–7, 532–542, 2016.
8. Law, Y.T.T., Doh, W.H.H., Luo, W., Zafeiratos, S., A comparative study of ethanol reactivity over Ni, Co and NiCo-ZnO model catalysts. *J. Mol. Catal. A: Chemical*, 381, 89–98, 2014.

9. Jacobs, G., Das, T.K., Zhang, Y., Li, J., Racoillet, G., Davis, B.H., Fischer-Tropsch synthesis: Support, loading, and promoter effects on the reducibility of cobalt catalysts. *Appl. Catal. A Gen.*, 233, 1–2, 263–281, 2002.
10. Subramanian, V., Ordomsky, V., Legras, B., Cheng, K., Cordier, C., Chernavskii, P., Khodakov, A., Design of iron carbon-silica composite catalysts with enhanced catalytic performance in high–temperature Fischer-Tropsch synthesis. *Catal. Sci. Technol.*, 2016.
11. Fu, T. and Li, Z., Review of recent development in Co-based catalysts supported on carbon materials for Fischer–Tropsch synthesis. *Chem. Eng. Sci.*, 135, 3–20, 2015.
12. Singh, V., Joung, D., Zhai, L., Das, S., Khondaker, S.I., Seal, S., Graphene based materials: Past, present and future. *Prog. Mater. Sci.*, 56, 8, 1178–1271, 2011.
13. Luo, W., Doh, W.H., Law, Y.T., Aweke, F., Ksiazek-Sobieszek, A., Sobieszek, A., Salamacha, L., Skrzypiec, K., Normand, F. Le, Machocki, A., Zafeiratos, S., Single-layer graphene as an effective mediator of the metal–support interaction. *J. Phys. Chem. Lett.*, 5, 11, 1837–1844, 2014.
14. Luo, W. and Zafeiratos, S., Tuning morphology and redox properties of cobalt particles supported on oxides by an in between graphene layer. *J. Phys. Chem. C*, 120, 26, 14130–14139, 2016.
15. Luo, W., Baaziz, W., Cao, Q., Ba, H., Baati, R., Ersen, O., Pham-Huu, C., Zafeiratos, S., Design and fabrication of highly reducible PtCo particles supported on graphene-coated ZnO. *ACS Appl. Mater. Interfaces*, 9, 39, 34256–34268, 2017.
16. Luo, W. and Zafeiratos, S., Graphene-coated ZnO and SiO_2 as supports for CoO nanoparticles with enhanced reducibility. *Chem. Phys. Chem.*, 17, 29, 3055–3061, 2016.
17. Luo, W., Mélart, C., Rach, A., Sutter, C., Zafeiratos, S., Interaction of bimetallic PtCo layers with bare and graphene-covered ZnO(0001) supports. *Surf. Sci.*, 669, 64–70, 2018.
18. Luo, W., *Tuning the redox properties of cobalt particles supported on oxides by an in-between graphene layer*, University of Strasbourg, 2016.
19. Suk, J.W., Magnuson, C.W., Hao, Y., Ahmed, S., An, J., Swan, A.K., Goldberg, B.B., Ruoff, R.S., Transfer of CVD-grown monolayer graphene onto arbitrary substrates. *ACS Nano*, 5, 9, 6916–6924.
20. Muñoz, R. and Gómez-Aleixandre, C., Review of CVD synthesis of graphene. *Chem. Vap. Depos.*, 19, 10-11–12, 297–322, 2013.
21. Lotya, M., King, P.J., Smith, R.J., Nicolosi, V., Karlsson, L.S., Blighe, F.M., De, S., Wang, Z., McGovern, I.T., Duesberg, G.S., Coleman, J.N., Liquid phase production of graphene by exfoliation of graphite in surfactant/water solutions. *J. Am. Chem. Soc.*, 131, 3611–3620, 2009.
22. Ferrari, A.C. and Basko, D.M., Raman spectroscopy as a versatile tool for studying the properties of graphene. *Nat. Nanotechnol.*, 8, 4, 235–246, 2013.
23. Casiraghi, C., Pisana, S., Novoselov, K.S., Geim, A.K., Ferrari, A.C., Raman fingerprint of charged impurities in graphene. *Appl. Phys. Lett.*, 91, 23, 2007.
24. Hyman, M.P., Martono, E., Vohs, J.M., Studies of the structure and interfacial chemistry of Co layers on ZnO(0001). *J. Phys. Chem. C*, 114, 40, 16892–16899, 2010.
25. Law, Y.T., Skála, T., Píš, I., Nehasil, V., Vondráček, M., Zafeiratos, S., Skala, T., Pis, I., Nehasil, V., Vondracek, M., Zafeiratos, S., Bimetallic nickel–cobalt nanosized layers supported on polar ZnO surfaces: Metal–support interaction and alloy effects studied by synchrotron radiation x-ray photoelectron spectroscopy. *J. Phys. Chem. C*, 116, 18, 10048–10056, 2012.
26. Luo, W., Jing, F.-L., Yu, X.-P., Sun, S., Luo, S.-Z., Chu, W., Synthesis of 2-methylpyrazine over highly dispersed copper catalysts. *Catal. Letters*, 142, 4, 492–500, 2012.
27. Matsuyama, T. and Ignatiev, A., LEED-AES study of the temperature dependent oxidation of the cobalt (0001) surface. *Surf. Sci.*, 102, 18–28, 1981.
28. Benitez, G., Carelli, J.L., Heras, J.M., Viscido, L., Interaction of oxygen with thin cobalt films. *Langmuir*, 12, 16, 57–60, 1996.
29. Baer, D.R. and Engelhard, M.H.J., XPS analysis of nanostructured materials and biological surfaces. *Electron Spectros. Relat. Phenomena*, 178, 415–432, 2010.

30. Jablonski, A. and Powell, C.J., The electron attenuation length revisited. *Surf. Sci. Rep.*, 47, 2–3, 33–91, 2002.
31. Yu, W., Porosoff, M.D., Chen, J.G., Review of Pt-based bimetallic catalysis: From model surfaces to supported catalysts. *Chem. Rev.*, 112, 11, 5780–5817, 2012.
32. Papaefthimiou, V., Dintzer, T., Lebedeva, M., Teschner, D., Hävecker, M., Knop-Gericke, A., Schlögl, R., Pierron-Bohnes, V., Savinova, E., Zafeiratos, S., Probing metal–support interaction in reactive environments: An *in situ* study of PtCo bimetallic nanoparticles supported on TiO_2. *J. Phys. Chem. C*, 116, 27, 14342–14349, 2012.
33. Ko, E.Y., Park, E.D., Lee, H.C., Lee, D., Kim, S., Supported Pt-Co catalysts for selective CO oxidation in a hydrogen-rich stream. *Angew. Chem. Int. Ed.*, 46, 5, 734–737, 2007.
34. Xu, X., Fu, Q., Wei, M., Wu, X., Bao, X., Comparative studies of redox behaviors of Pt–Co/SiO_2 and Au–Co/SiO_2 catalysts and their activities in CO oxidation. *Catal. Sci. Technol.*, 4, 9, 3151, 2014.
35. Jiang, Z.-Z., Wang, Z.-B., Chu, Y.-Y., Gu, D.-M., Yin, G.-P., Ultrahigh stable carbon riveted Pt/TiO_2–C catalyst prepared by *in situ* carbonized glucose for proton exchange membrane fuel cell. *Energy Environ. Sci.*, 4, 3, 728, 2011.
36. Naitabdi, A., Fagiewicz, R., Boucly, A., Olivieri, G., Bournel, F., Tissot, H., Xu, Y., Benbalagh, R., Silly, M.G., Sirotti, F., Gallet, J.J., Rochet, F., Oxidation of small supported platinum-based nanoparticles under near-ambient pressure exposure to oxygen. *Top. Catal.*, 59, 5–7, 550–563, 2016.
37. Werdinius, C., Österlund, L., Kasemo, B., Nanofabrication of planar model catalysts by colloidal lithography: Pt/Ceria and Pt/Alumina. *Langmuir*, 19, 2, 458–468, 2003.
38. Schalow, T., Laurin, M., Brandt, B., Schauermann, S., Guimond, S., Kuhlenbeck, H., Starr, D.E., Shaikhutdinov, S.K., Libuda, J., Freund, H.-J., Oxygen storage at the metal/oxide interface of catalyst nanoparticles. *Angew. Chem. Int. Ed. Engl.*, 44, 46, 7601–7605, 2005.
39. Hess, C. and Schlögl, R., Ruthenium active catalytic states: Oxidation states and methanol oxidation reactions. *Nanostructured Catalysts: Selective Oxidations*, pp. 248–265, 2011.
40. Son, D.I., Kwon, B.W., Park, D.H., Seo, W.-S., Yi, Y., Angadi, B., Lee, C.-L., Choi, W.K., Emissive ZnO–graphene quantum dots for white-light-emitting diodes. *Nat. Nanotechnol.*, 7, 7, 465–471, 2012.
41. Gong, C., Mcdonnell, S., Qin, X., Azcatl, A., Dong, H., Chabal, Y.J., Cho, K., Wallace, R.M., Realistic metal–graphene contact structures. *ACS Nano*, 8, 1, 642–649, 2014.
42. Giovannetti, G., Khomyakov, P.A., Brocks, G., Karpan, V.M., Van Den Brink, J., Kelly, P.J., Doping graphene with metal contacts. *Phys. Rev. Lett.*, 101, 026803, 2008.
43. Cancado, L.G., Jorio, A., Martins Ferreira, E.H., Stavale, F., Achete, C.A., Capaz, R.B., Moutinho, M.V.O., Lombardo, A., Kulmala, T.S., Ferrari, A.C., Quantifying defects in graphene via raman spectroscopy at different excitation energies. *Nano Lett.*, 11, 8, 3190–3196, 2011.
44. Liu, L., Ryu, S., Tomasik, M.R., Stolyarova, E., Jung, N., Hybertsen, M.S., Steigerwald, M.L., Brus, L.E., Flynn, G.W., Graphene oxidation: Thickness-dependent etching and strong chemical doping. *Nano Lett.*, 8, 7, 1965–1970, 2008.
45. Ryu, S., Han, M.Y., Maultzsch, J., Heinz, T.F., Kim, P., Steigerwald, M.L., Brus, L.E., Reversible basal plane hydrogenation of graphene. *Nano Lett.*, 8, 12, 4597–4602, 2008.
46. Mun, D.H., Lee, H.J., Bae, S., Kim, T.W., Lee, S.H., Photocatalytic decomposition of graphene over a ZnO surface under UV irradiation. *Phys. Chem. Chem. Phys.*, 17, 24, 15683–15686, 2015.
47. Liu, Y., Luo, J., Girleanu, M., Ersen, O., Pham-Huu, C., Meny, C., Efficient hierarchically structured composites containing cobalt catalyst for clean synthetic fuel production from Fischer–Tropsch synthesis. *J. Catal.*, 318, 179–192, 2014.
48. Luo, W. and Zafeiratos, S., A brief review of synthesis and catalytic applications of graphene-coated oxides. *Chem. Cat. Chem.*, 9, 13, 2432–2442, 2017.
49. Zhang, N., Yang, M., Liu, S., Sun, Y., Xu, Y., Waltzing with the versatile platform of graphene to synthesize composite photocatalysts. *Chem. Rev.*, 115, 10307–10377, 2015.

3

The Combinatorial Structure of Graphene

J.E. Graver[2*] and E.J. Hartung[1†]

[1]*Department of Mathematics, Massachusetts College of Liberal Arts, North Adams, MA*
[2]*Department of Mathematics, Syracuse University, Syracuse, NY*

Abstract
We think of graphene patches as regions of the hexagonal tessellation of the plane. In Section 3.1, we introduce basic parameters of graphene patches and describe the relationships between them. In Section 3.2, we discuss results about Kekulé structures (double bond structures) for graphene patches. In Section 3.3, we generalize the results about Kekulé structures for general patches, including disordered graphene patches (graphene with some non-hexagonal faces) as well as more general structures. Section 3.4 describes the topological structure of disordered graphene patches.

Keywords: Clar number, Fries number, graphene patch, benzenoid, Kekulé structure

3.1 Basic Definitions and Results

In this chapter, we consider mathematical properties of plane graphs that model graphene. Specifically, $G = (V, E, F)$ is a 2-connected plane graph that can be embedded in the hexagonal tessellation of the plane where all vertices are of degree 2 or 3, all faces are hexagonal except one face referred to as the "outside face," and all degree-2 vertices are on the boundary of the outside face (also called the boundary of G). We refer to G as a *graphene patch*; small graphene patches are often referred to as *benzenoids*.

3.1.1 Relations Among the Basic Parameters

Let $G = (V, E, F)$ be a graphene patch. We denote the basic parameters of a graphene patch by $\mathbf{b}$ for the length of the boundary, $\mathbf{b}_2$ for the number of degree-2 vertices on the boundary, $\mathbf{b}_3$ for the number of degree-3 vertices on the boundary, $\mathbf{v}$ for the number of vertices, $\mathbf{i}$ for the number of internal vertices (vertices not on the boundary), $\mathbf{h}$ for the number of hexagonal faces, and $\mathbf{e}$ for the number of edges. By definition, $\mathbf{b} = \mathbf{b}_2 + \mathbf{b}_3$. If we move around the boundary of G clockwise and keep track of the orientations of the edges, we note that the orientation of the second edge of a degree-2 vertex is rotated 60 degrees clockwise from its predecessor while the second edge of a degree-3 vertex is rotated 60 degrees counterclockwise from its predecessor. Since the boundary closes, there must be exactly six more 60-degree clockwise

**Corresponding author*: jegraver@syr.edu
†*Corresponding author*: lizhartung@gmail.com

Tobias Stauber (ed.) Handbook of Graphene: Volume 2, (73–94) © 2019 Scrivener Publishing LLC

rotations than 60-degree counterclockwise rotations. Hence, $\mathbf{b}_2 = \mathbf{b}_3 + 6$ (a formal proof of this formula is included in the proof of Theorem 6). From this, it follows that:

$$\mathbf{b}_2 = \frac{\mathbf{b}+6}{2}, \mathbf{b}_3 = \frac{\mathbf{b}-6}{2} \text{ and } \mathbf{b} = 2\mathbf{b}_3 + 6 = 2\mathbf{b}_2 - 6.$$

Concentrating on **b**, **i**, **v**, **h**, and **e**, we see they are related by three linear equations:

(i) $\mathbf{v} = \mathbf{b} + \mathbf{i}$
(ii) $\mathbf{v} - \mathbf{e} + \mathbf{h} = 1$ (Euler's formula)
(iii) $6\mathbf{i} + 5\mathbf{b} = 4\mathbf{e} + 6$ (Sum all vertex degrees to get $3(\mathbf{b}_3 + \mathbf{i}) + 2\mathbf{b}_2 = 2\mathbf{e}$. Then write $\mathbf{b}_2$ and $\mathbf{b}_3$ in terms of **b** and simplify.)

Using these three equations and those relating **b**, $\mathbf{b}_2$, and $\mathbf{b}_3$, we may write all seven of the parameters as functions of any two of these parameters (except $\mathbf{b}_2$ and $\mathbf{b}_3$). As a resource, we list four particularly useful cases:

$\mathbf{b}_2$, **h**	$\mathbf{b}_2$, **v**	**h**, **v**	**h**, **i**
$\mathbf{i} = 2\mathbf{h} - \mathbf{b}_2 + 4$	$\mathbf{i} = \mathbf{v} - 2\mathbf{b}_2 + 6$	$\mathbf{b} = 2\mathbf{v} - 4\mathbf{h} - 2$	$\mathbf{b} = 4\mathbf{h} - 2\mathbf{i} + 2$
$\mathbf{v} = 2\mathbf{h} + \mathbf{b}_2 - 2$	$\mathbf{h} = \frac{\mathbf{v} - \mathbf{b}_2 + 2}{2}$	$\mathbf{e} = \mathbf{v} + \mathbf{h} - 1$	$\mathbf{v} = 4\mathbf{h} - \mathbf{i} + 2$
$\mathbf{e} = 3\mathbf{h} + \mathbf{b}_2 - 2$	$\mathbf{e} = \frac{3\mathbf{v} - \mathbf{b}_2}{2}$	$\mathbf{i} = 4\mathbf{h} - \mathbf{v} + 2$	$\mathbf{e} = 5\mathbf{h} - \mathbf{i} + 1$
$\mathbf{b} = 2\mathbf{b}_2 - 6$ $\mathbf{b}_3 = \mathbf{b}_2 - 6$	$\mathbf{b} = 2\mathbf{b}_2 - 6$ $\mathbf{b}_3 = \mathbf{b}_2 - 6$	$\mathbf{b}_2 = \mathbf{v} - 2\mathbf{h} + 2$ $\mathbf{b}_3 = \mathbf{v} - 2\mathbf{h} - 4$	$\mathbf{b}_2 = 2\mathbf{h} - \mathbf{i} + 4$ $\mathbf{b}_3 = 2\mathbf{h} - \mathbf{i} - 2$

The parameters $\mathbf{b}_2$ and **h** are probably the easiest to count. For the graphene patch in Figure 3.1, we have $\mathbf{b}_2 = 27$ and $\mathbf{h} = 27$; giving $\mathbf{i} = 31$, $\mathbf{v} = 79$, $\mathbf{e} = 106$, $\mathbf{b} = 48$, and $\mathbf{b}_3 = 21$.

3.1.2 Kekulé Structures and the Clar and Fries Numbers

Carbon atoms have valence 4 while the vertices in *G*, the graphene patch model, have degree 3 or 2. The convention is to increase the degree at each vertex by doubling the edges in a perfect matching for *G*. Such a perfect matching is called a *Kekulé* structure and corresponds to the double bond structure of the patch.

There are two problems to consider: first, *G* may not admit a perfect matching, and second, doubling the edges of a perfect matching still only brings the original degree-2 vertices up to degree 3. In Section 3.1.3, we deal with both of these problems. Graphene patches that do admit a Kekulé structure generally admit very many of them; the number of Kekulé structures of a graphene patch is denoted by *K*(*G*) and has been studied extensively [4].

Given a Kekulé structure *K* for the graphene patch *G*, a hexagonal face bounded by three (doubled) edges from *K* is called a *benzene ring*, a *conjugated 6-cycle*, or a *conjugated face*. The maximum of the number of benzene rings over all Kekulé structures for *G* is called the *Fries number* of *G* and denoted *F*(*G*); the maximum of the number of independent (pairwise disjoint) benzene rings over all Kekulé structures for G is called the *Clar number* of G and denoted $C(G)$. The stability of a benzenoid has been positively correlated with a higher Fries number, Clar number, and number of Kekulé structures. For more background on these parameters, see [4], [8], [10], and [12].

3.1.3 Coloring Structures

An important tool for visualizing the structures within a graphene patch is a particular coloring of its vertices, faces, and edges. Since all faces, including the outside face, have even degree, the graph is bipartite and admits a vertex 2-coloring that is unique up to a reversal of colors. We will use black and white for the vertex coloring, as pictured in the upper left of Figure 3.1. A graphene patch G may be thought of as a finite piece of Λ, the hexagonal tessellation of the plane. Λ admits a unique (up to permutation of colors) 3-coloring of its faces and a unique (up to permutation of colors) 3-coloring of its edges. A graphene patch G inherits these colorings as illustrated in the upper right and lower left of Figure 3.1 with the colors red, blue, and green.

On the bottom left in the figure, we have integrated all three colorings. This coloring structure for a graphene patch G inherited by the hexagonal tessellation will always have the following properties:

- (i) The three face and edge colors (here red, blue, and green) appear in clockwise order around black vertices and counterclockwise order around white vertices;
- (ii) the face colors on either side of an internal edge are assigned the two colors different from the color of the edge;
- (iii) the color of the third (not bounded by the edge) face at a degree-3 endpoint of an edge is assigned same color as that edge;
- (iv) The edges bounding a hexagon are assigned different colors from the hexagon, and these two edge colors alternate around the face.

In Section 3.2.2, we will see that a color class of edges can be used as a basis for a Kekulé structure that is densely packed with benzene rings, and that the face color classes can often largely correspond with the benzene rings and independent benzene rings in the graphene patch.

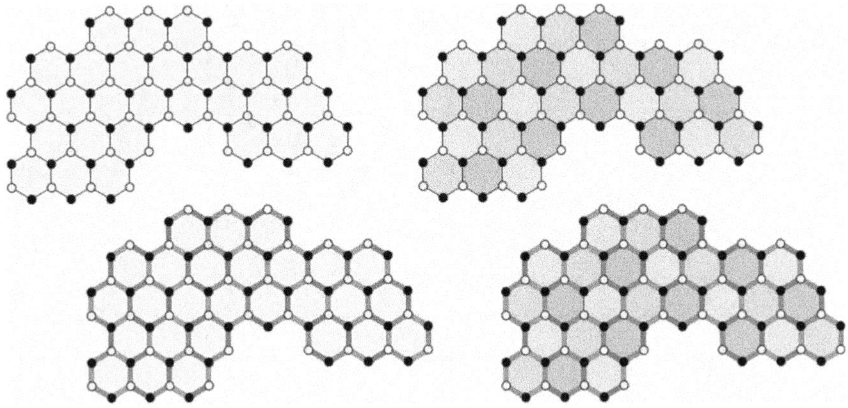

Figure 3.1 Colorings of a graphene patch.

3.2 Kekulé Structures

Given a graphene patch $G = (V, E, F)$, the first question concerning Kekulé Structures is: Does it have one? Since in any perfect matching, white and black vertices are matched in pairs, the obvious necessary condition for the existence of a Kekulé structure is that the number of white vertices equals the number of black vertices.

3.2.1 Sachs Approach

In [11], Sachs approached the problem by orienting the graphene patch so that the edges in one of the parallelism families of edges are vertical. The patch in Figure 3.1 is orientated this way. Consider the first patch in that figure. Sachs identified the degree-2 vertices "*pointing up*" as *peaks* and the degree-2 vertices "pointing down" as *valleys.* He then made three observations:

(i) Peaks all belong to the same color class (white in Figure 3.1);
(ii) valleys all belong to the other color class (black in Figure 3.1);
(iii) the vertical edges give a matching between all of the vertices that are not peaks or valleys (top right in Figure 3.2 below).

It follows at once from these observations that the number of white vertices equals the number of black vertices if and only if the number of peaks equals the number of valleys. Hence, a necessary condition for the existence of a Kekulé structure is that, in any orientation, the number of peaks equals the number of valleys.

Considering the graphene patch G in Figure 3.1, we see that it has six peaks and seven valleys; we conclude that G does not admit a Kekulé structure. In Figure 3.2, we have the patch G' obtained by deleting the rightmost hexagon from G. G' now has the same number of peaks and valleys (six of each). Hence, G' could have a Kekulé structure. The last two diagrams in Figure 3.2 illustrate the technique employed by Sachs to construct a Kekulé structure. The first step is to build the partial matching given by the vertical

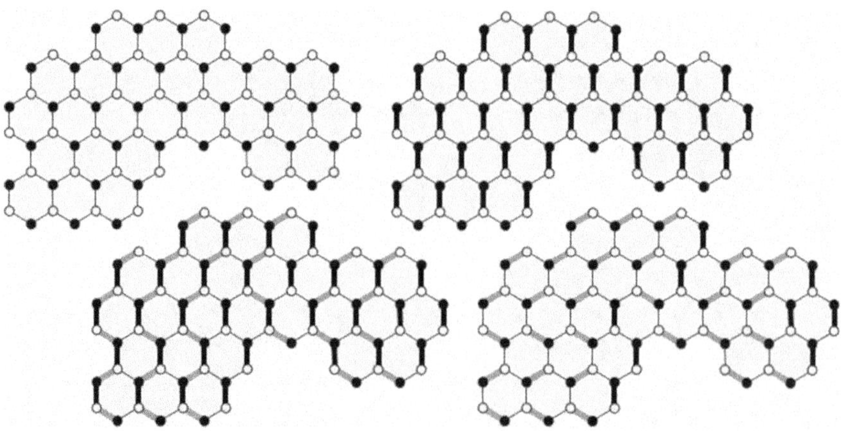

Figure 3.2 Kekulé structures in a graphene patch.

edges. Next, construct a set of disjoint alternating paths connecting the peaks to the valleys: by an alternating path, we mean a path that alternates between edges that belong to the partial matching (thick black edges) and edges that do not belong to the partial matching (colored red in the figure). Constructing such a family is easy for our example as is illustrated by the lower left diagram in Figure 3.2. Finally, deleting the black edges of these alternating paths from the partial matching and including the red edges of these alternating paths into the partial matching gives a perfect matching, as illustrated in the final diagram of the figure.

Theorem 1 (Sachs [11]). *A graphene patch admits a perfect matching if and only if in one orientation (and hence all three orientations) the number of peaks equals the number of valleys and the peaks can be joined to the valleys by a set of disjoint alternating paths.*

An algorithm for finding a set of disjoint alternating paths or showing that none exists can be found in [9]. The problem of finding a set of disjoint alternating paths or showing that none exists can be interpreted as a network flow problem where the alternating paths correspond to a maximum integer flow. In this interpretation, the nonexistence of a flow matching all peaks would correspond to a cut with capacity less than the number of peaks. In Figure 3.3, we give a simple example of a graphene patch with the same number of peaks and valleys, three of each in each orientation, but with no Kekulé structure. In all three orientations, the cut is indicated by the dashed red line.

In the first and third diagram, there is an alternating path joining one peak to one valley on each side of the cut, but the remaining peak and valley on different sides cannot be joined by an alternating path. In the center diagram, only two alternating paths can pass through the cut—we need three. We can see directly that there can be no perfect matching using Hall's theorem: consider the nine white vertices above the cut in the center diagram and note that all of the edges leaving those nine vertices go to just eight black vertices. Hence, at most, eight of those nine white vertices can be paired with a black vertex.

Since the Sachs construction of a Kekulé structure starts with a partial matching that has no benzene rings, it is not surprising that the constructed Kekulé structure contains relatively few benzene rings—exactly four in the example in Figure 3.2. So this construction, while giving a Kekulé structure, does not give one that could be used to compute the Fries and Clar numbers. Next, we consider an approach that produces more benzene rings.

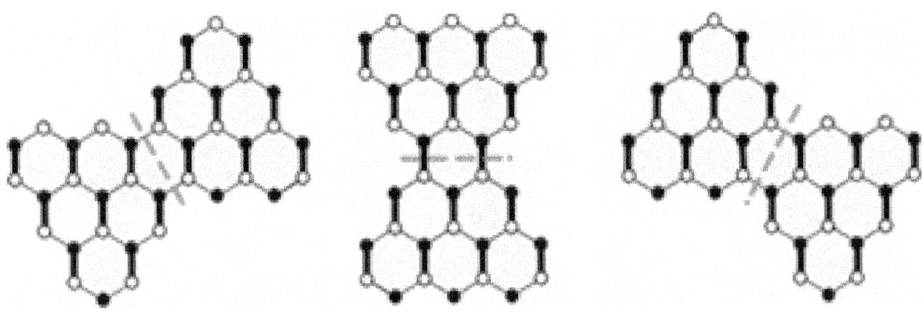

Figure 3.3 # peaks = # valleys but no Kekulé structure.

3.2.2 Kekulé Structures Giving the Clar and Fries Numbers

In Figure 3.4, we consider the same graphene patch with the (unique) face 3-coloring described in Section 3.1.3. In its first diagram, we have included the partial Kekulé structure given by the red edges. This approach from [6] leaves the degree-2 boundary vertices bounding red faces unpaired; we have labeled these *a,b,...,j*. Again, we note that these unmatched vertices have to be half black and half white. As in the Sachs construction, we use alternating paths to pair these unmatched vertices and we would like to do so in a way that disrupts as few benzene rings as possible. When a bounding red face has just two degree-2 vertices, those vertices can be paired without disrupting any benzene rings.

So, in this example, we can pair *a* and *b*, *d* and *e*, and *f* and *g*, leaving only *c*, *h*, *i* and *j* to be paired. However, matching c would require a path cutting across the patch, and alternating paths that cut across the patch will be rather convoluted and destroy many benzene rings. In the top-right diagram, we have chosen paths that connect these degree-2 vertices consecutively around the boundary. The alternating paths are shown as (new) black edges and dotted (deleted) red edges. The Clar faces that remain are indicated by orange rings and the original benzene rings that have been destroyed by these alternating paths are indicated by X's.

In the bottom two diagrams of Figure 3.4, we have extended the green and blue partial Kekulé structures. Here, it is easy to see that the Fries number is 17 and is achieved using the extended green Kekulé structure. The Clar number is 9 and it can be given by the red faces in either the green or blue extended Kekulé structures. In the green extended Kekulé structure, the six leftmost blue benzene rings and the rightmost red benzene rings also form an independent set of nine benzene rings. As we will see later, in some cases, none of the color classes of benzene rings can give the Clar number.

Consider the green Kekulé structure for the graphene patch. Unlike the red and blue Kekulé structures, all of the unpaired vertices may be paired by single edges with the clear advantage that none of the red and blue benzene rings are destroyed. Graphene patches for which the unpaired vertices can be paired by single edges in all three of the color class Kekulé structures were studied in [7]. Looking closer at the red Kekulé structure pictured

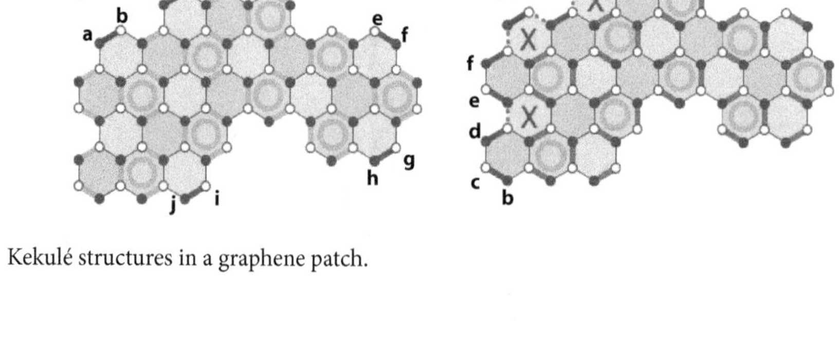

Figure 3.4 Kekulé structures in a graphene patch.

in the upper left of Figure 3.4, we can make several observations: the red boundary faces whose unpaired vertices can be paired by a single edge all have boundary paths of length 3 while the red boundary faces containing vertices that cannot be paired by a single edge all have boundary paths of length 2. We also note that all of the green boundary faces have boundary paths of length 1 or 3. In [7], the parity of the boundary paths is proved to be the key property. Combining several results from that paper we have:

Theorem 2. *Let G = (V, E, F) be a properly colored graphene patch with red, blue, and green color classes. The following four conditions are equivalent.*

(i) *All boundary faces have boundary paths of odd length.*
(ii) *Each of the edge color classes can be extended to a Kekulé structure using only single-edge pairings.*
(iii) *For each of these color class Kekulé structures, all faces in the other two color classes are benzene rings.*
(iv) *Around the boundary, the boundary faces alternate between just two colors.*

This last condition is a bit surprising until, looking at Figure 3.4, we note that when the boundary path of a boundary face is odd, the boundary faces on either side have the same color and when the boundary path of a boundary face is even, the colors of the boundary faces on either side are different. In [7], the collection of graphene patches that satisfy any one and hence all of these conditions is denoted by $\mathcal{H}$.

This leads to some interesting and useful general formulas. Let R, B, and G denote the red, blue, and green faces, respectively; let $b(R)$ denote the number of red boundary faces with $b(B)$ and $b(G)$ defined similarly; and let $\ell(R)$, $\ell(B)$, and $\ell(G)$ denote the number of boundary vertices that are on red, blue, and green faces respectively. We note that every internal vertex lies on exactly one red face as do the $\ell(R)$ vertices and every vertex on a red face is either an internal vertex or on the boundary of a red face. Hence $i + \ell(R) = 6|R|$. We have several other formulas as well:

Lemma 1. *Let G = (V, E, F) be a properly colored graphene patch with red faces R, blue faces B, and green faces G. Then*

(i) $6|R| - \ell(R) = 6|B| - \ell(B) = 6|G| - \ell(G) = i$,
(ii) $b(R) + b(B) + b(G) = b_3$,
(iii) $\ell(R) + \ell(B) + \ell(G) = b + b_3$,

To complete the proof, we note that each boundary face has two degree-3 boundary vertices and each degree-3 boundary vertex is on the boundary of two boundary faces. Finally, summing the lengths of the paths counts each degree-2 boundary vertex once and each degree-3 boundary vertex twice.

For the graphene patches in $\mathcal{H}$, one of the face color classes does not appear on boundary, say $\ell(R) = 0$. Then, it is easy to show that choosing the red Kekulé structure gives the Fries number of the patch. Combining several of the formulas, we can show that the Fries number for a graphene patch in $\mathcal{H}$ is given by $\frac{v}{3} - \frac{b_2}{6}$. In many cases, the Clar number will be simply the size of the largest face color class—in all cases, the size of the largest face color class will be a lower bound on the Clar number. Much of [7] is devoted to techniques for computing the Clar number for graphene patches in $\mathcal{H}$.

3.2.3 Pairwise Incompatibility of the Kekulé, Fries, and Clar Numbers for Benzenoids

The Clar number, Fries number, and number of Kekulé structures for a graphene patch or benzenoid B are three classic parameters that have been positively correlated with the stability of graphene patches. It was generally assumed that if two graphene patches B_1 and B_2 had the same number of vertices (atoms) and hexagons, and B_1 had a higher Clar number than B_2, it would follow that B_1 had more Kekulé structures and a higher Fries number than B_2. However, in [3], this is shown not to be the case. Figure 3.5 shows a smallest pair of graphene patches, B_1 and B_2, that has the same number of hexagons and vertices but disagrees on their largest Clar number versus highest number of Kekulé structures. B_1 and B_2 each have 7 hexagons and 28 vertices (a smaller incompatible pair with 6 hexagons exists, but these pairs of graphene patches do not have the same number of vertices). B_1 has 13 Kekulé structures and a Clar number of 3; B_2 has 16 Kekulé structures and a Clar number of 2. On the right of Figure 3.5 is a smallest pair of graphene patches, B_3 and B_4, that disagrees on the largest Clar versus Fries number, and is also a smallest pair that disagrees on their Fries number versus number of Kekule structures. These graphene patches each have 6 hexagons and 26 vertices. B_3 has Fries number 5, Clar number 4, and 22 Kekulé structures; B_4 has Fries number 6, Clar number 3, and 21 Kekulé structures.

Two natural questions arise: (1) How large of a gap between the disagreeing parameters can be created? (2) How often do pairwise disagreements occur? In [3], the authors showed that the gaps in these parameters could grow arbitrarily large for specified classes of graphene patches. They constructed a pair of classes Z_n and Y_n on n hexagons and 4n + 2 vertices such that $C(Z_n) < C(Y_n)$ but $F(Z_n) > F(Y_n)$, and both of these inequalities grow linearly with the number of hexagons. Similarly, they constructed a pair of classes W_n and V_n such that $C(V_n) < C(W_n)$ and $F(V_n) < F(W_n)$, but $K(V_n) > K(W_n)$. In this case, the growth of the gap is logarithmic in terms of the number of hexagons.

The authors have investigated the second question through the help of a computer search. This paper finds the proportion of disagreements for each pair of parameters for graphene patches on up to 13 hexagons [3]. This data was generated programmatically using C, Python, and awk, and the computer search was based on a method of generation established by Brinkmann, Caporossi, and Hansen [1].

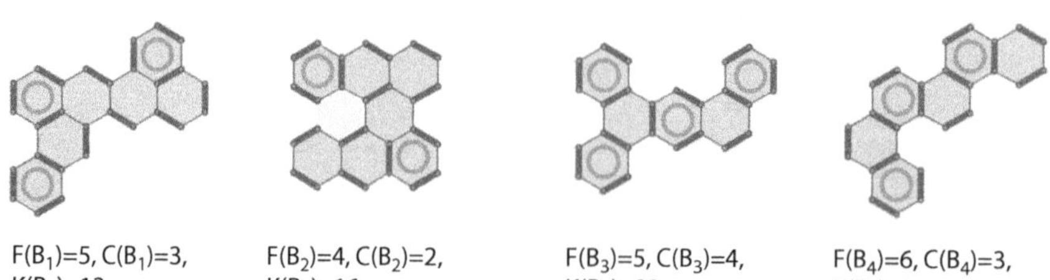

$F(B_1)=5, C(B_1)=3, K(B_1)=13$ $F(B_2)=4, C(B_2)=2, K(B_2)=16$ $F(B_3)=5, C(B_3)=4, K(B_3)=22$ $F(B_4)=6, C(B_4)=3, K(B_4)=21$

Figure 3.5 Smallest incompatible pairs of graphene patches.

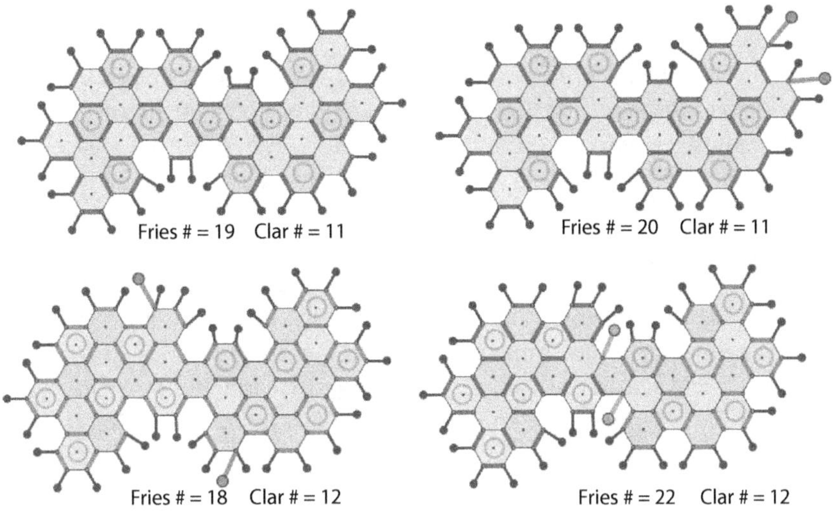

Figure 3.6 Doping patterns.

3.2.4 Doping and Kekulé Structures

Doping, attaching non-carbon atoms to a graphene patch, is done for a variety of reasons. Here, we consider doping for the single the purpose of increasing the Clar or Fries number. A Kekulé structure for a graphene patch brings the valence of every degree-3 vertex up to 4, but the b_2 degree-2 vertices become degree-3 vertices. Doping by attaching an atom of hydrogen to the original degree-2 vertices brings the valence of every vertex up to 4. However, different doping patterns could be used to change the Clar and Fries numbers. In Figure 3.6, we illustrate this with several doping patterns for the same graphene patch. In the upper-left diagram, we have the standard doping pattern along with the blue Kekule structure. This arrangement yields a Fries number of 19 and a Clar number of 11. In the upper-right diagram, we demonstrate that, by simply double doping two adjacent degree-2 vertices, we can increase the Fries number to 20. On the lower left, we have double doped two boundary vertices and increased the Clar number to 12. Finally, in the lower right, we have doped two degree-3 vertices to achieve a Fries number of 22 and a Clar number of 12.

3.3 Internal Defects

3.3.1 Internal Kekulé Structures

In Section 3.2.2, we describe extending a partial Kekulé structure given by a color class of edges to the boundary as a full Kekulé structure that disturbs as few of the benzene rings as possible. As described in Section 3.2.1, not all graphene patches have a Kekulé structure; some vertices on the boundary may be unmatched. In this section, we avoid dealing with unmatched vertices on the boundary by concentrating on *internal Kekulé structures*, that is, matchings that match all degree-3 vertices but not necessarily all degree-2 vertices. We apply this concept to graphene patches as well as *general patches*.

3.3.2 General Patches

We consider *general patches*, which will include disordered patches of graphene (graphene with some defective non-hexagonal faces) as well as more general structures. We define a general patch or simply a *patch* to be a plane graph $\Pi = (V, E, F)$ with one face $f_0 \in F$ designated as the *outside face* such that all vertices have degree 2 or 3, with all degree-2 vertices restricted to the boundary of f_0. The boundary of f_0 must be an elementary circuit. In the case where all faces except f_0 are hexagons, Π models a *graphene patch*. If all of the faces in F are of even degree, then the outside face f_0 will be as well, and we call this an *even patch*. Note that even patches will be bipartite and that a graphene patch is an even patch.

Even patches have many of the same properties as graphene patches. Even patches have a face 3-coloring [5], and also admit a vertex 2-coloring since they are bipartite. Furthermore, even patches will have a unique (up to permutation) face-edge 3-coloring that follows the same coloring rules given in Section 3.1.3, namely that

(i) If a face is assigned color c_1, the edges bounding it are alternately assigned colors c_2 and c_3.
(ii) Each edge and its bounding face or faces are all assigned different colors.

These two properties can be used to show that faces and edge colors appear in clockwise order around one of the partite vertex sets (say, black) and in counterclockwise order around vertices from the other partite set (white) [5].

We use this face-edge 3-coloring to construct three *perfect internal Kekulé structures* for any even patch. We define these to be internal Kekulé structures where every face is either conjugated or void. These internal Kekulé structures are optimal in terms of their number of benzene rings, because at any degree-3 vertex, at most two faces can be conjugated and at most, one face can be void [5]. It is further shown that there are exactly three perfect internal Kekulé structures for even patches, and that these are exactly the three different edge color classes in the unique face-edge 3-coloring of Π. To summarize from [5]:

Theorem 3. *Let* Π *be an even patch. Then*

(i) Π *admits an edge-face 3-coloring that is unique up to a permutation of the colors.*
(ii) *The edge color classes of this edge-face 3-coloring are three distinct perfect internal Kekulé structures for* Π *and these are the only perfect internal Kekulé structures for* Π.

Thus, if Π is a general patch with only faces of even degree, or a graphene patch with a few disordered faces of even degree, the perfect internal Kekulé structures for Π are essentially the same as for pure graphene.

3.3.3 Clusters

We now consider patches with disordered faces in their interiors, some of which are of odd degree. To study these patches, we choose a simply connected interior subpatch including all faces of odd degree and delete it. The boundary of the interior patch we delete must be an elementary circuit, and we delete all vertices, edges, and faces in its interior, leaving a single face f_I. What remains is an *annular patch* $\Theta = (V, E, F)$, where $f_O \in F$ is the *outside face and $f_I \in F$ is the inside face.* All vertices of the annular patch have degree 2 or 3, with all degree-2 vertices on the boundaries of f_O and f_I. Again, a matching that includes all degree-3 vertices but not necessarily all degree-2 vertices is called an *internal Kekulé structure* for the annular patch and a *perfect internal structure* for the annular patch if all faces are either conjugated or void. We restrict our attention to situations in which all faces of odd degree can be included in the interior patch, and so the faces of the annular patch all have even degree. The left figure in Figure 3.7 shows an example of a patch with three interior faces of odd degree in its interior.

One question is whether the annulus has perfect internal Kekulé structures. For example, the right figure in Figure 3.7 shows a perfect internal Kekulé structure for the annulus. A second question is: if the annulus has perfect internal Kekulé structures, can these structures be extended to the interior patch with relatively few changes to the matching around the annulus? The approach depends on whether the number of odd degree faces in the interior is even or odd. The details of this are worked out in [5], and we summarize the main results here.

Theorem 4. *Let Π be a patch with some faces of odd degree in its interior, and let Θ be the annular patch surrounding a subpatch containing all odd degree faces.*

(i) *If the number of odd degree faces in the interior is odd, then the annulus is not bipartite, and the annulus will have exactly one perfect internal Kekulé structure.*
(ii) *If the number of odd degree faces in the interior is even, then the annulus is bipartite, and the annulus will either have exactly three perfect internal Kekulé structures (the three edge color classes) or no perfect internal Kekulé structures.*

Further, it is shown in [5] that if an annular patch has a perfect internal Kekulé structure, the structure can be extended to the interior of the patch with a few small alterations on faces bounding the interior patch, as illustrated in Figure 3.8. This will be a Kekulé structure on the interior subpatch, but not necessarily "perfect" on the interior in the sense that there may be some faces on the interior subpatch that are neither void nor conjugated.

Theorem 5. *Let Π be a patch with a central subpatch containing all faces of odd degree, and let $\mathcal{K}$ be a perfect internal Kekulé structure for the annular patch surrounding this central patch. Then, $\mathcal{K}$, with a few possible alterations on faces bounding the central patch, may be extended to an internal Kekulé structure for Π.*

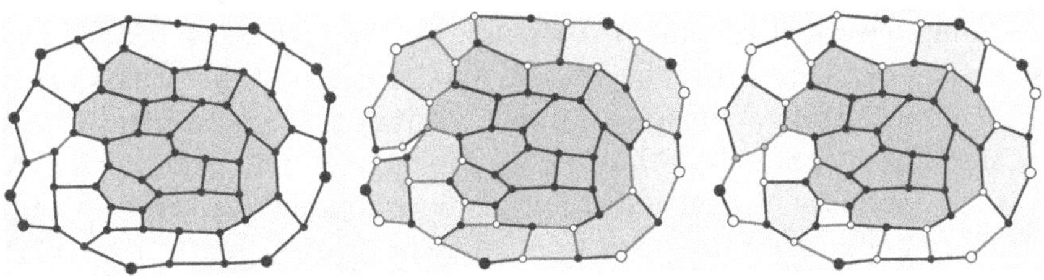

Figure 3.7 Odd faces: one edge color class around the annulus matching.

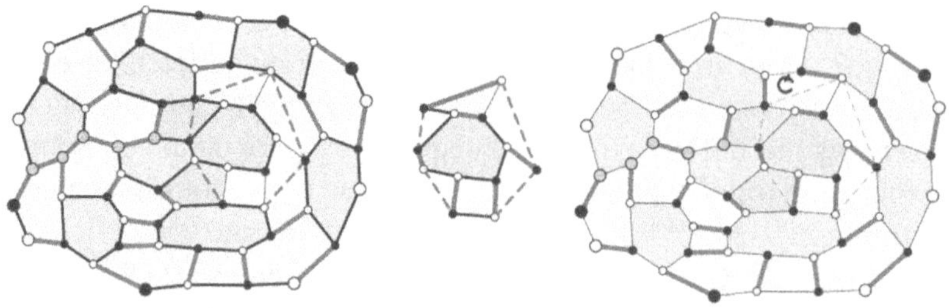

Figure 3.8 An extension for the example in Figure 3.7.

One way to determine whether the annulus will have three perfect internal Kekulé structures, one perfect internal Kekulé structure, or none is to examine the boundary. Given a patch, add a pendant vertex to every degree-2 vertex on the boundary. Now, select vertex on its rim, color that vertex white, and color its edges red, blue, and green in clockwise order. Moving clockwise around the rim, color the next vertex black and complete the coloring of the edges at that vertex using the counterclockwise orientation; the next vertex on the rim will be colored white and its edges colored using the clockwise orientation. We continue in this way until we return to the initial vertex.

(i) If the vertex colors match and the edge colors match, then the number of odd faces in the patch is even (perhaps 0) and, if they can be isolated in a subpatch in the interior, the resulting annular patch admits three perfect internal Kekulé structures, each of which extends to an internal Kekulé structure for the entire patch.

(ii) If the vertex colors match but the edge colors do not match, then the number of odd faces in the patch is even, but not 0, and there will be no perfect internal Kekulé structures for the annular patch obtained by deleting any internal subpatch containing the odd faces.

(iii) If the vertex colors do not match, then one of the edge colors must match, the number of odd faces in the patch is odd, and, if they can be isolated in a subpatch in the interior, the resulting annular patch admits exactly one perfect internal Kekulé structure (of the matching edge color), which then extends to an internal Kekulé structure for the entire patch.

3.4 Curvature

In this section, we consider *general graphene patches*: graphene-like patches that contain some defects but nevertheless retain the feature that they can be embedded in a hexagonal tessellation—not necessarily the hexagonal tessellation of the plane. Specifically, if $G = (V, E, F)$ is a general graphene patch, then we may add a border of hexagons to its boundary and do so repeatedly without being forced to include additional defects. For example, if the boundary of G included five or more consecutive degree-3 vertices, the face added to the boundary at this point would have to have a degree of at least 7; hence, G would not be a general graphene patch as we have defined them. Since we can add borders to a general graphene patch G indefinitely, we are constructing a hexagonal tessellation T outside of the patch G. The annular patch obtained by deleting G from T can easily be seen to be like the hexagonal tessellation of the plane locally, that is, all of its finite (but not annular) subpatches can be embedded in the hexagonal tessellation of the plane. However, if G has curvature other than 0, the entire annular patch cannot be embedded in the hexagonal tessellation of the plane. We will discuss these conditions in detail after we have introduced some definitions and proved some basic results. Let $G = (V, E, F)$ be a general graphene patch and let D denote the collection of defects, that is, the faces other than the outside face with a degree other than 6. Let d(f) denote the degree of the face f. We define the *curvature* of G to be the parameter

$$c(\mathrm{G}) = \sum_{f \in D} (6 - \mathrm{d}(f)).$$

3.4.1 Curvature and Growth

As we noted at the very beginning of this chapter, the number of degree-2 vertices on the boundary, b_2, and the number of degree-3 vertices on the boundary, b_3, are related by the simple formula $b_2 = b_3 + 6$ for all graphene patches. That result generalizes to general graphene patches:

Theorem 6. *Let* $G = (V, E, F)$ *be a general graphene patch with curvature c. Then,* $b_2 = b_3 + 6 - \mathrm{c}$.

Proof. One easily checks that this result holds for all general graphene patches consisting of just one "patch-face" f and the outside face: $b_2 = d(f)$, $b_3 = 0$, and $c = 6 - d(f)$; so $b_2 = 6 - c = b_3 + 6 - c$. We now proceed by induction: Let $G = (V, E, F)$ be a general graphene patch with curvature c that includes $n > 1$ "patch-faces" and assume that the formula is valid for all general graphene patches with fewer faces. Let f be a face whose removal does not disconnect the graph—such a face must exist. Finally, let G' denote the patch obtained by deleting f; see the leftmost patch in Figure 3.9. Note that G' and f intersect in a path, the red path joining vertices x and y in the figure. Let k denote its length. Then, the $k - 1$ vertices interior to the red path from x to y are degree-3 boundary vertices of G' that become interior vertices of G, x and y are degree-2 boundary vertices of G' that become degree-3 boundary vertices

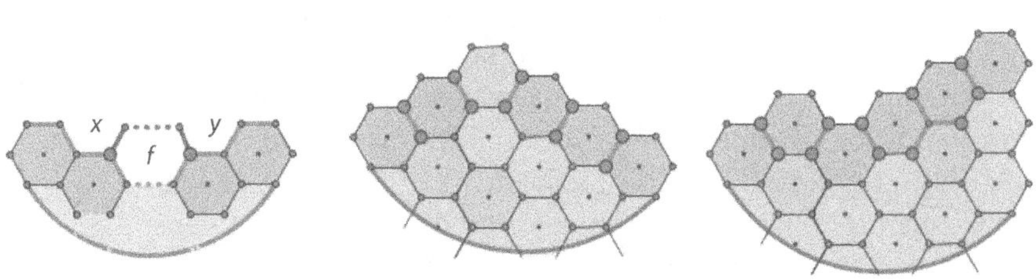

Figure 3.9 Curvature and the boundary.

of G, while the $d(f) - (k+1)$ vertices not shared with G' (the blue path from x to y) are new degree-2 boundary vertices of G. We have:

(i) $c = c' + (6 - d(f))$ or $c' = c - 6 + d(f)$;
(ii) $b_2 = b_2' - 2 + d(f) - (k+1)$ or $b_2' = b_2 + 3 + k - d(f)$
(iii) $b_3 = b_3' - (k-1) + 2$ or $b_3' = b_3 - 3 + k$.

Substituting these values in $0 = b_2' - b_3' - 6 + c'$ (the induction hypothesis):

$$0 = [b_2 + 3 + k - d(f)] - [b_3 - 3 + k] - 6 + [c - 6 + d(f)] = b_2 - b_3 - 6 + c \qquad \square$$

The formula $b_2 = b_3 + 6$ for graphene patches is just the special case of this result with $c = 0$.

Let $G = (V, E, F)$ be a general graphene patch with curvature c. In addition to the circuit that we have defined to be the boundary of G, the "circuit" of faces that share an edge with the outside face is also useful. We call this set of faces the *border* of G. These two notions are closely related. The length of the boundary is given by $d(f_O) = b_2 + b_3 = 2b_3 + 6 - c$. Working around the boundary, we note that the degree-3 vertices on the boundary alternate with the faces of the border. Hence b_3 counts the boundary faces with the caveat that a face f whose removal disconnects the patch is counted with multiplicity equal to the number of components of $f \cap f_O$. The patch consisting of just one face has $b_3 = 0$ and is also an exception. Therefore, when there are at least two faces and no face disconnects the patch upon its removal, the *border length* of G is simply b_3. Now, consider embedding G in a hexagonal tessellation by adding new borders, one border at a time, such that each edge on the boundary circuit of G must be adjacent to a new hexagon. This is illustrated in the second and third diagrams of Figure 3.9. We call the patch G', obtained by adding a border of new hexagonal faces to G, a *dilation* G. Here, we have added a new gray border to the blue patch; as illustrated on the right, this may involve filling in some deep indentations of the original boundary, and thus not all added faces are on the border of G'. We then track how the lengths of the borders grow in time. Informally, this is the *growth rate* of G. The next lemma is key to understanding the growth rate.

Theorem 7. *Let $G = (V, E, F)$ be a general graphene patch with all border faces hexagonal and let $G' = (V', E', F')$ denote the patch obtained by adding a border of new hexagonal faces. Then,*

(i) $b'_3 < b_3 + 6 - c$;
(ii) *the curvature of G is 6 or less;*
(iii) *if G has no two adjacent degree-3 vertices on its boundary, then G' also satisfies this condition and $b'_3 = b_3 + 6 - c$.*

Proof. In Figure 3.9, we see that each degree-3 boundary vertex of G is adjacent to a degree-2 vertex on the boundary of G' or is joined to a degree-2 vertex on G by the clockwise path around the new face to its left. It is also clear that each degree-2 vertex of G will be identified to at most one degree-3 vertex of G' in this way. Hence, $b'_3 \leq b_2 = b_3 + 6 - c$, giving the first inequality. It follows from this inequality that if c is greater than 6, each additional border will be smaller, eventually forcing another defect. Hence, a general graphene patch must have curvature less than or equal to 6. Finally, the strict inequality can only occur when deeper indentations are filled in, as illustrated on the right in the figure. The central diagram illustrates that when G has no two adjacent degree-3 vertices on its boundary, then G' also satisfies this condition and the correspondence between degree-3 vertices of G' and degree-2 vertices on G is 1 to 1 and onto; hence, in that case, equality holds in the formula.

If G has no three consecutive degree-3 vertices on the boundary, then $b'_3 = b_3 + 6$. However, G' may not satisfy this condition and subsequent boundaries may continue to shrink. Hence, the condition "no two consecutive degree-3 vertices" will permit us to define the growth rate. Patches that satisfy the condition *"no two adjacent degree-3 vertices on the boundary"* and the additional condition *"no three adjacent degree-2 vertices on the boundary"* are particularly nice and we say that such a patch has a *polygonal border* and that such a patch is itself *polygonal.* Intuitively, we think of the border faces with two degree-2 vertices on the boundary as the *corners* of the border and the run of faces between consecutive corners plus half of each corner as the *sides* of the border. The *length* of a side is then the number of faces in the run plus 1. Hence, the border may be interpreted as a polygon with the sum of the side lengths being the number of faces in the border or its *perimeter.*

Lemma 2. *Let $G = (V, E, F)$ be a general graphene patch with curvature less than 6. Then, G can be embedded in a polygonal patch.*

Proof. It is not too difficult (just tedious and hence omitted) to show that, repeatedly filling in the cavities formed by consecutive degree-3 vertices with hexagons, terminates in a patch satisfying the *"no two adjacent degree-3 vertices on the boundary"* condition. Once this is done, a single dilation eliminates all sets of three or more consecutive degree-2 vertices yielding a polygonal border. □

General graphene patches with curvature 6 will be considered later. The next lemma is easily verified.

Lemma 3. *Let $G = (V, E, F)$ be a polygonal graphene patch with curvature $c<6$. Then,*

(i) *The number of corners of G is* $6 - c$;
(ii) *The dilation of G is also polygonal with the same number of corners, with corresponding side lengths increased by 1 and with its perimeter increased by* $6 - c$.

Since every general graphene patch can be embedded on a polygonal patch and all subsequent dilations increase the perimeter by $6 - c$, we define $6 - c$ to be the *growth rate* of the patch G. The curvature of a graphene patch is 0 and its growth rate is 6. If G is a general graphene patch with a single pentagonal defect, then its curvature is 1 and its growth rate is 5; if G is a patch with a single heptagonal defect, then its curvature is −1 and its growth rate is 7. In Sections 3.4.2 through 3.4.4, we consider the structure of general graphene patches with curvature different from 0 and the ways to visualize them; in Section 3.4.5, we consider general graphene patches with curvature 0 that nevertheless contain defects. We end with a consideration of how curvature and Kekulé structures are related.

3.4.2 Cones

A patch with a single pentagonal defect may be thought of as being embedded in a cone with the pentagon at its center. In the hexagonal tessellation, select one hexagon and construct a 60-degree wedge whose vertex is at the center of that face. Deleting the wedge and identifying the sides of the wedge results in a cone with a pentagon at its apex. This process is reversible:

on the left in Figure 3.10, we have embedded the border of a polygonal graphene patch with curvature 1 in the hexagonal tessellation of the plane. In the hexagonal tessellation, this border does not close; the dark-gray faces represent the same face and must be identified. To make this identification, we construct an equilateral triangle whose base is the segment joining the centers of the faces to be identified and whose third vertex is the apex of the cone. Note that the apex is the center of another face. In the second diagram, we have extended the sides of the triangle to form a wedge to be deleted. In the third diagram, we have deleted a slightly skewed 60-degree wedge with the same apex. Identifying the faces and closing this gap results in a general graphene patch with a single pentagonal face. In the second diagram of Figure 3.10, we have added (yellow) faces to the border to enlarge this polygonal patch to a polygonal patch with a regular border and the pentagon at its center. What about other general graphene patches with curvature 1? If we enlarge their border enough, we can carry out the same construction yielding a pentagonal cone with successive borders that are regular pentagons. On the right in the figure, we have a general graphene patch with two pentagons and a heptagon, but with the same border. So the basic cone structure away from the original patch containing the defects is the same in this case, and in fact is the same in all cases. We may think of all general graphene patches with curvature 1 as truncated pentagonal cones.

Cones have been studied extensively, see [2]. We will consider one more case before simply summarizing the remaining cases. Let G be a polygonal graphene patch with curvature 2. We have $b_2 = b_3 + 4$, so it takes four 60-degree turns to close the border instead of the usual six 60-degree turns to close the border of a polygon in the hexagonal tessellation of the plane. Hence, we must remove two 60-degree wedges or, equivalently, one 120-degree wedge. Again, we embed the border in the hexagonal tessellation of the plane and join the centers of the faces that must be superimposed. Next, we construct the 30-degree –120-degree–30-degree isosceles triangle on this base; see Figure 3.11. In this case there are two possibilities: The apex could be the center of (1) a face or (2) a vertex. If the apex is the center of a face, then removing

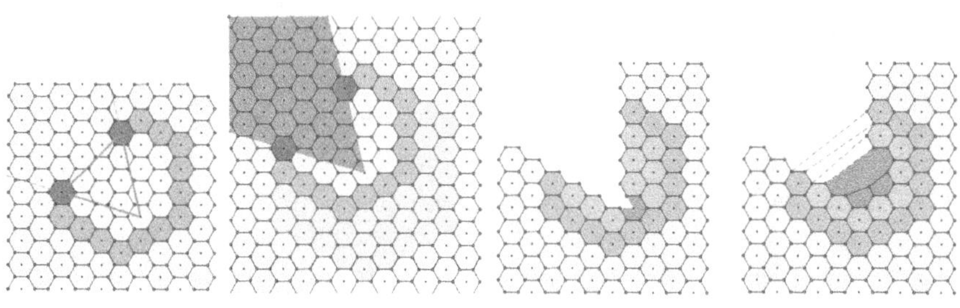

Figure 3.10 Curvature 1.

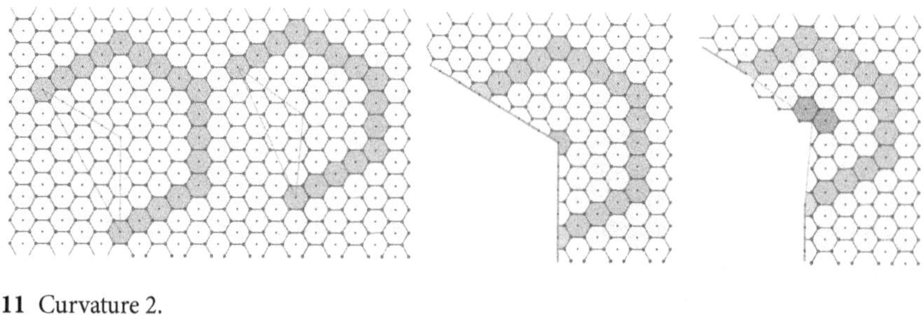

Figure 3.11 Curvature 2.

the 120-degree wedge and identifying its sides results in a cone with a quadrilateral face at the apex. If it is a vertex at the apex, then removing the 120-degree wedge and identifying its sides results in a cone with a degree-2 vertex at the apex. In this case, the degree-2 vertex is the center of a path of length 2 separating two hexagons. Replacing this path by a single edge yields a cone with two pentagons sharing an edge at the apex. Removing any two disjoint 60-degree wedges from different faces would produce a cone with two pentagonal faces. It is not obvious, but nevertheless true, that away from the defects, any general graphene patch with curvature 2 would correspond to one of the two cones constructed by removing a 120-degree wedge.

Why just two possible cones? Removing a θ-degree wedge and identifying its sides is possible only if the apex of the wedge is the center of a θ-degree rotation of the hexagonal tessellation—otherwise, the sides won't match. The rotations of the tessellation are: a 60-degree rotation about the center of a face; a 120-degree rotation about the center of a face or a vertex; a 180-degree rotation about the center of a face or the center of an edge; a 240-degree rotation about the center of a face or a vertex; and 300-degree rotation about the center of a face. For example, in the case of curvature 3, we are taking a half-plane and folding its side back on itself. If the folding point is the center of a face, that face becomes a triangle at the apex of a cone. If the folding point is the center of an edge, that edge folds back on itself. Removing that "half-edge" results in a path containing a degree-2 vertex separating a pentagon and a hexagon, and replacing the path by an edge results in a cone with a pentagon adjacent to a quadrilateral at its apex. Any cone containing a general graphene patch with curvature 4 is either a truncation of the cone with two adjacent quadrilaterals at its apex or a pentagon adjacent to a triangle at its apex. Curvature 5 admits just one possibility: the truncation of the cone with a quadrilateral adjacent to a triangle at its apex. Before we consider the cases of curvature 6, there is one more natural question to answer.

Are the two cones (say, for curvature 2) really different away from the defects? If so, how? Consider first the cone with a quadrilateral at the apex in Figure 3.11. The border in gray is a square of length 5 on each side. To its immediate right, the border of the cone with two pentagons at the center is a rectangle with one set of parallel sides of length 5 and the other of length 4. It is not too difficult to show that any rectangle enclosing the defects will have one pair of parallel sides of even length while the other side has odd length; a square is not possible. For curvature 3, one cone type will admit an equilateral triangle around the apex while, in the other cone type, none of the triangles around the apex can be equilateral. For curvature 4, the simple polygons enclosing the apex will be lunes: two straight paths with hexagons meeting at their ends. In one type of cone, the sides will have the same length, and, in the other, the side lengths will be different.

3.4.3 Curvature 6

The previous constructions can be summarized as follows: first, we filled in the cavities in the boundary, then we embedded the polygonal border in the hexagonal tessellation of the plane, constructed an isosceles triangle on the segment joining the centers of the repeated face, and, finally, we deleted the appropriate wedge defined by the apex angle of this triangle, identifying its sides. Two problems arise when trying to carry out this construction in the case of curvature 6. Note that $b_2 = b_3$. The first problem occurs in the process of filling in the cavities. This will either reduce the number of adjacent pairs of degree-3 vertices on the boundary to no such pairs or just one such pair. If there is a pair of adjacent degree-3 vertices on the boundary, then we have a straight path of hexagons, a 60-degree turn to the left (creating an adjacent pair of degree-3 vertices), a straight path of hexagons, a 60-degree turn to the right (creating an adjacent pair of degree-2 vertices), followed by a straight path of hexagons to the second copy of the repeated face. All subsequent borders will be congruent to this one, and the patch will not have a polygonal border. If there are no adjacent degree-3 vertices on the boundary, then the border is a straight path of hexagons between the two copies of the repeated face. In this case, the border is technically polygonal, but has no corners, as would follow from Lemma 3.

The second problem occurs in deleting the "wedge." Consider the wedges that were deleted under the construction in Section 3.4.2. Starting with the segment joining the centers of the two copies of the repeated face, we wish to construct the isosceles triangle with a 360-degree or 0-degree angle at its apex. This means the base angles of this "triangle" are 90 degrees, that is, the sides are parallel. Hence, we cut out the strip bounded by these two lines and identify them to get a cylinder! Another way of thinking about this case is to consider a patch containing six pentagonal faces: identify six hexagons in the tessellation, say, $f_1, \ldots f_6$ in clockwise order and, for illustration, assume that they are spread out roughly equidistant from some central point. Now, remove a 60-degree wedge at the center of f_1 opening away from the "center"; repeat this at f_2 and orient the wedge so that nearest sides of the two wedges are parallel. Continue all the way around and note that the sides of the first and last wedges that are closest to one another are also parallel. So when the wedges are removed and their sides are identified, the result is a cylinder consisting of six strips glued together.

This results in infinitely many possible borders to the cylinder that bound infinitely many "cylinder caps." The simplest possibility is a boundary consisting of a straight run of p hexagonal faces; each value for p gives a different cylinder. Slightly more complicated is the case with a 60-degree left turn at one face, f_1, and a 60-degree right turn at another face, f_2, with a straight run of p hexagons from f_1 to f_2 followed by a straight run of q hexagons from f_2 back to f_1. Again, each choice for p and q results in a different cylinder. With or without caps, these structures are called nanotubes.

3.4.4 Ruffles

Consider a defective patch G with curvature -1. In this case, the growth rate is 7 and the symmetric boundary that encloses the defects is a heptagon. On the left in Figure 3.12, we have embedded the boundary of a curvature -1 patch with an irregular boundary. As in the case of curvature 1, we construct the equilateral triangle from the segment joining the centers of the repeated face and the apex interior to the patch. But now, instead of excising the 60-degree wedge defined by the apex angle, we duplicate it, with one copy containing each end of the boundary path, and then identify the two edges of the wedge. As before, the two edges must match, so the apex must be at the center of a 60-degree rotation of the tessellation. In this

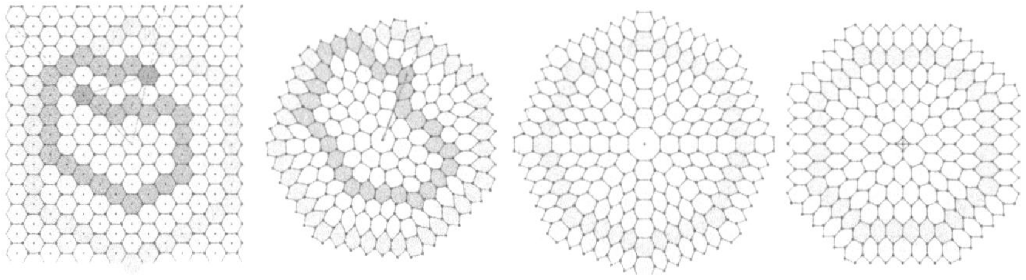

Figure 3.12 The left two images have curvature-1; the right two have curvature-2.

case, the face containing the apex is stretched into a heptagon (shown in the second diagram of Figure 3.12). We proceed in the same manner for all negative curvatures. Again, depending on the angle at the apex of the duplicated wedge, there are usually just two basic ruffles.

Introducing one defect, an $n + 6$-gon, at the center always yields a defective patch with curvature $-n$. These correspond to apex angles at the center of a face. Usually, there is a second option. In the case of curvature −2, the apex angle is 120 degrees and the apex could be at a vertex, which when the wedge is inserted becomes a degree-4 vertex. The third diagram in Figure 3.12 has an octagon at the center, resulting in a patch with curvature −2. The last diagram in the figure has a degree-4 vertex at its center. To make this a patch, we replace the degree-4 vertex by a small square. The resulting patch has a square at its center adjacent to four heptagons. Each heptagon contributes −1 to the curvature and the square contributes 2 for a total patch curvature of −2. Again, regardless of the actual set of defects of an $-n$ curvature patch, the type of the ruffle can be deduced by the shapes of the simple polygonal $n + 6$-gons around the defects. In the case corresponding to the prototype cone with an $n + 6$-gon at the center, a regular simple polygonal $n + 6$-gon is possible; in the other case with curvature −n, there are no regular simple polygonal $n + 6$-gons. In the case of curvature −3, the prototype “irregular ruffle” has three heptagons sharing a vertex at the center. For curvature −4, the prototype “irregular ruffle” has a degree-5 vertex at its center, which when converted to a patch has a pentagon bounded by five heptagons at its center. Curvature −5 has just one possible ruffle type, as was the case with curvature 5. As with curvature 6, there are infinitely many different ruffles with curvature −6.

3.4.5 0-Curvature Clusters and Flatness

One defining feature of the hexagonal tessellation of the plane is that every subpatch can be extended to a patch with a *regular* hexagonal border. The cones, cylinders, and ruffles that we have investigated satisfy this property for all patches away from the defects, but do not satisfy this property if the defects are included in the patch. So the tessellations in which they can be embedded are locally planar away from the defects but are quite distinct from the planar tessellation. Another way to see this is to note that, given the extended tessellation of a patch containing defects and having non-zero curvature, the annulus obtained by deleting a finite patch containing all of the defects cannot be embedded in the hexagonal tessellation of the plane. This leads us to the natural question: what if we have a patch with defects but with zero curvature?

We start by investigating the simplest 0-curvature patch with defects: one pentagon and one heptagon. With Diagram and Model **A** of Figure 3.13, we consider a specific example of this simplest case. To create a pentagon at face **f**, we delete a 60-degree wedge; then, to create the compensating heptagon at face **g**, we delete a 150-degree wedge from each of the two faces labeled g that we intend to identify. Once the deleted portion of the plane (shaded) is removed and the corresponding edges are identified, **f** becomes a pentagon and the two copies of **g** come together to form a heptagon. The curvature of this patch is 0, and so it can be embedded in a polygonal patch with a hexagonal-shaped border. However, no polygonal patch containing these defects will be regular and the annular patch obtained by deleting the interior of the patch with the blue hexagonal border cannot be embedded in the hexagonal tessellation of the plane. We say that a general patch with curvature 0 is *flat* if it can be embedded in a polygonal border that is a regular hexagon. In the next three examples, we illustrate that, by carefully arranging pentagon–heptagon pairs, we can construct 0-curvature general patches that also satisfy this flatness condition.

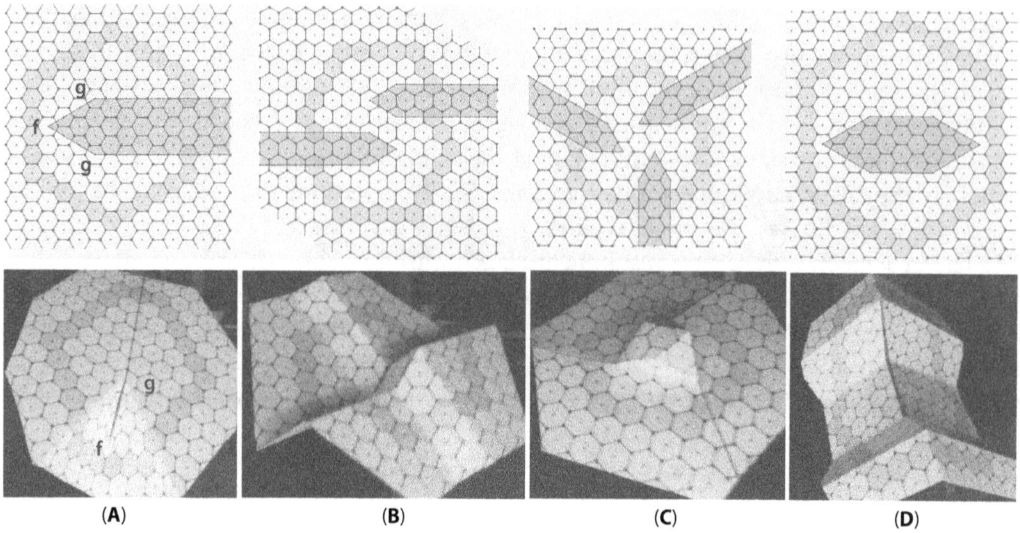

Figure 3.13 0 curvature and flatness.

In describing the shapes of these models, we find it convenient to use topographical terminology. Diagram and Model B, with two pentagon–heptagon pairs, yields a flat patch with two escarpments running perpendicular to one another. The sheets of the four quadrants away from the escarpments are all horizontal (parallel to one another but at different levels). Once the interior of the patch with the blue hexagonal border is deleted, the folds can be flattened out yielding a planar annular patch. Diagram and Model C, with three pentagon–heptagon pairs, yields an obviously flat patch with a bubble in the center. Diagram and Model D, again with two pentagon–heptagon pairs, yields a flat patch with two parallel escarpments perpendicular to a single ridge in the patch. The quadrants of this patch are not horizontal, so while it is technically flat, it introduces a permanent fold in the patch. But again, when the central patch containing the defects is deleted, the folds can be flatted out.

With carefully arranged pentagon–heptagon pairs, one can can construct flat patches with an infinite variety of shapes. For example, a repeated pattern of the diagram D pentagon–heptagon pair can be designed to give a flat general graphene patch with a family of parallel waves. Also, the possibilities using other defects are unbounded. We end this section with one such example. Consider a parallelogram ABCD embedded in the hexagonal tessellation of the plane:

(i) The vertices A, B, C, and D are at the centers of faces;
(ii) the angles at the opposite vertices A and C are 60 degrees;
(iii) the angles at the opposite vertices B and D are 120 degrees.

Now, excise this parallelogram. If we collapse this parallelogram by identifying vertices B and D, the faces containing vertices A and C become pentagons and the face resulting from identifying B and D becomes an octagon. The curvature is easily computed to be zero, and, since the parallelogram can be surrounded by a regular hexagonal-shaped border before it is excised, this patch is flat. By identifying A and C in this same parallelogram, we construct a flat general graphene patch with two quadrilateral faces and one 10-sided face.

3.4.6 Curvature and Perfect Kekulé Structures

In Section 3.3, we developed the conditions under which a general graphene patch admits three or one or no perfect internal Kekulé structures away from the defects. (recall that internal Kekulé structures may not match all degree-2 vertices on the boundary; perfect internal Kekulé structures are those in which every face is conjugated or void). In this last section, we wish to see just how those conditions relate to curvature. To start with, we note that the curvature of a general graphene patch will be odd (positive or negative) if and only if the patch has an odd number of odd defects and the curvature of a general graphene patch will be even (positive or negative) if and only if the patch has an even number of odd defects. So, by Theorem 4 (i), all general graphene patches with odd curvature will admit exactly one perfect internal Kekulé structure away from the defects which, by Theorem 5, can be extended to an internal Kekulé structure for the entire patch. For general graphene patches with even curvature, Theorem 4 (ii) tells us that either the annulus away from the defects admits no perfect internal Kekulé structure or it admits three perfect internal Kekulé structures, each of which can be extended to an internal Kekulé structure for the entire patch, by Theorem 5.

We consider flat general graphene patches first, and here the answer is easy to see. Since the annular patch can be embedded in the hexagonal tessellation of the plane, it admits a perfect face 3-coloring and hence three perfect internal Kekulé structures. In all of the remaining cases of even curvature, including 0-curvature but not flat, the question becomes: do the tessellations we construct admit a perfect face 3-coloring? If they do, then the polygonal border inherits this face 3-coloring. So the natural next question to ask is: can we tell from the polygonal border alone if the outer tessellation can be face 3-colored? The answer is yes. Before we can state the next result, we need a definition. A face 3-coloring of a polygonal border is *proper* if, as one progresses along a side of a polygonal border, the faces are colored in one of the two cyclic orders and the cyclic order changes at each corner. In other words, the cyclic order of colors alternates as it moves from one side to another side around the border.

Theorem 8. *Let G be a general graphene patch with even curvature; let B denote the border of a polygonal patch containing the defective faces; and let T denote the annular tessellation constructed by building out from B. Then, T admits a face 3-coloring if and only if B admits a proper face 3-coloring.*

Proof. Assume that T admits a face 3-coloring; we must show that the inherited face 3-coloring of B is proper. Start at a corner and move along a side towards the next corner. Let the corner face have color red and the next face on the side, blue. Now, rotating around the blue face in T the faces alternate red and green; hence, the next face along the side is green. Repeating this argument as we move along the side, we see that the face colors are in the red, blue, green cyclic order. Now, when we get to the next corner, the face just before the corner and the face just after the corner have the same color and the cyclic order is reversed.

Now assume that B admits a proper face 3-coloring. Each of the faces of the sides of the very next polygonal border inside of T is adjacent to two adjacent faces on B and hence can be assigned the third color. One easily checks that this coloring of the sides forces a coloring of the corners and that the result is a proper face 3-coloring of this next border. Inductively, we construct a face 3-coloring for T.

Corollary 1. *Let G be a general graphene patch with even curvature and a regular polygonal boundary. Then, the tessellation in which G can be embedded admits a face 3-coloring away from the defects.*

For the general graphene patches with even curvature but only non-regular polygonal boundaries, we must consider them individually. Consider first the case of curvature 2. Here, as we have seen, the polygonal border is a quadrilateral, say, with sides S, T, U, V in clockwise order. Assume that S is longer than U and add a new side S' at S. Note that S' is shorter than S by 1 and that both T and V have been increased in length by 1. We can repeat this process until S* and U have the same length. Now, apply this process to even up the lengths of T and V and note that the sides S* and U are lengthened the same amount in the process. Hence, we may always insist that our border is a rectangle, say, with side lengths p and q. Using tools not developed in this chapter, one can prove the next lemma.

Lemma 4. *Let G be a general graphene patch with curvature 2 and a rectangular boundary B having side lengths p and q. Then, B admits a proper face 3-coloring if and only if p and q are congruent modulo 3.*

For each of the cases of even curvature with no regular polygonal border, one can construct highly symmetric borders and give congruence conditions for the existence of a proper face 3-coloring of the border.

References

1. Brinkmann, G., Caporossi, G., Hansen, P., A constructive enumeration of fusenes and benzenoids. *J. Algorithms*, 45, 155–166, 2002.
2. Brinkmann, G. and Cleemput, N.V., Classification and generation of nanocones. *Discrete Appl. Math.*, 159, 1528–1539, 2011.
3. Chapman, J., Foos, J., Hartung, E.J., Nelson, A., Williams, A., Pairwise disagreements of the Kekule, Clar, and Fries numbers for benzenoids: A mathematical and computational investigation. *MATCH*, 80, 189–206, 2018.
4. Cyvin, S.J. and Gutman, I., *Kekulé Structures in Benzenoid Hydrocarbons*, Lecture Notes in Mathematical Chemistry (46) Springer-Verlag, 1988.
5. Graver, J.E. and Hartung, E.J., Internal Kekulé structures for graphene and general patches. *MATCH*, 76, 693–705, 2016.
6. Graver, J.E. and Hartung, E.J., Kekuléan benzenoids. *J. Math. Chem.*, 52, 977–989, 2014.
7. Graver, J.E., Hartung, E.J., Souid, A., Clar and Fries numbers for benzenoids. *J. Math. Chem.*, 51, 8, 1981–1989, 2013.
8. Gutman, I. and Bosanac, S., Quantitative approach to Huckel rule the relations between the cycles of a molecular graph and the thermodynamic stability of a conjugated molecule. *Tetrahedron*, 33, 1809–1812, 1977.
9. Hansen, P., Jaumard, B., Sachs, H., Zheng, M., Finding a Kekulé structure in a benzenoid system in linear time. *J. Chem. Inf. Comput. Sci.*, 35, 561–567, 1995.
10. Randic, M., Aromaticity of polycyclic conjugated hydrocarbons. *Chem. Rev.*, 103, 3349–3605, 2001.
11. Sachs, H., Perfect matchings in hexagonal systems. *Combinatorica*, 4, 89–99, 1984.
12. Zhang, F., Guo, X., Zhang, H., Advances of Clar's Aromatic Sextet Theory and Randic's Conjugated Circuit model. *Open Org. Chem. J.*, 5, 87–111, 2011.

4

Interacting Electrons in Graphene

T. Stauber[1*], P. Parida[2], M. Trushin[3], M. V. Ulybyshev[4], D. L. Boyda[5,6] and J. Schliemann[4]

[1]*Departamento de Teoría y Simulacíon de Materiales, Instituto de Ciencia de Materiales de Madrid, CSIC, Madrid, Spain*
[2]*Department of Physical Sciences, Central University of Punjab, City Campus, Mansa Road, Bathinda, Punjab, India*
[3]*Centre for Advanced 2D Materials, National University of Singapore, Singapore*
[4]*Institute for Theoretical Physics, University of Regensburg, Regensburg, Germany*
[5]*Far Eastern Federal University, Vladivostok, Russia*
[6]*ITEP, B., Moscow, Russia*

Abstract
In this chapter, we develop a Hartree–Fock theory for electrons on a honeycomb lattice aiming to solve the Fermi velocity renormalization problem for graphene. Our model employs no fitting parameters (like an unknown band cut-off) but relies on a topological invariant (crystal structure function) that makes the Hartree–Fock sublattice spinor independent of the electron–electron interaction. Agreement with the experimental data is obtained assuming static self-screening including local field effects. As an application of the model, we derive an explicit expression for the optical conductivity and discuss the renormalization of the Drude weight. The optical conductivity is also obtained via precise quantum Monte Carlo calculations, which compares well to our mean-field approach.

Keywords: Graphene, correlated electrons, Hartree-Fock, quantum Monte Carlo, Fermi velocity renormalization, optical conductivity, drude weight

4.1 Introduction

Graphene is a two-dimensional (2D) crystal of sp^2 bonded carbon atoms on a honeycomb lattice [1]. In the absence of electron–electron (e-e) interactions, valence and conduction bands touch only at two inequivalent Dirac points giving rise to a linear relativistic dispersion of the low-energy excitations $\epsilon(k) = v_F|k|$ with a Fermi velocity of $v_F \sim c/300$ [2–5]. The growth of research activities surrounding graphene-based materials and their applications has increased considerably in the past years due to their unique physical properties such as extremely high carrier mobilities [2, 6], fractional quantum Hall effect [7, 8], Klein tunneling [9, 10], long spin-relaxation length [11], and Andreev reflection [12]. However, the lack of a band gap limits the direct utilization of graphene in nanoelectronics and usual nanophotonic devices. To date, a variety of strategies has been explored in order to engineer the band structure of the

*Corresponding author: tobias.stauber@csic.es

Tobias Stauber (ed.) Handbook of Graphene: Volume 2, (95–126) © 2019 Scrivener Publishing LLC

graphene by surface adsorption [13, 14], chemical doping [15], interaction with a substrate [16–18], cutting into nanoribbons [19], or applying external field/strain [1, 20].

While the Dirac nature of the charge carriers is established, the issue of gap formation has remained far less clear. The generation of a gap in the quasiparticle spectrum of the suspended graphene is still illusive from experimental point of view. In QED, i.e., in 3+1 dimensions, the gap generation happens in the strong coupling regime [21, 22]. The chiral symmetry, which can break spontaneously at a large-enough Coulomb coupling, generates a gap in the quasiparticle spectrum. Suspended graphene provides a condensed-matter analog of strongly coupled quantum electrodynamics (QED) in 2+1 dimensions and many theoretical studies propose the possibility of broken symmetry gapped phases of graphene for sufficiently strong e-e interactions like the one that occurs in QED [23–28]. Despite the fact that this particular scenario seems to be not relevant for suspended graphene due to the balance between the Coulomb tail and the short-range repulsion [29, 30], other channels of spontaneous symmetry breaking can still generate the mass gap. For example, it was argued that the Dirac particles can acquire a mass due to the spontaneous breaking of the translational symmetry driven by the e-e interactions [31–33]. Also, time-reversal symmetry breaking can lead to a gap formation [34, 35].

Thus, the role of Coulomb interactions in graphene is still an open and important question [36], also in view of the regime of hydrodynamic electron liquids [37, 38] in which the e-e interaction represents the dominant scattering process [39, 40], see [41] for a recent review. The influence becomes especially crucial around the neutrality point where the Fermi surface shrinks to a single point and the Fermi liquid concept is not applicable directly. The role of e-e interactions in graphene has been revealed through measurements of the effective cyclotron mass [42], by scanning tunneling spectroscopy [43, 44], by direct angle-resolved photoemission spectroscopy (ARPES) of the Dirac cones [45], by quantum capacitance measurements [46], and also by Landau-level spectroscopy [47], that there is a Fermi velocity renormalization when lowering the electronic density close to half-filling. Moreover, in the absence of magnetic field, a spontaneous gap has still not been observed in a suspended monolayer graphene, experimentally, although a gap is found in case of a bilayer graphene [48, 49]. However, interaction-induced gaps have been observed in both single layer and bilayer graphene at many integer filling factors within the Hall ferromagnetism scenario [7, 8, 50].

A one-loop renormalization group (RG) and analogous Hartree–Fock (HF) analysis based on the continuous Dirac model predicts the following scaling behavior [42, 51]:

$$\frac{v_F^*}{v_F} = 1 + \frac{\alpha}{4}\ln\frac{\Lambda}{k}, \tag{4.1}$$

where $\alpha = \frac{1}{4\pi\epsilon_0\epsilon}\frac{e^2}{\hbar v_F} \approx 2.2/\epsilon$ is the fine-structure constant of graphene with the bare Fermi velocity v_F, ϵ characterizes the static dielectric environment, and Λ is the momentum cutoff. Equation 4.1 has been extended by several authors [27, 52–55], but a recent multi-loop expansion claims that perturbation theory may be inadequate particularly for suspended graphene [56]. Nonetheless, within a non-perturbative functional RG analysis, the perturbative series can be summed up to again yield Equation 4.1 with almost the same prefactor $\alpha/4$ as obtained from the HF approach [57, 58]. This suggests that a self-consistent mean-field theory will contain all the necessary ingredients to address interaction effects even close to the neutrality point.

The experimental data for the velocity renormalization can be fitted to Equation 4.1 by adjusting the band cut-off Λ as well as the effective dielectric screening constant ϵ [42]. Nevertheless, some ambiguities inherent to the RG approach can only be resolved by resorting to a realistic tight-binding Hamiltonian rather than working with an effective low-energy theory [59]. This especially holds for the optical conductivity, which has been the subject of persistent discussion regarding the constant C in the expansion $\sigma^*/\sigma_0 = 1 + C\alpha^* + O(\alpha^{*2})$, with $\sigma_0 = e^2/4\hbar$ the universal conductivity and $\alpha^*/\alpha = v_F/v_F^*$ [54, 60–67]. After almost 10 years of debate, one way to resolve this controversy could be an alternative, but well-defined numerical approach that still allows for analytical insight. We have therefore performed detailed HF calculations on the honeycomb lattice that hopefully will be able to shed some light on this issue from a different angle. We complement this with state-of-the-art quantum Monte Carlo (QMC) calculations, which now precisely determine the optical conductivity at energies of the order of the hopping parameter.

In contrast to earlier HF calculations preformed for a graphene quantum dot [68], the Dirac model [26], multilayer graphene [69], and graphene on a lattice [28], we now take into account self-screening and finite electronic densities, which are shown to be crucial to explain the experimental data *without* the need for a fitting parameter. In Section 4.2, we will show that the HF wave function is independent of the interaction strength even for a lattice model and relate this to a topological invariant that protects the chirality of the Dirac fermions around the nodal points. This reduces the numerical cost and further in Section 4.3 results in HF equations that in some limits are identical to the ones obtained from RG equations [70] and Hubbard–Stratonovich transformation [71]. Our solution is analyzed in Section 4.4 in connection with the energy dispersion and Fermi velocity renormalization, respectively.

The knowledge of the HF wave function further enables us to derive analytical expressions for the optical conductivity in Section 4.5, close to the value of [62] without screening (C = 1/4). Including self-screening, we obtain $C \approx 0.05$ for suspended samples in agreement with our Monte Carlo calculations described in Section 4.7 in great detail. In Section 4.6, we will focus on the Drude weight because studies of the electromagnetic response of various classes of correlated electron materials are often based on the f-sum rule [72]. Integrating the optical conductivity over the spectral range is then related to the Drude weight D, which is independent of the interaction in a Galilean invariant system. However, this is not the case for Dirac systems anymore, and e-e interactions modify the Drude weight in a nontrivial way, which is larger than the Drude weight of the non-interacting system [59, 73]. Sum rule analysis in Dirac systems [74, 75] thus has to be taken with care. Renormalization of the Drude weight is also of interest for plasmonics in Dirac systems as the plasmon energy scales as $\sqrt{D}$ [76]. Within our approach, we can analytically discuss the Drude weight for electronic densities close to half-filling.

4.2 The Model

We will model interacting Dirac fermions in graphene within a nearest-neighbor tight-binding model using the HF theory. We will first introduce the notation in the non-interacting case. We then outline the effective HF Hamiltonian. Special emphasis is placed on the screening and local field effects as well as on the derivation of the form factor.

4.2.1 The Non-Interacting Tight-Binding Model

The honeycomb lattice of graphene is a non-Bravais bipartite lattice, which means that it consists of two interpenetrating sublattices, each of them forming a triangular Bravais lattice. The lattice vectors are chosen as follows:

$$\boldsymbol{a}_1 = \frac{a}{2}\left(3,\sqrt{3}\right), \quad \boldsymbol{a}_2 = \frac{a}{2}\left(3,-\sqrt{3}\right), \tag{4.2}$$

where a ~ 1.42 Å is the C-C covalent bond length. The three nearest-neighbor vectors in real space are given by

$$\boldsymbol{\delta}_1 = \frac{a}{2}\left(-1,\sqrt{3}\right), \boldsymbol{\delta}_2 = \frac{a}{2}\left(-1,-\sqrt{3}\right), \boldsymbol{\delta}_3 = a(1,0). \tag{4.3}$$

The reciprocal lattice vectors $\boldsymbol{b}_1$ and $\boldsymbol{b}_2$ defined by the condition $\boldsymbol{a}_i \bullet \boldsymbol{b}_j = 2\pi\delta_{ij}$ are then given by

$$\boldsymbol{b}_1 = \frac{2\pi}{3a}\left(1,\sqrt{3}\right), \boldsymbol{b}_2 = \frac{2\pi}{3a}\left(1,-\sqrt{3}\right). \tag{4.4}$$

Within the nearest-neighbor tight-binding model, the general Bloch basis state is given by

$$\boldsymbol{\Psi}_{\mathbf{k}\lambda}(r) = \frac{1}{\sqrt{N_c}}\sum_j e^{ik\cdot(R_j+\eta_\nu)}\zeta(\boldsymbol{r}-\boldsymbol{R}_j-\boldsymbol{\eta}_\nu)\xi_\sigma, \tag{4.5}$$

where N_c is the number of unit cells, ξ_σ denotes the spin part of the wave function, ζ the one-electron atomic wave function (p_z orbital of carbon) at $\boldsymbol{R}_j + \boldsymbol{\eta}_\nu$, $\boldsymbol{\eta}_\nu$ is the position of sublattice ν in the crystallographic basis, and $\lambda = (\nu, \sigma)$ shall include sublattice and spin degrees of freedom. In the following, we will mostly choose $\boldsymbol{\eta}_a = (0,0)$, $\boldsymbol{\eta}_b = (a, 0)$.

The free tight-binding Hamiltonian can then be written as

$$H_{\mathbf{k}}^0 = -t\,|\phi_{\mathbf{k}}| \begin{pmatrix} m_{\mathbf{k}}^0 & e^{i\varphi_{\mathbf{k}}} \\ e^{-i\varphi_{\mathbf{k}}} & -m_{\mathbf{k}}^0 \end{pmatrix} + E_{\mathbf{k}}^0 1_{2\times 2}, \tag{4.6}$$

where $t = 3.1\text{eV}$, $\phi_{\mathbf{k}} = \sum_{i=1,2,3} e^{ik\cdot\delta_i}$ the structure factor, and $e^{i\varphi_{\mathbf{k}}} = \phi_{\mathbf{k}}/|\phi_{\mathbf{k}}|$. We also included a mass term and a constant energy term for sake of generality. The eigenenergies read

$$\varepsilon_{\mathbf{k}}^{0,\pm} = E_{\mathbf{k}}^0 \pm t\,|\phi_{\mathbf{k}}|\sqrt{1+{m_{\mathbf{k}}^0}^2}. \tag{4.7}$$

The eigenvectors are given by

$$|\psi_{\mathbf{k}}^{-}\rangle=\begin{pmatrix}\cos\frac{\vartheta_{\mathbf{k}}^{0}}{2}\\ \sin\frac{\vartheta_{\mathbf{k}}^{0}}{2}e^{-i\varphi_{\mathbf{k}}}\end{pmatrix},|\psi_{\mathbf{k}}^{+}\rangle=\begin{pmatrix}\sin\frac{\vartheta_{\mathbf{k}}^{0}}{2}\\ -\cos\frac{\vartheta_{\mathbf{k}}^{0}}{2}e^{-i\varphi_{\mathbf{k}}}\end{pmatrix}, \tag{4.8}$$

with $\cos\vartheta_{\mathbf{k}}^{0}=m_{\mathbf{k}}^{0}/\sqrt{1+m_{\mathbf{k}}^{0\,2}}$ and $\sin\vartheta_{\mathbf{k}}^{0}=1/\sqrt{1+m_{\mathbf{k}}^{0\,2}}$. We note that by defining the Bloch state as usual, i.e., $\mathbf{\Psi}_{\mathbf{k}\lambda}(\boldsymbol{r}+\mathbf{R}_i)=e^{i\mathbf{k}\cdot\mathbf{R}i}\,\mathbf{\Psi}_{\mathbf{k}\lambda}(\boldsymbol{r})$, we arrive at the eigenvectors of the Hamiltonian with a relative phase $e^{-i\varphi\mathbf{k}}$ between the first and second spinor component.

4.2.2 Mean-Field Theory

Let us introduce the Hamiltonian for the e-e interaction as

$$V=\frac{1}{2}\sum_{i,j;\lambda,\lambda'}c_{i\lambda}^{\dagger}c_{j\lambda'}^{\dagger}<i\lambda,j\lambda'|V|i\lambda,j\lambda'>c_{j\lambda'}c_{i\lambda}. \tag{4.9}$$

We define the Fourier transformation as follows:

$$c_{i\lambda}^{\dagger}=\frac{1}{N_c}\sum_{k\in 1.BZ}e^{-i\mathbf{k}\cdot(\mathbf{R}_i+\eta_v)}c_{k\lambda}^{\dagger},\ c_{k\lambda}^{\dagger}=\frac{1}{N_c}\sum_{i}e^{i\mathbf{k}\cdot(\mathbf{R}_i+\eta_v)}c_{i\lambda}^{\dagger}. \tag{4.10}$$

This is consistent with Equation 4.5, i.e., $\Psi_{\mathbf{k}\lambda}(\boldsymbol{r})=\langle\boldsymbol{r}|c_{k\lambda}^{\dagger}|0\rangle$. With the Coulomb propagator $U_{\mathbf{q}}^{\lambda,\lambda'}$ defined in the next section, we then have

$$V=\frac{1}{2A}\sum_{\mathbf{k},\mathbf{k}',\mathbf{q}}\sum_{\lambda,\lambda'}U_{\mathbf{q}}^{\lambda,\lambda'}c_{\mathbf{k}+\mathbf{q}\lambda}^{\dagger}c_{\mathbf{k}'-\mathbf{q}\lambda'}^{\dagger}c_{\mathbf{k}'\lambda'}c_{\mathbf{k}\lambda}, \tag{4.11}$$

where $A=A_CN_C$ is the system area with $A_c=3\sqrt{3}a^2/2$.

The interaction term within the mean-field approximation reads

$$\begin{aligned}H_{\mathbf{k}}^{ee}&=\sum_{\lambda,\lambda'}U_0^{\lambda\lambda'}\frac{1}{A}\sum_{\mathbf{k}'}\langle c_{\mathbf{k}'\lambda'}^{\dagger}c_{\mathbf{k}'\lambda'}\rangle c_{\mathbf{k}\lambda}^{\dagger}c_{\mathbf{k}\lambda}\\&-\frac{1}{A}\sum_{\mathbf{k}'\lambda,\lambda'}U_{\mathbf{k}-\mathbf{k}'}^{\lambda\lambda'}<c_{\mathbf{k}'\lambda'}^{\dagger}c_{\mathbf{k}'\lambda}>c_{\mathbf{k}\lambda}^{\dagger}c_{\mathbf{k}\lambda'}.\end{aligned} \tag{4.12}$$

The two terms are the Hartree and Fock (exchange) term, respectively. In the following, we will only consider the exchange interaction of the HF Hamiltonian as the Hartree term is neutralized by a positive background.

We numerically solved the total Hamiltonian $H = \Sigma_k H_k$, $H_k = H_k^0 + H_k^{ee}$ and found that the in-plane (azimuthal) angle of the pseudospin phase ϕ is not changed by the interaction strength. This can also be shown analytically, see below.

Here, we show that the phase between the components of the spinor is invariant with respect to the Coulomb interaction. This property is due to the boundary condition $c_{k+G,\lambda} = e^{-iG\cdot\eta_\nu} c_{k,\lambda}$ and the underlying three-fold rotational symmetry at the Dirac point.

Let us define $\phi_k^{ee} = \langle c_{k,b,\sigma}^\dagger c_{k,a,\sigma} \rangle /2$. The boundary condition implies

$$\phi_{k+G}^{ee} = e^{-iG\cdot(\eta_a - \eta_b)} \phi_k^{ee} = e^{iG_x a} \phi_k^{ee}, \tag{4.13}$$

where we choose $\eta_a = (0, 0)$, $\eta_b = (a, 0)$.

Due to the three-fold symmetry, we can write $\phi_k^{ee} = \sum_{i=1..3} f_i(k)$, where f_1 denotes an arbitrary function and $f_2(f_3)$ are related by a $\phi = 2\pi/3(\phi = 4\pi/3)$-rotation. At the Dirac point, the system is further rotationally invariant and we can set $f_i(k) = f(v_i \cdot k)$, with an arbitrary function f and v_i being the three vectors related by a rotation of $\phi = 2\pi/3$. In order to obtain the boundary condition (Equation 4.13), the arbitrary function must be related to the exponential and we thus have $f(x) = Ae^{ix}$ and $v_i = \delta_i$ the nearest-neighbor vectors with an arbitrary constant A.

We thus have $\phi_k^{ee} \propto \sum_i e^{i\delta_i \cdot k} = \phi_k =_0 \langle c_{k,b,\sigma}^\dagger c_{k,a,\sigma} \rangle_0 /2$. The structure factor in the case of the non-interacting system can thus be obtained from general symmetry arguments with the same boundary conditions as in Equation 4.13 and it is independent of the interaction or other perturbations that preserve the basic symmetries.

Finally, let us note that Equation 4.21 of [70] comes to the same conclusion based on the Feynman graphs with two external fermionic lines with the same wave vector k and a more formal proof of this relation can be found in App. C of the same reference.

The relative phase between the spinor components can thus be seen as a topological invariant and with $\phi_k^{ee} = < c_{k,b,\sigma}^\dagger c_{k,a,\sigma} > /2 \propto \phi_k$, the single particle H_k for any interaction strength can be written as

$$H_k = -t\,|\phi_k^{ee}| \begin{pmatrix} m_k & e^{i\varphi_k} \\ e^{-i\varphi_k} & -m_k \end{pmatrix} + E_k 1_{2\times 2}. \tag{4.14}$$

This parameterization will allow for a more efficient numerical analysis, which is crucial in order to address the physics close to the neutrality point. The eigenenergies of the interacting system now read

$$\varepsilon_k^{\pm} = E_k \pm t\,|\phi_k^{ee}|\sqrt{1+m_k^2}, \tag{4.15}$$

and the eigenvectors are again given by Equation 4.8 after replacing $\vartheta_k^0 \to \vartheta_k$ where $\cos\vartheta_k = m_k/\sqrt{1+m_k^2}$ and $\sin\vartheta_k = 1/\sqrt{1+m_k^2}$.

In the above equations, we have introduced the renormalized energy dispersion $|\phi_{\mathbf{k}}^{ee}|$ if $m_{\mathbf{k}} = 0$ and $E_{\mathbf{k}} = 0$, the total dimensionless, **k**-dependent mass $m_{\mathbf{k}}$, and the total **k**-dependent energy shift $E_{\mathbf{k}}$. These quantities are determined self-consistently by the following relations:

$$t|\phi_{\mathbf{k}}^{ee}| = t|\phi_{\mathbf{k}}| + \frac{1}{2A}\sum_{\mathbf{k}'} U_{\mathbf{k}-\mathbf{k}'}^{12} \frac{e^{i(\varphi_{\mathbf{k}'}-\varphi_{\mathbf{k}})}}{\sqrt{1+m_{\mathbf{k}'}^2}} F_{\mathbf{k}'}^{-}, \tag{4.16}$$

$$t|\phi_{\mathbf{k}}^{ee}| m_{\mathbf{k}} = t|\phi_{\mathbf{k}}| m_{\mathbf{k}}^0 + \frac{1}{2A}\sum_{\mathbf{k}'} U_{\mathbf{k}-\mathbf{k}'}^{11} \frac{m_{\mathbf{k}'}}{\sqrt{1+m_{\mathbf{k}'}^2}} F_{\mathbf{k}'}^{-}, \tag{4.17}$$

$$E_{\mathbf{k}} = E_{\mathbf{k}}^0 - \frac{1}{2A}\sum_{\mathbf{k}'} U_{\mathbf{k}-\mathbf{k}'}^{11} F_{\mathbf{k}'}^{+}. \tag{4.18}$$

Above, we defined $F_{\mathbf{k}}^{\pm} = n_F(\varepsilon_{\mathbf{k}}^{-}) \pm n_F(\varepsilon_{\mathbf{k}}^{+})$ with $n_F(\in) = (e^{\beta(\in-\mu)} + 1)^{-1}$ the Fermi distribution function and μ the chemical potential at a finite temperature $\beta = 1/(k_B T)$.

During the iteration process to obtain self-consistency, the chemical potential relative to the neutrality point will be kept constant. For $\mu = 0$, Equation 4.18 becomes trivial because $\Sigma_{\mathbf{k}'} U_{\mathbf{k}-\mathbf{k}'}^{11} = 0$. If we also neglect the self-consistency, we arrive at the same set of equation if one-loop corrections to the electron propagator and to the interaction potential are considered [71].

4.2.2.1 *Screening and Local Field Effects*

We will now specify the Coulomb propagator $U_{\mathbf{q}}^{\nu,\nu'}$ defined in Equation 4.11. If one incorporates screening due to tight-binding electrons, the interaction potential is not translationally invariant anymore and we have instead ($\lambda = (y,\sigma)$)

$$\langle i\lambda, j\lambda'|V|i\lambda, j\lambda'\rangle = \int d^2 r_1 \int d^2 r_2 \, |\zeta(\mathbf{r}_1 - \mathbf{R}_i - \boldsymbol{\eta}_\nu)|^2 \, V(\mathbf{r}_1, \mathbf{r}_2) |\zeta(\mathbf{r}_2 - \mathbf{R}_j - \boldsymbol{\eta}_{\nu'})|^2. \tag{4.19}$$

With the condition $V(\mathbf{r}_1, \mathbf{r}_2) = V(\mathbf{r}_1 + \mathbf{R}_i, \mathbf{r}_2 + \mathbf{R}_i)$, we obtain the following representation [77]:

$$V(\mathbf{r}_1, \mathbf{r}_2) = \frac{1}{A^2} \sum_{\mathbf{q}\in 1.BZ, \mathbf{G}_1, \mathbf{G}_2} e^{i(\mathbf{q}+\mathbf{G}_1)\mathbf{r}_1} e^{-i(\mathbf{q}+\mathbf{G}_2)\mathbf{r}_2} V(\mathbf{q}+\mathbf{G}_1, \mathbf{q}+\mathbf{G}_2), \tag{4.20}$$

where the sum runs over all reciprocal lattice vectors $\mathbf{G} = n\boldsymbol{b}_1 + m\boldsymbol{b}_2$ with integers $n, m \in \mathbb{Z}$. The interaction potential H^{ee} of Equation 4.11 is then defined with the Coulomb propagator given by

$$U_{\mathbf{q}}^{\nu,\nu'} = \frac{1}{A}\sum_{\mathbf{G},\mathbf{G}'} e^{i\mathbf{G}\eta_\nu} e^{-i\mathbf{G}'\eta_{\nu'}} f^*(\mathbf{q}+\mathbf{G}) f(\mathbf{q}+\mathbf{G}') V(\mathbf{q}+\mathbf{G},\mathbf{q}+\mathbf{G}') \tag{4.21}$$

with $f(\mathbf{q}) = \int d^2r|\zeta(r)|^2 e^{-i\mathbf{q}\cdot\boldsymbol{r}}$ and where we used $c_{\boldsymbol{k}+\mathbf{G},\lambda} = e^{i\mathbf{G}\cdot\boldsymbol{\eta}_\nu} c_{\boldsymbol{k},\lambda}$. We note that the definition is independent of the choice of the crystallographic basis η_ν and that $U_{\mathbf{q}}^{\nu,\nu'} = U_{-\mathbf{q}}^{\nu',\nu}$. It can also be shown that $U_{\mathbf{q}}^{1,1} = U_{\mathbf{q}}^{2,2}$ expected from equivalence of the two sublattices.

The screened potential shall be calculated within the RPA approximation:

$$V(\mathbf{q}+\mathbf{G},\mathbf{q}+\mathbf{G}') = A[\delta_{\mathbf{G},\mathbf{G}'} - v_{\mathbf{G}}(\mathbf{q})\chi_{\mathbf{G},\mathbf{G}'}(\mathbf{q})]^{-1} v_{\mathbf{G}'}(\mathbf{q}), \tag{4.22}$$

with $v_{\mathbf{G}}(\mathbf{q}) = v(\mathbf{q}+\mathbf{G}) = \dfrac{e^2}{2\epsilon_0\epsilon|\mathbf{q}+\mathbf{G}|}$. The (dynamical) response function is given by

$$\chi_{\mathbf{G},\mathbf{G}'}(\mathbf{q},\omega) = \frac{1}{A}\sum_{\boldsymbol{k},s,s'} \frac{n_F(\varepsilon_{\boldsymbol{k}}^s) - n_F(\varepsilon_{\boldsymbol{k}+\mathbf{q}}^{s'})}{\varepsilon_{\boldsymbol{k}}^s - \varepsilon_{\boldsymbol{k}+\mathbf{q}}^{s'} + \hbar\omega + i0} \langle \boldsymbol{k},s|e^{-i(\mathbf{q}+\mathbf{G})\hat{r}}|\boldsymbol{k}+\mathbf{q},s'\rangle\langle \boldsymbol{k}+\mathbf{q},s'|e^{i(\mathbf{q}+\mathbf{G}')\hat{r}}|\boldsymbol{k},s\rangle. \tag{4.23}$$

The eigenfunctions are given by

$$\langle \boldsymbol{r}|\boldsymbol{k},s\rangle = \sum_{\nu=a,b} \psi_{\nu,\boldsymbol{k},s} \Psi_{\boldsymbol{k},\lambda}(\boldsymbol{r}) \tag{4.24}$$

with the general Bloch eigenstate $\Psi_{\boldsymbol{k},\lambda}(\boldsymbol{r})$ given in Equation 4.5 and $\psi_{a,\boldsymbol{k},-} = \cos\frac{\vartheta_{\mathbf{k}}}{2}$, $\psi_{b,\boldsymbol{k},-} = \sin\frac{\vartheta_{\mathbf{k}}}{2} e^{-i\varphi_{\mathbf{k}}}$, $\psi_{a,\boldsymbol{k},+} = \sin\frac{\vartheta_{\mathbf{k}}}{2}$ and $\psi_{b,\boldsymbol{k},+} = -\cos\frac{\vartheta_{\mathbf{k}}}{2} e^{-i\varphi_{\mathbf{k}}}$. The matrix overlap function is given by

$$\left\langle \boldsymbol{k},s|e^{-i(\mathbf{q}+\mathbf{G})\hat{r}}|\boldsymbol{k}+\mathbf{q},s'\right\rangle = f(\mathbf{q}+\mathbf{G})\sum_{\nu=a,b} \psi_{\nu,\boldsymbol{k},s}^* \psi_{\nu,\boldsymbol{k}+\mathbf{q},s'} e^{-i\mathbf{G}\cdot\eta_\nu}. \tag{4.25}$$

Finally, we note that due to time-reversal symmetry, we have $\chi_{\mathbf{G},\mathbf{G}'}(\mathbf{q},\omega) = \chi_{\mathbf{G}',\mathbf{G}}(-\mathbf{q},\omega)$.

We note that the condition

$$U_{\mathbf{q}+\mathbf{G}}^{\nu,\nu'} = e^{-i\mathbf{G}\cdot(\eta_\nu - \eta_{\nu'})} U_{\mathbf{q}}^{\nu,\nu'} \tag{4.26}$$

is fulfilled in all cases, which is needed to fulfill the self-consistency equations.

For small in-plane momentum q, we can use the Dirac cone approximation for the evaluation of the sum over the Brillouin zone. This can be done analytically and we get for the doped case

$$\chi_{\mathbf{G},\mathbf{G}'}(q) = -\frac{g_s g_v k_F}{2\pi v_F \hbar} f(\mathbf{q}+\mathbf{G}) f(\mathbf{q}+\mathbf{G}') \times \left[\gamma_1 + \left(\frac{q}{4k_F}\gamma_2 \arccos(\frac{q}{2k_F}) - \gamma_3 \frac{1}{2}\sqrt{1-(\frac{2k_F}{q})^2} \right) \Theta(\frac{q}{2k_F}-1) + \frac{q}{4k_F}\gamma_2 \left(\arccos(\frac{q}{2\Lambda}) - \frac{\pi}{2} \right) + \gamma_4 \sqrt{(2\Lambda)^2 - q^2}/k_F \right] \tag{4.27}$$

with $\gamma_1 = (e^{-i\mathbf{G}\cdot\eta} + e^{i\mathbf{G}'\cdot\eta})/2$, $\gamma_2 = (e^{-i\mathbf{G}\cdot\eta} + e^{i\mathbf{G}'\cdot\eta})/2 - (1 - e^{-i\mathbf{G}\cdot\eta})(1 - e^{i\mathbf{G}'\cdot\eta})/4$, $\gamma_3 = (1 + e^{-i\mathbf{G}\cdot\eta})(1 + e^{i\mathbf{G}'\cdot\eta})/4$, $\gamma_4 = (1 - e^{-i\mathbf{G}\cdot\eta})(1 - e^{i\mathbf{G}'\cdot\eta})/8$. For the undoped case, this reduces to

$$\chi_{\mathbf{G},\mathbf{G}'}(q) = -\frac{g_s g_v}{2\pi v_F \hbar} f(\mathbf{q}+\mathbf{G}) f(\mathbf{q}+\mathbf{G}') \left[\gamma_2 \frac{q}{4} \arccos(\frac{q}{2\Lambda}) + \gamma_4 \sqrt{(2\Lambda)^2 - q^2} \right]. \tag{4.28}$$

The band cut-off Λ can be obtained from the exact numerical solution and we get $\Lambda = 2.29$ Å^{-1}.

4.2.2.2 Derivation of the Form Factor

Including lattice-scale effects, the explicit form of $f(q)$ can be calculated considering $\zeta(r)$ to be a hydrogenic $2p_z$ orbital type, $\zeta(r,\vartheta) = \frac{1}{4\sqrt{2\pi}} \left(\frac{Z}{a_0} \right)^{3/2} \frac{Zr}{a_0} e^{\frac{-Zr}{2a_0}} \cos(\vartheta)$ with Z the effective (screened) atomic charge. By taking the Fourier transformation of the charge distribution, the form factor is defined by

$$f(\mathbf{q}) = \int d\mathbf{r} e^{-i\mathbf{q}.\mathbf{r}} |\zeta(\mathbf{r})|^2. \tag{4.29}$$

We choose $\mathbf{q}$ to be in xy-plane $\mathbf{q}$ = q(1,0,0) and define $\tilde{a}_0 = \frac{a_0}{Z}$. This yields

$$\begin{aligned} f(\mathbf{q}) &= \frac{1}{32\pi \tilde{a}_0^3} \int d\mathbf{r} \left(\frac{r}{\tilde{a}_0} \right)^2 e^{-i\mathbf{q}\cdot\mathbf{r}} e^{-\frac{r}{\tilde{a}_0}} \cos^2\vartheta \\ &= \frac{1}{32\pi} \int_0^\infty dx x^4 e^{-x} \int_0^\pi d\vartheta \sin\vartheta \cos^2\vartheta \int_0^{2\pi} d\phi \sum_{n=0}^\infty \frac{(-iq\tilde{a}_0 \sin\vartheta \cos\phi)^n}{n!} \\ &= \sum_{n=0}^\infty (q\tilde{a}_0)^{2n} (-1)^n \frac{(n+1)(n+2)}{2}. \end{aligned} \tag{4.30}$$

We thus obtain the final result:

$$f(q)=\frac{1}{(1+q^2\tilde{a}_0^2)^3}. \tag{4.31}$$

With $<r^2>=30\tilde{a}_0^2$, we can reproduce the covalent bond radius of carbon by choosing $\tilde{a}_0=a/2/\sqrt{30}$. A larger effective radius could also consider screening effects from the sp^2-orbitals and we will choose sss $\tilde{a}_0=3a/2/\sqrt{30}$. The short-distance cut-off only has little influence on the scaling behavior of the Fermi velocity, i.e., the cut-off parameter only slightly increases with increasing short-distance screening corresponding to the absolute bandwidth with Λ = 1.75 Å^{-1} and Λ = 1.82 Å^{-1} for the two parameters mentioned above.

4.3 Numerical Implementation

As mentioned above, the relative phase between the spinor components is a topological invariant and the single-particle HF Hamiltonian $H_{\mathbf{k}}$ for any interaction strength can be written as

$$H_{\mathbf{k}}=-E_{\mathbf{k}}[\cos(\varphi_{\mathbf{k}})\sigma_x-\sin(\varphi_{\mathbf{k}})\sigma_y], \tag{4.32}$$

The above Hamiltonian is characterized by the renormalized energy dispersion $\varepsilon_{\mathbf{k}}$, which is determined self-consistently by the following equation:

$$\varepsilon_{\mathbf{k}}=\varepsilon_{\mathbf{k}}^0+\frac{1}{2A}\sum_{\mathbf{k}'\in 1.BZ}U(\mathbf{k}-\mathbf{k}')e^{i(\varphi_{\mathbf{k}'}-\varphi_{\mathbf{k}})}F_{\mathbf{k}'}, \tag{4.33}$$

where we introduced $\varepsilon_{\mathbf{k}}^0=t|\phi_{\mathbf{k}}|$ as the non-interacting dispersion relation, t is the tunnel-matrix element between nearest carbon atoms, and A denotes the sample area. Further, we have $F_{\mathbf{k}}=n_F(-\varepsilon_{\mathbf{k}})-n_F(\varepsilon_{\mathbf{k}})$ with the Fermi distribution function $n_F(\in)=(e^{\beta(\in-\mu)}+1)^{-1}$ and the chemical potential μ at a finite temperature $\beta=1/(k_BT)$. The Coulomb potential conserving the lattice symmetry and including the local field effects reads

$$U(\mathbf{q})=\sum_{\mathbf{G},\mathbf{G}'}e^{-i\mathbf{G}'a}f^*(\mathbf{q}+\mathbf{G})f(\mathbf{q}+\mathbf{G}')\times[\delta_{\mathbf{G},\mathbf{G}'}-v_{\mathbf{G}}(\mathbf{q})\chi_{\mathbf{G},\mathbf{G}'}(\mathbf{q})]^{-1}v_{\mathbf{G}'}(\mathbf{q}), \tag{4.34}$$

where G, G′ are the reciprocal lattice vectors, $f(\mathbf{q})$ is the form factor, $v_{\mathbf{G}}(\mathbf{q})=\frac{e^2}{2\varepsilon_0\varepsilon}\frac{1}{|\mathbf{q}+\mathbf{G}|}$ is the Fourier-transformed screened Coulomb potential, and $\chi_{\mathbf{G},\mathbf{G}'}(\mathbf{q})$ is the static polarizability matrix with local field effects. Neglecting the self-consistency by replacing $\varepsilon_{\mathbf{k}}\rightarrow\varepsilon_{\mathbf{k}}^0$ on the right-hand side of Equation 4.33, we arrive at the same equation that was obtained from a Hubbard–Stratonovich transformation on the lattice [71].

The Equation 4.34 is numerically solved using a grid with up to $N_c^2 = 15000^2$ lattice sites. This allows us to accurately discuss the scaling behavior for the undoped system as well as the Fermi-velocity renormalization for small finite electronic densities down to $n = 10^{10}\text{cm}^{-2}$.

In order to match the experimental results, it is crucial to incorporate self-screening effects. For this, the static polarization function χ has been calculated for momenta on a small grid $N_c = 300$, and later, we used bilinear interpolation to obtain the polarization on a larger grid $N_c = 3000$. The momentum summation for the polarization function has also been performed on a larger grid $N_c = 600$. Finally, we approximated x for small momenta by the analytical formula of the Dirac cone approximation, see Equation 4.27. Finally, the sum and matrix over the reciprocal lattice vectors $\mathbf{G} = n\mathbf{b}_1 + m\mathbf{b}_2$ is truncated by $|n|, |m| \le n_{max}$ with $n_{max} = 4, 6$.

For the self-consistent solution, we approximated the static polarization by the bare polarization divided by the renormalized Fermi velocity. This is justified following the work by Sodemann and Fogler [63] who found that $\epsilon(q) = 1 + \frac{\pi}{2}\alpha^* + O(\alpha^{*2})$ where $\alpha^* = \frac{1}{4\pi\epsilon_0\epsilon}\frac{e^2}{\hbar v_F^*}$ is the fine-structure constant with respect to the renormalized velocity. The self-consistent dispersion is scaled by a factor 1.4 compared to the dispersion coming from the bare polarization and is usually larger than the experimental data.

4.4 Fermi Velocity Renormalization

In Figure 4.1, the renormalized band structure of neutral graphene is shown for $t = 3.1\text{eV}$ ($\upsilon_F = 10^6\text{m/s}$) between the high symmetry points of the Brillouin zone for various coupling constant α, i.e., for different dielectric environments ϵ. By this, we can discuss suspended

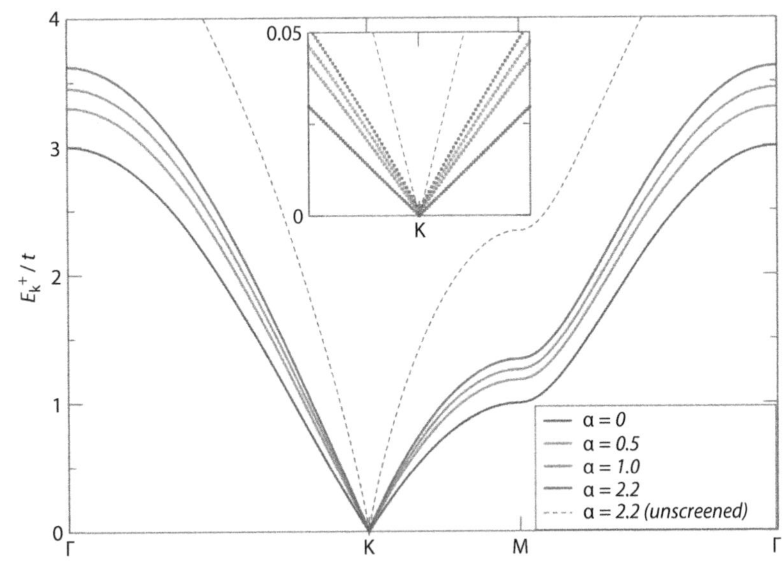

Figure 4.1 Band structure (undoped) along the high symmetry directions in the first Brillouin zone for various fine-structure constants α and self-screening (solid lines). For suspended graphene ($\alpha = 2.2$, $t = 3.1\text{eV}$), also, the unscreened dispersion is shown (dashed line). Inset: Close in of the dispersion around the Dirac cone.

graphene ($\epsilon = 1, \alpha = 2.2$), graphene on top of silicon ($\epsilon = 2.45, \alpha = 0.9$) or hexagonal boron nitride (hBN)-encapsulated graphene ($\epsilon = 4.9, \alpha = 0.45$). The solid lines refer to self-screened interaction, which is compared to the dispersion due to bare interaction for $\alpha = 2.2$ (dashed line). The inset shows the region close to the Dirac point where only slight deviations from the linear behavior can be seen.

At half-filling $\mu = 0$, $T = 0$ and no self-screening, Equation 4.33 is an explicit equation. This yields the analytical expression of Equation (4.1) when assuming $U(\mathbf{q}) = v_{G=0}(\mathbf{q})$ and converting the summation over the Brillouin zone by the Dirac cone approximation. Solving Equation 4.33 numerically, we obtain a fit for the cut-off parameter with $\Lambda \approx 1.75$ Å^{-1}. In Figure 4.2, we show the universal scaling behavior $(\frac{v_F^*}{v_F} - 1)/\alpha$ for different grid sizes $N_c =$ 3000, 15000. A fit to

$$v_F^* = v_F(1 + B\ln(\Lambda/k)) \tag{4.35}$$

with $B = \pi/4$ yields $\Lambda = 1.75$ Å^{-1}. Screening can be incorporated by $\alpha \to \alpha/\epsilon^{RPA}$, which agrees well with the full calculation. This is in contrast to [28], where $\Lambda \approx 20$ Å^{-1} was obtained, but agrees well with the usual argument of fixing Λ by conserving the total number of states in the Brillouin zone when compared to the tight-binding model, yielding $\Lambda \approx 1.58$ Å^{-1}. The precise value of Λ depends only weakly on the non-universal short-ranged Coulomb interaction.

For finite densities, we observe a kink in sss v_F^* at k_F, which is smeared out at finite temperatures. This kink is clearly an artefact of the unscreened interaction and screening cannot be incorporated afterwards. It is thus crucial to calculate the Fermi velocity renormalization including a self-screened interaction. The results are shown in Figure 4.3 for the two orthogonal directions. As expected, finite doping acts as a cut-off of the scaling law, but sizeable effects are still observed if we trace the Fermi velocity at the Fermi level k_F. We also obtain a larger prefactor of the logarithm (black dashed line) compared to the undoped case (magenta dotted-dashed line) along the KM-direction.

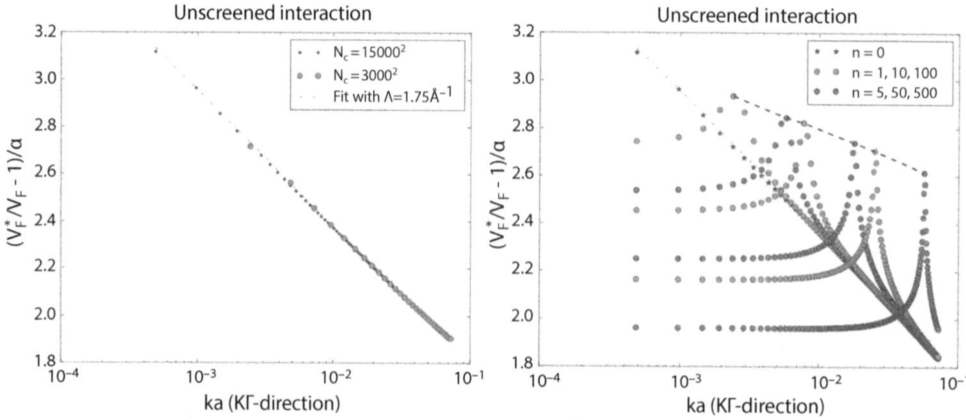

Figure 4.2 Left: Universal expression of the renormalised Fermi velocity screening as a function of the momentum for two different grids $N_c = 3000, 15000$. Also shown the fit to Equation 4.35 with $B = \alpha/4$. Right: Universal expression of the renormalized Fermi velocity screening as a function of the momentum for different electronic densities n in units of 10^{10}cm^{-2}. Also shown the fit for $n = 0$ (magenta dashed-dotted line) and the Fermi velocity at k_F (black dashed line).

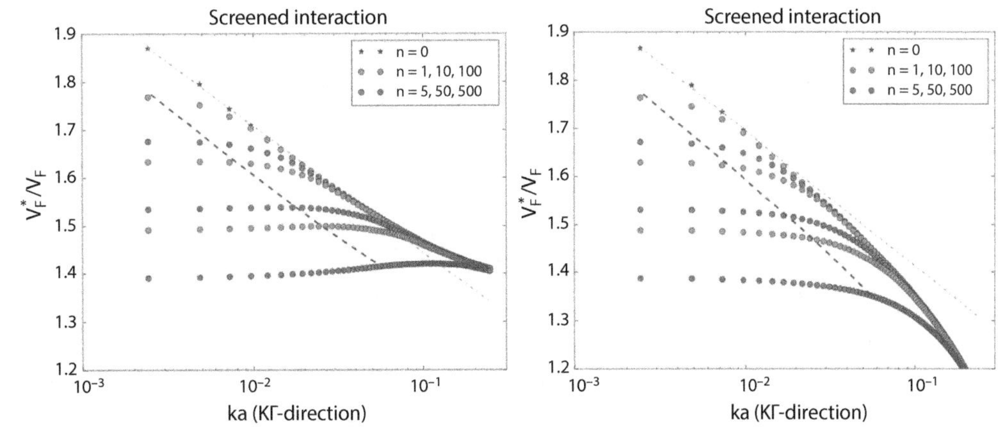

Figure 4.3 Renormalized Fermi velocity for suspended graphene ($\alpha = 2.2$ with $t = 3.1$eV) screening as a function of the momentum for the two orthogonal directions and for different electronic densities n in units of 10^{10}cm^{-2}. Also shown the fit for $n = 0$ (magenta dashed-dotted line) and the Fermi velocity at k_F (black dashed line).

In Figure 4.4, we present the numerical curves for the renormalized Fermi velocity at the neutrality point for various coupling strength α. We used $t = 3.1$eV, but the asymptotic results do not depend on the hopping parameter. The fits to Equation 4.35 are shown as blue lines. They are slightly different for the two orthogonal directions, and we will consider the one along the $K\Gamma$-direction since it is more closely related to Dirac cone physics [78, 79]. For unscreened interaction, the correction to the Fermi velocity $v_F^*/v_F - 1$ is proportional to α and the result is scale-independent of the hopping parameter t. At zero doping, self-screening can be incorporated by $\alpha \to \alpha/\epsilon^{RPA}$ because $\epsilon^{RPA} = 1 + \frac{\pi}{2}\alpha$ is momentum independent within the Dirac cone approximation. This yields good agreement with the experimental data of [42] for small densities $n \lesssim 20 \times 10^{10}\,\text{cm}^{-2}$ *without* the need for any fitting parameter, see magenta curve of Figure 4.5.

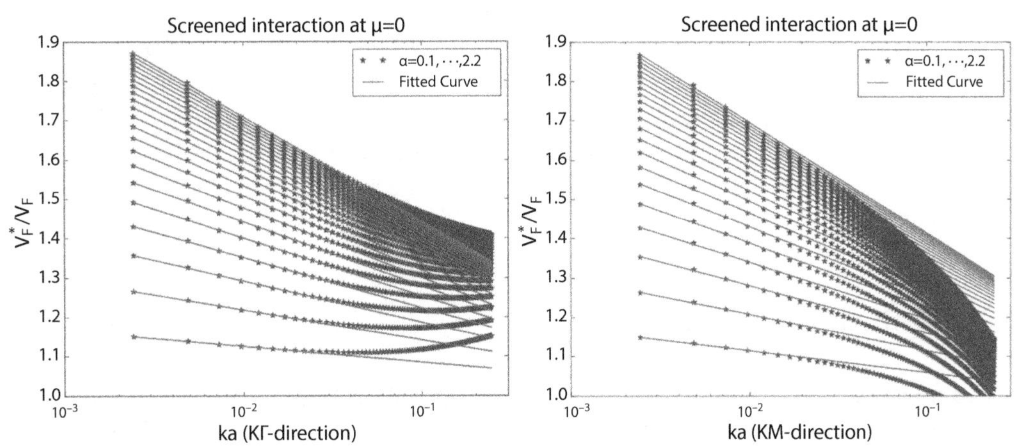

Figure 4.4 Renormalized Fermi velocity at the neutrality point ($\mu = 0$) as a function of the momentum for the two orthogonal directions and for various coupling constants $\alpha = 0.1, \ldots 2.2$ with $t = 3.1$eV. Also shown the fit functions to Equation 4.35.

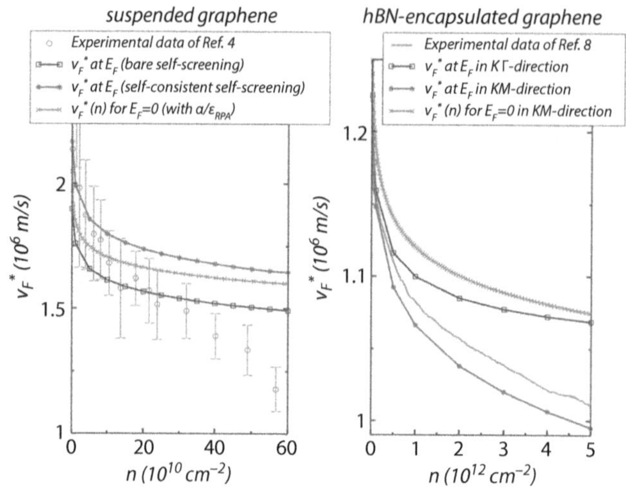

Figure 4.5 The renormalized Fermi velocity v_F^* for suspended ($\epsilon = 1$ and $t = 3.1$eV) and hBN-encapsulated ($\epsilon = 4.9$ and $t = 2.6$eV) graphene. Left panel is for suspended graphene: The experimental data of [42] compared to v_F^* at the Fermi surface based on the bare and self-consistent self-screened Coulomb interaction. Right panel is for hBN-encapsulated graphene: The experimental data of [46] compared to v_F^* at the Fermi surface in *KΓ* (black squares) and *KM* (blue stars) direction based on the bare self-screened Coulomb interaction. In both cases, the result for v_F^* at the neutrality point as function of the electronic density n is also shown based on the unscreened interaction with $\alpha = \alpha/\epsilon^{RPA}$.

For densities $n \lesssim 20-60\times10^{10}\,\text{cm}^{-2}$, there is a decrease of the Fermi velocity that cannot be accounted for by the results for neutral graphene. Since it might be due to screening at finite densities, we incorporated the momentum-dependent polarization function as outlined above. Even though the renormalized Fermi velocity now depends on α as well as on t in a nontrivial way, in the asymptotic limit, it becomes independent of t for $\mu = 0$.

In the left panel of Figure 4.5, we show the solution of Equation 4.33 using the bare (black squares) and the self-consistent (blue stars) polarization function. The bare solution agrees well with the experimental data for suspended graphene up to $n \lesssim 20-40\times10^{10}\,\text{cm}^{-2}$, but at higher densities, the experimental data drops whereas the theoretical value remains approximately constant (on a logarithmic scale). This has to be contrasted with the experimental data for hBN-encapsulated graphene [46], where good agreement is obtained over the whole density range up to s, see Figure 4.5 (right panel). A possible explanation for the failure of our theory could involve deformations of the suspended graphene sheet due to the applied gate potential.

4.5 Optical Response

Let us now discuss the interaction effects on the optical response. For this, we will couple the gauge field employing the Peierls substitution by replacing $\mathbf{k} \to \mathbf{k} + \frac{e}{\hbar}\boldsymbol{A}$ in the mean-field Hamiltonian $H_{\mathbf{k}}$. This procedure provides the correct vertex correction such that the optical f-sum rule is satisfied.

Assuming a linearly polarized light along the i-direction with $i = x,y$, the current operator reads

$$j_i = -\frac{\partial H}{\partial A_i} = j_i^P + j_i^D A_i + O(A_i^2), \tag{4.36}$$

where j_i^P and j_i^D are the paramagnetic and diamagnetic contribution, respectively.

The real part of the conductivity $\sigma_{ii}(\omega)$ represents the optical absorption of the light. Using the Kubo formula within the linear response formalism, $\Re\sigma_{ii}$ can be split into two terms containing a regular part and a delta singularity:

$$\Re\sigma_{ii}(\omega) = \pi D_{ii}\delta(\omega) + \sigma_{ii}^{reg}(\omega), \tag{4.37}$$

where D_{ii} is the Drude weight corresponding to the charge stiffness. The regular part of the conductivity reads for $\omega > 0$

$$\sigma_{ii}^{reg}(\omega) = \left(\frac{e}{\hbar}\right)^2 \frac{g_s\pi}{A\hbar\omega}\sum_{\mathbf{k}} \left|P_{\mathbf{k}}^i\right|^2 F_{\mathbf{k}}^- \delta(\varepsilon_{\mathbf{k}}^- - \varepsilon_{\mathbf{k}}^+ + \hbar\omega), \tag{4.38}$$

where $P_{\mathbf{k}}^i = \left\langle \psi_{\mathbf{k}}^+ \,|\, \hbar v_{\mathbf{k}}^i \,|\, \psi_{\mathbf{k}}^- \right\rangle$ is the interband momentum matrix element with the velocity operator along i-direction defined as $\hbar v_{\mathbf{k}}^i = \partial_{k_i} H_{\mathbf{k}}$, see Equation 4.36. For the mean-field Hamiltonian $H_{\mathbf{k}}$ (Equation 4.14), we finally obtain for $P_{\mathbf{k}}^i = \left\langle \psi_{\mathbf{k}}^+ \,|\, \hbar v_{\mathbf{k}}^i \,|\, \psi_{\mathbf{k}}^- \right\rangle$:

$$P_{\mathbf{k}}^i = -t\left|\phi_{\mathbf{k}}^{ee}\right|\left[i\partial_{k_i}\varphi_{\mathbf{k}} + \sin\vartheta_{\mathbf{k}}\partial_{k_i} m_{\mathbf{k}}\right]. \tag{4.39}$$

If $m_{\mathbf{k}} = 0$,

$$|P_{\mathbf{k}}^i|^2 = \frac{|\Phi_{\mathbf{k}}^{ee}|^2}{|\Phi_{\mathbf{k}}|^2}|P_k^{i,0}|^2, \tag{4.40}$$

where $|P_{\mathbf{k}}^{i,0}|^2 = \frac{t^2a^2}{16} g_{\mathbf{k}}$ corresponds to the value in non-interacting case and

$$g_{\mathbf{k}} = 18 + 4|\phi_{\mathbf{k}}|^2 - 24\Re\tilde{\phi}_{\mathbf{k}} + 18\frac{[\Re\tilde{\phi}_{\mathbf{k}}]^2 - [\Im\tilde{\phi}_{\mathbf{k}}]^2}{|\phi_{\mathbf{k}}|^2}, \tag{4.41}$$

with $\tilde{\phi}_{\mathbf{k}} = e^{-ik\cdot\delta_3}\phi_{\mathbf{k}}$ [80].

There has been considerable work on the renormalization of the conductivity due to e-e interaction. To leading order in $\alpha^* = \alpha\frac{v_F{}^*}{v_F}$, one gets via an RG analysis

$$\frac{\sigma}{\sigma_0} = 1 + C\alpha^* + O(\alpha^{*2}) \tag{4.42}$$

with $C_1 = 25/12 - \pi/2 \approx 0.51$ (hard cut-off) [55], $C_2 = 19/\alpha^*12 - \pi/2 \approx 0.01$ (soft cut-off) [60, 61] and $C_3 = 11/6 - \pi/2 \approx 0.26$ (dimensional regularization) [62]. Chiral anomalies have been claimed to be responsible for these discrepancies, and a perturbative analysis based on the tight-binding model yields $C = C_3$ [64].

Since for particle-hole symmetric Hamiltonian, there is no sign problem, the optical conductivity of graphene can also be studied using QMC calculations and values $C \approx 0.05$ were obtained for different interaction strengths [67]. Here, we obtain the analytical result $C = 0.25$ for unscreened interaction, but numerical results suggest lower values for self-screened interaction with $C \approx 0.05$ for suspended graphene ($\alpha = 2.5$ for $t = 2.7$eV).

For $m_k = 0$ and momenta close to the Dirac point, we have $|\phi_k| = \frac{3}{2}ak$ and an isotropic dispersion. The regular part of the real optical conductivity then reads

$$\sigma(\omega = \varepsilon_k^+) = \sigma_0 \frac{\varepsilon_k^+}{k\partial_k \varepsilon_k^+}. \tag{4.43}$$

As shown by Sharma and Kopietz [58], close to the neutrality point, we have the following functional behavior valid to all orders in α:

$$v_F^* = v_F \left[A + B \ln\left(\frac{\Lambda}{k}\right)\right]. \tag{4.44}$$

From Equation 4.16, we get $A = 1$ and integrating the above scaling law, we then have $\varepsilon_k^+ = \hbar v_F k[1 + B\ln(\Lambda/k) + B]$. For small ω, we then get the following result:

$$\sigma = \sigma_0 \left(1 + \frac{B v_F}{v_F^*}\right), \tag{4.45}$$

with $\sigma_0 = \frac{g_s g_v e^2}{16\hbar}$ universal conductivity. For unscreened interaction, we have the analytical result $B = \alpha/4$ and thus $C = 0.25$. For screened interaction, a good approximation is given by $B = \alpha/4\epsilon^{RPA}$ and thus $C(\alpha) = 0.25/(1 + \frac{\pi}{2}\alpha)$. These results also hold in the Dirac cone approximation. Thus, for momenta close to the Dirac point, the dispersion is isotropic with $v_F^*/v_F = 1 + C(\alpha)\alpha \ln \Lambda/k$. We then obtain for small frequencies ω the following result:

$$\frac{\sigma^*}{\sigma_0} = 1 + C(\alpha)\alpha \frac{v_F}{v_F^*}. \tag{4.46}$$

For unscreened interaction, we obtain *explicitly* $C = 1/4$ as mentioned in the introduction. This compares well with $C \approx 0.26$ obtained in [62] and [64].

For the self-screened interaction, $C \rightarrow C(\alpha)$ becomes a function with $C(\alpha \rightarrow 0) \rightarrow 0.25$ (unscreened limit) and $C(\alpha = 2.5) \approx 0.05$ for suspended graphene with $t = 2.7$eV. It is interesting to note that the universal factor of the scaling law $C(\alpha)$ is independent of all considered hopping matrix elements from $t = 2.6$eV to $t = 3.1$eV, and it compares well if self-screening is incorporated via RPA within the Dirac cone approximation, i.e., $C(\alpha) = [4(1+\frac{\pi}{2}\alpha)]^{-1}$. This is shown on the left of Figure 4.6; assuming a small, but finite electronic density would further decrease the constant $C(\alpha)$.

On the right-hand side, we plot the same for the conductivity of Equation 4.46 with $v_F^*/v_F = 1$. This is compared to the conductivity at $\hbar\omega = 0.7t$ as obtained from HF and QMC calculations. We obtain good agreement between the two approaches for suspended graphene, which justifies our mean-field approach. The Monte Carlo approach will be outlined below.

In Figure 4.7, we show the optical conductivity for suspended graphene with $t = 2.7$eV in the vicinity of the Dirac point, based on a system with $N_c = 3000^2$ lattice sites. Convergence is reached for energies up to $\hbar\omega \approx 1.2t$ for $N_{DeltaK} = 513^2$ grid points of the Brillouin zone around one Dirac cone (due to time-reversal symmetry, it is sufficient to only consider one Dirac cone).

The interacting curve crosses the non-interacting curve at $\hbar\omega \approx 1.12t$. Thus, one has to be careful to extract information from Monte Carlo simulations that are based on a comparison at this energy scale and compare them with the constant C of Equation 4.46, which is based on the Dirac cone approximation. Nevertheless, there is excellent agreement between the HF and Monte Carlo approach for $\hbar\omega \sim t$.

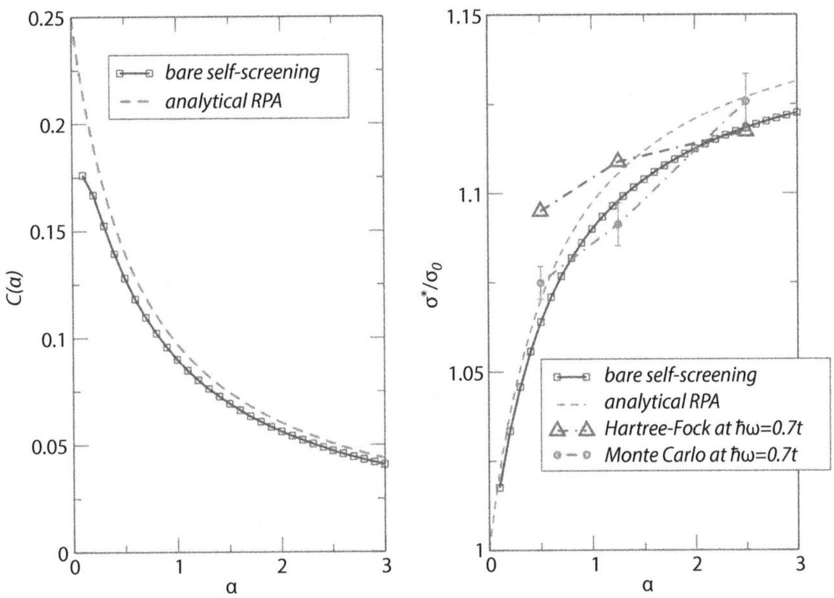

Figure 4.6 Left: The correction to the optical conductivity $C(\alpha)$ of Equation 4.46 compared to the result expected from the RPA, i.e., $[4\varepsilon^{RPA}]^{-1} = [4(1+\frac{\pi}{2}\alpha)]^{-1}$. Right: The same for the conductivity of Equation 4.46 with $v_F^*/v_F = 1$ compared to the conductivity at $\hbar\omega = 0.7t$ obtained from HF (blue triangle) and QMC (red circle) calculations.

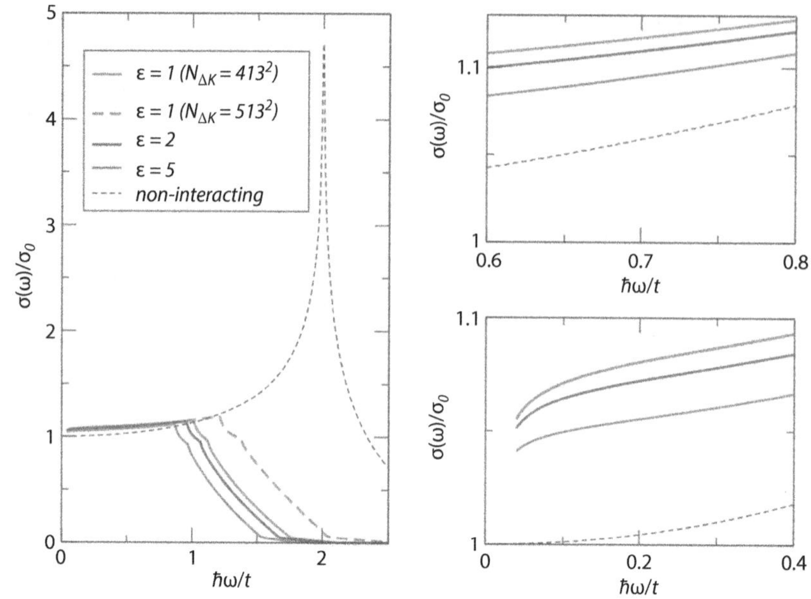

Figure 4.7 The optical conductivity for suspended graphene with $t = 2.7$eV calculated around one Dirac point including $N_{\Delta K} = 413^2$ (red) and $N_{\Delta K} = 513^2$ (blue) grid points based on a system with $N_c = 3000^2$ lattice sites. Also shown the optical conductivity without Coulomb interactions (dashed line).

In the lower-right panel, we show the behavior of σ for small energies. Noticeably, there is a large renormalization due to Coulomb interaction, which does not agree with the experimental results. We are, therefore, led to believe that suspended, neutral graphene contains electron-hole puddles with an average electronic density of $n \sim 5 \times 10^{11}\text{cm}^{-2}$. This leads to more effective screening in agreement with the results for the Fermi velocity renormalization.

In order to consistently include e-e interactions, the Coulomb potential needs to be periodic as discussed by Jung and MacDonald [28]. This introduces screening of the Coulomb interaction at small distances. If one further incorporates screening at large distances due to tight-binding electrons, the interaction potential is not translationally invariant anymore and additional non-diagonal local field effects need to be considered. To calculate the atomic orbital form factor $f(\mathbf{q}) = \int d\mathbf{r} e^{-i\mathbf{q}\cdot\mathbf{r}} |\zeta(\mathbf{r})|^2$ with $\zeta(\mathbf{r})$ being the one-electron atomic wave function, we take the full angular dependence of the wave function into account in contrary to [28] or [66].

4.6 Drude Weight

In a Galilean invariant system, the Drude weight D is independent of the interaction. However, this is not the case for Dirac systems and e-e interactions modify the Drude weight in a nontrivial way that is larger than the Drude weight of the non-interacting system [59, 73].

Making use of the fact that the HF wave function is given by the non-interacting wave function, one can obtain an analytical expression for the Drude weight from the optical f-sum rule.

The optical f-sum rule reads

$$\int_0^\infty \Re\sigma_{ii}(\omega)d\omega = -\frac{\pi}{2}\frac{\langle j_D^i \rangle}{A}, \tag{4.47}$$

where $j_D^i \equiv -\frac{e^2}{\hbar^2}\sum_{\mathbf{k}} \partial_{k_i}^2 H_{\mathbf{k}}$ is the diamagnetic current operator. From the above f-sum rule, we obtain the expression for the Drude weight:

$$D_{ii} = \left(\frac{e}{\hbar}\right)^2 \frac{g_s}{A} \sum_{\mathbf{k}\in 1.BZ, s=\pm} s[\partial_{k_i}^2 \varepsilon_{\mathbf{k}}] n_F(s\varepsilon_{\mathbf{k}}), \tag{4.48}$$

with $g_s = 2$ the spin degeneracy, and $i = x,y$. This is a general result, valid also in the presence of an arbitrary mass term.

$$D_{ii} = \left(\frac{e}{\hbar}\right)^2 \frac{g_s}{A} \sum_{\mathbf{k},s} [\partial_{k_i}^2 E_{\mathbf{k}}^s] f(E_{\mathbf{k}}^s). \tag{4.49}$$

Note that the expectation value of the diamagnetic current operator, $\langle j_D^i \rangle$, is not equal to the Drude weight since the Hellmann–Feynman theorem does not apply for the second derivative. For particles with mass m and parabolic dispersion, the well-known Drude weight $D = e^2 n/m$ with n the particle density is obtained [76]. The above formula generalizes this result to two bands, and we expect it to hold also for general multi-band systems (index $s = 1...n$).

Using partial integration, one gets at $T = 0$ in the thermodynamic limit $A_s \to \infty$

$$D_{ii} = g_s \left(\frac{e}{h}\right)^2 \int dk_{\bar{i}} \sum_{k_x | E_{\mathbf{k}}^s = \mu, s} |\partial_{k_i} E_{\mathbf{k}}^s| \mapsto \left(\frac{e}{h}\right)^2 \pi\mu, \tag{4.50}$$

where the last equation holds for isotropic Fermi surfaces. We thus obtain a direct link between the Drude weight and the Fermi energy μ. As it might have been expected, within our mean-field theory, the interacting electrons behave as independent quasi-particles with renormalized dispersion $\lambda\varepsilon_{\mathbf{k}}$ and we obtain

$$\frac{D^*}{D} = \frac{v_F^*}{v_F}. \tag{4.51}$$

A similar relation was obtained in [59] in the case of unscreened interactions. Changes due to trigonal warping and finite temperature can also be discussed by our approach.

4.7 Precise QMC Study of Graphene Conductivity

To back up our semi-analytical HF approach, we also perform QMC calculations for the following interacting tight-binding model for electrons at π-orbitals:

$$\hat{H} = -\sum_{\langle x,y\rangle,\sigma} t\left(\hat{a}^{\dagger}_{y,\sigma}\hat{a}_{x,\sigma} + h.c.\right) + \frac{1}{2}\sum_{x,y} V_{xy}\hat{q}_x\hat{q}_y. \tag{4.52}$$

Here, $\sum_{\langle x,y\rangle}$ means the summation over all pairs of nearest-neighbor sites, $\hat{a}^{\dagger}_{x,\sigma}$ ($\hat{a}_{x,\sigma}$) is the electron creation (annihilation) operator with spin σ, $t = 2.7$ eV is the hopping amplitude, $\hat{q}_x = \sum_\sigma \hat{a}^{\dagger}_{x,\sigma}\hat{a}_{x,\sigma} - 1$ is the charge operator at site x, and V_{xy} is the e-e interaction potential. We use the potentials calculated with the constrained RPA method [81] for suspended graphene (see [29] for details). These potentials correspond to the case $\epsilon = 1.0$ in our calculations. For $\epsilon \neq 1$, we uniformly rescale potentials at all distances by a factor of $1/\epsilon$. Since we can treat only finite sample, we impose periodic spatial boundary conditions as in [29], [82], and [83].

The calculation of graphene conductivity is performed with the same numerical procedure as was already described in [67]. We calculate Euclidean current–current correlator $G(\tau)$ using hybrid Monte Carlo algorithm, which is especially advantageous for systems of large spatial extent in the presence of long-range interaction. Since this particular Hamiltonian Equation 4.52 doesn't lead to sign problem in QMC calculations, we obtain numerically exact results without any further physical assumptions. Optical conductivity $\sigma(\omega)$ is extracted from the Green–Kubo relation:

$$G(\tau) = \int_0^\infty \sigma(\omega)K(\tau,\omega)d\omega, \tag{4.53}$$

$$K(\tau,\omega) = \frac{\omega\cosh(\omega(\beta/2-\tau))}{\pi\sinh(\omega\beta/2)}. \tag{4.54}$$

This integral equation is unstable; thus, we need a special numerical algorithm with some kind of regularization to find the stable solution. It's especially important in our case when the input data for correlator $G(\tau)$ is defined up to some statistical errors because it was obtained in statistical Monte Carlo procedure. In general, we rely on the Backus–Gilbert (BG) method proposed in [84], which we already employed in [67].

In order to improve accuracy and reduce statistical and systematic errors, we switched to substantially lower temperatures in comparison with our previous study (0.0625 and 0.125 eV in comparison with 0.5 eV in [67]). According to the general idea of the Euclidean time formalism, it leads to increased number of time slices. Unfortunately, it's impossible to rely on the old version of the BG method in this case, since the regularization scheme used in [84] and [67] doesn't work well for lattices with a large (~100) number of Euclidean time slices. Significant modifications of the algorithm were introduced in order to increase its stability and to improve the resolution in frequency. These modifications are discussed below alongside with the careful study of possible systematic errors.

We start from the brief description of the BG method. The estimator of the spectral function $\sigma(\omega)$ is defined as the convolution of the exact spectral function $\tilde{\sigma}(\tilde{\omega})$ with the resolution function $\delta(\omega,\tilde{\omega})$

$$\sigma(\omega)=\int_0^\infty d\tilde{\omega}\delta(\omega,\tilde{\omega})\tilde{\sigma}(\tilde{\omega}). \tag{4.55}$$

The resolution function is defined as a linear combination of the kernel profiles:

$$\delta(\omega,\tilde{\omega})=\sum_j q_j(\omega)K(\tau_j,\tilde{\omega}). \tag{4.56}$$

The coefficients $q_j(\omega)$ are determined by minimizing the width of the corresponding resolution function, concentrated around ω: $\partial_{q_j}D=0$, where D is defined as

$$D\equiv\int_0^\infty d\tilde{\omega}(\tilde{\omega}-\omega)^2\delta(\omega,\tilde{\omega}). \tag{4.57}$$

In practice, we minimize the width imposing the normalization condition:

$$\int_0^\infty d\tilde{\omega}\delta(\omega,\tilde{\omega})=1, \tag{4.58}$$

and sometimes the additional condition $\delta(\omega, 0) = 0$. The second condition is introduced in order to eliminate the influence of Drude peak on the estimator for optical conductivity. The result of this minimization yields

$$q_j(\omega)=\frac{W^{-1}(\omega)_{j,k}R_k}{R_nW^{-1}(\omega)_{n,m}R_m}, \tag{4.59}$$

where

$$W(\omega)_{j,k}=\int_0^\infty d\tilde{\omega}(\tilde{\omega}-\omega)^2K(\tau_j,\tilde{\omega})K(\tau_k,\tilde{\omega}),\quad R_n=\int_0^\infty d\tilde{\omega}K(\tau_n,\tilde{\omega}). \tag{4.60}$$

The matrix W is extremely ill-conditioned, with $C(W) \equiv \frac{\lambda_{\max}}{\lambda_{\min}} \approx O(10^{20})$. Thus, one needs to regularize the method in order to obtain stable results for a given set of data $G(\tau_i)$.

Previous studies employing the BG algorithm [67, 84] have used the regularization scheme based on the addition of the covariance matrix $C_{j,k}$ of the Euclidean correlator G to the kernel (Equation 4.60):

$$W(\omega)_{j,k} \to (1-\lambda)W(\omega)_{j,k} + \lambda C_{j,k}. \tag{4.61}$$

where λ is a small regularization parameter. The method worked well in [67] for lattices with 20 Euclidean time steps, but we failed with this approach in the calculations at small temperatures where the typical number of time steps is of the order of 100. Moreover, there is an important type of calculations where we apply the analytical continuation to formally exact correlator computed for the free fermions. It should be done in order to check the validity of the method and to study the systematic errors. Despite the fact that the numerical values of exact data contain only round-off errors, it's still impossible to use them in the Green–Kubo relation without regularization. Unfortunately, the covariance matrix is not defined for this kind of data; thus, we again need some alternative regularization methods. In this work, we used the combination of the so-called Tikhonov regularization and averaging of the correlator over intervals in Euclidean time. The Tikhonov regularization is widely used for the ill-posed problems of the form $Ax = b$. In this method, one seeks a solution to the modified least-squares function:

$$\min\left(\| Ax - b \|_2^2 + \| \Gamma x \|_2^2 \right), \tag{4.62}$$

where Γ is an appropriately chosen matrix. In the standard Tikhonov regularization, we choose $\Gamma = \lambda I$, where λ is again a small regularization parameter. Usually, for small λ, the solutions fit the data well but are oscillatory, while at large λ the solutions are smooth but do not fit the data.

In practice, the Tikhonov regularization is introduced during the inversion of the kernel matrix W, where we use the singular value decomposition (SVD):

$$W = U\Sigma V^{\top}, UU^{\top} = VV^{\top} = 1, \tag{4.63}$$

where $\Sigma = \text{diag}\,(\sigma_1, \sigma_2, \ldots, \sigma_N), \sigma_1 \geq \sigma_2 \geq \cdots \geq \sigma_N$. The inverse is thus easily expressed as

$$W^{-1} = V\Sigma^{-1}U^{\top}, \Sigma^{-1} = \text{diag}(\sigma_1^{-1}, \sigma_2^{-1}, \ldots, \sigma_N^{-1}). \tag{4.64}$$

Applying standard Tikhonov regularization simply modifies the matrix Σ in the following way:

$$\Sigma_{i,j}^{-1} \to \tilde{\Sigma}_{i,j}^{-1} = \delta_{ij} \frac{\sigma_i}{\sigma_i^2 + \lambda^2}. \tag{4.65}$$

Thus, one can see that the small eigenvalues that satisfy $\lambda >> \sigma_i$ are smoothly cut off.

The comparison of the Tikhonov and the covariance matrix regularization schemes is shown in Figure 4.8. We plot the width of the resolution function with center at $\omega = 0$ versus the statistical error of the reconstructed spectral function (at the same point $\omega = 0$) for both regularization methods. This data is taken from the recent paper where the BG method was used to reconstruct the dispersion relation of plasmons [85]. As one can see, the Tikhonov regularization works better in the sense that it provides better resolution in frequency being equally efficient in suppressing the statistical error. Or, in other words, for the Tikhonov regularization, the statistical error is smaller for the same resolution in frequencies.

Now let's turn to the second regularization technique, namely, the averaging of the correlator over some intervals in time. In this procedure, we take the correlator data $\{G(\tau_i); i = 0,1,\ldots, N_\tau - 1\}$ and map this to a new set $\{\tilde{G}(\tilde{\tau}_j); j=1,\ldots,N_{\text{int}}\}$ where

$$\bar{G}(\bar{\tau}_j) \equiv \frac{1}{\bar{N}_j}\sum_{i=1}^{\bar{N}_j} G(\tau_i^{(j)}), \tag{4.66}$$

$$N_\tau = \sum_{j=1}^{N_{\text{int}}} \bar{N}_j, 1 \leq \bar{N}_j < N_\tau. \tag{4.67}$$

Due to the linearity of the Green–Kubo relation, one can construct $\{\bar{K}(\bar{\tau}_j); j=1,\ldots,N_{\text{int}}\}$ in an analogous manner and use the new kernel and the new correlator in the BG method without any further modifications of the algorithm. The averaging increases the signal-to-noise ratio in the input data. From the point of view of the ill-defined inversion problem for the kernel matrix W, we reduce the matrix size, thus reducing the number of extremely

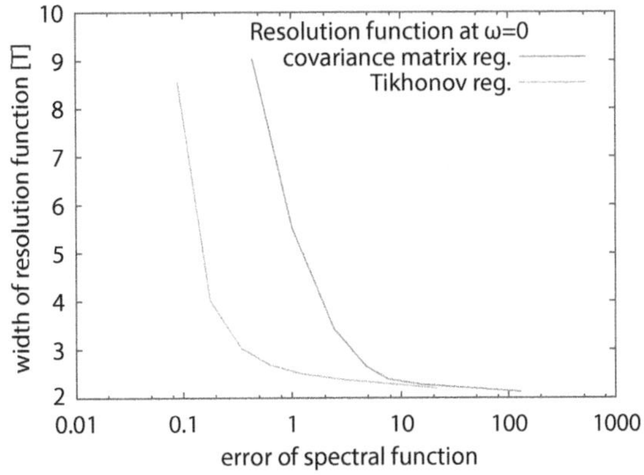

Figure 4.8 The half-peak width of the resolution function centered around $\omega = 0$ versus the relative statistical error of the reconstructed spectral function at the same point. The calculation is made for both covariance matrix and Tikhonov regularization. The width of the spectral function is plotted in the units of temperature. This example is taken from the paper devoted to the study of the dispersion relation of plasmons in strongly correlated models [85].

small numbers σ_i in Σ (see Equations 4.63 and 4.64) and making the SVD decomposition more stable.

We use lattices with spatial sizes 24 × 24, 36 × 36, 48 × 48, 72 × 72 and 96 × 96 (the two largest sizes only in the test calculations for the free fermions). In the most of our QMC runs, the temperature is equal to 0.125 eV with 80 steps in Euclidean time. As it was shown in [86], this discretization is good enough if we are far from the antiferromagnetic phase transition. In analytical continuation, we use Tikhonov regularization with constant $\lambda = 10^{-12}...10^{-11}$ and additional averaging over Euclidean time starting from the 10th step in time. The length of averaging intervals is equal to 10 time slices.

In the beginning, we should study the systematic errors appeared due to finite lattice size, non-zero temperature, and regularization during analytical continuation. We start from the check that the method applied to the free current–current correlator really reproduces the analytic profile of conductivity for the free tight-binding model. Figure 4.9 shows the results of this study. First of all, one can see that the profiles become stable in the limit of large lattices. Special attention should be paid to the plateau with center around $\omega = 0.7t$ (the point is defined as a position corresponding to the minimal value of the derivative $d\sigma/d\omega$). In the limit of large lattices, this plateau reproduces the correct value of non-interacting conductivity at $\omega = 0.7t$ ($\sigma_{NI} = 1.058\sigma$) with systematic error less than 1%, see Figure 4.10.

Another argument why we should look at this plateau is that it evolves into the plateau corresponding to the Dirac fermions in the low temperature limit. This feature is demonstrated in Figure 4.11. Indeed, once the temperature is decreased, the plateau shifts toward lower frequencies. The value of $\sigma(\omega)$ at the plateau simultaneously goes toward the standard limit of 2D Dirac fermions σ_0. Thus, we can conclude that the correct value of conductivity for Dirac fermions can be reproduced in our approach once we simultaneously increase the lattice size and decrease the temperature.

However, in practice, it's impossible to make QMC simulation at very large lattices and very small temperature (the latter needs increased number of steps in Euclidean time).

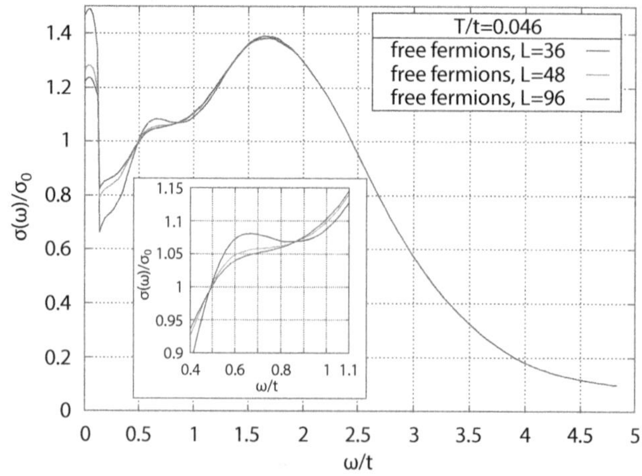

Figure 4.9 Finite size effects in the full profile $\sigma(\omega)$ in the case of free fermions. Discontinuities at small frequency are the consequence of the presence of the Drude peak (delta-function for the free fermions) at $\omega = 0$.

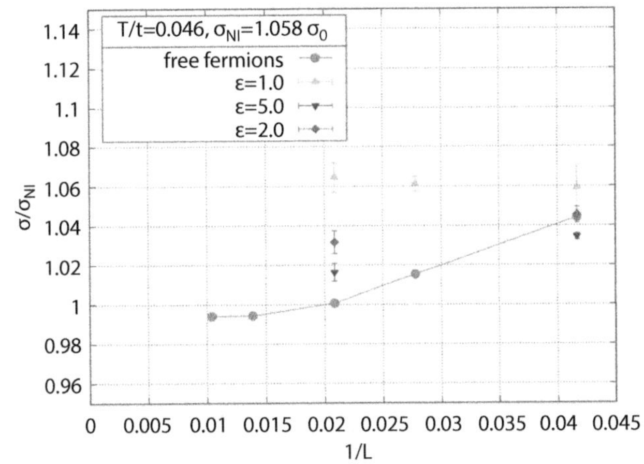

Figure 4.10 Dependence of optical conductivity $\sigma|_{\omega=0.7t}$ on the lattice size in the free case and in interacting systems. T = 0.125 eV in all calculations.

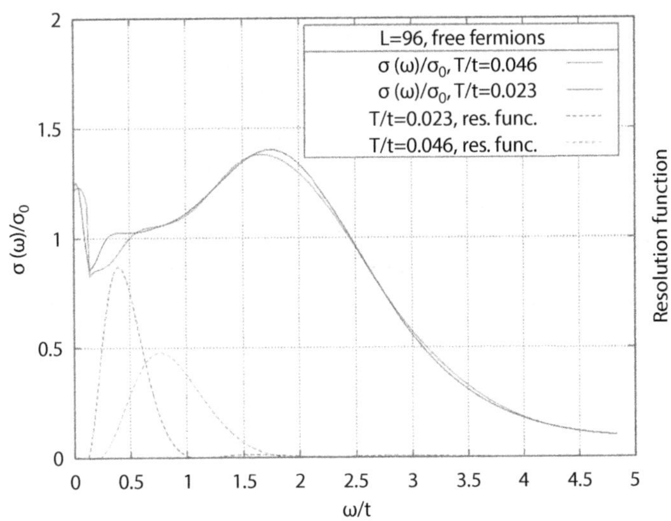

Figure 4.11 Comparison of $\sigma(\omega)$ profiles obtained from analytical continuation of free current–current correlator at two temperatures: T = 0.0625 eV and T = 0.125 eV. The resolution functions $\delta(\omega,\tilde{\omega})$ with centers (ω) at the plateau are plotted to illustrate the improvement of the resolution once we decrease the temperature. The plateau tends to correct value $\sigma = \sigma_0$ when temperature goes down.

Thus, we'll stop at lattice size equal to 48 × 48, where the correct value of $\sigma|_{\omega=0.7t}$ for the free fermions is already reproduced with good precision once we perform calculations at the temperature T = 0.125 eV. This frequency is used in all further calculations where we compare QMC results with analytic calculations.

The last potential source of systematic errors is the regularization. We checked the dependence of conductivity on the constant λ in the whole interval $\lambda = 10^{-12}...10^{-11}$ and the relative

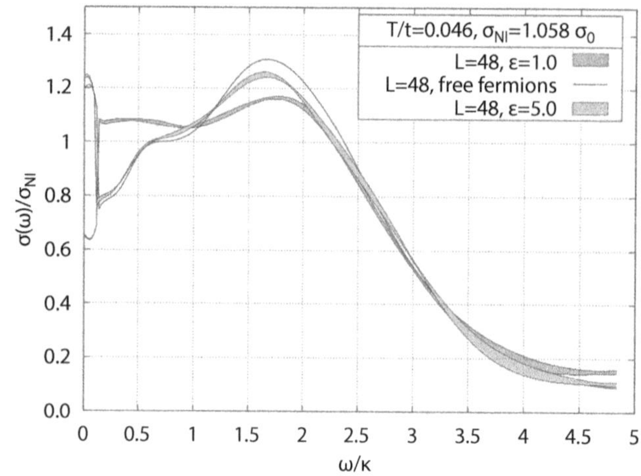

Figure 4.12 Comparison of interacting case (suspended graphene) with the results for the free fermions. Temperature is equal to 0.125 eV and lattice size $L = 48$ in both cases.

variation of conductivity $\sigma\,|_{\omega=0.7t}$ is smaller than 1%. Thus, we can conclude that systematic errors are under control since all of them are actually smaller than statistical uncertainties.

Conductivity profiles obtained in real QMC calculations are shown in Figure 4.12. Once can see that the general feature of $\sigma(\omega)$ profiles: the plateau around $\omega = 0.7t$, is preserved. Actually, for suspended graphene the plateau is even substantially wider. Results for conductivity at $\omega = 0.7t$ are summarized in Figure 4.10. Interestingly, the finite size effects for interacting systems become milder with increased interaction strength. This feature provides additional support for the statement that the finite size effects are under control, because the case of free fermions shows the upper limit for the finite size effects. Results for the largest lattice accessible in QMC (48 × 48) are used in the main text for the comparison with other approaches.

4.8 Conclusion

We presented a realistic tight-binding approach to the electron band structure of graphene renormalized by the Coulomb interaction. We identified a topological invariant that leaves the HF wave function unchanged even in the presence of the Coulomb interaction and found analytical expressions for the optical conductivity as well as for the Drude weight. By this, we were able to link our findings to the measured optical conductivity, which shows only little renormalization due to self-screened interactions. Precise Monte Carlo calculations yield good agreement for suspended samples and support our mean-field approach. We have also shown that the Fermi velocity and Drude weight renormalization are the same according to the expectations of a mean-field theory.

Our results compare well with experiments for suspended as well as for hBN-encapsulated graphene without invoking any fitting parameters. But in the case of suspended graphene, we were not able to account for the velocity renormalization in the case of larger densities $n \gtrsim 40 \times 10^{10}\text{cm}^{-2}$. This effect cannot be explained by our tight-binding model, and we expect the influence of ripples and corrugations, partially due to the applied gate, to be responsible for this effective screening of the long-ranged Coulomb interaction. This would also imply the absence of interaction renormalization of (high-density) plasmons in suspended graphene.

4.9 Acknowledgments

The authors acknowledge the interesting discussions with José González, Igor Herbut, Jeil Jung, Peter Kopietz, Vieri Mastropietro, and Anand Sharma. We also thank Kostya Novoselov for providing us with the experimental data. This work has been supported by Spain's Ministerio de Economía y Empresa (MINECO) under Grants No. FIS2017-82260-P and No. FIS2014-57432-P, by the Comunidad de Madrid under Grant No. S2013/MIT-3007 MAD2D-CM, and by Germany's Deutsche Forschungs-gemeinschaft (DFG) via SFB 689. M. Trushin is supported by the Director's Senior Research Fellowship from the Centre for Advanced 2D Materials at the National University of Singapore (NRF Medium Sized Centre Programme R-723-000-001-281). The work of M. Ulybyshev was supported by DFG grant BU 2626/2-1. D. Boyda acknowledges the support by RFBR Grant No. 16-32-00362-mol-a.

References

1. Novoselov, K.S., Geim, A.K., Morozov, S.V., Jiang, D., Zhang, Y., Dubonos, S.V., Grigorieva, I.V., Firsov, A.A., Electric Field Effect in Atomically Thin Carbon Films. *Science*, 306, 666–669, 2004.
2. Geim, A.K. and Novoselov, K.S., The rise of graphene. *Nat. Mater.*, 6, 183–191, 2007.
3. Novoselov, K.S., Geim, A.K., Morozov, S.V., Jiang, D., Katsnelson, M.I., Grigorieva, I.V., Dubonos, S.V., Firsov, A.A., Two-dimensional gas of massless Dirac fermions in graphene. *Nature*, 438, 197–200, 2005.
4. Castro Neto, A.H., Guinea, F., Peres, N.M.R., Novoselov, K.S., Geim, A.K., The electronic properties of graphene. *Rev. Mod. Phys.*, 81, 109–162, 2009.
5. Wallace, P.R., The band theory of graphite. *Phys. Rev.*, 71, 622–634, 1947.
6. Obradovic, B., Kotlyar, R., Heinz, F., Matagne, P., Rakshit, T., Giles, M.D., Stettler, M.A., Nikonov, D.E., Analysis of graphene nanoribbons as a channel material for field-effect transistors. *Applied Physics Letters*, 88, 142102, 2006.
7. Du, X., Skachko, I., Duerr, F., Luican, A., Andrei, E.Y., Fractional quantum Hall effect and insulating phase of Dirac electrons in graphene. *Nature*, 462, 192–195, 2009.
8. Bolotin, K.I., Ghahari, F., Shulman, M.D., Stormer, H.L., Kim, P., Observation of the fractional quantum Hall effect in graphene. *Nature*, 462, 196–199, 2009.
9. Katsnelson, M.I., Novoselov, K.S., Geim, A.K., Chiral tunnelling and the Klein paradox in graphene. *Nat. Phys.*, 2, 620–625, 2006.
10. Cheianov, V.V., Fal'ko, V., Altshuler, B.L., The focusing of electron flow and a veselago lens in graphene p-n junctions. *Science*, 315, 1252–1255, 2007.
11. Tombros, N., Jozsa, C., Popinciuc, M., Jonkman, H.T., van Wees, B.J., Electronic spin transport and spin precession in single graphene layers at room temperature. *Nature*, 448, 571–574, 2007.
12. Heersche, H.B., Jarillo-Herrero, P., Oostinga, J.B., Vandersypen, L.M.K., Morpurgo, A.F., Bipolar supercurrent in graphene. *Nature*, 446, 56–59, 2007.

13. Elias, D.C., Nair, R.R., Mohiuddin, T.M.G., Morozov, S.V., Blake, P., Halsall, M.P., Ferrari, A.C., Boukhvalov, D.W., Katsnelson, M.I., Geim, A.K., Novoselov, K.S., Control of graphene's properties by reversible hydrogenation: Evidence for graphane. *Science*, 323, 610–613, 2009.
14. Zhou, J., Wang, Q., Sun, Q., Chen, X.S., Kawazoe, Y., Jena, P., Ferromagnetism in semihydrogenated graphene sheet. *Nano Letters*, 9, 3867–3870, 2009.
15. Biel, B., Blase, X., Triozon, F.M.C., Roche, S., Anomalous doping effects on charge transport in graphene nanoribbons. *Phys. Rev. Lett*, 102, 096803, 2009.
16. Hunt, B., Sanchez-Yamagishi, J.D., Young, A.F., Yankowitz, M., LeRoy, B.J., Watanabe, K., Taniguchi, T., Moon, P., Koshino, M., Jarillo-Herrero, P., Ashoori, R.C., Massive Dirac Fermions and Hofstadter Butterfly in a van der Waals Heterostructure. *Science*, 340, 1427–1430, 2013.
17. Kharche, N. and Nayak, S.K., Quasiparticle band gap engineering of graphene and graphone on hexagonal boron nitride substrate. *Nano Letters*, 11, 5274–5278, 2011.
18. Chen, Z.-G., Shi, Z., Yang, W., Lu, X., Lai, Y., Yan, H., Wang, F., Zhang, G., Li, Z., Observation of an intrinsic bandgap and Landau level renormalization in graphene/boronnitride heterostructures. *Nature Comm.*, 5, 4461, 2014.
19. Han, M.Y., Özyilmaz, B., Zhang, Y., Kim, P., Energy band-gap engineering of graphene nanoribbons. *Phys. Rev. Lett.*, 798, 206805, 2007.
20. Pereira, V.M. and Castro Neto, A.H., Strain engineering of graphene's electronic structure. *Phys. Rev.*, 103, 046801, 2009.
21. Kogut, J.B., Dagotto, E., Kocic, A., On the existence of quantum electrodynamics. *Phys. Rev. Lett.*, 61, 2416–2419, 1988.
22. Fomin, P., Gusynin, V., Miransky, V., Vacuum instability of massless electrodynamics and the Gell-Mann-Low eigenvalue condition for the bare coupling constant. *Physics Letters B*, 78, 136–139, 1978.
23. Drut, J.E. and Lähde, T.A., Is graphene in vacuum an insulator? *Phys. Rev. Lett.*, 102, 026802, 2009.
24. Khveshchenko, D.V., Ghost excitonic insulator transition in layered graphite. *Phys. Rev. Lett.*, 87, 246802, 2001.
25. Herbut, I.F., Interactions and phase transitions on graphene's honeycomb lattice. *Phys. Rev. Lett.*, 97, 146401, 2006.
26. Trushin, M. and Schliemann, J., Pseudospin in optical and transport properties of graphene. *Phys. Rev. Lett.*, 107, 156801, 2011.
27. Stauber, T., Guinea, F., Vozmediano, M.A.H., Disorder and interaction effects in two-dimensional graphene sheets. *Phys. Rev. B*, 71, 041406, 2005.
28. Jung, J. and MacDonald, A.H., Enhancement of nonlocal exchange near isolated band crossings in graphene. *Phys. Rev. B*, 84, 085446, 2011.
29. Ulybyshev, M.V., Buividovich, P.V., Katsnelson, M.I., Polikarpov, M.I., Monte Carlo study of the semimetal-insulator phase transition in monolayer graphene with a realistic interelectron interaction potential. *Phys. Rev. Lett.*, 111, 056801, 2013.
30. Tupitsyn, I.S. and Prokof'ev, N.V., Stability of dirac liquids with strong coulomb interaction. *Phys. Rev. Lett.*, 118, 026403, 2017.
31. Gusynin, V.P. and Sharapov, S.G., Transport of Dirac quasiparticles in graphene: Hall and optical conductivities. *Phys. Rev. B*, 73, 245411, 2006.
32. Herbut, I.F., Theory of integer quantum Hall effect in graphene. *Phys. Rev. B*, 75, 165411, 2007.
33. Ezawa, M., Supersymmetric structure of quantum Hall effects in graphene. *Phys. Lett. A.*, 372, 924–929, 2008.
34. González, J., Dynamical breakdown of parity and time-reversal invariance in the manybody theory of graphene. *J. High Energy Phys.*, 175, 2013.

35. Marino, E.C., Nascimento, L.O., Alves, V.S., Smith, C.M., Interaction induced quantum valley Hall effect in *graphene. Phys. Rev. X*, 5, 011040, 2015.
36. Kotov, V.N., Uchoa, B., Pereira, V.M., Guinea, F., Castro Neto, A.H., Electron-electron interactions in graphene: Current status and perspectives. *Rev. Mod. Phys.*, 84, 1067–1125, 2012.
37. Torre, I., Tomadin, A., Geim, A.K., Polini, M., Nonlocal transport and the hydrodynamic shear viscosity in graphene. *Physical Review B*, 92, 165433, 2015.
38. Ho, D.Y., Yudhistira, I., Chakraborty, N., Adam, S., Microscopic theory for electron hydrodynamics in monolayer and bilayer graphene. *Phys. Rev. B*, 97, 121404(R), 2018.
39. Bandurin, D.A., Torre, I., Kumar, R.K., Ben Shalom, M., Tomadin, A., Principi, A., Auton, G.H., Khestanova, E., Novoselov, K.S., Grigorieva, I.V., Ponomarenko, L.A., Geim, A.K., Polini, M., Negative local resistance caused by viscous electron backflow in graphene. *Science*, 351, 1055–1058, 2016.
40. Crossno, J., Shi, J.K., Wang, K., Liu, X., Harzheim, A., Lucas, A., Sachdev, S., Kim, P., Taniguchi, T., Watanabe, K., Ohki, T.A., Fong, K.C., Observation of the Dirac fluid and the breakdown of the Wiedemann-Franz law in graphene. *Science*, 1058–1061, 351, 2016.
41. Lucas, A. and Fong, K.C., Hydrodynamics of electrons in graphene. *J. Physics: Condensed Matter*, 30, 053001, 2018.
42. Elias, D.C., Gorbachev, R.V., Mayorov, A.S., Morozov, S.V., Zhukov, A.A., Blake, P., Ponomarenko, L.A., Grigorieva, I.V., Novoselov, K.S., Guinea, F., Geim, A.K., Dirac cones reshaped by interaction effects in suspended graphene. *Nat. Phys.*, 7, 701–704, 2011.
43. Li, G., Luican, A., Andrei, E.Y., Scanning tunneling spectroscopy of graphene on graphite. *Phys. Rev. Lett.*, 2, 102, 176804, 2009.
44. Chae, J., Jung, S., Young, A.F., Dean, C.R., Wang, L., Gao, Y., Watanabe, K., Taniguchi, T., Hone, J., Shepard, K.L., Kim, P., Zhitenev, N.B., Stroscio, J.A., Renormalization of the graphene dispersion velocity determined from scanning tunneling spectroscopy. *Phys. Rev. Lett.*, 109, 116802, 2012.
45. Siegel, D.A., Park, C.-H., Hwang, C., Deslippe, J., Fedorov, A.V., Louie, S.G., Lanzara, A., Many-body interactions in quasi-freestanding graphene. *Proc. Nat. Acad. Sci.*, 108, 11365–11369, 2011.
46. Yu, G.L., Jalil, R., Belle, B., Mayorov, A.S., Blake, P., Schedin, F., Morozov, S.V., Ponomarenko, L.A., Chiappini, F., Wiedmann, S., Zeitler, U., Katsnelson, M.I., Geim, A.K., Novoselov, K.S., Elias, D.C., Interaction phenomena in graphene seen through quantum capacitance. *Proc. Nat. Acad. Sci.*, 110, 3282–3286, 2013.
47. Faugeras, C., Berciaud, S., Leszczynski, P., Henni, Y., Nogajewski, K., Orlita, M., Taniguchi, T., Watanabe, K., Forsythe, C., Kim, P., Jalil, R., Geim, A.K., Basko, D.M., Potemski, M., Landau Level spectroscopy of electron-electron interactions in graphene. *Phys. Rev. Lett.*, 114, 126804, 2015.
48. Martin, J., Feldman, B.E., Weitz, R.T., Allen, M.T., Yacoby, A., Local compressibility measurements of correlated states in suspended bilayer graphene. *Phys. Rev. Lett.*, 105, 256806, 2010.
49. Weitz, R.T., Allen, M.T., Feldman, B.E., Martin, J., Yacoby, A., Broken-symmetry states in doubly gated suspended bilayer graphene. *Science*, 330, 812–816, 2010.
50. Dean, C.R., Young, A.F., Cadden-Zimansky, P., Wang, L., Ren, H., Watanabe, K., Taniguchi, T., Kim, P., Hone, J., Shepard, K.L., Multicomponent fractional quantum Hall effect in graphene. *Nat. Phys.*, 7, 693–696, 2011.
51. González, J., Guinea, F., Vozmediano, M., Non-Fermi liquid behavior of electrons in the half-filled honeycomb lattice (A renormalization group approach). *Nuclear Physics B*, 424, 595–618, 1994.
52. Mishchenko, E.G., Effect of Electron-electron interactions on the conductivity of clean graphene. *Phys. Rev. Lett.*, 98, 216801, 2007.

53. Sheehy, D.E. and Schmalian, J., Quantum critical scaling in graphene. *Phys. Rev. Lett.*, 99, 226803, 2007.
54. Herbut, I.F., Juricic, V., Vafek, O., Coulomb interaction, Ripples, and the minimal conductivity of graphene. *Phys. Rev. Lett.*, 100, 046403, 2008.
55. Vafek, O. and Case, M.J., Renormalization group approach to two-dimensional Coulomb interacting Dirac fermions with random gauge potential. *Phys. Rev. B*, 77, 033410, 2008.
56. Barnes, E., Hwang, E.H., Throckmorton, R.E., Das Sarma, S., Effective field theory, three-loop perturbative expansion, and their experimental implications in graphene many-body effects. *Phys. Rev. B*, 89, 235431, 2014.
57. Bauer, C., Rückriegel, A., Sharma, A., Kopietz, P., Nonperturbative renormalization group calculation of quasiparticle velocity and dielectric function of graphene. *Phys. Rev. B*, 92, 121409, 2015.
58. Sharma, A. and Kopietz, P., Multilogarithmic velocity renormalization in graphene. *Phys. Rev. B*, 93, 235425, 2016.
59. Abedinpour, S.H., Vignale, G., Principi, A., Polini, M., Tse, W.-K., MacDonald, A.H., Drude weight, plasmon dispersion, and ac conductivity in doped graphene sheets. *Phys. Rev. B*, 84, 045429, 2011.
60. Mishchenko, E.G., Minimal conductivity in graphene: Interaction corrections and ultraviolet anomaly. *EPL (Europhysics Letters)*, 83, 17005, 2008.
61. Sheehy, D.E. and Schmalian, J., Optical transparency of graphene as determined by the fine-structure constant. *Phys. Rev. B*, 80, 193411, 2009.
62. Juricic, V., Vafek, O., Herbut, I.F., Conductivity of interacting massless Dirac particles in graphene: Collisionless regime. *Phys. Rev. B*, 82, 235402, 2010.
63. Sodemann, I. and Fogler, M.M., Interaction corrections to the polarization function of graphene. *Phys. Rev. B*, 86, 115408, 2012.
64. Rosenstein, B., Lewkowicz, M., Maniv, T., Chiral anomaly and strength of the electron-electron interaction in graphene. *Phys. Rev. Lett.*, 110, 066602, 2013.
65. Teber, S. and Kotikov, A.V., Interaction corrections to the minimal conductivity of graphene via dimensional regularization. *EPL (Europhysics Letters)*, 107, 57001, 2014.
66. Link, J.M., Orth, P.P., Sheehy, D.E., Schmalian, J., Universal collisionless transport of graphene. *Phys. Rev. B*, 93, 235447, 2016.
67. Boyda, D.L., Braguta, V.V., Katsnelson, M.I., Ulybyshev, M.V., Many-body effectson graphene conductivity: Quantum Monte Carlo calculations. *Phys. Rev. B*, 94, 085421, 2016.
68. Ozfidan, I., Korkusinski, M., Hawrylak, P., Theory of biexcitons and biexciton-exciton cascade in graphene quantum dots. *Phys. Rev. B*, 91, 115314, 2015.
69. Trushin, M. and Schliemann, J., Polarization-sensitive absorption of THz radiation by interacting electrons in chirally stacked multilayer graphene. *New J. Phys.*, 14, 095005, 2012.
70. Giuliani, A., Mastropietro, V., Porta, M., Lattice quantum electrodynamics for graphene. *Ann. Phys.*, 327, 461–511, 2012.
71. Astrakhantsev, N.Y., Braguta, V.V., Katsnelson, M.I., Many-body effects in graphene beyond the Dirac model with Coulomb interaction. *Phys. Rev. B*, 92, 245105, 2015.
72. Basov, D.N., Averitt, R.D., van der Marel, D., Dressel, M., Haule, K., Electrodynamics of correlated electron materials. *Rev. Mod. Phys.*, 83, 471–541, 2011.
73. Levitov, L.S., Shtyk, A.V., Feigelman, M.V., Electron-electron interactions and plasmon dispersion in graphene. *Phys. Rev. B*, 88, 235403, 2013.
74. Wu, L., Tse, W.-K., Brahlek, M., Morris, C.M., Aguilar, R.V., Koirala, N., Oh, S., Armitage, N.P., High-resolution faraday rotation and electron-phonon coupling in surface states of the bulk-insulating topological insulator Cu0:02Bi2Se3. *Phys. Rev. Lett.*, 115, 217602, 2015.
75. Post, K.W., Chapler, B.C., Liu, M.K., Wu, J.S., Stinson, H.T., Goldflam, M.D., Richardella, A.R., Lee, J.S., Reijnders, A.A., Burch, K.S., Fogler, M.M., Samarth, N., Basov, D.N., Sum-Rule

Constraints on the surface state conductance of topological insulators. *Phys. Rev. Lett.*, 115, 116804, 2015.
76. Stauber, T., Plasmonics in Dirac systems: From graphene to topological insulators. *J. Phys.: Condens. Matter*, 26, 123201, 2014.
77. Giuliani, G. and Vignale, G., *Quantum Theory of the Electron Liquid*, Cambridge Cambridge University Press, Cambridge, 2005.
78. Stauber, T., Schliemann, J., Peres, N.M.R., Dynamical polarizability of graphene beyond the Dirac cone approximation. *Phys. Rev. B*, 81, 085409, 2010.
79. Stauber, T., Analytical expressions for the polarizability of the honeycomb lattice. *Phys. Rev. B*, 82, 201404, 2010.
80. Peres, N.M.R. and Stauber, T., Transport in a clean graphene sheet at finite temperature and frequency. *Int. J. Mod. Phys. B*, 22, 2529–2536, 2008.
81. Wehling, T.O., Sasıoglu, E., Friedrich, C., Lichtenstein, A.I., Katsnelson, M.I., Blügel, S., Strength of effective coulomb interactions in graphene and graphite. *Phys. Rev. Lett.*, 106, 236805, 2011.
82. Buividovich, P.V. and Polikarpov, M.I., Monte Carlo study of the electron transport properties of monolayer graphene within the tight-binding model. *Phys. Rev. B*, 86, 245117, 2012.
83. Smith, D. and von Smekal, L., Monte Carlo simulation of the tight-binding model of graphene with partially screened Coulomb interactions. *Phys. Rev. B*, 89, 195429, 2014.
84. Brandt, B.B., Francis, A., Meyer, H.B., Robaina, D., Pion quasiparticle in the low temperature phase of QCD. *Phys. Rev. D*, 92, 094510, 2015.
85. Ulybyshev, M., Winterowd, C., Zafeiropoulos, S., Collective charge excitations and the metal-insulator transition in the square lattice Hubbard-Coulomb model. *Phys. Rev. B*, 96, 205115, 2017.
86. Buividovich, P., Smith, D., Ulybyshev, M., von Smekal, L., Competing order in the fermionic Hubbard model on the hexagonal graphene lattice. *PoS, LATTICE2016*, 244, 2016.

5

Computational Determination of the Properties of Graphene Nanoribbons

Frank J. Owens

Department of Physics, Hunter College, City University of New York, New York, N.Y.

Abstract

Computational Material Science has become an important tool in materials research. It has been used to identify new materials of technological interest and predict properties of existing materials such as geometric, vibrational and electronic structure. This paper reviews applications of computational methods to graphene nanoribbons. Graphene has enormous potential to increase the speed of electronic devices such as use in field effect transistors. One of the draw backs of graphene sheets is the absence of a band gap at the center of the Brillouin zone which is necessary for its function in electronic devices. However, it has been shown, both experimentally and theoretically, that narrow strips of graphene have an energy gap at the zone center. For this reason this review focuses on the properties of graphene nanoribbons. Examples of the application of theoretical methods, primarily density functional theory to armchair and zigzag graphene nanoribbons is presented. Calculations of the geometric, electronic and vibrational structure of pristine graphene ribbons as well as doped and defected graphene ribbons is discussed. Also calculations of potential applications is presented such as the possibility of graphene as a catalyst in fuel cells.

Keywords: Density functional theory, armchair nanoribbons, zigzag nanoribbons, electronic structure, vibrational frequencies, vacancies, catalysts

5.1 Computational Material Science

5.1.1 Application to Low-Dimensional Carbon Nanostructures

The field of computational material science has grown in the last 30 years to become an important part of materials research. There have been many applications to graphene. Computational materials science refers to theoretically predicting the properties of materials such as the structural, vibrational, and electronic properties. It can and has been used to predict the existence of new materials. For example, the possibility of the existence of C_{60} and its properties such as the vibrational frequencies were predicted long before its discovery. Theoretical modeling can also be used to determine the properties of existing materials and is particularly useful in predicting

Email: Owensfj@gmail.com

Tobias Stauber (ed.) Handbook of Graphene: Volume 2, (127–146) © 2019 Scrivener Publishing LLC

properties that are difficult to determine experimentally. Calculations of elastic constants give results quite close to experimental values and are much easier to do compared to experimental methods. The development of computational material science as an important tool in materials research is largely a result of the development of high-performance computers and theoretical models such as density functional theory (DFT). A critical issue in modeling solids is choice of the size of the solids, meaning the number of atoms or ions in the system. The number of substituents treated must be such that the results give values of properties corresponding to macroscopic systems. A cubic micron of copper has approximately 10^4 atoms, which is near the limit of numbers that can be handled by present-day computers using the most sophisticated theoretical models. Another limitation is the time scale, which should be less than 10^{-15} seconds in order to deal with atomic vibrations in the materials.

More recently, theoretical modeling has been applied to carbon nanotubes and graphene. Perhaps because of the smaller size of these nanostructures, which typically have less than 10^6 atoms, these limitations may be less of a problem. The objective of this paper is to review examples of computational predictions of new and interesting modifications of graphene ribbons that have application potential. This can provide guidance for identifying interesting new modifications to synthesize. For example, boron nitride nanoribbons, which have a structure analogous to graphene, have band gaps greater than 5 eV, limiting their electronic application potential. However, theoretical modeling has predicted that increasing the boron content relative to the nitrogen content or applying an electric field can significantly reduce the band gap to a value that makes them semiconductors. This may allow for the possibility of electronic applications. This paper will discuss the application of computational methods, particularly DFT, to graphene nanoribbons. The reason for concentration on graphene nanoribbons is that they have a band gap at the zone center whereas graphene sheets do not. Generally, for electronic applications, the materials have to be semiconductors.

5.1.2 DFT

Various modifications of molecular orbital theory have been used to calculate the electronic and vibrational properties as well as geometric conformation of solids. Here, a brief overview of molecular orbital theory is presented, which is needed to understand the approximations in DFT. The object of molecular orbital theory is to obtain approximate solutions of the nonrelativistic Schrodinger wave equation for a system of many atoms and electrons, which is given by,

$$\left[-\Sigma(\mathrm{h}/\pi)^2/2\mathrm{m}\left(\nabla_i^2\right)+\Sigma_{1<j}\,\mathrm{e}^2/\mathrm{r}_{\mathrm{ij}}-\Sigma_{\mathrm{iA}}\,\mathrm{Z}_{\mathrm{A}}\mathrm{e}^2/\mathrm{r}_{\mathrm{Ai}}+\Sigma_{\mathrm{A<B}}\,\mathrm{Z}_{\mathrm{A}}\mathrm{Z}_{\mathrm{B}}\mathrm{e}^2/\mathrm{R}_{\mathrm{AB}}\right]\psi_e=E\psi_e \quad (5.1)$$

The first term is the kinetic energy given by,

$$\nabla_i^2=\mathrm{d}^2/\mathrm{dx}_i^2+\mathrm{d}^2/\mathrm{dy}_i^2+\mathrm{d}^2/\mathrm{dz}_i^2 \quad (5.2)$$

The second term represents the electrostatic repulsion between the electrons. The third term is the electrostatic attraction between the electrons and the nuclei, and the fourth term is the electrostatic repulsion between the nuclei. E is the energy and ψ_e is the wave function.

Equation 5.1 omits the kinetic energy of the nuclei because the electrons move much more rapidly than the nuclei, which are much heavier. In effect, the motion of the nuclei is ignored. This is referred to as the Born–Oppenheimer approximation. The first successful solution of the Schrodinger equation of the hydrogen molecule, H_2, was by Heitler and London in 1927 [1]. An exact solution of the Schrodinger wave equation for systems having a large number of atoms and electrons is not possible. Theorists, however, have made a number of approximations to treat solids. One of these approximations is DFT, which has been proven to predict the properties of materials with a large number of atoms with reasonable accuracy.

Density functional theory is an extension of the Thomas–Fermi model of a many-electron atom. The theory treats the electrons as a gas of free electrons confined to a volume V by a spherically symmetric potential .In the context of the free electron model of metals for the electrons confined to a cubic volume, V, the Fermi level, the highest-filled level for a three-dimensional solid, is,

$$E_f = [(h/\pi)^2/2m]\ [3\pi^2 N/V]^{2/3} \tag{5.3}$$

where N is the number of electrons and M is their mass.

The depth of the potential at any value of r can be related to the density of electrons for that value of r by assuming that the depth of the potential is such that the energy levels are filled to the top, i.e., $E_f = -V[r]$. This yields a relationship between the potential and the density of electrons, $\rho = N/V$, given by,

$$-V[r] = [(h/\pi)^2/2m][3\ \pi^2\rho]^{2/3} \tag{5.4}$$

The model assumes that the V(r) does not change significantly in lengths compared to the wavelengths of the electrons. This means that a number of electrons can be localized in a volume in which V(r) is constant. The importance of this result is that it transforms the electrostatic interaction of the electrons with the nuclei and the other electrons (the second and third terms in Equation 5.1) to a form where the interaction between each electron and every other electron can be represented by the interaction of the electron and nuclei with a charge density. For a molecule having many nuclei and electrons, the electrostatic interaction of the electrons with each other and the nucleus becomes that of non-interacting electrons in an effective potential that is a function of the charge density. The Thomas–Fermi model of the electronic structure of atoms did not predict the energy levels of atoms very well primarily because it did not include the exchange interaction. However, it did provide a conceptual basis for the development of DFT by proposing that the electron-repulsive interaction between electrons could be replaced by a charge density.

In DFT, the ground state energy of a many-electron system is a function only of the electron density $\rho[r]^2$. The wave function must satisfy a Schrodinger-like equation having the form,

$$\left[-\Sigma(h/\pi)^2/2m(\nabla_i^2)+V_N[\mathbf{r}]+\int \rho[\mathbf{r}']/\mathrm{l}\mathbf{r}-\mathbf{r}']d^3r'+\in(\rho[\mathbf{r}])\right]\Psi_i(\mathbf{r})=E_i\,\Psi_i(\mathbf{r}) \tag{5.5}$$

All terms, except the electrostatic repulsion between the nuclei, are a function of the charge density. The first term is the kinetic energy. The kinetic energy density at each point is assumed to correspond to the kinetic energy density of a homogeneous non-interacting electron gas, which has been shown to be proportional to $\rho^{5/3}$. The second term is the interaction of the electrons with the nucleus. The third term is the electrostatic repulsion of the electrons with each other in terms of an electron density. The last term represents the exchange and correlation interactions. In DFT, accurate formulas for $\in$ have been developed from simulations of a uniform non-interacting electron gas. The model used to calculate the exchange interaction in DFT is based on the idea to treat the electronic structure of a simple metal such as lithium as a homogenous gas of electrons around a lattice of positive charges. This model is used in DFT because it is the only one that yields an exact and accurate form of the exchange interaction. The model is referred to as the local density approximation (LDA), and the exchange term has the form,

$$E_{XC}^{LDA}[\rho] = \int \rho[r] \in_{XC}(\rho[r])dr \tag{5.6}$$

where $\in_{XC}(\rho[r])$ is the energy density, i.e., the exchange plus correlation energy per electron. The use of the variational method to find the energy of a many-electron system always leads to a larger energy than the exact energy. The difference between the two energies is called the correlation energy. In a homogeneous electron gas with electron density "ρ," it is assumed that the electron density "ρ" varies slowly with position. The specific form of the exchange energy was obtained by Slater and is given by,

$$\in_{XC} = (-3/4)[3\,\rho[r]/\pi]^{1/3} \tag{5.7}$$

Substituting into Equation 5.6 gives the following form for the exchange energy:

$$\in_{XC}^{LDA}[\rho] = (-3/4)[3/\pi]^{1/3} \int \rho[r]^{4/3}\,dr \tag{5.8}$$

A brief overview of DFT approach within the LDA approximation has been presented without going into the details and proofs of the Hohenberg–Kohn existence and variational theorems [2]. The basic understanding that the reader should be left with is that DFT transforms the complex many-body problem of interacting electrons in the external potential of the nucleus to a tractable problem of non-interacting electrons moving in an effective potential that is a function of the electron charge density. With the DFT approach, it became possible to perform calculations on quite large structures with good to excellent accuracies.

5.1.3 An Example Application of DFT

Using the example of C_{60}, this section will examine how well DFT predicts the properties of this structure [3]. Figure 5.1 shows the structure of C_{60}. It consists of 12 pentagonal (5-sided) and 20 hexagonal (6-sided) carbon rings symmetrically arranged to form a soccer

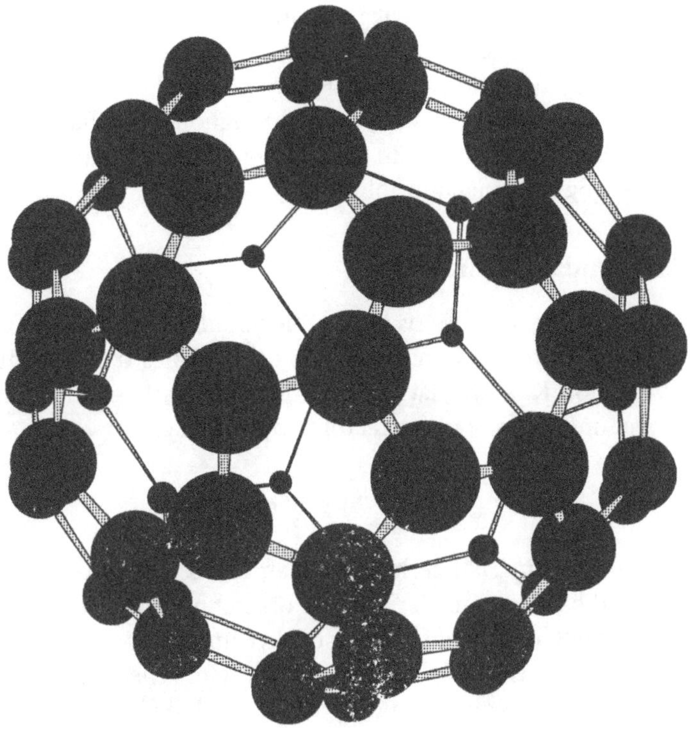

Figure 5.1 Structure of the C_{60} molecule calculated by DFT.

ball-like structure. Table 5.1 compares the bond lengths of the double- and single-carbon bonds of C_{60} calculated by DFT with experimental values. The DFT method predicts bond lengths in good agreement with experiment. The most intense observed Raman vibrational mode of C_{60} is the pentagonal pinch mode, which involves an expansion and contraction of the carbon pentagonal ring parallel to the surface of the ball. The DFT calculation predicts the frequency of this mode to within 1.6% of the experimental value. In order to examine the ability of the models to predict electronic structure, the results of calculating the vertical ionization energy are shown. This is calculated by taking the difference in the total energy of the minimum energy structure of C_{60} and the ion C_{60^+} having the same structure. Again, the DFT model gives reasonable agreement with the experimental value. Results such as these

Table 5.1 Comparison of some experimental and calculated properties of C_{60} using DFT [3].

Property	Experimental	DFT
C-C (A)	1.46	1.46
C = C (A)	1.40	1.40
IP (eV)	7.6 ± 0.2	7.22
$\dot{\omega}$(cm^{-1})	1493	1455

and the ability to treat large structures with reasonable computer time are the reason that the DFT theory is widely used to predict the properties of materials. It is also possible to employ periodic boundary conditions using DFT, which enables a calculation of the energy levels of solids as a function of the wave vector. Throughout this chapter, examples of the application of DFT will be used to obtain the energy levels and vibrational properties of various modifications of graphene nanoribbons.

5.1.4 Periodic Boundary Conditions

The use of periodic boundary conditions is possible because of Bloch's theorem.

The theorem states that for any wave function that satisfies the Schrodinger equation, there exists a wave vector, K, such that a translation by a lattice vector a is equivalent to multiplying the wave function by a phase factor exp(iK· a), i.e.,

$$\Psi_K(r+a) = \exp(iK\cdot a)\Psi_K(r) \tag{5.9}$$

Consider a one-dimensional lattice having a lattice vector a. To correctly obtain the energy levels, we would have to consider a lattice of length Na where N is infinite. This, of course, is not computationally possible. Now, if N were finite, it would be necessary that the wave function be zero at the ends of the lattice. Because of reflections from the ends, this would lead to standing waves, which would have to be included in the solution and do not exist in large crystals.

Periodic boundary conditions or Born–von Karman boundary conditions provide a mathematical device to get around the physical effects of boundaries. In one dimension, the device forms the lattice into a circle of N cells. To ensure that there is no discontinuity of the wave function, it is required that,

$$\Psi(X+Na) = \Psi(X) \tag{5.10}$$

where a, is the lattice parameter and N is the number of cells in the lattice. The Bloch condition in one dimension is

$$\Psi_K(X+Na) = \exp(iKNa)\Psi_K(X) \tag{5.11}$$

Thus, exp(iKNa) = 1, which means $K = 2\pi p/Na$ where p is an integer. In the reduced zone in one dimension, K must have values as $-\pi/a < K > \pi/a$. The integers p go from −(1/2)N to (1/2)N. The number of allowed wave vectors in the Brillouin zone equals the number of unit cells in the crystal.

5.1.5 Polyacenes, an Example of a Low-Dimensional Carbon Compound

Acenes or polyacenes are a class of two-dimensional carbon compounds consisting of a sequence of bonded benzene rings forming a two-dimensional plane. The acenes are essentially narrow graphene nanoribbons, and their existence long preceded the discovery of graphene. It is therefore of interest to examine theoretical calculations of their properties. Graphene is a two-dimensional structure of carbon atoms arranged in the same structure as the carbon atoms in the planes of graphite. Its properties will be discussed in detail in later sections. Naphthalene, $C_{10}H_8$, consists of two benzene rings, and the longest of the series, heptacene, $C_{30}H_{18}$, has seven benzene rings. The possibility of even longer chains is being pursued. Pentacene, which consists of five benzene rings, has application in organic field effect transistors (FETs). Here, the interest is in discussing their properties as an example of a covalently bonded two-dimensional carbon material. Figure 5.2 shows the minimum energy structure of pentacene calculated by DFT at the B3LYP/6-31G * level [4]. The calculation indicates that the structure is two-dimensional in agreement with experiment. Calculation of the frequencies of the structure yielded no imaginary values indicating the structure is at a minimum on the potential energy surface. The calculated energy difference between the highest occupied molecular orbital (HOMO) and the lowest unoccupied molecular orbital (LUMO), known as the energy gap, is plotted in Figure 5.3 versus the number of carbon atoms in polyacene chains showing that the energy gap approaches zero as the length of the chain increases [4]. This is a general result that has been noted in a number of narrow two-dimensional materials such as polyacetylene. Figure 5.4 is a plot

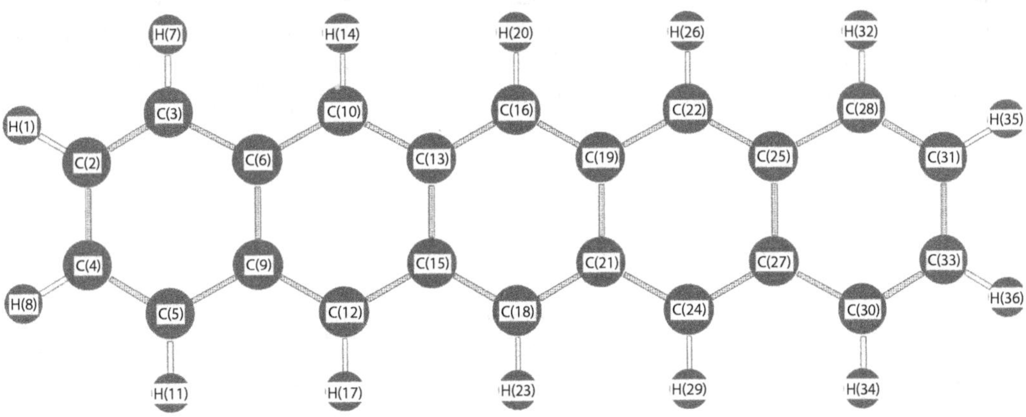

Figure 5.2 Structure of the pentacene molecule [4].

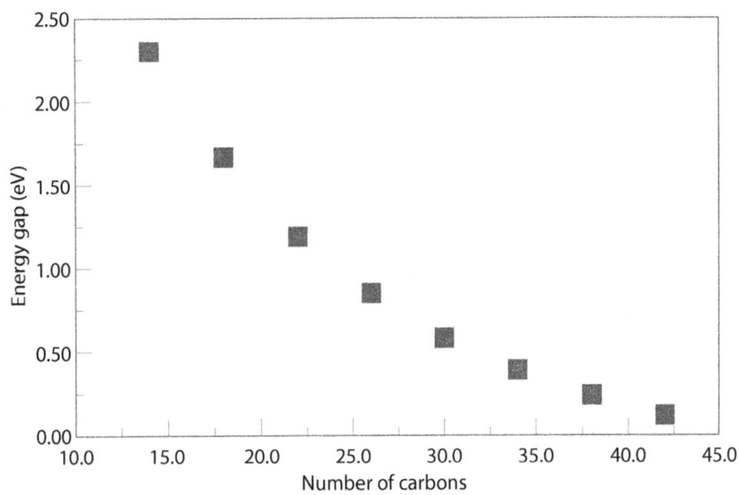

Figure 5.3 Energy gap at the center of the Brillouin zone versus the length of an acene chain [4].

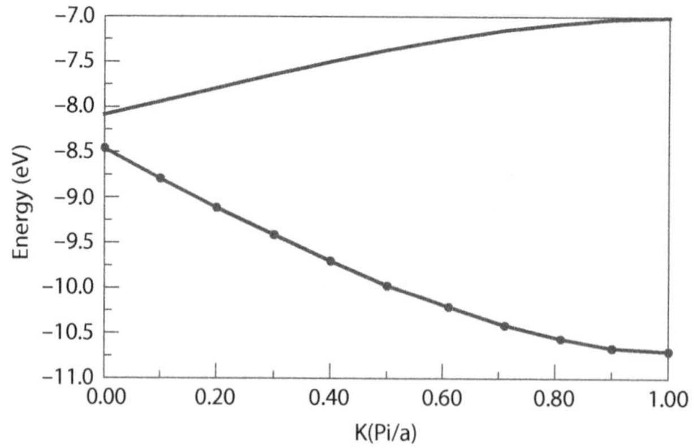

Figure 5.4 Highest occupied energy level and lowest unoccupied level versus the wave vector of acenes [4].

of a calculation of the HOMO and LUMO energy levels of polyacene, as a function of the wave vector K, using periodic boundary conditions [4]. The dependence of the HOMO and LUMO energy levels is predicted to be linear near K = 0 similar to graphene, and thus the conduction electrons should have zero effective mass. This implies that a polyacene should have similar electronic properties to graphene and similar application potential.

5.2 Graphene

5.2.1 Structure and Fabrication

Graphene is a two-dimensional array of carbon atoms where the carbon atoms have the same arrangement as the atoms in the planes of graphite. Its structure is illustrated in Figure 5.5. It had been predicted that two-dimensional solids such as graphene are thermodynamically unstable and would likely distort from planarity [5, 6]. The invalidity of this idea was demonstrated when

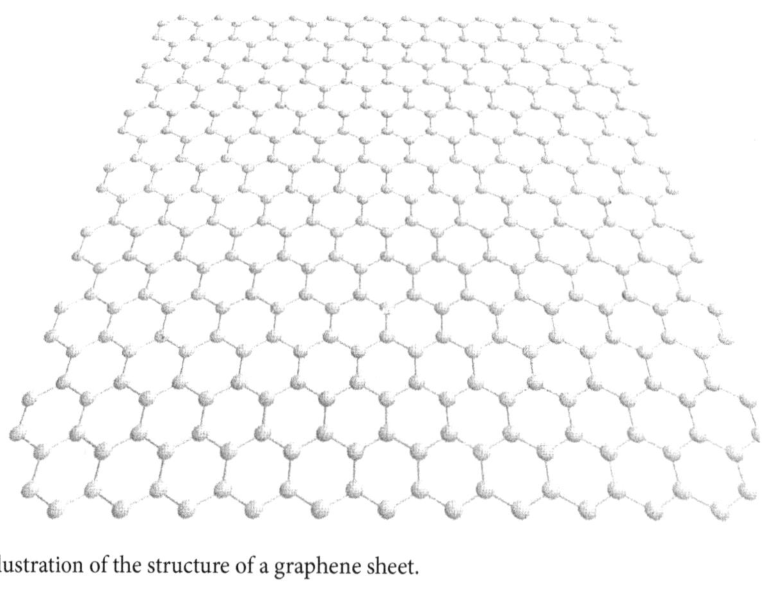

Figure 5.5 Illustration of the structure of a graphene sheet.

scientists in England and Russia produced single graphene sheets. The discovery generated much research activity in the materials science community because the unique electronic properties of graphene suggested the potential for many applications. Properties, such as its high strength and high electron mobility in the order of a million times more than copper, suggested applications such as much higher speed FETs and enhancement of the strength of composites. The first fabrication of graphene was quite simple [7]. Scotch tape was pressed on the large surface of single crystals of graphite. The graphite planes are parallel to the large facet of the crystal. The tape was carefully peeled off and then gently rubbed onto a silicon oxide substrate. The flakes on the substrate were of varying thicknesses. The thicknesses were determined by examining the flakes in a microscope using white light, which consists of many wavelengths, and examining the interference patterns of light reflected from the front and back surfaces of the flakes.

5.2.2 Calculation of Electronic Structure

Figure 5.6 shows the results of the calculation of the energy of the valence band and conduction band on the wave vector K for a sheet of graphene [8]. This calculation was done long before the discovery of graphene on a single plane of graphite, which was then considered a hypothetical material. The cones of energy versus wave vector in three dimensions are referred to as Dirac cones. There are two points in the Brillouin zone, K and K′, where the energy at the top of the valence band is the same as the energy of the conduction band. At these points, the material is metallic. At other points in the zone, this is not the case and there is an energy gap between the conduction and valence band. Such a material is referred to as a semi-metal. Another interesting feature of the dependence of the energies on the K vector is that near the points K and K', the energy is linearly dependent on the K vector. In a semi-conductor, typically the energy bands near K = 0 depend quadratically on the K vector given by,

$$E = (h/2\pi)K^2/2m^* \tag{5.12}$$

where m* is the effective mass of the conduction electrons or holes. When the electrons or holes are subjected to an applied electric field, they are accelerated with respect to the lattice as if they had a mass m*. This mass reflects the interaction of the electrons and holes with the lattice. In the mid 80s, theorists were examining the electronic structure of isolated graphite planes, essentially graphene, even though such a one-dimensional structure did not exist.

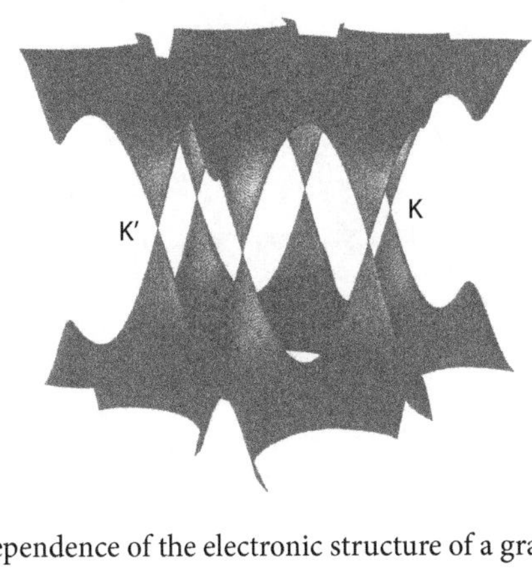

Figure 5.6 Calculation of the dependence of the electronic structure of a graphene sheet on the K vector [8].

The work resulted in an unusual prediction that at the contact points between the bands, K and K′, the effective mass of the conduction electrons is zero and that they would behave as though they are massless relativistic fermions [9, 10]. As a result, the electron dynamics in graphene is treated using the Dirac equation, which is a modification of the Schrodinger equation that deals with massless relativistic particles such as photons. This means that the conduction electrons move very fast in graphene having mobilities in the order of 50,000 cm^2/Vs, which is an order of magnitude greater than in present-day silicon transistors. This makes graphene a promising candidate for electronic devices. However, the fact that the band gap at the center of the Brillouin zone is zero is a problem that may limit application potential.

5.2.3 Graphene Nanoribbons

There has been much interest in graphene nanoribbons, long narrow strips of graphene, because such structures are likely to be constituents of graphene electronic devices. This is because the ribbons have a small band gap at the center of the Brillouin zone that is required for use in electronic devices such as FETs. There are two types of graphene nanoribbons referred to as armchair and zigzag, whose structure is illustrated in Figure 5.7. Figure 5.8

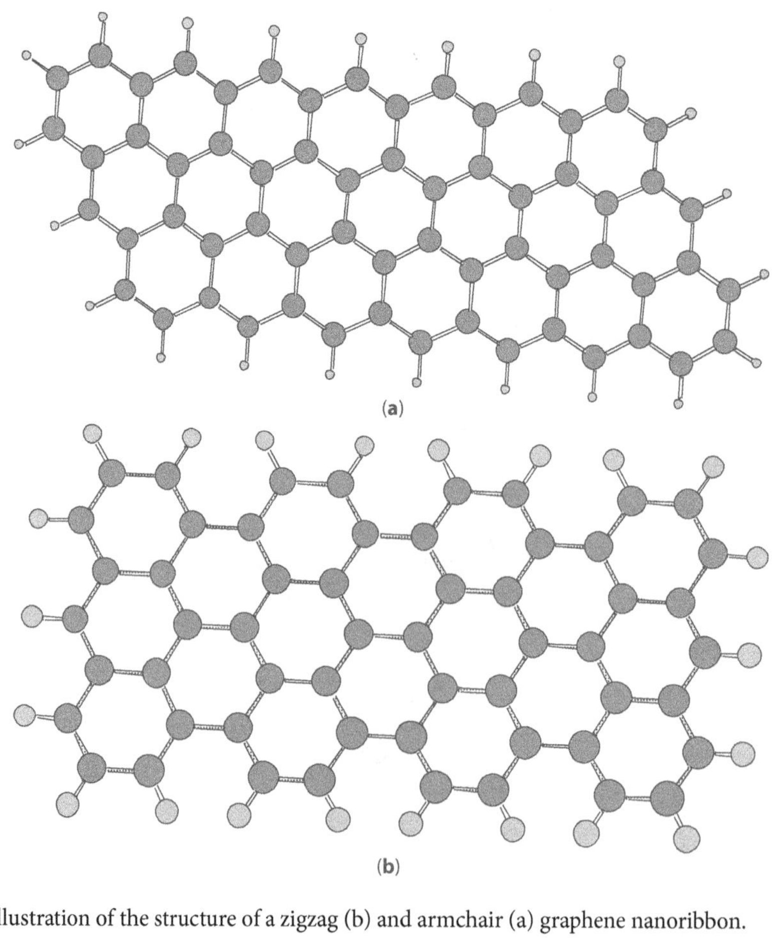

Figure 5.7 Illustration of the structure of a zigzag (b) and armchair (a) graphene nanoribbon.

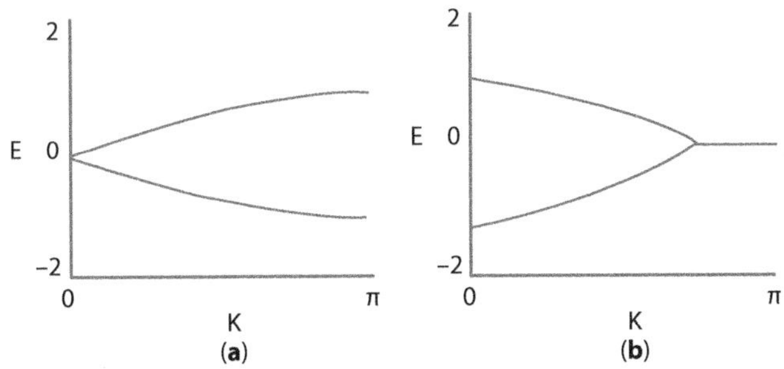

Figure 5.8 Calculation of the dependence of the energy of the highest occupied orbital and the lowest unoccupied orbital on the wave vector for a zigzag (b) and armchair (a) graphene nanoribbon [11].

shows the results of calculation of the dependence of the HOMO and the LUMO on the wave vector for the armchair and zigzag ribbons [11]. As discussed above, large graphene sheets are metallic at the center of Brillouin zone. However, it turns out that graphene nanoribbons have a band gap that depends on the length of the ribbon. Figure 5.9 shows the results of a DFT calculation of the band gap at the zone center of an armchair ribbon, three carbon rings wide and H terminated, as a function of the number of carbons in the ribbon (effectively the length) [12]. The calculation shows that as the length of the ribbon increases, the band gap approaches zero. This means that relatively short ribbons will be necessary for graphene-based electronic devices. Interestingly, the calculated dependence of the band gap of zigzag ribbons is quite different as it depends on whether the ribbons have an odd or even number of electrons. Figure 5.10 shows the dependence for the two cases. Figure 5.11 illustrates one concept of an FET using graphene nanoribbons [13]. Two graphene nanoribbons are connected by a small chain of benzene rings. In the same plane is another graphene nanoribbon slightly separated from the other two. This ribbon serves as a gate, and when a voltage is applied to it, current flows across the bridge between the two other ribbons.

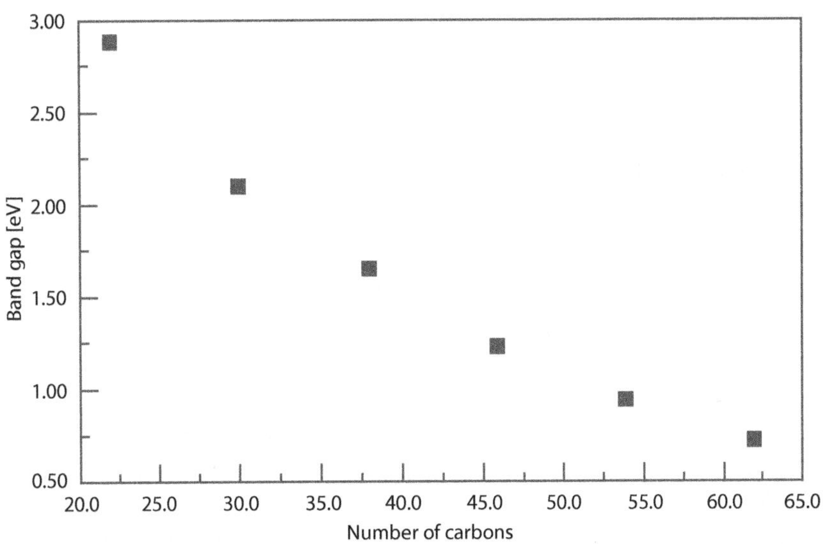

Figure 5.9 A DFT calculation of the band gap at K = 0 of an armchair graphene nanoribbon versus the number of carbons in the ribbon [12].

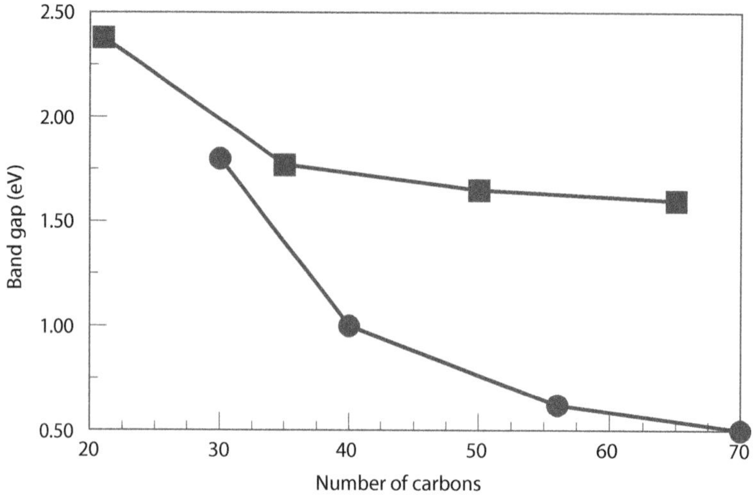

Figure 5.10 Plot of the dependence of the calculated band gap versus the number of carbons in a zigzag carbon nanoribbon depending on whether the number of electrons is even (bottom curve) or odd (top curve) [12].

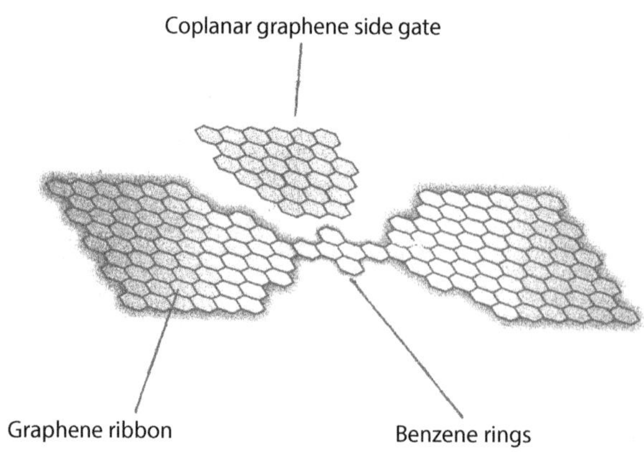

Figure 5.11 Concept of an FET using graphene ribbons [13].

5.2.4 Defected Graphene Ribbons

Developing graphene electronic devices based on graphene ribbons will likely require ribbons of 10 nm or less. However, their large-scale fabrication will have to produce ribbons that are defect-free, which may be a challenge. DFT calculations of the minimum energy structure of graphene ribbons having over 100 carbon atoms and containing one carbon vacancy indicate that they are significantly distorted from two dimensions [14]. Figure 5.12 shows the calculated minimum energy structure of an armchair ribbon with a carbon vacancy. This distortion from two dimensions, which is a critical determinant of the unique electronic properties of graphene, may cause the deterioration of the performance of graphene electronic devices when defects are present.

Thus, for use in electronic devices, the graphene ribbons have to be defect-free.

In order to be used in FETs, a material has to be a semiconductor having a small band gap at the zone center. This allows the application of a gate voltage to turn on and off the flow

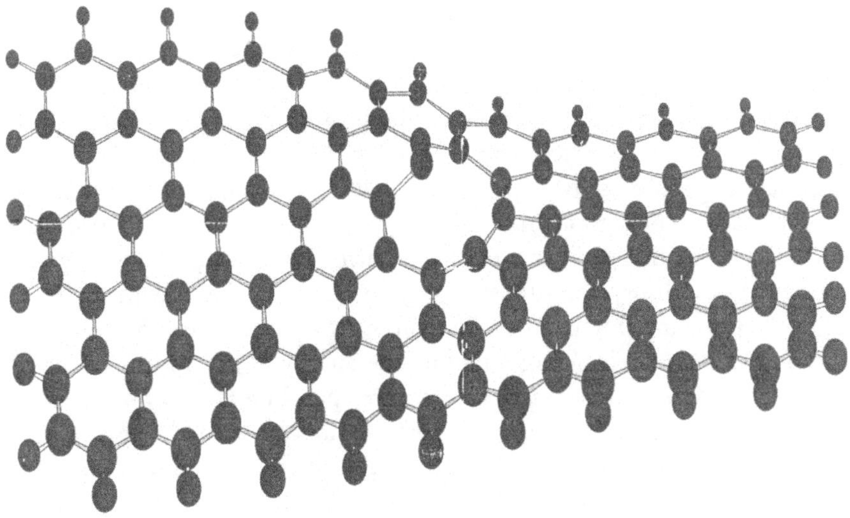

Figure 5.12 A DFT calculation of the minimum energy structure of an armchair graphene nanoribbon having one carbon vacancy [14].

of current in the semiconductor. As discussed above, graphene does not have a band gap at K = 0. Thus, there has been research aimed at finding ways to open the gap at the center of the Brillouin zone. One possibility is to use short nanoribbons, which, as shown in Figure 5.9, have a band gap. However, large-scale fabrication of such short narrow ribbons could be a challenge. Recently, it has been shown that by thermally decomposing a silicon carbide substrate, a carbon layer having the structure of graphene can be formed on the surface, and depending on the temperature used to decompose it, the layer had a band gap centered around the Fermi level [15]. Angle-resolved photo electron spectroscopy measurements of SiC heated at 1360°C indicated a band gap of 0.5 eV. Samples grown at 20 degrees lower did not have a band gap. It was also shown that the effective mass and the electron velocity near the top of the valence band depended on the direction of propagation. This was the first observation of such anisotropy in graphene. It was suggested that the opening of the gap and the anisotropy resulted from periodic bonding between the graphene layer and the SiC substrate.

5.2.5 Magnetism in Graphene Nanoribbons

For ferromagnetism to occur in a solid, the individual constituents of the solid must have a magnetic moment and the magnetic moments must be coupled so as to all align in the same direction. The major interaction that accomplishes this is the exchange interaction, which is a short-range interaction between nearest magnetic neighbors. These considerations clearly indicate that graphene ribbons or sheets cannot display ferromagnetism because the carbon atom is not paramagnetic. Further doping will not accomplish this. However, calculations have predicted the possibility of spin ordering in graphene ribbons. The calculations of the dependence of the band gap on the K vector for zigzag and armchair ribbons shown in Figure 5.8 show some unusual features that suggest spin ordering. Figure 5.8a and b shows the calculated dependence of the energy of the HOMO and the LUMO for a zigzag and armchair ribbon having a width of four carbon rings. The results for the zigzag ribbon show an unusual coincidence of the energy levels of the HOMO and the LUMO near $K = \pi$. This degeneracy is not predicted for graphite or the armchair ribbons. It is found that in this region, where the energies are coincident, charge is localized on the zigzag edges. These states are referred to as edge states. The calculation also predicts that there are magnetic moments associated with these edge states and that these moments are ordered ferromagnetically on one side of the ribbon and antiferromagnetically on the opposite side.

There is some preliminary experimental evidence for the edge states in graphene ribbons. Near edge x-ray absorption fine structure spectroscopy has been used to investigate this issue [16]. In this experiment, an x-ray photon excites an electron from the core carbon 1s level, producing a photoelectron emission. The energy of the emitted electron is measured. In graphite, there is a peak at 285.5 eV, corresponding to a transition from the carbon 1s level to the LUMO state. In zigzag graphene ribbons, an additional small peak was observed on the low-energy side of the graphite peak and attributed to spins at the edges.

The graphene was synthesized by a chemical vapor deposition method, and the samples were then annealed at different temperatures. The new peak was only observed in the samples annealed at 1000°C and 1500°C. A narrow electron spin resonance signal was also observed in the same samples and attributed to spins at the edges. Both the line width and g value decreased as the temperature was lowered, which was attributed to a strong coupling of the spins to the conduction carriers. These effects could also be due to the onset of ferromagnetic order of the edge spins.

5.2.6 Doped Graphene Ribbons as Catalysts for Oxygen Reduction Reaction in Fuel Cells

There have been a number of theoretical studies of boron- and nitrogen-doped graphene sheets and ribbons aimed at predicting their properties. Such doping can make the ribbons N or P semiconductors. The doping introduces an atom that has a spin and generally affects the size of the band gap. Thus, the doping can be used to engineer the band gap. Another area of interest is to assess whether doped graphene could be a catalyst for the oxygen reduction reaction in fuel cells [17, 18]. There is a need to find less-expensive replacements for the presently used platinum catalyst. Likely reactions that produce H_2O at the cathodes are the dissociation of O_2 bonded to the catalyst followed by atomic oxygen undergoing the following reaction:

$$2O + 4H^{+} + 4e \rightarrow 2H_2O \tag{5.13}$$

Another possibility is the formation of HO_2, which bonds to the catalyst, followed by the removal of OH, which could then undergo the following reaction:

$$OH + H^{+} + e \rightarrow H_2O \tag{5.14}$$

For the catalyst to be effective, the energy needed to dissociate O and OH from O_2 and HO_2 bonded to the catalyst should be significantly lower than that needed to dissociate free

O_2 and HO_2. The modeling approach seeks to identify a material X that bonds to O_2 or HO_2 to form X-O_2 or X-O_2H such that the bond dissociation energy (BDE) to remove O or OH is less than the BDE to dissociate free O_2 or O_2H. Because there are no available bonds on graphene for O_2 and HO_2 to bond to, it is necessary to dope the structures with an atom such as boron, which has one less electron than carbon, to provide an available bond.

As an example of such a theoretical prediction of catalytic activity, consider a boron-doped graphene nanoribbon and the reaction of Equation 5.14. The minimum energy structures of XHO_2 where X is a boron-doped graphene ribbon are calculated using DFT. The BDE is defined as,

$$BDE = [E(XO) + E(Z)] - [E(XY)] \quad (5.15)$$

where Y is HO_2 and Z is OH. E is the total electronic energy plus the zero point energy (ZPE) of the minimum energy structure. The ZPE is the total ZPE of all of the normal modes of vibration given by,

$$E_{zpe} = (1/2)h\sum_{i}^{3N-6} f_i \quad (5.16)$$

where f_I are the vibrational frequencies of the normal modes and N is the number of atoms in the molecule. The calculated BDE, given by Equation 5.15, is compared with that to dissociate free HO_2. If it is significantly less, it can be concluded that X may be a good catalyst for the HO_2 dissociation.

Another issue to be considered in assessing whether the nanostructures can catalyze the reactions at the cathode is whether HO_2 can bond to the boron-doped graphene ribbon. This can be evaluated by calculating the adsorption energy, E_{ads}, given by,

$$E_{ads} = E(XY) - E(X) - E(Y) \quad (5.17)$$

where X is a boron-doped graphene nanoribbon and Y is HO_2 and the Es are the total electronic energy of the structures at minimum energy. If the result is a negative value, it indicates that HO_2 can form a stable bond at the boron site of the doped carbon graphene nanoribbon.

Figure 5.13 shows the calculated minimum energy structure of a boron-doped armchair graphene ribbon [19]. The adsorption energies calculated using Equation 5.17 for HO_2 is negative. This indicates that HO_2 can bond to the armchair ribbon at the boron site. Figure 5.14 shows the calculated minimum energy structure of HO_2 bonded to the boron site of the armchair ribbon. The calculated BDE to remove OH from this structure is 1.4 eV less than needed to dissociate free HO_2 molecule into O and OH, which is 5.0 eV. This suggests that armchair boron ribbons may have the potential to be catalysts for the dissociation of O_2H. There is experimental evidence to support this prediction. Interestingly, boron-doped graphene has been synthesized and demonstrated experimentally using cyclic voltammetry to be excellent catalysts for the oxygen reduction reaction comparable to platinum [20].

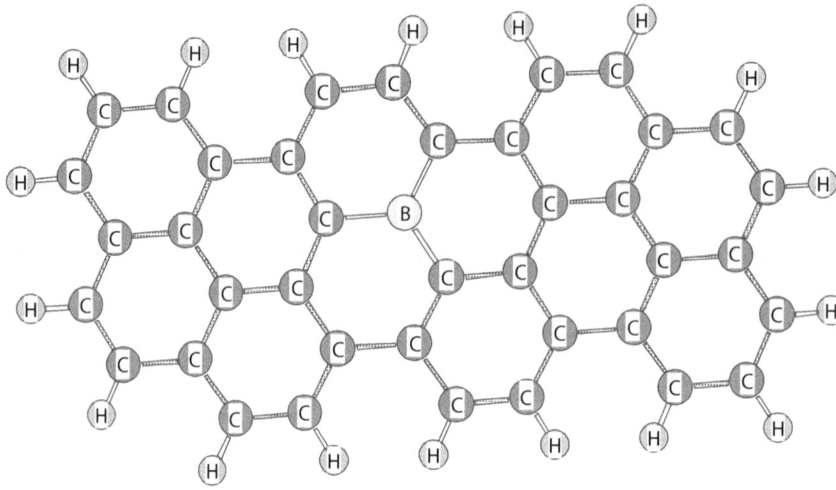

Figure 5.13 Calculated minimum energy structure of a boron-doped graphene ribbon [19].

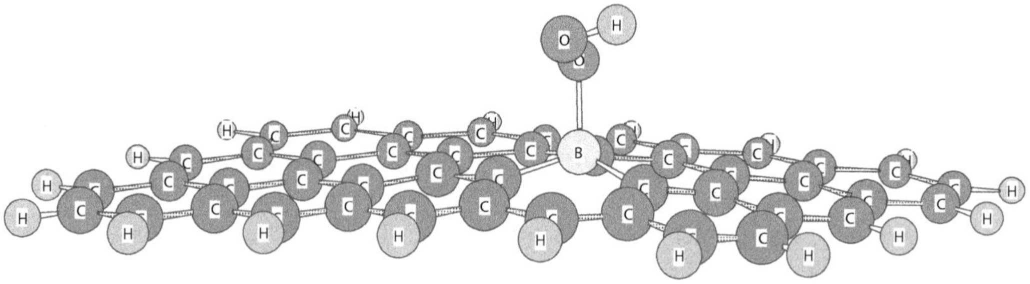

Figure 5.14 Calculated minimum energy structure of HO_2 bonded to the boron site of a boron-doped graphene ribbon [19].

5.2.7 Doped Graphene Ribbons as Catalysts for Hydrogen Production

Combustion of fossil fuels, such as coal and gasoline, account for more than half of the greenhouse gas emissions globally. One possible alternative energy source that would have no greenhouse gas emissions is a fuel cell. However, fuel cells require hydrogen in order to generate electricity, and thus a reliable source of hydrogen is needed. One approach is to develop catalysts that could remove hydrogen from hydrogen-rich molecules, such as water or formic acid, which have two H atoms. Presently, platinum and related metals are used as catalysts and these metals are approximately half the cost of fuel cells. There is a need to identify other potential catalysts that would be less costly. Theoretical modeling has been used as an approach to identify possible catalysts that could abstract hydrogen from various molecules. Metal-embedded nitrogen-doped graphene has also been investigated as a potential catalyst for H_2O dissociation [21, 22] Recently, boron-doped Pd metal was shown to catalyze the removal of H from formic acid [23, 24]. This latter result suggests the possibility that compounds containing boron may have the potential to be catalysts for H dissociation from H-containing molecules. DFT has been used to assess whether boron-doped

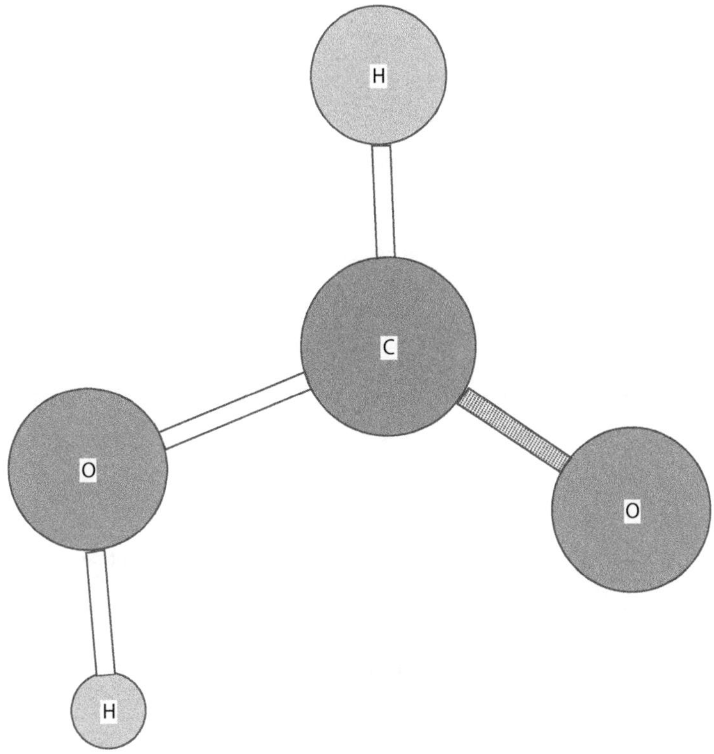

Figure 5.15 Structure of formic acid molecule.

graphene nanoribbons can catalyze the removal of hydrogen from formic acid. The structure of formic acid, HCOOH, is shown in Figure 5.15. Figure 5.16 shows the calculated structure of formic acid bonded to the boron site of a doped graphene nanoribbon [3]. The calculation yielded no imaginary frequencies indicating that it is at a minimum on the potential energy surface. The calculation of the adsorption energy, given by Equation 5.17, yields a negative value indicating that the formic acid molecule can bond to the boron-doped graphene nanoribbon. Figure 5.17 shows the calculated minimum energy structure of formic acid less one H atom bonded to the boron-doped nanoribbon. A calculation of the bond dissociation energy, using Equation 5.15, to remove an H atom from the oxygen of formic acid bonded to the doped graphene nanoribbon yields 2.7 eV, which is substantially lower than the calculated value of 4.3 eV needed to remove an H atom from the free formic acid molecule.

5.3 Conclusion

In this chapter, theoretical calculations using DFT of the properties of graphene nanoribbons have been discussed. The focus on nanoribbons is because they have a band gap at the zone center whereas graphene sheets do not. A band gap at the center of the Brillouin zone is needed for use in devices such as FETs. Calculations of the dependence of the band gap

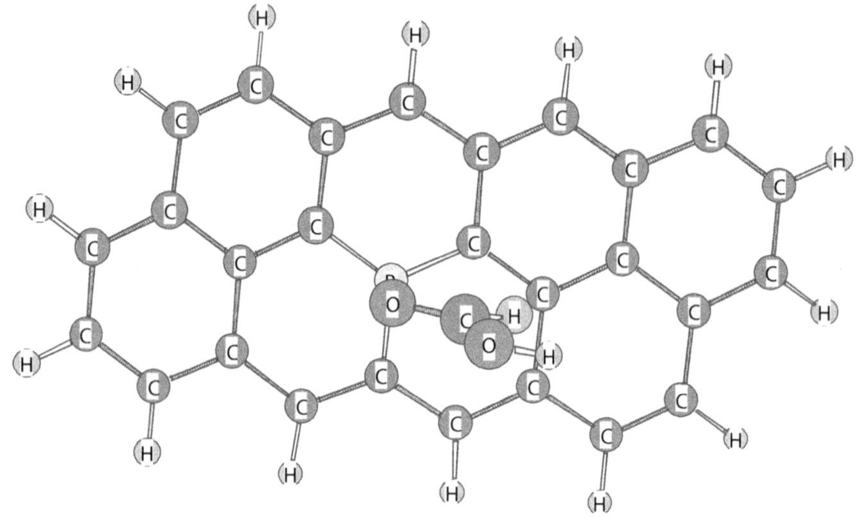

Figure 5.16 DDFT-calculated minimum energy structure of formic acid bonded to a boron-doped graphene ribbon.

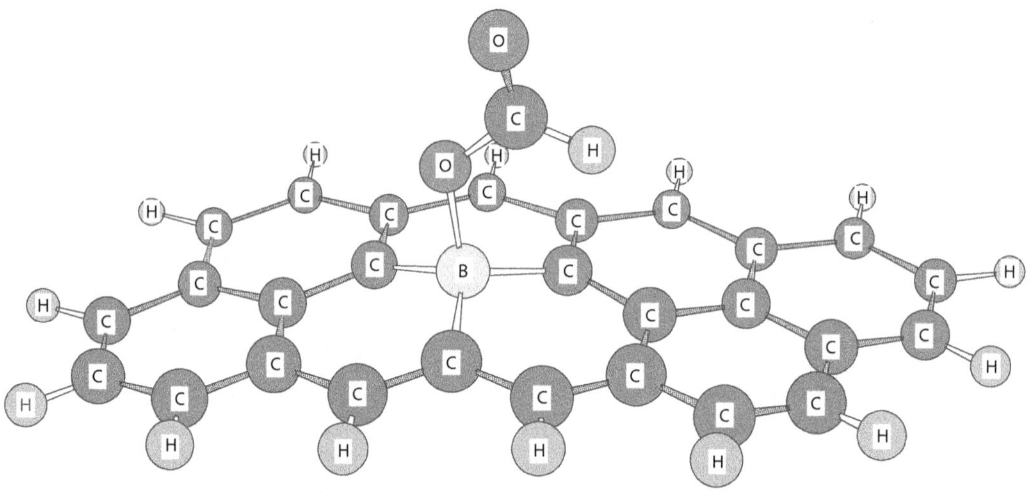

Figure 5.17 DFT-calculated minimum energy structure of formic acid less one H atom bonded to a boron-doped graphene ribbon.

on the wave vector, on the number of carbons in the ribbons (effectively the length), and on doping with boron or nitrogen have been discussed. Calculations of the effect of vacancies on the structure of the ribbons have shown distortion from planarity, which would be detrimental to the performance in electronic devices. The significant conclusion from these results is that if nanoribbons are to be used in electronic devices, they must be quite short, approximately less than 10 nm, and have no defects such as vacancies.

Also discussed are potential applications of B- and N-doped graphene nanoribbons as catalysts for reaction in fuel cells. The calculations have shown that the doped nanoribbons have the potential to catalyze the oxygen reduction reaction in fuel cells and for removal of hydrogen, needed for fuel cells, from hydrogen-rich molecules.

References

1. Heitler, W. and London, F., Wechselwirkung neutraler atome und homoopolare bindung nach der Quantummechanic. *Z. Phys.*, 44, 455, 1927.
2. Hohenberg, P. and Kohn, W., Inhomogeneous electron gas. *Phys. Rev.*, 136, B864, 1964.
3. Owens, F.J., unpublished.
4. Owens, F.J., Pentacene and poly-pentacene as graphene nanoribbons. *Solid State Comm.*, 185, 58, 2014.
5. Peierls, R.E., Quelques propriets typiques der corpse solides. *Ann. I. H. Poincare.*, 5, 177, 1935.
6. Landau, L.D., Zur Theorie der phasenumwandlungen. *Phys. Z. Sovjetunion.*, 11, 26, 1937.
7. Novoselov, K.S. *et al.*, Electric field effect in atomically thin carbon films. *Science*, 306, 666, 2004.
8. Wallace, P.R., The band theory of graphite. *Phys. Rev.*, 71, 622, 1974.
9. Semenoff, G.W., Condensed matter simulation of a three dimensional anomaly. *Phys. Rev. Lett.*, 53, 2449, 1984.
10. Haldane, F.D.M., Model for quantum Hall effect without Landu levels. *Phys. Rev. Lett.*, 61, 2015, 1988.
11. Nakada, K., Fujita, M., Dresselhaus, G., Dresselhaus, M.S., Edge states in graphene ribbons. *Phys. Rev.*, B54, 1795, 1996.
12. Owens, F.J., Electronic and magnetic properties of armchair and zigzag graphene nanoribbons. *J. Chem. Phys.*, 128, 194701, 2008.
13. Ozyilmaz, B. *et al.*, Electronic transport and quantum Hall effect in bipolar graphene p-n-p junctions. *Phys. Rev. Lett.*, 99, 166804, 2007.
14. Miller, M. and Owens, F.J., Defect induced distortion of armchair and zigzag graphene and boron nitride nanoribbons. *Chem. Phys. Lett.*, 570, 42, 2013.
15. Nevius, M.S. *et al.*, Semiconducting graphene from highly ordered substrate interactions. *Phys. Rev. Lett.*, 115, 136802, 2015.
16. Joly, V.L. *et al.*, Observation of magnetic edge states in graphene nanoribbons. *Phys. Rev.*, B81, 245428, 2010.
17. Zhang, L. and Xia, Z., Mechanism of oxygen reduction reaction on nitrogen doped graphene for fuel cells. *J. Phys. Chem. C*, 115, 11170, 2011.
18. Zhang, L., Niu, J., Dai, L., Xia, Z., Effect of microstructure of nitrogen doped graphene on oxygen reduction activity in fuel cells. *Langmuir*, 128, 7542, 2012.
19. Owens, F.J., Modeling boron doped carbon and boron nitride nanostructures as catalysts at the cathode of fuel cells, in: *Boron Nitride Properties and Synthesis*, E. Morgan (Ed.), Nova Science Publishers, 2017.
20. Sheng, Z. *et al.*, Synthesis of boron doped graphene for oxygen reduction reaction in fuel cells. *J. Mater. Chem.*, 22, 390, 2012.
21. Liu, L.L., Chen, C., Zhao, L., Wang, Y., Wang, X., Metal-embedded nitrogen-doped graphene for H_2O dissociation. *Carbon*, 115, 773–780, 2017.
22. Ghuman, K.K., Yadau, S., Singh, C.V., Absorption and dissociation of H_2O on monolayered MoS_2 edges: Energetic and mechanism from ab initio simulations. *J. Phys. Chem.*, C119, 6518–6529, 2015.
23. Jiang, K., Xu, K., Zou, S., Cai, W.B., B doped Pd catalyst: Boosting room temperature hydrogen production from formic acid solutions. *J. Am. Chem. Soc.*, 136, 4861–4864, 2014.
24. Yoo, J.S., Zhao, Z., Norskov, J.K., Studt, F., Effect of boron modifications of palladium catalysts for the production of hydrogen from formic acid. *ACS Catal.*, 5, 6579–6586, 2015.

6

Synthetic Electric Fields Influence the Non-Stationary Processes in Graphene

N.E. Firsova[1,2*‡] and Yu. A. Firsov[†]

[1]*A.F. Ioffe Physical-Technical Institute, the Russian Academy of Sciences, St. Petersburg, Russia*
[2]*St. Petersburg Polytechnic University, St. Petersburg, Russia*

Abstract

We consider two examples of synthetic electric fields influence non-stationary processes in graphene. The first one is a graphene nanoresonator new loss mechanism caused by dissipative intra-valley currents stipulated by these fields that are generated by external periodic electromotive force and intrinsic out-of-plane distortions. These fields arise as a time derivative of time-dependent gauge fields. The obtained formula for quality factor has a quantum mechanical origin. This new mechanism accounts for an essential part (about 30%) of loss. The ways of minimization of this kind of dissipation are discussed. The second example is a graphene electromagnetic response. It was earlier shown that in the framework of flatland model-induced currents, paths were straight lines. Our taking into account synthetic electric fields led to their curvature, which is consistent with experiments. This is not yet studied in detail. If the difference between the theory and experiment were found to be insignificant, we could consider the inverse problem, i.e., to determine the surface corrugations form if induced current paths are known. Then, we could develop a new imaging method. Also, it is very important to study synthetic electric fields that influence other non-stationary processes in graphene. These problems are awaiting researchers and promise to award them with interesting results.

Keywords: Graphene, nanoresonator, nonstationary processes, synthetic electric fields, out-of-plane displacements

6.1 Introduction

The recent successful preparation of one-atomic carbon sheet, i.e., graphene, was made in 2004 (see [1]) and the first two-dimensional (2D) crystal in 2005 (see [2]) by K.S. Novoselov, A.K. Geim *et al.* The first nontrivial experimental observations ("Dirac-like" charge carriers spectrum and anomalous quantum Hall effect) were published in 2005 [2–3]. A recent development of solution-phase exfoliation technique (in contrast to micromechanical

**Corresponding author*: nef2@mail.ru
†*Deceased*
‡*Dedication: To my son and granddaughters*

Tobias Stauber (ed.) Handbook of Graphene: Volume 2, (147–194) © 2019 Scrivener Publishing LLC

cleavage and epitaxial growth) provides a low-cost, high-yield method for mass production of unoxidized, defect-free graphene [4]. The new exciting facts concerning graphene gave rise to the development of physics; many unusual graphene properties were elucidated. It was found that graphene exhibits a variety of interesting transport phenomena [5–7]. Graphene also shows remarkable optical properties, which makes it an ideal photonic and optoelectronic material [8].

However, the question whether a 2D membrane can be stable was a subject of a long-standing theoretical debate. The question whether a strictly crystal can exist was first raised theoretically more than 70 years ago by Peierls [9–10] and Landau [11–12]. They showed that in the standard harmonic approximation, long wavelength fluctuations should destroy long-range order, essentially resulting in "melting" of a lattice at any finite temperature. Mermin and Wagner proved that a magnetic long-range order could not exist in one and two dimensions, [13] and, later, the proof was extended to the crystalline order in 2D [14].

Then, in [15–17], the phenomenological theory for flexible membranes was derived by Nelson D.R., Pelity L., Radzihovssky L., Le Doussal *et al.*, which showed that these dangerous fluctuations can, however, be suppressed by anharmonic coupling between bending and stretching modes making that a two-dimensional membrane can exist but should present strong height fluctuations. We shall review the main ideas and results of this theory following the way it was described in [18]. The primary quantities are two-component displacement vector in the plane $\vec{u}$, the height of out-of-plane deformation h and the normal unit vector $\boldsymbol{n}$ with in-plane component $-\nabla h/\sqrt{1+(\nabla h)^2}$. The elastic energy of the membrane is given by

$$E=\int d^2x\left[\frac{\kappa}{2}\left(\nabla^2 h\right)^2+\mu\bar{u}_{\alpha\beta}^2+\frac{\lambda}{2}\bar{u}_{\alpha\alpha}^2\right] \tag{6.1}$$

where κ is the bending rigidity, μ and λ are Lame parameters, and $\bar{u}_{\alpha\beta}$ is the deformation tensor:

$$\bar{u}_{\alpha\beta}=\frac{1}{2}\left(\frac{\partial u_\alpha}{\partial x_\beta}+\frac{\partial u_\beta}{\partial x_\alpha}+\frac{\partial h}{\partial x_\alpha}\frac{\partial h}{\partial x_\beta}\right) \tag{6.2}$$

In harmonic approximation, by neglecting the last, nonlinear term in the deformation tensor (6.2), the bending (h) and stretching ($\boldsymbol{n}$) modes are decoupled. In this approximation, the Fourier components of the bending correlation function with wavevector $\vec{q}$ are

$$\left\langle\left|h_q\right|\right\rangle^2=\frac{TN}{\kappa S_0 q^4} \tag{6.3}$$

where N is the number of atoms, $S_0 = L_xL_y/N$ is the area per atom, and T is the temperature in units of energy. In this approximation, the mean displacement in the direction normal to the layer is

$$\langle h^2 \rangle = \sum_q \left\langle \left| h_q \right|^2 \right\rangle \propto \frac{T}{\kappa} L^2 \tag{6.4}$$

where L is a typical linear sample size. The correlation function of the normals

$$G(q) = \left\langle \left| n_q \right|^2 \right\rangle = q^2 \left\langle \left| h_q \right|^2 \right\rangle \tag{6.5}$$

in this approximation becomes

$$G_0(q) = \frac{TN}{\kappa S_0 q^2} \tag{6.6}$$

which implies that the mean square angle between the normals is logarithmically divergent as $L \to \infty$. This behavior indicates the tendency to crumpling of membranes due to thermal fluctuations. Deviations from this harmonic behavior, namely, anharmonic coupling between bending and stretching modes, can stabilize the flat phase by suppressing the long wavelength fluctuations [15–17]. In this case, the corresponding correlation function of the normals is given by the Dyson equation:

$$G_a^{-1}(q) = G_0^{-1}(q) + \Sigma(q) \tag{6.7}$$

with self-energy

$$\Sigma(q) = \frac{AS_0}{Nq^2} \left(\frac{q}{q_0} \right)^{\eta} \tag{6.8}$$

where $q_0 = 2\pi\sqrt{\frac{B}{\kappa}}$, B is the two-dimensional bulk modulus, η is the anomalous rigidity exponent, and A is a numerical factor.

The simplest way to derive this expression is to use the self-consistent perturbation [17], which gives $\eta \approx 0.8$.

As a result of this anharmonic coupling, the typical height of fluctuations in the direction normal to the membrane is much smaller than the one given by Equation 6.4 and scales with the sample size as L^{ζ} with $\zeta = 1 - \eta/2$. Nevertheless, the fluctuations are still anomalously large and they can be much larger than the interatomic distance for large samples. Thus, the theory predicts an intrinsic tendency to ripple formation. At the same time, the amplitude $\bar{h} \propto L^{\zeta}$ of these transverse fluctuations remains much smaller than the sample size and preserves the long-range order of the normals so that the membrane can be considered as approximately flat and not crumpled.

So the theory developed in [15–17] predicted that graphene samples were stable but they should absolutely have out-of-plane distortions. And really among very interesting graphene properties, it was opened that one of the most important of them was that strictly speaking graphene membrane is not "flatland" (2D system) as it was theoretically predicted. Really, all the observed mono-atomic graphene samples have inherent stable out-of-plane corrugations, i.e., out-of-plane deformations (ripples, bubbles, wrinkles, etc.). For the first time, it was observed by Meyer Jannic C. *et al.* (see [19–20]), where it was written, that "....graphene sheets are not perfectly flat, but exhibit intrinsic microscopic roughening..." and also "...the observed corrugations in the third dimension may shed light on subtle reasons behind the stability of 2*D*-crystals").

The phenomenological theories for flexible membranes [15–17] were derived in the continuum limit without including any microscopic feature, and their applicability to graphene in the interesting range of temperatures, sample sizes, etc. is not evident. In [18], Fasolino A., Los J.H. and Katsnelson M.I. presented Monte Carlo simulations of the equilibrium structure of single-layer graphene. By monitoring the normal–normal correlation functions, they directly compared their results to predictions of earlier theories. The effect of interest was cruelly dependent on acoustical phonons and interactions between them, and therefore the simulations require samples much larger than interatomic distances, typically many thousands of atoms at thermal equilibrium, making prohibitive ab-initio simulations a la Car-Partinello [21]. However, for carbon, a very accurate description of energetic and thermodynamic properties of different phases was provided by the effective many-body potential long-range carbon bond-order potential (LCBOPII) [22–23]. This bond potential was constructed in such a way as to provide a unified description of the energetic and elastic constants of all carbon phases as well as the energy characteristics of different defects with accuracy comparable to the experimental one. They found clear deviations from harmonic behavior for long wavelength fluctuations, but, instead of the expected power-law scaling at long wavelengths, they found a marked maximum of fluctuations with wavelength of about $70\AA$. Against the expectations, they also found a stiffening of the bending rigidity with increasing temperature. The authors in [18] related these features to fluctuations in bond length that, in carbon, signal a partial change from conjugated to single/double bonds, with consequent deviations from planarity.

To obtain a quantitative comparison with theoretical predictions [15–17] for the spatial distribution of ripples the authors in [18] have calculated numerically the Fourier components of the correlation function of the normals $G(q)$ for q_x and q_y multiples of $2\pi/L_y$ and $2\pi/L_y$, respectively. To their surprise, their numerical results were not described by this general theory.

At small q, the behavior of $G(q)$ was not described by the harmonic approximation $G_0(q)$ nor by the anharmonic expression $G_a(q)$. The most remarkable feature of $G(q)$ was a maximum instead of the power law dependence $G_a(q)$ that implies the absence of any relevant length scale in the system. The presence of this maximum, instead, means that there is a preferred average value of about $70\AA$ (at $300K$). This length is also recognizable in real space images, as shown by arrows in [18], Figure 6.1. Indeed, the two samples that are smaller than this length do not show this decrease of $G(q)$ at low q. The results at $T = 3500K$ confirm this picture but with the maximum shifted to a larger q corresponding to a length of roughly $30\AA$. This temperature dependence of the typical ripple length scale should be investigated.

The described results are relevant not only for better understanding of the stability and structure of graphene but also of electronic transport. The fluctuations of normals lead to a modulations of the hopping integrals and are bound to affect the electronic structure [24–25]. Knowledge of the normal–normal correlation functions is necessary for calculations of the electron scattering by ripples.

It is also very important that it was shown in [18] that even fluctuations at the scale of tens of interatomic distances cannot be described by continuum medium theory. It would be very instructive to carry out systematic experimental and theoretical investigations of other two-dimensional crystals to understand which properties are common to flexible membranes and which ones are consequences of particular features of the chemical bonding and interatomic interactions.

In result, a two-dimensional graphene membrane can exist and be stable, but strong height thermal fluctuations (about 7nm) should be present and ripples spontaneously appear. So considering graphene membrane with out-of-plane distortions, we study not a specific case, but a general one.

Due to static disorder of static lattice out-of-plane distortions, the synthetic vector potential $\boldsymbol{A}(\boldsymbol{r})$ arises. Its presence affects the electronic dynamics through an effective pseudomagnetic field $e\mathbf{B} = \nabla \times \boldsymbol{A}(\boldsymbol{r})$, which, by time reversal symmetry, points in opposite directions at the two Dirac points of the electronic dispersion. Consequences of this effective pseudomagnetic (gauge) field have been considered by different authors. The influence of these static pseudomagnetic fields (gauge fields) in the physical properties of graphene was studied in detail: their connection with minimal conductivity [26], weak antilocalization [24], topological insulator state [27], pseudomagnetic quantum Hall effect [28], and so on. All these results as well as other ones of the sort are discussed in the review [29]. It is interesting that these inevitably existing gauge fields in graphene have a magnitude of about several Tesla!

So the graphene surface form deformations couple directly to the electrons making graphene a unique example of a metallic membrane. We shall write what happens to electrons when the graphene is deformed either by strain or bending following the description given by A.N. Castro-Neto and E.-A. Kim in [30–31]. One effect is the change in the distance between atoms; the other is the change in the overlap between the different orbitals. In both cases, the hopping energy between different carbon atoms is affected. The perturbed low-energy effective Hamiltonian in the vicinity of a Dirac point $\mathbf{K} = \left(\frac{4\pi}{3\sqrt{3}a}, 0\right)$ will be as follows:

$$H = \int d^2 r \psi^+(\boldsymbol{r})\left\{\sigma \cdot \left[iv_F \nabla + \boldsymbol{A}(\boldsymbol{r})\right] + \Phi(\boldsymbol{r}) - \mu\right\}\psi(\boldsymbol{r}) \tag{6.9}$$

Here, μ is the chemical potential measured away from Dirac point, $v_F = 3t_0 a/2$ is the Fermi velocity, bare nearest hopping integral $t_0 = 2.7\text{eV}$, the nearest neighbor distance a = $2.5\dot{A}$ and $v_F = 10^8\text{cm/sec}$; $\sigma = (\sigma_x, \sigma_y)$ are Pauli matrices in the sublattice basis. In the case of the nearest-neighbor hopping that changes t_0 to $t_0 + \delta t_i(\boldsymbol{\delta})$ at site R_i in the direction $\boldsymbol{\delta}$, we have $\boldsymbol{A}(\boldsymbol{r}) = (A_x(\boldsymbol{r}), A_y(\boldsymbol{r}))$,

$$A_x(\boldsymbol{r}) + iA_y(\boldsymbol{r}) = -\sum_j \delta t_j(\boldsymbol{r}) e^{iu_j\mathbf{K}} \tag{6.10}$$

Hence, these changes in the nearest neighbor hopping couple like a gauge field. In the case of hopping between next nearest neighbor sites, the scalar potential is generated:

$$\Phi(r) = -3\sum_{\vec{v}} \delta t_j(\mathrm{v}) e^{i\,\mathrm{v}\,K} \tag{6.11}$$

which describes essentially hopping on a triangular lattice with lattice spacing $\sqrt{3}a$ with nearest neighbor vectors

$$\mathrm{v}_1 = \sqrt{3}a(0,1), \quad \mathrm{v}_2 = \sqrt{3}a\left(\sqrt{3}/2,-1/2\right), \quad \mathrm{v}_3 = \sqrt{3}a\left(-\sqrt{3}/2,-1/2\right)$$

Given the local changes in the hopping energies, the vector and scalar potentials can be readily computed through Equations 6.10 and 6.11, respectively.

These changes in hopping energies can be connected to changes in the structures of the lattice. Let us consider first the case of in-plane distortions. In this case, the only change in the hopping energy is due to the change in the distance between the p_z orbitals. Consider the case where the distance between sublattices changes by a distance δl in the direction of $\vec{\delta}$. In the first order, we have

$$\delta t \approx (\partial t/\partial a)\delta l, \tag{6.12}$$

$$\delta l \approx (\boldsymbol{\delta} \cdot \nabla)\boldsymbol{u}, \tag{6.13}$$

where $\boldsymbol{u}(\mathbf{r})$ is the lattice displacement. Replacing the above expression (Equation 6.10), we get

$$A_x^{(u)}(\mathbf{r}) \approx \alpha a\left(u_{xx} - u_{yy}\right) \tag{6.14}$$

$$A_y^{(u)}(\mathbf{r}) \approx \alpha a u_{xy}, \tag{6.15}$$

where α is a constant with dimensions of energy, and we have used the standard definition of the strain tensor:

$$u_{ij} = \frac{1}{2}\left(\frac{\partial u_i}{\partial x_j} + \frac{\partial u_j}{\partial x_i}\right). \tag{6.16}$$

An analogous calculation for the change of next-to-nearest neighbor hopping energy leads to

$$\Phi^{(u)}(\mathbf{r}) \approx g(u_{xx} + u_{yy}). \tag{6.17}$$

Equations 6.14, 6.15, and 6.17 relate the strain tensor to potentials that couple directly to the Dirac particles.

Less trivial is the coupling to the out-of-plane modes since those involve rotations of the orbitals. In Figure 12 in [31], a rotation of two orbitals by an angle θ is shown. The rotation mixes π and σ states, and for small angles, the change in the hybridization energy is given by:

$$t \approx t_0 + \delta V \cdot \theta^2 \tag{6.18}$$

where δV is the energy mixing between π and σ states. Notice that $\theta = 2\pi/R$ where R is the curvature radius. In terms of the height variable, it was calculated in [31] for this case that the change in the hopping amplitude due to the bending in the direction **u** is given by

$$\delta t \approx \delta V[(\mathbf{u} \cdot \nabla)\nabla h]^2 \tag{6.19}$$

where the function $h = h(x,y)$ describes the graphene membrane surface form that it has due to out-of-plane displacements.

On the other hand, if **u** is a nearest-neighbor vector $\boldsymbol{\delta}$, we get from Equation 6.10

$$A_x^{(h)}(\mathbf{r}) = -\frac{3E_{ab}a^2}{8}\left[\left(\partial_x^2 h\right)^2 - \left(\partial_x^2 h\right)^2\right], \tag{6.20}$$

$$A_y^{(h)}(\mathbf{r}) = \frac{3E_{ab}a^2}{4}\left(\partial_x^2 h + \partial_y^2 h\right)\partial_x h \partial_x h, \tag{6.21}$$

where the coupling constant E_{ab} depends on microscopic details. On the other hand, if **u** is the next-to-nearest neighbor vector, we find in accordance with Equation 6.11

$$\Phi^{(h)}(\mathbf{r}) \approx -E_{aa}a^2[\nabla^2 h(\mathbf{r})]^2, \tag{6.22}$$

where E_{aa} is an energy scale associated with the mixing between orbitals.

So one can consider pseudomagnetic vector potential caused by in-plane strain and use the formulae (6.14) and (6.15) (see for instance [24], [32]) or the case of out-of-plane deformations using Equations 6.20 and 6.21). We shall investigate below the last case.

We formulate the main conclusion citing Castro Neto ([31]): “For smooth distortions of graphene due to strain or bending the Dirac particles are subject to scalar or vector potentials leading to an “electrodynamics” that is purely geometrical (there is no electric charge associated with the “electric” and “magnetic” fields created by structural deformations). This structural “electrodynamics” has strong consequences for the electronic motion in graphene leading to many unusual effects that cannot be found in ordinary materials. In particular one can manipulate the electrons by “constructing” appropriate deformations of the lattice that mimic electric and magnetic fields. This is the so-called strain-engineering and is a field of research that is still in its infancy."

Additional physics arises when the synthetic vector potential is caused by *time-dependent* distortions. In this case, vector potential becomes time dependent, i.e., we have $\mathbf{A} = \mathbf{A}(x,y,t)$ and it will generate not only a magnetic but also an effective electric field (synthetic electric field) $\mathbf{E}_{syn}(x,y,t) = -\frac{1}{c}\frac{\partial \mathbf{A}}{\partial t}(x,y,t)$. As far as we know, the influence of $\mathbf{E}_{syn}(x,y,t)$ in graphene was first considered by Von Oppen *et al.* in [33]. These authors in [33] studied the case when pseudomagnetic gauge field is generated by time-dependent distortions such as flexural phonons.

It is interesting that such fields were artificially created not long ago in another nontrivial system. It was done in rubidium Bose–Einstein condensate (BEC). This field was produced as a time-dependent one that led to the appearance of the so-called synthetic electric fields [34]. In [34], the effective time-dependent vector potential for neutral atoms was created via interaction with the laser light generating a synthetic electric field-simulating charged condensed matter system in the array of neutral atoms.

Let us describe the results of the paper [33] more in detail. Within the Dirac theory of the electronic properties of graphene, smoothly varying lattice strain affects the Dirac carriers through a synthetic gauge field. For static lattice strain, the gauge field induces a synthetic magnetic field, which is known to suppress weak localization corrections by a dynamical breaking of time reversal symmetry. When the lattice strain is time dependent, in connection with phononic excitations, the gauge field becomes time dependent and the synthetic vector potential is also associated with an electric field. In [33], the authors show that this synthetic electric field has observable consequences. They find that the Joule heating associated with the currents driven by the synthetic electric field dominates the intrinsic damping, caused by the electron–phonon interaction, of many acoustic phonon modes of graphene and metallic carbon nanotubes (CNTs) when including the effects of disorder and Coulomb interactions. Several important consequences follow from the observation that by time-reversal symmetry, the synthetic electric field associated with the vector potential has opposite signs for the two valleys. First, this implies that the synthetic electric field drives charge-neutral valley currants and therefore unaffected by screening. This frequently makes the effects of synthetic vector potential more relevant than a competing effect of the scalar deformation potential which has a much larger bare coupling constant. Second, valley currents decay by electron–electron (e-e) scattering (valley Coulomb drag), which causes interesting temperature dependence of the damping rates.

The authors in [33] considered the damping of low-energy phonons in CNTs and graphene originating from the electron–phonon interaction. For most phonon modes, this damping is closely related to a synthetic electric field associated with a strain-induced vector potential in the Dirac equation for the electronic properties of graphene. They found that it was very instructive to analyze phonon damping in terms of these synthetic electric fields: (i) Within this approach, phonon damping is a direct consequence of Joule-type heating. (ii) This establishes a close relation between phonon damping and the dynamic conductivity which they exploit to derive damping rates in the presence of disorder and e-e interactions. (iii) They find rich physics emerging from the fact that the synthetic electric field has opposite signs in the two valleys. They identify valley Coulomb drag as an important dissipation channel to unconventional temperature dependence of the damping rate.

The results obtained in [33] should be of direct relevance to the intense ongoing experimental efforts to build and explore nanomechanical as well as nanoelectromechanical devices based on graphene nanostructures. For instance, the approach introducing a new Joule-type loss mechanism for graphene nanoresonators, its influence the quality factor and optimization strategies is considered below in Section 6.2.

In [33], the authors show that these synthetic electric fields have observable consequences. They consider a low-energy phonon mode of graphene or CNTs. The phonon is associated with a synthetic electric field which, when the system is metallic, drives currents. The dissipation (Joule-type heating) associated with these currents causes damping of the phonon mode. They found that frequently the synthetic electric fields are responsible for the damping of phonon modes in metallic CNTs and graphene. In the clean limit, this damping mechanism is equivalent to dissipation by electron-hole pair creation. Approaching the problem from the point of view of the synthetic electric fields allows us to calculate damping rates including the effects of disorder and e-e interactions which we find to be significant.

The second problem we consider in this chapter is graphene electromagnetic response. This problem was first studied by Mikhailov and Ziegler in [35]. In the framework of flatland model they considered this question developing a quasi-classical kinetic theory approach, taking into account the self-consistent-field effects. Response of the system to both harmonic and pulse excitation was studied. Possible applications of graphene in terahertz electronics were discussed.

More in detail, the results obtained in [35] are as follows. First of all, it is the equation for nonlinear electromagnetic response of graphene found in approximation with taking into account the self-consistent field effect, collisions of electrons with impurities, photons and other lattice imperfections having been ignored. We shall describe the most important steps of the self-consistent procedure made in [35].

Assuming that the chemical potential μ lies in the upper band $E_{p_2} = v_F p$, the temperature is small, $T \ll \mu$ and the system is subjected to the external time-dependent periodic electric field $\mathbf{E}^{ext}(t)$. Then, the distribution function of electrons $f_{\mathbf{p}}(t)$ is described by the Boltzmann equation

$$\frac{\partial f_{\mathbf{p}}(t)}{\partial t} - e\mathbf{E}^{ext}(t)\frac{\partial f_{\mathbf{p}}(t)}{\partial \mathbf{p}} = 0 \tag{6.23}$$

in which we ignored collisions of electrons with impurities, phonons, and other lattice imperfections. Equation 6.23 has the exact solution

$$f_p(t) = F_0(p - \mathbf{p}_0(t)) \tag{6.24}$$

where

$$F_0(\boldsymbol{p}) = \left[1 + exp\left(\frac{v_F p - \mu}{T}\right)\right]^{-1} \tag{6.25}$$

is a Fermi–Dirac function, and

$$\mathbf{p}_0(t) = -e\int_{-\infty}^{t}\mathbf{E}^{ext}(t')dt' \tag{6.26}$$

is the solution of a single-particle classical equation of motion. The electrical current then assumes the form

$$\mathbf{j}(t) = -\frac{g_s\, g_v e v_F}{(2\pi h)^2}\int\frac{\mathbf{p}d\mathbf{p}}{\mathrm{p}}F_0(\mathbf{p}-\mathbf{p}_0(t)), \tag{6.27}$$

where $g_s = 2$ is the spin degeneracy and $g_v = 2$ is the valley-degeneracy factor. If the temperature is zero, $T = 0$ and the chemical potential is finite, $\mu > 0$, the current $\boldsymbol{j}(t)$ can be rewritten in the form

$$\mathbf{j}(t) = en_s v_F\frac{\mathbf{P}}{\sqrt{1+P^2}}G(Q), \tag{6.28}$$

where $\mathbf{P} \equiv \mathbf{P}(t) = -\mathbf{p}_0(t)/p_F, P(t) = :\mathbf{P}(t):, p_F = \mu/v_F$ is the Fermi momentum and the formula for the density of electrons n_s in the upper band may be written as follows:

$$n_s = n_e = \frac{g_s g_v p_F^2}{4\pi h^2} = \frac{g_s g_v \mu^2}{4\pi h^2 v_F^2} \tag{6.29}$$

The function $G(Q)$ is defined as

$$G(Q) = \frac{4}{\pi Q}\int_0^{\pi/2}\cos x dx\left(\sqrt{1+Q\cos x}-\sqrt{1-Q\cos x}\right),\quad Q \equiv 2P(t)/(1+P^2) \tag{6.30}$$

So far, we did not consider the effects of radiative decay, which can be important under realistic experimental conditions in graphene. It was assumed that the graphene electrons move in the sample under the action of the external electric field $\mathbf{E}^{ext}(t)$, and this directly leads to the time-dependent electric current $\mathbf{j}(t)$ (Equation 6.18). In general, however, the time-dependent electric current creates, in its turn, a secondary (induced) electric field $\mathbf{E}^{ind}(t)$, which acts back on the electrons and should be added to the external field. Calculating the response of the system, one should take into account that electrons respond not to the external, but to the total self-consistent electric field $\mathbf{E}^{tot}(t) = \mathbf{E}^{ext}(t) + \mathbf{E}^{ind}(t)$. These results in the effects of electromagnetic reaction of the medium (graphene) to the external field and can reduce the frequency upconversion efficiency. Now, we shall study how substantially the radiative decay suppresses the efficiency of the frequency multiplication.

Consider an infinite $2D$ electron system with the graphene sheet lying at the plane $z = 0$. We assume that the external electromagnetic wave is incident upon the graphene layer along the axis and induces the ac in the layer. This current produces the induced electric field $\mathbf{E}^{ind}(t)$, which is added to the external one. The Boltzmann equation for the momentum distribution function of the electrons should then be written as

$$\frac{\partial f_{\mathbf{p}}(t)}{\partial t} - e\mathbf{E}_{z=0}^{tot}(t)\frac{\partial f_{\mathbf{p}}(t)}{\partial p} = 0 \tag{6.31}$$

$$\mathbf{E}_{z=0}^{tot}(t) = \mathbf{E}_{z=0}^{ext}(t) + \mathbf{E}_{z=0}^{ind}(t), \tag{6.32}$$

instead of Equation 6.23. The solution of this equation as well as the electric current can again be written in the form [35] (Equations 6.24, 6.25, and 6.27), respectively, but the classical momentum $\mathbf{p}_0(t)$ now satisfies the equation

$$\mathbf{p}_0(\mathrm{t}) = -\mathrm{e}\int_{-\infty}^{\mathrm{t}} \mathbf{E}_{z=0}^{tot}(t')dt' = -e\int_{-\infty}^{\mathrm{t}} \left[\mathbf{E}_{\mathrm{z=0}}^{\mathrm{ind}}(t') + \mathbf{E}_{\mathrm{z=0}}^{\mathrm{ext}}(t')\right]dt' \tag{6.33}$$

The field $\mathbf{E}_{z=0}^{tot}(t) = \left[\mathbf{E}_{\mathrm{z=0}}^{\mathrm{ind}}(t) + \mathbf{E}_{\mathrm{z=0}}^{\mathrm{ext}}(t)\right]$ is not known and should be calculated self-consistently. To do this, the authors use the Maxwell equations relating to the current and induced electric field:

$$\mathbf{E}_{z=0}^{ind}(t) = -2\pi\mathbf{j}(t)/c \tag{6.34}$$

Here, $z = z(x,y) = 0$ is the equation of the plane graphene sheet. Combining Equations 6.33, 6.34, and 6.27, they get [35] the following self-consistent equation of motion for the momentum:

$$\frac{d\mathbf{p}_0(t)}{dt} + \frac{e^2 g_s g_v v_F}{2\pi h^2 c}\int \frac{\mathbf{p}d\mathbf{p}}{\mathrm{p}} F_0(\mathbf{p} - \mathbf{p}_0(t)) = -e\mathbf{E}_{\mathrm{z=0}}^{\mathrm{ext}}(\mathrm{t}) \tag{6.35}$$

After the nonlinear equation (Equation 6.35) is resolved with respect to the momentum $\mathbf{p}_0(t)$, the current $\mathbf{j}(t)$ can be found from the formula (6.27).

$$\mathbf{j}(t) = -\frac{g_s g_v e v_F}{(2\pi h)^2}\int \frac{\mathbf{p}d\mathbf{p}}{\mathrm{p}} F_0(\mathbf{p} - \mathbf{p}_0(t)), \tag{6.36}$$

Equations 6.35 and 6.36 describe the nonlinear self-consistent response of graphene to an arbitrary external electric field $\mathbf{E}_{z=0}^{\text{ext}}(t)$. The second term in the left-hand side of Equation 6.35 describes the radiative decay effects in the infinite $2D$ graphene layer.

In conventional $2D$ electron systems with the parabolic energy dispersion and the effective mass m^* of $2D$ electrons, the self-consistent equation for $\mathbf{p}_0(t)$ and $\mathbf{j}(t)$ similar to Equations 6.35 and 6.36 has the form

$$\frac{d\mathbf{p}_0(t)}{dt}+\frac{2\pi n_s e^2}{m^* c}\mathbf{p}_0(t)=-e\mathbf{E}_{z=0}^{\text{ext}}(t),\quad \mathbf{j}(t)=-\frac{en_s}{m^*}\mathbf{p}_0(t). \tag{6.37}$$

Here, $\Gamma_{par}\equiv\frac{2\pi n_s e^2}{m^* c}$ is the radiative decay rate [29] in the conventional (parabolic) $2D$ electron system. In Equation 6.37, we have ignored the scattering due to impurities and phonons [the corresponding term $\gamma\mathbf{p}_0(t)$ can be added to the left-hand side of the first equation (Equation 6.37)]. In high electron mobility GaAs/AlGaAs quantum-well samples, the radiative decay Γ_{par} substantially exceeds the scattering rate γ, $\Gamma_{par}\gg\gamma$ and determines the linewidth of the cyclotron, plasmon, and magnetoplasmon resonances [36–37]. As the graphene mobility is also very high, one can expect that, at high frequencies, the radiative effects are more important in graphene than the scattering effects. This justifies ignoring the scattering terms in Equation 6.37.

Returning back to the non-particle graphene system, we rewrite Equation 6.35 (at $T = 0$) in terms of the dimensionless momentum $\mathbf{P}(t) = -\mathbf{p}_0(t)/p_F$:

$$\frac{d\mathbf{P}(t)}{dt}+\Gamma\frac{\mathbf{P}}{\sqrt{1+P^2}}G(Q)=\frac{e}{p_F}\mathbf{E}_{z=0}^{ext}(t) \tag{6.38}$$

$$\Gamma=\frac{g_s g_v}{4}\frac{e^2}{\text{hc}}\frac{2\mu}{h}=v_F\frac{e^2}{\text{hc}}\sqrt{g_s g_v \pi n_s} \tag{6.39}$$

The current is determined again by Equation 6.28.

In the linear-response regime, when $|P|\ll 1$ and $G\approx 1$, Equation 6.38 gives

$$\frac{d\mathbf{P}(t)}{dt}+\Gamma\mathbf{P}(t)=\frac{e}{p_F}\mathbf{E}_{z=0}^{ext}(t) \tag{6.40}$$

From here, one can see that the quantity Γ has the physical meaning of the radiative decay rate in graphene in the linear-response regime. In contrast to Γ_{par}, Γ is proportional to the square root of the charge carrier density. For experimentally relevant densities n_s, the value of Γ lies in the subterahertz range.

Now consider the response of graphene to a harmonic excitation:

$$\boldsymbol{E}_{ext}(t) = \boldsymbol{E}_0 \cos \Omega t \tag{6.41}$$

It is convenient to rewrite Equations 6.38 and 6.28 in the form

$$\frac{d\boldsymbol{P}(t)}{dt} + \frac{\Gamma}{\Omega}\frac{\boldsymbol{P}}{\sqrt{1+P^2}}G(Q) = \frac{e\boldsymbol{E}_0 v_F}{\mu\Omega}\cos\tau, \quad \tau = \Omega t \tag{6.42}$$

$$\frac{\boldsymbol{j}}{en_s v_F} = \frac{\boldsymbol{P}}{\sqrt{1+P^2}}G(Q) \tag{6.43}$$

from which one sees that the solutions depend on the dimensionless parameters $\varepsilon = eE_0v_F/\mu\Omega$ and Γ/Ω. If the field parameter ε is small, the response is linear, $|P| \ll 1$ and

$$\boldsymbol{P}(t) = \frac{e\boldsymbol{E}_0}{p_F\sqrt{\Omega^2+\Gamma^2}}\cos\left(\Omega t - arctan\frac{\Omega}{\Gamma}\right), \tag{6.44}$$

$$\mathbf{j}(t) = en_s v_F \boldsymbol{P}(\mathrm{t}). \tag{6.45}$$

In [35], the case $\varepsilon \gg 1$ is also discussed. Figure 5 in [35] shows the time dependence of the momentum P(t) and of the dimensionless electric current $j(t)/en_s$ at $\varepsilon = 10$ and several values of the radiative decay parameter Γ/Ω. If Γ/Ω does not exceed the value of about $\varepsilon/2$, the self-consistent field effects lead only to the phase shift of the current, not influencing the shape of the current–time curves and hence not reducing the amplitudes of the higher harmonics. At higher values of Γ/Ω (between $\Gamma/\Omega \sim \varepsilon/2$ and $\Gamma/\Omega \sim \varepsilon$), the shape of the current–time curves smoothly modifies from the step-like form to the sinusoidal form, and at $\Gamma/\Omega \geq \varepsilon$, the higher harmonics are fully suppressed.

All the graphene electromagnetic response studies in [35] were done in the framework of flatland model. However, graphene always has out-of-plane deformations, which play an essential role. We shall consider their influence on this process in Section 6.3 neglecting radiation decay, and in Section 6.4, we shall take into account the radiation decay. Such correction of the results will allow to do the measurements in graphene-including devices more exact, which is very important for applications.

6.2 New Loss Mechanism in Graphene Nanoresonators Due to the Synthetic Electric Fields Caused by Inherent Out-of-Plane Membrane Corrugations

New loss mechanism in monolayer graphene nanoresonators caused by dissipative intra-valley currents stipulated by the synthetic electric fields is considered. These fields are generated by time-dependent gauge fields arising in graphene membrane due to its intrinsic out-of-plane distortions and the influence of the external periodic electromotive force. The corresponding formula for quality factor is obtained using quantum mechanical approach and includes quantum mechanical parameters. This loss mechanism accounts for essential part (about 40%) of loss in graphene nanoresonator and is specific just for graphene. The ways of the minimization of this kind of dissipation (increase of the quality factor of the electromechanical system) are discussed. It is explained why one can increase the quality factor by correctly choosing a combination of strains (by strain engineering). Besides, it is shown that the quality factor can be increased by switching on a magnetic field perpendicular to the graphene membrane.

6.2.1 Preliminaries

Nanoresonators proved to be very useful in a great number of applications in different spheres of activity. In a series of new small-size devices named nanoelectromechanical systems (NEMS) (see [38–39]), the nanoresonators seem to be especially perspective.

At first, materials such as piezoelectrics, silicon, metallic nanowires, and CNTs were used for fabrication of nanoresonators. The best dynamic characteristics may be achieved as the resonator size and mass scaled down (which is assumed in classical linear-elastic Bernoulli–Euler beam theory). Resonance frequency may be essentially increased, while the quality factor Q will not become much worse (see [40–41], for instance). This allows the sensitive detection of many microscopic physical phenomena and measuring parameters such as spin, force, and molecular mass. These possibilities opened new investigations in biology: virus, protein, and deoxyribonucleic acid (DNA) detection, detection of enzymatic activity, etc.

New opportunities arise if we use such 2D material as graphene—with one carbon atom layer. For instance, recently, a new especially precise method was suggested for mass detection (with zg sensitivity) based on NEM mass spectrometer [42] exploiting the advantage of graphene membranes.

Different modifications of graphene nanoresonators were studied, for instance, in [43–45]. It was shown that the damping rate increases linearly with resonance frequency. Different kinds of loss mechanisms are discussed in [43–48]. Some of them are common to all experimental setups: attachment losses, thermoelastic dissipation, etc. The others depend on actuation scheme, for instance, magnetomotive actuation scheme, capacitive coupling, etc. Surface-relative losses can usually be modeled by distribution of effective two-level systems. All these possibilities were considered in detail in [47]. Authors of [47] pointed out that in graphene nanoresonators, dissipation is dominated by electrostatically coupled graphene layer and doped metallic backgate, the energy being dissipated by increasing electron-hole excitations and due to interaction of charge fluctuation with lower-energy flexural phonons.

However, these approaches do not take into account very specific properties of the systems mentioned above. Namely they do not consider the significant influence of the so called gauge pseudomagnetic fields [29] created due to the sponteneous generations of large-scale stable distortions of graphene surface (ripples, wrinkles, etc.) responsible for its high bending rigidity. There exist expectations that these pseudomagnetic fields can be used for the creation of new graphene nanoelectromechanics. Later, it was discovered that these gauge fields might be varied by applying of external strains (see [49–50], [27]) (strain engineering). Below, we use analytical formulae for pseudo-vector potential $\vec{A}$ for monolayer graphene sheet obtained by authors of paper [30].

However, only in [33] was it pointed out that in graphene, one should also take into account that so-called synthetic electric fields should arise if pseudomagnetic gauge fields turn to be time-dependent. Having this idea in mind, authors of [33] calculated the damping rate (see (14) in [33]) for flexural phonons. Comparing it with the Kubo formula for conductivity (see (13) in [33]), they were able to interpret the dissipation caused by synthetic electric fields and associated them with currents as Joule heating.

Our problem essentially differs from the one analyzed in [33]. We consider synthetic electric fields that inevitably arise during nanoresonator vibrations driven by external electromotive force. While in [33], an amplitude of vibrations in the long wave limit tends to zero, in our case of external electromotive force, it is not so. Therefore, the formula (14) from [33] cannot be used in our case of the long wave limit ($q \approx 1/L$, is the characteristic length scale of a membrane). Nevertheless, the necessary procedures demonstrating that physics of damping is of a Joule-type heating can be done in our case as well. Below, we also estimate the resonator intrinsic losses (quality factor Q) caused by the synthetic electric fields. We show that the corresponding contribution to $1/Q$ is very essential and leads to rather large Joule-type loss in graphene nanoresonators.

Of course, the role of synthetic electric fields in other NEMS may be also important.

In the last subsection of this section we discuss the methods of graphene nanoresonators Joule-type loss possible reduction.

6.2.2 The Model

We consider graphene nanoresonator (see for instance Figure 1 in [44] or Figure 1 in [45]). For the monolayer graphene membrane described by the equation of surface $z = h(x,y)$, for any atom the vectors directed to three nearest neighbors have the form (see for instance [30])

$$\mathbf{u}_1 = a\left(\sqrt{3}/2, 1/2\right), \quad \mathbf{u}_2 = a\left(-\sqrt{3}/2, 1/2\right), \quad \mathbf{u}_3 = a\left(0, -1\right).$$

Here, $a = 2{,}5A$ is a distance between the nearest neighbors in the lattice; $h = h(x,y)$ is a distance from a point on membrane with projection (x,y) till the plane XOY.

The following formulae for gauge field vector potential $\vec{A}$ are obtained in paper [30] (see also [31]):

$$A_x\left(\mathbf{r}\right) + iA_y\left(\mathbf{r}\right) = -\sum_j \delta t_j\left(\mathbf{r}\right) e^{i\mathbf{u}_j K} = -\frac{\varepsilon_{\pi\pi}}{2} \sum_j \left[\left(\mathbf{u}_j \cdot \nabla\right) \nabla h\right]^2 e^{i\mathbf{u}_j K}, \tag{6.46}$$

$$A_x = -\frac{1}{2} A^0 \left[\left(h_{xx} \right)^2 - \left(h_{yy} \right)^2 \right] a^2, \quad A_y = A^0 \left[h_{xy} \left(h_{xx} + h_{yy} \right) \right] a^2, \tag{6.47}$$

$$A^0 = 3/4 \cdot \varepsilon_{\pi\pi}/e \cdot c/V_F. \tag{6.48}$$

Here $V_F = 10^8$ *cm/sec* is Fermi velocity, the energy $\varepsilon_{\pi\pi}$ characterizes strength of valence bonds, $\varepsilon_{\pi\pi}$=2,89 *eV*, $\mathbf{K} = a^{-1}\left(4\pi/3\sqrt{3},0\right)$ is the Dirac point position and t_j is the exchange integral with the *j*-th nearest nearest neighbor j = 1,2,3 and A^0 has the same dimension as the vector potential. Products of the expressions in square brackets in formulae for A_x,A_y in (6.47) by a^2 are dimensionless, i.e., they are numerical coefficients, their magnitudes being dependent on the deflection depth of the graphene membrane (we take into consideration large-scale deformations such as ripples, wrinkles, etc.) and also on the lattice constant value for the current moment of time. Note that in [24, 32] (see also [29]), a different formula for pseudomagnetic vector potential was considered. The difference between these two formulae is discussed in [31] and [29].

When periodic electromotive field directed along the *OZ* axis is switched on, the vectors $\mathbf{u}_j$ should get the time-depending variation $\Delta\mathbf{u}_j(t)$, which is proportional to $E_0 \sin \omega t$, i.e., we have in the linear approximation

$$a(t) = a_0 + \Delta a(t), \tag{6.49}$$

where a_0 = 2,5 Å is the initial value of the parameter "*a*" at $t = 0$ and

$$\Delta a(t) = \eta_1 E_0 \sin \omega t = a_{00} \sin \omega t \tag{6.50}$$

Here, the coefficient η_1 has dimensionality [cm²/V].

Similarly, we assume

$$h(x,y,t) = h_0(x,y) + \Delta h(t) \tag{6.51}$$

$$\Delta h(t) = \eta_2 E_0 \sin \omega t \cdot \cos(\pi x/2L) = h_{00} \sin \omega t \cdot \cos(\pi x/2L) \tag{6.52}$$

where $z = h_0(x,y)$ is the equation of the initial membrane surface form and η_2 has the same dimensionality as η_1. Both of them describe interaction with actuating field on the microscopic level. The coefficients η_1, η_2 may, generally speaking, depend on *x,y*, but it does not influence the main results of our paper. The last factor in Equation 6.52) is connected with the clumping of the opposite membrane edges by $x = \pm L$ (doubly clumped).

Notice that as it is shown in [44], the linear approximation is reasonable if the deflection amplitude of the graphene nanoresonator vibrations does not exceed 1.1nm. As we assume in our calculations below, it equals 1nm really. Therefore, our assumption about linearity is quite reasonable. By the way, nonlinear problem was studied in a number of works (see [51] and references therein) as well. However, we restrict ourselves here by the linear case.

In presence of the external actuating periodic electric field $E_0 \sin \omega t$ the gauge field vector potential $\vec{A}$ will depend on time, i.e., in the monolayer graphene membrane the so-called synthetic electric field arises

$$\vec{\mathrm{E}}_{\mathrm{syn}} = -\mathrm{c}^{-1}\vec{\mathrm{A}}_{\mathrm{t}} \tag{6.53}$$

Let $\omega \approx \omega_{res}$ where ω_{res} is the eigenfrequency of our resonator. Then, substituting Equations 6.47 through 6.52 into 6.53, we find

$$\left(\vec{E}_{syn}\right)_x = -c^{-1}\left(\vec{A}_x\right)_t = A^0 / c \cdot \left\{\left[\left(h_{xx}^2 - h_{yy}^2\right)\left(\Delta a\right)_t + a h_{xx}\left(\Delta h\right)_{xxt}\right]a\right\}, \tag{6.54}$$

$$\left(\vec{E}_{syn}\right)_y = -c^{-1}\left(\vec{A}_y\right)_t = -A^0 / c \cdot \left\{h_{xy}\left[2\left(h_{xx} + h_{yy}\right)\left(\Delta a\right)_t + a\left(\Delta h\right)_{xxt}\right]a\right\} \tag{6.55}$$

We can write formulae (6.54) and (6.55) in the form

$$\left(\vec{E}_{syn}\right)_x = E^0\left(\omega\right)h_{00}I_x \cos\omega t, \quad \left(\vec{E}_{syn}\right)_y = E^0\left(\omega\right)h_{00}I_y \cos\omega t \tag{6.56}$$

where $h_{00} = (E_0\eta_2)$ is a resonator oscillation amplitude (deflection) and

$$I_x = \left\{\left[\left(\eta_1/\eta_2\right)\left(h_{xx}^2 - h_{yy}^2\right) - a h_{xx}\left(\pi/2L\right)^2 \cos\left(\pi x/2L\right)\right]a\right\} \tag{6.57}$$

$$\mathrm{I}_y = \{h_{xy}[2(\eta_1/\eta_2)(h_{xx} + h_{yy}) - a(\pi/2L)^2 \cos(\pi x/2L)]a\} \tag{6.58}$$

$$E^0(\omega) = 3/4 \cdot \epsilon_{\pi\pi}/\mathrm{e} \cdot \omega/V_F \tag{6.59}$$

Remark that dimensionless quantities I_x, I_y do not turn to zero even by zero deflection because of the presence in graphene of deformations such as ripples, wrinkles, and so on. It follows from Equations 6.56 through 6.59 that after time averaging, we have

$$\left(\vec{E}_{syn}\right)^2 = \left(\vec{E}_{syn}\right)_x^2 + \left(\vec{E}_{syn}\right)_y^2 = \left(E^0\left(\omega\right)\right)^2 h_{00}^2\left(I_x^2 + I_y^2\right)/2 \tag{6.60}$$

Note that our problem very much differs from one considered in [33]. In particular, in [33], where flexural phonons damping rate due to Joule-type heating is calculated, their vibrational amplitude tends to zero in the long wave limit. However, in our case, the amplitude of the membrane vibration does not tend to zero because of the external electromotive force. Nevertheless, the corresponding procedure can be done in our case as well. As it was explained, dissipation mechanism in monolayer graphene resonator driven by time-dependent electromotive force may be considered as Joule-type loss (like the problem of finding flexural phonons damping rate in [33]). So we can write for the Joule-type loss in our problem $\Delta\epsilon_J$ for the period $T = 2\pi/\omega$ in the form

$$\Delta\epsilon_J \approx 2\pi\left(\vec{E}_{syn}\right)^2 \sigma L_x L_y/\omega \tag{6.61}$$

where L_x, L_y are membrane sizes and conductivity σ is two dimensional conductivity (for more details, see below). From Equations 6.60 and 6.61, we obtain

$$\Delta\epsilon_J \approx \pi\left(E^0(\omega)\right)^2 h_{00}^2\left(I_x^2 + I_y^2\right)\sigma L_x L_y/\omega \tag{6.62}$$

Remark that in Equations 6.61 and 6.62, we took into consideration a contribution only from one Dirac cone K (only one valley, i.e., only one sublattice). We will discuss the correction connected with the role of another valley K^* in Section 6.2.4 "Summary."

From formulae (6.59) and (6.62), we see that the damping rate linearly depends on frequency. It is interesting that in nanoresonators on the basis of CNTs, the dissipation mechanism connected with the electron tunneling through a vibrating nanotube also gives damping rate linearly depending on frequency [48].

Graphene resonators general loss includes parts of a different nature:

$$Q^{-1} = Q_0^{-1} + Q_J^{-1} \tag{6.63}$$

Here, Q_0^{-1} is connected with the dissipation mechanisms studied earlier by other authors (see for instance [43–48]), and Q_J^{-1} is connected with the mechanism considered and analyzed in the present paper for the first time.

We introduce quality factor Q_j connected with Joule-type loss as follows

$$Q_J^{-1} = \Delta\epsilon_J/\epsilon_{total} \tag{6.64}$$

Here $\Delta\epsilon_J$ was found in Equation 6.62, and the total energy is defined as follows:

$$\epsilon_{total} = N\cdot m_{at}\cdot\omega^2\cdot h_{00}^2, \qquad N = L_x L_y/\left(a^2 3\sqrt{3}/2\right),$$

where N is the number of atoms in graphene membrane, m_{at} is the atom mass and h_{00} is the membrane oscillation amplitude. So we obtain

$$Q_J^{-1} = \pi\frac{3\sqrt{3}}{2}\cdot\frac{\left(E^0(\omega)\right)^2\cdot\sigma\cdot\left[a^2\left(I_x^2 + I_y^2\right)\right]}{\omega^3 m_{at}}. \tag{6.65}$$

Note that using for $E^0(\omega)$ Equation 6.59 and for conductivity the general formula in the form

$$\sigma \cong \frac{e^2}{h}\frac{\epsilon_F \tau}{h} \cong \frac{e^2}{h} k_F l,$$

we see that quality factor (Equation 6.65) does not depend on the charge It is connected with the fact that synthetic fields are chargeless ($div\mathbf{E}_{syn} = div\, \mathbf{A}_t = 0$, see [31]).

Note also that our formula (6.65) for Joule-type loss quality factor contains the Plank constant h and the energy $\epsilon_{\pi\pi}$ characterizing strength of valence bonds. So considered loss mechanism due to intrinsic synthetic fields has quantum origin. Besides, it was shown in [33] (see formula (45) in [33] and discussion thereabout) that 2D conductivity σ does not (or weakly) depend on activating field frequency in graphene. But estimating in the next point below the approximate value of Joule-type loss, we shall take the measured value for σ using experimental data.

6.2.3 Joule-Type Loss Estimation and the Ways of Their Minimization

Let us estimate the value of Joule-type loss found in formula (6.65) and compare the calculated value with experimental data. We consider graphene nanoresonator with frequency $\omega_{res} \approx 130MHz$ investigated in paper [44]. As for the case $m_{at} = 12 \cdot 1{,}67 \cdot 10^{-24}$ g, we have $m_{at} \cdot \omega^3 \approx 42$ [g/s^3].

From formula (6.59), we obtain

$$E^0(\omega) = 3/4 \cdot \epsilon_{\pi\pi}/e \cdot \omega/V_F \approx 3/4 \cdot 3 \cdot 1{,}3/3\ volt/cm = 3{,}9/4 \cdot 1/300 \text{ CGSE} \qquad (6.66)$$

The conductivity value for graphene sample mentioned above was not given in [44] and we take it from another paper [53] where the parameters of the experiment are close to those in [44, 21]. From paper [53], for concentration value $n = 2{,}5 \cdot 10^{11}[\text{cm}^{-2}]$ we find from Figure 1 in [53, 35] that $\sigma \approx 1{,}2 \cdot 10^9$ [cm/s] (for the sample of good quality).

Estimate now the factor $a^2\left(I_x^2+I_y^2\right)$ in (6.65). In [44], it is demonstrated that membrane oscillation critical amplitude after which nonlinearity becomes essential is equal to 1,5nm. We assume it to be $h_{00} \approx 1nm$. It is natural to think that $\Delta a/a \approx h_{00}/h \approx 0{,}1$), i.e., $\eta_1/\eta_2 \approx \Delta a/h_{00} \approx (\Delta a/a) \cdot (a/h_{00}) \approx 2{,}5 \cdot 10^{-2}$.

Let us estimate the first term in the expression for $a^2\left(I_x^2+I_y^2\right)$ using formulae (6.57) and (6.58). Taking into consideration that graphene membrane surface has corrugations and assuming for simplicity the deformation height (depth) and the basis (length, width) to have close sizes, we find $a^2 \cdot I_x^2 \approx \left(6{,}25 \cdot 10^{-4} a^4 \cdot \left(h_{xx}^2\right)^2 + \ldots\right) \approx \left(6{,}25/81\right) \cdot 10^{-8}$. When estimating, we assumed the deformation radius to be $\delta_x \approx 15\ nm$, and $\delta h/\delta_x \approx 2$. Other terms in formula for $a^2\left(I_x^2+I_y^2\right)$ can be estimated similarly. Therefore, we obtain $a^2\left(I_x^2+I_y^2\right) \approx 0{,}7 \cdot 10^{-8}$. Hence, we find from Equation 6.65, (6.66) the approximate theoretical numerical value for Joule-type loss in the sample 6.mentioned above $Q_J^{-1} = \Delta\epsilon_J/\epsilon_{total} \approx 3 \cdot 10^{-5}$. Since the experiment in [44] gives the result $Q \approx 14000$, we see that the Joule-type loss are responsible for about 40% of the total loss and our model gives the reasonable magnitude of damping rate.

It is interesting that in paper [45] for the sample with about the same resonance frequency, they obtained the quality factor $Q \approx 100\,000$. The measured increase of the quality factor to our point of view was obtained by authors as they used tension. From our formula (6.65), it is well seen that in this case, the factor $(I_x^2 + I_y^2)$ decreases, which enhances the quality factor, i.e., the measured increase of quality factor is in accordance with our theory.

Now consider the question, how can one minimize the Joule loss Q_J^{-1}. It is clear that the expressions I_x, I_y in Equations 6.57 and 6.58, and consequently the loss (Equation 6.64) can be reduced by varying the form of the function $h(x,y)$ with the help of strains of different kinds. The fact that one can increase the quality factor by such actions was opened experimentally, and it has become a subject of a new special field of activity that was called strain engineering. From the formula (6.65), the reason of this phenomenon is obvious.

One can decrease Joule loss also by switching on a magnetic field perpendicular to the graphene membrane plane. Indeed, in paper [54], Figure 4, we see that for temperature $T = 300\ K$, the magnetic fields less than 8 Tesla are not quantizing (are classical) and longitudinal resistivity is increasing function of field (in [55], Figure 3a, $T = 5$, nonquantizing magnetic field is much less and resistivity is again increasing function). So, for example, for $H = 6$ Tesla, we have ([54], Figure 4) that Joule loss Q_J^{-1} are about six times less. We saw above that the percentage of these loss is about 40%. Consequently, the quality factor would be one-and-a-half times more if magnetic field $H = 6$ Tesla were switched on.

Since the graphene membrane surface has corrugations, the external magnetic field components parallel to vibrating membrane can arise. These components play the role of magnetomotive force. Hence, as it is shown in [56–57], we can obtain extra damping and the increase of Q may be less.

Note that the formulae obtained using Boltzmann equation stop to be correct when quantization in magnetic field of Landau type starts. Nevertheless, a tendency of loss decreasing remains valid though the form of dependence may be different. By the way, the experimental investigation of the dependence $Q(H_\perp)$ appeared just now (see [58], Figure 3c) for the temperature $T = 5K$, i.e., the fields about 6 Tesla are quantizing. However, for magnetic field 6 Tesla, they measured an increase of Q of about 30%. The role of synthetic fields wasn't discussed in [54].

6.2.4 Summary

In this section, we considered new quantum dissipation mechanism for graphene nanoresonators, i.e., "Joule-type" loss caused by synthetic electric fields. For the linear case (i.e., electromotive periodic force is rather weak), the formulae for Joule-type loss are obtained. However, though graphene lattice consists of two sublattices in formula (6.65) for quality factor derivation, we took into consideration the contribution only from one valley K (one sublattice). Note that in the ideal case, i.e., if there is the time-reversal symmetry [52], the gauge fields in K and K^* have opposite directions and equal magnitudes, and the two valley currents compensate each other. However, this question was analyzed in [33], where it was shown that the two corresponding valley currents do not compensate each other if we take into account intervalley Coulomb drag effect and intervalley scattering on short-range impurities. So, really we should have in (6.65) additionally also dimensionless factor, which depends on relaxation times τ_D and τ_v. We meant that τ_D^{-1} is the intervalley scattering rate responsible for the drag effect and τ_v^{-1} is the intervalley scattering rate due to the short-range impurities. But such a theory is beyond the scope of this section.

We would like to stress especially that while in the major part of papers dedicated to nanoresonators, the phenomenological approach within framework of continuum nonlinear elastic model (see [59] and last review-like paper [51]) was used (nonlinear Duffing oscillator), our results for Joule-type loss are obtained on the basis of microscopic theory taking into account the specific features of graphene. The membrane vibration is supposed to be classical, but the mechanism of loss in graphene nanoresonator is described within the framework of quantum solid-state physics. In particular, our main formulae for Joule loss and quality factor include quantum mechanical parameters $\varepsilon_{\pi\pi}, V_F$.

Using the formulae obtained for the Joule-type loss, we calculated approximately their magnitudes. This estimate shows that their contribution to the general dissipation appears to be about 40%.

The possible methods of lowering down of Joule loss are as follows:

- Application of strain engineering methods to minimize quantities I_x,I_y, and consequently Q
- Switching on the magnetic field perpendicular to the graphene membrane

Note that just now, a paper [60] has appeared where it was stated that in polycrystalline graphene, sheets of multilayer graphene nanoresonator-measured loss proved to be much larger than calculated for monolayer graphene. The authors in [60] mentioned that it could have been due to grain boundaries angle misorientations, which generate out-of-plain buckling (see Figure 2 in [60]). This type of corrugations should also cause artificial gauge fields and lead to the microscopic mechanism of loss suggested in our paper. In [60], it was also mentioned the significant Q-factor increasing when reduction in out-of-plane buckling heights by using tensile strain. These strain-engineering method to reduce loss were theoretically found by us at first in [61]. It would be also interesting to study CVP-grown graphene Q-factors increasing, when external perpendicular magnetic field is switched, its magnitude and temperature being varied.

If the prediction of the quality factor increases due to the magnetic field switching on perpendicular to the membrane we published at first in [61], the experimental investigation of the dependence of Q on $B_\perp$ is done just now (see Figure 3(c) in [58]) in quantum limit for the temperature $T = 5K$. The increase of Q in quantizing magnetic field $B_\perp$ was really observed and its value proved to be about 30%. Note that the authors of [54] think that in standard 2D model of graphene (without out-of-plane corrugations), this effect may be due to the magnetization of graphene in quantum Hall limit, which changes electromechanics.

Note also that the synthetic currents we considered in this section not only lead to Joule-type loss but also cause dissipation due to their interaction with currents arising on gate. We hope to analyze these losses in the next section.

In this section, we found that the taking into account the inevitably existing in graphene membrane various corrugations gives essential contribution into the magnitude of the quality factor in graphene nanoresonator in megahertz and gigahertz frequency range. Obviously, this mechanism should influence also a nonlinear electromagnetic response of graphene in terahertz and optical frequency range. In transport phenomena, we should also take it into consideration. So for exact estimates of losses while constructing different kinds of devices where graphene is used in nonstationary regime in branches such as photonics, optoelectronics, etc., we should also take into account the generation of synthetic electric fields and investigate their influence. Some ideas to minimize its negative action were also outlined.

The results described in this section were published in [62].

6.3 Surface Corrugations Influence on Monolayer Graphene Electromagnetic Response

The known quasiclassical self-consistent equations for graphene nonlinear electromagnetic response are generalized by taking into account so-called synthetic electric fields arising due to the presence in graphene of inherent out-of-plane nanodeformations (ripples, etc.). The modified equations are discussed.

6.3.1 Preliminaries

In 2007, S.A. Mikhailov [63] has first predicted that graphene's linear dispersion properties should lead to a strongly nonlinear optical behavior at microwave and terahertz frequencies. In [63, 35] the kinetic (quasiclassical) transport MZ (Mikhailov-Ziegler) equation was obtained to describe first the nonlinear graphene electromagnetic response. Note that in [63, 27], only intraband transitions were taken into account which was reasonable as they considered radiation in terahertz frequency range. For the case of higher optical frequencies, Ishikawa [64] used the time-dependent Dirac equation cast into the form of extended optical Bloch equation, which allowed him to show that for $\hbar\omega > E_F$ due to the interplay of interband and intraband transitions, total nonlinear optical response grows more slowly. The theory of H. Dong *et al.* [65] is consistent with the more precise quantum approach of Ishikawa and gives the same result for $\hbar\omega > E_F$. An interesting new result obtained in [65] is a moving Townes-like spatial solitary waves, i.e., soliton (see Figures 2 and 3 in [65]), which arises due to inclusion of the term $\Delta f_p(x,y,t)$ into the MZ quasiclassical transport equation. Exellent photophysical properties and large optical nonlinearities with ultrafast response times including fast optical communications promise many potential applications (see review [66]).

Most of previous theoretical studies and experimental observations were devoted to very peculiar properties of the second- or third-order harmonics excited in graphene by external time-dependent radiation. But there exist situations when one cannot confine themselves only to these first harmonics, for instance, for description of strong terahertz emission stimulated from optically pumped graphene etc. Therefore, it is necessary to develop the general theory of nonlinear electromagnetic response for graphene. In the papers [63–65] mentioned above, the first studies of the nonlinear optical response in graphene were published, which were not based on the investigation of separate harmonics. Now, there appeared a series of papers based on the more-developed Floque theory approach. Very interesting new results for the nonlinear optical response in the presence of intense circularly and linearly polarized terahertz fields were obtained in [67] on the basis of Floque theory (see for instance [68]). It was shown that a gap was opened in the quasi-energy spectrum due to the single-photon (or multi-photon) resonances. An effective adiabatic perturbation theory for quantum systems responding to short laser pulses was presented in [68].

All the mentioned methods are based on the one-electron approximation and do not take into account many-body effect. They also do not consider the influence of the inherent out-of-plane corrugations. To begin with, we'll make use of a rather simple approach formulated by Mikhailov in [63] and elaborated in [35] for the description of monolayer graphene nonlinear optics and based on the model of non-interacting Dirac 2D electrons. We'll generalize this approach additionally taking into account "internal" time-dependent "synthetic" electric fields induced by radiation influence on the inherent always existing out-of-plane corrugations (ripples, wrinkles, etc.). We'll explain in this subsection

1. Why it is important to take into account the influence of the out-of-plane corrugations on the monolayer graphene electromagnetic response, and
2. Why it is reasonable to begin the investigation of nonlinear optical response with simplified MZ model of the noninteracting 2D Dirac electrons?

(1) More than 70 years ago, Peierls [9, 10] and Landau [11, 12] showed that in standard harmonic approximation, strictly 2D crystals cannot exist because thermal fluctuations should destroy long-range order resulting in "melting" of a lattice at any finite temperature. Really (see [18]), all the observed monolayer graphene samples have inherent stable out-of-plane deformations (ripples, bubbles, wrinkles, etc.). These corrugations lead to the arising of pseudomagnetic fields (gauge fields) (see for instance the review in [29]). These inevitably existing fields in graphene are about several Tesla. Not long ago, such fields were artificially created in another nontrivial system—in rubidium Bose–Einstein condensate (BEC). This pseudomagnetic field was created time-dependent via interaction of BEC atoms with laser light. This led to the appearance of the so-called synthetic electric fields [34].

Time-dependent (due to the presence of flexural phonons (f-ph)) gauge field and its time-derivative, i.e., synthetic electric fields for monolayer graphene, was first considered in [33] where it was shown that (f-ph) damping rate is determined by interaction with these fields.

In graphene nanoresonators, vector potential depends on time due to the influence of the external time-dependent electromotive force. This generates synthetic electric fields and currents associated with them. It was shown in [62] and in Section 6.2 that this leads to a new mechanism of the dissipation ("heating") and determines the loss that strongly influence the quality factor (about 40%).

In this subsection, we consider the graphene out-of-plane corrugations influencing the nonlinear electromagnetic response. If in [62] and in Section 6.2, considered frequencies were not more than 1GHz in the present investigation, we analyze the corrugations influencing monolayer graphene response to electromagnetic radiation of terahertz [300GHz–3THz] and infrared [201THz–790THz] frequency range.

(2) As it is well known now (see for instance the review in [69]), in the 2D system of interacting Dirac electrons, many-body effects should be taken into account. It was shown that not only quasiparticles ("massless electrons" and "holes") but also plasmons (collective charge density excitations) and plasmarons [70] ("dressed" quasiparticle excitations coupled to plasmons) play an important role. These predictions are based on a very specific form of

quasiparticle spectral function obtained theoretically [71–72] and observed experimentally by a variety of direct and indirect methods: electron energy-loss spectroscopy (EELS) (see in [71]), angle-resolved photoemission spectroscopy (ARPES) [73], and scanning tunneling spectroscopy (STS) [74] on exfoliated graphene sheets and on epitaxial graphene samples.

Large efforts were made to observe these predictions in optical measurements. As a rule, indirect methods are used “by engineering plasmons coupling to infrared light in a number of intriguing ways” [69], by making more strong light–matter interaction and alleviating the momentum mismatch between plasmon modes and the incident radiation. To illustrate this idea, we give as an example Mikhailov’s formula for the plasmon-enhanced second harmonic generation in graphene [75]:

$$I_{2q,2\omega}^{graph} \sim \left(I_{q,\omega}^{ext}\right)^2 / \left\{ \left[\left(\omega^2 - \frac{\omega_p^2(q)}{2} \right)^2 + \frac{\omega^2\gamma^2}{4} \right] \left[\left(\omega^2 - \omega_p^2(q)\right)^2 + \omega^2\gamma^2 \right]^2 \right\}$$

Here, $I_{q,\omega}^{ext}$ is the intensity of the incident light, $I_{2q,2\omega}^{graph}$ is the intensity of the second harmonic wave in graphene, $\omega_p(q) = \sqrt{\frac{8E_F}{h\varepsilon}\frac{\pi}{2}\frac{e^2}{h}q}$ is the plasmon frequency, q is the plasmon wave vector, and γ is the momentum scattering rate for plasmon. The denominator in the last formula strongly decreases as $\omega \to \omega_p(q)$ and $I_{2q,2\omega}^{graph}$ strongly increases. However, one should not forget about the momentum conservation law: $q_{light} = q_{plasm}$. Let us estimate the ratio q_{pl}/q_{light} for $\omega = \omega_{light} \approx \omega_{pl}$. As $\omega_{light} = cq_{light}$ for external field we obtain $q_{pl} / q_{light} = \frac{h\omega_{light}}{E_F}\left(\frac{e^2}{hc}\right)^{-1}$ which shows that it is very difficult to satisfy simultaneously the both conditions $\omega_{light} \approx \omega_{pl}$ and $q_{light} = q_{pl}$. However, in 2011, Basov *et al.* [76] managed to overcome these difficulties creating stronger light–matter interaction by confining mid-IR radiation at the apex of a nanoscale tip of atomic force microscopy. This turned to be a very effective method to alleviate the momentum mismatch between plasmon modes and incident radiation. Through infrared nanoimaging, the authors of [76] have shown that graphene/SiO_2/Si back-gated structures may support a propagation of surface plasmons.

Independently, Chen, J. *et al.* [77] have made a similar discovery using graphene films deposited by a gas rather than peeled from graphite. They used near field produced by a permanent dipole. As far as we know, up to now, it was not shown experimentally that homogeneous light radiation of a single-layer graphene membrane may excite plasmons.

It is interesting to note that L. Crasse *et al.* [78] have shown that in graphene epitaxially grown on SiC, the Drude absorption is transformed into a strong terahertz plasmon peak due to natural nanoscale inhomogeneities such as substrate terraces and wrinkles. This is a natural confinement potential and does not require special lithography patterning.

Thus, we have come to the conclusion that under incident homogenous radiation, surface plasmons (SPs) cannot be driven without using special efforts: or by using AFM tip to confine the incident radiation to q nanoscale region around the tip as Fei *et al.* [74] or by exciting SP in graphene micro-ribbon array or by using stacked graphene microdiscs to realize electromagnetic radiation shield with 97.5% effectiveness [69].

Measurement of the linear optical conductivity of graphene by Mak *et al.* [79] are in agreement with theoretical results predicted within a model of non-interacting massless Dirac fermions. Taking into account e-e interactions (see [80]) with the renormalization group techniques results to the notable modifications but does not reveal plasmon excitation. At present, we are not aware of any experimental investigations of graphene nonlinear electromagnetic response to homogenous radiation in which SP were displayed. Theoretical investigations of the single-layer graphene transmittance (T) in nonlinear case shows very small (~2%) changes of T (see [81]). Therefore, we shall use the model of noninteracting Dirac electrons as it was done in [63, 35]. Using our method we essentially modify the known Mikhailov's photonics equations ([63, 35]) in terahertz range. Our model allows us to calculate nonlinear electromagnetic response of graphene with synthetic electric fields taken into account on the basis of the kinetic (quasiclassical) transport equation with generated by corrugations synthetic electric fields taken into account. The modified equations are studied. Radiation decay rate and response of graphene to pulse excitations are discussed. The obtained results can be applied to the analysis of different devices in terahertz optics and optoelectronics.

6.3.2 Generalization of MZ Equation

Below, we shall generalize MZ equation for nonlinear electromagnetic response of graphene obtained in approximation with taking into account the self-consistent field-effect, collisions of electrons with impurities, photons and other lattice imperfections having been ignored. At first we shall describe the most important steps of the self-consistent procedure made in [63, 35]. We shall stress also that the real model of the single-layer graphene (with taking into account corrugations) leads to the essential difference from Mikhailov's flatland model to make it easier for the reader to understand our self-consistent procedure and it generalizes the flatland approach. So the MZ equation was obtained in [63, 35] on the basis of kinetic transport equation for Fermi momentum distribution function $f(\mathbf{p},t)$ in the presence of the external time-dependent field $\mathbf{E}_{z=0}^{ext}(t)$ incident on the noncorrugated graphene 2D surface $z = 0$,

$$\frac{\partial f_{\mathbf{p}}(t)}{\partial t} - e\mathbf{E}_{z=0}^{tot}(t)\frac{\partial f_{\mathbf{p}}(t)}{\partial \mathbf{p}} = 0, \tag{6.67}$$

$$\mathbf{E}_{z=0}^{tot}(t) = \mathbf{E}_{z=0}^{ext}(t) + \mathbf{E}_{z=0}^{ind}(t), \tag{6.68}$$

where the induced electric field $\mathbf{E}_{z=0}^{ind}(t)$ and a current are related by the so-called 2*d* Maxwell equation (see [63, 35])

$$\mathbf{E}_{z=0}^{ind}(t) = -2\pi\mathbf{j}(t)/c \tag{6.69}$$

Here, $z = z(x,y) = 0$ is the equation of the plane graphene sheet. It was shown that electric current

$$\mathbf{j}(t) = -\frac{g_s\, g_v eV}{(2\pi h)^2}\int \frac{\mathbf{p}d\mathbf{p}}{\mathrm{p}} F_0\left(\boldsymbol{p}-\mathbf{p}_0(t)\right), \tag{6.70}$$

where $g_s = 2$ is the spin degeneracy, $g_v = 2$ is the valley-degeneracy factor, $f_p(t) = F_0(\boldsymbol{p}-\boldsymbol{p}_0(t))$ is the exact solution of the equation (6.67) for $\mathbf{E}^{tot}_{z=0}(t) = \mathbf{E}^{ext}_{z=0}(t)$

$$F_0(\boldsymbol{p}) = \left[1 + exp\left(\frac{Vp-\mu}{T}\right)\right]^{-1}, \tag{6.71}$$

$$\mathbf{p}_0(\mathrm{t}) = -\mathrm{e}\int_{-\infty}^{\mathrm{t}} \mathbf{E}^{\mathrm{tot}}(t')dt' = -e\int_{-\infty}^{\mathrm{t}}\left[\mathbf{E}^{\mathrm{ind}}_{\mathrm{z=0}}(t') + \mathbf{E}^{\mathrm{ext}}_{\mathrm{z=0}}(t')\right]dt' \tag{6.72}$$

If the temperature $T = 0$ and the chemical potential is finite, $\mu > 0$ the current $\boldsymbol{j}(\mathrm{t})$ can be rewritten in the form (see [63, 35])

$$\mathbf{j}(t) = en_s V\frac{\mathbf{P}}{\sqrt{1+P^2}}G(Q), \tag{6.73}$$

where $\mathbf{P} \equiv \mathbf{P}(t) = -\mathbf{p}_0(t)/p_F, P(t) = \,\vdots\, \mathbf{P}(t) \,\vdots, p_F = \mu/V$ is the Fermi momentum and the formula for the density of electrons n_s in the upper band may be written as follows:

$$n_s = n_e = \frac{g_s g_v p_F^2}{4\pi h^2} = \frac{g_s g_v \mu^2}{4\pi h^2 V^2}$$

The function $G(Q)$ is defined as

$$G(Q) = \frac{4}{\pi Q}\int_0^{\pi/2} \cos x dx\left(\sqrt{1+Q\cos x} - \sqrt{1-Q\cos x}\right),\quad Q \equiv 2P(t)/\left(1+P^2\right) \tag{6.74}$$

From Equations 6.67 through 6.69 and 6.73, one gets the following equation (obtained in [63, 35]):

$$\frac{d\mathbf{P}(t)}{dt} + \gamma\frac{\mathbf{P}}{\sqrt{1+P^2}}G(Q) = \frac{e}{p_F}\mathbf{E}^{ext}_{z=0}(t), \tag{6.75}$$

$$\gamma = \frac{g_s g_v}{4} \frac{e^2}{hc} \frac{\mu}{h} = V \frac{e^2}{hc} \sqrt{\pi g_s g_v n_s} \tag{6.76}$$

Coefficient γ in Equation 6.75 plays the role of natural decay, thanks to radiation. As it was explained above, the Mikhailov's equation (Equations 6.75 and 6.76) was obtained in [63, 35] on the basis of quasiclassical kinetic theory. It allows to describe graphene plane sheet nonlinear electromagnetic response. However, this equation is written for an absolutely plane monolayer graphene membrane and doesn't take into account a specific graphene property: the existence of the inherent out-of-plane deformations (bubbles, ripples, etc.), which leads to the arising of quantum effects due to the changing of overlap for different valence bonds having quantum origin. Namely, it leads to the arising of the pseudo-vector potential $\mathbf{A}(x,y)$ due to out-of-plane deformations (see [10]). Under the action of the external electromagnetic field, the vector potential $\mathbf{A}(x,y,t)$ begins to depend on time, which leads to the arising of so-called synthetic electric field

$$\mathbf{E}_{syn}(x,y,t) = -c^{-1}\partial\mathbf{A}(\mathrm{x,y,t})/\partial t \tag{6.77}$$

Our approach is to take into account a corrugated graphene surface form $z = h(x,y,t)$; that is why instead of Equation 6.69 considered in [35], we write

$$\mathbf{E}^{ind}_{z=0}(x,y,t) = -2\pi\frac{\mathbf{j}(x,y,t)}{c} + \mathbf{E}_{syn}(x,y,t) \tag{6.78}$$

As it follows from Equation 6.78 in formula (6.72) in integral over t' a new term depending on x,y should arise, and in Boltzmann equation gradient terms ($v_{Fx}\frac{\partial f}{\partial x}, v_{Fy}\frac{\partial f}{\partial y}$) must be taken into account, which complicates the analysis of the solution of this equation.

In $2D$ Dirac model, the above-mentioned gradient term does not change the magnitude of velocity modulus ($\vdots\, v \,\vdots = v_F$) and describes only the rotation of velocity vector $\mathbf{v}$ in p-space. These terms do not depend on electron energy, and that is why they are not so important as both terms in Equation 6.67 for satisfactory description of absorption and emission of electromagnetic radiation and the current generation in graphene membrane. Absorption of electromagnetic radiation stipulates the increase of momentum $p(t)$.

However, the preliminary study shows that these terms may be neglected if $E_{syn} < E_0$ (see the discussion after the formula (6.88)).

So using Equations 6.77 and 6.78 instead of 6.69, in Equation 6.72, we find the nonlinear electromagnetic response, taking into consideration the quantum effects related with specific microscopic features of graphene:

$$\frac{d\mathbf{P}(x,y,t)}{dt} + \gamma\frac{\mathbf{P}(\mathrm{x,y,t})}{\sqrt{1+P^2}} G(\mathrm{Q}) = \frac{e}{p_F}\left[\mathbf{E}^{ext}_{z=0}(t) - c^{-1}\partial\mathbf{A}(\mathrm{x,y,t})/\partial t\right] \tag{6.79}$$

For the gauge field $\boldsymbol{A}(x,y)$, the formulae were obtained in [30] (see also [82–83]):

$$A_x(\mathbf{r})+iA_y(\mathbf{r})=-\sum_j \delta t_j(\mathbf{r})e^{i\mathbf{u}_j\mathbf{K}}=-\frac{\epsilon_1}{2}\sum_j\left[(\mathbf{u}_j\cdot\nabla)\nabla h\right]^2 e^{i\mathbf{u}_j\mathbf{K}} \tag{6.80}$$

$$A_x=-\frac{1}{2}A^0\left[(h_{xx})^2-(h_{yy})^2\right]a^2,\quad A_y=A^0\left[h_{xy}(h_{xx}+h_{yy})\right]a^2 \tag{6.81}$$

$$A^0 = 3/4\cdot\epsilon_1/e\cdot c/V_F \tag{6.82}$$

Here, $\epsilon_1=2{,}89ev$, $\mathbf{K}=a^{-1}\left(4\pi/3\sqrt{3},0\right)$ is a Dirac point and t_j-exchange integral with the j-th nearest neighbor, j = 1,2,3, and A^0 has the same dimensionality as vector potential. In the formulae for A_x,A_y in (6.81), the products of the expressions in square brackets multiplied by a^2 are dimensionless, i.e., they are numerical coefficients, their magnitudes being dependent on the deflection depth of the graphene membrane (we take into consideration large-scale deformations such as ripples, wrinkles, etc.) and also on the lattice constant value for the current moment of time.

Using Equation 6.77, we can rewrite Equation 6.79 in the form

$$\frac{d\mathbf{P}(x,y,\tau)}{d\tau}+\frac{\gamma}{\omega}\frac{\mathbf{P}(x,y,\tau)}{\sqrt{1+P^2(x,y,\tau)}}G(Q)=\frac{e}{p_F}\left[\frac{\mathbf{E}_0}{\omega}\cos\tau+\mathbf{E}_{syn}(x,y,\tau)\right],\quad \tau=\omega t \tag{6.83}$$

i.e., in terms of synthetic electric field. The influence of synthetic electric fields $\mathbf{E}_{syn}$ on the Joule-type loss in graphene nanoresonator was studied in [65]. The formulae for $\mathbf{E}_{syn}$ were obtained for electromotive force $\boldsymbol{E}^{ext}=\boldsymbol{E}_0\cos\omega t$ with small amplitude E_0 under assumption that the membrane is moving as a whole without a notable changing of a corrugated surface form. Namely,

$$\left(\mathbf{E}_{syn}\right)_x(x,y,t)=-E^0(\omega)h_{00}I_x(x,y)\sin\omega t, \tag{6.84}$$

$$\left(\mathbf{E}_{syn}\right)_y(x,y,t)=-E^0(\omega)h_{00}I_y(x,y)\sin\omega t \tag{6.85}$$

where $h_{00}=(E_0\eta_2)$ is a resonator oscillation amplitude (deflection) and

$$E^0(\omega) = 3/4\cdot\epsilon_1/e\cdot\omega/V_F \tag{6.86}$$

$$I_x=\left\{\left[(\eta_1/\eta_2)\left(h_{xx}^2-h_{yy}^2\right)-ah_{xx}(\pi/2L)^2\cos(\pi x/2L)\right]a\right\} \tag{6.87}$$

$$I_y = \{h_{xy}[2(\eta_1/\eta_2)(h_{xx} + h_{yy}) - a(\pi/2L)^2 \cos(\pi x/2L)]a\} \tag{6.88}$$

Coefficients η_1,η_2 determine time corrections to the lattice constant $\Delta a(t)$ and the variation of the corrugation height $\Delta h(t)$, i.e., $a(t) = a_0 + \Delta a(t), h(t) = h_0(x,y) + \Delta h(t)$,

$$a(t) = \eta_1 E_0 \sin \omega t, \qquad h(t) = \eta_2 E_0 \sin \omega t$$

Coefficients $\eta_{1,2}$ have the dimension cm^2V^{-1}. Note that dimensionless quantities I_x, I_y do not turn to zero even by zero deflection because of the presence in graphene of deformations such as ripples, wrinkles and so on. In [62], it was shown that taking into account the new loss mechanism, i.e., Joule-like loss (heating) caused by $\mathbf{E}_{syn}$, in linear case is rather essential (about 40%).

Consider now Equation 6.81 for linear case $P \ll 1$. Then, we have

$$\frac{d\mathbf{P}(\mathrm{x,y},\tau)}{d\tau} + \frac{\gamma}{\omega}\mathbf{P}(\mathrm{x,y},\tau) = \frac{e}{\omega p_F}\left[\mathbf{E}_0 \cos\tau + \mathbf{E}_{syn}(x,y,\tau)\right], \quad \tau = \omega t, \tag{6.89}$$

$$\mathbf{j}(x,y,\tau) = en_s V\, \mathbf{P}(x,y,\tau) \tag{6.90}$$

We may rewrite Equation 6.89 in the form

$$\frac{d\mathrm{P}_{x,y}(\mathrm{x,y},\tau)}{d\tau} + \frac{\gamma}{\omega} P_{x,y}(\mathrm{x,y},\tau) = C_{x,y}(x,y)\cos\left(\tau + \tan^{-1} D_{x,y}(x,y)\right), \tag{6.91}$$

$$C_{x,y}(x,y) = \frac{e}{\omega p_F}\sqrt{E_0^2 + \left(E^0(\omega)\mathrm{h}_{00} I_{x,y}(x,y)\right)^2}, \quad D_{x,y}(x,y) = \frac{E^0(\omega)\mathrm{h}_{00} I_{x,y}(x,y)}{E_0} \tag{6.92}$$

So we get the solution for the case of the weak intensity with out-of-plane corrugations taken into account

$$\mathrm{P}_{x,y}(\mathrm{x,y},t) = \frac{C_{x,y}(x,y)}{\sqrt{\omega^2 + \gamma^2}} \cos\left[\left(\omega t + \tan^{-1} D_{x,y}(x,y)\right) - \tan^{-1}\frac{\omega}{\gamma}\right] \tag{6.93}$$

For the plane graphene sheet (without corrugations), from here follows the formula obtained in [35]:

$$\mathbf{P}(\mathrm{t}) = \frac{e\mathbf{E}_0}{p_F\sqrt{\omega^2 + \gamma^2}} \cos\left(\omega t - \tan^{-1}\frac{\omega}{\gamma}\right)$$

From formulae (6.86) and (6.89), we see that in case of weak intensity of the incident electromagnetic radiation and corrugations influence taken into account, the current depends on the point (x,y), i.e., the current lines are curved and the electromagnetic response process is now inhomogenous while it were homogenous if graphene were plane and the current lines were direct (see [63, 35]). It would be interesting to study the current lines of the grapheme membrane under electromagnetic radiation experimentally and to investigate the problem of determining of the form of corrugations if the form of the current curves is known (i.e., to solve the inverse problem). From the formulae (6.89), one can also see that the electromagnetic response is elliptically polarized compared to the linearly polarized incident radiation.

In case of the strong radiation $P \gg 1$, we have Equations 6.72 and 6.77 or 6.81. We have nonlinear electromagnetic response, which is inhomogenous because of the inevitable existence of corrugations. The current lines are curved. However, we cannot calculate their form as we do not know the value of the phenomenological constants $\eta_{1,2}$. For this analysis, experiments should be made that would allow to determine their values.

6.3.3 Summary

The quasiclassical kinetic theory of the nonlinear electromagnetic response of graphene was developed in [63, 35]. The first experimental investigation of nonlinear optical properties of graphene was presented in [64] where a remarkably high third-order optical nonlinearity of graphene was demonstrated. It was noted there that this graphene property is of great use for applications. For instance, graphene can be utilized for imaging purposes with its image contrasts, which is orders of magnitude higher than those obtained using linear microscopy. The irradiated graphene property to emit multiplied harmonics also gives exciting opportunities.

In [65] and [64], intraband and interband transitions were studied, which were ignored in [63, 35], where quasiclassical kinetic approach was used.

In the present subsection, MZ equation describing nonlinear self-consistent electromagnetic response of graphene was generalized taking into account the specific graphene property to have a corrugated surface, which creates synthetic electric fields in irradiated graphene.

The generalized equation shows that these fields reduce nonlinear effects, which is negative to applications. We also discuss minimization of this negative influence by strain engineering methods, which can be useful in calculations for different optoelectronic devices.

The results described in this section were published in [84].

6.4 Radiative Decay Effects Influencing the Local Electromagnetic Response of the Monolayer Graphene with Surface Corrugations in Terahertz Range

In this section, we continue the study of out-of-plane surface corrugations' influence on the monolayer graphene local electromagnetic response in terahertz range, which we started in Section 6.3. The effects of radiative decay, double-valley structure of charge carriers spectrum in graphene, and the breathing surface curvature form induced synthetic electric fields are taken into account. To fulfill this program, the generalized nonlinear self-consistent equation is obtained. In case of weak external alternating electric field $\vec{E}^{ext}(t)=\vec{E}_0\cos\omega t$ for the obtained equation in linear approximation on the external electric field, the exact solution is found. It shows that in this case, we get local dephasing in induced current paths depending on the surface form $z = h(x,y)$. This result could become the basis of the new method of the imaging of surface corrugations form for the given picture of local current paths. The obtained results allow to study the way the mechanical behavior of materials at nanoscale deviates from macroscoping established concepts, and this is of particular importance for graphene.

6.4.1 Preliminaries

As it was already mentioned above, any graphene membrane always has out-of-plane corrugations, i.e., considering graphene membrane with out-of-plane distortions, we do not study a specific case, but rather a general one. Let us introduce the function $z = h(x,y)$ describing graphene membrane surface form, which it has due to out-of-plane displacements. It was shown that these corrugations together with graphene membrane in-plane strains lead to arising of the pseudomagnetic field (gauge field) $\vec{A}(x,y)$, which was studied in many papers (see for instance the review [29]).

If out-of-plane corrugations are excited by an external periodic electric field, they begin to move ("breathe") and an additional ("extra") synthetic electric field $\mathbf{E}_{syn}(x,y,t)=-\frac{1}{c}\frac{\partial\mathbf{A}}{\partial t}(x,y,t)$ will arise (here $\mathbf{A}(x,y,t)$, which is a vector potential corresponding to "internal" gauge field (see above)). As far as we know, the influence of $\mathbf{E}_{syn}(x,y,t)$ in graphene was considered first in [33] and then in [62], Section 6.2 and in [84], Section 6.3 by the authors of this chapter. In [33], it was shown that (f-ph) damping rate was determined by interaction with these fields due to surface deformation by moving flexural phonons and its time derivative. In [62], we have shown that synthetic fields excited by time-dependent electromotive force in graphene nanoresonator and generated by their currents lead to new loss mechanism, which essentially influences the quality factor value. In [84], we studied the total induced current (from both valleys) in irradiated corrugated graphene membrane taking into account synthetic electric fields generated by activating external periodic electric fields with due regard of the inversion and time-reversal symmetry breaking [53]. The "breathing" surface form $z = h(x,y,t)$ determines the synthetic electric field dependence $\mathbf{E}_{syn}(x,y,t)$ on the point (x,y) in every moment. The formula for $\mathbf{E}_{syn}(x,y,t)$ was obtained under assumption that the external electric field is weak enough to consider the spectrum to be continuous (Stark levels are washed out due to current carriers scattering and thermal fluctuation) and suggesting the "breathing" membrane linear dependence on the external electric field. Under these assumptions, the current paths depending on the point were found for a given surface form and its curvature. So the synthetic field $\mathbf{E}_{syn}(x,y,t)$ differs by strength, direction, and phase, and consequently, current paths are different and their directions and phases change in the neighborhood of every ripple.

Actually, it means that we have another model of disordered system where the curvature also depends on time, i.e., disorder is not static (see Section 6.4.4 "Summary and Discussion"). In the present section, we are going to study in addition the role of radiative decay of charge carriers "overheated" by terahertz irradiation, which may be more effective than standard relaxation mechanism and make the picture of special alternating current paths more exact.

In [84], we did not take into account e-e interaction considering graphene response to homogeneous radiation in terahertz range. There are many cases (see for instance the review [69] and discussion below), when in the 2D system of interacting Dirac electrons, many-body effects should be taken into account. However, sometimes, this interaction does not play a significant role (see below). For instance, it was shown that not only quasiparticles ("massless electrons" and "holes") but also plasmons (collective charge density excitations) and plasmarons [70] ("dressed" quasiparticle excitations coupled to plasmons) play an important role. These predictions are based on a very specific form of quasiparticle spectral function obtained theoretically [71–72] and observed experimentally by a variety of direct and indirect methods: electron energy-loss spectroscopy (EELS) (see in [73]), angle-resolved photoemission spectroscopy (ARPES) [70], and scanning tunneling spectroscopy (STS) [73] on exfoliated graphene sheets and on epitaxial graphene samples.

Large efforts were made to observe these predictions in optical measurements. As a rule, indirect methods are used "by engineering plasmons coupling to infrared light in a number of intriguing ways" [69], by making more strong light–matter interaction and alleviating the momentum mismatch between plasmon modes and of the incident radiation.

Thus, we have come to the conclusion that under incident homogenous radiation, surface plasmons (SP) cannot be driven without using special efforts: either by using AFM tip to confine the incident radiation to nanoscale region around the tip as Z. Fei, D.N. Basov *et al.* [76] or by exciting SP in graphene micro-ribbon array or by using stacked graphene microdiscs to realize electromagnetic radiation shield with 97.5% effectiveness [69].

The manifestation of e-e interaction in optical and other properties in graphene is considered in more detail in Section 6.3 (see also [84]).

To conclude the discussion about the strong or weak influence of e-e interactions in optical experiments in monolayer graphene, we also refer to three papers [85–87] (here [85] is a theoretical article and [86–87] are experimental ones sent in 2014). In [86], it is written that the negative dynamical conductivity discovered before in experiment one can explain only taking into account e-e interactions. In [86], it is suggested with reasonable arguments that the mentioned phenomenon should be explained by defect-mediated collision of the hot carriers with the acoustic phonons. The mentioned experimental data once again confirm our opinion described above that without special serious tricks, e-e interactions may not manifest (display) itself.

In [87], a transient decrease in graphene conductivity was observed. The THz frequency-dependence of graphene photoconductive response increases in the Drude weight.

It should be noted that recently, the study of graphene current patterns under the action of the electric field became a vogue. The case of the constant electric field and the flatland model was theoretically analyzed on the basis of quantum description and non-equilibrium Keldysh–Green formalism in [88] and using the trajectory-based semiclassical analysis in [89].

In Section 6.3 (see also [84]), we were the first to consider the case with time-dependent electric field and the monolayer graphene membrane having corrugations. We also took into account the double valley spectrum. We suggested that the external electric field is weak enough to consider the spectrum as continuous (see above) and the formula for synthetic electric field was obtained suggesting the "breathing" linear dependence on the external electric field. Under these assumptions, the local current paths depending on the point were found for a given surface form and its curvature but radiation rate was not taken into account.

Summarizing all above mentioned, it may be told that we carried on our studies of out-of-plane corrugations' (ripples) influence on monolayer graphene electromagnetic response based on the kinetic Boltzmann equation for Fermi–Dirac momentum distribution function and linear Dirac energy spectrum (~*k*) for current carriers, but taking into account double-valleys picture of energy spectrum in first Brillouin zone for Dirac electrons. We took into account inversion and time-reversal symmetry broken by ripples in graphene membrane. The same way as in Section 6.3 (see also [84]), we did not take into account "direct" e-e interactions (a detailed explanation is given in [84]) but important terms, for instance, radiation rate that is proportional to e^2 (here, e is the electron charge), arise due to the use of the very specific "self-consistent-field approach" developed in [63] and [35], which takes into account the mutual influence of excited and induced currents.

Note that experimental research of local charge transport visualized with atomic resolution using scanning tunneling microscopy (STM) [90] showed the current patterns dependence on the point. We think that this dependence is connected with corrugations that are always present, and this experiment proves the correctness of our theory. There are also other experimental results of the sort (see [91]) that used a new technique that allows one to probe transport by creating a movable scatter with an scanning probe microscopy (SPM) tip.

Our theoretical description is based on the kinetic equation and quantum description of Dirac charge carriers with double valley spectrum.

In this subsection, we get the nonlinear self-consistent equation with corrugations taken into account and we take into consideration radiation loss.

The time-dependent electric current $j(x,y,t)$ excited in the sample under the action of external periodic electric field $\mathbf{E}_{ext}(t)$ creates in its turn a secondary (induced) electric field $\mathbf{E}_{ind}(t)$ which acts back on electrons and should be added to the external field. Our results obtained below for corrugated (rippled) graphene membrane are compared with the results obtained in [63], [35] for flatland model of the graphene membrane.

We also consider the case when the external field is weak enough so that this nonlinear equation turns to be linear. In this case, we obtain its exact solution that shows that in this case, the currents get dephasing, which depends on the point. This dephasing is determined by the surface curvature in the point. So observing the dephasing map, we can get information about the surface form of the membrane. It gives the idea of the new method of imaging of ripples, which could be useful for many applications.

Note that the experimental research of local charge transport visualized with atomic resolution using STM [90] showed the current pattern's dependence on the point. We think that this dependence is connected with corrugations that are always present and this experiment proves the correctness of our theory. There are also other experimental results of the sort (see [91]) that used a new technique that allows one to probe transport by creating a movable scatter with an SPM tip.

The adequate methods for theoretical description of the current path's structure for the given surface form and experimental efforts to find a more precise (exact) methods for current paths imaging is concerned to be one of the most actual problems to be solved for graphene. It would be of utmost importance also for the further development of graphene-based nanoelectronics and DNA sequencing [92].

A more detailed discussion of all these items will be given below in Subsection 6.4.4 “Summary and Discussion.”

6.4.2 Generalized Self-Consistent Equation

To describe the electromagnetic response or total current in corrugated graphene membrane in linear approximation taking into account the radiative loss, we should find the self-consistent equation generalizing the one obtained in [35] for flatland model. In [84], we obtained the kinetic equation in the form

$$\frac{\partial f(\mathbf{p},\mathbf{r},t)}{\partial t} - e\left[\mathbf{E}^{ext}(t) + \mathbf{E}^{syn}(\mathbf{r},t)\right]\frac{\partial f(\mathbf{p},\mathbf{r},t)}{\partial \mathbf{p}} = 0 \tag{6.94}$$

where synthetic electric field $\mathbf{E}^{syn}(\mathbf{r},t)$ appears as a consequence of the presence of out-of-plane deformations $z = h(x,y)$. Here,

$$\mathbf{E}^{syn} = -c^{-1}\partial\mathbf{A}(x,y,t)/\partial t \tag{6.95}$$

and $\mathbf{A}(x,y,t)$ is a gauge field. We used for it the formula obtained in [30]:

$$A_x = -\frac{1}{2}A^0\left[(h_{xx})^2 - (h_{yy})^2\right]a^2, \quad A_y = A^0\left[h_{xy}(h_{xx} + h_{yy})\right]a^2 \tag{6.96}$$

$$A^0 = 3/4 \cdot \epsilon_1/e \cdot c/v_F \tag{6.97}$$

Here, $\epsilon_1 = 2{,}89ev$, $\vec{K} = a^{-1}\left(4\pi/3\sqrt{3},0\right)$ corresponds to a Dirac point

Assuming that the external electric field is weak enough to use the linear approximation $\Delta a(t) = \eta(x,y)E_0 \sin \omega t$ (we consider for simplicity that $\eta(x,y) = const$) it was obtained in [84]. To simplify the form of the equations we obtain, we shall assume below that $\eta(x,y) = const$

$$(\mathbf{E}^{syn})_x(x,y,t) = -E^0(\omega)E_{0x}I_x(x,y)\sin \omega t, \tag{6.98}$$

$$(\mathbf{E}^{syn})_y(x,y,t) = -E^0(\omega)E_{0y}I_y(x,y)\sin \omega t \tag{6.99}$$

where

$$E^0(\omega) = 3/4 \cdot \epsilon_1/e \cdot \omega/V_F, \quad I_x = a\eta\left(h_{xx}^2 - h_{yy}^2\right), \quad I_y = 2a\eta\left(h_{xx} + h_{yy}\right)h_{xy}, \tag{6.100}$$

The equation (6.94) for the temperature close to zero has the exact solution:

$$f(\mathbf{p},\mathbf{r},t) = F_0\left(\left|\mathbf{p} - \mathbf{p}_0(\mathbf{r},t)\right|\right), \quad F_0(\mathbf{p}) = \left[1 + exp\left(\frac{vp - \mu}{kT}\right)\right]^{-1} \tag{6.101}$$

where $F_0(\mathbf{p})$ is the Fermi–Dirac distribution function, and

$$\mathbf{p}_0(\vec{r},t) = -e\int_{-\infty}^{t}\left[\mathbf{E}^{ext}(t') + \mathbf{E}^{syn}(\mathbf{r},t')\right]dt' \tag{6.102}$$

If the temperature is zero, $T = 0$, and the chemical potential is finite, $\mu > 0$, the current $\mathbf{j}(\mathbf{r},t)$ can be presented in the form

$$j_x = \frac{g_s e v_F}{(2\pi\hbar)^2} p_F^2 G_x(Q_x,Q_y)\frac{2P_{0x}}{\sqrt{1+P_{0x}^2+P_{0y}^2}},\ j_y = \frac{g_s e v_F}{(2\pi\hbar)^2} p_F^2 G_y(Q_x,Q_y)\frac{2P_{0y}}{\sqrt{1+P_{0x}^2+P_{0y}^2}} \tag{6.103}$$

where

$$G_x(Q_x,Q_y) = \frac{1}{Q_x}\int_0^{\pi/2}\left[\sqrt{1+Q_y\sin\varphi+Q_x cos\ \varphi} - \sqrt{1+Q_y\sin\varphi-Q_x cos\ \varphi}\right]\cos\varphi d\varphi$$

$$G_y(Q_x,Q_y) = \frac{1}{Q_y}\int_0^{\pi/2}\left[\sqrt{1+Q_x cos\ \varphi+Q_y\sin\varphi} - \sqrt{1+Q_x cos\ \varphi-Q_y\sin\varphi}\right]\sin\varphi d\varphi$$

$$Q_{x,y} = \frac{2P_{0x,0y}}{1+P_{0x}^2+P_{0y}^2}, \qquad P_{ox,oy} = \frac{p_{0x,0y}}{p_F} \tag{6.104}$$

The equation (6.94) and its solution (6.101 and 6.102) is the generalization of kinetic equation considered in [35] for flatland. Note that according to the Landau–Peierls theorem demonstrated in [9–11], these corrugations exist always and consequently our generalization is essential.

All the formulae written above were obtained without taking into account the radiative loss. Now, we should find the effect produced by this circumstance. We assume that the external electromagnetic field is incident upon graphene membrane along z-axis and induces the periodic current in the layer. In general, however, the time-dependent electric current creates in its turn a secondary (induced) electric field $\mathbf{E}^{ind}$, which acts back on the electrons and should be added to the external field. Calculating the response of the system, one should take into account that electrons respond not to the external but to the total self-consistent electric field:

$$\mathbf{E}^{tot} = \mathbf{E}_{ext}^{tot}(t) + \mathbf{E}_{syn}^{tot}(\mathbf{r},t),\quad \mathbf{E}_{ext}^{tot}(t) = \mathbf{E}_{ext} + \mathbf{E}_{ext}^{ind},\quad \mathbf{E}_{syn}^{tot}(t) = \mathbf{E}_{syn} + \mathbf{E}_{syn}^{ind}$$

So instead of the equation (6.94) for the momentum distribution function of the electron, we obtain

$$\frac{\partial f(\mathbf{p},\mathbf{r},t)}{\partial t} - e\left[\mathbf{E}_{ext}^{tot}(t) + \mathbf{E}_{syn}^{tot}(\mathbf{r},t)\right]\frac{\partial f(\mathbf{p},\mathbf{r},t)}{\partial \mathbf{p}} = 0 \tag{6.105}$$

The solution of this equation can be written again in the form

$$\mathbf{p}_0 = -e\int_{-\infty}^{t} \mathbf{E}_{z=0}^{tot}(t')dt' \tag{6.106}$$

where the field $\mathbf{E}_{z=0}^{tot}$ is not known and should be calculated self-consistently. To do this, we recall that the current and the electric field are related by the Maxwell equations [93]:

$$\mathbf{E}_{z=0}^{ind} = -\frac{2\pi\mathbf{j}}{c} \tag{6.107}$$

Using Equations 6.106, 6.107, and 6.105, we get the following self-consistent equation of motion for the momentum $\mathbf{p}_0$:

$$\frac{d}{dt}\mathbf{p}_0 + e\frac{2\pi}{c}\left[\mathbf{j}^{ext} + \mathbf{j}^{syn}\right] = e\left[E^{ext}(t) + E^{syn}(\mathbf{r},t)\right] \tag{6.108}$$

Introducing $\mathbf{P} = -\mathbf{p}/p_F$ and $\tau = \omega t$, we can rewrite (6.108) in the form

$$\frac{d}{d\tau}P_{0x} + \frac{\gamma}{\omega}\frac{P_{0x}}{\sqrt{1+P_{0x}^2+P_{0y}^2}}G_x(Q_x,Q_y) = \frac{e}{\omega p_F}\left[E_x^{ext}(t) + E_x^{syn}(\mathbf{r},t)\right] \tag{6.109}$$

$$\frac{d}{d\tau}P_{0y} + \frac{\gamma}{\omega}\frac{P_{0y}}{\sqrt{1+P_{0x}^2+P_{0y}^2}}G_y(Q_x,Q_y) = \frac{e}{\omega p_F}\left[E_y^{ext}(t) + E_y^{syn}(\mathbf{r},t)\right] \tag{6.110}$$

$$\gamma = \frac{g_s}{\pi}\frac{e^2}{hc}\frac{v_F p_F}{h} = v_F\frac{e^2}{hc}\sqrt{g_s \pi n_s} \tag{6.111}$$

Here, γ has a physical meaning of the radiative decay rate. For experimentally relevant densities $n_s \approx 10^{12}$, we have $\gamma \approx 7THz$. After this, the nonlinear equation should be resolved with respect to the momentum $\mathbf{p}_0$ and the current can be found from Equations 6.103 and 6.104.

Equations 6.109 through 6.111 describe the nonlinear self-consistent response of graphene to an arbitrary external time-dependent electric field $\mathbf{E}^{ext}_{z=0}$ and the synthetic electric field determined by the given membrane form.

6.4.3 Induced Current Pattern as Graphene Electromagnetic Response for Weak Fields

If the external field $\mathbf{E}^{ext}(t)$ is weak (and consequently $\mathbf{E}^{syn}(\mathbf{r},t)$ is weak) and $\boldsymbol{P}_{0x,0y} \ll 1$, and $G_{x,y}(Q_x,Q_y) \approx 1$, we can consider instead of the nonlinear self-consistent equations (Equations 6.109 through 6.111) their linear approximations

$$\frac{d}{d\tau}P_{0x} + \frac{\gamma}{\omega}P_{0x} = \frac{e}{\omega p_F}\left[E_x^{ext}(\tau) + E_x^{syn}(\mathbf{r},\tau)\right] \tag{6.112}$$

$$\frac{d}{d\tau}P_{0y} + \frac{\gamma}{\omega}P_{0y} = \frac{e}{\omega p_F}\left[E_y^{ext}(\tau) + E_y^{syn}(\mathbf{r},\tau)\right] \tag{6.113}$$

where $\tau = \omega t$. It is clear (see Equations 6.98 and 6.99) that

$$E^{ext}_{x,y}(\tau) + E^{syn}_{x,y}(\vec{r},\tau) = E_{0x,0y}\sqrt{1+\left(D^{(x,y)}(x,y,\omega)\right)^2}\cos\left[\tau - \delta_{x,y}(x,y,\omega)\right], \tag{6.114}$$

$$\delta_{x,y}(x,y,\omega) = \arctan D^{(x,y)}(x,y,\omega),\ D^{(x,y)}(x,y,\omega) = E^0(\omega)I_{x,y}(x,y), \tag{6.115}$$

So we can rewrite Equations 6.112 and 6.113 in the form

$$\frac{d}{d\tau}P_{0x,0y} + \frac{\gamma}{\omega}P_{0x,0y} = \frac{e}{\omega p_F}E_{0x,0y}\sqrt{1+\left(D^{(x,y)}(x,y,\omega)\right)^2}\cos\left[\tau - \delta_{x,y}(x,y,\omega)\right], \tag{6.116}$$

Solving this equation, we obtain

$$P_{0x,0y}(x,y,t) = \frac{e}{p_F}\frac{E_{0x,0y}}{\sqrt{\omega^2+\gamma^2}}\sqrt{1+\left(D^{(x,y)}(x,y,\omega)\right)^2}\cos\left[\tau - \delta_0(\omega) - \delta_{x,y}(x,y,\omega)\right],$$

$$\delta_0(\omega) = \arctan\frac{\gamma}{\omega} \tag{6.117}$$

And consequently (see Equation 6.103)

$$j_{x,y} = 2\frac{g_s e v_F}{\left(2\pi\hbar\right)^2} p_F^2 P_{0x,0y} =$$

$$= g_s \frac{e^2}{2\pi h}\frac{v_F p_F}{\pi h}\frac{E_{0x,0y}}{\sqrt{\omega^2+\gamma^2}}\sqrt{1+\left(D^{(x,y)}\left(x,y,\omega\right)\right)^2}\cos\left[\omega t-\delta_0\left(\omega\right)-\delta_{x,y}\left(x,y,\omega\right)\right] \tag{6.118}$$

Neglecting corrugation's influence, i.e., for flatland model, we get from Equation 6.118 the formula obtained in [35].

$$j_{x,y} = g_s \frac{e^2}{2\pi h}\frac{v_F p_F}{\pi h}\frac{E_{0x,0y}}{\sqrt{\omega^2+\gamma^2}}\cos\left[\omega t-\delta_0\left(\omega\right)\right] \tag{6.119}$$

Note that our formula (6.118) was calculated taking into account corrugations influence as well as (6.119) calculated without it looking different compared to the commonly used Drude formula:

$$\sigma\left(\omega\right)=\frac{\sigma\left(0\right)}{1+\left(\omega\tau\right)^2}$$

However, when $(\omega/\gamma) << 1$ (which is true in terahertz range, see Equation 6.111) these formulae weakly differ since then $\sqrt{1+\left(\omega/\gamma\right)^2}\approx 1+2^{-1}\left(\omega/\gamma\right)^2$. But in terahertz range where the main mechanism of decay is radiation loss (not because of weak nonelastic scattering with phonons emission) our theory is in agreement with the experiment [94] where the intra conductivity, i.e., its real part in terahertz range was measured to be a constant not depending on ω. Also, it was measured in [94] that inter part, i.e., actually its imaginary part, is close to zero in terahertz range. So we see that our quasiclassic approximation neglecting the imaginary part of conductivity in terahertz range does not contradict the experiment [94].

If we assume that there is no radiation loss (i.e., $\gamma = 0$) we get the formula for currents found in Section 6.2 ([84]).

$$\mathbf{j}_x \approx \sigma_0 E_{0x}[\cos\omega t + D^{(x)}(x,y)\sin\omega t], \qquad \mathbf{j}_y \approx \sigma_0 E_{0y}[\cos\omega t + D^{(y)}(x,y)\sin\omega t] \tag{6.120}$$

$$\sigma_0 = \frac{n_s e^2 v_F}{p_F\omega}$$

The formulae (6.119), (6.118), and (6.120) for the current density in the linear approximation describe current paths that are straight lines in flatland model (see Equation 6.119) and are curved lines depending on the point (x,y) (see Equations 6.118 and 6.120) when corrugations are taken into account. It means that the graphene membrane curvature leads to the curving of the current paths, the total current through the section conserving the constant value. So there is an important difference from the usual percolation description where the origin of the curved current lines is due to the random potential. Formulae (6.118) describe current patterns when radiation loss is taken into account. The formulae (6.118) and (6.120) show that the induced current paths are curved which is the influence of corrugations always present. It leads to the fact that actually the current density depends on the point that was observed in a number of experiments (see [95]).

Note that the flatland model formula (6.119) for a valley current taken with the opposite sign gives us the valley current in another valley (see [53]) so that total current is equal to zero. However, if we take into consideration the corrugations influence (formulae (6.118) or (6.120)), we have symmetry breaking and the total current is not equal to zero. The corresponding formula was obtained in Section 6.3 ([84]) without taking into account radiative loss. If radiation loss is taken into consideration (see (3.25)), we obtain for the total current

$$j_{x,y}^{tot} = 2\sigma_0\ E_{0x,0y} D^{(x,y)}(x,y,\omega)\sin\omega t,\ \sigma_0 = g_s \frac{e^2}{2\pi h}\frac{v_F p_F}{\pi h\sqrt{\omega^2+\gamma^2}} \tag{6.121}$$

In other words, we have

$$\sigma^{tot} = 2\sigma_0 D^{(x,y)}(x,y,\omega)$$

So the presence of corrugations makes the conductivity a function of the point (x,y). So if we know the surface form $z = h(x,y)$, we can predict the form of current pattern and vice versa; if we know from the experiment the current pattern, we can restore the surface form. This result could be used for creating new method of corrugations imaging when we have the experimentally measured current patterns.

6.4.4 Summary and Discussion

In this subsection, we obtained the generalized nonlinear self-consistent equation describing the corrugated graphene membrane electromagnetic response in linear on the field approximation taking into account the radiative loss. We did not take into consideration the direct e-e interaction between particles but considered the electrodynamics interaction between currents using self-consistent field method [63, 35]. Obtaining this equation, we assumed that the graphene membrane surface form changes under the action of the external field, i.e., we have "breathing" corrugations. This assumption was justified by experimental works (see [95], Figures 2 and 3). Assuming the external field to be weak enough to use the linear approximation, we simplified this equation, which allowed finding exact solution. The current pattern found for the given surface form $z = h(x,y)$ proved to depend on the point (x,y), which we predicted in Section 6.3 ([84]) as the influence of the always-existing ripples. Note that after the publishing of [84], we found experimental papers (see for instance [91], Figure 3) that showed that this dependence really exists. However, we could not compare our formula with experimental current patterns as we do not know the value of phenomenological constants we introduced. Note that according to our solution, the induced currents, which are the graphene electromagnetic response, the amplitude is less due to the radiation loss and the extra phase appears, which depends on the point. It is important that morphology and current maps, generally speaking, do not coincide. Figuratively, one may say that charge current in such graphene membrane is not homogeneous in any cross-section and reminds the "turbulent stream" of a very thin layer of charged liquid intermixed by many dephased and unequal electro mixers. In the present paper, we studied in addition the role of radiative decay of charge carriers "overheated" by terahertz irradiation, which may be more effective than standard relaxation mechanism and make the picture of spatial alternating current paths more exact.

Note that we determined the gauge field $\mathbf{A}(x,y)$ on the basis of the formula [30] that was obtained using atomic-scale quantum approach. It is especially important because it was experimentally found that there are scanning probe microscopy (SPM) in graphene nanoscale-wavelength size ripples [96, 97].

For qualitative comparison between our formula and experimental current patterns, more in-detail experimental investigation using methods developed in [91] should be made to study the values of the phenomenological constants we introduced. However, one can determine these parameters experimentally using methods developed in [91].

In case new more sensitive experimental investigation allows to show that our formula describes the phenomenon well enough, we could try to solve the inverse problem, i.e., the imaging of the surface form on the basis of the observed current patterns. It could be interesting for applications.

The problem of theoretical description of current patterns in different experiments is very important. The experimental study of spatially resolved photocurrents has already been done in monolayer graphene [98], and in several graphene nanoribbon devices [99], (the last is especially important for understanding of the functioning of ultrafast photodetectors and mode-locked lasers where graphene is used as an important component). Also it was considered by the description of coherent transport in graphene heterojunctions [100] and in graphene quantum dots [101], in a number of cases quantum mechanical approach being absolutely naturally used.

Imaging experiments show that current profiles in low dimensional and mesoscopic systems may be very complicated and in the current density may arise streamlines. Theory of such nontrivial nonstationary phenomenon is awaiting the attentive investigator.

The results described in this section were published in [102].

6.5 Conclusion

In this chapter, we analyzed how graphene out-of-plane deformations influence nonstationary processes due to the synthetic electric fields. We studied two examples. The first one was devoted to the calculation of the corresponding additional loss (Joule-type loss) for graphene nanoresonators due to the generation of the induced currents in the graphene membrane. The formula for the quality factor related to this loss was found for the given graphene membrane surface form. Using this formula, the dissipation of the sort was estimated. It proved to be rather significant, about 30–40%, depending the curvature of out-of-plane deformations.

So we see that in researching graphene nanoresonators and their applications as components of nanoelectromechanical and optomechanical systems for the highly sensitive detection and visualization of material structure, we should take into account the obtained results that show that there is the additional loss and consequently the additional mistake in measurement.

We also see (and it is very important) that the mentioned additional mistake depends on the graphene membrane surface form, which is different for different devices. And consequently, the results of measurements done by different graphene nanoresonators will lead to slightly different values for the measured magnitude. And as far as we cannot make identical the two graphene membrane surface forms, we cannot get two identical values measuring by two different devices. As far as for a measurement we need different copies of devices which we might think identical; we see here the boundary for the possible sensitivity of measurement.

It is clear that to reach a high exactitude of measurement, we should have devices we can think to be identical enough. But the obtained results show the limit of possible sensitivity of measurement when we cannot consider devices to be identical because of different graphene membrane surface forms. So the obtained results show the limit of possible sensitivity of measurement.

The second example considered in this chapter is a problem of theoretical description of induced current paths arising in the irradiated graphene membrane. It was shown that the value and direction of induced current are the functions of a point, the direction being determined by the direction of the synthetic electric field, which in turn is determined by the out-of-plane deformations, i.e., graphene surface curvature at the point. The corresponding formulae for induced current paths for a given surface form were found.

The obtained results allow to estimate the scale of harmful influence of these induced currents on the useful processes in the device containing graphene membrane. The problem to minimize this drawback using strain engineering is discussed in the chapter.

Note that this harmful influence will be different in different devices as it depends on the value and curvature of induced current paths, which is determined by out-of-plane graphene membrane deformations, which are different for different devices, and they cannot be done identically for different devices within the needed sensitivity. The problem to minimize this drawback using strain engineering is discussed in the chapter.

A number of experiments where induced current paths were measured show that they are complicated curved lines that are different for different graphene samples. These experiments are consistent with our results while the flatland model theory says that induced current trajectories should be straight lines. However, the more detailed experiments that would allow to determine the phenomenological parameters we introduced have not yet been done. So our theory is preliminary and should be developed together with the experimental part.

It would be very useful to calculate the current paths for a given graphene membrane with known surface form using the obtained formulae, then to measure the current paths for this sample and then to compare the theoretical calculations and the results of the measurement. If the theoretical and experimental paths do not differ much, it could be possible to consider the inverse problem, i.e., for given experimental current paths to determine the out-of-plane deformations of the graphene sample using the obtained formulae. That could give a new method of visualization of the unknown surface form of a graphene flake.

If comparison of the theoretical prediction based on the obtained formulae for a given sample and experimental measurement for it does not demonstrate good consistency, then it is clear that the theoretical formulae should be corrected by taking into account the factors and parameters that were neglected for simplicity in our preliminary consideration. If the corrected formula gave predictions close enough to experimental data, it could be considered as a basis to state the inverse problem and to suggest a new method of graphene surface form visualization.

The considered examples show how important it is to study the influence of the synthetic electric fields generated by out-of-plane deformations the nonstationary processes in graphene. This research is a first step in this direction. For instance, we analyzed the simplest linear case, which was reasonable as we considered weak activating external electric field. It would be interesting to study also the nonlinear case.

It is clear also that synthetic electric fields should exist as well in other low-dimensional materials discovered recently as out-of-plane deformations exist there too. So we anticipate the fruitful development of the preliminary investigations described in this chapter and their applications in the near future.

References

1. Novoselov, K.S., Geim, A.K., Morozov, S.V., Jiang, D., Zhang, Y., Dubonos, S.V., Grigorieva, I.V., Firsov, A.A., Electric field effect in atomically thin carbon films. *Science*, 306, 666, 2004.
2. Novoselov, K.S., Geim, A.K., Morozov, S.V., Jiang, D., Katsenelson, M.I., Grigorieva, I.V., Dubonos, S.V., Firsov, A.A., Two-dimensional gas of massless Dirac fermions in graphene. *Nature (London)*, 438, 197, 2005.
3. Zhang, Y., Tan, Y.-W., Stormer, H.L., Kim, P., Experimental observation of quantum Hall Effect and Berry's phase in graphene. *Nature*, 438, 201, 2005.
4. Hernandes, Y., Nicolosi, V., Lotya, M., Blighe, F.M., Sun, Z., De, S., McGovern, I.T., Holland, B., Byrne, M., Gun'Ko, Yu.K., Boland, J.J., Niraj, P., Duesberg, G., Krishnamurty, S., Goodhue, R., Hutchison, J., Scardaci, V., Ferrari, A.C., Coleman, J.N., High-yield production of graphene by liquid-phase exfoliation of graphite. *Nat. Nanotechnol.*, 3, 563, 2008.
5. Castro Neto, A.H., Guinea, F., Peres, N.M.R., Novoselov, K.S., Geim, A.K., The electronic properties of graphene. *Rev. Mod. Phys.*, 81, 109–162, 2009.
6. Firsova, N.E. and Ktitorov, S.A., Electron scattering in the monolayer graphene with the short-range impurities. *Phys. Lett. A*, 374, 1270–1273, 2010.
7. Firsova, N.E. and Ktitorov, S.A., Electron scattering and conductivity in the monolayer graphene. *Appl. Surf. Sci.*, 267, 189–191, 2013.
8. Bonaccorso, F., Sun, Z., Hasan, T., Ferrari, A.C., Graphene photonics and optoelectronics. *Nat. Photonics*, 4, 611, 2010.
9. Peierls, R.E., Bemerkungen uber Umwandlungtemperaturen. *Helv. Phys. Acta*, 7, 81–83, 1934.
10. Peierls, R.E., Quelques proprieties typiques des corpes solides. *Ann. I. E. Poincare*, 5, 177–222, 1935.

11. Landau, L.D., Zur Theorie der Phasenumwandlungen II. *Phys. Z. Sowietunion*, 11, 26–35, 1937.
12. Landau, L.D. and Lifshitz, E.M., *Statistical Physics, Part I*, Pergamon Press, Oxford, 1981.
13. Mermin, N.D. and Wagner, H., Absence of ferromagnetism or antiferromagnetizm in one- or two-dimensional isotropic Heisenberg models. *Phys. Rev. Lett.*, 17, 1133–1136, 1966.
14. Mermin, N.D., Crystalline order in two dimensions. *Phys. Rev.*, 176, 250–254, 1968.
15. Nelson, D.R., Piran, T., Weinberg, S. (Eds.), *Statistical Mechanics of Membranes and Surfaces*, World Scientific, Singapore, 2004.
16. Nelson, D.R. and Peliti, L., Fluctuations in membranes with crystalline and hexatic order. *J. Physique*, 48, 1085–1092, 1987.
17. Radzihovsky, L. and Le Doussal, P., Self-consistent theory of polymerized membranes. *Phys. Rev. Lett.*, 69, 1209–1212, 1992.
18. Fasolino, A., Los, J.H., Katsnelson, M.I., Intrinsic ripples in graphene. *Nat. Mater.*, 6, 858–861, 2007.
19. Meyer, J.C., Geim, A.K., Katsnelson, M.I., Novoselov, K.S., Booth, T.J., Roth, S., The structure of suspended graphene sheets. *Nature*, 446, 60–63, 2007.
20. Meyer, J.C., Geim, A.K., Katsnelson, M.I., Novoselov, K.S., Obergfell, D., Roth, S., Girit, C., Zettl, A., On the roughness of single- and bi-layer graphene membranes. *Solid State Commun.*, 143, 101–109, 2007.
21. Car, R. and Parrinello, M., Unified approach for molecular dynamics and density-functional theory. *Phys, Rev. Lett.*, 55, 2471–2474, 1985.
22. Los, J.H., Ghiringhelli, L.M., Meijer, E.J., Fasolino, A., Improved long-range reactive bond-order potential for carbon. I Construction. *Phys. Rev. B*, 72, 214102, 2005.
23. Ghiringhelli, L.M., Los, J.H., Meijer, E.J., Fasolino, A., Frenkel, D., Modelling the phase diagram of carbon. *Phys. Rev. Lett.*, 94, 145701, 2005.
24. Morozov, S.V., Novoselov, K.S., Katsnelson, M.I., Schedin, F., Jiang, D., Geim, A.K., Strong suppression of weak localization in graphene. *Phys. Rev. Lett.*, 97, 016801, 2006.
25. Castro Neto, A.H. and Kim, E.A., Charge inhomogeneity and the structure of graphene sheets. *E-print at arxiv.org:cond-mat/0702562.*
26. Herbut, I.F., Juricic, V., Vafek, O., Coulomb interaction, ripples, and the minimal conductivity of graphene. *Phys. Rev. Lett.*, 100, 046403, 2008.
27. Guinea, F., Katsnelson, M.I., Geim, A.K., Energy gaps, topological insulator state and zero-field quantum Hall effect in graphene by strain engineering. *Nat. Phys.*, 6, 30, 2010.
28. Geim, A.K. and Novoselov, K.S., The rise of graphene. *Nat. Mater.*, 6, 183–191, 2007.
29. Vozmediano, M.A., Katsnelson, M.I., Guinea, F., Gauge fields in graphene. *Phys. Rep.*, 496, 109, 2010.
30. Kim, E.-A. and Castro Neto, A.N., Graphene as an electronic membrane. *EPL*, 84, 57007, 2008.
31. Castro Neto, A.H., *Lecture notes on graphene for the Les Houches School on Modern theory of correlation electron systems*, (May 11–29), Les Houches, 2009 (arXiv:1004.3682,(2009)).
32. Katsnelson, M.I. and Novoselov, K.S., Graphene: New bridge between condensed matter physics and quantum electrodynamics. *Solid State Commun.*, 143, 3, 2007.
33. Von Oppen, F., Guinea, F., Martini, E., Synthetic electric fields and phonon damping in carbon nanotubes and graphene. *Phys. Rev. B*, 80, 075420, 2009.
34. Lin, Y.-J., Compton, R.L., Jimenez-Garcia, K., Phillips, W.D., Spielman, I.B., A synthetic electric force acting on neutral atoms. *Nat. Phys.*, 7, 531–534, 2011.
35. Mikhailov, S.A. and Ziegler, K., Nonlinear electromagnetic response of graphene: Frequency multiplication and self-consistent-fields effects. *J. Phys. Condens. Matter*, 20, 384204, 2008.
36. Mikhailov, S.A., Radiative decay of collective excitations in an array of quantum dots. *Phys. Rev. B*, 54, 10335, 1996.

37. Mikhailov, S.A., Microwave-induced magnetotransport phenomena in two-dimensional electron systems: Importance of electrodynamic effects. *Phys. Rev. B*, 70, 165311, 2004.
38. Cleland, A.N., *Foundation of Nanomechanics*, Springer, Berlin, 2003.
39. Ekinci, K.L. and Roukes, M.L., Nanoelectromechanical systems. *Rev. Sci. Instrum.*, 76, 061101, 2005.
40. Gaidarzhy, A., High quality gigahertz frequencies in nanomechanical diamond resonators. *Appl. Phys. Lett.*, 91, 203503, 2007.
41. Eom, K., Park, H.S., Yoon, D.S., Kwon, T., Nanomechanical resonators and their applications in biological/chemical detection: Nanomechanics principles. *Phys. Rep.*, 503, 4–5, 115–163, 2011.
42. Atalaya, J., Kinaret, J.M., Isacsson, A., Nanomechanical mass measurement using nonlinear response of a graphene membrane. *EPL*, 91, 48001, 2010.
43. Bunch, J.S., van der Zande, A.M., Verbridge, S.S., Frank, I.W., Tanenbaum, D.M., Parpia, J.M., Craighead, H.G., McEuen, P.L., Electromechanical resonators from graphene sheets. *Science*, 315, 490–493, 2007.
44. Chen, C., Rosenblatt, S., Bolotin, K.I., Kalb, W., Kim, P., Kumissis, I., Stormer, H.L., Heinz, T.F., Hone, J., Performance of monolayer graphene nanomechanical resonators with electrical readout. *Nat. Nanotechnol.*, 4, 861, 2009.
45. Eichler, A., Moser, J., Chaste, J., Zdrojek, M., Wilson-Rae, J., Bachtold, A., Nonlinear damping in mechanical resonators based on graphene and carbon nanotubes. *Nat. Nanotechnol.*, 6, 339, 2011.
46. Sazonova, V., Yaish, Y., Ustinet, H., Roundy, D., Arias, T.A., A tunable carbon nanotube electromechanical oscillator. *Nature*, 431, 284, 2004.
47. Seoanez, C., Guinea, F., Castro Neto, A.N., Dissipation in graphene and nanotube resonators. *Phys. Rev. B*, 76, 125427, 2007.
48. Lassagne, B., Tarakanov, Y., Kinaret, J., Sanchez, D.-G., Bachtold, A., Coupling mechanics to charge transport in carbon nanotube mechanical resonators. *Science*, 325, 1107, 2009.
49. Pereira, V.P. and Castro Neto, A.N., All-graphene integrated circuits via strain engineering. *Phys. Rev. Lett.*, 103, 046801, 2009.
50. Low, T. and Guinea, F., Strain-induced pseudo-magnetic field for novel graphene electronics. *Nano Lett.*, 10, 3551, 2010.
51. Moser, J., Eichler, A., Lassagne, B., Chaste, B., Tarakanov, Y., Kinaret, J., Wilson-Rae, J., Bachtold, A., Dissipative and conservative nonlinearity in carbon nanotube and graphene mechanical resonators, arXiv:1110.1234v1, [cond-mat. Mes-nan], 6 Oct, 2011.
52. Morpurgo, A.F. and Guinea, F., Intervalley scattering, long-range disorder, and effective time reversal symmetry breaking in graphene. *Phys. Rev. Lett.*, 97, 196804, 2006.
53. Bolotin, K.I., Sikes., K.J., Home, J., Stormer, H.L., Kim, P., Temperature dependent transport in suspended graphene. *Phys. Rev. Lett.*, 101, 096802, 2008.
54. Cho, S. and Fuhrer, M.S., Charge transport and inhomogeneity near the charge neutrality point in graphene. *Phys. Rev. B*, 77, 08140(R), 2008.
55. Bolotin, K.I., Sikes, K.J., Jiang, Z., Klima, M., Fudenberg, G., Hone, J., Kim, P., Stormer, H.L., Ultrahigh electron mobility in suspended graphene. *Solid State Commun.*, 146, 351, 2008.
56. Cleland, A.N. and Roukes, M.L., External control of dissipation in a nanometer-scale radiofrequency mechanical resonator. *Sens. Actuators*, 72, 256–261, 1999.
57. Feng, X.L., Zorman, C.A., Mehregany, M., Roukes, M.L., Dissipation in single-crystal 3C-SIC ultra-high frequency nanomechanical resonators. *Nano Lett.*, 7, 1953, 2007.
58. Singh, V., Irfan, B., Subramanian, C., Solanki, H.S., Sengupta, S., Dubey, S., Kumar, A., Ramakrishna, S., Deshmuch, M.M., Coupling between quantum Hall state and electromechanics in suspended graphene resonator. *Appl. Phys. Lett.*, 100, 233103, 2012.

59. Landau, L.D. and Lifshitz, E.M., *Theory of Elasticity*, 3rd ed., Butter-Heinemann and Oxford, 1986.
60. Qi, Z. and Park, H.S., Intrinsic energy dissipation in CVD-grown graphene nanoresonators. *Nanoscale*, 11, 11, 2012.
61. Firsova, N.E. and Firsov, Y.A., Losses in monolayer graphene nanoresonators due to Joule dissipation caused by synthetic electric fields and the ways of Joule losses minimizations, arXiv: 1110.5742v., 2011.
62. Firsova, N.E. and Firsov, Y.A., A new loss mechanism in graphene nanoresonators due to the synthetic electric fields caused by inherent out-of-plane membrane corrugations. *J. Phys. D: Appl. Phys.*, 45, 435102 (7pp), 2012.
63. Mikhailov, S.A., Non-linear self-consistent response of graphene in time domain. *Europhys. Lett.*, 79, 27002, 2007. arXiv:0709.3024v1 [cond-mat.mes-hall] 19 Sep 2007.
64. Kenichi, L. and Ishikawa, M., Nonlinear optical response of graphene in time domain. *Phys. Rev. B*, 82, 201402, 2010.
65. Dong, H., Contu, C., Marini, A., Bianealana, F., Terahertz relativistic special solitons in doped graphene metamaterials. *J. Phys. B: At. Mol. Opt. Phys.*, 46, 15, 15, 2013.
66. Bo, L., Graphene photophysical properties and large optical nonlinearities. *Chin. Sci. Bull.*, 57, 23, 2971–2982, 2012.
67. Zhou, Y. and Wu, M.W., Optical response of graphene under intense terahertz fields. *Phys. Rev. B*, 83, 245436, 2011.
68. Grifone, M. and Hanggi, P., Driven quantum tunneling. *Phys. Rep.*, 304, 229–354, 1998.
69. Grigorenko, A.M., Polini, M., Novoselov, K.S., Graphene plasmonics—Optics in flatland, Arxiv: 1301.4241 [cond-mat], 2013.
70. Bostwick, A., Speck, F., Seyller, T., Horn, K., Polini, M., Asgari, R., Macdonald, A.H., Rotenberg, E., Observation of plasmarons in quasi-freestanding doped graphene. *Science*, 328, 999–1002, 2010.
71. Hwang, E.H. and Das Sarma, S., Dielectric function screening, and plasmons in two-dimensional graphene. *Phys. Rev. B*, 75, 205418, 2007.
72. Polini, M., Asgari, R., Barghi, G., Pereg-Barnea, T., MacDonald, A.H., Plasmons and the spectral function of graphene. *Phys. Rev. B*, 77, 081411, 2008.
73. Walter, A.I., Bostwick, A., Jeon, K.-J., Speck, F., Ostler, M., Seyller, T., Moreshini, L., Chang, Y.J., Polini, M., Asgari, R., MacDonald, A.H., Horn, K., Rotenberg, E., (ARPES) Effective screening and plasmaron bands in graphene. *Phys. Rev. B*, 84, 085410, 2011.
74. Brar, V.W., Wickenburg, S., Palasigui, M., Park, C.-H., Wehling, T.O., Zhang, Y., Decker, R., Gerit, C., Balatsky, A.V., Loule, S.G., Zettl, A., Crommie, M.F., Observation of carrier-density-dependent many-body effects in graphene via tunneling spectrosopy. *Phys. Rev. Lett.*, 104, 036805, 2010.
75. Mikhailov, S.A., Theory of the giant plasmon enhanced second harmonic generation in graphene and semiconductor two-dimensional electron system. *Phys. Rev. B*, 84, 045432, 2011.
76. Fei, Z., Andreev, G.O., Bao, W., Zhang, L.M., McLeo, A.S., Wang, C., Stewart, M.K., Zhao, Z., Dominguez, G., Thiesmens, M., Fogeert, M.M., Tauber, M.J., Casnhj-Beto, A.H., Lau, C.N., Keilmen, F., Basov, D.N., Infrared nanoscopy of Dirac plasmons at the graphene-SiO_2 interface. *Nano Lett.*, 11, 11, 4701–4705, 2011.
77. Chen, J., Badioli, M., Alonso-Gonzalez, P., Thongrattanasiri, S., Huth, F., Osmond, J., Spasenovic, M., Centano, A., Pesquera, A., Godignon, P., Elorza, A.Z., Camara, N., Garda de Abajo, F.J., Hillerbrand, R., Koppens, F.H.L., Optical nano-imaging of gate-tunable graphene plasmons. *Nature*, 487, 77, 2012.

78. Crassee, L., Orlita, M., Potemski, M., Walter, A.L., Ostler, M., Seyller, Th., Gaponenko, J., Chen, J., Kuzmenko, A.B., Intrinsic terrahertz plasmons and magnetoplasmons in large scale monolayer graphene. *Nano Lett.*, 12, 2470–2474, 2012.
79. Mak, K.F., Sfeir, M.V., Wu, Y., Lui, C.H., Misevich, J.A., Heinz, T.F., Measurement of the optical conductivity of graphene. *Phys. Rev. Lett.*, 101, 196405, 2008.
80. Grushin, A.G., Valenzuele, B., Vozmediano, M.A.H., Effect of Coulomb interaction on the optical properties of doped graphene. *Phys. Rev. B*, 80, 155417, 2009.
81. Mishenko, E.G., Dynamic conductivity in graphene beyond linear response. *Phys. Rev. Lett.*, 103, 246802, 2009.
82. Trif, M., Upadhyaya, P., Tserkovnyak, Y., Theory of electromechanical coupling in dynamical graphene. *Phys. Rev. B*, 88, 245423, 2013.
83. Note that in [82] the more generalized formulae for vector potential coordinates were obtained than in [30] taking into account that the rotated orbitals can lie out of the same plane. Assuming that the rotated orbitals are lying in the same plane as it was done in [30], one gets (2.2.15), (2.2.16) from the generalized formulae [82]. We consider for simplicity this case and hence use the formulae (2.2.15), (2.2.16).
84. Firsov, Yu.A. and Firsova, N.E., Surface corrugations influence monolayer graphene electromagnetic response. *Physica E Low Dimens. Syst. Nanostruct.*, 62, 36–42, 2014.
85. Svintsov, D., Ryzhii, V., Saton, A., Otsuji, T., Carrier-carrier scattering and negative dynamic conductivity in pumped graphene. *Optics Express*, 22, 17, 19873, 2014.
86. Kar, S., Mohapatra, D.R., Ereysz, E., Sood, A.K., Tuning photoinduced terahertz probe spectroscopy. *Phys. Rev. B*, 90, 165420, 2014.
87. Giriray, J., Yi, R., Hugen, Y., Tony, F.H., Observation of a transient decrease in terahertz conductivity of single-layer graphene induced by ultrafast optical excitations. *Nano Lett.*, 13, 524–530, 2013.
88. Van Leeuwen, R., Dahlen, N.E., Stefanucci, G., Almbladh, C.-O., Von Barth, U., Introduction to Keldysh formalism, in: *Time-Dependent Density Functional Theory*, Marques, M.A.L., Ullrich, C., Noguera, F., Rubio, A., Burke, K. (Eds.), pp. 36–59, Springer, Berlin, Heidelberg, 2006.
89. Schneider, M. and Brouwer, P.W., Quantum corrections to transport in graphene: Trajectory based semiclassical analysis. *New J. Phys.*, 16, 063015, 2014.
90. Beresovsky, J., Borunda, M.F., Heller, E.J., Westervelt, R.M., Imaging coherent transport in graphene (part 1): Mapping universal conductance fluctuations. *Nanotechnology*, 21, 274014, 2010.
91. Borunda, M.F., Beresovsky, J., Heller, E.J., Imaging conductance fluctuations in graphene. *ACS Nano*, 5, 6, 3622, 2011.
92. Joel, M., Tankut, C., Dirk, K.M., Spatial current patterns, dephasing and current imaging in graphene nanoribbons. *New J. Phys.*, 16, open access, 2014.
93. Considering response of charge carriers to the external electromagnetic field, one should take into account self-consistent field effect studied in [89] for flatland model. The external time-dependent electric field of the wave induces in the electron gas lying in the plane the electric current . According to the Maxwell equations, this time-dependent current produces in its turn a secondary (induced) electric field.
94. Dawlaty, J.M., Shivaraman, S., Strait, J., George, P., Chandrashekhar, M., Rana, F., Spencer, M.G., Veksler, D., Chen, Y., Measurement of the optical absorption spectra of epitaxial graphene from terahertz to visible. *Appl. Phys. Lett.*, 93, 131905, 2008.
95. Osvath, Z., Lefloch, F., Bouchiat, V., Chapelier, C., Electric field-controlled rippling of graphene. *Nanoscale*, 5, 22, 10996, 2013.
96. Levente, T., Dumitrica, T., Kim, S.J., Nemes-Incze, P., Hwang, C., Biro, L.P., Breakdown of continuum mechanics for nanometer-wavelength rippling of graphene. *Nat. Phys.*, 8, 739, 2012.

97. Wang, W.L., Bhandari, S., Yi, W., Bell, D.C., Westervelt, R., Kaxiras, E., Direct imaging of atomic-scale ripples in few-layer graphene. *Nano Lett.*, 12, 5, 2278–2282, 2012.
98. Park, J., Ahn, Y.H., Ruiz-Vargas, C., Imaging of photocurrents generation and collection on single-layer graphene. *Nano Lett.*, 9, 5, 1742–1746, 2009.
99. Stutzel, E.U., Dufaux, T., Sagar, A., Rauschenbach, S., Balasubramanian, K., Burghard, M., Kern, K., Spatially resolved photocurrents in graphene nanoribbon devices. *Appl. Phys. Lett.*, 102, 043106, 2013.
100. Herbschleb, E.D., Puddy, R.K., Marconcini, P., Griffiths, J.P., Jones, G.A.C., Macucci, M., Smith, C.G., Conolly, M.R., Direct imaging of coherent quantum transport in graphene heterojunctions. *Phys. Rev. B*, 22, 125414, 2015.
101. Puddy, R.K., Chua, C.J., Buitelaar, M.R., Transport spectroscopy of a graphene quantum dot fabricated by atomic force microscope nanolithography. *Appl. Phys. Lett.*, 103, 18, 101063, 2013.
102. Firsov, Yu.A. and Firsova, N.E., Radiative decay effects influence the local electromagnetic response of the monolayer graphene with surface corrugations in terahertz range. *Physica E Low Dimens. Syst. Nanostruct.*, 71, 134, 2015.

7

Interaction and Manipulation of Bi Adatoms on Monolayer Epitaxial Graphene

Shu Hsuan Su[1], Shih Yang Lin[1], Jung Chun Andrew Huang[1,2,3*] and Min Fa Lin[1†]

[1]Department of Physics, National Cheng Kung University, Tainan, Taiwan (R.O.C.)
[2]Advanced Optoelectronic Technology Center, National Cheng Kung University, Tainan, Taiwan (R.O.C.)
[3]Taiwan Consortium of Emergent Crystalline Materials, Ministry of Science and Technology, Taipei, Taiwan (R.O.C.)

Abstract

The metal–graphene interaction can determine to what extent the graphene properties might be modified. Semimetal bismuth (Bi), being one of the most extensively studied heavy elements, has received considerable attention because of its unique electronic properties in reduced length scale. In this chapter, long-range electronic interaction between Bi adatoms deposited on monolayer epitaxial graphene (MEG) formed on a 4H-SiC (0001) substrate at room temperature (RT) has been studied. The introduction of the motivation and purpose of this work will be described in the first part of this chapter, followed by the detailed preparation methods of the MEG. The provision of large-area MEG makes the characterization and application of graphene-based devices become easier. The interaction and manipulation of Bi adatoms on MEG studied by scanning tunneling microscopy (STM) will be described in the second part of this chapter. We elucidate that the oscillatory interaction among Bi adatoms results from both the mediation of graphene Dirac-like electrons and the effect of the corrugated surface of SiC substrate. A series of structural transition of Bi adatoms, adsorbed on MEG at RT, is explored with Bi coverage-variation. Bi adatoms undergo a structural transition from one-dimensional (1D) linear structures to two-dimensional (2D) triangular islands and such growth modes are strongly affected by the corrugated substrate. Additionally, upon Bi deposition, some charge transfers between Bi and graphene occurs and a characteristic peak can be observed in the scanning tunneling spectrum (STS), reflecting by the distinctive electronic structure of Bi adatoms. With annealing to ~500K, 2D triangular Bi islands aggregate into Bi nanoclusters of uniform size of three to four adatoms. A well-controlled fabrication and manipulation method is demonstrated. Finally, first-principles calculations are conducted to investigate the substrate effect of Bi-adsorbed monolayer (ML) graphene. The slightly deformed ML graphene, corrugated buffer layer, and six-layered substrate are all simulated. Interactions between the buffer layer and graphene dominate the hexagonal Bi adsorption patterns. Specially, we demonstrate that the room temperature stability comes from a buffer-layer-induced energy barrier. The density of states exhibits low-lying dip and peak, as observed in the experimental measurements. The approaches adopted herein provide perspectives for fabricating and characterizing periodic networks on graphene and related systems, which are useful in realizing graphene-based electronic, energy, sensor, and spintronic devices.

**Corresponding author*: jcahuang@mail.ncku.edu.tw
†*Corresponding author*: mflin@mail.ncku.edu.tw

Tobias Stauber (ed.) Handbook of Graphene: Volume 2, (195–218) © 2019 Scrivener Publishing LLC

Keywords: Bismuth, graphene, scanning tunneling microscopy, scanning tunneling spectrum, long-range interaction, first-principles calculations

7.1 Introduction

A single sheet of graphene is made of a two-dimensional (2D) honeycomb structure of carbon atoms [1, 2], with two atoms per unit cell and a unit cell vector of $a_G = 2.46$ Å [3]. The symmetry of the graphene lattice has subtle effects on conduction. There are two distinct (triangular) sublattices corresponding to the two equivalent atoms in the unit cell. The two sublattices cause some special properties of graphene such as the pseudospin. The carbon atoms are bonded to each other by sp^2 bonds in plane. The delocalized π-electrons, out of plane, show the interesting electronic properties of graphene. There are two high symmetry points, Γ-point in the center and the K-point in the corner, in the hexagonal Brillouin zone. Although nearly everything in condensed matter physics is described by the Schrödinger equation, graphene reveals a totally different behavior compared to above. In contrast to the ordinary parabolic dispersion relation of a free electron, graphene exhibits a linear dispersion relation in the vicinity of the Fermi surface [4]. These conical bands suggest comparison of charge carriers in graphene to photons, which travel at a constant speed (c, the speed of light) and almost have no mass. Electrons in graphene may be considered effectively massless particles, and travel at constant speed ~ c/300. That is, same as photons, they may be treated mathematically as relativistic particles. As consequences, the band structure can best be described by the Dirac equation, as the electrons considered into massless Dirac fermions. The crossing point is then called the Dirac point E_D. For undoped (freestanding) graphene, the energy of the Dirac point is identical with the Fermi level E_F. Therefore, freestanding graphene is called a zero-gap semiconductor or semimetal.

Understanding how adsorbates interact with each other on an atomically well-defined surface and controlling their structures are essential to the fabrication of low-dimensional materials with novel electronic structures. In recent decades, adsorbate–adsorbate interactions have been extensively studied both theoretically [5, 6] and experimentally [7–9], but can almost be observed only at rather low temperatures. These interactions have various origins that can be generally categorized according to their adsorbate separation statistics, resulting in several possible interaction behaviors as a function of distance between two adsorbates [10]. In particular, the electronic indirect interaction mediated by surface electrons exhibits a characteristic feature of oscillatory interaction energy. These oscillatory interactions can be observed directly by scanning tunneling microscopy (STM) [11, 12]. The inability to clarify the interactions between adsorbates prevents the clear identification of the nucleation and growth of adatoms and limits the development of advanced devices with nanometer dimensions. Despite many of the concerns paid to interactions between adsorbate and a well-defined surface, most studies have focused on metal surfaces. An atomically well-defined surface with unique properties for the fundamental study of nucleation and growth of adatoms and invention of high-performance devices is of our great interest. Graphene is a model candidate, owing to its remarkable physical properties [13, 14] such as high carrier mobility and Dirac-like electrons. Additionally, doping of graphene with heavy metals can control the number of carriers and spin–orbit coupling can be exploited for fundamental electronic studies and spintronic applications. Charge transfer between metal adatoms and carbon atoms are also worthy of investigation and may lead to a distinct graphene electronic structure. To date, many theoretical [15–19] and experimental studies [20–23] of adsorbate–graphene systems have been performed. Almost all related experiments have focused on either the adsorbate–graphene interactions or the modified graphene properties following the adsorption of atoms and molecules; interactions between adatoms had been seldom addressed until S. M. Binz *et al.* and Xiaojie Liu *et al.* reported the deposition of Fe clusters on graphene [24, 25] and Can-Li Song *et al.* explored the adsorption of Cs adatoms on graphene [26].

Owing to the flourishing development of nanotechnology, the techniques of STM become crucial for examining the intriguing properties originated from nano-effects. Notably, STM is extremely sensitive to the investigation of surface morphology and surface electronic structures. Recently, material graphene, a single layer of carbon atoms, has incited a massive attention and lead to the development of the various new research field. Importantly, the applications of graphene require connections to metallic elements such as Bi used in this chapter. The metal–graphene interaction will determine to what extent the graphene properties might be modified. Nevertheless, to the best of our knowledge, Bi adatoms grown on graphene, has not yet been described to understand the adatom–adatom interactions as well as the nucleation and growth of adatoms on graphene using STM measurement. We think that this is a crucial key to fill the gap in the graphene-related literature, which heavily consists of studies on clean graphene. Using advantages of STM, the related issue mentioned above can be further investigated and elucidated, as presented in the following.

7.1.1 Long-Range Interactions of Bismuth Adatoms at Room Temperature

Long-range electronic interaction between Bi adatoms on graphene formed on a 4H-SiC (0001) substrate are clearly observed at room temperature (T = 300K). Using STM and density functional theory (DFT) calculations, we have demonstrated that such oscillatory interaction results mainly from the mediation of graphene Dirac-like electrons and the effect of the corrugated surface of SiC substrate. These two factors cause the observed oscillatory interaction with characteristic distribution distances and linear arrangements of Bi adatoms. The present study sheds light on understanding and controlling the nucleation of adatoms and subsequent growth of nanostructures on graphene surface. This part of the work will be introduced in Chapter 7.2.

7.1.2 Low-Dimensional Structures of Bismuth Adatoms and Temperature Effect

A series of structural transition of Bi adatoms, adsorbed on monolayer epitaxial graphene (MEG), is explored at room temperature. Bi adatoms undergo a structural transition from one-dimensional (1D) linear structure to 2D triangular islands, and such 2D growth mode is affected by the corrugated substrate. Upon Bi deposition, a little charge transfer occurs and a characteristic peak can be observed in the tunneling spectrum, reflecting the distinctive electronic structure of the Bi adatoms. When annealed to ~500K, 2D triangular Bi islands aggregate into Bi nanoclusters (NCs) of uniform size. A well-controlled fabrication method is thus demonstrated. The approaches adopted herein provide perspectives for fabricating and characterizing periodic networks on MEG and related systems, which are useful in realizing graphene-based electronic, energy, sensor, and spintronic devices. This part of the work will be introduced in Chapter 7.3.

7.1.3 The Energetically Favorable Distribution of Bi Adatoms Using First-Principles Calculations

There are some theoretical studies about Bi-adsorbed and Bi-intercalated graphenes that focus on the geometric structures and energy bands [19, 27]. The study of Bi-adsorbed graphene has been performed on ML graphene without simulation of the buffer layer graphene and substrate; hence, the deformed graphene surface structure may be unreliable [19]. The Bi-intercalated graphene is calculated for Bi and/or Sb as a buffer layer on the four-layer SiC substrate, which can be an energetically unfavorable environment for the metal atoms to be adsorbed on the graphene sheet [27]. A systematic study on the critical roles played by the configuration of the Bi adatoms, buffer layer, and substrate is needed to date.

The energetically favorable configuration of Bi adatoms can be simulated under the substrate effect and the buffer layer by the first-principles calculations. This section shows that the dependence of adsorption energy on the various atomic sites and hexagonal positions. The atomic sites include bridge, hollow, and top sites, and the different hexagonal positions are closely related to the non-uniform van der Walls interactions. Two different bismuth distributions could be obtained, which is uniform or aggregated adatom arrangements with the buffer-layer-induced energy barriers. The aggregated arrangements of adatom are as associated with the large-scale hexagonal symmetry, Bi coverage, and relatively few vacancies. The optimal geometric structures were validated by our STM measurements [28, 29]. Moreover, the main influences of the Bi adsorption on the density of states (DOS) could be understood through a detailed comparison with tunneling conductance measurements [28, 29]. This part of the work will be introduced in Chapter 7.4.

7.2 Long-Range Interactions of Bismuth Growth on MEG

7.2.1 As-Prepared MEG Surface

The Si-terminated 4H-SiC (0001) substrate was used and cleaned ultrasonically in acetone and isopropanol before *in situ* degassing at ~600°C in UHV for several hours. The substrate was then annealed under a low Si flux at ~1000°C to remove the native oxide before annealing to higher temperatures without Si flux. Graphene was subsequently formed by thermal annealing in the temperature range of 1200–1300°C [30]. An atomically clean surface of MEG is presented in Figure 7.1a. The image reveals a well-identified honeycomb lattice of the graphene superimposed onto the 66 superstructure (bright hexagon) [31, 32], as well as the corresponding fast Fourier transform (FFT), in Figure 7.1b. A white solid arrow points to a 1×1-G spot, and a green dashed arrow assigns to a 6×6-SiC spot [33, 34]. Figure 7.1c schematically depicts a cross-section view of the MEG/4H-SiC (0001) structure using ball-and-stick representations. The dI/dV spectrum of the as-grown MEG in Figure 7.1d exhibits a local minimum ~–0.37eV, representing the Dirac point of graphene [35]. Although some packets are observed within the range –0.2 eV to 0.4 eV, these are likely deemed to be a

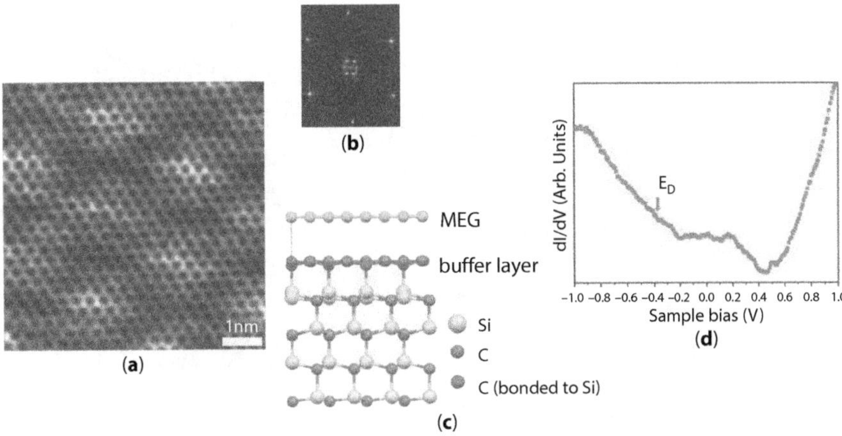

Figure 7.1 Clean MEG surface. (a) Atomically resolved STM image of MEG on 4H-SiC (0001). (b) (c) Side-view ball-and-stick representations of MEG/4H-SiC (0001). (d) dI/dV-V curve obtained on clean surface of MEG.

result of many-body effects in graphene [36] or the effect of corrugated substrate. The main feature (Dirac point) can still be identified.

7.2.2 Low Coverage of Bismuth Growth on MEG

Immediately after preparing a clean surface of MEG, Bi was deposited on the graphene surface from a Bi rod (with a purity of 99.9999%) by a commercial evaporator (Omicrometer EFM 3) at room temperature. The atomic density of the deposited metal was represented in the ML form, corresponding to a Bi packing density of 1.1×10^{15} Bi atom/cm^{-2} when proceeding along the growth plane of graphene [37]. The growth rate was determined to be approximately 0.0013 ML per minute.

At 0.0013 ML, the surface reveals dispersed protrusions, as illustrated in Figure 7.2a. The interactions of Bi adatoms are pairwise and additive with increasing Bi coverage. As coverage increased to 0.0078 ML, the arrangement of the protrusions exhibits linear-like ordering (Figure 7.2b). Moreover, at 0.0092 ML, large-scale linear structures can be detected in STM image, as illustrated in Figure 7.2c, indicating possibly 1D growth mode. The protrusions of approximately 0.24 nm in height are observed within different coverages of Bi, as shown in Figure 7.2d.

7.2.3 The Interaction Potential between the Bi Adatoms Using Pair Distance Distribution Analysis

In order to clarify the Bi-Bi interactions of such structures, the pair distance distribution analysis between Bi adatoms, by analyzing Figure 7.2c, is introduced by firstly evaluating over 18,800 inter-adatom separations from a series of non-overlapping STM images and then plotting a statistical histogram $N(r)$ of these separations, as illustrated in Figure 7.3a. By comparing the theoretical function with the random distribution $N'(r)$ of interadatom separations r in the absence of Bi-Bi interactions (blue dashed line in Figure 7.3a) [38],

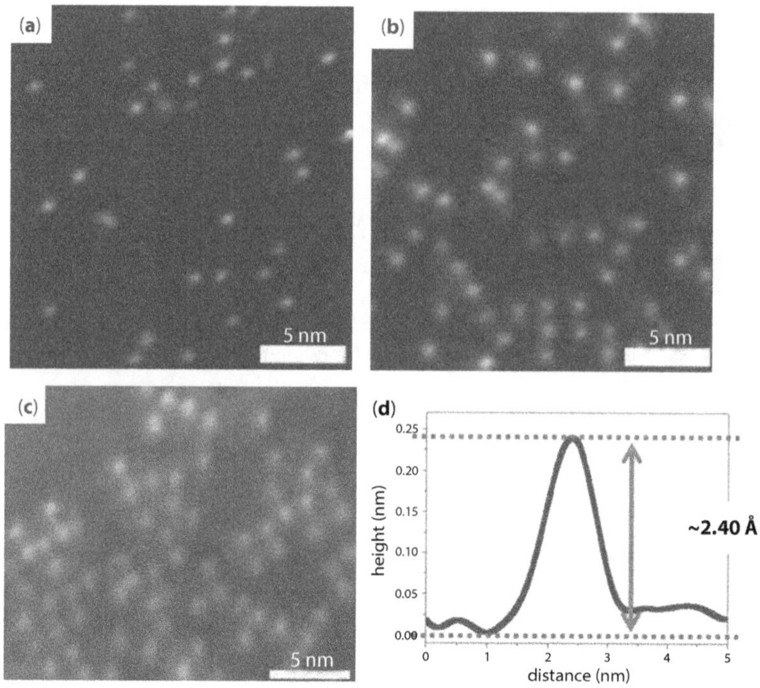

Figure 7.2 STM images of Bi adatoms on MEG with (a) 0.0013 ML, (b) 0.0078 ML, and (c) 0.0092 ML. (d) The height of Bi adatom.

the pair-correlation function, in terms of the two-body interaction, can be described as $g(r)=\frac{N(r)}{N'(r)}$. Next, the interaction potential $E(r)$ between the Bi adatoms is determined by $E(r) = -k_B T \ln [(g(r))]$ [38], where T denotes the temperature (T = 300K), and k_B represents the Boltzmann constant.

Figure 7.3b plots the interaction potential between Bi adatoms, E(r), as a function of distance between the interadatom separations r. The interaction potential, E(r), reveals a clearly oscillatory long-range interaction. This oscillatory feature is very likely to originate from Friedel oscillations [8]. To verify the origin of the oscillatory feature, the interaction energy is theoretically fitted as follows. The experimental interaction energy (solid line, Figure 7.3b) is overlaid with two theoretical energy plots (red and blue dashed lines in Figure 7.3b) adapted from Hyldgaard [6] and V. V. Mkhitaryan [18], respectively, with adequate fitting. The first theoretical energy plot (red dashed line in Figure 7.3b) resembles a Shockley surface state fit of a (111) noble metal surface and can be formulated as follows [6]:

$$E(r)=-A(\delta_F,r)\left(\frac{4\varepsilon}{\pi^2}\right)\frac{\sin(2k_F r+2\delta_F)}{(k_F r)^2} \tag{7.1}$$

where δ_F represents the phase shift, ε is the carrier energy, $k_F \frac{2\pi}{\lambda_F}$ is the Fermi wave vector, r is the interatomic distance, and A is a dimensionless value representing the interaction

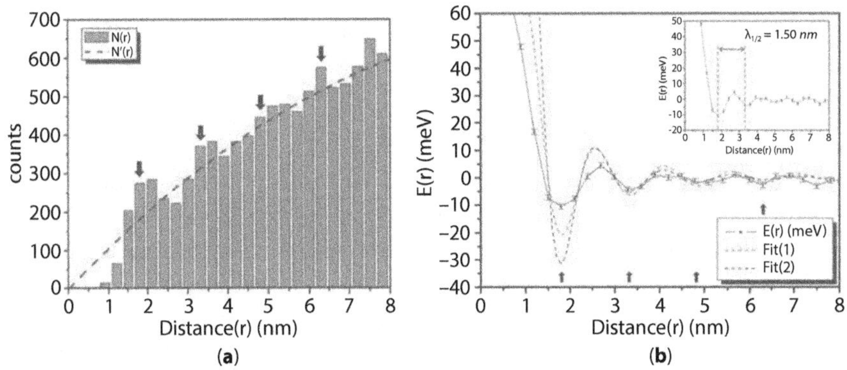

Figure 7.3 (a) Statistical histogram $N(r)$ of separations between Bi adatoms obtained from a series of STM images recorded at room temperature. Blue dashed line $N'(r)$ represents the expected distribution in the absence of any interatomic interactions. Black arrows correspond to the energy minima in Figure 7.3b. (b) The experimentally derived potential energy (solid line) is overlaid with two theoretically derived plots (red and blue dashed lines). The standard deviation of the mean is shown as the error bar. Inset: $\lambda_{1/2}$ = 1.50 nm measured from experimental interaction potential. Arrows indicate four potential energy minima at 1.8, 3.3, 4.8, and 6.3 nm.

strength. Hence, the first observed position of minimum potential energy, at a distance of 1.8 nm, corresponds to the first favorable distance between the adatoms, and the following minima are observed at intervals of ~1.5 nm (which is, $\lambda_F/2$). Additionally, the amplitude of the interaction strength decays as $1/r^2$. The theoretical energy plot is based on Equation 7.1 with $A = 0.78$ and $\delta = 0$, and the measured A is larger than previous values of noble metal surface in the literature [7]. This difference probably results from the fact that Bi is a strong scattering element. The phase shift of Fit (1) is zero, which is smaller than that for the noble metal surface case [39]. This discrepancy may be attributed to fundamental property differences between graphene Dirac-like electrons and the electrons of the Shockley surface state of a (111) noble metal surface. Notably, Shockley surface-state electrons have a finite Fermi wave vector [8] but Dirac-like electrons of intrinsic graphene have a very small (almost zero) Fermi wave vector. Only by gating or doping in graphene can finite Fermi wave vector be achieved. This issue will be addressed in the following discussions of Fit (2). Furthermore, the first minimum of potential energy in the theoretical energy plot is less than 25meV, indicating that the observed distribution of Bi-Bi distances described by Equation 7.1 should not be stable at room temperature. Therefore, the description of a conventional 2D electron gas description cannot be fully applied to our study. The second theoretical energy fit (blue dashed line in Figure 7.3b describes a system with the Berry phase and nontrivial chiral spectrum, if the states with momenta k and -k are mutually orthogonal and then backscattering is suppressed [39]. This situation could happen in the case of doped or gated graphene [14], when the Fermi level is shifted away from the Dirac points, such that $k_F \neq 0$. The doping level reveals non-zero Fermi momentum, and this phenomenon can be roughly described as [8]

$$E(r) = \alpha \times 2\sin(\delta_F)\frac{\cos\left(2k_F r + \delta_F\right)}{\pi^2 k_F r^3} \tag{7.2}$$

where δ_F denotes the phase shift, $k_F = \frac{2\pi}{\lambda_F}$ is the Fermi wave vector, r is the distance, and α is a constant value that represents the energy scale. The theoretical energy plot is based on Equation 7.2 above with $\alpha = 2000$ and $\delta = \frac{\pi}{2}$, and the same oscillatory period and similar potential energy minimum features are determined. Moreover, the potential energy minima fit better than those in Fit (1) and the experimentally observed first energy maximum is about 11 meV, likely originating from a strong scattering effect of Bi adatom [7]. Notably, the deviation of experiment plot from two theoretical plots is large at $r < 1.5$ nm. With a closer examination, experimental E(r) approximately decays as 1/r at $r < 1.5$ nm, suggesting that a possibly electrostatic interaction exists between Bi adatoms. Such 1/r behavior could be originated from the interaction between Bi adatom and graphene, and it is similar with previous literature [26]. The two theoretical formulae employed herein do not include the electrostatic interaction, which could be the reason of large deviation at $r < 1.5$ nm. Meanwhile, two interesting features can be observed in experimental plot. First, the potential wells in the empirical E(r) in Figure 7.3b and the first potential energy minimum, ~32 meV in Fit (2), are seemingly not completely sufficient to keep the Bi atoms at a certain site. We then speculate that the observed order structure could be a metastable phase. Second, compared with previous report [35], one finds that the Fermi wave vector obtained from Figure 7.3b is quite large, indicating heavily doped graphene. This conflict with the evaluated Dirac point measured in Figure 7.1d. That is, there are other factors existing in the formation of the observed oscillatory interaction and it will be addressed in the following discussions.

Based on the above findings, we suggest that not only the Dirac-like electrons in graphene but also some factors are responsible for the room-temperature oscillatory interaction between Bi adatoms. It is noticed that Dirac-like electrons mainly originate from the doping effect of the underlying SiC substrate and the charge transfer between graphene and Bi adatom. Previous literature has verified very weak interactions between Bi nanoribbon and MEG [37, 40] from photoemission spectroscopy (PES) data and STM studies, combined with DFT calculations. Therefore, we suggest that the charge transfer between graphene and Bi should be a minor effect and the doping effect of SiC substrate is very likely to play a major role in the contribution of the Dirac-like electrons [37, 40]. Remarkably, the induced energy difference follows the usual Friedel $\cos(2k_F r)/r^3$ dependence, with the resonance behavior entering via the phase shift, as shown in Equation 7.2. For such a reason, the measured $\frac{\pi}{2}$ phase shift indicates that the deposition of Bi adatoms on graphene is in the resonance region [18]. In contrast, Fe clusters that are deposited on epitaxial graphene [24, 25] exhibit Fe-Fe interactions that involve short-range attraction and long-range repulsion. Such interactions are attributed to sum of dipole–dipole, elastic (substrate), and indirect interactions. On the other hand, Cs adatoms that are deposited on epitaxial graphene [26] exhibit Cs-Cs long-range electrostatic interaction mainly caused by significant charge transfer from Cs to epitaxial graphene, which is also coupled with the corrugated substrate. Same as previous literature [26], MEG exhibits the unambiguously characteristic 6 × 6-SiC periodicity in Figure 7.1a, indicating the existence of influence of the corrugated substrate. However, the Bi-Bi interactions herein differ significantly from the behaviors described above and an unexpected Fermi wave vector is extracted from Figure 7.3. Therefore, the effect of corrugated substrate should need a careful investigation for the formation of oscillatory interaction between Bi adatoms.

7.2.4 The Relation between the Linear Bi Structures and Buffer Layer of SiC

In general, corrugated substrate can be most likely represented as exhibiting the curvature effect, which noticeably changes the chemical activity and binding energy of graphene [31]. A closer examination of the STM image, as illustrated in Figure 7.4a, reveals linear arrangements of Bi adatoms rather than short-range ordered hexagonal structures with an average interadatom separation of 6 × 6 superstructure [26], indicating weak influence of 6 × 6 superstructure. Therefore, the structure of the underlying buffer layer should also be charily examined instead of a simple consideration of the 6 × 6 superstructure. Additionally, most linear Bi structures are indicated by red lines (oriented at ~60° with respect to the green lines and ~30° with respect to the blue lines); the blue lines are oriented at ~90° with respect to green lines in Figure 7.4a. The linear Bi distribution in three different orientations is analyzed by means of statistical histogram, as shown in Figure 7.4b. The observation of linear arrangements of Bi adatoms along the three different orientations is adequately consistent with the fact that the deposition of Bi nanoribbons grown on MEG [40].

DFT calculation is introduced and ball-and-stick representations of MEG/4H-SiC (0001) structures are schematically presented in Figure 7.5a. As in the literature [33, 41], DFT calculations demonstrate reliable structure features. In the structure of MEG/4H-SiC (0001), some bonding length variations are observed in the buffer layer and such C atoms are marked in red color in Figure 7.5a. That is, the buffer layer reveals bonding length variations around the C atom due to C atom bonded to Si atom or not. Moreover, the average Bi-Bi interatomic distance obtained from Figure 7.2 is ~1.8 nm, which is almost four times the lattice constant along the [11–20] direction of Bi nanoribbon [40].

Furthermore, DFT calculations reveal that the buffer layer clearly exhibits two different bonding length variation units, d_1 and d_2 of Unit 1 and Unit 2, respectively. The length ratio d_1/d_2 is approximately 0.866 obtained from DFT calculations and the statistical length ratio D_{Red}/D_{Blue} is roughly 0.878 ± 0.008 extracted from experimental data, as shown in Figures 7.4a and 7.5b. As described above, the structure of buffer layer should be an important reason for the formation of the linear Bi structure and causing small variations in the interatomic distances between Bi adatoms. With a closer examination of the underlying buffer layer of SiC, Unit 1 or 2 is oriented at 60° with respect to itself and 90° with respect to the other, as illustrated in Figure 7.5c. Evidently, the underlying buffer layer of SiC determines

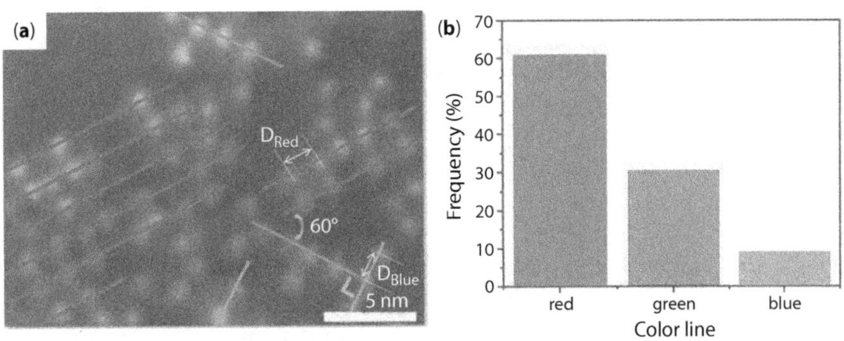

Figure 7.4 (a) Linear Bi structures are represented by several colored lines (red, green, and blue), representing different orientations. (b) Statistical histogram of different directional arrangements of Bi linear structures.

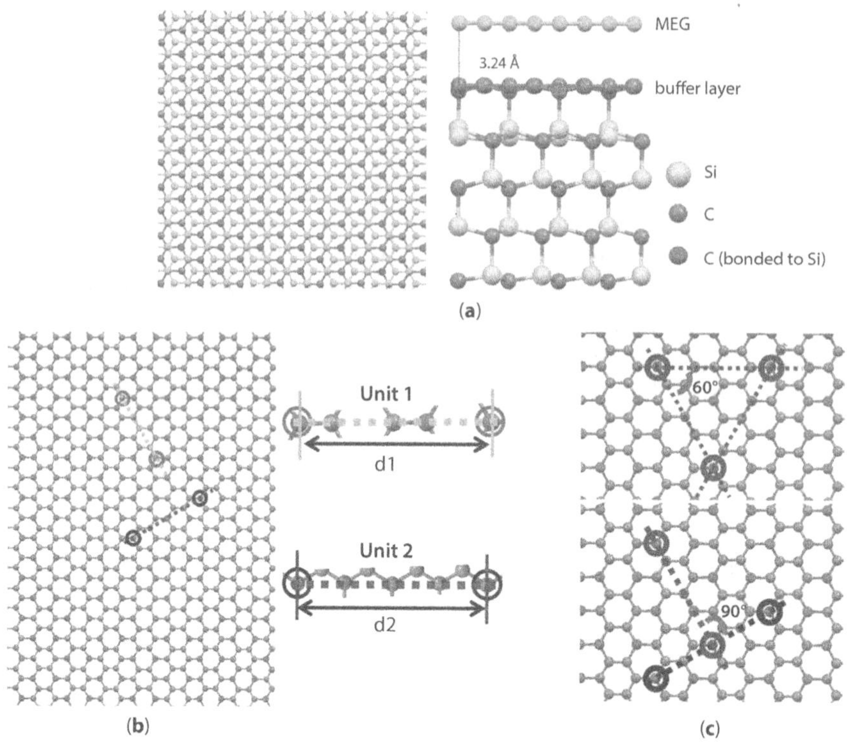

Figure 7.5 (a) Top- and side-view ball-and-stick representations of MEG/4H-SiC (0001). MEG, C, and Si atoms are shown in different colors. (b) DFT results of two different length variation units on buffer layer structure. (c) Possible orientation with respect to each Unit 1 due to bond length variation and angle of structure of 90° between Unit 1 and Unit 2.

the effect of the corrugated substrate, which causes the linear arrangements of Bi adatoms. Furthermore, the different orientations of the linear structures of Bi adatoms are consistent with Bi nanoribbons grown on MEG [40]. This detailed investigation provides a route to elucidating the initial nucleation of Bi adatoms and the subsequent growth of Bi nanoribbons.

7.3 Low-Dimensional Structures of Bismuth on MEG

7.3.1 Coverage-Dependent Structural Transition of Bi Adatoms Adsorbed on MEG

Upon the deposition of Bi, the transition of surface structural is observed as a function of the Co coverage in Figure 7.6a–d. The interatomic distance distribution in different coverage is analyzed by means of statistical histogram, as shown in Figure 7.6e–h. At 0.0013 ML, dispersed protrusions and local Bi nearest-neighbor distances, 1.3 nm and 1.5 nm, are obtained from the histogram in Figure 7.6e. The protrusions exhibit linear ordering when the coverage is increased to 0.0078 ML, as shown in Figure 7.6b. The 1D linear structure appears (green dashed line) in accompaniment with a Bi-Bi nearest-neighbor distance of

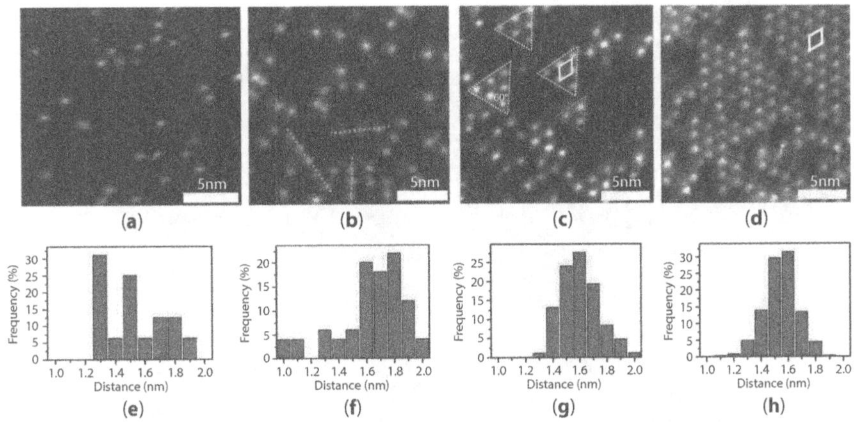

Figure 7.6 STM images of different Bi coverage on 4H-SiC (0001) surface. (a) 0.0013 ML, (b) 0.0078 ML, (c) 0.039 ML, and (d) 0.078 ML. Panels (e)–(f) present the histogram of Bi-Bi interatomic distance.

around 1.8 nm (Figure 7.6f), reflecting a characteristic 1D growth mode at low coverage. Notably, the Bi–Bi distance of 1.8 nm is almost four times the lattice spacing along the [11–20] direction of Bi nanoribbon [40]. At 0.039 ML, 2D triangular islands (indicated by yellow and green dashed lines) are observed and linear structures are still detected, as presented in Figure 7.6c, revealing the transition from 1D linear to 2D triangular structures. Such 2D triangular islands consist of Bi atoms with equally interatomic distance and form regular triangles, as displayed in Figure 7.6g. In triangular islands, the Bi-Bi distance is about 1.6 nm, which is very close to $2\sqrt{3}\times a = 1.57$ nm {a = 4.54 Å, which is the lattice constant along the [11–20] direction of Bi nanoribbon [40]}. The white solid line in Figure 7.6c delineates a unit cell of Bi adatoms. A large-scale hexagonal array of Bi atoms is formed at 0.078 ML, as presented in Figure 7.6d. The interatomic distance of Bi atoms is maintained at 1.6 nm (in Figure 7.6h), and the same unit cell is still observed (white solid line in Figure 7.6d) in the STM images. Therefore, the large-scale 2D hexagonal array is unambiguously identified as coverage-induced $2\sqrt{3}\times 2\sqrt{3}$ Bi reconstruction on the MEG surface. No ordered structures have been observed in previous literature at such low Bi coverage [42–44]. The results indicate the existence of some particular interactions between Bi adatoms and MEG. This issue will be addressed in the following discussion.

7.3.2 Structural Analysis of Bi Hexagonal Array

The MEG and the Bi hexagonal array are analyzed by zooming in the FFT of two STM topographic images with same size, as respectively depicted in Figure 7.7. The six outer spots originate from the hexagonal graphene lattice and the six inner spots correspond to the observed 6 × 6-SiC moiré pattern, as shown in Figure 7.7b. The moiré pattern of the MEG is due to the $(6\sqrt{3}\times 6\sqrt{3})$ R30° reconstruction of the buffer layer, which occurs as a result of the large difference between lattice parameters of SiC (3.08 Å) and graphene (2.46Å). Figure 7.7d, which is the FFT of Figure 7.7c, shows a similar sixfold symmetry pattern to that of the inner spots in Figure 7.7b. The upper inset, which zooms in the FFT of the inset in Figure 7.1b, shows clearer pattern. That is, the large-scale hexagonal array of Bi adatoms evidences a moiré-like superstructure and such arrangement is strongly affected by the corrugated

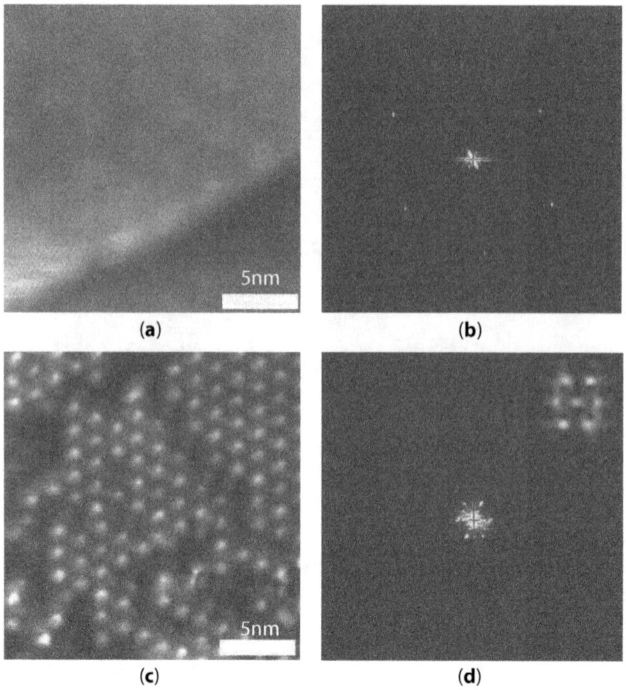

Figure 7.7 (a) The atomically resolved MEG image. (b) Room-in FFT of (a) image. (c) The image of 0.078 ML Bi coverage on MEG. (d) Room-in FFT of (c) image and top-right panel corresponds to zooming in the FFT of inset in Figure 7.1b.

substrate. As stated above, the interatomic distance between Bi atoms changes from 1.8 nm for the 1D linear chain to 1.6 nm for the 2D triangular island structure, equivalently from four times to $2\sqrt{3}$ times the lattice spacing of the Bi nanoribbon, indicating a coverage-dependent 1D → 2D growth mode transition in this system.

According to the DFT calculations, combined with previous literature [28, 45], Bi adatom is preferentially adsorbed at the bridge (B) site, as shown in Figure 7.8a. The theoretical arrangements of 1D and 2D structures are then illustrated in Figure 7.8b and c. Left and

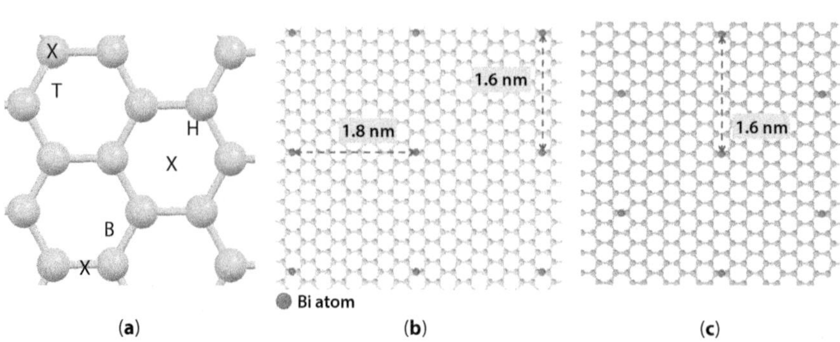

Figure 7.8 (a) The three adsorption sites considered: hollow (*H*), bridge (*B*), and top (*T*). (b) and (c) The theoretically atomic structure of the 1D chain and 2D triangular structures.

right panels reveal the atomic arrangements of Bi adatoms at different coverage, respectively. From the DFT calculations, the interatomic distances between Bi atoms, 1.8 nm and 1.6 nm, coexist at 1D-structure dominated configuration while only one characteristic distance of 1.6 nm appears at 2D structure. The DFT results are consistent with the STM observations. This structural transition could be related with coverage-dependent phenomenon and the effect of corrugated substrate.

Correspondingly, the electronic properties of Bi adatom in hexagonal array are probed using tunneling spectroscopy, as shown in Figure 7.9. The dI/dV spectrum of the as-grown MEG exhibits a characteristic minimum at -0.37eV. This is attributed to Dirac point, indicative of n-type doping by the SiC substrate, while some packets exist around Fermi level (E_F) [36]. Upon the deposition of Bi adatoms on MEG, the Dirac point shifts to near E_F, appearing at -0.32eV, indicating a little charge transfer (~50meV) from MEG to Bi adatom.

The observed coverage-dependent structural transition and small charge transfer are basically consistent with the STM and synchrotron-based PES measurements [40]. The tunneling spectrum also exhibits one peak at −0.72eV (black arrow in Figure 7.9), which is similar to one of the four characteristic peaks of the Bi (110) nanoribbon, based on rhombohedral indexing [37, 46, 47]. As recorded in the literature [46, 48], the peak at ~ −0.75 eV is mainly attributable to the p-states that are localized at the topmost layer of Bi (110) nanoribbon. Thus, we speculate that the peak in Figure 7.9 results from the contribution of the p band of 2D Bi hexagonal array. Additionally, the characteristic feature (−0.72eV) suggests that Bi adatom forms a bound state on MEG due likely to the Bi adatom–MEG interaction.

7.3.3 Temperature Effect of Bi Adatoms

To investigate the effect of temperature, the sample with 0.039 ML of Bi atoms was annealed to an elevated temperature of 500 K. Figure 7.10a and b displays the STM images obtained before and after annealing, respectively. After annealing at 500 K for 10 minutes, 1D linear arrangement of Bi NCs, instead of the Bi adatoms in 2D triangular islands, is clearly observed. Interestingly, the Bi NCs are very uniform in size. Within line profile measurements (red and gray dashed lines in Figure 7.10a and b), the Bi adatoms are ~0.24 nm

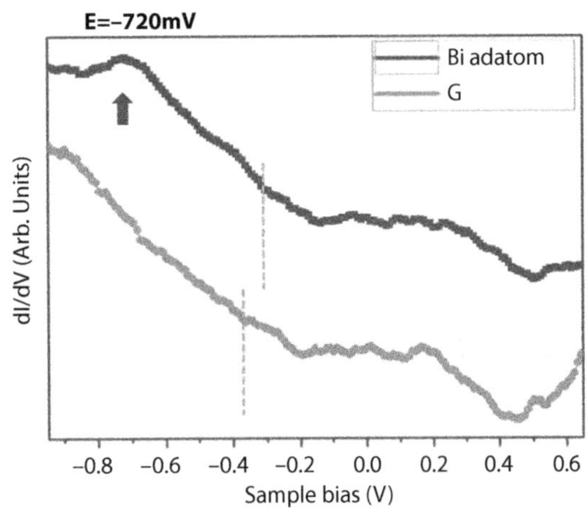

Figure 7.9 dI/dV spectra of as-grown graphene and Bi adatom in hexagonal array.

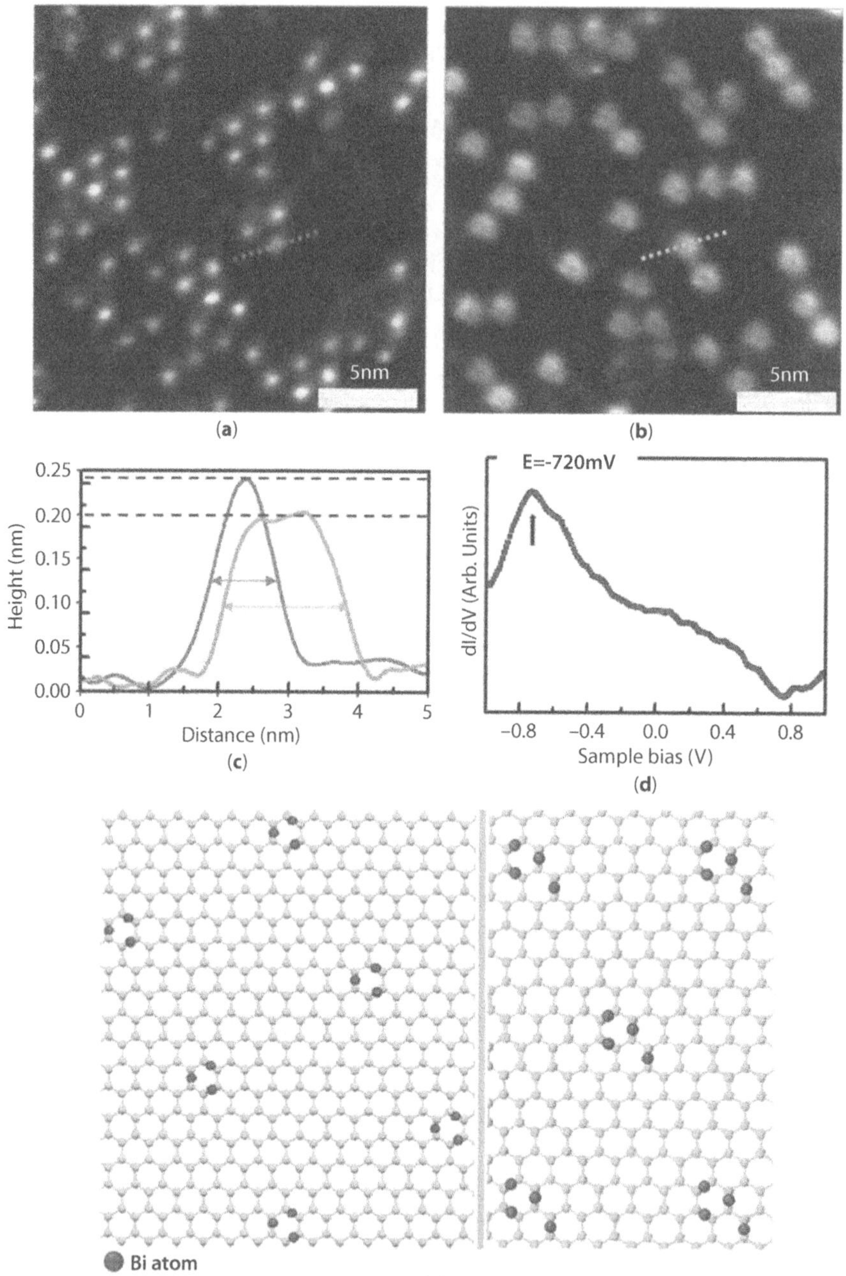

Figure 7.10 STM images of temperature effect. (a) STM image of 0.039 ML Bi coverage. (b) Annealed STM image of Figure 7.10a. (c) Line profile measurements of (a) and (b). (d) dI/dV spectrum of Bi NCs at room temperature. (e) The atomic model of 1D linear Bi NCs. NCs are consisted of 3 or 4 Bi adatoms, as displayed in left and right panels, respectively. The interatomic distance between NCs of 3 Bi adatoms is approximately 2.65 Å.

height and the full width at half maximum (FWHM) ~1 nm in lateral size while Bi NCs are ~0.20 nm height and the FWHM ~2 nm in lateral size, as shown in Figure 7.10c. These results reveal that the Bi NCs are located closer to MEG than are the Bi adatoms. Each Bi NC is likely an aggregation of 3~4 Bi adatoms. Hence, the 2D triangular islands are transformed into the 1D Bi NCs by annealing. Based on the DFT calculations, the atomic model of 1D linear arrangement of Bi NCs are displayed in Figure 7.10e. Notably, Bi adatoms in NCs remain located at B site. Moreover, both tri-atom and quad-atom configurations of the Bi NCs are possible according to the DFT results. Annealing at a higher elevated temperature (>600K) causes Bi NCs to dissociate into Bi atoms, most of which evaporate from MEG. The spectrum of NCs, as shown in Figure 7.10d, reveals a similar characteristic peak located at ~−0.72 eV (black arrow) to that in the dI/dV profile of the Bi adatom in hexagonal array, but with a larger FWHM, indicating a stronger feature of the p-state of Bi atoms in NCs. Based on the above, a well-controlled method for forming Bi-based low-dimensional structures can be developed. However, further DFT calculations and experimental works to study the effect of corrugated substrate are needed to carry out in the future.

7.4 The Energetically Favorable Distribution of Bi Adatoms Using First-Principles Calculations

7.4.1 Adsorption Energies of Various Bi Adsorption Sites on MEG

The atomic structure of buffer layer is a controversial subject. A pioneer work done by van Bommel *et al.* [49] shows that the graphite-like layers are bonded to the surface of SiC (0001). Another study introduces the 6 × 6 hexagonal reconstruction of the buffer layer observed in STM images [50–52]. This is different from the $(6\sqrt{3}\times 6\sqrt{3})$ R30° reconstruction revealed by the low-energy electron diffraction (LEED) patterns. [50, 53, 54] The geometric structure of these works is determined by the evidences in STM/LEED measurements and numerical calculations. Based on the experimental analysis [28, 29] and a detailed comparison with other theoretical results [33, 55], the long-range ripple structure due to the $(4\sqrt{3}\times 4\sqrt{3})$R30° reconstruction is confirmed in our study. The in-plane lattice constants are 3.06 Å, 2.30 Å, and 2.65 Å in our calculation for SiC, buffer layer, and graphene, respectively. The buffer layer is identified to be a rippled shape. They play an important role in the optimal geometric structures. This further induces the non-uniform van der Walls interactions between buffer and ML graphene, dominating the adsorption sites of Bi adatom (discussed in detail later).

Because of different stacking configurations, there exist a large number of silicon carbide polytypes [56–58]. The three commonly studied hexagonal polytypes include 2H-, 4H-, and 6H-SiC (0001), respectively, with AB, ABCB, and ABCACB stacking sequences. The six-layer 4H-SiC (0001) substrate in this calculation has a CBABCB stacking sequence, being the most energetically favorable structure [57]. The optimized structures show that the four-layer substrate (ABCB) exhibit has almost identical geometric properties compared with the six-layer (CBABCB) one, which largely reduces the computational resources. A buffer layer is in a periodic ripple shape after relaxation, as shown in region II of Figure 7.11. The troughs of the buffer layer bond with the silicon atoms of the substrate. The 11 Si atoms in the unit cell bond with the C atoms near the trough of the buffer layer have their Si-C

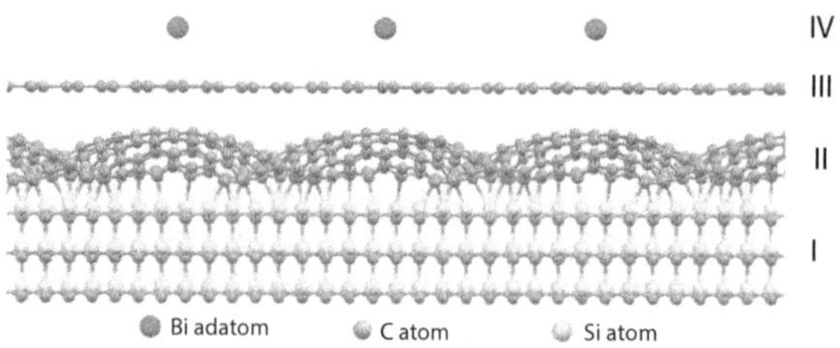

Figure 7.11 Geometric structures of silicon carbide substrate, buffer layer, monolayer graphene, and bismuth adatoms. C, Si, and Bi are, respectively, indicated by the gray, yellow, and red solid circles.

bond length in the range of 1.932 to 2.244 Å, while the other 16 Si atoms with long Si-C distances almost have no bonding with near C atom. Such covalent bonds mainly come from the strong hybridizations of the Si- and C-sp3 orbitals. The periodic ripple structure is almost identical to that revealed by the STM measurements [28]. The ML graphene, as indicated in region III, is almost with an extended C-C bond length of 1.50 Å. The interlayer distance between the buffer layer and the ML graphene varies from 3.21 Å to 5.45 Å at the crests and troughs, respectively. This clearly illustrates the non-uniform van der Waals interactions between them, and further dominates the distribution of the Bi adatoms. Bismuth atoms can be adsorbed on ML graphene in self-consistent calculations, as shown in region IV in Figure 7.11. Based on the different experimental environments, two kinds of adatom distributions have been observed, explicitly a uniform hexagonal distribution and bismuth NCs, which are dependent on the adsorption energies. The adsorption energy ΔE is calculated from the reduced energy due to the bismuth adsorption on graphene, which is very useful for understanding the optimal geometric structure. It is defined as

$$\Delta E = E_t - E_g - E_{bl} - E_{bi} \tag{7.3}$$

where E_t, E_g, E_{bl}, and E_{bi} are the total energies of the composite system, pristine ML graphene, buffer layer, and isolated bismuth atoms, respectively. Three kinds of most commonly studied adsorption sites with higher geometric symmetry are investigated, that is, the hollow, bridge, and top sites in Table 7.1. The bismuth atoms on the bridge and top sites have comparable adsorption energies with the former possessing the largest adsorption energy. The hollow site adsorption has the smallest ΔE, indicating that this configuration is less stable. This suggests that bismuth atoms are most likely to be observed at bridge sites. For the different adsorption sites, the optimal distance h between the adatoms and

Table 7.1 Adsorption energies and heights of various Bi adsorption sites.

Site	ΔE (eV)	h(Å)
B	1.3450	2.32
T	1.2904	2.34
H	0.6602	2.51

graphene surface is different. The bridge sites with shorter distance h indicate stronger interactions between adatoms and graphene, which account for the structural stability. The adatom height of 2.32 Å is consistent with that of STM measurements (Figure 7.2d).

The bismuth atoms distribution can be clearly explored by the ground-state energy. According to the above-mentioned results, the bridge site is a much more stable site within a hexagonal ring. The ground-state energies for the bridge sites in different hexagons are all calculated along the periodic armchair direction. These energies provide the information to determine the most stable position in ML graphene, as shown in Figure 7.12. These positions can be further categorized into three regions: gray, green, and blue regions. The blue hexagonal is closest to the crest of buffer layer with a distance of 3.21 Å; it possesses the lowest ground-state energy among all bridge sites. In order to compare it with the other bridge sites, the lowest ground-state energy is set to zero. Away from the blue region, the bridge sites between blue and green sticks have higher ground-state energies about 17–23 meV. The gray hexagons, which possess comparable ground-state energies to each other, reveal the highest energy difference in the range of 48–52 meV. Such energy variations strongly indicate that the Bi adatoms are hardly transported from the blue to the other regions at room temperature. The van der Walls interactions between Bi adatoms and the crests of the buffer layer are strongest at the blue hexagonal rings, which is responsible for the stability of Bi adatoms. The energy barrier of ~50 meV creates a potential well for Bi adatoms, which plays a critical role in the dramatic change of adatom distribution during the variation of temperature.

The ground-state energies and the adsorption energies are used to understand the large-scale pattern of Bi adatoms identified by experimental measurements. Bi-adsorbed ML graphene in a periodic hexagon pattern is shown in Figure 7.13a, where the most stable and unstable positions are represented by the blue and gray regions, respectively. The longest, shortest, and average distances between two Bi adatoms are, respectively, 18.1Å, 14.2Å, and 15.9Å. The former two are indicated by brown and black arrows between each blue region.

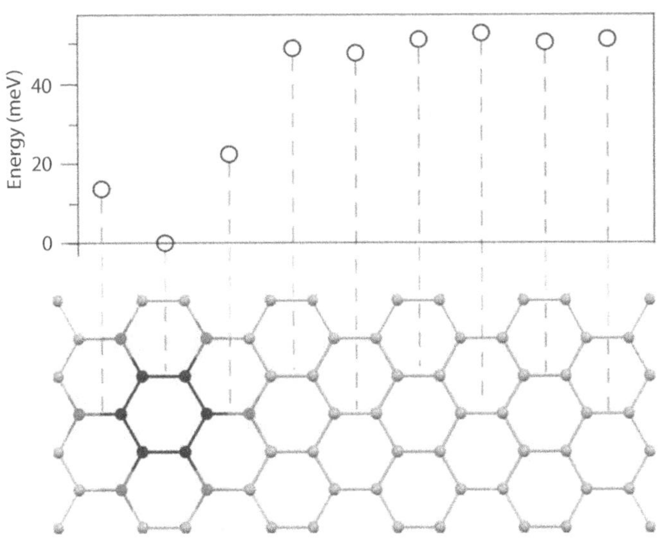

Figure 7.12 Ground-state energies of bismuth adsorption on different sites above monolayer graphene.

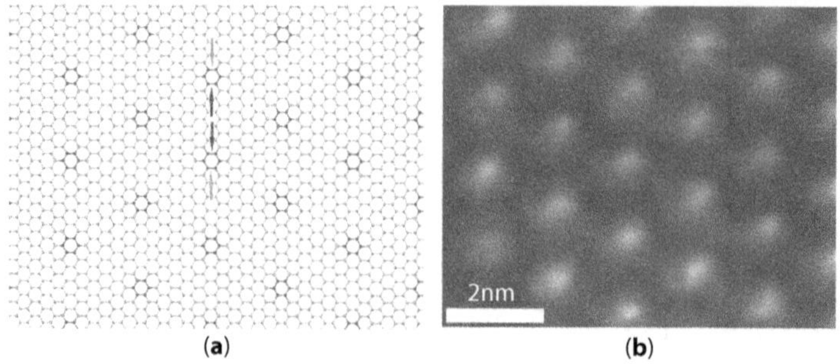

Figure 7.13 Geometric structures of hexagonal Bi array by the (a) DFT calculation and (b) STM measurement.

This indicates that the most possible interatomic distance lies in 14.2–18.1 Å. The large-scale hexagonal array of Bi atoms is clearly identified in Figure 7.13b with the STM measurement [28, 29]. It is determined that the interatomic distance is 15 Å and 16 Å for most of Bi adatoms via the statistical analysis, but 14 Å and 17 Å for some others. All the simulation results agree with the STM measurements. This illustrates that the periodic corrugation of the buffer layer influences the arrangement of the Bi adatoms.

7.4.2 The Interaction Energies and DOS of Various Bi NCs with Annealing Treatment

In addition to the large-scale hexagonal Bi pattern, there are metastable nanostructures induced by the annealing treatment [29]. During the treatment, many Bi-adatom NCs are observed in Figure 7.14a. Most of these are distributed in triangular and rectangular form. The NC configurations can be studied by Bi-Bi interaction energies and the ground-state energies of optimized structures, as shown in Table 7.2. All the Bi adatoms are at the bridge sites with the stronger Bi-C interactions, as indicated in Figure 7.14b for various adatom numbers (from 1 to 6). The higher number of adatoms results in the lower E_t. The stronger attractive Bi-Bi interactions are responsible for this result. The Bi-C interaction energy is

Table 7.2 The Bi-Bi interaction energies and total energies of various Bi NCs.

# Bi atoms	E_t **(eV)**	ΔE_{bi-bi} **(eV)**
1	-2317.86	x
2	-2322.14	-1.13
3	-2326.77	-1.62
4	-2330.60	-1.67
5	-2334.19	-1.65
6	-2337.86	-1.65

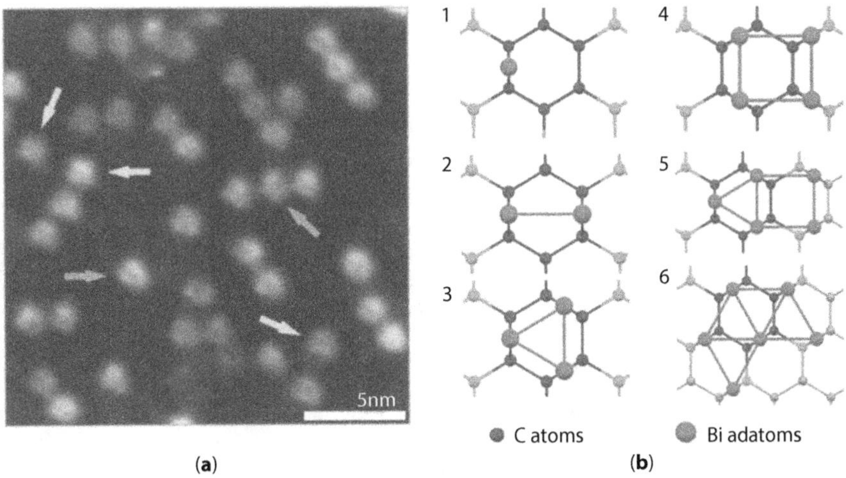

Figure 7.14 Geometric structures of Bi NCs by the (a) STM measurement and (b) DFT calculations. 3- and 4-adatoms are, respectively, indicated by the green and yellow arrows.

$E_{bi-c} = -1.01$ eV, which is obtained by subtracting the E_t of graphene from that of single-Bi-adsorbed graphene. The reduced energy due to the Bi-Bi interactions is

$$\Delta E_{bi-bi} = \left(E_t - E_g - \mathrm{E_{bl}} - nE_{bi} - \sum_{i=1}^{n} E_{bi-c} \right) / \mathrm{n} \tag{7.4}$$

ΔE_{bi-bi}'s in the 3-, 4-, 5-, and 6-adatom NCs are much lower than that of the 2-adatom one. This suggests that the former four are energetic favorable metastable structures. The experimental measurements reveal that most of the NCs are composed of 3 and 4 adatoms (green and yellow arrows in Figure 7.14a). The absence of 5- and 6-adatom NCs possibly originates from the insufficient bismuth coverage. It is noticed that the nearest distance between two Bi clusters is about 16 Å, which indicates that the energetic favorable adsorption positions of them correspond to the blue hexagonal rings in Figure 7.12.

The DOS can directly reflect the primary electronic properties, as shown in Figure 7.15a and b. DOS is finite at E = 0 for a hexagonal array of Bi-adsorbed graphene, and it exhibits a dip structure at low energy, and a peak at E ~ −0.6 eV (Figure 7.15a). The finite DOS at E = 0 indicates that there exists a certain amount of free carriers. The dip structure at −0.2 eV is ascribed to the Dirac point of graphene. The peak structure comes from the bismuth atoms, which is a critical factor to identify the existence of bismuth atoms. The STS measurements provide an accurate and prompt way to examine theoretical calculation, where the tunneling differential conductance of the dI/dV-V curve is proportional to the DOS. STS measurements exhibit the finite DOS at E = 0 and the small dip at low energy shows good agreement with the theoretical results, as shown in Figure 7.15b. The peak at ~ −0.7 V is similar to the peak structure in our calculation, which originates from the bismuth atoms.

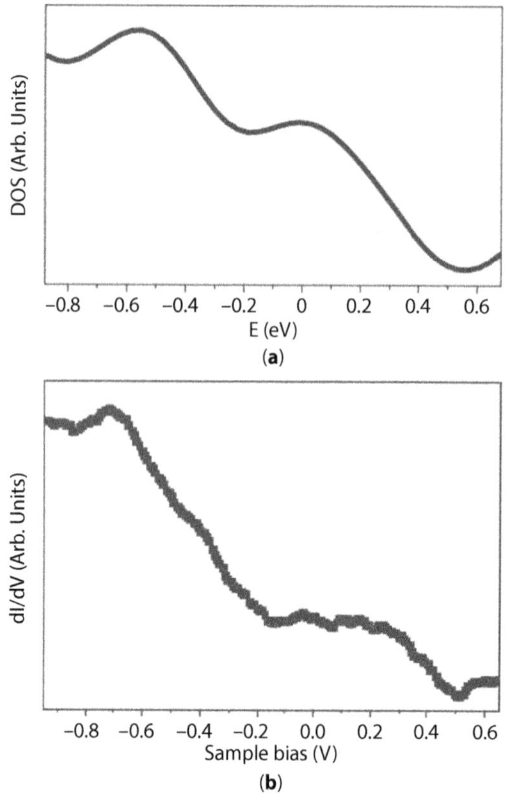

Figure 7.15 DOS for hexagonal Bi array by the (a) DFT calculation and (b) STS measurement.

7.5 Conclusion

This study investigated fundamental issues of nanotechnology and surface science in graphene-related field. The first part of the thesis explores the behaviors of Bi adatoms adsorbed on MEG formed on 4H-SiC (0001) at room temperature. At 0.0092 ML, large-scale linear structures can be detected in STM image, indicating possibly 1D growth mode. In order to clarify the Bi-Bi interactions of such structures, the pair distance distribution analysis between Bi adatoms is introduced by evaluating over 18,800 interadatom separations from a series of non-overlapping STM images and then transforming such distribution into the interaction potential between Bi adatoms, $E(r)$, as a function of distance between the interadatom separations r. The interaction potential reveals a clearly oscillatory long-range interaction. The oscillatory feature resembles much as Friedel oscillations. Within our STM experiments and DFT calculations, such interaction is explained by the mediation of graphene Dirac-like electrons and the effect of the corrugated substrate. The mediation of graphene Dirac-like electrons and the corrugated substrate are responsible to the observed oscillatory interaction with characteristic distribution distances and linear arrangements of Bi adatoms. Notably, the ratio of these two factors cannot be identified at present stage. Bi adatoms' growth on different substrates could be a crucial way to test these two factors. However, an interesting question that comes to our mind is, how about the different coverage and temperature effects of such system?

The second part of this work is then focused on the detailed physical insight of Bi/MEG system, which has elucidated the coverage and temperature-dependent structural transition of Bi adatoms adsorbed on MEG that is formed on 4H-SiC (0001) at room temperature. Within our STM and DFT calculations, Bi adatoms reveal an interestingly structural transition from 1D linear chains to 2D triangular islands, from 4 times to $2\sqrt{3}$ times the lattice spacing of Bi nanoribbon, as coverage increased from below 0.01 ML to above 0.03 ML. Such 2D growth mode is clearly affected by the corrugated substrate by FFT analysis. Tunneling spectroscopy measurements show a little charge transfer from MEG to Bi adatom, which is also confirmed in interaction energy analysis in the first part. Additionally, a characteristic peak (–0.72eV), corresponding to the p band of Bi, is detected upon deposition of Bi adatoms, suggesting that the Bi adatom forms a bound state at the bridge (B) site on MEG due likely to the Bi adatom–MEG interaction. After annealing to 500K, triangular Bi islands aggregate into uniform Bi NCs and reveal a stronger feature of the p-state of Bi atoms in NCs. Annealing at a higher elevated temperature (> 600K) causes Bi NCs to dissociate into Bi atoms, most of which evaporate from MEG.

In addition, the geometric and electronic properties of bismuth adsorption on ML graphene are studied by first-principles calculations. The six-layered substrate, periodically corrugated buffer layer, slightly deformed ML graphene, and adatom arrangements are calculated. The bismuth adsorption energies, ground-state energies, and Bi-Bi interaction energies for various Bi adatom configurations are investigated in detail. Importantly, the buffer layer has periodic non-uniform van der Walls interactions with the ML graphene, enriching the adatom distributions. The optimized structure presents a large-scale hexagonal array with an inter-adatom distance of 14 ~ 17 Å. The 50 meV energy barrier accounts for the room temperature stability. The stable and metastable structures are consistent with STM measurements after the annealing processes. The DOS exhibits a finite value at the Fermi level, a dip structure at E ~ –0.2 eV, and peak at E ~ –0.6 eV, originating from the free conduction electrons, the shifted Dirac-cone, and the bismuth-dominated electronic states, respectively. These features are consistent with those observed by the STS measurements.

In our work of Bi/MEG system, Bi-Bi interaction, Bi-MEG interaction, and Bi-substrate interaction coexist with different influences. The inability to clarify these three interactions will prevent the clear identification of the nucleation and growth of adatoms on MEG and limits the development of advanced devices with nanometer dimensions. Within our work, we observed fascinating phenomena such as the possibly metastable distribution in first part and a series of structure- and temperature-dependent transformation in second part, which is experimentally observed with some important features: (1) the mediation of graphene Dirac-like electron (Bi-Bi interaction); (2) weak charge transfer between Bi adatom and MEG (Bi-MEG interaction); (3) a bound state formed, between Bi adatom and MEG (Bi-MEG interaction), at B site; and (4) complex effect of corrugated substrate (Bi-substrate interaction). To our present understanding, when inert atoms are deposited on the inert substrate, many tiny and interesting effects can be enlarged and observed. Finally, Bi/MEG is a very interesting system, which can open many possibilities in various research and industrial fields. The approaches adopted in this study demonstrate the opportunity to engineer the nucleation and growth of Bi adatoms on MEG. The results also access to a new stepwise route for preparing self-assembled Bi-based low-dimensional structures. Most importantly, it demonstrates a reliable process toward multifunctional hybrid architectures for use in graphene-based devices at room temperature.

References

1. Geim, A.K. and Novoselov, K.S., The rise of graphene. *Nat. Mater.*, 6, 183–191, 2007.
2. Seyller, Th., Bostwick, A., Emtsev, K.V., Horn, K., Ley, L., McChesney, J.L., Ohta, T., Riley, J.D., Rotenberg, E., Speck, F., Epitaxial graphene: A new material. *Phys. Status Solidi. B*, 245, 1436–1446, 2008.
3. Weiss, N.O., Zhou, H., Liao, L., Liu, Y., Jiang, S., Huang, Y., Duan, X., Graphene: An emerging electronic material. *Adv. Mater.*, 24, 5782–5825, 2012.
4. Novoselov, K.S., Geim, A.K., Morozov, S.V., Jiang, D., Katsnelson, M.I., Grigorieva, I.V., Dubonos, S.V., Firsov, A.A., Two-dimensional gas of massless Dirac fermions in graphene. *Nature*, 438, 197–200, 2005.
5. Bogicevic, A., Ovesson, S., Hyldgaard, P., Lundqvist, B.I., Brune, H., Jennison, D.R., Nature, strength, and consequences of indirect adsorbate interactions on metals. *Phys. Rev. Lett.*, 85, 1910–1913, 2000.
6. Hyldgaard, P. and Persson, M., Long-ranged adsorbate–adsorbate interactions mediated by a surface-state band. *J. Phys: Condens. Matter*, 12, L13–L19, 2000.
7. Nanayakkara, S.U., Sykes, E.C., Fernández-Torres, L.C., Blake, M.M., Weiss, P.S., Long-range electronic interactions at a high temperature: Bromine adatom islands on Cu(111). *Phys. Rev. Lett.*, 98, 206108, 2007.
8. Brune, H., Epple, M., Hirstein, A., Schneider, M.A., Kern, K., Long-range adsorbate interactions mediated by a two-dimensional electron gas. *Phys. Rev. B*, 65, 115420, 2002.
9. Stepanyuk, V.S., Baranov, A.N., Tsivlin, D.V., Hergert, W., Bruno, P., Knorr, N., Schneider, M.A., Kern, K., Quantum interference and long-range adsorbate-adsorbate interactions. *Phys. Rev. B*, 68, 205410, 2003.
10. Ternes, M., Pivetta, M., Patthey, F., Schneider, W.D., Creation, electronic properties, disorder, and melting of two-dimensional surface-state-mediated adatom superlattices. *Prog. Surf. Sci.*, 85, 1–27, 2010.
11. Kamna, M.M., Stranick, S.J., Weiss, P.S., Imaging substrate-mediated interactions. *Science*, 274, 118–119, 1996.
12. Merrick, M.L., Luo, W.W., Fichthorn, K.A., Substrate-mediated interactions on solid surfaces: Theory, experiment, and consequences for thin-film morphology. *Prog. Surf. Sci.*, 72, 117–134, 2003.
13. Kim, K., Choi, J.Y., Kim, T., Cho, S.H., Chung, H.J., A role for graphene in silicon-based semiconductor devices. *Nature*, 479, 338–344, 2011.
14. Neto, A.H.C., Guinea, F., Peres, N.M.R., Novoselov, K.S., Geim, A.K., The electronic properties of graphene. *Rev. Mod. Phys.*, 81, 109, 2009.
15. LeBohec, S., Talbot, J., Mishchenko, E.G., Attraction-repulsion transition in the interaction of adatoms and vacancies in graphene. *Phys. Rev. B*, 89, 045433, 2014.
16. Shytov, A.V., Abanin, D.A., Levitov, L.S., Long-range interaction between adatoms in graphene. *Phys. Rev. Lett.*, 103, 016806, 2009.
17. de Laissardiere, G.T. and Mayou, D., Conductivity of graphene with resonant and nonresonant adsorbates. *Phys. Rev. Lett.*, 111, 146601, 2013.
18. Mkhitaryan, V.V. and Mishchenko, E.G., Resonant finite-size impurities in graphene, unitary limit, and Friedel oscillations. *Phys. Rev. B*, 86, 115442, 2012.
19. Aktürk, O.U. and Tomak, M., Bismuth doping of graphene. *Appl. Phys. Lett.*, 96, 081914, 2010.
20. Eelbo, T., Waśniowska, M., Thakur, P., Gyamfi, M., Sachs, B., Wehling, T.O., Forti, S., Starke, U., Tieg, C., Lichtenstein, A.I., Wiesendanger, R., Adatoms and clusters of 3d transition metals on graphene: Electronic and magnetic configurations. *Phys. Rev. Lett.*, 110, 136804, 2013.

21. Gyamfi, M., Eelbo, T., Waśniowska, M., Wehling, T.O., Forti, S., Starke, U., Lichtenstein, A.I., Katsnelson, M.I., Wiesendanger, R., Orbital selective coupling between Ni adatoms and graphene Dirac electrons. *Phys. Rev. B*, 85, 161406(R), 2012.
22. Liu, X., Wang, C.Z., Hupalo, M., Lin, H.Q., Ho, K.M., Tringides, M.C., Metals on graphene: Interactions, growth morphology, and thermal stability. *Crystals*, 3, 79, 2013.
23. Marchenko, D., Varykhalov, A., Scholz, M.R., Bihlmayer, G., Rashba, E.I., Rybkin, A., Shikin, A.M., Rader, O., Giant Rashba splitting in graphene due to hybridization with gold. *Nat. Commun.*, 3, 1232, 2012.
24. Binz, S.M., Hupalo, M., Liu, X., Wang, C.Z., Lu, W.C., Thiel, P.A., Ho, K.M., Conrad, E.H., Tringides, M.C., High island densities and long range repulsive interactions: Fe on epitaxial graphene. *Phys. Rev. Lett.*, 109, 026103, 2012.
25. Liu, X., Wang, C.Z., Hupalo, M., Lu, W.C., Thiel, P.A., Ho, K.M., Tringides, M.C., Fe-Fe adatom interaction and growth morphology on graphene. *Phys. Rev. B*, 84, 235446, 2011.
26. Song, C.L., Sun, B., Wang, Y.L., Jiang, Y.P., Wang, L., He, K., Chen, X., Zhang, P., Ma, X.C., Xue, Q.K., Charge-transfer-induced cesium superlattices on graphene. *Phys. Rev. Lett.*, 108, 156803, 2012.
27. Hsu, C.H., Ozolins, V., Chuang, F.C., First-principles study of Bi and Sb intercalated graphene on SiC (0001) substrate. *Surf. Sci.*, 616, 149–154, 2013.
28. Chen, H.H., Su, S.H., Chang, S.L., Cheng, B.Y., Chen, S.W., Chong, C.W., Huang, J.C.A., Lin, M.F., Long-range interactions of bismuth growth on monolayer epitaxial graphene at room temperature. *Carbon*, 93, 180–186, 2015.
29. Chen, H.H., Su, S.H., Chang, S.L., Cheng, B.Y., Chen, S.W., Chen, H.Y., Lin, M.F., Huang, J.C.A., Tailoring low-dimensional structures of bismuth on monolayer epitaxial graphene. *Sci. Rep.*, 5, 11623, 2015.
30. Poon, S.W., Chen, W., Tok, E.S., Wee, A.T.S., Probing epitaxial growth of graphene on silicon carbide by metal decoration. *Appl. Phys. Lett.*, 92, 104102, 2008.
31. Balog, R., Jørgensen, B., Wells, J., Laegsgaard, E., Hofmann, P., Besenbacher, F., Hornekaer, L., Atomic hydrogen adsorbate structures on graphene. *J. Am. Chem. Soc.*, 131, 8744–8745, 2009.
32. Merino, P., Švec, M., Martínez, J.I., Mutombo, P., Gonzalez, C., Martín-Gago, J.A., de Andres, P.L., Jelinek, P., Ortho and para hydrogen dimers on G/SiC(0001): Combined STM and DFT study. *Langmuir*, 31, 233–239, 2015.
33. Varchon, F., Mallet, P., Veuillen, J., Magaud, L., Ripples in epitaxial graphene on the Si-terminated SiC (0001) surface. *Phys. Rev. B*, 77, 235412, 2008.
34. Joucken, F., Frising, F., Sporken, R., Fourier transform analysis of STM images of multilayer graphene moiré patterns. *Carbon*, 83, 48–52, 2015.
35. Mallet, P., Brihuega, I., Bose, S., Ugeda, M.M., Gómez-Rodríguez, J.M., Kern, K., Veuillen, J.Y., Role of pseudospin in quasiparticle interferences in epitaxial graphene probed by high-resolution scanning tunneling microscopy. *Phys. Rev. B*, 86, 045444, 2012.
36. Brar, V.W., Wickenburg, S., Panlasigui., M., Park, C.H., Wehling, T.O., Zhang, Y., Decker, R., Girit, C., Balatsky, A.V., Louie, S.G., Zettl, A., Crommie, M.F., Observation of carrier-density-dependent many-body effects in graphene via tunneling spectroscopy. *Phys. Rev. Lett.*, 104, 036805, 2010.
37. Sun, J.T., Huang, H., Wong, S.L., Gao, H.J., Feng, Y.P., Wee, A.T., Energy-gap opening in a Bi (110) nanoribbon induced by edge reconstruction. *Phys. Rev. Lett.*, 109, 246804, 2012.
38. Lai, J.H., Su, S.H., Chen, H.H., Huang, J.C.A., Wu, C.L., Stabilization of ZnO polar plane with charged surface nanodefects. *Phys. Rev. B*, 82, 155406, 2010.
39. Chen, G.H. and Raikh, M.E., Exchange-induced enhancement of spin-orbit coupling in two-dimensional electronic systems. *Phys. Rev. B*, 60, 4826, 1999.

40. Huang, H., Wong, S.L., Wang, Y., Sun, J.T., Gao, X., Wee, A.T.S., Scanning tunneling microscope and photoemission spectroscopy investigations of bismuth on epitaxial graphene on SiC (0001). *J. Phys. Chem. C*, 118, 24995–24999, 2014.
41. Varchon, F., Feng, R., Hass, J., Li, X., Nguyen, B.N., Naud, C., Mallet, P., Veuillen, J.Y., Berger, C., Conrad, E.H., Magaud, L., Electronic structure of epitaxial graphene layers on SiC: Effect of the substrate. *Phys. Rev. Lett.*, 99, 126805, 2007.
42. Girard, Y., Chacon, C., de Abreu, G., Lagoute, J., Repain, V., Rousset, S., Growth of Bi on Cu (111): Alloying and dealloying transitions. *Surf. Sci.*, 617, 118–123, 2013.
43. Zhang, K.H.L., McLeod, I.M., Lu, Y.H., Dhanak, V.R., Matilainen, A., Lahti, M., Pussi, K., Egdell, R.G., Wang, X.S., Wee, A.T.S., Chen, W., Observation of a surface alloying-to-dealloying transition during growth of Bi on Ag (111). *Phys. Rev. B*, 83, 235418, 2011.
44. Kato, C., Aoki, Y., Hirayama, H., Scanning tunneling microscopy of Bi-induced Ag (111) surface structures. *Phys. Rev. B*, 82, 165407, 2010.
45. Chan, K.T., Neaton, J.B., Cohen, M.L., First-principles study of metal adatom adsorption on graphene. *Phys. Rev. B*, 77, 235430, 2008.
46. Yaginuma, S., Nagaoka, K., Nagao, T., Bihlmayer, G., Koroteev, Y.M., Chulkov, E.V., Nakayama, T., Electronic structure of ultrathin bismuth films with A7 and black-phosphorus-like structures. *J. Phys. Soc. Jpn.*, 77, 014701, 2008.
47. Bobaru, S., Gaudry, É., de Weerd, M.C., Ledieu, J., Fournée, V., Competing allotropes of Bi deposited on the Al13Co4 (100) alloy surface. *Phys. Rev. B*, 86, 214201, 2012.
48. Koroteev, Y.M., Bihlmayer, G., Chulkov, E.V., Blügel, S., First-principles investigation of structural and electronic properties of ultrathin Bi films. *Phys. Rev. B*, 77, 045428, 2008.
49. Van Bommel, A.J., Crombeen, J.E., Van Tooren, A., LEED and Auger electron observations of the SiC(0001) surface. *Surf. Sci.*, 48, 463–472, 1975.
50. Berger, C., Song, Z., Li, X., Wu, X., Brown, N., Naud, C., Mayou, D., Li, T., Hass, J., Marchenkov, A.N., Conrad, E.H., First, P.N., de Heer, W.A., Electronic confinement and coherence in patterned epitaxial graphene. *Science*, 312, 1191–1196, 2006.
51. Brar, V.W., Zhang, Y., Yayon, Y., Ohta, T., McChesney, J.L., Bostwick, A., Rotenberg, E., Horn, K., Crommie, M.F., Scanning tunneling spectroscopy of inhomogeneous electronic structure in monolayer and bilayer graphene on SiC. *Appl. Phys. Lett.*, 91, 122102, 2007.
52. Mallet, P., Varchon, F., Naud, C., Magaud, L., Berger, C., Veuillen, J.Y., Electron states of mono- and bilayer graphene on SiC probed by scanning-tunneling microscopy. *Phys. Rev. B*, 76, 041403(R), 2007.
53. Chen, W., Xu, H., Liu, L., Gao, X., Qi, D., Peng, G., Tan, S.C., Feng, Y., Loh, K.P., Wee, A.T.S., Atomic structure of the 6H-SiC nanomesh. *Surf. Sci.*, 596, 176–186, 2005.
54. De Heer, W.A., Berger, C., Wu, X., First, P.N., Conrad, E.H., Li, X., Li, T., Sprinkle, M., Hass, J., Sadowski, M.L., Potemski, M., Martinez, G., Epitaxial graphene. *Solid State Commun.*, 143, 92–100, 2007.
55. Mattausch, A. and Pankratov, O., *Ab initio* study of graphene on SiC. *Phys. Rev. Lett.*, 99, 076802, 2007.
56. Käckell, P., Furthmüller, J., Bechstedt, F., Polytypic transformations in SiC: An ab initio study. *Phys. Rev. B*, 60, 13261, 1999.
57. Righi, M.C., Pignedoli, C.A., Borghi, G., Di Felice, R., Bertoni, C.M., Catellani, A., Surface-induced stacking transition at SiC (0001). *Phys. Rev. B*, 66, 045320, 2002.
58. Sołtys, J., Piechota, J., Łopuszyński, M., Krukowski, S., A comparative DFT study of electronic properties of 2H-, 4H- and 6H-SiC (0001) and SiC (000-1) clean surfaces: Significance of the surface Stark effect. *New J. Phys.*, 12, 043024, 2010.

8

Strain Engineering: Electromechanical Properties of Graphene

Shuze Zhu

Department of Chemical Engineering, Massachusetts Institute of Technology, Cambridge, MA, USA

Abstract

The charge carriers in graphene behave like massless Dirac fermions, and graphene shows ballistic charge transport, turning it into an ideal material for circuit fabrication. However, graphene lacks a bandgap around the Fermi level, which causes challenge to controlling the conductivity by electronic means and realization of graphene-based electronic devices. Since graphene's electronic structure is strongly tied to the symmetry of its hexagonal crystal lattice, the change or breaking of its crystal lattice symmetry via mechanical deformations, can alter its electronic properties. This opens up fertile opportunities to leverage the electromechanical properties of graphene, which is also known as strain engineering. Finite bandgap could be engineered from the confinement of charge carriers induced by strain-induced pseudomagnetic fields in graphene, which has been explored as a potential approach to engineering the electronic states of graphene and enabling the design of next-generation graphene-based nanoelectronics. In this chapter, the role of mechanical strain in electromechanical properties of graphene is discussed, in the context of theoretical derivation, computational insights, experimental results, and application perspectives.

Keywords: Graphene, strain engineering, electronic properties, bandgap, pseudomagnetic field

8.1 The Era of Strain Engineering of Graphene

Carbon atoms are able to densely pack into a honeycomb lattice in a plane with an interatomic distance approximately 0.142 nm. They form a two-dimensional crystal that has inspired an outburst of research endeavors in various fields for over a decade. This groundbreaking material is called graphene, as epitomized by the 2010 Nobel Prize in Physics, opening up a new era in condensed matter physics, materials science, and many other engineering fields. Without a doubt, it is the unprecedented combination of a variety of superior material properties in one that makes graphene superior in advancing the boundary of human knowledge. The amazing material properties are mostly due to the unique nature of graphene's two-dimensional crystal structure.

Email: shuzezhu@mit.edu

Tobias Stauber (ed.) Handbook of Graphene: Volume 2, (219–244) © 2019 Scrivener Publishing LLC

Graphene is the strongest material ever discovered [1]. Its fracture strain is up to 25% [1, 2], making graphene one of the highly deformable crystals, in the sense that it can avoid inelastic relaxation up to a significant fraction of their ideal strength.

Besides the exceptional mechanical properties, graphene has the highest electron mobility ever measured, which is 100 times higher than that of silicon at room temperature [3]. Graphene also has the lowest known resistivity at room temperature, less than that of silver [4].

However, intrinsic graphene is a zero-gap semiconductor [5], which poses challenges for their application in electronics, where typically bandgaps are essential. To create gapped electronic states, it is found that graphene nanoribbon with a finite width shows a completely different band structure [6, 7]. Experimentally a 200 meV energy gap can be achieved by narrowing graphene nanoribbon to a width of 15 nm [8]. Therefore, cutting graphene into ribbons remains an effective method for engineering graphene's electronic states.

In graphene, the electrons are confined in a two-dimensional space, an ideal platform to experimentally probe the quantum Hall effect [9, 10], which is a typical quantum-mechanically enhanced transport phenomena when electrons are confined in two-dimensional materials. Such effect results from the influence of external magnetic field on the states of charge carriers in graphene. In two dimensions, when classical electrons are subjected to a magnetic field, they follow circular cyclotron orbits. Treated quantum mechanically, these orbits are quantized. The energy levels of these quantized orbitals take on discrete values, known as the Landau levels. Essentially, energy gaps are created between Landau levels, which could be utilized toward graphene-based electronic and magnetoelectronic devices. Nevertheless, creating strong magnetic field at nanoscale spatial resolution is not a trivial task.

Because graphene is a highly deformable crystal, and its electronic properties are strongly tied to its lattice structure, a tempting prospect of engineering graphene's electronic properties is through mechanical deformation.

In 2010, N. Levy *et al.* reported the creation of pseudomagnetic fields far stronger than the strongest magnetic fields ever sustained in a laboratory [11], in highly strained graphene nanobubbles. This was about the first time that the behaviors of electrons as if subjected to extreme magnetic fields in excess of 300 T were accessed. Such an intensity is far beyond the capacity of normal laboratory environments. Testing materials in magnetic fields to investigate how electrons behave has been ongoing for many decades, but it's prohibitive to sustain tremendously strong magnetic fields in a laboratory setting. When stronger fields are created, the magnets would experience extremely strong forces, which could fracture themselves. Pseudomagnetic field in graphene is a brand new physical effect. The ability to make electrons in graphene behave as if they were in magnetic fields of extreme intensity, unattainable in laboratory condition, just from mechanical deformation, offers a new path on a source of important electronics applications and fundamental scientific discoveries.

8.2 Electronic Dispersion of Graphene and Dirac Fermions

To lay out the physical picture of strain-induced pseudomagnetic field in graphene, it is essential to revisit one of the most intriguing signatures of the electronic dispersion spectrum of graphene, the Dirac cone.

The unit cell of graphene lattice contains two sublattice sites (Figure 8.1a). Each site has three nearest neighbors, and the distances to each neighbor are equal. In the tight binding approximation, if we consider only the first nearest neighbor, and only the out-of-plane p_z orbitals, the Hamiltonian of graphene can be expressed as

$$\boldsymbol{H}=\begin{bmatrix} 0 & f(\vec{k}) \\ f^*(\vec{k}) & 0 \end{bmatrix} \tag{8.1}$$

where

$$f(\vec{k})=-\sum_{j=1}^{3} t_j \cdot e^{i\,\vec{k}\cdot\vec{\delta}_j}$$

where $t_j=(pp\pi)_j=\langle p_z^A|\hat{H}|p_z^B\rangle_j$, is a Slater–Koster parameter [12] denoting the energy integral between the *jth* (*j=1,2,3*) first nearest-neighboring p_z orbitals of carbon atoms.

It is important to note that such energy integrals for all the three nearest neighbors are the same for graphene in undeformed state, and can therefore be denoted simply as *t*. Therefore, Hamiltonian of graphene in undeformed state can be simplified as

$$\boldsymbol{H}=\begin{bmatrix} 0 & -t\cdot g(\vec{k}) \\ -t\cdot g^*(\vec{k}) & 0 \end{bmatrix} \tag{8.2}$$

where

$$g(\vec{k})=\sum_{j=1}^{3} e^{i\,\vec{k}\cdot\vec{\delta}_j}$$

Hereafter, the hopping vectors $\vec{\delta}_j$ are defined as (Figure 8.1a),

$$\vec{\delta_1}=\frac{a}{2}(\sqrt{3},1)$$

$$\vec{\delta_2}=\frac{a}{2}(-\sqrt{3},1)$$

$$\vec{\delta_3}=a(0,-1) \tag{8.3}$$

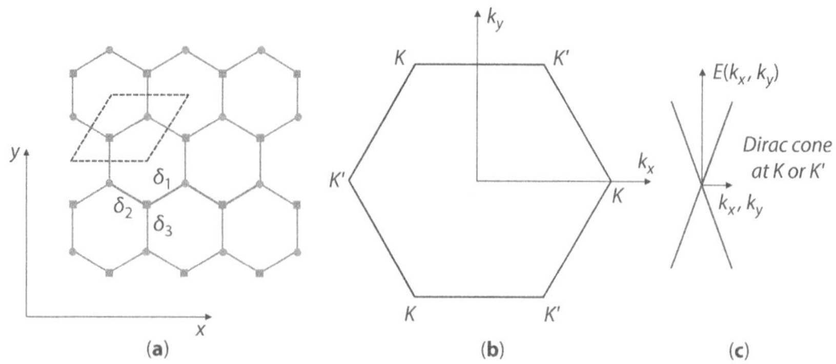

Figure 8.1 (a) Graphene lattice in real space showing two sublattices denoted by red square and green circle. The dashed box denotes the unit cell. δ_1, δ_2, δ_3 are three hopping vectors. The x-axis is along zigzag direction. (b) Schematic showing the six Dirac points located at the corner of the hexagonal Brillouin zone in reciprocal space. These six points can be classified into two inequivalent points, denoted as K and K'. (c) Schematic showing the electronic dispersion relation of graphene around Dirac cone.

where the x-axis is chosen along zigzag direction, and a is the equilibrium nearest-neighbor distance.

The energy dispersion can be calculated from the secular equation

$$\begin{vmatrix} -E & -t\, g(\vec{k}) \\ -t\, g^*(\vec{k}) & -E \end{vmatrix} = 0 \tag{8.4}$$

We can obtain

$$E(\vec{k}) = \pm t \left| g(\vec{k}) \right| \tag{8.5}$$

The Dirac cone can be visualized by plotting this result in reciprocal space. The Dirac points are located at the corner of the hexagonal Brillouin zone (Figure 8.1b), where the energy is assigned to zero (Fermi energy). It can be seen that the electronic dispersion is approximately linear around the Dirac points (Figure 8.1c). Due to this property, we can introduce a linear approximation, which is called the Dirac approximation, to simplify the electronic dispersion. The basic idea is that the $g(\vec{k})$ can be rewritten as $g_D(\vec{q})$, where $\vec{q}$ is a vector in reciprocal space, measured relative to the Dirac point (**K** point, or $\vec{K}$), so that $\vec{q} = \vec{k} - \vec{K}$, we can have

$$g_D(\vec{q}) = g(\vec{q} + \vec{K}) = \sum_{j=1}^{3} e^{i(\vec{q}+\vec{K})\cdot\vec{\delta}_j} = \sum_{j=1}^{3} e^{i\,\vec{q}\cdot\vec{\delta}_j} e^{i\vec{K}\cdot\vec{\delta}_j} \tag{8.6}$$

There are in fact two inequivalent Dirac points [13], which we denote as $\vec{K}$ (**K** point) and $\vec{K}'$ (**K′** point).

Using $\vec{K}=\left(\frac{4\pi}{3\sqrt{3}a},0\right)$, and expanding $e^{i\,\vec{q}\cdot\vec{\delta}_j}$ as a linear approximation near $\vec{q}=\vec{0}$, one can arrive at the low-energy effective Dirac Hamiltonian around $\vec{K}$ for strain-free graphene as [13]

$$\boldsymbol{H}_{\vec{K}}=\hbar v_F\begin{bmatrix}0 & q_x-iq_y\\ q_x+iq_y & 0\end{bmatrix} \tag{8.7}$$

where $v_F=\frac{3ta}{2\hbar}$ is the pristine Fermi velocity.

The energy dissipation becomes

$$E_{\vec{K}}=\pm\hbar v_F\,\vec{q} \tag{8.8}$$

Similarly, using $\vec{K}'=\left(-\frac{4\pi}{3\sqrt{3}a},0\right)$, we can have [13]

$$\boldsymbol{H}_{\vec{K}'}=\hbar v_F\begin{bmatrix}0 & q_x+iq_y\\ q_x-iq_y & 0\end{bmatrix} \tag{8.9}$$

$$E_{\vec{K}'}=\pm\hbar v_F\,\vec{q} \tag{8.10}$$

8.3 Dirac Fermions in External Magnetic Field and Landau Levels

In classical physics, electrons in a magnetic field travel in circles called cyclotron orbits. Quantum mechanically, cyclotron orbits become quantized and exhibit discrete energy levels, which are called Landau levels. They correspond to energies where constructive interference occurs in an orbiting electron's quantum wave function.

To show this in graphene, we can rewrite the Dirac Hamiltonian around **K** point in the following more compact format

$$\boldsymbol{H}_{\vec{K}}=\hbar v_F\,\vec{\sigma}\cdot\vec{q}=v_F\,\vec{\sigma}\cdot\vec{p} \tag{8.11}$$

where $\vec{\sigma} = (\sigma_x, \ \sigma_y)$, $\boldsymbol{\sigma}_x$ and $\boldsymbol{\sigma}_y$ are the Pauli matrices, and $\vec{p}$ is the momentum measured from ***K*** point.

In the presence of a magnetic field, in momentum space, we can make the substitution $\vec{p} \to \vec{P} = \vec{p} - e\vec{A}$, where e is the electron charge, and $\vec{A}$ is the vector potential such that $\vec{\nabla} \times \vec{A} = B\vec{e}_z$ is the magnetic field, which is considered to be constant. The corresponding vector potential is $\vec{A} = B(-y, 0)$. Please note that in real space, this substitution is done by using the minimal coupling (i.e., replacing $\vec{p} = -i\hbar\vec{\nabla}$ by $\vec{P} = -i\hbar\vec{\nabla} - e\vec{A}$, taking effective mass approximation).

Then the Dirac Hamiltonian will change to

$$\boldsymbol{H}_{\vec{K}} = v_F \ \vec{\sigma} \cdot \vec{P} = v_F \begin{bmatrix} 0 & P_x - iP_y \\ P_x + iP_y & 0 \end{bmatrix} \tag{8.12}$$

The two-component wave function is denoted as

$$\Psi = \begin{pmatrix} \Phi_A^K \\ \Phi_B^K \end{pmatrix}$$

Φ_A^K and Φ_B^K are the wave functions of sublattice A and B for the valley at ***K*** point.

The stationary equation to be solved is

$$\boldsymbol{H}_{\vec{K}} \ \Psi = \mathrm{E}\Psi$$

The eigen energies for Φ_B^K can be solved as [13–15]

$$E_{n|\Phi_B^K} = sgn(n) v_F \sqrt{2e\hbar B|n|} \tag{8.13}$$

where $n = 0, \pm 1, \pm 2, \ldots$ is the Landau level index. The positive values correspond to electrons (conduction band), while the negative values correspond to holes (valence band).

The eigen energies for Φ_A^K can be solved as

$$E_{n|\Phi_A^K} = sgn(n+1) v_F \sqrt{2e\hbar B|n+1|} \tag{8.14}$$

where the Landau level indexes satisfy $|n + 1| \geq 1$.

Similarly, for the valley at ***K'*** point, the eigenenergies for $\Phi_A^{K'}$ can be solved as

$$E_{n|\Phi_A^{K'}} = sgn(n) v_F \sqrt{2e\hbar B|n|} \tag{8.15}$$

where $n = 0, \pm 1, \pm 2, \ldots$ is the Landau level index.

The eigen energies for $\Phi_B^{K'}$ can be solved as

$$E_{n|\Phi_B^{K'}} = sgn(n+1)v_F\sqrt{2e\hbar B|n+1|} \tag{8.16}$$

where the Landau level indexes satisfy $|n + 1| \geq 1$.

There are several interesting observations from these results. First, the gap between neighboring Landau energies is index dependent. Considering the linear scaling of Landau level energy with respect to $sgn(n)\sqrt{|n|}$, the largest energy gap is between the zero and the first Landau level. This large spacing facilitates the observation of the quantum Hall effect in graphene. Apparently, the number of electrons occupying each Landau level depends on the strength of the magnetic field. The stronger the field, the more energy spacing between Landau levels, and the denser the electron states become at each level. Second, by characterizing the scaling of the Landau level energies, it is quite straightforward to calculate the magnitude of applied magnetic field. As to be shown later, these two observations also hold for the strain-induced pseudomagnetic fields in graphene.

From the symmetry relation $E_{n|\Phi_B^{K'}} \overset{sym}{=} E_{n|\Phi_A^{K}}$ and $E_{n|\Phi_A^{K'}} \overset{sym}{=} E_{n|\Phi_B^{K}}$, we can see that although different valleys contribute different amount to the local density of states of a particular sublattice, the overall sublattice equivalence is preserved when the contributions from both valleys are combined (i.e., $E_{n|\Phi_B^{K'}} + E_{n|\Phi_B^{K}} \overset{sym}{=} E_{n|\Phi_A^{K}} + E_{n|\Phi_A^{K'}}$). Nevertheless, as to be shown later, this overall sublattice equivalence does not hold for the strain-induced pseudomagnetic fields, which is coupled to its nature of time reversal symmetry.

8.4 Dirac Hamiltonian of Graphene in Strain Field and Pseudomagnetic Field

To generalize the above discussion, the Hamiltonian in the tight binding approximation of graphene considering only the first nearest neighbor is given by

$$H = -\sum_{i,j} t_0 \, a_i^\dagger b_j + h.c., \tag{8.17}$$

where t_0 is the hopping parameter, a_i and $a_i^\dagger$ (b_i and $b_i^\dagger$) are the annihilation and creation operators for an electron on sublattice A (B).

To offer the most simplified picture of the strain effect, we assume that when the mechanical deformation is in play, only the hopping energy integral changes due to the changing interatomic distance. Let's first assume the modification is linear, given by [16]

$$t_n = t_0 (1 + \Delta_n) \tag{8.18}$$

Next, applying the Fourier transformation of the creation and annihilation operators, we have

$$a_i = \frac{1}{\sqrt{N}} \sum_{\vec{k}} e^{i\,\vec{k}\cdot\vec{R_i}} a_{\vec{k}}$$

$$b_i = \frac{1}{\sqrt{N}} \sum_{\vec{k}} e^{i\,\vec{k}\cdot\vec{R_i}} b_{\vec{k}} \tag{8.19}$$

Then, we obtain the strained Hamiltonian as

$$H = -\sum_{n,\vec{k}} t_n e^{-i\,\vec{k}\cdot\vec{\delta_n}}\, a_{\vec{k}}^{\dagger} b_{\vec{k}} + h.c., \tag{8.20}$$

Expanding around $\boldsymbol{K}$ point to the linear order of strain and wave vector, and collecting terms, we have [17]

$$\begin{aligned} H &= -\sum_{n,\vec{k}} t_n e^{-i\vec{k}\cdot\vec{\delta_n}} a_{\vec{k}}^{\dagger} b_{\vec{k}} + h.c. = -\sum_{n=1}^{3} t_n \begin{pmatrix} 0 & e^{-i(\vec{K}+\vec{q})\cdot\vec{\delta_n}} \\ e^{i(\vec{K}+\vec{q})\cdot\vec{\delta_n}} & 0 \end{pmatrix} \\ &= -\sum_{n=1}^{3} t_0(1+\Delta_n) \begin{pmatrix} 0 & e^{-i(\vec{K})\cdot\vec{\delta_n}} \\ e^{i(\vec{K})\cdot\vec{\delta_n}} & 0 \end{pmatrix} \times (\boldsymbol{I} + i\sigma_z \vec{q}\cdot\vec{\delta_n}) \\ &= -\sum_{n=1}^{3} t_0(1+\Delta_n) \left(i\frac{\vec{\sigma}\cdot\vec{\delta_n}}{a}\sigma_z \right) \times (\boldsymbol{I} + i\sigma_z \vec{q}\cdot\vec{\delta_n}) \\ &= -\sum_{n=1}^{3} t_0 \left(i\frac{\vec{\sigma}\cdot\vec{\delta_n}}{a}\sigma_z \right) \times [\boldsymbol{I} + i\sigma_z \vec{q}\cdot\vec{\delta_n} + \Delta_n \boldsymbol{I}] \\ &= \hbar v_F \vec{\sigma}\cdot\vec{q} - \hbar v_F \vec{\sigma}\cdot\overrightarrow{A_{sk}} \end{aligned} \tag{8.21}$$

We arrive at the following Hamiltonian

$$H = \hbar v_F\, \vec{\sigma}\cdot(\vec{q} - \overrightarrow{A_{sk}}) \tag{8.22}$$

where we define $\overrightarrow{A_{sk}}$ as a vector in the reciprocal space associated with the term Δ_n.

The above equation suggests that if Δ_n is not zero, then the mechanical deformation is possible to shift the $\boldsymbol{K}$ point with the vector $\overrightarrow{A_{sk}}$ (Figure 8.2).

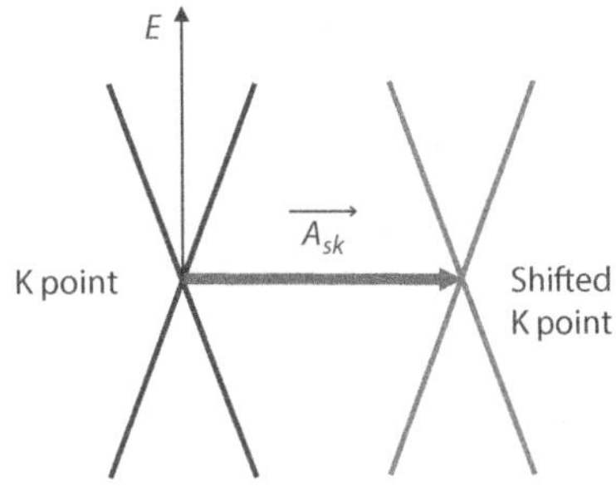

Figure 8.2 Dirac cone shift at K point before (left, black) and after (right, red) deformation.

We can also rewrite the above equation as

$$H = v_F\ \vec{\sigma}\cdot(\hbar\vec{q} - \hbar\overrightarrow{A_{sk}}) = v_F\ \vec{\sigma}\cdot(\vec{p} - \hbar\overrightarrow{A_{sk}}) = v_F\ \vec{\sigma}\cdot(\vec{p} - e\overrightarrow{A_{ps}}) \tag{8.23}$$

where

$$\overrightarrow{A_{ps}} = \frac{\hbar\overrightarrow{A_{sk}}}{e}$$

Now let's recall the modification of Dirac Hamiltonian in the presence of a magnetic field. We make the substitution $\vec{p} \to \vec{P} = \vec{p} - e\vec{A}$ and we obtain

$$\boldsymbol{H}_{\vec{K}} = v_F\ \vec{\sigma}\cdot(\vec{p} - e\vec{A}) \tag{8.24}$$

The resemblance between the Dirac Hamiltonian in a magnetic field and the Dirac Hamiltonian in a strain field has led to the rise of strain-induced pseudomagnetic field.

Similarly, the strained Hamiltonian at the $\boldsymbol{K}'$ point is obtained by using the transformation $\vec{\sigma} = (\sigma_x,\ \sigma_y) \to \vec{\sigma} = (\sigma_x,\ -\sigma_y)$ and $\vec{A} \to -\vec{A}$, so that the time reversal symmetry is conserved. This shows that the shifting vectors at $\boldsymbol{K}$ and $\boldsymbol{K}'$ points have opposite signs.

In the presence of strain field, the unit cell of graphene lattice may be distorted. The length of the three nearest hopping vectors may no longer be the same, introducing a bond-dependent nearest-neighbor hopping amplitude. The electrons in strained graphene are still governed by a Dirac equation, but one in which the strain effect is accommodated by the shifting of the Dirac cones [18]. This conclusion is obtained in the context of the conventional minimal coupling scenario. It says the charge carriers can possibly respond to the strain field as if they are in a real external magnetic field. Information on the strength of this pseudomagnetic field can be directly gathered from the shift of the Dirac cones. The nature of cone shape energy dispersion is preserved, unless the deformation is too strong that the tight binding assumption is no longer reliable.

8.5 The Coupling between Strain Field and Hopping Energy

As mentioned above, the modification in hopping energy integral due to the changing interatomic distance is given linearly as $t_n = t_0\,(1 + \Delta_n)$. Apparently, before we can proceed to further quantify the shift of $\boldsymbol{K}$ point in response to a certain strain field, we need to explicitly correlate the term Δ_n with strain field.

A classic empirical expression for the simplified modification of the hopping parameter due to strain is given by [16]

$$t_n = t_0 e^{-\beta\left(\frac{\delta_n}{a}-1\right)} \tag{8.25}$$

where a is the unstrained nearest neighbor distance, and β can be approximated by measuring the change in the hopping parameter as follows.

$$\beta = \left| \frac{\log(t_n) - \log(t_0)}{\frac{\delta_n}{a} - 1} \right| \tag{8.26}$$

In the limit of $\delta_n \to a$, $t_n \to t_0$, we have

$$\beta = \left| \frac{\partial \log(t_0)}{\partial \log(a)} \right| \tag{8.27}$$

Apparently β is an approximation. In fact, $\beta \cong 2 - 3.37$ [16].

The information of strain field is now apparently coupled with the term $\frac{\delta_n}{a}$, which is the ratio of the length of the hopping vector over its original length.

The Cauchy strain tensor is given by

$$\epsilon_{ij} = \frac{1}{2}\left(\frac{\partial u_i}{\partial x_j} + \frac{\partial u_j}{\partial x_i} \right) \tag{8.28}$$

where $\vec{u}$ is the displacement vector and $\vec{x}$ denotes in-plane coordinate system axes.

From the theory of linear elasticity, the ratio of the length change of the a line element before and after deformation in such strain field is given by

$$\lambda_v = \sqrt{1 + 2\epsilon_{ij} v_i v_j} \tag{8.29}$$

where v_i is the directional cosine of the line element. Due to the small strain assumption, v_i can be calculated either from the undeformed line element, or from the deformed line element, and we assume their difference is trivial. Further, λ_v can be expanded to the linear order of strain tensor as

$$\lambda_v \cong 1 + \epsilon_{ij} v_i v_j \tag{8.30}$$

Therefore,

$$\frac{\delta_n}{a} = 1 + \frac{\delta_n^i}{a} \cdot \epsilon_{ij} \cdot \frac{\delta_n^j}{a} \tag{8.31}$$

Finally, expand t_n to the linear order of strain tensor

$$t_n = t_0 e^{-\beta\left(\frac{\delta_n}{a}-1\right)} = t_0\left(1-\beta\left(\frac{\delta_n}{a}-1\right)\right) = t_0\left(1-\beta\frac{\delta_n^i}{a}\cdot\epsilon_{ij}\cdot\frac{\delta_n^j}{a}\right)$$

Therefore, we have arrived at a more detailed presentation about the modification in hopping energy integral due to the strain field as [17]

$$\Delta_n = \frac{-\beta\delta_n^i \cdot \epsilon_{ij} \cdot \delta_n^j}{a^2} \tag{8.32}$$

8.6 The Coupling between Strain Field and Pseudomagnetic Field

We denote the shifted $\boldsymbol{K}$ point as $\overrightarrow{K_{ps}}$, and it satisfies the condition $E(\overrightarrow{K_{ps}})=0$, which can be equivalently rewritten as

$$\sum_{n=1}^{3} t_n e^{i\,\overrightarrow{K_{ps}}\cdot\overrightarrow{\delta_n}} = 0 \tag{8.33}$$

If the strain tensor ϵ_{ij} is representing an isotropic strain field, $\Delta_n = 0$. Therefore, $t_n = t_0$, which leads to the conclusion that the Dirac points do not shift. For example, $\overrightarrow{K_{ps}} = \vec{K} = \left(\frac{4\pi}{3\sqrt{3}a}, 0\right)$.

If the strain tensor ϵ_{ij} is representing an anisotropic strain field, $\overrightarrow{K_{ps}} \neq \vec{K}$. We assume that $\overrightarrow{K_{ps}} = \vec{K} + \overrightarrow{A_{sk}}$. Then from $E(\overrightarrow{K_{ps}}) = 0$, we have

$$\sum_{n=1}^{3} t_0(1+\Delta_n)e^{i(\vec{K}+\overrightarrow{A_{sk}})\cdot\vec{\delta_n}} = 0 \tag{8.34}$$

Expand to the linear order of $\overrightarrow{A_{sk}}$, we have

$$\sum_{n=1}^{3} (1+\Delta_n)(1+i\ \overrightarrow{A_{sk}}\cdot\overrightarrow{\delta_n})e^{i\ \vec{K}\cdot\vec{\delta_n}} = 0 \tag{8.35}$$

Denote the strain tensor as

$$[\epsilon_{ij}] = \begin{bmatrix} \epsilon_{xx} & \epsilon_{xy} \\ \epsilon_{yx} & \epsilon_{yy} \end{bmatrix} \tag{8.36}$$

We can also calculate from Equation 8.32 (note $\epsilon_{xy} = \epsilon_{yx}$)

$$\Delta_1 = -\beta\left(\frac{3}{4}\epsilon_{xx} + \frac{\sqrt{3}}{2}\epsilon_{xy} + \frac{1}{4}\epsilon_{yy}\right)$$

$$\Delta_2 = -\beta\left(\frac{3}{4}\epsilon_{xx} - \frac{\sqrt{3}}{2}\epsilon_{xy} + \frac{1}{4}\epsilon_{yy}\right)$$

$$\Delta_3 = -\beta\epsilon_{yy} \tag{8.37}$$

Finally, the shift vector for ***K*** point can be solved from Equation 8.35 as

$$\overrightarrow{A_{sk}} = (A_{skx}, A_{sky})$$

$$A_{skx} = \frac{1}{2}\frac{\beta}{a}(\epsilon_{xx} - \epsilon_{yy})$$

$$A_{sky} = \frac{1}{2}\frac{\beta}{a}(-2\epsilon_{xy}) \tag{8.38}$$

Recall the notation in Equation 8.22

$$\overrightarrow{A_{ps}} = \frac{\hbar\overrightarrow{A_{sk}}}{e}$$

Then we can rewrite

$$\overrightarrow{A_{ps}} = \frac{\hbar\beta}{2ae}\begin{pmatrix} \epsilon_{xx} - \epsilon_{yy} \\ -2\epsilon_{xy} \end{pmatrix} = const \cdot \begin{pmatrix} \epsilon_{xx} - \epsilon_{yy} \\ -2\epsilon_{xy} \end{pmatrix} \tag{8.39}$$

The pseudomagnetic field can be calculated from

$$\vec{\nabla} \times \overrightarrow{A_{ps}} = B_{ps}\,\vec{e}_z \tag{8.40}$$

Apparently the vector $\begin{pmatrix} \epsilon_{xx} - \epsilon_{yy} \\ -2\epsilon_{xy} \end{pmatrix}$ is the quantity of interest and $\frac{\hbar\beta}{2ae}$ is only a constant scaling factor. Also, $\overrightarrow{A_{ps}}$ and the corresponding B_{ps} for $\boldsymbol{K}'$ point carries an opposite sign.

The above equation is the most fundamental result. Its mathematic beauty is stunning! The pseudomagnetic field is linearly proportional to a linear combination of the gradient of a two-dimensional strain tensor!

Considering $\beta \cong 2$–3.37 [16] and $v_F = \frac{3ta}{2\hbar}$, a practical and physically significant form of the pseudomagnetic field is [19, 20]

$$B_{ps} = \frac{t\beta}{ev_F}\left[-\frac{2\partial\epsilon_{xy}}{\partial x} - \frac{\partial}{\partial y}(\epsilon_{xx} - \epsilon_{yy})\right] \tag{8.41}$$

It is important to note that we have taken very simplified assumptions in arriving at the above first-order picture. There are a bunch of elaborated studies available regarding more refined treatments [21–26]. For example, we do not consider the length change in the hopping vector $\vec{\delta}_n$ in response to the strain field. This is essentially related to the reference frame with which the shift of Dirac cones is calculated, and it is tempting to study such lattice correction effect. Nevertheless, it turns out that, although lattice corrections can affect the description of the shift in the positions of the Dirac cones, they do not contribute to the pseudomagnetic field [21–26].

8.7 Pseudo Landau Levels and Pseudospin Polarization

The vector potential enters the Dirac Hamiltonian in the same way as a real magnetic field does. We can therefore compare the strain-induced pseudomagnetic field to a real magnetic field (Section 8.3).

For example, if we consider the valley at $\boldsymbol{K}$ point, the real magnetic field has the following Landau levels:

$$E_n = sgn(n)v_F\sqrt{2e\hbar B|n|}$$

If we switch the B with the B_{ps}, we can immediately obtain

$$E_n = sgn(n)v_F\sqrt{2e\hbar B_{ps}|n|} \tag{8.42}$$

where n = 0, ±1, ±2,... is the pseudo Landau level index.

The above equation is the foundation for probing the intensity of pseudomagnetic field in the scanning tunneling microscopy (STM) experiments, where the spatial distribution of local density of states can be accessed, noticeable as distinct peaks in the curve of dI/dV versus sample voltage, and the corresponding peak energies should follow the scaling law with respect to pseudo Landau level index. These principles were applied in characterizing pseudomagnetic fields of 300 T [11]. The observation is also called strain-induced Landau quantization, possibly opening energy gaps that are even observable at room temperature [27]. A good measurement requires that the pseudomagnetic field is homogeneous on the cyclotron radius scale and can be directly probed by the STM tip. An STM experiment works by using a sharp needle probe that skims along the surface of the nanobubble to measure the differential conductance (dI/dV), proportional to the local density of states at each point in the scan while building an image of the surface. This is also called the scanning tunneling spectroscopy (STS) measurements.

Nevertheless, unlike the real magnetic field (Section 8.3), the pseudomagnetic field has opposite signs for charge carriers in valleys at ***K*** and ***K′*** points where the vector potentials have opposite signs. For example, the electrons in K valley are circling in a direction opposite to the electrons in ***K′*** valley. This is a consequence of the time reversal symmetry nature of pseudomagnetic field.

Following the discussion in Section 8.3, the time reversal symmetry nature of pseudomagnetic field gives rise to a different set of relations $E_{n|\Phi_B^{K'}} \overset{sym}{=} E_{n|\Phi_B^{K}}$ and $E_{n|\Phi_A^{K'}} \overset{sym}{=} E_{n|\Phi_A^{K}}$. These relations indicate that for each sublattice, both valleys contribute the same amount to the local density of states associated with this sublattice, and overall sublattice equivalence is thus broken when the contributions from both valleys are combined (i.e., $E_{n|\Phi_B^{K'}} + E_{n|\Phi_B^{K}} \overset{sym}{\neq} E_{n|\Phi_A^{K}} + E_{n|\Phi_A^{K'}}$). Such a redistribution of the local density of states associated with pseudomagnetic field recently has been directly observed in an STM/STS experiment [28]. The pseudomagnetic field shifts the energy of the respective states at sublattices A and B in opposite directions, thereby giving rise to a sublattice symmetry breaking. The local density of states in one sublattice is increased while that of the other sublattice is decreased. This property is fundamentally different from a real magnetic field. The sublattice symmetry breaking is another signature of pseudomagnetic field, in addition to the Landau quantization. In addition, sublattice symmetry breaking implies a valley filtering property that can be employed for valleytronic applications, which will be briefly discussed later.

In the Dirac description, a sublattice symmetry breaking is analogous to a pseudospin polarization, where the sublattice degree of freedom is represented by a pseudospin [29]. In the presence of a pseudomagnetic field, we can thus compare the sublattice symmetry breaking to the alignment of the pseudospin in response to the pseudomagnetic field.

8.8 Strain-Induced Pseudomagnetic Field Greater Than 300 T

Let's revisit the experiments on strain-induced pseudomagnetic field greater than 300 T [11]. Highly strained graphene nanobubbles, on the order of a few nanometers in size, were formed on a platinum substrate during the cooling down to a few degrees above absolute zero. A scanning tunneling microscope was used to study these monolayer graphene nanobubbles. It is important to note that the pseudomagnetic field was measured much more uniform over the nanobubbles with triangular shape (Figure 8.3).

Distinct peaks in the local density of states were observed when scanned on the nanobubble surface [11]. These peaks corresponded to pseudo Landau levels, as they follow the linear scaling behavior between the peak energies. It was also found that the larger the curvature of the bubbles, the greater the strength of the pseudomagnetic field. This could be correlated to the fact that larger curvature usually indicated larger strain gradient, on which the intensity of pseudomagnetic field is linearly dependent. The intensity also increased at the edges of the nanobubble, where the strain gradient profile was expected to reach an extreme, since further away from the edges the graphene was rather flat on substrate. The intensity of the pseudomagnetic field calculated from the pseudo Landau levels in some cases was on the order of 300 T or more.

This experiment confirms that strain engineering is a promising approach to controlling the electronic structure of graphene, even at room temperature. It provides a new perspective for the study of extreme high magnetic field regimes in a condensed matter environment.

While it is exciting to demonstrate pseudomagnetic field of extreme intensity on graphene nanobubbles, one has to realize that these nanobubbles are formed naturally during cooling down of graphene grown by chemical vapor deposition on a platinum substrate to cryogenic temperatures, due to a relatively large difference between the thermal expansion coefficients of graphene and platinum. Therefore, there is limited degree of control from engineering perspectives. First, the locations where the nanobubbles might appear are not known in advance. Second, the size and shape of the nanobubble, closely related to its strain field, are not in good control. After all, the strain field is governing pseudomagnetic field.

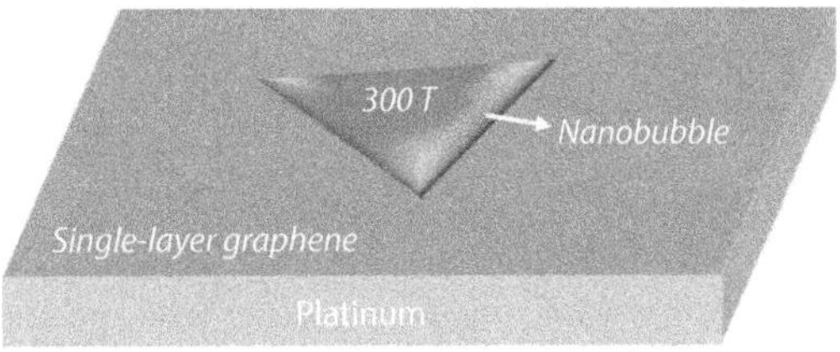

Figure 8.3 Schematic showing rather uniform pseudomagnetic field on the order of 300 T measured over triangular graphene nanobubble naturally formed on a platinum substrate during the cooling down to a few degrees above absolute zero [11].

8.9 Graphene Drumheads and On-Demand Activation of Pseudomagnetic Field

A more controllable demonstration is using the graphene drumheads. In 2012, N. N. Klimov reported that they could tune the strain in monolayer graphene drumhead structure using the STM tip [19]. Tuning the strain enabled the creation of semiconducting areas in the graphene where pseudomagnetic fields emerged.

Monolayer graphene was placed over shallow holes in a substrate of silicon dioxide, making an array of graphene drumheads (Figure 8.4a). As the tip scanned over, the graphene rose up to meet the tip. This is a result of the van der Waals interaction. The tent-like deformation created a localized strain field beneath the tip. STS measurements on the deformed suspended graphene displayed an electronic spectrum indicating the formation of a spatially confined quantum dot in the region at the apex of the deformed morphology. Quantum dot [30] is a type of semiconductor in which electrons are confined to a small region of space. Normally, to make a graphene quantum dot, one has to cut out a nanosize piece of graphene. But in this experiment, semiconducting quantum dot regions were created in graphene by deliberately engineering its morphology and strain field using the tip. Further measurement and computational analysis confirmed that the quantum dot behavior is caused by a three-fold-symmetry pseudomagnetic field (Figure 8.4b).

The on-demand activation of pseudomagnetic field can be achieved using tip–graphene interaction, although the intensity of pseudomagnetic fields in this drumhead experiment is not as strong as those in the nanobubble experiment [11]. The strain field is highly localized beneath the tip so that the pseudomagnetic field region follows the movement of the tip. The size of the quantum dot and the intensity of the pseudomagnetic field are independent on the tip location, but are related to the radius of the tip and the diameter of the drumhead [31]. The on-demand activation of the pseudomagnetic field is fully controlled by the tip–graphene contact.

Nevertheless, the energy spacing of the individual levels from STS measurement follows the classic energy spectrum of a quantum dot, instead of a set of quantized Landau levels as appeared in the nanobubble experiment [11]. This indicates that the pseudomagnetic field in the drumhead experiment is not a uniform field. In fact, the location where the

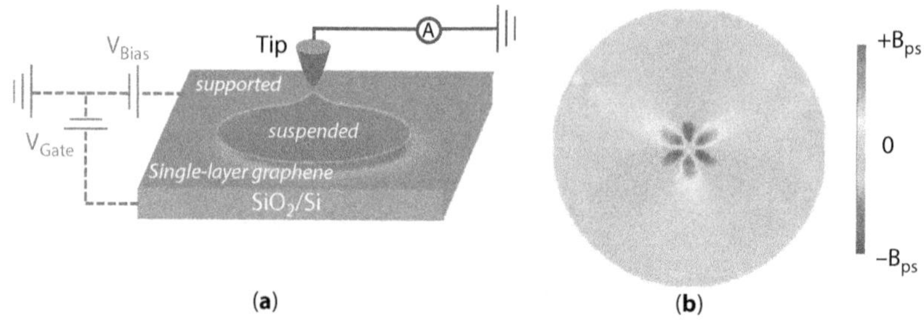

Figure 8.4 (a) Schematic showing STM tip-induced deformation of graphene drumhead structure, which is formed by depositing a monolayer graphene on a substrate patterned with a pit. (b) Typical distribution of threefold symmetry pseudomagnetic field for STM tip-induced deformation of suspended graphene. The pseudomagnetic field presented here is reproduced from a molecular dynamics simulation as described in Ref. [19], where the intensity B_{ps} is around 10 T, for a scaled-down structural model.

pseudomagnetic field is maximum is some distance away from the region directly under the probe tip. Based on the calculation on the strain field of the deformed graphene, the pseudomagnetic field is found as a spatially alternating field with threefold symmetry. Such a pseudomagnetic field confines the graphene carriers around the alternating peaks of the pseudomagnetic field. However, some electronic states corresponding to classic snake orbits that propagate along the lines where the pseudomagnetic field changes sign will not be confined. Upon application of an external magnetic field with the intensity matching one component of the pseudomagnetic field that opposes the applied magnetic field, the leaky confinement disappeared [19]. Both such measurement and computational estimation suggest the peak intensity of the spatial alternating pseudomagnetic field is around 10 T.

The above distinctions between the drumhead [19] and nanobubble [11] suggest a wide range of factors to modify electronic states of graphene through pseudomagnetic fields. An immediate question is that, why in the drumhead case, the pseudomagnetic field is a spatially alternating field with threefold symmetry, while in the nanobubble case, the pseudomagnetic field is rather uniform?

First, let's understand the drumhead case. The tip-induced deformation can be regarded as rotational symmetric. With this assumption and in a cylindrical coordinate system (r,θ), the pseudomagnetic field can be expressed as [31]

$$B_{ps} = \frac{t\beta}{ev_F} sin3\theta \left(-\frac{\partial(\epsilon_{rr} - \epsilon_{\theta\theta})}{\partial r} + \frac{2(\epsilon_{rr} - \epsilon_{\theta\theta})}{r} \right) \tag{8.43}$$

This suggests that a strain field in graphene having rotational symmetry can lead to a pseudomagnetic field with a threefold (triangular) symmetry (as a result of the prefactor $sin3\theta$).

Then, think in another way, in order to achieve a uniform pseudomagnetic field, the strain field has to cancel out the prefactor $sin3\theta$, which seems to suggest that the strain field should be somewhat triangular symmetric. Could this be an explanation for the uniform pseudomagnetic field measured over triangular nanobubbles [11]? Can we engineer the pseudomagnetic field via actively controlling the strain field?

8.10 Strain Engineering of Pseudomagnetic Field: Triaxial Stretching

The experimental evidence of enormous pseudomagnetic fields in locally strained graphene nanobubbles [11] and graphene drumheads [19] inspires enthusiasm in exploring the abundant potential of strain engineering of graphene, as well as charge carrier behaviors under extreme magnetic fields that otherwise do not exist in normal laboratory environments. Nevertheless, both graphene nanobubble and drumhead experiments demonstrated pseudomagnetic fields in highly localized regions of graphene, but usually a large area field is desirable for characterization and practical control. In addition, the nature of strain gradient dependence of pseudomagnetic field is troublesome. For example, no energy gaps were found for uniaxial strain [16] simply because there was no strain gradient. A fundamental question is that, what kind of strain field can render a large area and uniform pseudomagnetic field?

In 2009, F. Guinea *et al.* reported that a uniform pseudomagnetic field on order of tens of Tesla [27] can be achieved by introducing a strain field of threefold symmetry in graphene. The theory successfully explained the rather uniform pseudomagnetic field measured over triangle-shaped nanobubbles [11], where the strain field was also expected to carry triangular symmetry.

Using polar coordinates in (r,θ), denote u_r and u_θ as displacement fields. If $u_r = const \cdot r^2 sin3\theta$ and $u_\theta = const \cdot r^2 cos3\theta$, we can calculate a uniform pseudomagnetic field as [27]

$$B_{ps} = const \cdot \frac{8t\beta}{ev_F} \tag{8.44}$$

The prescribed displacement field (u_r and u_θ) can correspond to a regular hexagon with normal stresses applied evenly at its three nonadjacent sides, oriented perpendicular to equivalent crystallographic directions ⟨100⟩ of graphene lattice (Figure 8.5). Denote the side length of this hexagon as L, the maximum strain at its perimeter as ϵ_{pmax}, then $const = \epsilon_{\text{pmax}}/L$ and

$$B_{ps} = \frac{8\epsilon_{\text{pmax}} t\beta}{ev_F L} \tag{8.45}$$

If L = 30 nm, ϵ_{pmax} = 1%, we can obtain rather uniform $B_{ps} \cong$ 7T, near the central region of the hexagon [27]. The peaks in local density of states are also observable as pseudo Landau levels [27]. There is a linear scaling relationship $B_{ps} \propto \epsilon_{\text{pmax}}/L$.

This triaxial loading scheme can immediately explain the nanobubble experiments [11] discussed earlier. The highly strained triangle-shaped nanobubble represented a triaxial stress state, and therefore the pseudomagnetic field was measured rather uniform over the nanobubbles.

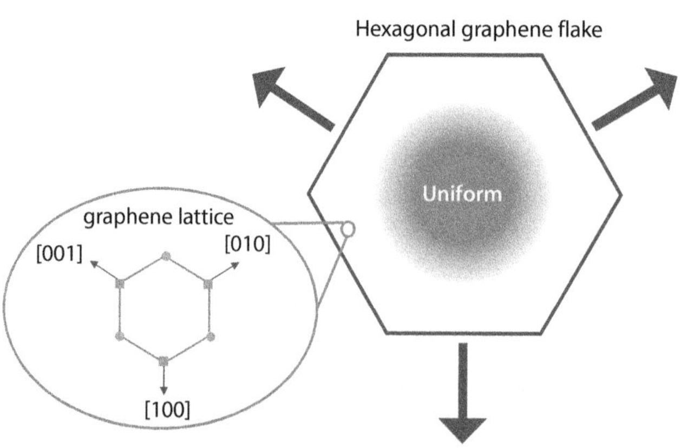

Figure 8.5 Schematic showing pseudomagnetic field for a regular hexagon graphene stretched by its three alternating edges. The field is rather uniform at the central region.

One can further imagine the application of the strain graphene superlattices, which can be created by depositing graphene on engineered substrate surfaces with profiled features that enable such triaxial stress state [27]. The resulting pseudomagnetic fields superlattice opens the energy gaps in the band of electronic states of graphene.

8.11 Strain Engineering of Pseudomagnetic Field: Uniaxial Stretching

From previous sections, it is quite interesting to obverse the following. An axisymmetric strain field in graphene leads to a pseudomagnetic field of rotational threefold symmetry, while a strain field of rotational threefold symmetry in graphene leads to a uniform pseudomagnetic field (which can be considered as axisymmetric to some extent). This is a clear demonstration of the nature of the dependence of the symmetry of pseudomagnetic field on the strain gradient in graphene, as embedded in the simple equation to calculate the pseudomagnetic field.

Nevertheless, are there other solutions to this simple equation for a uniform pseudomagnetic field?

In 2015, S. Zhu *et al.* reported a "simple uniaxial stretch" solution to achieve programmable extreme pseudomagnetic fields with uniform distributions in a planar graphene sheet over a large area [32]. This is achieved by patterning the planar graphene geometry with a shape function to engineer a desired strain gradient. The method is geometrical, offering new fertile opportunities of strain engineering of electronic properties of two-dimensional materials in general.

In two-dimensional materials, we have the constitutive law as $\sigma_{xx} = \frac{E}{1-\nu^2}(\epsilon_{xx} + \nu\epsilon_{yy})$, $\sigma_{yy} = \frac{E}{1-\nu^2}(\epsilon_{yy} + \nu\epsilon_{xx})$, $\sigma_{xy} = 2G\,\epsilon_{xy}$. E is the Young's modulus, ν is Poisson's ratio, and G is the shear modulus, which follow the relation $G = \frac{E}{2(1+\nu)}$.

We also have the stress equilibrium equations (without body force) as $\frac{\partial\sigma_{xx}}{\partial x} + \frac{\partial\sigma_{yx}}{\partial y} = 0$, $\frac{\partial\sigma_{xy}}{\partial x} + \frac{\partial\sigma_{yy}}{\partial y} = 0$.

Combining the constitutive law and stress equilibrium equations gives $\frac{E}{1-\nu^2}\left(\frac{\partial\epsilon_{xx}}{\partial x} + \frac{\nu\partial\epsilon_{yy}}{\partial x}\right) + 2G\frac{\partial\epsilon_{xy}}{\partial y} = 0$, $\frac{E}{1-\nu^2}\left(\frac{\partial\epsilon_{yy}}{\partial y} + \frac{\nu\partial\epsilon_{xx}}{\partial y}\right) + 2G\frac{\partial\epsilon_{xy}}{\partial x} = 0$.

If we assume a unidirectional tensile loading ($\epsilon_{xx} = -\nu\epsilon_{xy}$), we obtain $\frac{\partial\epsilon_{xy}}{\partial y} = 0$ and $\frac{\partial\epsilon_{xy}}{\partial x} = -(1+\nu)\frac{\partial\epsilon_{yy}}{\partial y}$. Therefore, the pseudomagnetic field can be rewritten as

$$B_{ps} = \frac{t\beta}{ev_F}\left(-\frac{2\partial\epsilon_{xy}}{\partial x} - \frac{\partial\epsilon_{xx}}{\partial y} + \frac{\partial\epsilon_{yy}}{\partial y}\right) = \frac{3t\beta}{ev_F}(1+\nu)\frac{\partial\epsilon_{yy}}{\partial y} \tag{8.46}$$

The above analysis points out that if $\frac{\partial\epsilon_{yy}}{\partial y}$ can be engineered to be a constant, we should be able to achieve a uniform pseudomagnetic field by a simple uniaxial stretch.

The key to engineer this strain gradient lies in the shape function of the graphene (Figure 8.6a)

$$W(y) = \frac{f_r \, LW_0}{f_r \, (L - y) + y} \tag{8.47}$$

where f_r is the ratio between the widths of the top and bottom ends of the graphene nanoribbon, L is the initial length of the nanoribbon, and W_0 is the basal initial width of the nanoribbon.

As a result,

$$\frac{\partial \epsilon_{yy}}{\partial y} = \frac{2\,\epsilon_{\text{app}}}{L} \frac{(1 - f_r)}{(1 + f_r)} \tag{8.48}$$

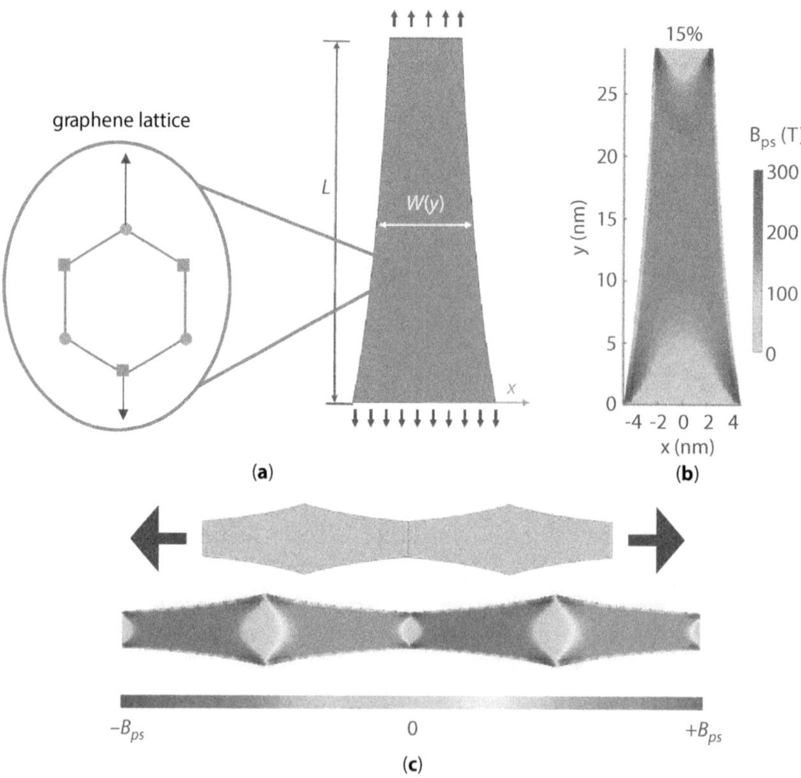

Figure 8.6 (a) Schematic showing a graphene nanoribbon of varying width under a uniaxial stretch producing a pseudomagnetic field. The inset shows the lattice orientation and a different degree of deformation is enforced as a result of the uniaxial stretch combined with the shape function W(y). (b) Resulting rather uniform pseudomagnetic fields in a graphene nanoribbon as shaped in (a) under 15% uniaxial stretch. The pseudomagnetic field presented here is reproduced from a finite element simulation as described in Ref. [32]. (c) Schematic showing patterned pseudomagnetic fields enabled by stretching in graphene nanoribbon patterned with repeating optimized geometry.

$$B_{ps} = \frac{6t\beta}{ev_F}\frac{\epsilon_{app}}{L}\frac{(1-f_r)}{(1+f_r)}(1+\nu) \tag{8.49}$$

where ϵ_{app} is the globally applied strain.

For a proper setting, the resulting intensity could be extreme (Figure 8.6b). Similar to the triaxial loading scheme, we have a linear scaling relationship $B_{ps} \propto \epsilon_{app}/L$.

This method can be readily extended to create the pseudomagnetic fields superlattice [32]. Let's consider a graphene nanoribbon that has been patterned with repeating optimized geometry (Figure 8.6c). Due to the symmetry, uniform pseudomagnetic fields with alternating signs are expected to appear under a uniaxial stretch. Such concept can also be applied to a graphene nanomesh structure.

8.12 Strain Engineering towards Topological Insulators and Valleytronics

In the reciprocal space, the conduction band and the valence band of graphene touch in six points in a hexagonal pattern (Figure 8.1b). These six points can be classified into two inequivalent points, which are usually denoted as $\boldsymbol{K}$ and $\boldsymbol{K}'$ points. Since pseudomagnetic field carries opposite signs for charge carriers in valleys at $\boldsymbol{K}$ and $\boldsymbol{K}'$ points where the vector potentials have opposite signs, strain engineering of graphene can thus offer alternative opportunities to achieve valley polarization as well as topological valley currents.

Two-dimensional topological insulators are a class of atomically thin layered materials that exhibit unique symmetry-protected helical metallic edge states with an insulating interior [33]. These states are also referred to as quantum spin Hall states. Such effect gives rise to edge states at the boundary that are topologically protected from backscattering, offering a potential application to electronic devices (e.g., spintronics [34]) to transport current without dissipation.

The quantum spin Hall effect is characterized by a full insulating gap in the bulk and helical gapless edge states where opposite spins counterpropagate at each boundary protected by time reversal symmetry [33]. In fact, quantum spin Hall effect was originally predicted in semiconductors in the presence of strain gradient, where the degenerate quantum Landau levels are created by the spin–orbit coupling in the absence of any magnetic field [35]. Considering the spin–orbit coupling, which couples the spin and the momentum degrees of freedom of the carriers, the strain in semiconductor heterostructures can lead to quasi Landau quantization with opposite effective magnetic field acting on two spins [27, 35]. However, weak spin–orbit coupling in graphene only allows tiny Landau gaps that would be hard to access even at a few kelvins, unless a proper magnetic field is present to create accessible gaps and thereby grant access to the quantum spin Hall states in graphene [36].

Pseudomagnetic field takes the advantage of pseudospin–orbit coupling in graphene, and resulting large gaps can be accessible at room temperature. Therefore, pseudomagnetic field in graphene is of potential significance to exploit its quantum spin Hall states. A strong pseudomagnetic field leads to Landau quantization, opening energy gaps that are even observable at room temperature. This is also called zero-field quantum Hall effect [27]. Unlike the standard quantum Hall effect in real magnetic field, the pseudomagnetic field has opposite signs for charge carriers in valleys at **K** and **K′** points. Therefore, there are edge states that circulate in opposite directions. In other words, without breaking the time reversal symmetry, strain can induce not only energy gap in the bulk, but also the counterpropagating edge states at the boundaries of the region where pseudomagnetic field emerges.

For a graphene nanoribbon, a strong pseudomagnetic field can lead to a quantum spin Hall state [37, 38]. Pseudomagnetic field enhances the carrier localization and pushes the edge states much closer to the boundaries of the nanoribbon, and stabilize the quantum spin Hall state. Further, the topological quantum phase transition between quantum spin Hall states and quantum anomalous Hall states were shown to be triggered by strain [37].

The sign difference of pseudomagnetic field for charge carriers in **K** and **K′** valleys can also be employed to separate electrons from different valleys, which can be exploited for valleytronics application [39–41]. Similar with spintronics [34], which is about manipulating electrons with different spin, valleytronics concerns the manipulating of electrons associated with different valleys. To this end, it is crucial to generate the valley-polarized current. In other words, valley filtering is needed.

The effect of pseudomagnetic field on valley filtering is promising. Due to the changing sign of pseudomagetic field at different valleys, electrons associated with **K** valley are forced to circle (classic circular trajectories) in a direction opposite to that with **K′** valley. This suggests that it might be possible to engineer an exclusive transport path for the electrons only in one valley by taking advantage of patterned pseudomagnetic fields. One possible design [42, 43] is to impose a dome geometry in graphene, which can be realized in experiments

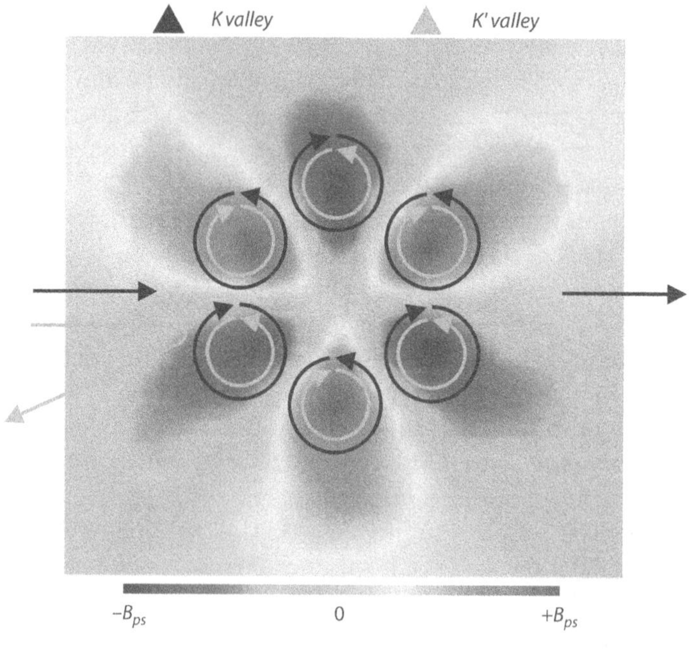

Figure 8.7 Schematic showing the valley filtering properties of a pseudomagnetic field with threefold symmetry. This is a zoomed view of Figure 8.4b. The vortices denote the classic counterpropagating electron trajectories in response to the pseudomagnetic field in the K (black) and K' valleys (gray). Considering an incoming current containing both K and K' electrons flows through this region from left to right. With a proper incident direction relative to the field, the K' valley electrons are scattered, while the K valley electrons are transmitted.

by STM tip pulling on suspended graphene [19], where the pseudomagnetic field intensity could rise up to 1000 T [28]. The resulting pseudomagnetic field carries threefold symmetry. When an incoming electron wave contains both valleys, with a proper incident direction, the electrons in one valley will be guided through the sign-alternating pseudomagnetic field regions along classic snake orbits, while the electrons in the other valley are to be backscattered (Figure 8.7). Such a design could stimulate the future development of effective THz valley filter, as a basic element of valleytronics [28, 42, 43].

8.13 Summary

In this chapter, the most essential picture of mechanical strain in modifying the electronic states of graphene via pseudomagnetic field is presented, in the context of theoretical derivation, computational insights, experimental results, and application perspectives.

When the graphene lattice is strained, the hopping energies between the two graphene sublattices change. The modified hopping energies add a term to the momentum operators in the low-energy Dirac Hamiltonian, in the same way a vector potential is added in real magnetic fields. This gives a very useful way to correlate the mechanical deformation in graphene with its electronic structure, ushering in the era of strain engineering. The signature of pseudomagnetic field is the pseudo Landau level peaks, and the sublattice symmetry breaking in the local density of states. These features are accessible in STM/STS measurements.

Uniform pseudomagnetic fields as strong as 300 T had been measured over highly strained graphene nanobubbles with triangular shape, which formed naturally during cooling down of graphene on a platinum substrate to cryogenic temperatures. The graphene was grown by chemical vapor deposition, and the large difference between the thermal expansion coefficients of graphene and platinum causes the lattice mismatch between two materials during the cooling process. This lattice mismatch governs the natural formation of high strained graphene nanobubbles.

Another scanning tunneling microscopy/spectroscopy (STM/STS) studies of suspended graphene drumheads and related calculations revealed that an STM probe tip interacting with a suspended graphene drumhead generated a highly localized radially symmetric strain field in graphene right under the STM probe tip. This strain field, in turn, leads to a pseudomagnetic field of threefold symmetry. This deformation method offers more controllability.

To take advantage of strain-induced pseudomagnetic fields to the full potential, a large area uniform field is desired. The main challenge comes from the fact that the pseudomagnetic field is dependent on the strain gradient in graphene. Two solutions to obtaining uniform pseudomagnetic fields are triaxial stretching and uniaxial stretching, and the resulting intensity can range from tens to hundreds of tesla. Both methods can be applied to engineer pseudomagnetic superlattices.

Being able to influence the motion of charge carriers, strain-induced pseudomagnetic fields in graphene are a useful tool to engineer the electronic states of graphene, vitally important in electronics application and condensed matter physics. The fundamental application is to create energy gaps in bulk graphene, so that its nature of zero-gap semiconductor can be tuned for electronics application. In addition, pseudomagnetic fields act oppositely on the two inequivalent valleys of the energy band structure of graphene, giving rise to zero-field quantum Hall effect. Patterning the pseudomagnetic fields can thus possibly render graphene as a two-dimensional topological insulator accessible in experiments, as well as a valley-filtering component in valleytronic applications.

References

1. Lee, C., Wei, X., Kysar, J., Hone, J., Measurement of the elastic properties and intrinsic strength of monolayer graphene. *Science*, 321, 385, 2008.
2. Lu, Q., Gao, W., Huang, R., Atomistic simulation and continuum modeling of graphene nanoribbons under uniaxial tension. *Modell. Simul. Mater. Sci. Eng.*, 19, 054006, 2011.
3. Morozov, S., Novoselov, K., Schedin, F., Jiang, D., Firsov, A., Geim, A., Two-dimensional electron and hole gases at the surface of graphite. *Phys. Rev. B*, 72, 201401, 2005.
4. Chen, J., Jang, C., Xiao, S., Ishigami, M., Fuhrer, M., Intrinsic and extrinsic performance limits of graphene devices on SiO2. *Nat. Nanotechnol.*, 3, 206, 2008.
5. Novoselov, K., Geim, A., Morozov, S., Jiang, D., Zhang, Y., Dubonos, S., Grigorieva, I., Firsov, A., Electric field effect in atomically thin carbon films. *Science*, 306, 666, 2004.
6. Barone, V., Hod, O., Scuseria, G., Electronic structure and stability of semiconducting graphene nanoribbons. *Nano Lett.*, 6, 2748, 2006.
7. Tapaszto, L., Dobrik, G., Lambin, P., Biro, L., Tailoring the atomic structure of graphene nanoribbons by scanning tunnelling microscope lithography. *Nat. Nanotechnol.*, 3, 397, 2008.
8. Han, M., Ozyilmaz, B., Zhang, Y., Kim, P., Energy band-gap engineering of graphene nanoribbons. *Phys. Rev. Lett.*, 98, 206805, 2007.
9. Zhang, Y., Tan, Y., Stormer, H., Kim, P., Experimental observation of the quantum Hall effect and Berry's phase in graphene. *Nature*, 438, 201, 2005.
10. Novoselov, K., Jiang, Z., Zhang, Y., Morozov, S., Stormer, H., Zeitler, U., Maan, J., Boebinger, G., Kim, P., Geim, A., Room-temperature quantum Hall effect in graphene. *Science*, 315, 1379, 2007.
11. Levy, N., Burke, S., Meaker, K., Panlasigui, M., Zettl, A., Guinea, F., Neto, A., Crommie, M., Strain-induced pseudo-magnetic fields greater than 300 tesla in graphene nanobubbles. *Science*, 329, 544, 2010.
12. Slater, J. and Koster, G., Simplified LCAO method for the periodic potential problem. *Phys. Rev.*, 94, 1498, 1954.
13. Castro Neto, A., Guinea, F., Peres, N., Novoselov, K., Geim, A., The electronic properties of graphene. *Rev. Mod. Phys.*, 81, 109, 2009.
14. Das Sarma, S., Adam, S., Hwang, E., Rossi, E., Electronic transport in two-dimensional graphene. *Rev. Mod. Phys.*, 83, 407, 2011.
15. Goerbig, M., Electronic properties of graphene in a strong magnetic field. *Rev. Mod. Phys.*, 83, 1193, 2011.
16. Pereira, V., Castro Neto, A., Peres, N., Tight-binding approach to uniaxial strain in graphene. *Phys. Rev. B*, 80, 045401, 2009.
17. de Juan, F., Sturla, M., Vozmediano, M., Space dependent Fermi velocity in strained graphene. *Phys. Rev. Lett.*, 108, 227205, 2012.
18. Hasegawa, Y., Konno, R., Nakano, H., Kohmoto, M., Zero modes of tight-binding electrons on the honeycomb lattice. *Phys. Rev. B*, 74, 033413, 2006.
19. Klimov, N., Jung, S., Zhu, S., Li, T., Wright, C., Solares, S., Newell, D., Zhitenev, N., Stroscio, J., Electromechanical properties of graphene drumheads. *Science*, 336, 1557, 2012.
20. Kim, K., Blanter, Y., Ahn, K., Interplay between real and pseudomagnetic field in graphene with strain. *Phys. Rev. B*, 84, 081401, 2011.
21. de Juan, F., Manes, J., Vozmediano, M., Gauge fields from strain in graphene. *Phys. Rev. B*, 87, 165131, 2013.
22. Masir, M., Moldovan, D., Peeters, P., Pseudo magnetic field in strained graphene: Revisited. *Solid State Commun.*, 175, 76, 2013.

23. Kitt, A., Pereira, V., Swan, A., Goldberg, B., Lattice-corrected strain-induced vector potentials in graphene. *Phys. Rev. B*, 87, 115432, 2013.
24. Oliva-Leyva, M. and Naumis, G., Understanding electron behavior in strained graphene as a reciprocal space distortion. *Phys. Rev. B*, 88, 085430, 2013.
25. Sloan, J., Sanjuan, A., Wang, Z., Horvath, C., Barraza-Lopez, S., Strain gauge fields for rippled graphene membranes under central mechanical load: An approach beyond first-order continuum elasticity. *Phys. Rev. B*, 87, 155436, 2013.
26. Barraza-Lopez, S., Sanjuan, A., Wang, Z., Vanevic, M., Strain-engineering of graphene's electronic structure beyond continuum elasticity. *Solid State Commun.*, 166, 70, 2013.
27. Guinea, F., Katsnelson, M., Geim, A., Energy gaps and a zero-field quantum Hall effect in graphene by strain engineering. *Nat. Phys.*, 6, 30, 2010.
28. Georgi, A., Nemes-Incze, P., Carrillo-Bastos, R., Faria, D., Kusminskiy, S., Zhai, D., Schneider, M., Subramaniam, D., Mashoff, T., Freitag, N., Liebmann, M., Pratzer, M., Wirtz, L., Woods, C., Gorbachey, R., Cao, Y., Novoselov, K., Sandier, N., Morgenstern, M., Tuning the pseudospin polarization of graphene by a pseudomagnetic field. *Nano Lett.*, 17, 2240, 2017.
29. Sasaki, K. and Saito, R., Pseudospin and deformation-induced gauge field in graphene. *Prog. Theor. Phys. Suppl.*, 176, 253, 2008.
30. Ponomarenko, L., Schedin, F., Katsnelson, M., Yang, R., Hill, E., Novoselov, K., Geim, A., Chaotic Dirac billiard in graphene quantum dots. *Science*, 320, 356, 2008.
31. Zhu, S., Huang, Y., Klimov, N., Newell, D., Zhitenev, N., Stroscio, J., Solares, S., Li, T., Pseudomagnetic fields in a locally strained graphene drumhead. *Phys. Rev. B*, 90, 075426, 2014.
32. Zhu, S., Stroscio, J., Li, T., Programmable extreme pseudomagnetic fields in graphene by a uniaxial stretch. *Phys. Rev. Lett.*, 115, 245501, 2015.
33. Hasan, M. and Kane, C., Colloquium: Topological insulators. *Rev. Mod. Phys.*, 82, 3045, 2010.
34. Felser, C., Fecher, G., Balke, B., Spintronics: A challenge for materials science and solid-state chemistry. *Angew. Chem., Int. Ed.*, 46, 668, 2007.
35. Bernevig, B. and Zhang, S., Quantum spin Hall effect. *Phys. Rev. Lett.*, 96, 226801, 2006.
36. Young, A., Sanchez-Yamagishi, J., Hunt, B., Choi, S., Watanabe, K., Taniguchi, T., Ashoori, R., Jarillo-Herrero, P., Tunable symmetry breaking and helical edge transport in a graphene quantum spin Hall state. *Nature*, 505, 528, 2014.
37. Guassi, M., Diniz, G., Sandler, N., Qu, F., Zero-field and time-reversal-symmetry-broken topological phase transitions in graphene. *Phys. Rev. B*, 92, 075426, 2015.
38. Liu, Z., Wu, Q., Chen, A., Xiao, X., Liu, N., Miao, G., Helical edge states and edge-state transport in strained armchair graphene nanoribbons. *Sci. Rep.*, 7, 8854, 2017.
39. Xiao, D., Yao, W., Niu, Q., Valley-contrasting physics in graphene: Magnetic moment and topological transport. *Phys. Rev. Lett.*, 99, 236809, 2007.
40. Rycerz, A., Tworzydlo, J., Beenakker, C., Valley filter and valley valve in graphene. *Nat. Phys.*, 3, 172, 2007.
41. Jiang, Y., Low, T., Chang, K., Katsnelson, M., Guinea, F., Generation of pure bulk valley current in graphene. *Phys. Rev. Lett.*, 110, 046601, 2013.
42. Milovanovic, S. and Peeters, F., Strain controlled valley filtering in multi-terminal graphene structures. *Appl. Phys. Lett.*, 109, 203108, 2016.
43. Settnes, M., Power, S., Brandbyge, M., Jauho, A., Graphene nanobubbles as valley filters and beam splitters. *Phys. Rev. Lett.*, 117, 276801, 2016.

9

Characteristic Mechanical Responses of Graphene Membranes

Young In Jhon

Sensor System Research Center, Korea Institute of Science and Technology, Seoul, Republic of Korea

Abstract

Graphene exhibits a superb mechanical strength with an ultrahigh stretchability. However, its detailed mechanical behavior has been less known than the electronic properties due to the absence of proper experimental tools for the relevant investigation. This chapter summarizes recent progress in computational studies on the mechanical response of graphene membrane under tensile and/or compressive loading, which was shown to be distinctly different from common ductile and/or brittle materials. In its four sections, the first section described the tensile fracture behavior of polycrystalline graphene in which unique plastic deformation and fracture were observed at the grain boundaries and noticeably monoatomic carbon chains (MACCs) were spontaneously formed between two side graphene domains upon the tensile failure. This process induced remarkable toughness even after quasi-mechanical fracture as well as providing a facile route for the fabrication of graphene–MACC–graphene modules. These modules have been considered to be promising nanoscale platforms for various electro-optic applications but their manufacturing has been severely hindered from technical challenges. The second and third sections investigated the compressive behavior of polycrystalline graphene and the grain boundary orientation effect on its tensile mechanical behavior, respectively. The fourth section described the orientation-dependent tensile behavior of monocrystalline graphene in one- and/or two-dimensional regimes by using systematic molecular dynamic simulations in which nanoindentation was employed for the two-dimensional study. Based on the nanoindentation simulation results, a new experimental scheme was suggested that can manifest the anisotropic mechanical response of monocrystalline graphene, which has never been validated experimentally so far.

Keywords: Graphene, mechanical behavior, tensile fracture, monoatomic carbon chain, grain boundaries, orientational dependence, nanoindentation, molecular dynamics simulation

9.1 Characteristic Tensile Fracture of Polycrystalline Graphene

Graphene, a single atomic carbon film with covalently bonded hexagonal lattice, has gained considerable attention due to its extraordinary properties such as ultrahigh electronic mobility [1, 2], excellent thermal conductivity [3, 4], and superb mechanical strength with

Email: yijhon@kaist.ac.kr

Tobias Stauber (ed.) Handbook of Graphene: Volume 2, (245–272) © 2019 Scrivener Publishing LLC

exceptional stretchability [5–7]. These properties in defect-free graphene can be easily obtained due to high formation energies of defects and strong bonding of carbon atoms [8]. There are two important lattice directions in monocrystalline graphene, which are zigzag (ZZ) and armchair (AC) directions as shown in Figure 9.1a. Graphene exhibits the strongest tensile strength along the ZZ direction while it yields the weakest tensile strength in the AC direction as shown in Figure 9.1b as obtained from molecular dynamic simulation.

With intensive continuing research, a breakthrough has occurred in the synthesis of large-scale graphene film by using the chemical vapor deposition technique, bringing it closer to practical application [9, 10]. However, the large-scale production has inevitably generated a polycrystalline form of graphene owing to crystal imperfection of substrate materials and kinetic influence in the growth process [11–14], and grain boundaries (GBs) have become one of the most important intrinsic defects. Different from any other material, graphene can host lattice defects through the reconstruction of atomic arrangement resulting in nonhexagonal ring formation in graphene, which include Stone–Thrower–Wales (STW) defects [15, 16] and a variety of GBs. While STW and/or mono-vacancy defects are point defects, GBs are line defects [17] and thus they can significantly influence the properties of a two-dimensional material [18–23]. In this context, a full understanding of graphene mechanics in the presence of GBs is very important for the development of advanced graphene technology.

The present section and the next section provide a systematic review of computational studies on the mechanical response of polycrystalline graphene as it is elongated and/or compressed, respectively [24], in directions perpendicular and/or parallel to GBs and they are denoted by filled and open arrows, respectively, in Figure 9.2a. Graphene membranes with mismatched ZZ-oriented (tilted) GBs and a perfectly matched AC-oriented (nontilted) GB [18, 19] were considered, and their representative structures were depicted in Figure 9.2. For tilted GBs, six cases were examined by gradually varying the misorientation angle (θ in Figure 9.1a), and the corresponding systems are denoted by T_i (i = 1–6, sequentially numbered in a descending order of the misorientation angle, i.e., 21.10°, 12.87°, 9.26°, 7.23°, 5.93°, and 5.03°). Meanwhile, a defect-free graphene and a graphene with perfectly matched AC-oriented (nontilted, that is, θ equals 0°) GBs were denoted as PR and TZ, respectively.

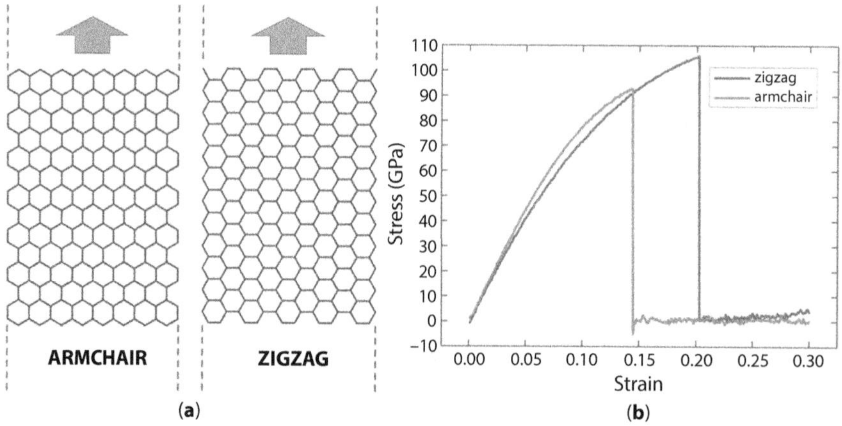

Figure 9.1 (a) The AC and ZZ directions in graphene; (b) The tensile stress–strain plots in monocrystalline graphene membrane along AC and ZZ directions.

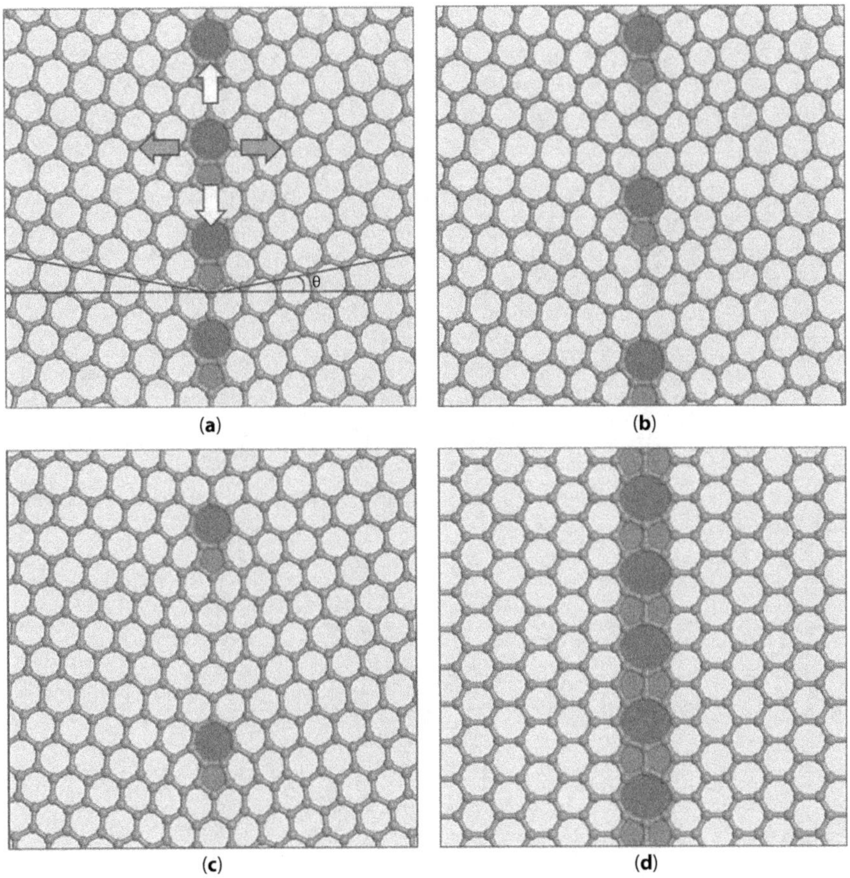

Figure 9.2 The structures of GBs (a–c) in tilted ZZ-oriented graphene membranes (T_1, T_2, and T_3 systems, respectively) and (d) nontilted AC-oriented graphene membrane (TZ system). The misorientation angle θ is defined as the angle between the direction normal to GBs and the ZZ and/or AC direction. Whether the direction is ZZ or AC depends on which is closer to the normal direction to GBs and it is the ZZ direction in the cases of (a–c).

This kind of nanomechanical investigation requires highly controlled sample fabrication and very subtle stress–strain measurement and unfortunately, modern experimental techniques cannot manage it. Hence, instead of it, molecular dynamics simulations have been largely perfomed for such researches in which a force field of "adaptive intermolecular reactive empirical bond order" was employed to account for fracture phenomenon properly [24].

For the reasonable description of the polycrystalline graphene membranes, the graphene simulation systems were constructed to meet the periodic boundary condition by embedding two directionally opposite tilted GBs into graphene as shown in Figure 9.3. The simulation showed that, when the graphene membrane was elongated parallel to the GBs, it sustained a considerable amount of the tensile strength being comparable to defect-free graphene (Figure 9.4). On the contrary, elongation perpendicular to the GBs sustained much less tensile strength than that of pristine graphene. Interestingly, for the tilted GBs, the tensile strength decreased as the misorientation angle decreased [22]. This seemingly counterintuitive phenomenon can

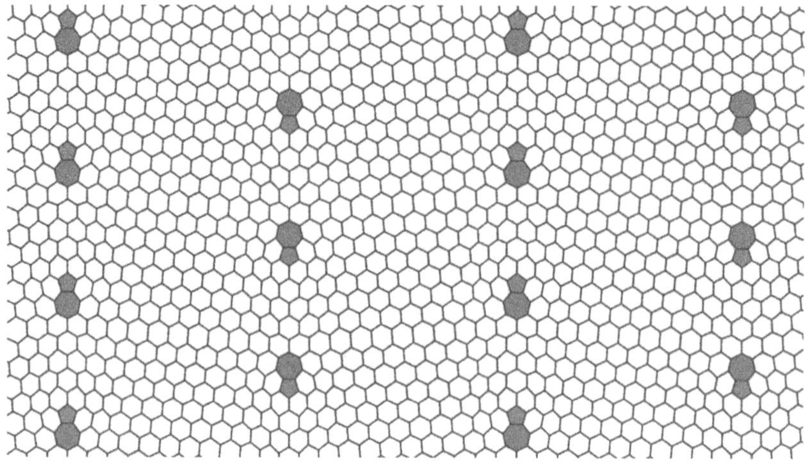

Figure 9.3 The structure of graphene membrane in which two directionally opposite GBs are embedded to meet periodic boundary condition.

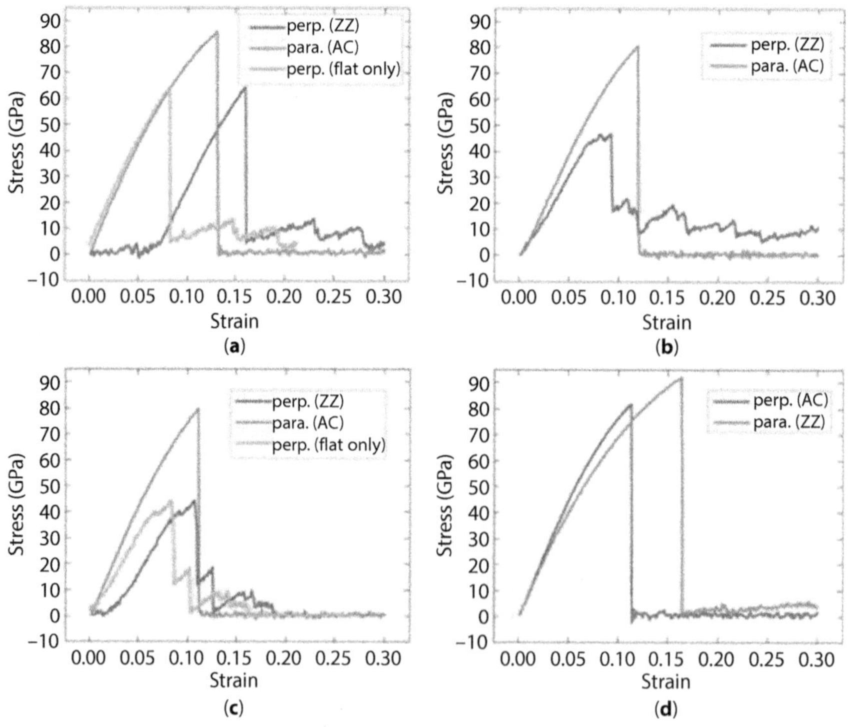

Figure 9.4 The stress–strain plots of (a–c) T_1, T_2, and T_3 systems, and TZ system, respectively, and (d) TZ system under tensile elongation perpendicular (perp.) and/or parallel (para.) to GBs. Certain polycrystalline graphene systems are initially folded along GBs and they become flattened with a negligible tensile energy consumption at the early stage of the tensile process. Green curves denote the stress–strain plots without the period for such a flattening process.

be explained by the fact that the weakest bond in these tilted GBs is initially more highly strained as the misorientation angle decreases, resulting in earlier fracture under tensile elongation [22]. More importantly, it was revealed that the tensile evolution of graphene with tilted GBs is remarkably different from that of TZ and/or defect-free graphene. It exhibited an incomplete fracture behavior for a long period of time even after the quasi-tensile failure, creating rugged tails in the stress–strain curves (Figure 9.4a–c). It is also noted that the slope of the stress–strain curves decreased rapidly just before the point of quasi tensile failure for T_2 and T_3 systems in sharp contrast to defect-free graphene.

Meanwhile, the GB with nontilted side domains (TZ), in which the GB is composed of carbon octagonal and pentagonal rings, does not show such an abnormal fracture (Figure 9.4d) similarly to that in defect-free graphene. Incidentally, for all graphene systems, the elongation parallel to GBs did not show such abnormal fracture response. The in-depth investigation on

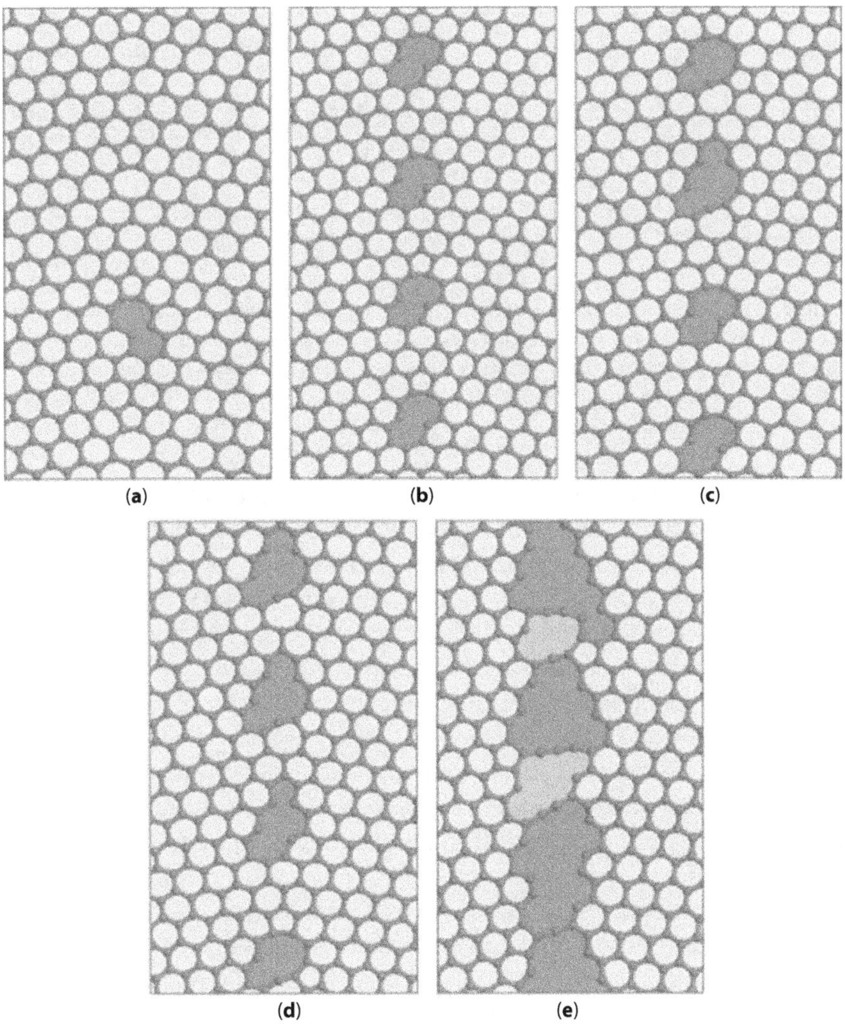

Figure 9.5 (a–e) The structures of T_2 system corresponding to A–E in Figure 9.6a, respectively. Green denotes crack regions that were formed and grown during tensile elongation and fluorescent denotes void regions newly created at the stage of tensile failure.

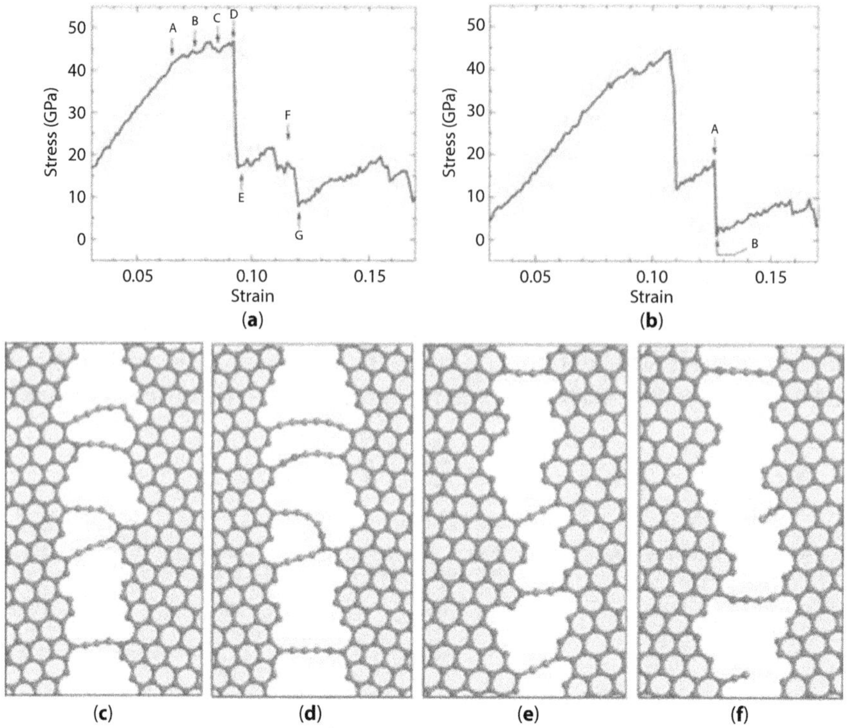

Figure 9.6 Characteristic stress–strain plots of (a) T_2 system and (b) T_3 system with indication of important stages marked with arrows. (c) and (d) The structures of T_2 system corresponding to stages of F and G in Figure 9.6a, respectively. (e) and (f) The structures of T_3 system corresponding to stages of A and B in Figure 9.6b, respectively. Green denotes strings before chemical extension of strings and red denotes carbon atoms to be used for the chemical extension.

the structural evolution of T_2 and T_3 systems elucidated the physical mechanism behind this extraordinary phenomenon as shown in Figures 9.5 and 9.6, in which several critical stages in the stress–strain curves of T_2 and T_3 systems were linked to their corresponding structures.

The starting point of decreasing the slope of the curve, which occurred before quasi-tensile failure, coincided with the stage at which the first bond breaking occurred. In all these systems, the first breaking took place at the heptagon structure in heptagon–pentagon pairs (Figure 9.5a) since this bond was initially most heavily strained. Once the bond had been broken at the heptagon, the crack began to propagate to other parts as shown in Figure 9.5b–d. During the crack propagation, the stress required for elongation was significantly alleviated, which was expressed as a rapid decrease of the slope in the stress–strain curve (Figure 9.6a).

However, such a pattern did not appear in the case of the largest misorientation angle (T_1) because the distance of crack propagation was very short, leading to a rapid completion of the crack propagation as shown in Figure 9.4a. The quasi-tensile failure occurred by the combination of a rapid growth of existing cracks and a creation of new voids as demonstrated in Figure 9.5e.

As described above, graphene membranes with tilted GBs did not exhibit complete fracture immediately even after quasi-tensile failure. Instead, they produced long-lasting strings bridged between the two separate parts, accompanied by a large persistent strength being tracked as rugged tails in the stress–strain plots (Figure 9.6). This trend was more striking in the systems with a large misorientation angle. The strings were sustained for a long period of time in T_i (i = 1–3) systems. From the structural analysis for this period, it was observed that tensile stress decreased remarkably either when a string was extended chemically using carbon atoms supplied from the side grain domains through bond-breaking/reforming process (Figure 9.6c and d) or when a string was disconnected permanently due to elongation. Particularly, the tensile stress precipitously dropped when these processes occurred simultaneously (Figure 9.6e and f).

It is supposed that the occurrence and propagation of cracks in these tilted GBs are attributed to uneven bond strength distribution along tilted GBs. Specifically, the ZZ-oriented tilted GBs in graphene are constructed by aligning 5–7 defects, each of which exhibits asymmetric stress distribution. Furthermore, linkage regions between the 5–7 defects, which are composed of hexagonal bonded structures, offer additional stress nonhomogeneity, and such stress distributions lead to the crack generation and propagation prior to tensile failure, which allowed the efficient production of MACC subsequently.

This carbon string, termed as MACC and traditionally known as carbyne or linear acetylenic carbon in chemistry, has long been the subject of considerable debate as a hypothetical carbon allotrope in the scientific community [25, 26]. MACC is a one-dimensional infinite chain molecule that entirely consists of sp-hybridized carbon atoms. Hence, in contrast to other carbon allotropes such as diamond, fullerene, and graphite, it possesses extremely high physical fragility and reactivity, which has severely thwarted attempts to isolate this intriguing material and to perform a full characterization of it.

The existence of MACC was actually observed in the most inner side of multi-walled carbon nanotubes (MWCNTs) at the cathode deposits prepared by hydrogen arc-discharge evaporation of carbon rods. However, it was only available at limited conditions and the heavy carbon wall shielding has significantly prevented its application and/or characterization [27]. The fabrication of MACC from graphene was recently achieved by carefully controlled electron irradiation using a transmission electron microscope (TEM), and these experiments demonstrated the stability of chains with lengths of a few nanometers, setting the stage for a new methodology to produce MACC bridged between graphene nanoribbons [28, 29]. This technique enables the applicability of MACC but its feasible productivity still remains challenging since it required quite an elaborate protocol and an expensive equipment.

In this regard, it is important to investigate the feasible production of MACC (up to a few nanometers) through mechanical fracture of tilted GBs in graphene instead of a time- and cost-consuming electron beam etching technology. We can see the similarity between the results of these two methods in Figure 9.7. It is expected that the length of MACC obtained from the mechanical fracture-based method could be further increased if graphene is elongated along the direction inclined to the direction normal to GBs since it seems that more facilitated and efficient production could be possible under an inclined elongation avoiding excessive accumulation of local stresses.

The potential of "graphene–MACC–graphene" modules as nanoelectronic switches has been theoretically investigated using density functional theory (DFT) calculations [30]. It was observed that a noncollinear MACC with a reconstructed pentagon undergoes structural transformation under strain, changing the number of carbon atoms composed of the chain from even to odd or vice versa. It was shown that the change in the carbon atom number, arising from the formation and/or annihilation of the rings at the interface of the MACC and the side graphene, resulted in a large fluctuation in the output current (Figure 9.8), suggesting that these modules can be used as nanoswitching devices.

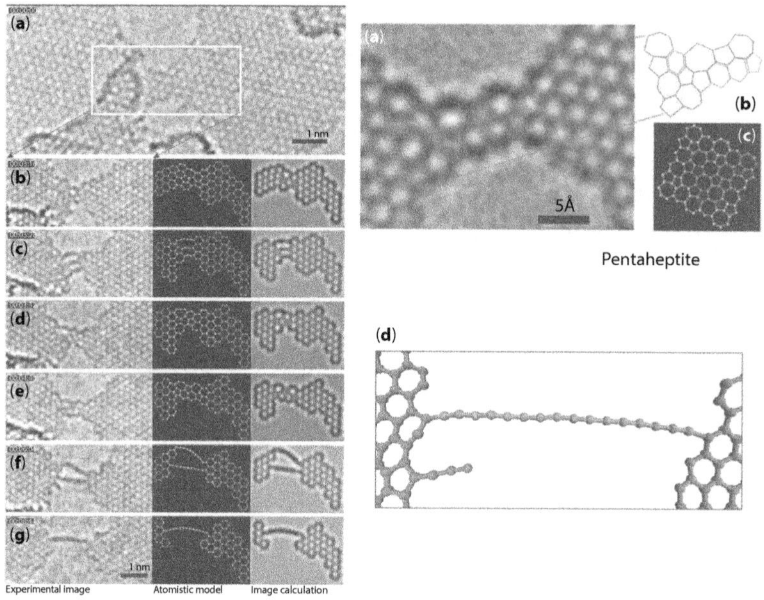

Figure 9.7 (left) (a–g) The structural evolution of graphene into MACC during electron irradiation using TEM [30]; (right) (a) The intermediate structure of narrowed graphene formed during electron irradiation of TEM, (b) the analyzed atomic structure of the intermediate structure, which is mainly composed of heptagonal and pentagonal carbon rings, (c) the similarity between the intermediate structure and a pentaheptite molecule, and (d) the MACC bridged between graphene domains fabricated by the tensile fracture, exhibiting the similarity to the structure shown in the left figure (g).

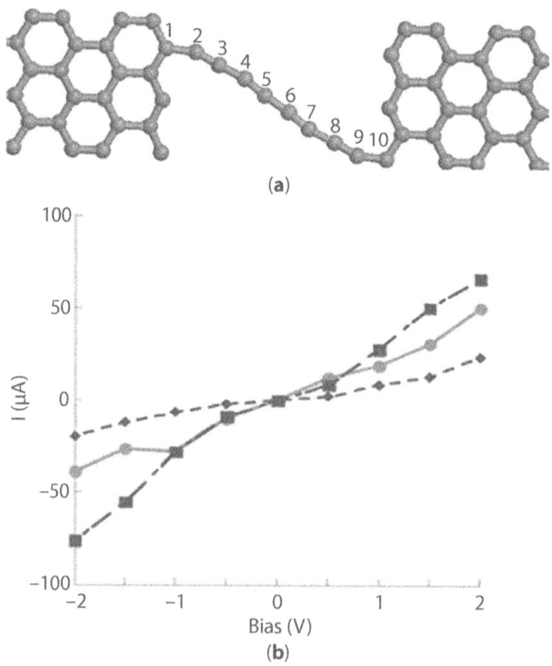

Figure 9.8 (a) The illustrated graphene–MACC–graphene modules in which the number of carbon atoms in MACC is eight; (b) I–V characteristics for graphene–MACC–graphene modules as the strain increases in the order of red circles, blue squares, and black diamonds, showing a large fluctuation in the output current.

The "graphene/p-type-doped MACC/graphene" modules under strain have also been theoretically studied to serve as highly tunable nanolasers [31]. While the graphene–MACC–graphene module is a two energy-level system, the calculation indicated that it can be a three energy-level system by doping MACC with B and/or Al as shown in Figure 9.9. In this three-level system, the pumping from the highest occupied state (HO) to the conduction band bottom (CB) was shown to be highly efficient. The band gap energy varied by only 7% (0.27–0.29 eV) upon the B doping with the composition ratio of 5%. In addition, the calculation showed that the radiationless decay rate from the valence band top (VT) to HO should be at least 10 times faster than the spontaneous rate from CB to VT, allowing a population inversion and thus enabling the lasing.

The production of MACC by the tensile fracture of graphene has been indeed realized later in experiments [32]. It should be noted that in the molecular dynamic simulation study showed that the MACC production can also be achieved for the tensile fracture of monocrystalline graphene as far as it is elongated along the ZZ direction although the extent was lower than that of polycrystalline graphene, as described in the third section. In the above experiment work, the cutting of the graphene flakes was first made by the moving Fe nanoparticles on the surface of the flakes, leading to the formation of thin graphene nanoribbons standing off the edges. These graphene ribbons were used as precursors for the formation of MACCs as summarized in Figure 9.10.

The process started by contacting the edge of a few-layer graphene sheet with an Au tip. That is, the first step was to ensure a good contact by gradually increasing the voltage

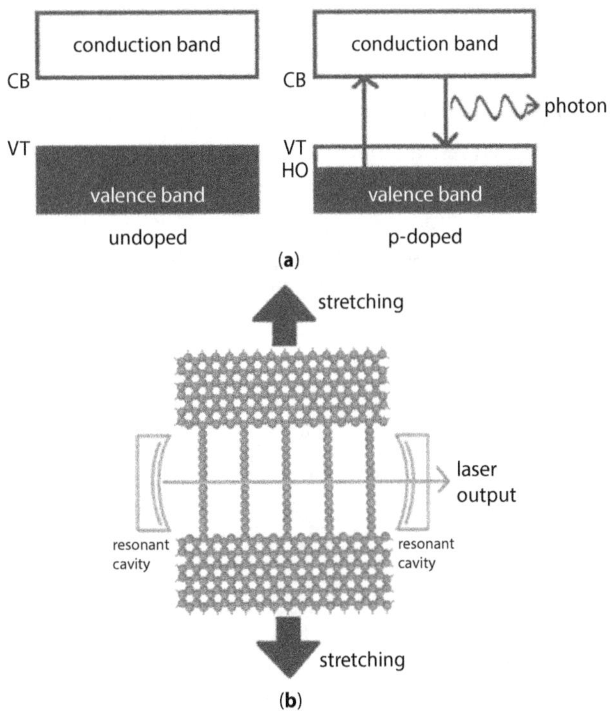

Figure 9.9 (a) The electronic band states of undoped and doped MACCs. (b) The illustrated laser system based on a graphene–MACC–graphene module [31].

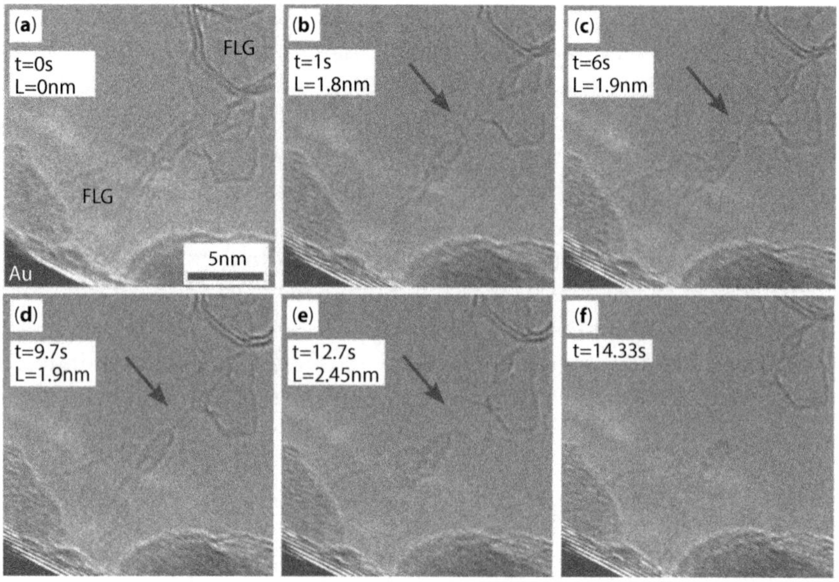

Figure 9.10 *In situ* synthesis of MACCs. (a–e) A few-layer graphene nanoribbon breaks and forms a carbon chain (arrowed) that is stable for a few seconds. (f) The chain eventually breaks forming the two separated few-layer graphene regions. The time scale and the measured length of the chain are indicated in the projection onto the image plane [32].

until the current through the circuit reached several 10^{-4} Å with typical voltages of 1–1.5 V. Alternatively, the voltage was increased abruptly to the value of 2–3 V while the current was constrained to a few 10^{-4} Å in order to prevent the destruction of the graphene. Once the good electrical contact had been achieved, the voltage was decreased down to ~1 V and the Au tip was retracted slowly. As a consequence, the graphene region in contact with the Au tip started to break from the flake, forming graphitic structures with the width less than 1 nm in between the two side regions (Figure 9.11a). Further separation of the tip from the sample resulted in the formation of stretched graphenic structures, eventually leading to the unraveling of the graphenic layer into MACC (Figure 9.11).

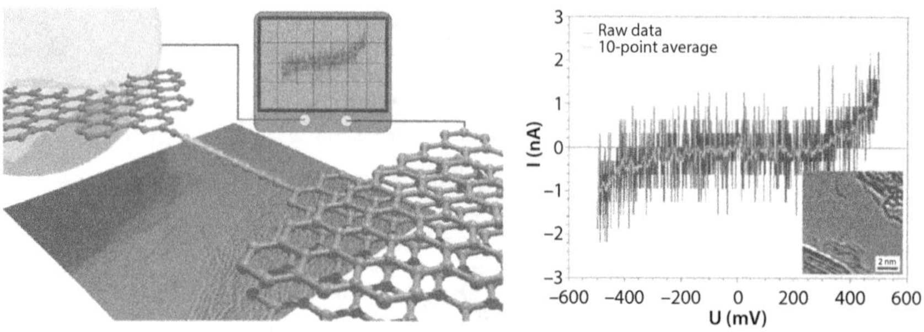

Figure 9.11 (a) The graphene–MACC–graphene module formed by mechanical stretching of graphene nanoribbon using an Au tip that was electrically controlled. (b) I–V measurements on MACC. The image of the MACCs is displayed in the inset [32].

9.2 Compressive Mechanical Response of Polycrystalline Graphene

The molecular dynamics simulations for the equilibration of polycrystalline graphene frequently lead to the bending along GBs, which suggests a feasible folding of graphene along GBs upon compression perpendicular to GBs. The simulations indeed showed the coincidence of the folding line and the GB line in polycrystalline graphene during the entire process of compression in the direction perpendicular to GBs while the folding line wandered randomly over the whole region of the graphene membrane in the case of monocrystalline graphene as shown in Figure 9.12.

This result indicates that the folding of graphene can be systematically tuned in a predictable manner using GBs, which has significant implication to the nanomechanic design of graphene devices. It should be noted that the GBs in this system were closely located each other with a separation of 24–26 Å and the compression was performed in the nanoscale. If the compression is performed in the macroscopic scale, the graphene could possibly bend at positions other than GBs.

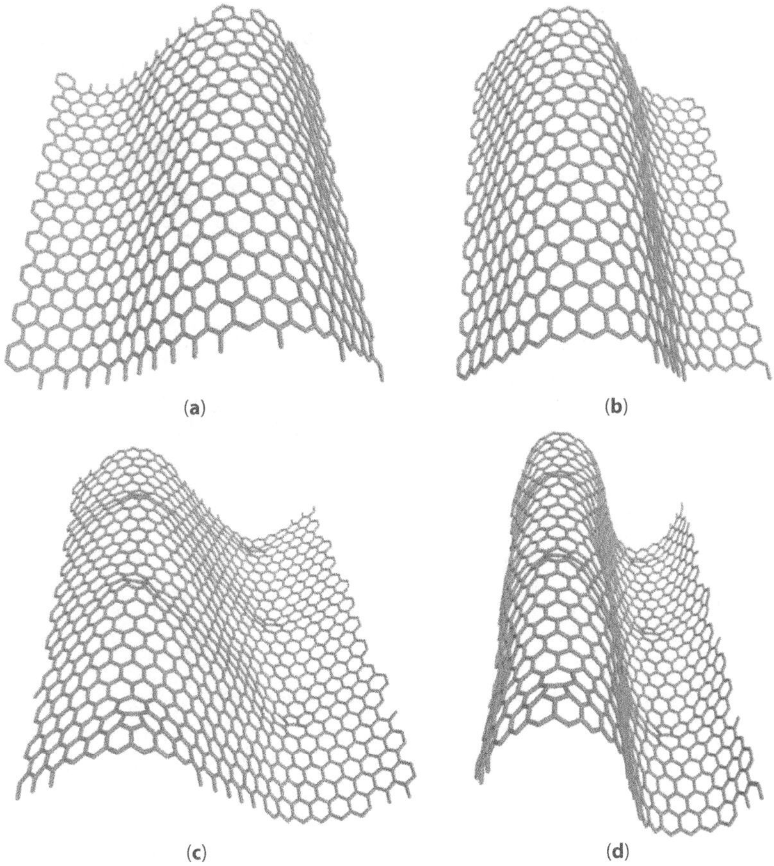

Figure 9.12 (a, b) The structural variation of monocrystalline graphene under compression at a compressive strain of 0.214 and 0.28, respectively. (c, d) The structures of T4 polycrystalline graphene under the compression perpendicular to GBs at a compressive strain of 0.214 and 0.28, respectively. GBs are highlighted in red for clarity.

The compression (which arises in the form of bending) in the perpendicular direction to GBs does not require as much energy as that of elongation. It is also the same under the compression parallel to GBs if the graphene system has a flat initial structure. However, when the T_1 graphene system is slightly folded along GBs, the compression parallel to GBs required five to six times greater energy than that of a flat initial structure (Figure 9.13), in which the folded graphene was assumed to have a sinusoidal shape from the side view in which the ratio of amplitude and wavelength was set to be ~0.186.

In this case, a rapid and remarkable decrease in the internal energy density (energy per unit area) was observed at a compressive strain of 0.172, and it resembled the critical buckling strength that can be found in the compression of graphene attached to a substrate [33].

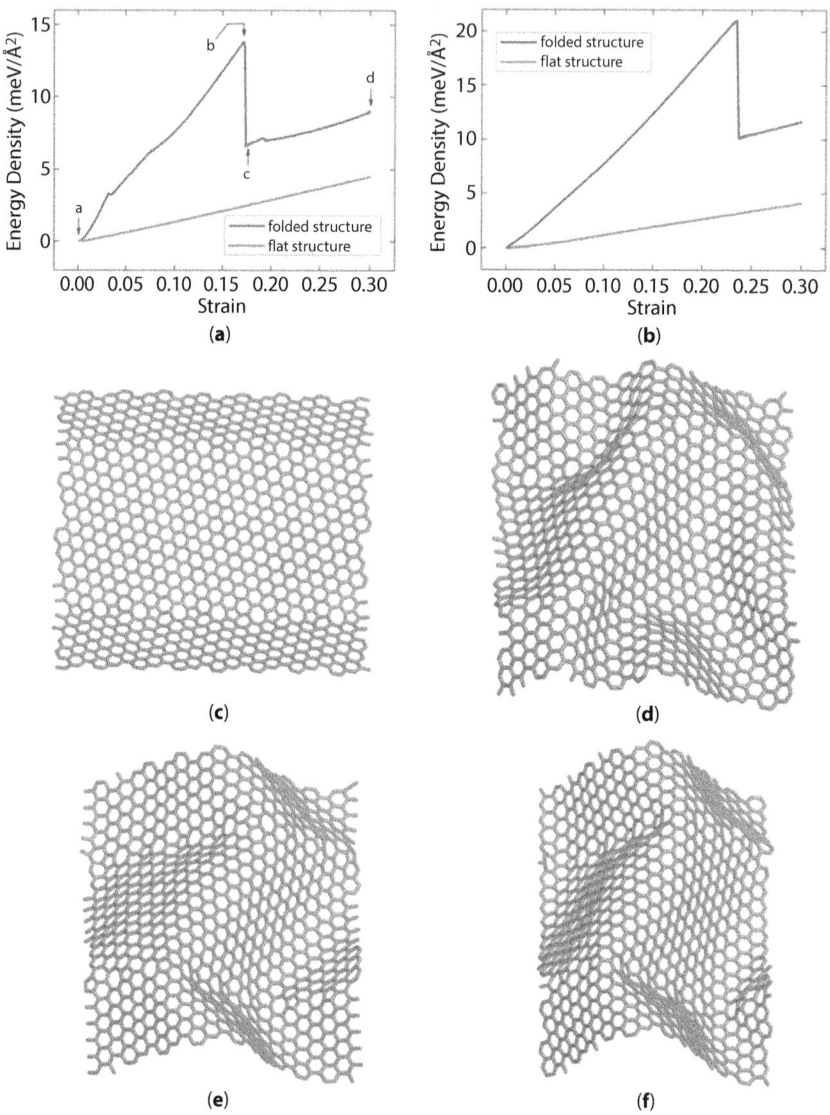

Figure 9.13 The variation of energy density under compression parallel to GBs for the folded and flat structures of (a) T1 system and (b) TZ system. (c–f) The structures of T1 system corresponding to stages of a–d in Figure 9.13a, respectively.

In addition, the slope of the energy density vs. strain curve also decreased considerably from 82 to 21 meV/Å^2 after the critical compressive strain of 0.172, allowing its easier movement. At the critical strain of 0.172, the large amount of energy was spent in changing the shape of the graphene drastically, indicating that the fold formed along the GBs can act as intrinsic reinforces in the graphene membrane.

9.3 The GB Orientation Effects on the Tensile Fracture

It is worth investigating how the GB orientational effect affects the tensile fracture mechanics of graphene by considering two representative GB groups, namely, AC- and ZZ-oriented tilted GBs as shown in Figure 9.14 [34]. We see that the closest direction to the normal direction to GB lines is ZZ and AC direction for ZZ- and AC-oriented GBs, respectively, as shown in magenta in Figure 9.14b and e. The degree of this orientation line (ZZ or AC direction) deviating from normal to GB lines is referred to as a misorientation angle. In other words, a misorientation angle denotes an intersection angle between the direction of orientation line and the direction normal to GB lines. For each of ZZ- and/or AC-oriented GB groups, we examined three specific cases by decreasing a misorientation angle from the largest magnitude and they were denoted by Ti (i = 1–3, in order of the largest to the smallest misorientation angle).

Following this nomenclature, we introduced symbols of ZZT$_i$ and ACT$_i$ to discern subgroups of ZZ- and AC-oriented GBs, respectively. The misorientation angles were 20.4°, 13.6°, and 11.1° for ZZT$_i$ (i = 1–3), respectively, and they were 27.5°, 22.5°, and 18.3° for ACT$_i$ (i = 1–3), respectively.

The simulation studies showed that polycrystalline graphene exhibited a remarkably different MACC production at the tensile fracture depending on the GB orientation, being

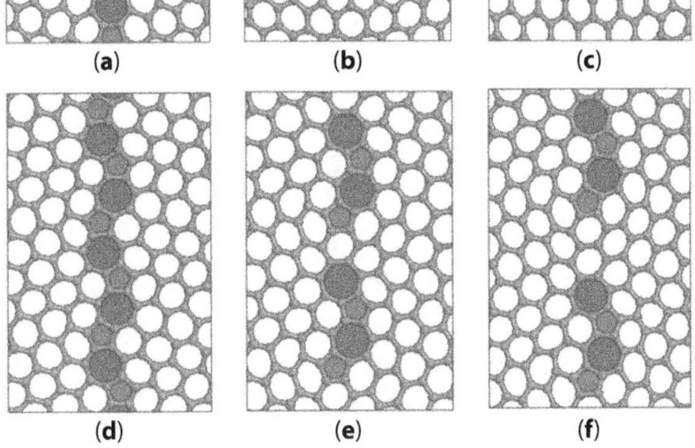

Figure 9.14 The structures of (a) ZZT$_1$ [(2,1)|(2,1)], (b) ZZT$_2$ [(3,2)|(3,2)], and (c) ZZT$_3$ (4,3)|(4,3)] for ZZ-oriented GBs and (d) ACT$_1$ [(3,1)|(3,1)], (e) ACT$_2$ [(4,1)|(4,1)], and (f) ACT$_3$ [(5,1)|(5,1)] for AC-oriented GBs. The bracket means the chiral directional notation of the GBs.

1.2–3.0 times denser and 1.6–5.0 times longer for ZZ-oriented GBs compared to those of AC-oriented GBs as shown in Figure 9.15. It is noteworthy that even GB-free, monocrystalline graphene shows a certain degree of the MACC production when it is elongated in the ZZ direction whereas it does not produce any MACC in the AC direction.

It is also worthy to note that polycrystalline graphene with ZZ-oriented GBs can yield considerably higher density and achievable length (4.51 nm^{-2} and 1.47 nm, maximally) of MACCs compared to monocrystalline graphene. The time-resolved analysis on the evolution of the MACC production provided informative pictures of this phenomenon (Figure 9.16). As the strain increased, the density of MACC decreased in a stepwise way, suggesting a series of the MACC breaking, and it lasted for a markedly longer time in the case of ZZ-oriented GBs compared to AC-oriented GBs.

As the misorientation angle increased, the tensile strength of graphene membranes with ZZ- and/or AC-oriented GBs (Figure 9.17a and b) decreased. This counterintuitive

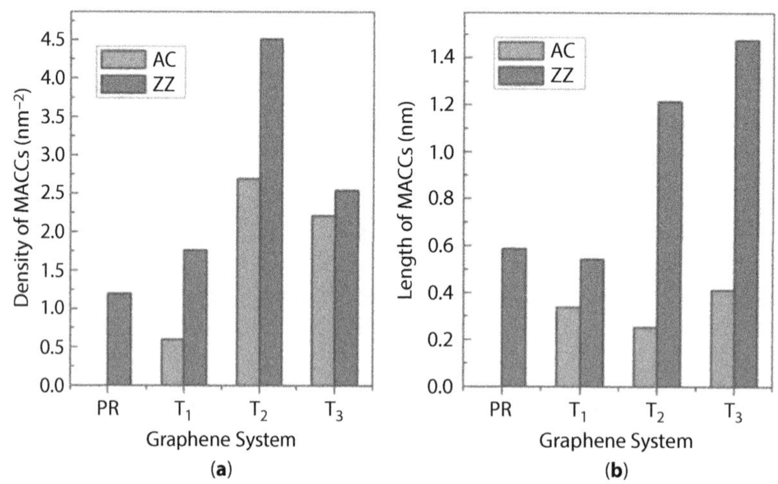

Figure 9.15 The achievable maximum values of (a) the density and (b) the length of MACC obtained from pristine graphene and polycrystalline graphenes with various AC- and/or ZZ-oriented GBs.

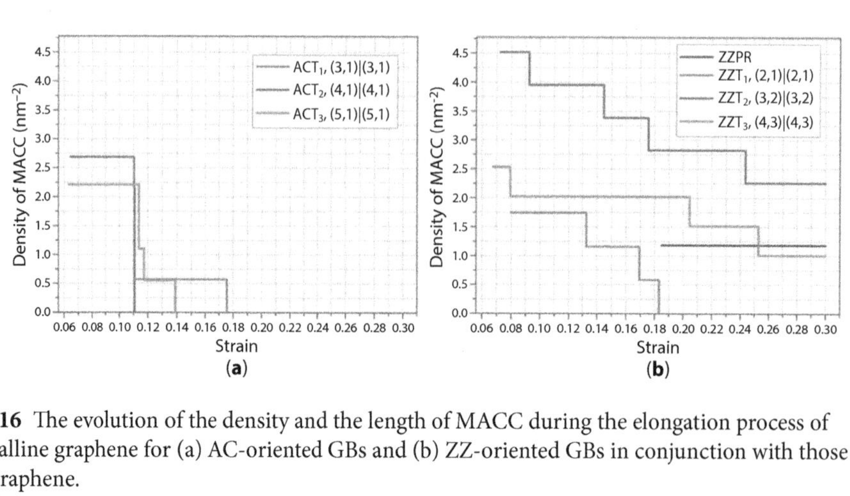

Figure 9.16 The evolution of the density and the length of MACC during the elongation process of polycrystalline graphene for (a) AC-oriented GBs and (b) ZZ-oriented GBs in conjunction with those of pristine graphene.

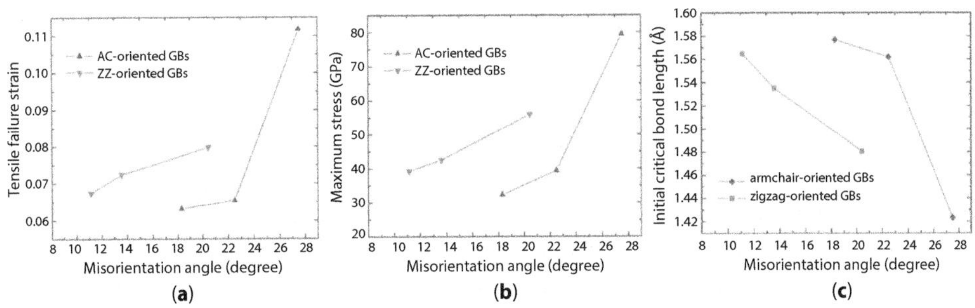

Figure 9.17 (a) The ultimate tensile strain and (b) its corresponding maximum stress as a function of a misorientation angle for the elongation process and (c) the initial critical bond lengths as a function of a misorientation angle for polycrystalline graphene composed of various AC- and ZZ-oriented GBs.

tendency can be explained by the degree of prestress in GBs, which is related to the maximum length of critical bonds at the initial stage (Figure 9.17c), as indicated in the case of graphene with ZZ-oriented GBs in the previous section. Specifically, the initial length of the (most highly strained) critical bond in GBs, which is one edge of the heptagonal ring, decreased as the misorientation angle increased and the crack initiated at this bond.

Atomic stress analyses clearly supported this fact for both cases of AC- and ZZ-oriented GBs as shown in Figure 9.18, and also showed that all key reactions emerging before tensile failure are not affected remarkably by the GB orientation (i.e., ZZ and/or AC orientation) of polycrystalline graphene since the reactions only occurred within the structure of GBs in contrast to the MACC dynamics occurring after tensile failure.

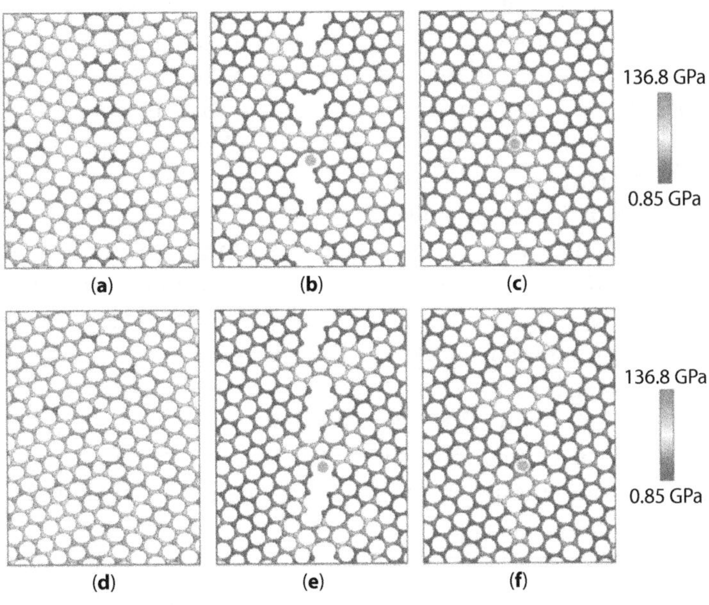

Figure 9.18 The distributions of atomic stresses for (a) ZZT_1 and (b) ZZT_2 just before tensile failure and for (c) ZZT_2 just before initial crack formation, and the distributions of atomic stresses for (d) ACT_1 and (e) ACT_2 just before tensile failure and for (f) ACT_2 just before initial crack formation. The green-filled circles in (b) and (c) and those in (e) and (f) denote the identical positions, respectively.

9.4 Orientation-Dependent Tensile Fracture in Monocrystalline Graphene

Graphene is simultaneously both very strong and stretchy, and such unique mechanical properties make it difficult to predict the orientation-dependent fracture strength and strain of graphene by using theories that have been commonly used to explain the mechanical properties of brittle and/or ductile materials. However, it is almost impossible to experimentally get detailed information about mechanical properties of graphene due to severe technical difficulties even using modern technologies. Despite such experimental technical challenges, by performing systematic molecular dynamics simulations about the tensile process of monocrystalline graphene in a wide range of tensile orientations (Figure 9.19), we are able to obtain the data about it and can also construct a numerical platform for its generic prediction based on the data [35]. Following this scheme, the tensile strength of graphene was plotted as a function of a tensile orientation angle θ, where θ is defined to be the angle made between the (reference) AC direction and the specific tensile direction (Figure 9.19). Consequently, for the AC and ZZ tensile directions, θ is 0° and 30°, respectively, and θ ranging between these values can denote an arbitrary tensile direction due to the six-fold symmetry of graphene.

The simulation study showed that as the uniaxial tensile direction rotates from AC (θ = 0°) to ZZ orientation (θ = 30°), both the tensile strength and strain remain almost constant up to an orientation angle (θ) of 12°, then rapidly increase (the quasi-exponential growth) at the temperature of 100 K as shown in Figure 9.20a. This unique fracture pattern was consistently observed at various temperatures, i.e., 100, 300, 500, and 700 K, although the tensile strength quantitatively decreases as the temperature increases due to the thermal softening (Figure 9.23). The tensile strain for fracture showed a very similar orientation dependence compared to that of the tensile strength (Figure 9.20b).

In order to investigate their orientation dependence excluding the thermal effect, the tensile strength and strain were plotted in a reduced form by dividing them with the values at an orientation angle of 0° (i.e., those obtained for the armchair tensile direction) at the temperatures of 100–700 K. These curves showed excellent coincidence with each other when they were put together (Figure 9.20c and d), which indicates that an identical physical origin exists regardless of the temperature.

As the orientation angle increases, the tensile strength increases quasi-exponentially in stark contrast to the strength obtained from the brittle counterpart in which the secant square growth is expected as shown in green dotted lines in Figure 9.20. To gain insight into the physics of this phenomena, the structural evolution of graphene was carefully monitored during the tensile process. The fracture of graphene dominantly occurred along the ZZ-lines of the hexagonal structure, regardless of the tensile orientation and temperature (Figure 9.21). Similar results were observed in the axial elongation of carbon nanotubes (CNTs) with various chiralities as well as in the longitudinal elongation of graphene nanoribbons with AC and/or ZZ edges [36, 37].

In these studies, Cauchy–Born equation was employed to describe homogenous deformation and predict the values of fracture strain at various tensile orientations. However, lattices with a basis like that of graphene lacks inversion symmetry and thus graphene nanoribbons and CNTs do not follow the Cauchy–Born rule precisely as demonstrated in Dumitrica *et al.*'s work [37]. More importantly, the orientation-dependent tensile strengths of CNTs

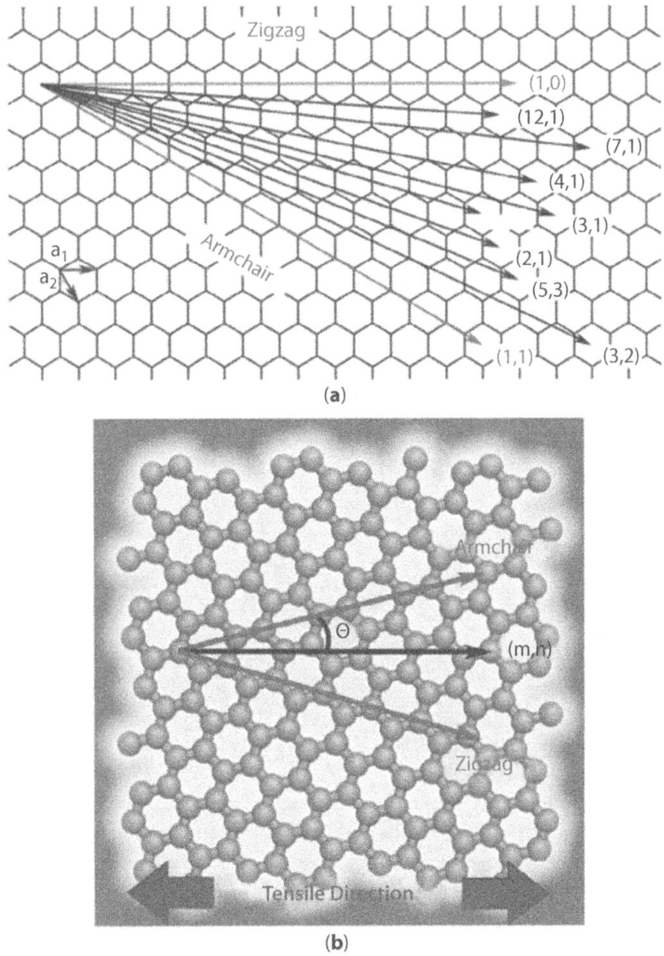

Figure 9.19 (a) The representative nine tensile orientations selected to investigate the orientation-dependent tensile mechanics of graphene, for which the corresponding chiral notations are given in parentheses. (b) The illustrated uniaxial tensile simulation system where the tensile direction is set to an arbitrary (m,n) chiral direction, and an orientation angle θ is defined as the angle made between the relevant tensile direction and the AC direction.

and graphene nanoribbons have not been theoretically addressed in these studies, although the tensile strength is a more important factor than the fracture strain in many cases.

Noticing the fact that the tensile failure of graphene always takes place along the ZZ-lines and it fairly resembles the slip or breaking of metal crystals occurring on the densest lattice plane under tensile deformation, stress transformation formalism [38] was adopted to evaluate the tensile strength of graphene for an arbitrary tensile direction, which is given by

$$S_n = \frac{1}{2}(S_x + S_y) + \frac{1}{2}(S_x - S_y)\cos 2\theta + T_{xy}\sin 2\theta \tag{9.1}$$

where S_n is the (engineering) stress normal to an arbitrary plane (P-plane) that is perpendicular to the xy-plane, θ is the angle made between the x direction and the normal direction of

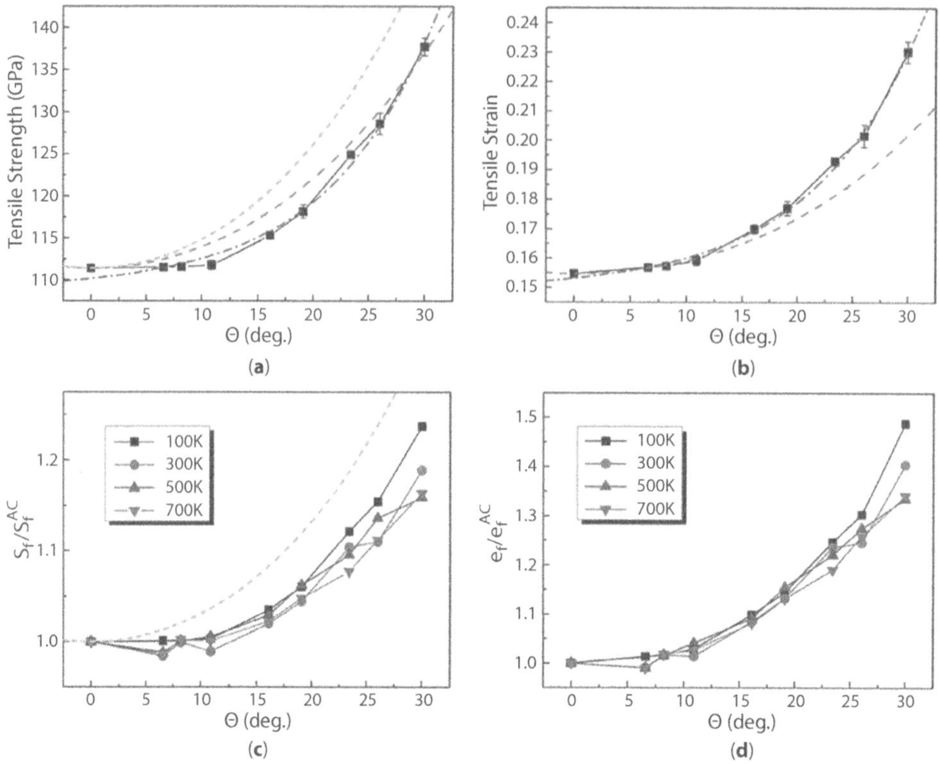

Figure 9.20 (a) The tensile strength and (b) the fracture strain of graphene plotted as a function of the orientation angle at 100 K. (c) The tensile strengths and (d) fracture strains of graphene plotted as a function of the orientation angle at various temperatures, where they are given in a reduced form such that they are divided by their respective values at the orientation angle of 0° (AC direction) for each temperature. In (a–c), the magenta dashed lines and violet dot-dashed lines indicate the values obtained from our fracture model and numerical platform, respectively, while the green short-dashed lines indicate the tensile strengths obtained from the brittle fracture model.

the P-plane, and S_x, S_y, and T_{xy} refer to the corresponding components of the Cauchy stress tensor. For simplicity, three assumptions were made in this equation for the application to the uniaxial tensile fracture of graphene. First, the x direction and normal direction correspond to the tensile direction and the AC direction that is closest to the tensile direction, respectively. Second, the values of S_y and T_{xy} are negligible during the tensile process. Third, the elongation is completed to the fracture point. The equation would then be written as

$$S_f^{AC} = \frac{S_f^{tensile}}{\cos^2\theta_f'} \tag{9.2}$$

where S_f^{AC} is the tensile strength in the AC direction, $S_f^{tensile}$ is the tensile strength in an arbitrary tensile direction, and θ_f' is the angle made between the AC direction and the tensile direction at the tensile fracture (Figure 9.22). If we assume the brittle fracture of graphene, θ_f' should approximate to θ_0 since the initial AC direction and the AC direction at tensile fracture are almost the same. However, graphene is not brittle and it stretches considerably

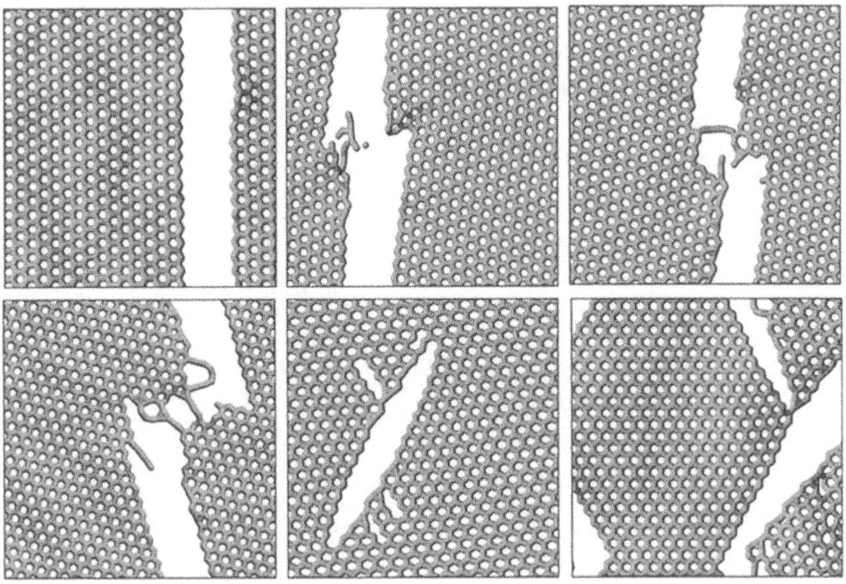

Figure 9.21 The graphene structures appearing at the fracture stage for the tensile directions of orientation angles of 0°, 6.6°, 8.2° (from left to right in the top panel), 16.1°, 23.4°, and 30° (from left to right in the bottom panel), respectively. Their equivalent chiral notations are (1, 1), (3, 2), (5, 3), (3, 1), (7, 1), and (1, 0), respectively.

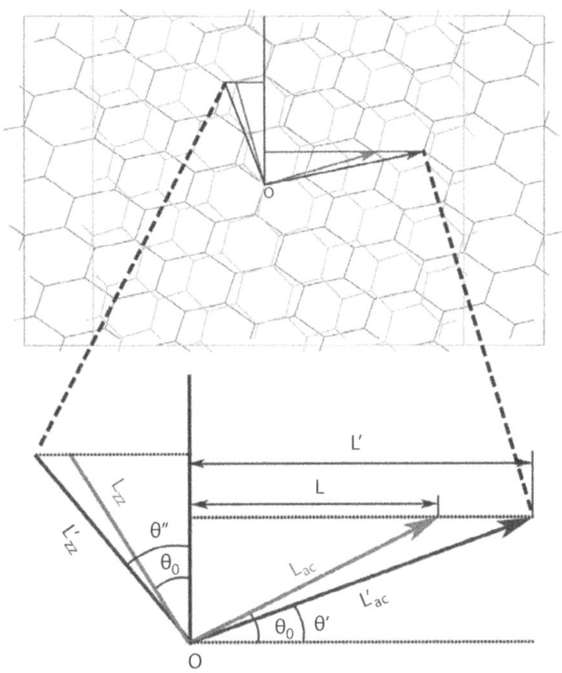

Figure 9.22 (top) The geometric variations of graphene occurring in the uniaxial tensile process and (bottom) the geometric diagram relating the angles of θ_0, θ′, and θ″ with the lengths along the AC, ZZ, and tensile directions of graphene. The red line and red arrow indicate the lengths along the AC and ZZ directions in the original graphene structure, respectively, while the blue line and blue arrow indicate the lengths in those directions in the elongated structure.

until the fracture point. Considering this factor and a possible distortion effect, it can be roughly assumed that θ_f' of Equation 9.2 should be the intersection angle measured at the tensile failure point rather than that of the initial structure (Figure 9.22).

From this geometric relationship appearing during the tensile process, several significant strains can be obtained as shown below:

$$e_{tensile} + 1 = \frac{L'}{L} = \frac{\tan\theta_0}{\tan\theta'} = \frac{\tan\theta''}{\tan\theta_0} \tag{9.3}$$

$$e_{AC} + 1 = \frac{L_{AC}}{L'_{AC}} = \frac{\sin\theta_0}{\sin\theta'} \tag{9.4}$$

$$e_{ZZ} + 1 = \frac{L'_{ZZ}}{L_{ZZ}} = \frac{\cos\theta_0}{\cos\theta''} \tag{9.5}$$

where $e_{tensile}$, e_{AC}, and e_{ZZ} are the tensile (engineering) strains along the tensile direction, AC direction, and ZZ direction, respectively. The meanings of the other variables are depicted in Figure 9.22. Given that the magnitudes of the tensile strength (S_f^{AC}) and fracture strain (e_f^{AC}) of graphene are known for the AC-directional elongation, we can evaluate the values of θ_f' and the fracture strain ($e_f^{tensile}$) of an arbitrary tensile direction using Equations 9.3–9.5. We are then able to know the value of the tensile strength ($S_f^{tensile}$) of an arbitrary tensile direction from the value of θ_f' using Equation 9.2. Here, we neglected a contraction perpendicular to the tensile direction during the tensile process as is the case in most tensile simulations.

This fracture model can reproduce the unique orientation dependence of the tensile strength of graphene at 100 K as well as explaining its physical origin well (as shown in the magenta dashed line in Figure 9.20). However, the predicted tensile strength deviated from the results obtained from the molecular dynamics simulations as the temperature increased, while the predicted fracture strain showed good agreement regardless of the temperature as shown in Figure 9.23. This implies that the structural relation (to obtain fracture strain) given in Figure 9.22 is reasonable regardless of the temperature, but the prediction using Equation 9.2 (to obtain fracture strength) is quantitatively inaccurate at the higher temperatures presumably due to the poor evaluation of a distortion effect on the tensile strength.

In order to achieve a best fitting to the fracture behavior of graphene for the entire range of tensile orientations and a wide range of operating temperatures, a numerical platform was constructed using an exponential growth template as given by

$$y(x) = A_0 \times \exp\left(\frac{x}{x_0}\right) + y_0 \tag{9.6}$$

where x is the tensile orientation angle and y is the corresponding tensile strength and/or fracture strain, while A_0, x_0, and y_0 are parameters to be determined by a fitting process.

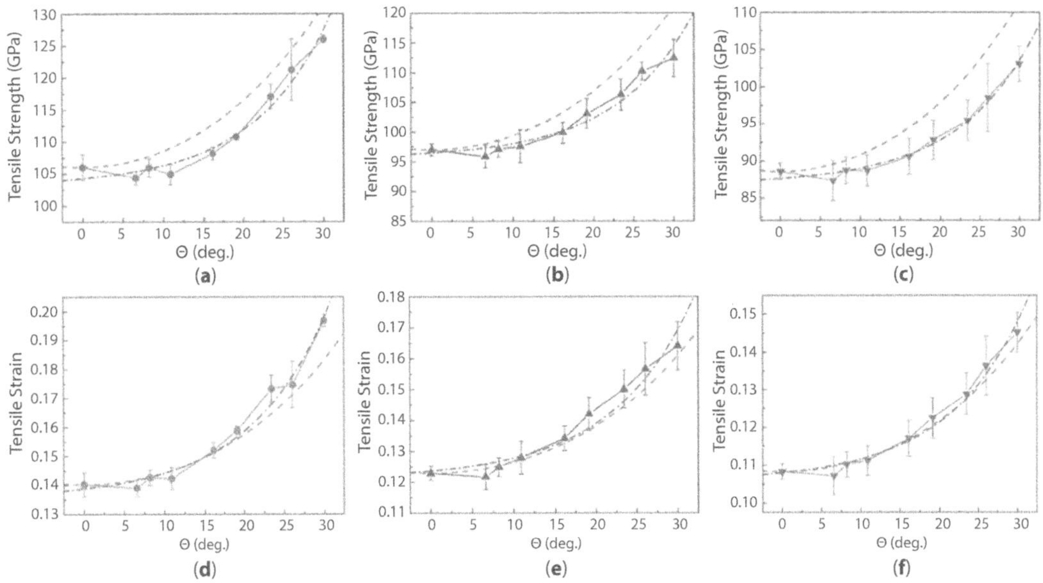

Figure 9.23 The tensile strength and the failure strain of graphene plotted as a function of the orientation angle at (a, d) 300, (b, e) 500, and (c, f) 700 K. Magenta dashed lines and violet dot-dashed lines indicate the values obtained from our fracture model and numerical platform, respectively.

Based on this regime, the parameters were first obtained so that the curves may be well fitted to the simulation results at an operating temperature of 100 K. The parameters were also obtained at 300, 500, and 700 K in a similar manner, while maintaining the value of x_0 to be that of 100 K. The fitting curves obtained using this method showed an excellent agreement with the simulation results (Figures 9.20 and 9.23). In order to cover all operating temperatures between 100 and 700 K (interpolation) as well as possibly covering the temperatures out of this range (extrapolation), quadratic polynomials of temperature were further fitted for A_0 and y_0 with a high precision [35].

Besides the orientation-dependent tensile fracture mechanics, the study on the orientation-dependent elastic response of monocrystalline graphene is also important in practical applications. Interestingly, the molecular dynamics simulations showed that graphene exhibits quasi-isotropic elastic behavior for all tensile orientations, in contrast to its distinct anisotropic tensile fracture behavior. In other words, almost the same magnitude of elastic stress is required for the same strain at all tensile orientations at 300 K as shown in Figure 9.24. Such an isotropic elastic behavior was also observed at other temperatures of 100, 500, and 700 K, indicating that this feature holds regardless of the temperature in monocrystalline graphene.

In Lee *et al.*'s indentation study [39], an isotropic mechanical response of graphene was assumed based on its six-fold rotation symmetry in deriving the equation that relates the tensile strength magnitudes and the measured values of the indenter force. Regarding this point, the above finding suggests that the isotropic mechanical response of graphene holds at a more fundamental level as far as it is concerned to the elastic deformation.

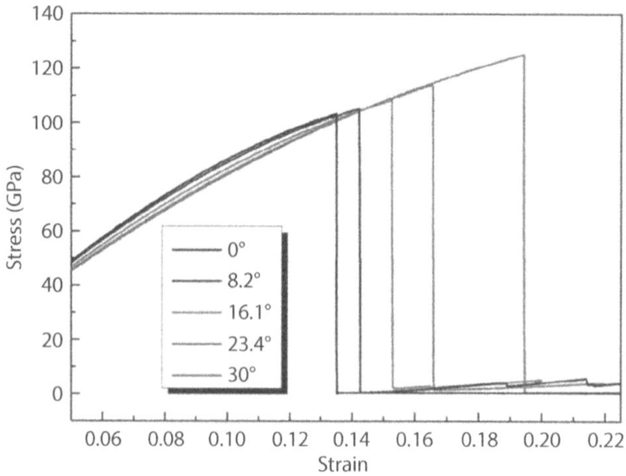

Figure 9.24 The tensile stress–strain curves of graphene obtained for a variety of tensile directions, the orientation angles of which are given in the box.

9.5 Two-Dimensional Tensile Systems: Nanoindention

It is not clear whether the anisotropic/isotropic features of the fracture/elasticity of graphene observed in the uniaxial tensile systems should be preserved in two-dimensional (2D) tensile systems and even if they are, to what extent the anisotropic performances of 1D and 2D tensile systems are quantitatively different from each other. To address these questions, a series of molecular dynamics simulations have been performed for nanoindentation of monocrystalline graphene membranes attached on SiO_2 substrates (Figure 9.25a), in which details of the scheme are similar to those of Lee *et al.*'s experimental study except for the shape of the wells patterned in the SiO_2 substrates. That is, the elliptical wells (where the lengths of the short and long axes are 30 and 60 Å, respectively) were used instead of circular one and the long-axis of the elliptical well was oriented in parallel with the AC and/or the ZZ direction of the graphene to investigate a possible anisotropic effect (Figure 9.25b).

The diameter of the indenter is around 20 Å, and the force exerted on each atom due to the indenter is given by

$$F(r) = K \times (r - R)^2 \tag{9.7}$$

where K is the specified force constant, r is the distance from each atom to the center of the indenter, and R is the radius of the indenter. The force is repulsive and F(r) = 0 for r > R. Here, K and R are set to be 1000 eV/Å^3 and 10 Å, respectively.

As the indenter gradually moved downward, the force exerted on the indenter was measured for these two systems in which the elliptical wells patterned in the silica substrates were differently oriented with each other. Virtually the same results were obtained as those observed in the uniaxial tensile process. These two indentation systems showed almost the same elastic behaviors quantitatively and qualitatively, regardless of the orientation of the elliptical wells (Figure 9.25c), indicating that the isotropic elastic behavior of graphene observed in the uniaxial (1D) tensile systems is still operative in the 2D tensile systems.

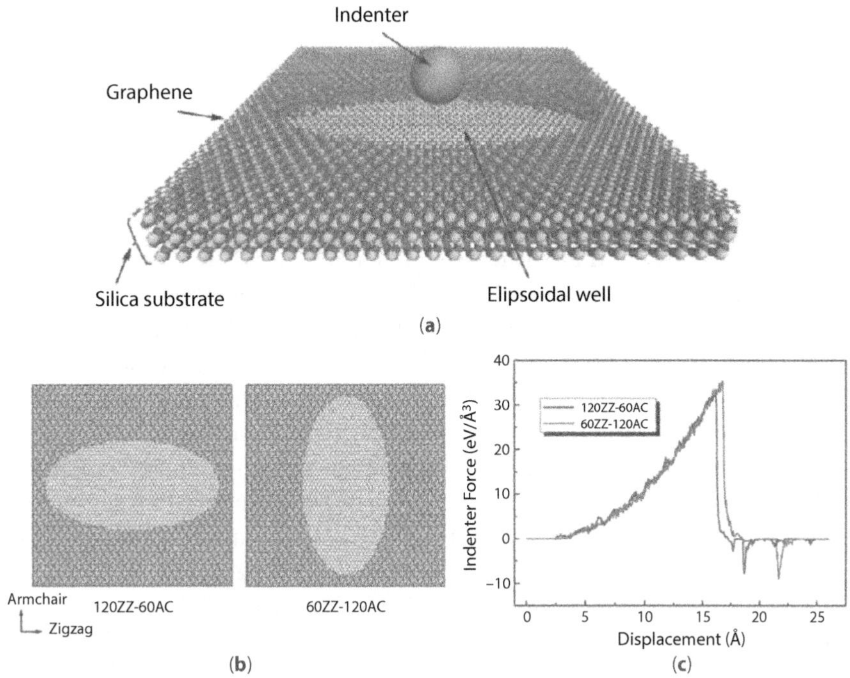

Figure 9.25 (a) The indentation simulations for the two-dimensional tensile systems in which carbon, silicon, and oxygen atoms are shown in gray, yellow, and red, respectively. (b) Two differently oriented elliptical wells that are patterned in the silica substrates. (c) The magnitudes of the force exerted on an indenter as it gradually moves down pushing and/or penetrating the graphene membrane.

In contrast to this, they yielded distinctly different values for the maximum indenter force (i.e., the breaking force) and the corresponding indenter position (Figure 9.25c). When the longer axis of the elliptical well was aligned in parallel with the AC direction (denoted as 60ZZ–120AC), it required a larger breaking force and greater movement of the indenter (34.71 ± 0.48 eV/Å and 16.60 ± 0.13 Å, respectively) compared to those (32.72 ± 0.59 eV/Å and 16.09 ± 0.11 Å, respectively) required in the other configuration (denoted as 120ZZ–60AC).

This anisotropic tensile strength can be explained as follows. The effective tensile strain of graphene of an elliptical well should be greater in the short axis direction (i.e., the shortest radius direction) than that in any other tensile directions, for the same amount of movement of the indenter.

In addition, the fracture of graphene will be critically affected by the deformation in the weakest tensile direction (that is, the AC direction). In this context, 120ZZ–60AC should require a weaker indenter force for breaking than that in 60ZZ–120AC because the short axis direction of an elliptical well coincides with the AC direction of the graphene membrane for 120ZZ–60AC while it does not in 60ZZ–120AC. This conjecture is in good agreement with the above simulation result.

In the studies using a continuum model for nanoindentation of a linear elastic graphene membrane with circular wells on the silica substrate, the tensile strength of graphene was estimated by measuring the breaking force and using the following equation [39, 40]:

$$\sigma_m^{2D} = \left(\frac{FE^{2D}}{4\pi R}\right)^{\frac{1}{2}} \tag{9.8}$$

where σ_m^{2D} is the maximum tensile stress, and E^{2D} is a fitting parameter that is set to 342 N m^{-1} as suggested by Lee *et al.* [39], R is the tip radius, and F is the breaking force.

Based on this relationship, the tensile strengths of 60ZZ–120AC and 120ZZ–60AC were roughly evaluated to be 116.15 and 112.78 GPa, respectively. They are within a reasonable range considering the tensile strengths obtained from the uniaxial tensile simulations. Although the above fracture behavior is qualitatively consistent with our understanding for the uniaxial tensile fracture of graphene, the anisotropic fracture behavior observed in the indentation process is remarkably weakened compared to the uniaxial tensile process.

This attenuation effect is attributed to the following two factors. First, as a certain tensile direction (indentation deformation is composed of many individual tensile deformations in various tensile directions) is oriented closer to the short axis direction of an elliptical well, both the *in-situ* strain (a current value) and the fracture strain (a threshold value) relevant to that tensile direction increase in the case of 60ZZ–120AC, while the *in-situ* strain and the fracture strains increase and decrease, respectively, in the case of 120ZZ–60AC. Due to such a cancelation effect, a critical tensile direction, which determines the fracture point, does not coincide with the short axis direction (the ZZ direction) of an elliptical well for 60ZZ–120AC. Instead, the critical tensile direction is located between the short axis (ZZ) direction and the AC direction that intersects the short axis direction with an angle of 30° (it should be noted that this AC direction is not the long axis AC direction of an elliptical well), lowering the magnitude of the breaking force compared to the value required in the case that the fracture occurs along the short axis (ZZ) direction (Figure 9.26). Such a decrease eventually weakens the anisotropic behavior of the fracture strength in the indentation process. This scenario is constructed assuming that the fracture occurs in the zigzag line in any cases.

The above assumption was indeed verified by investigating the fractured structures of graphene membranes under nanoindentation in which the fracture line was created along the longest axis direction (i.e., the ZZ lines that are perpendicular to the shortest axis

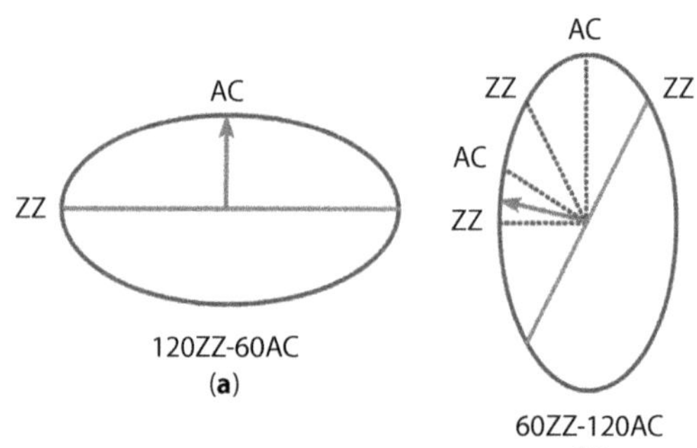

Figure 9.26 The underlying fracture mechanism in the indentation of (a) 120ZZ–60AC and (b) 60ZZ–120AC. The blue arrow denotes a critical tensile direction that determines the fracture point of the indentation process while the red lines denote a fracture line that agrees well with the simulation result shown in Figure 9.27.

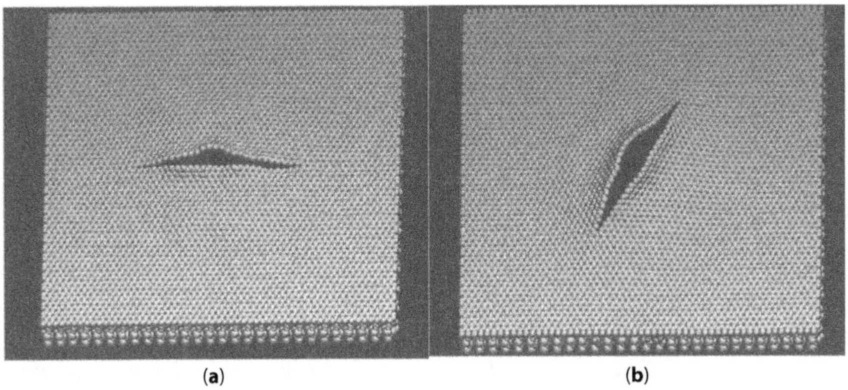

Figure 9.27 The fractured structures of graphene membranes under nanoindentation for (a) 120ZZ–60AC and (b) 60ZZ–120AC. The orientations of the graphene membranes and substrate elliptical wells are the same as those in Figure 9.25b.

direction) for 120ZZ–60AC, while the fracture line was formed along the ZZ lines of the direction that intersects the shortest axis direction with an angle of 60° for 60ZZ–120AC (Figure 9.27).

Second, the size allowed for the intersection angle between the critical tensile direction of the graphene membrane and the short axis direction of an elliptical well is at most as small as 30° (accordingly, the difference between the two radii of an elliptical well corresponding to these directions should also be small). As a result, the indentation process always brings severe impact on the tensile fracture of the graphene membrane, irrespective of the orientation of the elliptical well. This is an inevitable factor arising from the inherent graphene structure, different from the first factor, which can hardly be resolved and hence inherently suppresses the anisotropic degree of indentation fracture of graphene membranes.

It is noteworthy that the indentation scheme used in this study, which utilizes elliptical substrate wells, can be directly applied to experiments for the anisotropic mechanical study of graphene membranes, since an indentation scheme using circular substrate wells has already been well established experimentally [5]. This experimental application has great importance because the anisotropic tensile fracture behavior of graphene has never been validated experimentally. It should be noted that both cases of using an anisotropic shape of the indenter tips and an anisotropic shape of the substrate wells were examined computationally and it was concluded that the latter is far more efficient than the former for the anisotropic fracture study of graphene membranes.

References

1. Chen, J.H., Jang, C., Xiao, S., Ishigami, M., Fuhrer, M.S., Intrinsic and extrinsic performance limits of graphene devices on SiO_2. *Nat. Nanotechnol.*, 3, 206, 2008.
2. Bolotin, K.I., Sikes, K.J., Jiang, Z., Klima, M., Fudenberg, G., Hone, J., Kim, P., Stormer, H.L., Ultrahigh electron mobility in suspended graphene. *Solid State Commun.*, 146, 351, 2008.
3. Balandin, A.A., Ghosh, S., Bao, W., Calizo, I., Teweldebrhan, D., Miao, F., Lau, C.N., Superior thermal conductivity of single-layer graphene. *Nano Lett.*, 8, 902, 2008.

4. Seol, J.H., Jo, I., Moore, A.L., Lindsay, L., Aitken, Z.H., Pettes, M.T., Li, X., Yao, Z., Huang, R., Broido, D., Mingo, N., Ruoff, R.S., Shi, L., Two-dimensional phonon transport in supported graphene. *Science*, 328, 213, 2010.
5. Lee, C., Wei, X., Kysar, J.W., Hone, J., Measurement of the elastic properties and intrinsic strength of monolayer graphene. *Science*, 321, 385, 2008.
6. Zhao, H., Min, K., Aluru, N.R., Size and chirality dependent elastic properties of graphene nanoribbons under uniaxial tension. *Nano Lett.*, 9, 3012, 2009.
7. Liu, F., Ming, P.M., Li, J., *Ab initio* calculation of ideal strength and phonon instability of graphene under tension. *Phys. Rev. B*, 76, 064120, 2007.
8. Banhart, F., Kotakoski, J., Krasheninnikov, A.V., Structural defects in graphene. *ACS Nano*, 5, 26, 2011.
9. Li, X., Cai, W., An, J., Kim, S., Nah, J., Yang, D., Piner, R., Velamakanni, A., Jung, I., Tutuc, E., Banerjee, S.K., Colombo, L., Ruoff, R.S., Large-area synthesis of high-quality and uniform graphene films on copper foils. *Science*, 324, 1312, 2009.
10. Bae, S., Kim, H., Lee, Y., Xu, X., Park, J.S., Zheng, Y., Balakrishnan, J., Lei, T., Kim, H.R., Song, Y.I., Kim, Y.J., Kim, K.S., Özyilmaz, B., Ahn, J.H., Hong, B.H., Iijima, S., Roll-to-roll production of 30-inch graphene films for transparent electrodes. *Nat. Nanotechnol.*, 5, 574, 2010.
11. Coraux, J., N'Diaye, A.T., Engler, M., Busse, C., Wall, D., Buckanie, N., Heringdorf, F.M., Gastel, R., Poelsema, B., Michely, T., Growth of graphene on Ir(111). *New J. Phys.*, 11, 023006, 2009.
12. Miller, D.L., Kubista, K.D., Rutter, G.M., Ruan, M., Heer, W.A.D., First, P.N., Stroscio, J.A., Observing the quantization of zero mass carriers in graphene. *Science*, 324, 924, 2009.
13. Loginova, E., Nie, S., Thurmer, K., Bartelt, N.C., McCarty, K.F., Defects of graphene on Ir(111): Rotational domains and ridges. *Phys. Rev. B*, 80, 085430, 2009.
14. Park, H.J., Meyer, J., Roth, S., Skákalová, V., Growth and properties of few-layer graphene prepared by chemical vapor deposition. *Carbon*, 48, 1088, 2010.
15. Stone, A.J. and Wales, D.J., Theoretical studies of icosahedral C60 and some related species. *Chem. Phys. Lett.*, 128, 501, 1986.
16. Thrower, P.A., The study of defects in graphite by transmission electron microscopy. *Chem. Phys. Carbon*, 5, 217, 1969.
17. Malola, S., Hakkinen, H., Koskinen, P., Structural, chemical, and dynamical trends in graphene grain boundaries. *Phys. Rev. B*, 81, 165447, 2010.
18. Yazyev, O.V. and Louie, S.G., Topological defects in graphene: Dislocations and grain boundaries. *Phys. Rev. B*, 81, 195420, 2010.
19. Lahiri, J., Lin, Y., Bozkurt, P., Oleynik, I.I., Batzill, M., An extended defect in graphene as a metallic wire. *Nat. Nanotechnol.*, 5, 326, 2010.
20. Yazyev, O.V. and Louie, S.G., Electronic transport in polycrystalline graphene. *Nat. Mater.*, 9, 806, 2010.
21. Liu, Y. and Yakobson, B.I., Cones, pringles, and grain boundary landscapes in graphene topology. *Nano Lett.*, 10, 2178, 2010.
22. Grantab, R., Shenoy, V.B., Ruoff, R.S., Anomalous strength characteristics of tilt grain boundaries in graphene. *Science*, 330, 946, 2010.
23. Huang, P.Y., Ruiz-Vargas, C.S., van der Zande, A.M., Whitney, W.S., Levendorf, M.P., Kevek, J.W., Garg, S., Alden, J.S., Hustedt, C.J., Zhu, Y., Park, J., McEuen, P.L., Muller, D.A., Grains and grain boundaries in single-layer graphene atomic patchwork quilts. *Nature*, 469, 389, 2011.
24. Jhon, Y.I., Zhu, S., Ahn, J.H., Jhon, M.S., The mechanical responses of tilted and non-tilted grain boundaries in graphene. *Carbon*, 50, 3708, 2012.
25. Heimann, R.B., Evsyukov, R.B., Kavan, L., *Carbyne and carbynoid structures*, Springer, Berlin, 1999.
26. Cataldo, F., *Polyynes: Synthesis, properties and applications*, CRC Press, Boca Raton, 2005.

27. Liu, Y., Jones, R.O., Zhao, X.L., Ando, Y., Carbon species confined inside carbon nanotubes: A density functional study. *Phys. Rev. B*, 68, 125413, 2003.
28. Jin, C., Lan, H., Suenaga, K., Iijima, S., Deriving carbon atomic chains from graphene. *Phys. Rev. Lett.*, 102, 205501, 2009.
29. Chuvilin, A., Meyer, J.C., Algara-Siller, G., Kaiser, U., From graphene constrictions to single carbon chains. *New J. Phys.*, 11, 083019, 2009.
30. Akdim, B. and Pachter, R., Switching behavior of carbon chains bridging graphene nanoribbons: Effects of uniaxial strain. *ACS Nano*, 5, 1769, 2011.
31. Lin, Z.Z., Zhuang, J., Ning, X.J., High-efficient tunable infrared laser from MACCs between graphenes. *Europhys. Lett.*, 97, 27006, 2012.
32. Cretu, O., Botello-Mendez, A.R., Janowska, I., Pham-Huu, C., Charlier, J., Banhart, F., Electrical transport measured in atomic carbon chains. *Nano Lett.*, 13, 3487, 2013.
33. Frank, O., Tsoukleri, G., Parthenios, J., Papagelis, K., Riaz, I., Jalil, R., Novoselov, K.S., Galiotis, C., Compression behavior of single-layer graphenes. *ACS Nano*, 4, 3131, 2010.
34. Jhon, Y.I., Chung, P.S., Smith, R., Min, K.S., Yeom, G.Y., Jhon, M.S., Grain boundaries orientation effects on tensile mechanics of polycrystalline graphene. *RSC Adv.*, 3, 9897, 2013.
35. Jhon, Y.I., Jhon, Y.M., Yeom, G.Y., Jhon, M.S., Orientation dependence of the fracture behavior of graphene. *Carbon*, 66, 619, 2014.
36. Zhao, H., Min, K., Aluru, N.R., Size and chirality dependent elastic properties of graphene nanoribbons under uniaxial tension. *Nano Lett.*, 9, 3012, 2009.
37. Dumitrica, T., Hua, M., Yakobson, B.I., Symmetry-, time-, and temperature-dependent strength of carbon nanotubes. *Proc. Natl. Acad. Sci. USA*, 103, 6105, 2006.
38. Bucciarelli, L.L., *Engineering mechanics for structures*, Dover Publications, New York, 2009.
39. Lee, C., Wei, X., Kysar, J.W., Hone, J., Measurement of the elastic properties and intrinsic strength of monolayer graphene. *Science*, 321, 385, 2008.
40. Bhatia, N.M. and Nachbar, W., Finite indentation of an elastic membrane by a spherical indenter. *Int. J. Nonlinear Mech.*, 3, 307, 1968.

10

Graphene and Its Derivatives as Platforms for MALDI-MS

Hani Nasser Abdelhamid[1]* and Hui-Fen Wu[2,3,4,5†]

[1]*Advanced Funtional Materials Laboratory, Department of Chemistry, Faculty of Science, Assiut University, Assiut, Egypt*
[2]*Department of Chemistry and Center for Nanoscience and Nanotechnology, National Sun Yat-Sen University, Kaohsiung, Taiwan*
[3]*School of Pharmacy, College of Pharmacy, Kaohsiung Medical University, Kaohsiung, Taiwan*
[4]*Institute of Medical Science and Technology, National Sun Yat-Sen University, Kaohsiung, Taiwan*
[5]*Doctoral Degree Program in Marine Biotechnology, National Sun Yat-Sen University and Academia Sinica, Kaohsiung, Taiwan*

Abstract
Graphene (G) and its derivatives have advanced mass spectrometry (MS). Graphene-based nanomaterials offer several derivatives such as graphene (G), graphene oxide (GO), reduced graphene oxide (rGO), and others. These derivatives are a rich library for applications including MS. Applications of graphene and its derivatives for MS offer detection of proteins, peptides, carbohydrates, oligonucleotides, biomarkers, and small molecules. The large surface area and the simple modification of G or its derivatives provide high sensitivity and offer better selectivity. Graphene derivatives' advanced several techniques include surface-assisted laser desorption/ionization mass spectrometry (SALDI-MS) and graphene-assisted laser desorption/ionization mass spectrometry (GALDI-MS). These methods show no interferences at low mass range (<1000 Da) and can be applied for the detection of small molecules with a free background spectrum. GALDI-MS shows no fragmentation and can be applied for labile biomolecules. This book chapter summarizes the applications of G or its derivatives for LDI-MS.

Keywords: Graphene, mass spectrometry, matrix-assisted laser desorption/ionization mass spectrometry, surface-assisted laser desorption/ionization mass spectrometry

10.1 Introduction

Graphene is a fully sp^2-hybridized 2D carbon nanomaterial with an atomically carbon size thickness (thickness of ~1 nm) [1]. Graphene and its derivatives have been applied for separation and preconcentration of protein species [2], biosensors [3], bioapplications [4], spintronics [5], devices [5], photodetectors [6], sensing [7], environmental applications [8],

**Corresponding author*: hany.abdelhameed@science.au.edu.eg
†*Corresponding author*: hwu@faculty.nsysu.edu.tw

Tobias Stauber (ed.) Handbook of Graphene: Volume 2, (273–290) © 2019 Scrivener Publishing LLC

biomedical applications [8], and filtration [9]. Graphene or its derivatives have advanced several fields including mass spectrometry (MS).

Matrix-assisted laser desorption/ionization time-of-flight mass spectrometry (MALDI-TOF) offers a soft ionization method [10]. Organic compounds, known as matrices, assist laser desorption/ionization mass spectrometry (LDI-MS) process for the investigated analytes [11, 12]. The LDI-MS process takes place also using salts of the organic matrices (known also as ionic liquid matrices) [13–18], and nanoparticles [19]. Nanomaterials, including carbon-based nanomaterials [20], quantum dots [21–27], magnetic nanoparticles [28–36], gold nanoparticles [37, 38], and others [39], were applied as surface for LDI-MS [40]. Nanoparticles show low background compared to organic matrices. They have low or no fragmentation for thermal labile species. They offer high sensitivity and better selectivity. The large surface area of nanoparticles gives high adsorption capacity and leads to higher ionization without the requirement of crystallization.

Graphene and its derivatives, including graphite particles [41], graphite in silicone polymer [42], graphene oxide (GO) [43], and fluorographene (FG) [44], were used as matrix for MALDI-MS. They were used for polar compounds including amino acids, polyamines, anticancer drugs and nucleosides [45], proteins [43], carbohydrates, oligonucleotides [46], polymers [47], and small molecules. They offer high sensitivity, ionization efficiency, and selectivity, and can be used as probe for separation/extraction and high throughput analysis.

This book chapter introduces the applications of graphene and its derivatives for the applications of LDI-MS. A brief introduction of MALDI-MS using conventional organic matrices will be discussed. The applications of graphene-based nanoparticles as matrices, probe, surface, pseudo-stationary phase and thin film technologies will be reviewed.

10.2 Matrix-Assisted Laser Desorption/Ionization Mass Spectrometry (MALDI-MS)

The analyte with high absorption of laser energy can be directly desorbed/ionized. Thus, these species can be analyzed using LDI-MS. However, LDI-MS is only limited to the species with high laser absorption. For the species that lack of laser absorption, a small organic matrix is usually used to assist the LDI-MS process. There are certain requirements for the organic compounds to serve as an effective matrix. First, organic matrices should have high absorption at the wavelength of the used laser. There are several types of laser including N_2 laser (337 nm), Nd:YAG lasers (355 nm and 266 nm), Er:YAG laser (2.94 μm), and carbon dioxide laser (10.6 μm). For each laser, there are certain organic matrices. Second, organic matrices should be lacking of sublimation under vacuum. Derivatives of benzoic acids are usually sublime before the interactions with the laser radiation. Thus, they show low ionization efficiency and lack of high reproducibility. Third, organic matrices usually undergo proton transfer with the investigated analytes. Organic matrices may be having acidic or basic properties. The protonated ($[M+H]^+$, M refers to the analyte) or alkali adducts ($[M+Na]^+$, or $[M+K]^+$) are usually observed. Fourth, the organic matrices should have crystallized with the target analyte to transfer the laser energy. The crystallization offers sweet spots that show high ionization.

Conventional organic matrices show wide applicability and can be applied for analytes including proteins, peptides, carbohydrates, oligonucleotides, small molecules, and metal ions [11]. They can be applied for molecules with high and low molecular weight. MALDI-MS using organic matrices offer soft ionization and is complementary to other soft ionization methods including electrospray ionization mass spectrometry (ESI-MS) [48–51]. However, the organic matrices have low ionization and need a high amount in order to ionize the investigated analytes. The optimum ratio of analyte:organic matrices are 1:1000–1:100000. The acid/base characters of organic matrices sometimes cause denature or fragmentation of proteins or large biomolecules. The self-ionization of organic matrices shows interference and limit the application of MALDI-MS for the analysis of small molecules. Thus, inorganic matrices including nanoparticles have been investigated to circumvent most of the drawbacks of conventional organic matrices.

Graphene and its derivatives can be used as surface to assist LDI-MS. Schematic representation for LDI-MS using graphene is shown in Figure 10.1. The investigated analytes are adsorbed in the surface of graphene or its derivatives. The analyte is desorbed/ionized using laser irradiation (Figure 10.1). The ionized analytes are extracted and accelerated before the free field region (time of flight (TOF) tube). The ions are detected in reflector detector for high molecular weight analytes or linear detector for small molecules (Figure 10.1).

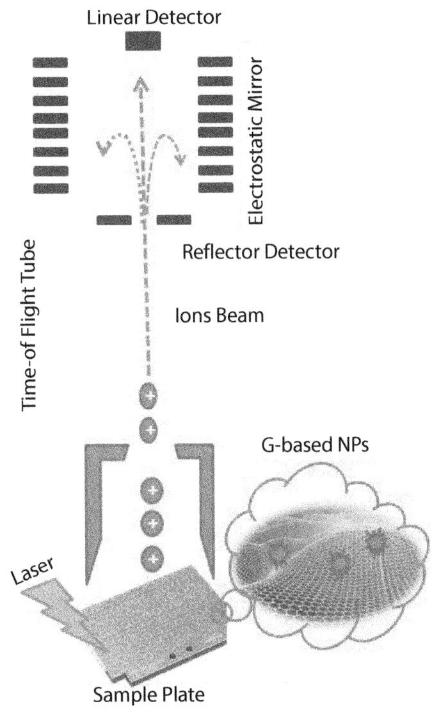

Figure 10.1 Schematic representation for LDI-MS using graphene and its derivatives as surface.

10.3 Application of Graphene and Its Derivatives for the Analysis of Large Biomolecules

Protein analysis is important for many disciplines including proteomics, biomedical, biomedicine, biotechnology, and others. The analysis of protein is crucial due to the low stability of protein species and the degradation tendency of the protein molecules. Thus, soft ionization and simple analysis of protein is highly desired.

The sample preparation for protein species is simple (Figure 10.2). The sample solution containing protein species is mixed with graphene (Figure 10.2). The solution was incubated for a few minutes for desorption of the analytes on the surface of graphene. Then, the sample was separated using centrifugation or other separation method. The separated species is rinsed and redispersed in small solvent amount and spotted in MALDI-MS plate (Figure 10.2). The spot after drying is analyzed using MALDI-MS.

There are several graphene derivatives that are applied as matrices or surface (Figure 10.3). The surface of graphene or its derivatives can be simply modified with biomolecules including aptamers that can offer selective detection of protein species. Protein mucin1 (MUC1), cancer biomarkers, was selectively detected using MUC1-binding aptamer (Apt_{MUC1})-conjugated gold nanoparticles immobilized graphene oxide (Apt_{MUC1}–Au NPs-GO) [52]. LDI-MS offered limit of detection (LOD) of as few as 100 MCF-7 cells [52]. The binding forces of graphene with biomolecules such as hydrophobic and π–π interactions enhance the efficiency of extraction and sensitivity of detection. Graphene provides an ultrahigh efficiency surface-enhanced laser desorption/ionization (SELDI) probe for single-stranded DNA and proteins [53]. The limit of detection (LOD) was 100 fM for DNA and 1 pM for protein [53]. Graphene derivatives have been applied for the analysis of proteins including N-linked glycans [54], lysozyme, α-lactalbumin, cellulase, and trypsin [43].

The surface of graphene derivatives can be simply tailored with other nanoparticles for multifunctional applications (Figure 10.3). Cu^{2+}-immobilized magnetic graphene@

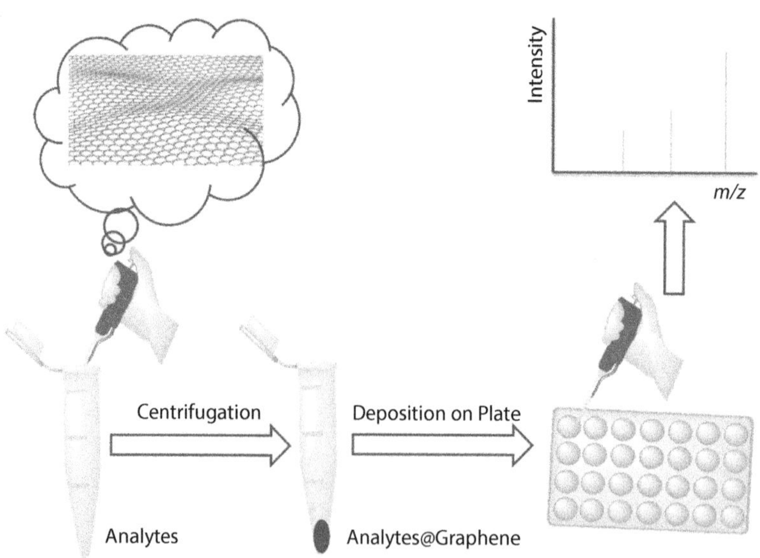

Figure 10.2 Sample preparation and analysis for proteins using graphene-assisted LDI-MS.

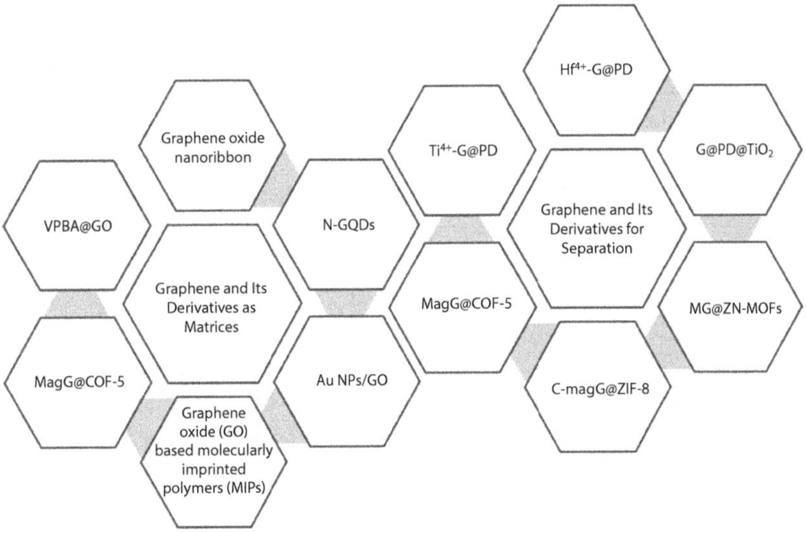

Figure 10.3 Graphene and its derivatives for matrices or separation.

polydopamine (magG@PDA@Cu^{2+}) composites were applied for the enrichment and identification of low-concentration peptides from biological samples including human urine and serum [55]. The magG@PDA@Cu^{2+} composites show great affinity to both hydrophobic and hydrophilic peptides with extremely low concentration of 10 pM. Peptide enrichment using external magnets provides a rapid and facile approach to the extraction of low-concentration peptides.

Graphene derivatives including graphene oxide (GO) were applied as probe for the detection of oligonucleotides using LDI-MS [46]. GO-assisted LDI-MS offers ultrasensitivity for oligonucleotides and can be applied for real sample analysis.

Graphene derivatives such as graphene nanoflakes were used as a matrix for the analysis of cancer cell and cancer stem cell lipids [56]. Graphene (G) and graphene oxide (GO) were utilized to enrich and ionize long-chain fatty acids including n-dodecanoic acid (C12), n-tetradecanoic acid (C14), n-hexadecanoic acid (C16), n-octadecanoic acid (C18), and n-eicosanoic acid (C20) [57].

Polymer analysis using graphene-based nanomaterials were reported. Few-layered graphene (FLG) was used as a matrix for the analysis of low molecular weight polymers, polar polyethylene glycol (PEG) of molecular weight 1000 Da, and nonpolar polymethyl methacrylate (PMMA) of molecular weight 650 Da [47]. Several polymers including polypropylene glycol, polystyrene and polymethyl methacrylate with average molecular weights from 425 to 3500 Da were investigated using graphene-based nanomaterials [58]. Graphene-assisted LDI-MS produced high-quality MS spectra with low background interference.

10.4 Application of Graphene and Its Derivatives for the Analysis of Small Molecules

Analysis of small molecules using organic matrices is limited due to their background. Thus, graphene derivatives including graphene oxide nanoribbons (GONRs) [59], N-doped graphene [60], aptamer-conjugated graphene oxide [61], converted graphene-like monolayer (CGM) plate [62], and graphite-coated paper [63] were applied as surface for the analysis of small molecules. Graphene-based nanomaterials show almost no background and offer high sensitivity compared to conventional organic matrices. They provide an extremely simple mass spectrometric method to analyze different kinds of small molecules.

A piece of graphite-coated paper offered a simple and sensitive method that is a great advantage in simplifying the analysis process of 51 compounds [63]. It showed universal applicability to detect different kinds of small molecules [63]. Core–shell structured gold@ graphitized mesoporous silica nanocomposite (Au@GMSN) was applied for small molecular weight analytes including amino acids, neutral saccharides, peptides, and traditional Chinese medicine (TCM) [64]. Au@GMSN offered effective surface for surface-assisted laser desorption/ionization mass spectrometry (SALDI-MS) with large surface area, good dispersibility, and strong ultraviolet (UV) absorption. Silver nanoparticles (AgNPs) and reduced graphene oxide (rGO) hybrid nanoporous structures were synthesized using layer-by-layer (LBL) electrostatic self-assembly [65]. AgNPs@rGO provided a simple platform for SALDI-MS for the rapid analysis of carboxyl-containing small molecules [65].

Graphene-based nanomaterials were applied for the analysis of endogenous caffeine and theanine of Chinese tea leaves [66], traditional Chinese medicine (TCM) herbs [67], octachlorodibenzo-p-dioxin (OCDD) [68], tetracycline residues in milk [69], polycyclic aromatic hydrocarbons (PAHs) and estrogen [70], flavonoids and phenylpropanoids, such as coumarin [71], dopamine in cerebrospinal fluid [72], metallodrugs [73], pollutants—organochlorine pesticide pentachlorophenol (PCP), endocrine disrupter estradiol (E2), brominated flame-retardant 2,2′,4,4′-tetrabromodiphenyl ether (BDE-47), and tetrabromobisphenol A (TBBPA) [59], amino acids, fatty acids, peptides, anabolic androgenic steroids as well as anticancer drugs [60], epi-androsterone, testosterone, and methyltestosterone [60], and cocaine and adenosine [61].

Graphene-based nanomaterials show high potential for device fabrication including thin film-based technologies. Hybrid films of gold nanoparticles and graphene oxide (GO) were synthesized on a glass slide and were applied for the analysis of small molecules using SALDI-MS [74]. Application of graphene derivatives for thin film technologies offer simple sample preparation and high sensitivity.

Fluorographene (FG) serves as an effective probe for high-throughput identification and screening of chemical contaminants in complex samples [44]. FG assisted LDI-MS showed higher sensitivity (detection limits at ppt or subppt levels), and better reproducibility compared to other graphene-based materials due to the unique chemical structure and self-assembly properties of FG [44].

10.5 Graphene Application for Extraction and Separation Prior to Analysis Using MALDI-MS

Graphene-based nanomaterials offers effective surface for extraction or separation of analytes before the analysis using MALDI-MS [45]. The extraction or separations are usually used for analytes with very low concentration in a complex sample. The process is required to improve the sensitivity or to avoid interference peaks from the remaining species in the sample. Graphene-based nanomaterials were applied for the extraction or separation of analytes including proteins, peptides, lipids, carbohydrates, and small molecules.

10.6 Extraction and Separation of Proteins and Peptides Using Graphene-Based Nanomaterials

Carbon-based nanomaterials can be applied as probe or stationary phase for separation of protein species from biological sample [75]. They are widely used for preconcentration of protein [2]. The π-rich skeleton of graphene-based nanomaterials offers high efficiency for the separation or enrichment of protein. They provide several forces to interact with protein biomolecules. Graphene derivatives require a simple extraction procedure that involves the vortexing of G-based nanomaterials in the mixture, followed by centrifugation in short time [53]. The separated species can be rinsed with a small amount of solvent and directly deposited in the MALDI-MS plate before the analysis. The sample preconcentration offers manyfold increase in the detection sensitivity in MALDI-MS [54].

The surface of graphene can be modified with other biomolecules that can serve as recognition biomolecules and magnetic nanoparticles that simplify the separation using external magnets. Aptamer-conjugated magnetic graphene–gold nanoparticle hybrid nanocomposites showed specific enrichment and rapid analysis of thrombin [76]. Fe_3O_4-chitosan@graphene (Fe_3O_4-CS@G) core–shell composite was applied to enrich low-abundance proteins from biological samples [77]. Fe_3O_4-CS@G can be also used as a magnetic adsorbent for the enrichment of protein analyzed by MALDI-MS without desorption [78].

Graphene and its derivatives can be used to immobilize enzyme for protein digestion. Trypsin-linked GO offers microwave-assisted on-plate digestion method for proteolysis of standard [79]. The process of protein digestion is simple, accelerated proteolysis, which reduces the time required for traditional procedures and offers high throughput protein identification. Graphene oxide-immobilized trypsin enzyme reactor (GO-IMER) offered simultaneous imaging mass spectrometry (IMS) for the proteome on tissue [80]. GO-IMER showed a superior proteolysis performance, which is sensitive to low protein concentration, and requires short processing time. The protein score and amino acid sequence coverage of on-plate digestion with free enzyme and GO-IMER are 44–62% (free enzyme) and 71–102% (GO-IMER) for casein and 60–83% (free enzyme) and 79–107% (GO-IMER) for BSA [80]. Dendrimer-grafted graphene oxide nanosheets (dGO) were synthesized via covalent bonding using glutaraldehyde as a coupling agent [81]. It efficiently digested proteins within 15 min, and showed high sequence coverages compared to those obtained by conventional overnight in-solution digestion [81].

Immobilized *N*-glycosidase F (PNGase F) on the surface of GO shortens the digestion time for the analysis of *N*-glycans [82]. GO-based immobilized PNGase F showed high reproducibility for *N*-glycan identification using MALDI-MS [82]. Au NPs-maltose-PDA-Fe_3O_4-RGO nanocomposite provided selective enrichment of the glycopeptides from a low concentration of horseradish peroxidase (HRP) tryptic digest (0.1 ng μL^{-1}) [83]. The presence of magnetic nanoparticles offered a simple separation method for the modified enzyme. Ultrahydrophilic dendrimer-modified magnetic graphene@polydopamine@poly(amidoamine) (magG@PDA@PAMAM) identified 15 glycopeptides from HRP digestion with limit of detection as low as 1 fmol μL^{-1} [84]. magG-PDA-Au-L-Cys offered high sensitivity (LOD, 0.1 fmol), and high selectivity (1:100), and repeatability (at least 10 times) for the enrichment of glycopeptides from human serum [85]. The superior hydrophilicity of these composites enhanced the analysis of glycoproteins.

Aminophenylboronic acid groups-conjugated graphene byphenolic-formaldehyde resin (magG@PF@APB) showed high selectivity in the presence of the background nonglycopeptides with a concentration 100-fold [86]. GO–polyethylenimine (PEI)–Au–L-Cysboronic acid composites can be applied as stationary phase for zwitterionic hydrophilic interaction chromatography (ZIC-HILIC) [87]. Boronic acid-functionalized graphene oxide (GO-APBA)-based molecularly imprinted polymers (MIPs) were used as a template for the analysis of glycoprotein (ovalbumin, OVA) [88]. It showed a loading capacity of 278 mg/g within 40 min [88]. Covalent organic framework (COF)-functionalized magnetic graphene composite (MagG@COF-5) was applied for the analysis of N-linked glycopeptide with a low detection limit (0.5 fmol μL^{-1}). The pore structure of COF is suitable for separation of the target analyte using size-exclusion effect (HRP digests/BSA, 1 : 600) [89]. 1-pyrenebutyryl chloride-functionalized graphene oxide (PCGO) (π–π stacking of 1-pyrenebutyryl chloride on the GO surface) offers a simple glycan enrichment [90]. The method is based on the reversible covalent bond formation between the hydroxyl groups of glycans and the acyl chloride groups on GO. Due to the presence of multiple hydroxyl groups of glycans, the cross-linking and self-assembly of free PCGO sheets and glycans require a very short time (within 30 s). This method is simple and improves the sensitivity and selectivity.

Graphene or its composites were widely investigated for phosphorylated proteins/peptides [91]. TiO_2–graphene composites were used for selective capture of phosphopeptides from peptide mixtures [92]. rGR–TiO_2–ZrO_2 (rGTZ) composite nanosheet selectively enrich phosphopeptides including α-casein, mixtures of β-casein, and bovine serum albumin (BSA) from semi-complex samples, nonfat milk, and mouse organs [93]. It offered the identification of 1980 phosphopeptides from 1769 proteins from mouse brain and 577 phosphopeptides from 1267 proteins from mouse liver [93]. TiO_2–graphene composites are an effective platform for the selective extraction of phosphopeptides from peptide mixtures [94]. It offered direct analysis of phosphopeptides using surface-enhanced laser desorption/ionization mass spectrometry (SELDI-MS). It also promotes the soft ionization of analytes and shows a detection limit in the attomole range, which is 10^2–10^5 more sensitive than conventional platforms. It is applicable for detecting phosphopeptides in cancer cells (HeLa cells) with high specificity (94%) [94]. Not only graphene–TiO_2 composite (GTOC) but also graphene–ZrO_2 composite (GZOC) showed selectivity enrichment for phosphopeptides including α- and β-casein, mixtures of β-casein and BSA, and nonfat milk tryptic digest [95]. Graphene–hafnium oxide composite (GHOC) also showed selective enrichment of phosphopeptides [95]. MoO_3–GO composite showed high detection sensitivity (1 fmol/mL) and well recovery (91.13%) for the enrichment of phosphopeptides using metal oxide affinity chromatography (MOAC) [96]. Other composites including Fe_3O_4–graphene–TiO_2 composites were also reported [97]. The presence of magnetic nanoparticles offered simple, fast, highly selective, and sensitive enrichment of phosphopeptides from biosamples [98]. Thus, they show multifunctionalities.

The surface of graphene derivatives can be decorated using magnetic nanoparticles and other reagents that ensure multifunctionalities and high sensitivity/selectivity. The nanocomposites show selective capture and rapid separation of low-abundant phosphopeptides from complex biological samples [99]. The surface can be modified with metal ions that offer highly selective and effective enrichment of phosphopeptides using immobilized metal ion affinity chromatography (IMAC). For instance, polydopamine coated on the surface of graphene and functionalized with titanium ions (denoted as Ti^{4+}-G@PD) was applied as a platform for phosphoproteome analysis [100]. Hf^{4+}-immobilized polydopamine-coated magnetic graphene (denoted as magG@PDA-Hf^{4+}) was applied for phosphopeptide enrichment [101]. The surface of PD-coated G can be also grafted using TiO_2 (denoted as G@PD@TiO_2) for phosphopeptide enrichment using metal oxide affinity chromatography (MOAC) [102]. G@PD@TiO_2 offered high sensitivity (detection limit of 5 fmol) and high selectivity for phosphopeptides at a molar ratio of 1:1000 (phosphopeptides/nonphosphopeptides) [102].

Selective separation can be achieved using pore size exclusion for graphene-based porous materials. Self-assembly of a thin layer of metal organic frameworks (MOFs) [103–106] on basal planes of Fe_3O_4–G nanosheets (denoted as MGMOF) showed high specific surface area (345.4 m^2 g^{-1}), and average pore size of 3.2 nm [107]. It offered selective extraction and separation of low-concentration biomolecules from biological samples via the size-selection property. The presence of magnetic nanoparticles offered simple separation using external magnets. Magnetic nanoparticles–graphene functionalized with MOFs (MG@Zn-MOFs) offered a simple method for recognition of glycopeptides [108]. It showed identification of 517 *N*-glycopeptides within 151 unique glycoproteins from human serum (1 μL) [108]. MG@Zn-MOFs have a large number of affinity sites, unique aperture-screen effect, strong magnetic responsiveness, and high specific surface area [108]. Zeolitic imidazolate frameworks (ZIF-8)-functionalized magnetic nanoporous carbon–G composites (C-magG@ZIF-8) offered excellent selectivity and sensitivity, good recyclability, and incredible size exclusion ability (roughly 2000 times) for the analysis of N-linked glycans from the normal human serum [109].

10.7 Extraction and Separation of Small Molecules Using Graphene-Based Nanomaterials

The separation or preconcentration of small molecules using graphene derivatives is paramount for the analysis using MALDI-MS. General extraction protocol requires the dispersion of graphene-based probe dispersion into the sample solution, and then the mixture was shaken before separation using centrifugation [59]. Selective separation and direct detection of glutathione (GSH) can be achieved using gold nanoparticles prior to the analysis using MALDI–MS with graphene as matrix [110]. N-doped graphene quantum dots (N-GQDs) were applied for solid-phase extraction (SPE) for the extraction of perfluoroalkyl sulfonates (PFSs) including perfluorooctane sulfonate (PFOS) and potassium salts of perfluorobutane sulfonate (PFBS) [111].

Microwave-assisted extraction (MAE) using graphene derivatives facilitated the analysis of two traditional Chinese medicine (TCM) herb models, *A. sinensis* and *S. barbata* D. Don [67]. Aptamer-conjugated GO offered high affinity for selective extraction and detection of cocaine and adenosine from plasma samples with greatly improved signal-to-noise ratios [61]. Antibody-functionalized graphene oxide nanoribbons (GONR-polyethylene glycol (PEG)-Ab) offered a probe for SELDI-MS for enrichment and detection of chloramphenicol (CAP) in complex media [112]. GONR-PEG-Ab-provided high sensitivity with detection limit (LOD, at S/N >3) of CAP in river water and serum samples was determined to be 10 pg mL^{-1}, which is better than that reported with amplified enzyme-linked immunosorbent assay (ELISA) and gas chromatography-MS [112]. Furthermore, it showed an enrichment factor of ~200, which is higher than that achieved on a commercial SPE cartridge with dynamic range of detection of 10–1000 pg mL^{-1} [112]. 4-vinylphenylboronic acid (VPBA)-functionalized GO showed selective enrichment and analysis of small-molecule compounds [113]. Boronic acid provided selective interactions with vicinal diols and showed a lower LOD (0.63 pmol mL^{-1}) compared to GO (73.0 pmol mL^{-1}) and DHB (83.0 pmol mL^{-1}) [113].

The surface of graphene-coated cobalt nanoparticles was functionalized with benzylamine groups (CoC–NH_2 nanomagnets) [114]. The material showed effective enrichment of small molecules including pentachlorophenol, bisphenol A, and polyfluorinated compounds (PFCs) [114]. The surface modification of CoC nanomagnets with benzylamine groups enhanced the yield of peptide ions and offered soft ionization method. It showed high sensitivity even at the sub-part-per-trillion level (~0.1 ng/L).

10.8 Conclusions and Outlook

Graphene and its derivatives offered sensitive and selective detection or enrichment of wide analytes. However, the key parameters, including particle sizes [115], degree of oxidation [116], morphology [112], thickness [117], and the number of layers [118, 119] for thin film, should be investigated further. This information will improve the analysis of current technologies and will advance the technique. Furthermore, it will open new venues for further applications. The mechanism of the interaction between the nanomaterials and the target species should be carried out.

Graphene derivatives offered several advantages including simple sample preparation, no or low fragmentation, i.e., soft ionization methods, effective analyte desorption/ionization, good reproducibility of peak intensities for analytes, and good salt-tolerance ability [57]. They can be used as enrichment absorbents and ionization matrices. They are wavelength independent and can be applied for several instruments equipped with different laser types with much higher desorption efficiency compared to conventional organic matrices [66]. It showed no ion suppression effect for low abundance molecules in the presence of other highly abundant biological molecules [54]. It can be used as probes for the purification, extraction, amplification, desorption, and ionization of the analytes.

Acknowledgments

H.N. Abdelhamid thanks the Ministry of Higher Education and Scientific Research (MHESR) in Egypt for support. H.F. Wu thanks the Ministry of Science and Technology, Taiwan for their financial support.

References

1. Novoselov, K.S., Geim, A.K., Morozov, S.V., Jiang, D., Zhang, Y., Dubonos, S.V., Grigorieva, I.V., Firsov, A.A., Electric field effect in atomically thin carbon films. *Science (80-)*, 306, 666–669, 2004.
2. Chen, X., Hai, X., Wang, J., Graphene/graphene oxide and their derivatives in the separation/isolation and preconcentration of protein species: A review. *Anal. Chim. Acta*, 922, 1–10, 2016.

3. Kim, J., Park, S.J., Min, D.H., Emerging approaches for graphene oxide biosensor. *Anal. Chem.*, 89, 232–248, 2017.
4. Yin, P.T., Shah, S., Chhowalla, M., Lee, K.B., Design, synthesis, and characterization of graphene-nanoparticle hybrid materials for bioapplications. *Chem. Rev.*, 115, 2483–2531, 2015.
5. Han, W., Kawakami, R.K., Gmitra, M., Fabian, J., Graphene spintronics. *Nat. Nanotechnol.*, 9, 794–807, 2014.
6. Koppens, F.H.L., Mueller, T., Avouris, P., Ferrari, A.C., Vitiello, M.S., Polini, M., Photodetectors based on graphene, other two-dimensional materials and hybrid systems. *Nat. Nanotechnol.*, 9, 780–793, 2014.
7. Liu, J., Liu, Z., Barrow, C.J., Yang, W., Molecularly engineered graphene surfaces for sensing applications: A review. *Anal. Chim. Acta*, 859, 1–19, 2015.
8. Perreault, F., Fonseca de Faria, A., Elimelech, M., Environmental applications of graphene-based nanomaterials. *Chem. Soc. Rev.*, 44, 5861–5896, 2015.
9. Aghigh, A., Alizadeh, V., Wong, H.Y., Islam, M.S., Amin, N., Zaman, M., Recent advances in utilization of graphene for filtration and desalination of water: A review. *Desalination*, 365, 389–397, 2015.
10. Karas, M. and Hillenkamp, F., Laser desorption ionization of proteins with molecular masses exceeding 10,000 daltons. *Anal. Chem.*, 60, 2299–2301, 1988.
11. Abdelhamid, H.N., Organic matrices, ionic liquids, and organic matrices@nanoparticles assisted laser desorption/ionization mass spectrometry. *TrAC - Trends Anal. Chem.*, 89, 68–98, 2017.
12. Nasser Abdelhamid, H. and Wu, H.F., Furoic and mefenamic acids as new matrices for matrix assisted laser desorption/ionization-(MALDI)-mass spectrometry. *Talanta*, 115, 442–450, 2013.
13. Abdelhamid, H.N., Ionic liquids for mass spectrometry: Matrices, separation and microextraction. *TrAC - Trends Anal. Chem.*, 77, 122–138, 2016.
14. Abdelhamid, H.N., Ionic liquids matrices for laser assisted desorption/ionization mass spectrometry. *Mass Spectrom. Purif. Tech.*, 1, 109–119, 2015.
15. Abdelhamid, H.N., Physicochemical properties of proteomic ionic liquids matrices for MALDI-MS. *J. Data Min. Genomics Proteomics*, 7, 2153–0602.1000, 2016.
16. Abdelhamid, H.N., Ionic liquids for mass spectrometry: Matrices, separation and microextraction. *TrAC - Trends Anal. Chem.*, 77, 122–138, 2016.
17. Abdelhamid, H.N., Gopal, J., Wu, H.F., Synthesis and application of ionic liquid matrices (ILMs) for effective pathogenic bacteria analysis in matrix assisted laser desorption/ionization (MALDI-MS). *Anal. Chim. Acta*, 767, 104–111, 2013.
18. Abdelhamid, H.N., Khan, M.S., Wu, H.F., Design, characterization and applications of new ionic liquid matrices for multifunctional analysis of biomolecules: A novel strategy for pathogenic bacteria biosensing. *Anal. Chim. Acta*, 823, 51–60, 2014.
19. Tanaka, K., Waki, H., Ido, Y., Akita, S., Yoshida, Y., Yoshida, T., Protein and polymer analyses up to m/z 100,000 by laser ionization TOF-MS. *Rapid Commun. Mass Spectrom.*, 2, 151, 1988.
20. Wang, J., Liu, Q., Liang, Y., Jiang, G., Recent progress in application of carbon nanomaterials in laser desorption/ionization mass spectrometry. *Anal. Bioanal. Chem.*, 408, 2861–2873, 2016.
21. Abdelhamid, H.N. and Wu, H.-F., Monitoring metallofulfenamic-bovine serum albumin interactions: A novel method for metallodrug analysis. *RSC Adv.*, 4, 2014.
22. Abdelhamid, H.N. and Wu, H.-F., Synthesis and multifunctional applications of quantum nanobeads for label-free and selective metal chemosensing. *RSC Adv.*, 5, 2015.
23. Abdelhamid, H.N. and Wu, H.-F., Synthesis and characterization of quantum dots for application in laser soft desorption/ionization mass spectrometry to detect labile metal–drug interactions and their antibacterial activity. *RSC Adv.*, 5, 76107–76115, 2015.

24. Wu, H.F., Gopal, J., Abdelhamid, H.N., Hasan, N., Quantum dot applications endowing novelty to analytical proteomics. *Proteomics*, 12, 2949–2961, 2012.
25. Abdelhamid, H.N. and Wu, H.-F., Probing the interactions of chitosan capped CdS quantum dots with pathogenic bacteria and their biosensing application. *J. Mater. Chem. B*, 1, 6094–6106, 2013.
26. Abdelhamid, H.N., Chen, Z.-Y., Wu, H.-F., Surface tuning laser desorption/ionization mass spectrometry (STLDI-MS) for the analysis of small molecules using quantum dots. *Anal. Bioanal. Chem.*, 409, 4943–4950, 2017.
27. Chen, Z.-Y., Abdelhamid, H.N., Wu, H.-F., Effect of surface capping of quantum dots (CdTe) on proteomics. *Rapid Commun. Mass Spectrom.*, 30, 1403–1412, 2016.
28. Abdelhamid, H.N., Lin, Y.C., Wu, H.F., Thymine chitosan nanomagnets for specific preconcentration of mercury (II) prior to analysis using SELDI-MS. *Microchim. Acta*, 184, 1517–1527, 2017.
29. Abdelhamid, H.N. and Wu, H.F., Thymine chitosan nanomagnets for specific preconcentration of mercury (II) prior to analysis using SELDI-MS. *Microchim. Acta*, 2017.
30. Abdelhamid, H.N., Kumaran, S., Wu, H.-F., One-pot synthesis of CuFeO2 nanoparticles capped with glycerol and proteomic analysis of their nanocytotoxicity against fungi. *RSC Adv.*, 6, 97629–97635, 2016.
31. Abdelhamid, H.N., Laser assisted synthesis, imaging and cancer therapy of magnetic nanoparticles. *Mater. Focus*, 5, 305–323, 2016.
32. Gopal, J., Abdelhamid, H.N., Hua, P.-Y., Wu, H.-F., Chitosan nanomagnets for effective extraction and sensitive mass spectrometric detection of pathogenic bacterial endotoxin from human urine. *J. Mater. Chem. B*, 1, 2463–2475, 2013.
33. Abdelhamid, H.N., Delafossite nanoparticle as new functional materials: Advances in energy, nanomedicine and environmental applications. *Mater. Sci. Forum.*, 832, 28–53, 2015.
34. Abdelhamid, H.N. and Wu, H.-F., Facile synthesis of nano silver ferrite (AgFeO2) modified with chitosan applied for biothiol separation. *Mater. Sci. Eng. C Mater. Biol. Appl.*, 45, 438–445, 2014.
35. Abdelhamid, H.N., Talib, A., Wu, H.-F., Facile synthesis of water soluble silver ferrite (AgFeO2) nanoparticles and their biological application as antibacterial agents. *RSC Adv.*, 5, 34594–34602, 2015.
36. Abdelhamid, H.N. and Wu, H.-F., Multifunctional graphene magnetic nanosheet decorated with chitosan for highly sensitive detection of pathogenic bacteria. *J. Mater. Chem. B*, 1, 3950–3961, 2013.
37. Abdelhamid, H.N. and Wu, H.-F., Gold nanoparticles assisted laser desorption/ionization mass spectrometry and applications: From simple molecules to intact cells. *Anal. Bioanal. Chem.*, 408, 4485–4502, 2016.
38. Abdelhamid, H.N., Talib, A., Wu, H.F., One pot synthesis of gold–carbon dots nanocomposite and its application for cytosensing of metals for cancer cells. *Talanta*, 166, 357–363, 2017.
39. Abdelhamid, H.N. and Wu, H.-F., Proteomics analysis of the mode of antibacterial action of nanoparticles and their interactions with proteins. *TrAC - Trends Anal. Chem.*, 65, 30–46, 2014.
40. Abdelhamid, H.N., *Applications of nanomaterials and organic semiconductors for bacteria & biomolecules analysis/biosensing using laser analytical spectroscopy*, National Sun-Yat Sen University, 2013.
41. Sunner, J., Dratz, E., Chen, Y.C., Graphite surface-assisted laser desorption/ionization time-of-flight mass spectrometry of peptides and proteins from liquid solutions. *Anal. Chem.*, 67, 4335–4342, 1995.
42. Li, X., Wilm, M., Franz, T., Silicone/graphite coating for on-target desalting and improved peptide mapping performance of matrix-assisted laser desorption/ionization-mass spectrometry targets in proteomic experiments. *Proteomics*, 5, 1460–1471, 2005.

43. Abdelhamid, H.N. and Wu, H.-F., Synthesis of a highly dispersive sinapinic acid@graphene oxide (SA@GO) and its applications as a novel surface assisted laser desorption/ionization mass spectrometry for proteomics and pathogenic bacteria biosensing. *Analyst*, 140, 1555–1565, 2015.
44. Huang, X., Liu, Q., Huang, X., Nie, Z., Ruan, T., Du, Y., Jiang, G., Fluorographene as a mass spectrometry probe for high-throughput identification and screening of emerging chemical contaminants in complex samples. *Anal. Chem.*, 89, 1307–1314, 2017.
45. Dong, X., Cheng, J., Li, J., Wang, Y., Graphene as a novel matrix for the analysis of small molecules by MALDI-TOF MS. *Anal. Chem.*, 82, 6208–6214, 2010.
46. Huang, R.C., Chiu, W.J., Li, Y.J., Huang, C.C., Detection of microRNA in tumor cells using exonuclease III and graphene oxide-regulated signal amplification. *ACS Appl. Mater. Interfaces*, 6, 21780–21787, 2014.
47. Cho, D., Hong, S., Shim, S., Few layer graphene matrix for matrix-assisted laser desorption/ionization time-of-flight mass spectrometry. *J. Nanosci. Nanotechnol.*, 13, 5811–5813, 2013.
48. Khan, N., Abdelhamid, H.N., Yan, J.-Y., Chung, F.-T., Wu, H.-F., Detection of flutamide in pharmaceutical dosage using higher electrospray ionization mass spectrometry (ESI-MS) tandem mass coupled with Soxhlet apparatus. *Anal. Chem. Res.*, 3, 89–97, 2015.
49. Sekar, R., Kailasa, S.K., Abdelhamid, H.N., Chen, Y.-C., Wu, H.-F., Electrospray ionization tandem mass spectrometric studies of copper and iron complexes with tobramycin. *Int. J. Mass Spectrom.*, 338, 23–29, 2013.
50. Abdelhamid, H.N. and Wu, H., Soft ionization of metallo-mefenamic using electrospray ionization mass spectrometry. *Mass Spectrom. Lett.*, 6, 43–47, 2015.
51. Abdelhamid, H.N. and Wu, H.-F., Soft ionization of metallo-mefenamic using electrospray ionization mass spectrometry. *Mass Spectrom. Lett.*, 6, 2015.
52. Huang, R.-C., Chiu, W.-J., Po-Jung Lai, I., Huang, C.-C., Multivalent aptamer/gold nanoparticle–modified graphene oxide for mass spectrometry–based tumor tissue imaging. *Sci. Rep.*, 5, 10292, 2015.
53. Tang, L.A.L., Wang, J., Loh, K.P., Graphene-based SELDI probe with ultrahigh extraction and sensitivity for DNA oligomer. *J. Am. Chem. Soc.*, 132, 10976–10977.
54. Bai, H., Pan, Y., Tong, W., Zhang, W., Ren, X., Tian, F., Peng, B., Wang, X., Zhang, Y., Deng, Y., Qin, W., Qian, X., Graphene based soft nanoreactors for facile "one-step" glycan enrichment and derivatization for MALDI-TOF-MS analysis. *Talanta*, 117, 1–7, 2013.
55. Zhao, M., Deng, C., Zhang, X., Synthesis of polydopamine-coated magnetic graphene for Cu2+ immobilization and application to the enrichment of low-concentration peptides for mass spectrometry analysis. *ACS Appl. Mater. Interfaces*, 5, 13104–13112, 2013.
56. Hua, P.-Y., Manikandan, M., Abdelhamid, H.N., Wu, H.-F., Graphene nanoflakes as an efficient ionizing matrix for MALDI-MS based lipidomics of cancer cells and cancer stem cells. *J. Mater. Chem. B*, 2, 7334–7343, 2014.
57. Liu, Y., Liu, J., Deng, C., Zhang, X., Graphene and graphene oxide: Two ideal choices for the enrichment and ionization of long-chain fatty acids free from matrix-assisted laser desorption/ionization matrix interference. *Rapid Commun. Mass Spectrom.*, 25, 3223–3234, 2011.
58. Lu, M., Lai, Y., Chen, G., Cai, Z., Laser desorption/ionization on the layer of graphene nanoparticles coupled with mass spectrometry for characterization of polymers. *Chem. Commun.*, 47, 12807, 2011.
59. Liu, Q., Cheng, M., Wang, J., Jiang, G., Graphene oxide nanoribbons: Improved synthesis and application in MALDI mass spectrometry. *Chem. - A Eur. J.*, 21, 5594–5599, 2015.
60. Min, Q., Zhang, X., Chen, X., Li, S., Zhu, J.J., N-Doped graphene: An alternative carbon-based matrix for highly efficient detection of small molecules by negative ion MALDI-TOF MS. *Anal. Chem.*, 86, 9122–9130, 2014.

61. Gulbakan, B., Yasun, E., Shukoor, M.I., Zhu, Z., You, M., Tan, X., Sanchez, H., Powell, D.H., Dai, H., Tan, W., A dual platform for selective analyte enrichment and ionization in mass spectrometry using aptamer-conjugated graphene oxide. *J. Am. Chem. Soc.*, 132, 17408–17410, 2010.
62. Kang, H., Yun, H., Lee, S.W., Yeo, W.S., Analysis of small biomolecules and xenobiotic metabolism using converted graphene-like monolayer plates and laser desorption/ionization time-of-flight mass spectrometry. *Talanta*, 168, 240–245, 2017.
63. Zhang, J., Li, Z., Zhang, C., Feng, B., Zhou, Z., Bai, Y., Liu, H., Graphite-coated paper as substrate for high sensitivity analysis in ambient surface-assisted laser desorption/ionization mass spectrometry. *Anal. Chem.*, 84, 3296–3301, 2012.
64. Xu, G., Liu, S., Peng, J., Lv, W., Wu, R., Facile synthesis of gold@graphitized mesoporous silica nanocomposite and its surface-assisted laser desorption/ionization for time-of-flight mass spectroscopy. *ACS Appl. Mater. Interfaces*, 7, 2032–2038, 2015.
65. Hong, M., Xu, L., Wang, F., Geng, Z., Li, H., Wang, H., Li, C., A direct assay of carboxyl-containing small molecules by SALDI-MS on a AgNP/rGO-based nanoporous hybrid film. *Analyst*, 141, 2712–2726, 2016.
66. Chang, C., Li, X., Bai, Y., Xu, G., Feng, B., Liao, Y., Liu, H., Graphene matrix for signal enhancement in ambient plasma assisted laser desorption ionization mass spectrometry. *Talanta*, 114, 54–59, 2013.
67. Liu, Y., Liu, J., Yin, P., Gao, M., Deng, C., Zhang, X., High throughput identification of components from traditional Chinese medicine herbs by utilizing graphene or graphene oxide as MALDI-TOF-MS matrix. *J. Mass Spectrom.*, 46, 804–815, 2011.
68. Zhou, X., Wei, Y., He, Q., Boey, F., Zhang, Q., Zhang, H., Reduced graphene oxide films used as matrix of MALDI-TOF-MS for detection of octachlorodibenzo-p-dioxin. *Chem. Commun.*, 46, 6974, 2010.
69. Liu, J., Liu, Y., Gao, M., Zhang, X., High throughput detection of tetracycline residues in milk using graphene or graphene oxide as MALDI-TOF MS matrix. *J. Am. Soc. Mass Spectrom.*, 23, 1424–1427, 2012.
70. Zhang, J., Dong, X., Cheng, J., Li, J., Wang, Y., Efficient analysis of non-polar environmental contaminants by MALDI-TOF MS with graphene as matrix. *J. Am. Soc. Mass Spectrom.*, 22, 1294–1298, 2011.
71. Liu, C.-W., Chien, M.-W., Su, C.-Y., Chen, H.-Y., Li, L.-J., Lai, C.-C., Analysis of flavonoids by graphene-based surface-assisted laser desorption/ionization time-of-flight mass spectrometry. *Analyst*, 137, 5809, 2012.
72. Zheng, X., Zhang, J., Wei, H., Chen, H., Tian, Y., Zhang, J., Determination of dopamine in cerebrospinal fluid by MALDI-TOF mass spectrometry with a functionalized graphene oxide matrix. *Anal. Lett.*, 49, 1847–1861, 2016.
73. Abdelhamid, H.N. and Wu, H.-F., A method to detect metal-drug complexes and their interactions with pathogenic bacteria via graphene nanosheet assist laser desorption/ionization mass spectrometry and biosensors. *Anal. Chim. Acta*, 751, 94–104, 2012.
74. Kim, Y.K. and Min, D.H., Preparation of the hybrid film of poly(allylamine hydrochloride)-functionalized graphene oxide and gold nanoparticle and its application for laser-induced desorption/ionization of small molecules. *Langmuir*, 28, 4453–4458, 2012.
75. Chen, X., Hu, L., Liu, J., Chen, S., Wang, J., Nanoscale carbon-based materials in protein isolation and preconcentration. *TrAC - Trends Anal. Chem.*, 48, 30–39, 2013.
76. Xiong, Y., Deng, C., Zhang, X., Development of aptamer-conjugated magnetic graphene/gold nanoparticle hybrid nanocomposites for specific enrichment and rapid analysis of thrombin by MALDI-TOF MS. *Talanta*, 129, 282–289, 2014.

77. Zhang, P., Fang, X., Yan, G., Gao, M., Zhang, X., Highly efficient enrichment of low-abundance intact proteins by core-shell structured Fe 3 O 4 -chitosan@graphene composites. *Talanta*, 174, 845–852, 2017.
78. Ye, N., Xie, Y., Shi, P., Gao, T., Ma, J., Synthesis of magnetite/graphene oxide/chitosan composite and its application for protein adsorption. *Mater. Sci. Eng. C*, 45, 8–14, 2014.
79. Xu, G., Chen, X., Hu, J., Yang, P., Yang, D., Wei, L., Immobilization of trypsin on graphene oxide for microwave-assisted on-plate proteolysis combined with MALDI-MS analysis. *Analyst*, 137, 2757, 2012.
80. Jiao, J., Miao, A., Zhang, X., Cai, Y., Lu, Y., Zhang, Y., Lu, H., Realization of on-tissue protein identification by highly efficient *in situ* digestion with graphene-immobilized trypsin for MALDI imaging analysis. *Analyst*, 138, 1645, 2013.
81. Jiang, B., Yang, K., Zhang, L., Liang, Z., Peng, X., Zhang, Y., Dendrimer-grafted graphene oxide nanosheets as novel support for trypsin immobilization to achieve fast on-plate digestion of proteins. *Talanta*, 122, 278–284, 2014.
82. Ren, X.J., Bai, H.H., Pan, Y.T., Tong, W., Qin, P.B., Yan, H., Deng, S.S., Zhong, R.G., Qin, W.J., Qian, X.H., A graphene oxide-based immobilized PNGase F reagent for highly efficient N-glycan release and MALDI-TOF MS profiling. *Anal. Methods*, 6, 2518–2525, 2014.
83. Bi, C., Jiang, R., He, X., Chen, L., Zhang, Y., Synthesis of a hydrophilic maltose functionalized Au NP/PDA/Fe3O4-RGO magnetic nanocomposite for the highly specific enrichment of glycopeptides. *RSC Adv.*, 5, 59408–59416, 2015.
84. Wang, Y., Wang, J., Gao, M., Zhang, X., An ultra hydrophilic dendrimer-modified magnetic graphene with a polydopamine coating for the selective enrichment of glycopeptides. *J. Mater. Chem. B*, 3, 8711–8716, 2015.
85. Wu, R., Li, L., Deng, C., Highly efficient and selective enrichment of glycopeptides using easily synthesized magG/PDA/Au/l-Cys composites. *Proteomics*, 16, 1311–1320, 2016.
86. Wang, J., Wang, Y., Gao, M., Zhang, X., Yang, P., Multilayer hydrophilic poly(phenol-formaldehyde resin)-coated magnetic graphene for boronic acid immobilization as a novel matrix for glycoproteome analysis. *ACS Appl. Mater. Interfaces*, 7, 16011–16017, 2015.
87. Jiang, B., Liang, Y., Wu, Q., Jiang, H., Yang, K., Zhang, L., Liang, Z., Peng, X., Zhang, Y., New GO–PEI–Au–L-Cys ZIC-HILIC composites: Synthesis and selective enrichment of glycopeptides. *Nanoscale*, 6, 5616–5619, 2014.
88. Luo, J., Huang, J., Cong, J., Wei, W., Liu, X., Double recognition and selective extraction of glycoprotein based on the molecular imprinted graphene oxide and boronate affinity. *ACS Appl. Mater. Interfaces*, 9, 7735–7744, 2017.
89. Wang, J., Li, J., Gao, M., Zhang, X., Self-assembling covalent organic framework functionalized magnetic graphene hydrophilic biocomposites as an ultrasensitive matrix for N-linked glycopeptide recognition. *Nanoscale*, 9, 10750–10756, 2017.
90. Zhang, W., Han, H., Bai, H., Tong, W., Zhang, Y., Ying, W., Qin, W., Qian, X., A highly efficient and visualized method for glycan enrichment by self-assembling pyrene derivative functionalized free graphene oxide. *Anal. Chem.*, 85, 2703–2709, 2013.
91. Wang, Z.G., Lv, N., Bi, W.Z., Zhang, J.L., Ni, J.Z., Development of the affinity materials for phosphorylated proteins/peptides enrichment in phosphoproteomics analysis. *ACS Appl. Mater. Interfaces*, 7, 8377–8392, 2015.
92. Lu, J., Wang, M., Li, Y., Deng, C., Facile synthesis of TiO2/graphene composites for selective enrichment of phosphopeptides. *Nanoscale*, 4, 1577, 2012.
93. Huang, X., Wang, J.P., Liu, C.C., Guo, T., Wang, S., A novel rGR-TiO2-ZrO2 composite nanosheet for capturing phosphopeptides from biosamples. *J. Mater. Chem. B*, 3, 2505–2515, 2015.

94. Tang, L.A.L., Wang, J., Lim, T.K., Bi, X., Lee, W.C., Lin, Q., Chang, Y.T., Lim, C.T., Loh, K.P., High-performance graphene-titania platform for detection of phosphopeptides in cancer cells. *Anal. Chem.*, 84, 6693–6700, 2012.
95. Huang, X., Wang, J.J., Wang, J.J., Liu, C., Wang, S., Preparation of graphene–hafnium oxide composite for selective enrichment and analysis of phosphopeptides. *RSC Adv.*, 5, 89644–89651, 2015.
96. Sun, H., Zhang, Q., Zhang, L., Zhang, W., Zhang, L., Facile preparation of molybdenum (VI) oxide—Modified graphene oxide nanocomposite for specific enrichment of phosphopeptides. *J. Chromatogr. A*, 1521, 36–43, 2017.
97. Lu, J., Deng, C., Zhang, X., Yang, P., Synthesis of Fe3O4/graphene/TiO2 composites for the highly selective enrichment of phosphopeptides from biological samples. *ACS Appl. Mater. Interfaces*, 5, 7330–7334, 2013.
98. Liang, Y., He, X., Chen, L., Zhang, Y., Facile preparation of graphene/Fe3O4/TiO2 multifunctional composite for highly selective and sensitive enrichment of phosphopeptides. *RSC Adv.*, 4, 18132–18135, 2014.
99. Cheng, G., Yu, X., Zhou, M.M.-D., Zheng, S.S.-Y., Preparation of magnetic graphene composites with hierarchical structure for selective capture of phosphopeptides. *J. Mater. Chem. B*, 2, 4711–4719, 2014.
100. Yan, Y., Zheng, Z., Deng, C., Li, Y., Zhang, X., Yang, P., Hydrophilic polydopamine-coated graphene for metal ion immobilization as a novel immobilized metal ion affinity chromatography platform for phosphoproteome analysis. *Anal. Chem.*, 85, 8483–8487, 2013.
101. Lin, H. and Deng, C., Development of Hf4+-immobilized polydopamine-coated magnetic graphene for highly selective enrichment of phosphopeptides. *Talanta*, 149, 91–97, 2016.
102. Yan, Y., Sun, X., Deng, C., Li, Y., Zhang, X., Metal oxide affinity chromatography platform-polydopamine coupled functional two-dimensional titania graphene nanohybrid for phosphoproteome research. *Anal. Chem.*, 86, 4327–4332, 2014.
103. Abdelhamid, H.N., *Lanthanide Metal-Organic Frameworks and Hierarchical Porous Zeolitic Imidazolate Frameworks: Synthesis, Properties, and Applications*, Stockholm University, Faculty of Science, 2017.
104. Abdelhamid, H.N., Huang, Z., El-Zohry, A.M., Zheng, H., Zou, X., A fast and scalable approach for synthesis of hierarchical porous zeolitic imidazolate frameworks and one-pot encapsulation of target molecules. *Inorg. Chem.*, 56, 9139–9146, 2017.
105. Abdelhamid, H.N., Bermejo-Gómez, A., Martín-Matute, B., Zou, X., A water-stable lanthanide metal-organic framework for fluorimetric detection of ferric ions and tryptophan. *Microchim. Acta*, 184, 3363–3371, 2017.
106. Yang, Y., Shen, K., Lin, J., Zhou, Y., Liu, Q., Hang, C., Abdelhamid, H.N., Zhang, Z., Chen, H., A Zn-MOF constructed from electron-rich π-conjugated ligands with an interpenetrated graphene-like net as an efficient nitroaromatic sensor. *RSC Adv.*, 6, 45475–45481, 2016.
107. Cheng, G., Wang, Z.G., Denagamage, S., Zheng, S.Y., Graphene-templated synthesis of magnetic metal organic framework nanocomposites for selective enrichment of biomolecules. *ACS Appl. Mater. Interfaces*, 8, 10234–10242, 2016.
108. Wang, J., Li, J., Wang, Y., Gao, M., Zhang, X., Yang, P., Development of versatile metal-organic framework functionalized magnetic graphene core-shell biocomposite for highly specific recognition of glycopeptides. *ACS Appl. Mater. Interfaces*, 8, 27482–27489, 2016.
109. Wang, J., Wang, Y., Gao, M., Zhang, X., Yang, P., Versatile metal-organic framework-functionalized magnetic graphene nanoporous composites: As deft matrix for high-effective extraction and purification of the N-linked glycans. *Anal. Chim. Acta*, 932, 41–48, 2016.

110. Wan, D., Gao, M., Wang, Y., Zhang, P., Zhang, X., A rapid and simple separation and direct detection of glutathione by gold nanoparticles and graphene-based MALDI-TOF-MS. *J. Sep. Sci.*, 36, 629–635, 2013.
111. Rao, Z., Geng, F., Zhou, Y., Dong, C., Kang, Y., N-doped graphene quantum dots as a novel matrix of high efficacy for the analysis of perfluoroalkyl sulfonates and other small molecules by MALDI-TOF MS. *Anal. Methods*, 9, 2014–2020, 2017.
112. Wang, J., Cheng, M., Zhang, Z., Guo, L., Liu, Q., Jiang, G., An antibody-graphene oxide nanoribbon conjugate as a surface enhanced laser desorption/ionization probe with high sensitivity and selectivity. *Chem. Commun.*, 51, 4619–4622, 2015.
113. Zhang, J., Zheng, X., Ni, Y., Selective enrichment and MALDI-TOF MS analysis of small molecule compounds with vicinal diols by boric acid-functionalized graphene oxide. *J. Am. Soc. Mass Spectrom.*, 26, 1291–1298, 2015.
114. Kawasaki, H., Nakai, K., Arakawa, R., Athanassiou, E.K., Grass, R.N., Stark, W.J., Functionalized graphene-coated cobalt nanoparticles for highly efficient surface-assisted laser desorption/ionization mass spectrometry analysis. *Anal. Chem.*, 84, 9268–9275, 2012.
115. Kim, Y.K. and Min, D.H., The structural influence of graphene oxide on its fragmentation during laser desorption/ionization mass spectrometry for efficient small-molecule analysis. *Chem. - A Eur. J.*, 21, 7217–7223, 2015.
116. Liu, Q., Cheng, M., Jiang, G., Mildly oxidized graphene: Facile synthesis, characterization, and application as a matrix in MALDI mass spectrometry. *Chem. - A Eur. J.*, 19, 5561–5565, 2013.
117. Kuo, T.R., Wang, D.Y., Chiu, Y.C., Yeh, Y.C., Chen, W.T., Chen, C.H., Chen, C.W., Chang, H.C., Hu, C.C., Chen, C.C., Layer-by-layer thin film of reduced graphene oxide and gold nanoparticles as an effective sample plate in laser-induced desorption/ionization mass spectrometry. *Anal. Chim. Acta*, 809, 97–103, 2014.
118. Kim, Y.-K. and Min, D.-H., Fabrication of alternating multilayer films of graphene oxide and carbon nanotube and its application in mechanistic study of laser desorption/ionization of small molecules. *ACS Appl. Mater. Interfaces*, 4, 2088–2095, 2012.
119. Anderson, P.W., Local moments and localized states, *Rev. Mod. Phys.* 50, 191, 1978.

11

Characterization and Dynamic Manipulation of Graphene by *In Situ* Transmission Electron Microscopy at Atomic Scale

Chaolun Wang, Chen Luo and Xing Wu*

Shanghai Key Laboratory of Multidimensional Information Processing, Department of Electronics Engineering, East China Normal University, Shanghai, China

Abstract

Graphene that possesses atomic thin geometry and remarkable properties is a star material for fundamental researches and advanced applications. The intrinsic and dynamic relationships among the crystal structures, chemical compositions, and electronic structures of graphene are critical and fundamental to understand its physical properties. With the advanced TEM technologies, such as aberration correction and low-voltage technologies, the morphology, crystal structure, chemical composition, and electronic structures of graphene can be directly characterized at atomic scale. In addition, the development of microelectromechanical systems enables the applications of external stimuli, such as thermal, electrical, mechanical, and environmental fields on graphene inside an *in situ* TEM. By applying the advanced *in situ* TEM, graphene could be manipulated under external fields and recorded in real time with atomic resolution. In this chapter, we introduce the applications of state-of-the-art TEM technologies on the studies of intrinsic and dynamic properties of graphene.

Keywords: Graphene, transmission electron microscopy, *in situ* manipulation, characterization, two-dimensional layered materials

11.1 Introduction

Transmission electron microscopy (TEM) is an advanced technology that is widely used for the characterization of morphology, structure, and composition of materials with atomic resolution. The incorporation of *in situ* technologies further extends the applications of TEM to manipulate materials under the stimulations of external fields, and record the dynamic process in real time [1], as shown in Figure 11.1.

The TEM with versatile characterization modes is a powerful tool that has been widely applied in materials science and life science. Here, we only focus on the applications of the TEM in nanomaterials, and Cryo-TEM technologies for life sciences are not included. Similar to optical microscope (OM), light beam and optical lens are used to probe samples

Corresponding author: xwu@ee.ecnu.edu.cn

Tobias Stauber (ed.) Handbook of Graphene: Volume 2, (291–314) © 2019 Scrivener Publishing LLC

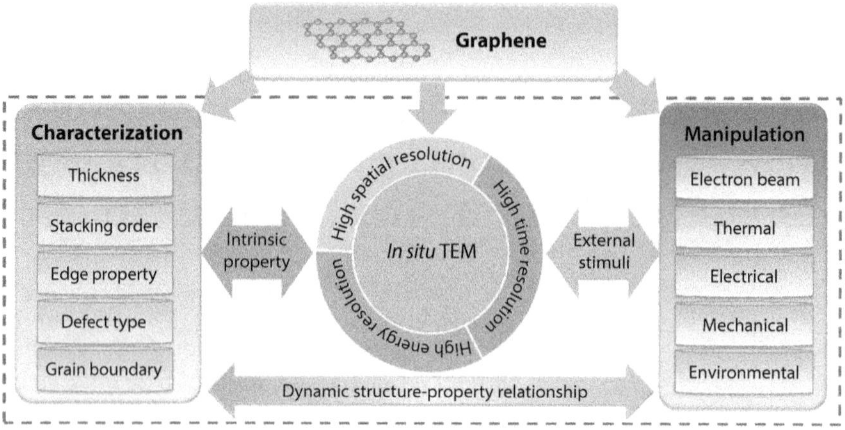

Figure 11.1 Overview of the advanced *in situ* TEM for versatile characterization and manipulation of graphene.

in an OM, while for TEM, electron beam and electromagnetic lens are the counterparts in a TEM. By taking advantage of the short wave length of an electron beam at 200~300 kV, about five orders smaller than that of the visible light, the resolution has been increased from submicron to subangstrom. TEM works under ultrahigh vacuum to increase the mean free path of electrons that are emitted from an electron gun by thermionic emission or cold field emission. Then the electrons are accelerated by 200~300 kV voltage and condensed into a parallel electron beam by a set of electromagnetic lenses. When the electron beam passes through a nano-thick sample, the election beam interacts with the electrostatic potentials of the sample atoms, and the transmitted electron beam carries the sample information. The transmitted electron beam is finally magnified and projected on a charge coupled device (CCD) camera for observation.

The interactions between electron beam and samples lead to three type of contrasts, the mass-thickness contrast, diffraction contrast, and phase contrast. The mass-thickness contrast generates from the incoherent elastic scattering of electrons, which is related with the atomic number Z and the thickness of the sample. Higher atomic number and thickness of the sample lead to less transmitted electrons and darker contrast. The diffraction contrast is induced by the electron diffraction that fulfill the Bragg's law in a crystalline sample. The diffracted electrons cannot reach the CCD camera, and display a dark contrast. Phase contrast results from the interaction of the electron waves with different phase. For an High Resolution Transmission Electron Microscope (HRTEM) image, the interactions of scattered and unscattered electron waves form the contrast of the crystal lattice.

Multifunctional TEM could probe a sample in different modes, such as diffraction mode, image mode, and scanning transmission electron microscopy (STEM) mode. The diffraction mode and image mode of TEM use the parallel electron beam and transmitted-electron signals to investigate a sample. In the diffraction mode, the diffraction patterns at the back focal plane of the objective lens are amplified by the intermediate lens, while in the image mode, the morphology of a sample at the image plane of the objective lens is amplified by the intermediate lens. The diffraction and image mode are used to study the structure information and morphology of a sample, respectively. The STEM mode uses the focused electron beam to scan the surface of the sample, and collects scattered-electron signals. The intensity of the STEM is proportional to the atomic number Z and the thickness of the sample, similar to that of the mass-thickness contrast. STEM could be used to study the chemistry and morphology of a sample. The analytical tools such as electron energy loss spectroscopy (EELS) and energy-dispersive X-ray spectroscopy (EDX) further extend the ability of TEM to electron state and chemical analysis, respectively.

11.2 The Development of TEM Technologies

11.2.1 Aberration Correction

After building the first demonstration of TEM in 1931, increasing the spatial resolution is the main effort in the TEM community. The main problems that limit the spatial resolution of TEM are spherical aberration (variation of focus ability along the radius of the lens), chromatic aberration (energy dispersion of the electron), and astigmation (asymmetry focus ability around the center of the lens). Chromatic aberration and astigmation could be corrected by the monochromator and stigmator. But spherical aberration generated form the uneven focus of the off-axis electron beams is the main obstacle for the further resolution improvement due to the difficulties of building a spherical aberration corrector [2]. With the advent of commercially available aberration-corrected TEM (Cs-corrected TEM) in 2004, the spatial resolution of TEM is pushed to subangstrom ~0.5 Å [3].

11.2.2 Low-Voltage TEM

Usually higher acceleration voltage leads to better spatial resolution, but high voltage also causes damages to samples due to knock-on effects (KOE) [4, 5], radiolysis [6, 7], and electron-stimulated desorption [8, 9]. To Probe electron-irradiation-sensitive materials such as graphene, low-voltage TEM (less than 80 kV) is indispensable [10]. Chromatic aberration resulting from nonuniform electrons energies becomes severe at low voltage. Therefore, monochromator is needed in low-voltage TEM.

11.2.3 Exit-Wave Reconstruction Technology

Besides the correction techniques of the intrinsic aberrations, reconstruction of electron exit waves is another technique to improve the spatial resolution. The data sets of the approach comprise a series of HRTEM images recorded at either varying defocus levels [11, 12] or with different illumination tilt directions [13, 14]. Each image contains independent information about the specimen exit plane wave function, which can be computationally recovered from the digital data [15]. Kisielowski *et al.* corrected the intrinsic aberrations and provided high-quality images of the detailed atoms arrangement in MoS_2 nanocatalysts at the single atom level by using the reconstruction of electron exit waves technique [16].

11.3 Characterization of the Intrinsic Properties of Graphene

This section introduces the represented applications of TEM for the characterization of intrinsic properties of graphene, such as thickness, stacking order, edge property, defect type, and grain boundary, that have important effects on the physical and chemical properties of graphene. The resolved structure and chemistry of graphene with atomic resolution provide a deep understanding of relations between structure and property, which is important for its preparation, processing, and application.

11.3.1 Characterization of the Layer Number of Graphene

The thickness of graphene influences its electronic properties [17]. TEM with high resolution and versatile working modes is a powerful tool for the determination of graphene layer number. Three working modes are applied for the determination of layer number, such as image mode (HRTEM and dark-field TEM (DFTEM)), STEM mode, and diffraction mode. The folded borders of layered graphene provide cross-sections to directly identify the layers by HRTEM. Figure 11.2a–c are examples of the HRTEM images of folded graphene borders with one, three, and four layers [18]. The layer number of the graphene could be identified by directly counting the dark lines in the folded borders. However, this method is only applicable to the homogeneous graphene that the thickness of the folded border can represent the whole graphene layer.

As discussed before, the proportional relation between the contrast of STEM and the thickness of the graphene can be applied for the evaluation of layer number of graphene. The intensity of the ADF-STEM can be written as in the following equation:

$$I \approx t \times Z^{1.7} \tag{11.1}$$

where I is the intensity of ADF-STEM signal; t is the thickness; and Z is the atomic number of scattering atoms. From Equation 11.1, it can be shown that the intensity of the ADF-STEM of graphene should be integral proportional to the layer number. As shown in Figure 11.2d, the layer numbers of the graphene in the middle hole, with the size of microscale, are determined by evaluating the intensity of the ADF-STEM image, and the corresponding layer numbers are labeled.

Another way to determine the layer number of graphene is to analyze the intensity of the HRTEM pattern [19]. Under the condition of −5 nm defocus value and negative C_s, the carbon atoms in 1-layer graphene present dark contrast. The HRTEM patterns of carbons in 2- and 3-layer graphene under the same condition show white and triangular (alternatively dark and white) contrast, respectively, as shown in Figure 11.2k–l. The simulated pattern contrast of HRTEM images of graphene with 1, 2, and 3 layers are consistent with the experiment as shown in Figure 11.2e–j.

DFTEM image generates from the diffracted electrons that satisfy the Bragg's law. The intensity of the diffracted electrons is proportional to the thickness of the graphene layer, which could be used to determine the layer number of the graphene [20, 21]. As shown in Figure 11.2m, the arrow-indicated regions are 1-layer graphene. The folded 2- and 3-layer graphenes are brighter than 1-layer graphene. The intensities of the line profile indicated by dashed lines are shown in Figure 11.2n, which presents integer multiples' increase.

The single- and bilayer Bernal (AB) stacked graphene can be distinguished directly by analyzing selected area electron diffraction (SAED) patterns. Figure 11.3a–b shows calculated 3D Fourier transforms of single- and bi-layer (AB-stacked) graphene atom positions;

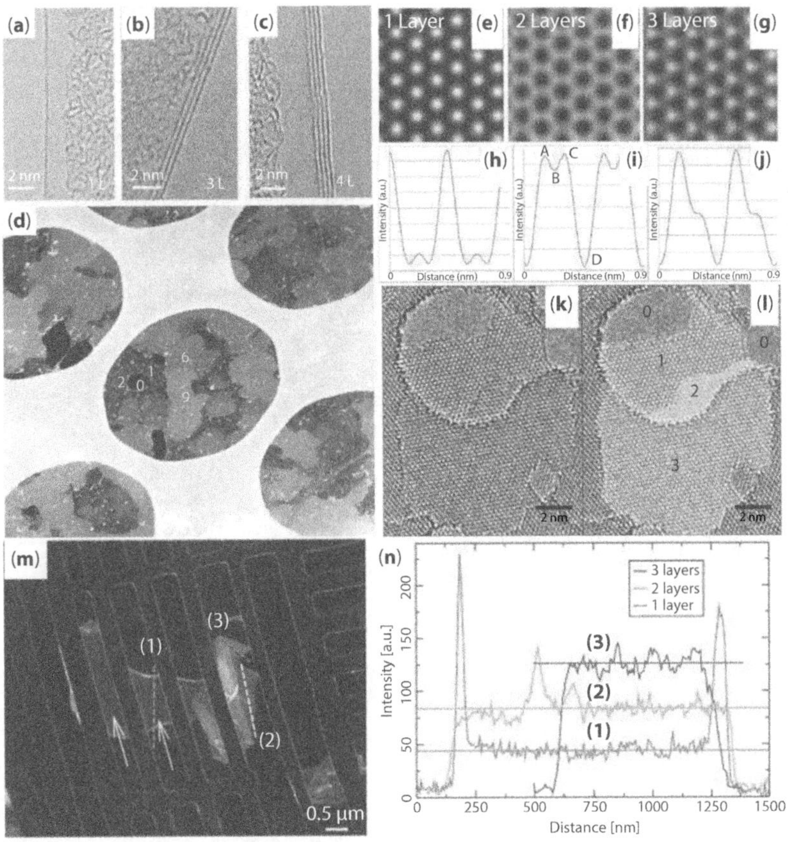

Figure 11.2 Thickness characterization of graphene. (a)–(c) HRTEM images showing the one, three, and four layers of the CVD-grown graphene films, by counting the dark lines of the folded border. (Reproduced with permission [18]. Copyright 2009, American Chemical Society.) (d) ADF-STEM image of a few-layer graphene located on a 3.5 μm hole in the center of the figure. The layer numbers of the graphene are labeled by numbers, which can be directly determined by evaluating the intensity of the ADF signal. (Reproduced with permission [22]. Copyright 2010, Elsevier.) (e)–(j) Simulations of HRTEM images of various layers of graphene, and the corresponding intensities of line profiles. (k)–(l) HRTEM images of a graphene sheet show different contrasts with the variation of thickness of graphene, which consists with the simulated results. The layer numbers of the graphene are indicated both by numbers and colors. (Reproduced with permission [19]. Copyright 2010 IOP Publishing Ltd.) (m) Small-angle DFTEM image of a graphene layer. The layer numbers of the graphene are determined by the intensity of the DFTEM signal similar to that of ADF-STEM. The signals of two and three layers of graphene are two and three integer multiples of that of the one-layer graphene as indicated by the corresponding line intensity profiles shown in (n). (Reproduced with permission [21]. Copyright 2007, Elsevier.)

the atomic scattering factors are not incorporated, and the intensities in Figure 11.3a–b are only qualitatively correct [21]. At normal incidence, the intensity in reciprocal space is not suppressed (blue planes) both in single- and bilayer graphene. The intensity of line profiles of single- and bilayer graphene along indices (0–110) (2.13 Å spacing) and (1–210) (1.23 Å spacing) is shown in Figure 11.3c and d. The layer number can be identified from the intensity ratios of the diffraction peaks. For the single-layer graphene, the intensity of inner spots {0–110} is stronger than the outer {1–210} spots, and the relative intensity is reversed in the bilayer. By varying incidence angles between the electron and graphene, diffraction peaks

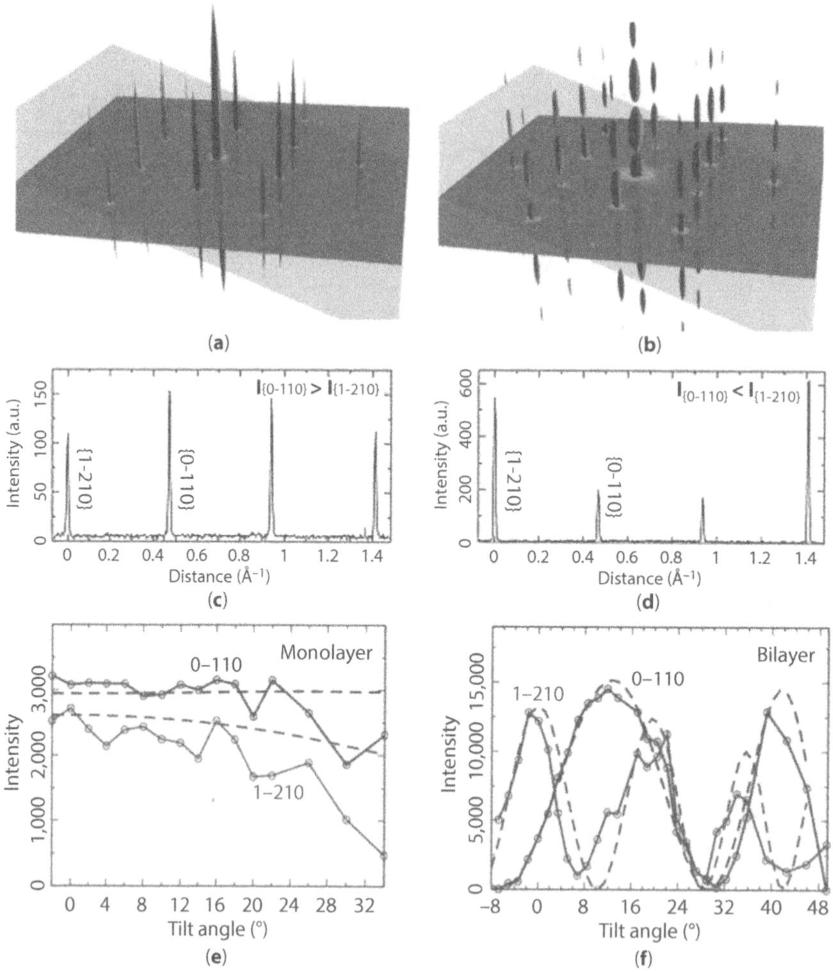

Figure 11.3 Layer-number determination of graphene by SAED. (a) and (b) are the calculated 3D Fourier transforms of single- and bilayer Bernal (AB)-stacked graphene atom positions. The intensities in reciprocal space are continuous rods with weak and monotonic intensities for single-layer graphene while they are suppressed at certain angles for bilayer graphene. The blue planes indicate a diffraction pattern that is obtained at normal incidence, and the red plane for tilt angle of 20°. (c) and (d) are the intensity of line profiles along indices {0–110} (2.13 Å spacing) and {1–210} (1.23 Å spacing), respectively. (Reproduced with permission [21]. Copyright 2007, Elsevier.) For the Bernal stacking of graphene, the layer number can be determined by intensity ratios of the inner and outer diffraction peaks in (c) and (d). (e) and (f) are the total intensity of single- and bilayer AB-stacked graphene as a function of tilt angle. The solid lines are the experimental data fitted by a Gaussian distribution; the dashed lines are numerical simulations. (Reproduced with permission [20]. Copyright 2007, Nature Publishing Group.)

are suppressed at certain angles (red planes). In this way, the whole 3D reciprocal space is probed. Figure 11.3e–f shows the total intensity of single- and bilayer AB-stacked graphene as a function of tilt angle [20]. For the single-layer graphene, the intensities in reciprocal space are continuous rods with weak and monotonic intensities (Figure 11.3a), and the changes in total intensity are relatively small when varying the tilt angle (Figure 11.3e). The key for the identification of single-layer graphene is that its reciprocal space has only the zero-order Laue zone and no dimming of the diffraction peaks should occur at any angle.

For bilayer graphene, the intensity strongly varies along the rods in 3D reciprocal space, and suppresses at certain tilt angles (Figure 11.3b). The total intensities vary so strongly that the same peaks become completely suppressed at some angles (Figure 11.3f). The method can also be used in distinguishing between single- and multilayer graphene. The layer number of other 2D materials such as MoS_2 can be also determined by this method [23].

11.3.2 Characterization of the Stacking State of Graphene

The stacking sequence and layer number of graphene have a great impact on the electronic properties, which leads a possible way for the band gap adjusting of graphene and its applications on wearable devices [24, 25]. Multilayer graphene composed of monolayer graphene has three possible configurations: AA stacking, AB stacking, and ABC stacking. The energy of AA and ABC stacking are 17.31 meV/atom and 0.11 meV/atom, respectively, larger than

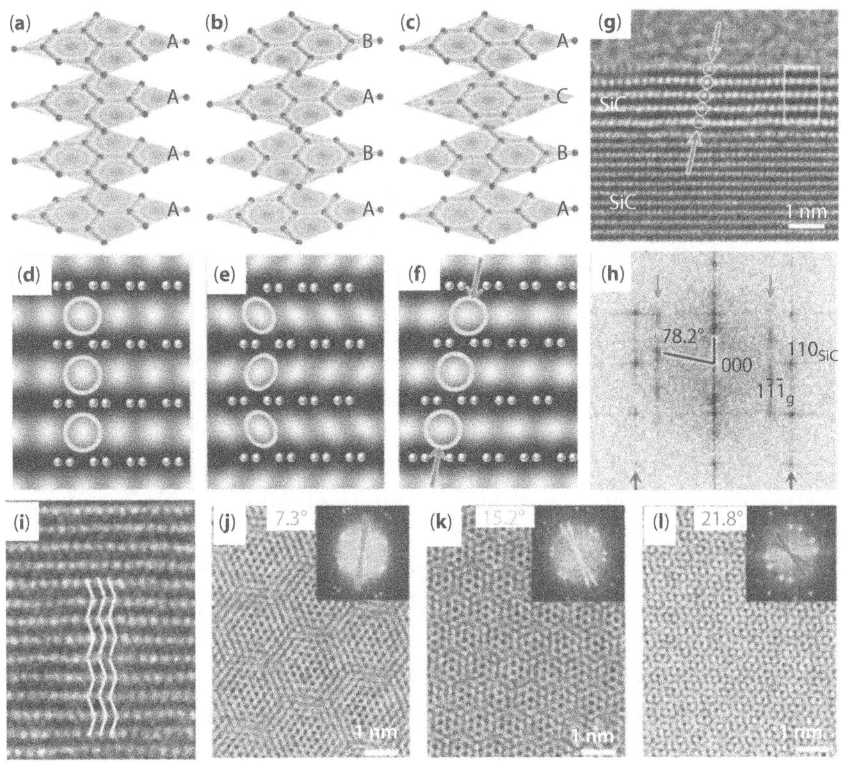

Figure 11.4 Characterization of the stacking modes of graphene. (a)–(f) The AA, AB, and ABC stacking models and the corresponding simulated HRTEM images of graphene. (g) The HRTEM image of the few-layer graphene grown on the SiC substrate. The stacked graphene layers indicated by the arrows have the same atomic configuration as (f) presenting the ABC stacking order. (h) The FFT of (g), the 78° angle between (003) and the ($\bar{1}$11) reflection points, is the feature of ABC stacking. (Reproduced with permission [27]. Copyright 2010, American Physical Society.) (i) The HRTEM image of the multilayer epitaxial graphene showing the AB stacking order as indicated by the folding lines. (Reproduced with permission [28]. Copyright 2012, AIP Publishing LLC.) (j)–(l) The HRTEM of bilayer graphene with different rotation angles. The Moiré patterns induced by the misorientation of the graphene layers show different periodicities with different rotation angles. Insets are the corresponding FFT patterns that can clearly determine the rotation angles of the two graphene layers. (Reproduced with permission [29]. Copyright 2013, American Chemical Society.)

that of AB stacking [26, 27]. Therefore, the AA stacking of graphene could hardly be found in nature, while the ABC stacking could be found in nature graphite. AA stacking can be found in the artificially folded single- and bilayer graphene. The stacking models of AA, AB, and ABC are presented in Figure 11.4a–c. Figure 11.4d–f are the simulated HRTEM images of AA-, AB-, and ABC-stacked graphite for a 40 nm underfocus and a sample thickness of 3 nm. The carbon atoms are indicated by the blue dots, and the graphene layers are dark lines. The white dots circled by the circles are main features to distinguish the stacking orders of the graphene. For the AA-stacked graphene, the white dots are round and vertically aligned. For the AB-stacked graphene, the white dots are vertically aligned but become elliptical and alternatively oriented. For the ABC stacking, the direction of white dots is not vertically aligned, but becomes 78° with the horizontal line. Figure 11.4g is an example of ABC-stacked graphite as indicated by the arrows, which consist of the simulated pattern. Figure 11.2h is the FFT of the HRTEM image graphite layer in Figure 11.4g and the 78.2° angle of the graphene patterns further confirm of ABC stacked order [27]. Figure 11.4i is an example of the normal AB-stacked graphite as confirmed by the alternative repeat of the graphene layers indicated by the guiding lines [28].

Except the translational stacking of the graphene layers, there are also rotational stacking of graphene layers, which leads to the formation of Moiré patterns. Figure 11.4j–l shows the examples of rotated graphene stacks with the degrees of 7.3°, 15.2°, and 21.8° as determined by the rotated angles between the two sets of SAED patterns. The HRTEM images clearly show that with increase of rotation angle, the period of the Moiré patterns becomes smaller [29].

11.3.3 Characterization of the Graphene Edge

The edge structure of graphene has a large impact on its electronic and chemical properties [30–32]. TEM equipped with EELS has high spatial and energy resolution, and is undoubtedly a versatile tool to characterize morphology, chemical composition, crystal structure, and electronic structure of graphene with atomic resolution. Armchair and zigzag edges of graphene are the preferred and simplest edge with few dangling bonds. Figure 11.5a–d shows the typical armchair and zigzag edge of a hole in graphene and the corresponding structure models [33]. The outlined region of the hole in Figure 11.5a is entirely "armchair," and that in Figure 11.5b is entirely "zigzag." The existence of such long-range order indicates the configurations are stable both in the experiment (Figure 11.5a–d) and simulation (Figure 11.5c–d). From the experiment and simulation, the armchair and zigzag edge are 6–6 rings. Kim *et al.* observe extended pentagon–heptagon (5–7) reconstruction at the zigzag edge (Figure 11.5e) [34]. They find that the zigzag edge frequently undergoes dramatic reversible transitions between a 5–7 reconstructed edge and a 6–6 zigzag edge, and armchair edge is relatively stable under the influence of the electron beam.

The atom at edge with different numbers of coordinated carbon atoms exhibits rich chemical information and local electronic structures. Except the long-range order structures like armchair and zigzag edge, some atoms with one or two coordinated carbon atoms exhibit different properties. Figure 11.5f shows a typical ADF image of the graphene edge of a single graphene layer [35]. The atom with single-, double-, and triple-coordinated carbon atoms is achieved with atomic resolution (Figure 11.5g); their corresponding atomic positions are shown in Figure 11.5f–j. From the energy-loss near-edge fine structure (ELNES)

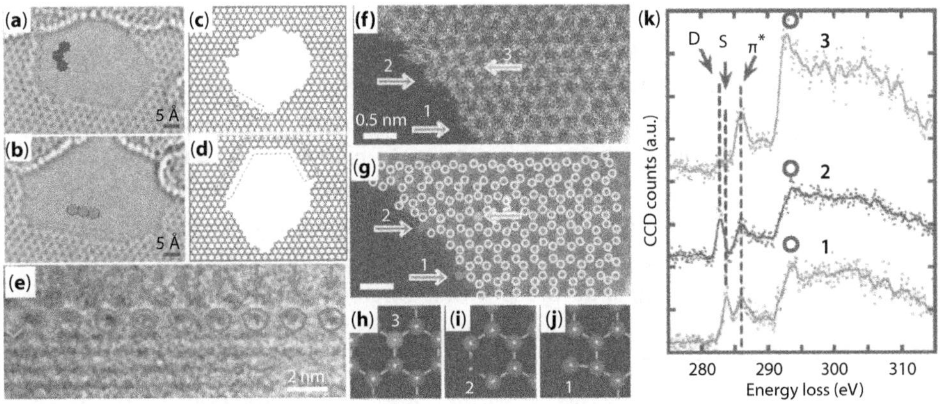

Figure 11.5 TEM characterization of the graphene edge. (a)–(d) The armchair and zigzag edges of a hole in graphene and the corresponding structure models. (Reproduced with permission [33]. Copyright 2009, American Association for the Advancement of Science.) (e) The HRTEM image of the 5–7 carbon-ring configuration of the zigzag edge of graphene. (Reproduced with permission [34]. Copyright 2013 Nature Publishing Group.) (f) ADF image of the graphene edge. (g) Carbon atoms near the graphene edge are marked by circles. The carbon atoms that are bonded with one, two, and three nearest neighbors are indicated by the dots pointed by the arrows. Figures on the arrows indicate the bond numbers. (h)–(j) Illustrations of the atom configurations with one, two, and three nearest neighbors shown in (g), and the corresponding EELS spectra are shown in (k). The atom states with various bonding conditions can be identified from the intensity of the D, S, and π peaks, as indicated by the dashed lines and arrows in (k). (Reproduced with permission [35]. Copyright 2010, Nature Publishing Group.)

spectra of carbon K (1s)-edge (Figure 11.5k), the carbon atom with different number of bonding neighbors can be distinguished. The spectrum of three bonding is recorded at an atomic position in bulk as a reference. This spectrum exhibits the features of typical sp^2-coordinated carbon atoms, the π^* peak around 286 eV and the excitation peak of σ^* at 292 eV. The spectrum in blue is recorded from an edge atom with two-coordination. The spectrum has an extra peak around 282.6 ± 0.2 eV (labeled D in Figure 11.5k), and the intensity of π^* peak reduced. The excitation peak intensity is reduced and broadened compared to the bulk spectrum (marked by open circles). The spectrum in red shows similar features, with weaker π^* peak and broadened σ^* peak. The extra peak occurs at a different energy position of 283.6 ± 0.2 eV (labeled S in Figure 11.5k). However, the structure of single-coordinated carbon atoms is very unstable under the incident electron beam, and the spectrum disappears quickly and is not fully reproducible.

11.3.4 Characterization of the Point Defects of Graphene

Defects including point defects such as vacancies, dopants, and adatoms, and line defects such as grain boundaries are common in graphene, which influences the properties of graphene especially the electric properties. Considering the potential applications of graphene in electronics, the types of defects and their properties need to be studied in detail. To achieve this target, probing the defects with atomic resolution is indispensable. TEM is a versatile tool with different working modes that can characterize the structure, chemistry, and electron states of graphene defects at atomic scale.

The interactions between illumination electron beam with the acceleration voltage higher than 80 kV and carbon atoms in graphene could lead to the sputtering of carbon atoms and generate vacancies at the original carbon position. Robertson points out that even at 80 kV, the threshold of the KOD of graphene, vacancy can also be generated when the electron beam current density is increased to ~10^8 $e^{-1}nm^{-2}s^{-1}$. Figure 11.6a–b is an example of the graphene double vacancy of 8 carbon ring with two adjacent 5 carbon rings that are created at 80 kV with current density of 10^8 $e^{-1}nm^{-2}s^{-1}$ after 30 seconds of electron beam exposure [36]. Except the electron irradiation, vacancies could also be created by the high-energy atom/ion bombardment. Wang *et al.* present that single vacancies of graphene can be created by Au atom bombing with the kinetic energy between 150 eV and 250 eV shown in Figure 11.6c–d [37]. Si atom is a common dopant in graphene due to the presence of Si element in the growth substrates, such as Si/SiO_2 and SiC. Figure 11.6e is the ADF-STEM image of the Si atoms in single- and double-vacancy sites and bond with three and four neighboring carbon atoms [38]. The Si atoms show brighter contrast than the C atoms due to the Z contrast of the ADF-STEM image. The ADF-STEM probing is operated at 60 kV to avoid damages induced by electron beam irradiation. Figure 11.6f and g shows illustrations of structure models around the Si dopants overlapped on the ADF image. In addition to the

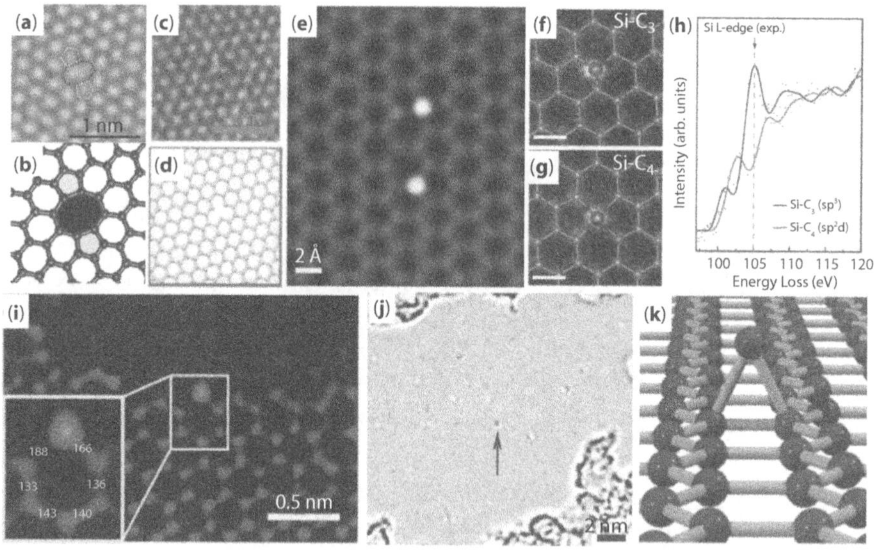

Figure 11.6 Characterization of point defects in graphene. (a)–(b) The vacancy created by electron beam irradiation and the corresponding atom model. (Reproduced with permission [36]. Copyright 2012, Nature Publishing Group. (c)–(d) The vacancy created by high-energy atom bombardment and the corresponding atom model. (Reproduced with permission [37]. Copyright 2012, American Chemical Society.) (e) The ADF-STEM image of Si atom substitutions in a monolayer graphene. (f)–(g) The illustration of structure models of the impurity Si atoms bonded with three or four nearest carbon atoms. (h) The EELS spectra of two types of Si-atom substitutions in (e) that can be separated by the signal intensity at the energy loss of 105 eV. (Reproduced with permission [38]. Copyright 2012, American Physical Society.) (i) STEM image of Si atom substitution in the zigzag edge of graphene. (Reproduced with permission [39]. Copyright 2016, American Chemical Society.) (j) Carbon adatom indicated by the arrow on a graphene membrane. (k) The structure model of a carbon adatom on monolayer graphene. (Reproduced with permission [40]. Copyright 2008, Nature Publishing Group.)

chemical analysis by ADF probe, the STEM-EELS could provide the bonding information between C and Si atoms. Figure 11.6h is the EELS of Si atoms with three and four bonded C atoms. Si spectra are extracted from a 0.16 × 0.16 nm^2 region around Si atom, and collected with a dwell time of 0.1 s per pixel and probe current of 100 pA. The energy-loss near-edge fine structure (ELNES) of Si atom bonded with three C atoms has a sharp peak at 105 eV and gradually increases above 107 eV, indicating sp^3 hybridization. The Si atom with sp^3 hybridization is not planar but occupies a site above the graphene plane. The ELNES of Si atom bonded with four C atoms does not have the sharp peak at 105 eV, but only has two small peaks at 102.6 and 107 eV. The absence of the sharp peak at 105 eV indicates stronger mixing of the 3*d* states with the 3*s* and 3*p* states forming sp^2d-like hybridization that makes up the four equal Si–C bonds. This hybridization leads to a planar configuration of Si atom in the graphene plane. The Si dopants could also be located at the edge of graphene as probed by HRTEM shown in Figure 11.6i [39]. Si atom at the zigzag edge is unstable as indicated by the uneven bond connected with the two neighbor C atoms, shown in the inset of Figure 11.6i. This 6–6 zigzag edge near the Si dopant atom tends to be reconstructed into 5–7 zigzag edge, which leaves a larger space to contain the Si atom.

Adatoms on a graphene even for light elements, such as C and H, can be directly observed by TEM, due to the ultrathin and highly transparent single layer of carbon atoms. Figure 11.6g is the TEM image of one carbon atom on a single-layer graphene as indicated by an arrow [40]. The atom type is determined by comparing the calculated contrast with the experimental result. A clear contrast could be observed by increasing the signal-to-noise ratio via summing multiple frames that are captured at the same location. Even hydrogen atoms can be clearly observed by this method. Figure 11.6k is the structure model of C adatom on the surface of graphene.

11.3.5 Characterization of the Grain Boundary of Graphene

The atomic arrangement of grain boundaries has a strong impact on the electronic, magnetic, chemical, and mechanical properties of graphene [41–44]. The grain boundary often forms in the growing process. Chemical vapor deposition (CVD) is a mature method to produce graphene with large scales up to meters, making the polycrystallinity almost unavoidable [45, 46]. The detailed structures of grain boundary and orientations of each grain can be characterized by TEM in imaging mode and diffraction mode. By ADF-STEM at 60 kV, the atomic structures of a tilted grain boundary between two graphene grains with a relative misorientation of 27° are clearly shown in Figure 11.7a [47]. The boundary consists of a series of pentagons, heptagons, and distorted hexagons (Figure 11.7b). Using atomic-resolution imaging with low voltage (60 kV), the location and every atom at a grain boundary are identified. However, tens of billions to hundreds of billions of pixels would be needed to image a single micrometer-scale grain by using atomic-resolution approaches. It will cost a day or more to finish it. Thus, diffraction-filtered imaging is used to rapidly map the location, orientation, and shape of several hundred grains and boundaries. By using an objective aperture filter in the back focal plane, the grains with specific orientation can be imaged (Figure 11.7c). The resulting real-space image shows only the grains that are selected and require only a few seconds to acquire. By repeating this process using several different aperture filters, maps of the graphene grain structure with all orientations are completed (Figure 11.7d).

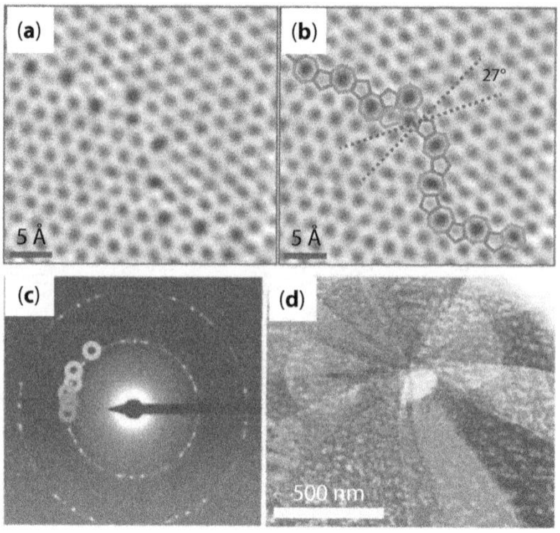

Figure 11.7 Grain boundary characterization of graphene. (a) The filtered ADF-STEM image of a monolayer-graphene grain boundary. (b) Structure of the grain boundary presented by the connected pentagons, heptagons, and distorted hexagons. The rotated angle between the two graphene crystals is 27°. (c) SAED of a polycrystalline graphene, each circle denotes one set of a diffraction pattern of a single crystal. (d) The shapes and orientations of graphene crystal grains are determined by the small-angle DFTEM, and coded by the same colors in (c). (Reproduced with permission [47]. Copyright 2011, Nature Publishing Group.)

11.3.6 Characterization of the Heterostructures of Graphene

Horizontally stitching and vertically staking of layered 2D materials with different properties can be used for the fabrication of heterostructure-based devices, such as electronics, optoelectronics, and energy convertors [48–51]. TEM is a powerful tool to characterize the structure and composition of heterostructures with atomic resolution. Unveiling the influences of stacking order, element distribution, and crystal structures at the border, to the properties of heterostructures by TEM, is critical for their preparation and application.

Figure 11.8a is the HRTEM image of an in-plane epitaxy growth of graphene indicated by the arrow from monolayer h-BN edge on a graphene substrate [52]. The B, N, and C atoms at the border could be directly determined by HRTEM as marked by dotted lines in the square. The clear BN/graphene structure could be extracted by processing the acquired HRTEM image, using the FFT and the inverse fast Fourier transform (IFFT) method. The contrast of the graphene substrate could be removed by erasing its corresponding "diffraction pattern" in the FFT image. The following IFFT shows the contrast of lateral BN/graphene heterostructure. By measuring the intensity of the atoms enclosed by the rectangle in Figure 11.8b, the atom positions of B, N, and C can be clearly identified as shown in Figure 11.8c. The STEM technology that is sensitive to the atomic number is another important way for heterostructure analysis. The structure, a vertical graphene/BN heterostructure on a BN substrate, is studied by the cross-sectional TEM as shown in Figure 11.8d [53]. Atomic layers of graphene are distinguished from the BN due to the lower STEM contrast. The STEM technology also provides a method for the evaluation of the thickness of single-layer BN by directly measuring the distance between BN layers.

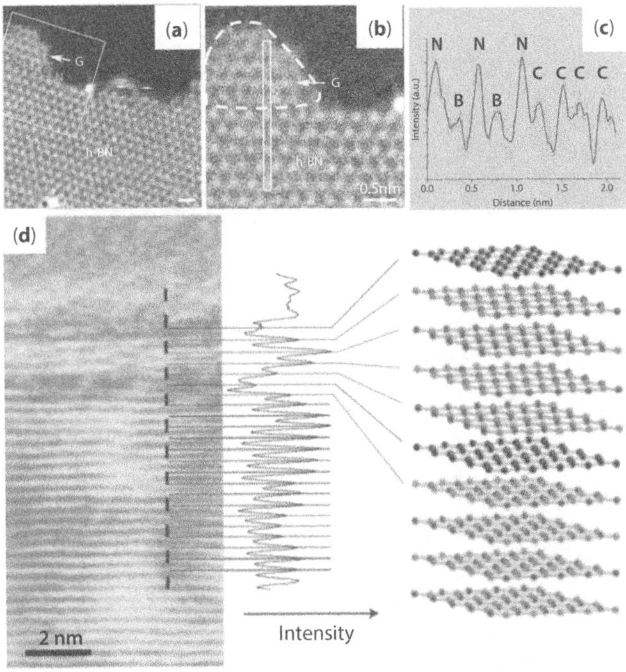

Figure 11.8 Heterostructure characterization of graphene. (a) Lateral heterostructure of graphene and h-BN characterized by ADF-STEM. The N, B, and C atoms at the border of the heterostructure are indicated by dots. (b) Zoom in the square area in (a). The dashed line shows the monolayer graphene epitaxially grown from the h-BN border. The intensity of the signal crossing the border in the rectangle is presented in (c). (Reproduced with permission [52]. Copyright 2016, Wiley.) The smoothed signal is illustrated by line. (d) The STEM image of the cross-section of the graphene/h-BN vertically stacked heterostructure. The graphene layer and h-BN layer can be directly identified by the signal intensity. The stacking order of the heterostructure is shown at the right side. (Reproduced with permission [53]. Copyright 2012, Nature Publishing Group.)

11.4 Dynamic Manipulation of Graphene

11.4.1 Fabrication of Graphene Nanostructures by Electron Beam Irradiation

By applying the high-energy electron beam, nanostructures such as nanopore and nanoribbon could be fabricated under control inside a TEM. The nanopore and nanoribbon of graphene are important components for the advanced DNA sequencing test and high-performance transistors.

The interaction between the electron beam and carbon atoms of graphene results in the energy transfer from high-energy electrons to carbon atoms. If the transferred energy is higher than the bonding energy between carbon atoms, the atoms will be knocked out leading to the KOE or sputtering effects. If the transferred energy is lower than the bonding energy, the atoms tend to repair the defects to reduce the surface energy. These two conditions are shown in Figure 11.9a [54]. The nanopore drilling on a few-layer graphene is demonstrated by Xu *et al.*, as shown in Figure 11.9b–c [55]. At the voltage of 300 kV and current density of 2.0×10^3 A/cm^2, a two-nanometer graphene nanopore is fabricated within 48 s. The formation of graphene nanopore is sensitive to the current density of the electron

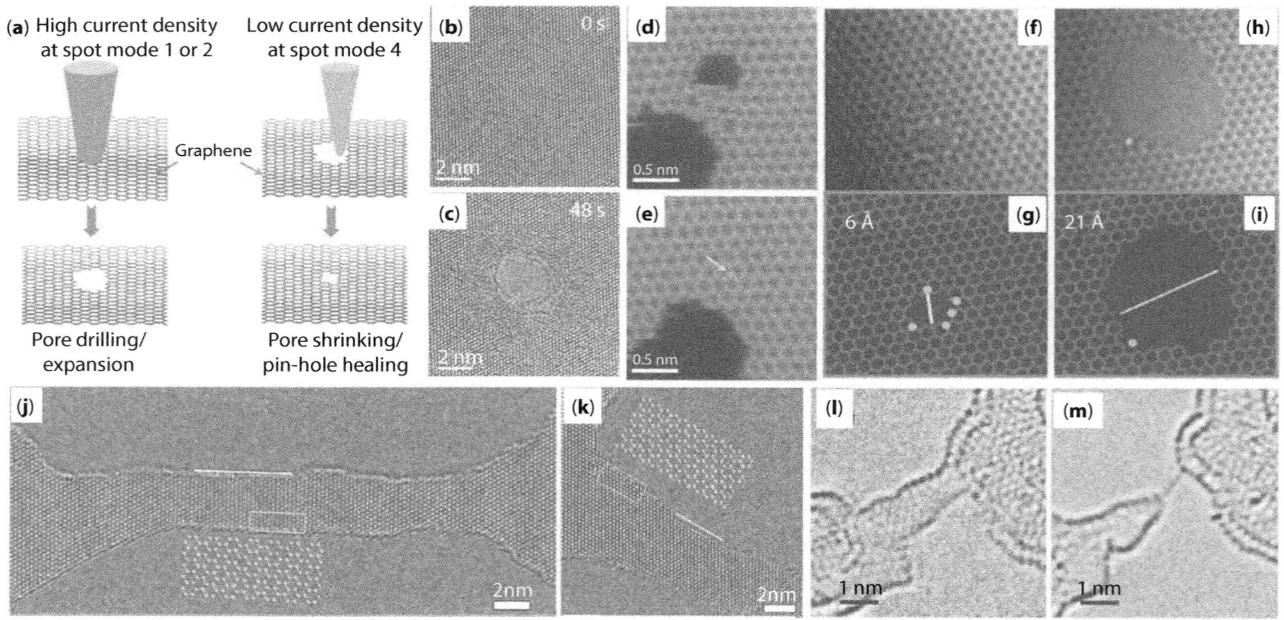

Figure 11.9 *In situ* fabrication of graphene nanostructures by electron beam irradiation. (a) The illustration of creating or healing graphene holes with different electron-beam parameter. High current density tends to create hole, while low current density tends to healing hole. (Reproduced with permission [54]. Copyright 2016, IOP Publishing Ltd.) (b)–(c) The HRTEM images of creating a graphene hole by electron beam irradiation. (Reproduced with permission [55]. Copyright 2014, Wiley.) (d)–(e) The HAADF images of a complete healing of a graphene hole. (Reproduced with permission [56]. Copyright 2012, American Chemical Society.) (f)–(h) The HRTEM image of a graphene pore opening with the help of catalytic Si atoms. (g)–(i) The corresponding atomic configurations at the graphene edge. (Reproduced with permission [57]. Copyright 2014, American Chemical Society.) (j)–(k) HRTEM images of the graphene nanoribbons terminated with the armchair and zigzag edges, respectively. (Reproduced with permission [58]. Copyright 2013, American Chemical Society.) The corresponding atomic structures of the two types of edges are shown nearby. The nanoribbons are fabricated by electron beam irradiation at 300 kV and 600°C. (l)–(m) The HRTEM image of fabricating a single carbon chain by further narrowing the graphene nanoribbon. (Reproduced with permission [59]. Copyright 2012, American Chemical Society.)

beam, which results in the nonlinear growth of the nanopore with time, due to the Gaussian distribution of the electron beam density from center to edge. They also point out that when the current density is larger than 500 A/cm^2, the electron beam-induced sputtering effect is dominant, while the electron beam-induced deposition is dominant when the current density is lower than 300 A/cm^2. Zan *et al.* report the electron beam-assisted healing of graphene nanopore using the high-angle annular dark field (HAADF) STEM, shown in Figure 11.9d–e [56]. Only within 10 s, the graphene nanopore of 0.5 nm is fully recovered with perfect hexagonal structures. There are two mechanisms for the recovery of the graphene nanopores. The recovery assisted by the transport of nearby hydrocarbon compounds leads to the reknits of graphene nanopore with defects, while the recovery through the reconstruction of graphene results in a perfectly mended hexagonal structure. The graphene nanopore could be fabricated by electron beam with the help of catalytic atoms such as Si and metal atoms. Wang *et al.* report the direct observation of Si-assisted nanopore opening by using the HRTEM at atomic scale [57]. The C atoms at the graphene edge are knocked out atom by atom, while catalytic Si atoms are reserved as shown in Figure 11.9f–i.

Similar to the fabrication of nanopore, if sculpting two parallel rectangular holes nearby, the narrow graphene in the middle forms a nanoribbon. The STEM is a desirable method for nanoribbon fabrication due to the ultrathin electron beam down to 0.1 nm, smaller than that of a carbon atom. And the observation and sculpting mode of STEM could be easily switched by adjusting the scanning dwell time. A high quality of graphene nanoribbons with selected type of edges has been fabricated by STEM. A 4 nm-thick graphene nanoribbon with armchair and zigzag edges is fabricated at 300 kV and imaged at 80 kV to reduce the KOE as presented in Figure 11.9j and k [58]. The whole process is at elevated temperature, 600°C, to reduce the generated defects by self-repair. The edges of the nanoribbons are atomically straight along $\langle 1\bar{1}00 \rangle$ (zigzag) and $\langle 1\bar{2}10 \rangle$ (armchair) directions. And the crystal lattices inside the graphene nanoribbons are defect free. Further sculpting the nanoribbons achieves the final widths of ~1.9 nm. Xu *et al.* summarized three primary process for the sculpting of graphene using high-energy electron beams: (1) the formation of vacancies by NOE; (2) self-healing of the graphene lattice by C adatoms or C-rich ad-molecules; (3) the electron-beam-induced formation of C-rich contamination. Boerrnert *et al.* report that by further sputtering the graphene nanoribbon with a broad electron beam at 80 kV, the edge atoms are gradually sputtered off, and finally form a single carbon chain that is stable under the electron beam for at least 24 s [59].

11.4.2 *In Situ* Heating Manipulation

In situ heating sample holder for TEM is designed by the MEMS microheater with four-point probe temperature measurement. With the high quality of Si_3N_4 thin film and improved MEMS-based heating chips, the controlled heating of samples up to 1200°C with minimized thermal drift has been achieved by *in situ* TEM and keeping the high resolution. This provides an advanced plant form to dynamically manipulate and probe graphene at high temperature.

The properties of graphene at high temperature, such as defect evolution and phase transition, are important issues for its preparation, processing, and application. The structure evolution of large closed grain boundary in graphene is studied by *in situ* heating TEM at 773 K as shown in Figure 11.10a–d [60]. The AC-TEM with monochromator operated

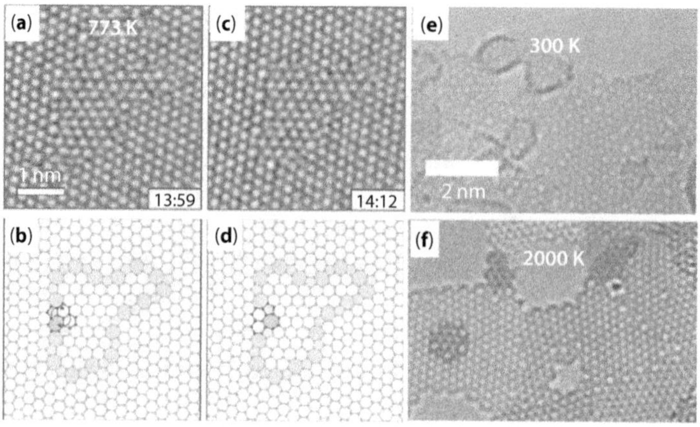

Figure 11.10 Characterization of the structure transition of graphene by *in situ* heating TEM. (a)–(d) HRTEM images of the grain boundary evolution of a monolayer graphene and the corresponding atomic structure model. The closed graphene boundary loops are created at 773 K under electron beam irradiation. Structural change of the grain boundary can be realized by bond rotation that is indicated by arrows shown in (b) and (d). (Reproduced with permission [60]. Copyright 2016, American Chemical Society.) (e)–(f) Crystallization of amorphous carbon monolayers into polycrystalline graphene at 2000 K with free edges of predominantly armchair configuration. The extremely high temperature is achieved by Joule heating of graphene, and measured by the melting of gold particles and the initial sublimation of SiN. (Reproduced with permission [61]. Copyright 2011, American Chemical Society.)

at 80 kV shows clear grain boundary loop with atomic resolution. The process of grain boundary evolution is accelerated at high temperature. The large closed grain boundary of graphene tends to relax into separated five to seven dislocations under electron beam irradiation. The structural change of the grain boundary can be realized by bond rotation and carbon evaporation. Westenfelder *et al.* present the transformations of hydrocarbon adsorbates on graphene substrates at extremely high temperature [61]. The high temperature is achieved by the Joule heating of graphene substrate and evaluated by the melting of gold nanoislands and SiN supporting film. The adsorbed hydrocarbon compounds transform into amorphous carbon monolayers, and start to crystallize. At 2000 K, the carbon monolayers form polycrystalline graphene with dominantly armchair edges. The contrast of amorphous carbon and polycrystalline graphene is clearly presented by subtracting the contrast of graphene substrate using FFT and IFFT methods as shown in Figure 11.10e–f.

11.4.3 *In Situ* Electrical Testing

The *in situ* electrical chip is similar to the *in situ* heating chip. By controlling the supplied voltage and measuring the responded electrical signals, the electric properties of the testing sample can be understudied. In this section, the properties of electron emission, stability, and charge storage of graphene with applied voltage studied by *in situ* TEM are discussed.

In situ electrical TEM equipped with scanning tunneling microscope sample holder that has two independent twin probes is applied for the *in situ* characterization of electron emission property of graphene nanoribbon, as shown in Figure 11.11a [62]. The emission current of graphene nanoribbon can be extracted out by a driving voltage less than 3 V and increases exponentially as shown in the inset of Figure 11.11a2. The electrons are emitted

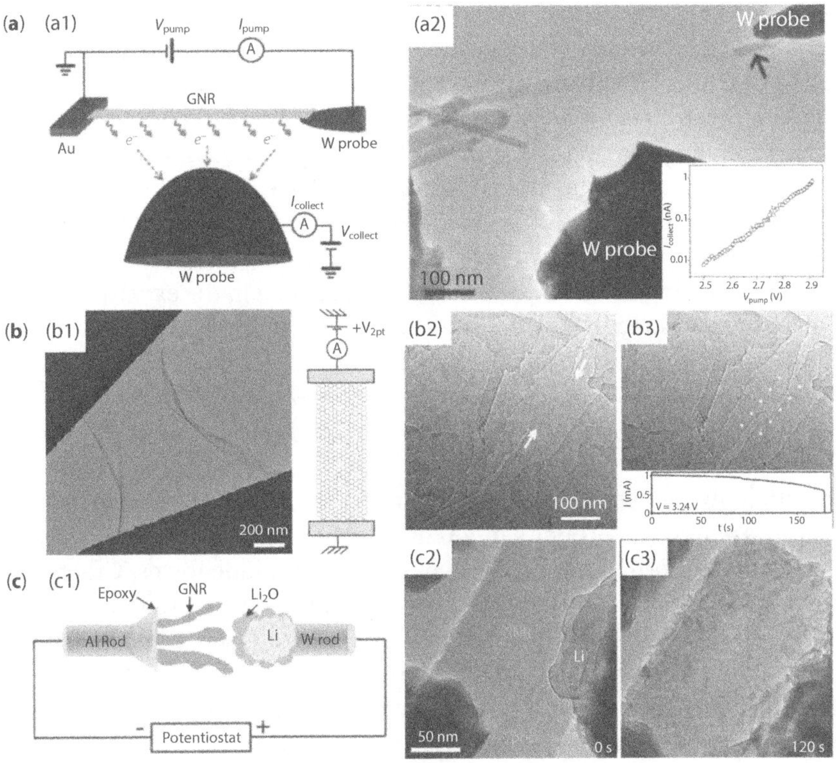

Figure 11.11 *In situ* electrical test of graphene. (a) Electron emission of graphene nanoribbon (a1)–(a2) Illustration and TEM image of *in situ* TEM setup for electron emission of graphene nanoribbon induced by internal electron field. The inset of (a2) shows the measured curve of $I_{collect}$ versus V_{pump} as indicated in (a1), which shows an exponentially increase. (Reproduced with permission [62]. Copyright 2012, American Chemical Society.) (b) Crack evolution of a graphene ribbon at high bias. (b1) TEM image and illustration of *in situ* TEM setup of graphene nanoribbon for high bias test. (b2)–(b3) TEM image of the propagation of two cracks toward each other indicated by arrows at high bias, and the final break of graphene marked by the dotted lines. The inset in (b3) is the recorded current versus time during the crack evolution. The current gradually decreases during the elongation of the crack, and becomes zero at the breakdown of graphene. (Reproduced with permission [63]. Copyright 2012, American Chemical Society.) (c) *In situ* electrochemical lithiation of a graphene nanoribbon. (c1) Schematic picture of the graphene lithium cell setup. (c2)–(c3) The TEM images of the pristine and lithiated graphene nanoribbon. The Li_2O compounds located at the surface of the graphene nanoribbon after lithiation can be clearly seen. (Reproduced with permission [64]. Copyright 2012, Elsevier.)

perpendicularly to the surface of graphene nanoribbon with high emission density of 12.7 A/cm^2, which is different form the conventional field-induced electron emission that happens at the edges. The stability of a graphene ribbon at high bias is directly observed by *in situ* TEM as shown in Figure 11.11b1 [63]. High bias-induced electroburning leads to the formation and propagation of cracks from the border of the graphene, and the layer-by-layer sublimation of few-layer graphene as shown in Figure 11.11b2–b3. The current continuously decreases during the narrowing of the graphene, and suddenly drops to zero at the breakdown. This effect provides a potential way for controlled preparation of single-layer graphene. On the contrary, two overlapped graphenes heal and form one continuous sheet at high bias. The charge storage properties of few-layer graphene used as the electrode material for Li ion battery are studied by *in situ* TEM that could track the dynamic structure and composition evolution during lithiation and delithiation. Graphene nanoribbon and Li metal on Al and W rods are used as the electrodes, and Li_2O layer covered on the Li metal is the solid electrolyte for Li^+ transport as shown in Figure 11.11c1 [64]. The lithiation process dominantly happens on the surface of the graphene nanoribbon with the formation of Li_2O nanocrystals as shown in Figure 11.11c2–c3. The SAED pattern of lithiated graphene confirms the formation of Li_2O nanocrystals. The variation of the layer distance of the graphene nanoribbon is directly examined by HRTEM from the cross-section side. The TEM-scanning tunneling microscopy (STM) holder applied in the experiment provides the ability of mechanical testing. By the *in situ* mechanical test, the lithiated graphene nanoribbon is robust, which is different from the lithiated carbon nanotubes.

11.4.4 *In Situ* Mechanical Manipulation

Graphene with high conductivity and excellent mechanical properties, such as high flexibility, intrinsic strength, and Young's modulus, is valuable for flexible electronics [45, 65]. In considering the uniform bond breaking in single-crystalline forms, Graphene is one of the strongest materials on Earth [66, 67]. The capability to resist uniform bond breaking is another important mechanical quantity of solids, which is described by fracture toughness. However, due to the ultrathin dimension of graphene, the mechanical test is difficult to carry out. Wei *et al.* report the fracture toughness testing on multilayer graphene sample having V-shaped premade single-edge notch, by using *in situ* TEM equipped with a side-entry AFM-TEM holder [68]. The fracture testing of a single-edge V-notched multilayer graphene nanosheet is shown in Figure 11.12. The graphene nanosheet is clamped

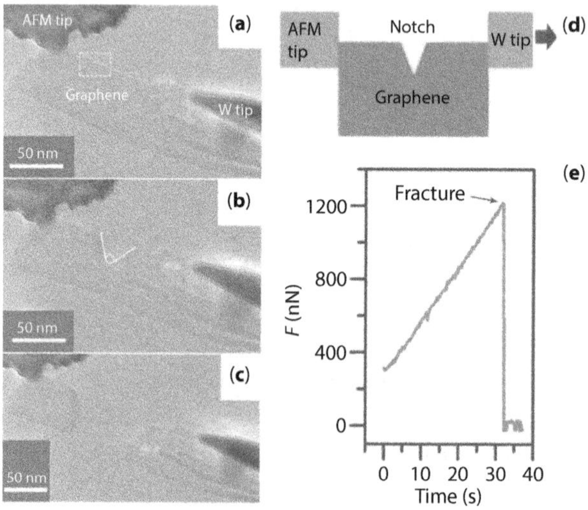

Figure 11.12 Mechanical property of graphene characterized by *in situ* AFM-TEM. (a) The TEM image of few-layer graphene attached between AFM tip and W tip. (b) TEM image of a notch created on a border of few-layer graphene for mechanical test, and the corresponding illustration of the device configuration is shown in (d). (c) TEM image of the fracture of graphene along the notch after mechanical test. (e) *In situ* recording of the mechanical force versus time during the mechanical test. The initial force is above zero due to the pre-existing tensile stress. (Reproduced with permission [68] Copyright 2015, American Chemical Society.)

between the AFM and W tips (Figure 11.12a). Then the V-shaped notch is fabricated by a 300 kV focused electron probe at its upper edge (Figure 11.12b). The graphene nanosheet is subjected to tension and fractured into two pieces with the fracture edges perpendicular to the tensile direction (Figure 11.12c). Tension is applied by keeping the W tip at a constant speed of ~2–4 nm/s with the direction perpendicular to the crack, and the force is simultaneously recorded through the deflection of the force-sensing cantilever. The tensile force increased linearly before reaching the critical value (1210 nN) and then abruptly dropped to zero, as shown in Figure 11.12e. The result indicates that the graphene nanosheet fractures in a brittle way.

11.4.5 Graphene Liquid Cell for *In Situ* TEM

Environmental TEM provides a unique opportunity for monitoring liquid phase reactions with high spatial resolution, and the materials of liquid-cell window play an important role in improving the spatial resolution [69, 70]. The resolution of the sample in the liquid cell is closely related to the thickness and the composition of the observation window, which cause unwanted electron scattering [71]. The conventional window is fabricated from Si_3N_4 or SiO_2. However, the relatively thick (tens to one hundred nanometers) and relatively high atomic number element windows will limit the resolution and perturb the natural state of the liquid or species suspended in the liquid. Graphene with a thickness of one carbon atom provides a high contrast (almost transparent to electron beam), an excellent electrical and thermal conductor, and displays minimal charging and heating effects under the electron beam. And the inert surface with fewer dangling bonds eliminates chemical and physical interference from the substrate. The excellent properties such as high flexibility, mechanical tensile strength, and permeability to small molecules make graphene a superior material as liquid-cell window. Figure 11.13a shows a TEM image of a graphene

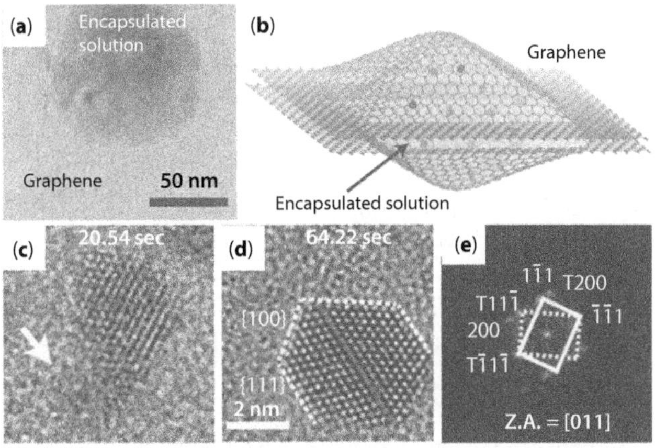

Figure 11.13 Graphene-based liquid cell for *in situ* TEM study. (a) TEM image of Pt colloidal solution encapsulated between two layers of graphene. (b) The illustration of graphene liquid cell shown in (a). (c)–(d) The growth of Pt nanoparticle by coalescence of small nanocrystal marked by arrow, and the following reshaping by straightening of the twin boundary indicated by the dotted line in (d). (e) The FFT of (d) that clearly presents two Pt grains. (Reproduced with permission [72]. Copyright 2012, American Association for the Advancement of Science.)

liquid cell. The area with darker contrast is the encapsulated liquid sample trapped between two suspended graphene sheets (lighter contrast). The graphene liquid cell remains intact under the electron beam (80 kV) during the whole TEM observation with high resolution. The critical steps in the evolution of colloidal platinum nanocrystal growth including site-selective coalescence, structural reshaping, and surface faceting are observed clearly (Figure 11.13c–e) [72].

11.5 Outlook and Challenges

In summary, we reviewed that the advanced TEM has played a critical role in the understanding of the relationships among the structures, properties, and functions of graphene, which can guide the preparation, processing, and application of graphene-based devices. However, there are a few challenges for the further improvement of TEM applications on graphene.

1. Improvement of time resolution. The dynamic analysis of the transition state of graphene by *in situ* TEM requires high time resolution to capture the fast evolution process. Two problems need to be solved. On the one hand, electron detector with high acquiring speed is indispensable. On the other hand, the massive data need to be stored and analyzed with the help of computer. The newly developed direct-electron-detection technology is a big step for the high-time-resolution TEM.
2. The spatial resolution of *in situ* TEM is usually decreased due to the undesired side effects caused by sophisticated sample holders, such as mechanical stability, thermal drifting, and electron scattering caused by additional supporting layers. Designing robust and stable *in situ* sample holders with improved thin-film technology is important.
3. The optical and magnetic *in situ* sample holders need to be developed to unveil the mechanism of their interactions with graphene.

References

1. Luo, C., Wang, C., Wu, X., Zhang, J., Chu, J., *In situ* transmission electron microscopy characterization and manipulation of two-dimensional layered materials beyond graphene. *Small*, 13, 1604259, 2017.
2. Williams, D.B. and Carter, C.B., *Transmission electron microscopy—A textbook for materials science*, Springer, Berlin, Germany, 2009.
3. Urban, K.W., Studying atomic structures by aberration-corrected transmission electron microscopy. *Science*, 321, 506, 2008.
4. Alem, N., Erni, R., Kisielowski, C., Rossell, M.D., Gannett, W., Zettl, A., Atomically thin hexagonal boron nitride probed by ultrahigh-resolution transmission electron microscopy. *Phys. Rev. B*, 80, 155425, 2009.
5. Kotakoski, J., Jin, C.H., Lehtinen, O., Suenaga, K., Krasheninnikov, A.V., Electron knock-on damage in hexagonal boron nitride monolayers. *Phys. Rev. B*, 82, 113404, 2010.

6. Egerton, R.F., Li, P., Malac, M., Radiation damage in the TEM and SEM. *Micron*, 35, 399, 2004.
7. Egerton, R.F., Control of radiation damage in the TEM. *Ultramicroscopy*, 127, 100, 2013.
8. Takeguchi, M., Furuya, K., Yoshihara, K., Electron energy loss spectroscopy study of the formation process of Si nanocrystals in SiO_2 due to electron stimulated desorption-decomposition. *Micron*, 30, 147, 1999.
9. Cazaux, J., Correlations between ionization radiation damage and charging effects in transmission electron microscopy. *Ultramicroscopy*, 60, 411, 1995.
10. Lee, Z., Meyer, J.C., Rose, H., Kaiser, U., Optimum HRTEM image contrast at 20 kV and 80 kV—Exemplified by graphene. *Ultramicroscopy*, 112, 39, 2012.
11. Coene, W.M.J., Thust, A., Beeck, M.O.D., Dyck, D.V., Maximum-likelihood method for focus-variation image reconstruction in high resolution transmission electron microscopy. *Ultramicroscopy*, 64, 109, 1996.
12. Beeck, M.O.D., Dyck, D.V., Coene, W., Wave function reconstruction in HRTEM: The parabola method. *Ultramicroscopy*, 64, 167, 1996.
13. Kirkland, A.I., Saxton, W.O., Chand, G., Multiple beam tilt microscopy for super resolved imaging. *Microscopy*, 46, 11, 1997.
14. Kirkland, A.I., Saxton, W.O., Chau, K.L., Tsuno, K., Kawasaki, M., Super-resolution by aperture synthesis: Tilt series reconstruction in CTEM. *Ultramicroscopy*, 57, 355, 1995.
15. Kirkland, A.I. and Meyer, R.R., "Indirect" high-resolution transmission electron microscopy: Aberration measurement and wavefunction reconstruction. *Microsc. Microanal.*, 10, 401, 2004.
16. Kisielowski, C., Ramasse, Q.M., Hansen, L.P., Brorson, M., Carlsson, A., Molenbroek, A.M., Topsøe, H., Helveg, S., Imaging MoS2 nanocatalysts with single-atom sensitivity. *Angew. Chem. Int. Ed.*, 49, 2708, 2010.
17. Castro Neto, A.H., Guinea, F., Peres, N.M.R., Novoselov, K.S., Geim, A.K., The electronic properties of graphene. *Rev. Mod. Phys.*, 81, 109, 2009.
18. Reina, A., Jia, X.T., Ho, J., Nezich, D., Son, H.B., Bulovic, V., Dresselhaus, M.S., Kong, J., Large area, few-layer graphene films on arbitrary substrates by chemical vapor deposition. *Nano Lett.*, 9, 30, 2009.
19. Warner, J.H., The influence of the number of graphene layers on the atomic resolution images obtained from aberration-corrected high resolution transmission electron microscopy. *Nanotechnology*, 21, 2010.
20. Meyer, J.C., Geim, A.K., Katsnelson, M.I., Novoselov, K.S., Booth, T.J., Roth, S., The structure of suspended graphene sheets. *Nature*, 446, 60, 2007.
21. Meyer, J.C., Geim, A.K., Katsnelson, M.I., Novoselov, K.S., Obergfell, D., Roth, S., Girit, C., Zettl, A., On the roughness of single- and bi-layer graphene membranes. *Solid State Commun.*, 143, 101, 2007.
22. Park, H.J., Meyer, J., Roth, S., Skakalova, V., Growth and properties of few-layer graphene prepared by chemical vapor deposition. *Carbon*, 48, 1088, 2010.
23. Brivio, J., Alexander, D.T.L., Kis, A., Ripples and layers in ultrathin MoS_2 membranes. *Nano Lett.*, 11, 5148, 2011.
24. Zhang, Y., Tang, T., Girit, C., Hao, Z., Martin, M.C., Zettl, A., Crommie, M.F., Shen, Y.R., Wang, F., Direct observation of a widely tunable bandgap in bilayer graphene. *Nature*, 459, 820, 2009.
25. Latil, S. and Henrard, L., Charge carriers in few-layer graphene films. *Phys. Rev. Lett.*, 97, 036803, 2006.
26. Charlier, J.C., Gonze, X., Michenaud, J.P., First-principles study of the stacking effect on the electronic properties of graphite(s). *Carbon*, 32, 289, 1994.
27. Norimatsu, W. and Kusunoki, M., Selective formation of ABC-stacked graphene layers on SiC(0001). *Phys. Rev. B*, 81, 2010.

28. Weng, X., Robinson, J.A., Trumbull, K., Cavalero, R., Fanton, M.A., Snyder, D., Epitaxial graphene on SiC(000(1)over-bar): Stacking order and interfacial structure. *Appl. Phys. Lett.*, 100, 2012.
29. Lu, C., Lin, Y., Liu, Z., Yeh, C., Suenaga, K., Chiu, P., Twisting bilayer graphene superlattices. *ACS Nano*, 7, 2587, 2013.
30. Son, Y., Cohen, M.L., Louie, S.G., Half-metallic graphene nanoribbons. *Nature*, 444, 347, 2006.
31. Kobayashi, Y., Fukui, K., Enoki, T., Kusakabe, K., Edge state on hydrogen-terminated graphite edges investigated by scanning tunneling microscopy. *Phys. Rev. B*, 73, 125415, 2006.
32. Yang, L., Park, C., Son, Y., Cohen, M.L., Louie, S.G., Quasiparticle energies and band gaps in graphene nanoribbons. *Phys. Rev. Lett.*, 99, 186801, 2007.
33. Girit, C.O., Meyer, J.C., Erni, R., Rossell, M.D., Kisielowski, C., Yang, L., Park, C., Crommie, M.F., Cohen, M.L., Louie, S.G., Zettl, A., Graphene at the edge: Stability and dynamics. *Science*, 323, 1705, 2009.
34. Kim, K., Coh, S., Kisielowski, C., Crommie, M.F., Louie, S.G., Cohen, M.L., Zettl, A., Atomically perfect torn graphene edges and their reversible reconstruction. *Nat. Commun.*, 4, 2013.
35. Suenaga, K. and Koshino, M., Atom-by-atom spectroscopy at graphene edge. *Nature*, 468, 1088, 2010.
36. Robertson, A.W., Allen, C.S., Wu, Y.A., He, K., Olivier, J., Neethling, J., Kirkland, A.I., Warner, J.H., Spatial control of defect creation in graphene at the nanoscale. *Nat. Commun.*, 3, 2012.
37. Wang, H., Wang, Q., Cheng, Y., Li, K., Yao, Y., Zhang, Q., Dong, C., Wang, P., Schwingenschloegl, U., Yang, W., Zhang, X.X., Doping monolayer gaphene with single atom substitutions. *Nano Lett.*, 12, 141, 2012.
38. Zhou, W., Kapetanakis, M.D., Prange, M.P., Pantelides, S.T., Pennycook, S.J., Idrobo, J., Direct determination of the chemical bonding of individual impurities in graphene. *Phys. Rev. Lett.*, 109, 2012.
39. Chen, Q., Robertson, A.W., He, K., Gong, C., Yoon, E., Kirkland, A.I., Lee, G., Warner, J.H., Elongated silicon-carbon bonds at graphene edges. *ACS Nano*, 10, 142, 2016.
40. Meyer, J.C., Girit, C.O., Crommie, M.F., Zettl, A., Imaging and dynamics of light atoms and molecules on graphene. *Nature*, 454, 319, 2008.
41. Yazyev, O.V. and Louie, S.G., Electronic transport in polycrystalline graphene. *Nat. Mater.*, 9, 806, 2010.
42. Červenka, J., Katsnelson, M.I., Flipse, C.F.J., Room-temperature ferromagnetism in graphite driven by two-dimensional networks of point defects. *Nat. Phys.*, 5, 840, 2009.
43. Malola, S., Häkkinen, H., Koskinen, P., Structural, chemical, and dynamical trends in graphene grain boundaries. *Phys. Rev. B*, 81, 165447, 2010.
44. Grantab, R., Shenoy, V.B., Ruoff, R.S., Anomalous strength characteristics of tilt grain boundaries in graphene. *Science*, 330, 946, 2010.
45. Bae, S., Kim, H., Lee, Y., Xu, X., Park, J., Zheng, Y., Balakrishnan, J., Lei, T., Ri Kim, H., Song, Y.I., Kim, Y., Kim, K.S., Özyilmaz, B., Ahn, J., Hong, B.H., Iijima, S., Roll-to-roll production of 30-inch graphene films for transparent electrodes. *Nat. Nanotechnol.*, 5, 574, 2010.
46. Xu, X., Zhang, Z., Qiu, L., Zhuang, J., Zhang, L., Wang, H., Liao, C., Song, H., Qiao, R., Gao, P., Hu, Z., Liao, L., Liao, Z., Yu, D., Wang, E., Ding, F., Peng, H., Liu, K., Ultrafast growth of single-crystal graphene assisted by a continuous oxygen supply. *Nat. Nanotechnol.*, 11, 930, 2016.
47. Huang, P.Y., Ruiz-Vargas, C.S., van der Zande, A.M., Whitney, W.S., Levendorf, M.P., Kevek, J.W., Garg, S., Alden, J.S., Hustedt, C.J., Zhu, Y., Park, J., McEuen, P.L., Muller, D.A., Grains and grain boundaries in single-layer graphene atomic patchwork quilts. *Nature*, 469, 389, 2011.
48. Shi, J., Ji, Q., Liu, Z., Zhang, Y., Recent advances in controlling syntheses and energy related applications of MX_2 and MX_2/graphene heterostructures. *Adv. Energy Mater.*, 2016. Ahead of Print.

49. Guo, Y. and Robertson, J., Band engineering in transition metal dichalcogenides: Stacked versus lateral heterostructures. *Appl. Phys. Lett.*, 108, 233104/1, 2016.
50. Novoselov, K.S., Mishchenko, A., Carvalho, A., Castro Neto, A.H., 2D materials and van der Waals heterostructures. *Science*, 353, 2016.
51. Roy, T., Tosun, M., Hettick, M., Ahn, G.H., Hu, C., Javey, A., 2D-2D tunneling field-effect transistors using $WSe_2/SnSe_2$ heterostructures. *Appl. Phys. Lett.*, 108, 083111/1, 2016.
52. Liu, Z., Tizei, L.H.G., Sato, Y., Lin, Y., Yeh, C., Chiu, P., Terauchi, M., Iijima, S., Suenaga, K., Postsynthesis of h-BN/Graphene heterostructures inside a STEM. *Small*, 12, 252, 2016.
53. Haigh, S.J., Gholinia, A., Jalil, R., Romani, S., Britnell, L., Elias, D.C., Novoselov, K.S., Ponomarenko, L.A., Geim, A.K., Gorbachev, R., Cross-sectional imaging of individual layers and buried interfaces of graphene-based heterostructures and superlattices. *Nat. Mater.*, 11, 764, 2012.
54. Goyal, G., Lee, Y.B., Darvish, A., Ahn, C.W., Kim, M.J., Hydrophilic and size-controlled graphene nanopores for protein detection. *Nanotechnology*, 27, 2016.
55. Xu, T., Xie, X., Yin, K., Sun, J., He, L., Sun, L., Controllable atomic-scale sculpting and deposition of carbon nanostructures on graphene. *Small*, 10, 1724, 2014.
56. Zan, R., Ramasse, Q.M., Bangert, U., Novoselov, K.S., Graphene reknits its holes. *Nano Lett.*, 12, 3936, 2012.
57. Wang, W.L., Santos, E.J.G., Jiang, B., Cubuk, E.D., Ophus, C., Centeno, A., Pesquera, A., Zurutuza, A., Ciston, J., Westervelt, R., Kaxiras, E., Direct observation of a long-lived single-atom catalyst chiseling atomic structures in graphene. *Nano Lett.*, 14, 450, 2014.
58. Xu, Q., Wu, M.-Y., Schneider, G.F., Houben, L., Malladi, S.K., Dekker, C., Yucelen, E., Duninborkowski, R.E., Zandbergen, H.W., Controllable atomic scale patterning of freestanding monolayer graphene at elevated temperature. *ACS Nano*, 7, 1566, 2013.
59. Boerrnert, F., Fu, L., Gorantla, S., Knupfer, M., Buechner, B., Ruemmeli, M.H., Programmable sub-nanometer sculpting of graphene with electron beams. *ACS Nano*, 6, 10327, 2012.
60. Gong, C., He, K., Chen, Q., Robertson, A.W., Warner, J.H., *In situ* high temperature atomic level studies of large closed grain boundary loops in graphene. *ACS Nano*, 10, 9165, 2016.
61. Westenfelder, B., Meyer, J.C., Biskupek, J., Kurasch, S., Scholz, F., Krill, C.E., III, U., Kaiser, Transformations of carbon adsorbates on graphene substrates under extreme heat. *Nano Lett.*, 11, 5123, 2011.
62. Wei, X., Bando, Y., Golberg, D., Electron emission from individual graphene nanoribbons driven by internal electric field. *ACS Nano*, 6, 705, 2012.
63. Barreiro, A., Boerrnert, F., Ruemmeli, M.H., Buechner, B., Vandersypen, L.M.K., Graphene at high bias: Cracking, layer by layer sublimation, and fusing. *Nano Lett.*, 12, 1873, 2012.
64. Liu, X.H., Wang, J.W., Liu, Y., Zheng, H., Kushima, A., Huang, S., Zhu, T., Mao, S.X., Li, J., Zhang, S., Lu, W., Tour, J.M., Huang, J.Y., *In situ* transmission electron microscopy of electrochemical lithiation, delithiation and deformation of individual graphene nanoribbons. *Carbon*, 50, 3836, 2012.
65. Novoselov, K.S., Falko, V.I., Colombo, L., Gellert, P.R., Schwab, M.G., Kim, K., A roadmap for graphene. *Nature*, 490, 192, 2012.
66. Lee, G., Cooper, R.C., An, S.J., Lee, S., van der Zande, A., Petrone, N., Hammerherg, A.G., Lee, C., Crawford, B., Oliver, W., Kysar, J.W., Hone, J., High-strength chemical-vapor deposited graphene and grain boundaries. *Science*, 340, 1073, 2013.
67. Lee, C., Wei, X., Kysar, J.W., Hone, J., Measurement of the elastic properties and intrinsic strength of monolayer graphene. *Science*, 321, 385, 2008.
68. Wei, X., Xiao, S., Li, F., Tang, D., Chen, Q., Bando, Y., Golberg, D., Comparative fracture toughness of multilayer graphenes and boronitrenes. *Nano Lett.*, 15, 689, 2015.

69. Wu, J., Shan, H., Chen, W., Gu, X., Tao, P., Song, C., Shang, W., Deng, T., *In situ* environmental TEM in imaging gas and liquid phase chemical reactions for materials research. *Adv. Mater.*, 28, 9686, 2016.
70. Chen, X., Li, C., Cao, H., Recent developments of the *in situ* wet cell technology for transmission electron microscopies. *Nanoscale*, 7, 4811, 2015.
71. de Jonge, N. and Ross, F.M., Electron microscopy of specimens in liquid. *Nat. Nanotechnol.*, 6, 695, 2011.
72. Yuk, J.M., Park, J., Ercius, P., Kim, K., Hellebusch, D.J., Crommie, M.F., Lee, J.Y., Zettl, A., Alivisatos, A.P., High-resolution EM of colloidal nanocrystal growth using graphene liquid cells. *Science*, 336, 61, 2012.

12

Peculiarities of Quasi-Particle Spectra in Graphene Nanostructures

E.S. Syrkin[1,2], V.A. Sirenko[1], S.B. Feodosyev[1*], I.A. Gospodarev[1] and K.A. Minakova[2]

[1]Verkin Institute for Low Temperature Physics and Engineering, Kharkiv, Ukraine
[2]National Technical University "Kharkiv Polytechnic Institute", Institute of Education and Science in Engineering and Physics, Kharkiv, Ukraine

Abstract

Phonon and electron spectra of graphene nanofilms and nanotubes with defects are analyzed. The electron and phonon local densities of states (LDOSs) are calculated for graphene with isolated vacancies and group of nearest-neighbor vacancies. Different geometries of graphene boundaries and step-edge imperfections on metallic bi- and trigraphene are considered. The dynamic planar stability of the considered structure is proved for temperatures above the ambient. A strong heterogeneity of behavior of electron and phonon LDOSs near Fermi level is demonstrated for atoms near defect.

It is shown that in electron spectra of graphene with zigzag boundary, there appear the waves, split off the bands of quasi-continuous spectrum, which propagate along the boundary and decay with increasing distance from it. They, moreover, propagate only via the atoms of sublattice, which contains atoms with dangling bond, appeared with a formation of boundary. Dispersion of these waves is determined by character of relaxation processes during formation of the boundary. Dispersion in electron spectrum is relativistic, but corresponds to the significantly less values of group velocity, compared to infinite monolayer of graphene.

The split gap waves lead to a formation on LDOS of the sharp resonances, which enrich significantly electron spectrum near Fermi level, as well as phonon spectrum near the point of intersection of acoustic and optical branches, polarized normally to the plane of graphene monolayer. These phonons within a considered frequency range do not practically interact with differently polarized phonons, and also possess high group velocities, which should provide a dominant contribution to electron–phonon coupling. The presented results demonstrate a possibility to facilitate superconductivity in a graphene matter by a controlled creation of defects like vacancy or zigzag boundary, which distort the atomic bonding in a particular sublattice of graphene monolayer. The similar behavior is demonstrated by electron spectra of graphene with isolated vacancy and vacancy groups as well as for zigzag step-edge imperfection on metallic bi- and trigraphene.

On the basis of calculations carried out at the microscopic level, there is a given quantitative description of the phonon heat capacity ultrathin graphene nanofilms as well as single-walled carbon nanotubes. The behavior of the flexural stiffness of graphene monolayers is analyzed. The temperature ranges are defined where the temperature dependence of the heat capacity is determined by the contributions of flexural vibrations. We show contributions to phonon heat capacity in both the

*Corresponding author: feodosiev@ilt.kharkov.ua

Tobias Stauber (ed.) Handbook of Graphene: Volume 2, (315–388) © 2019 Scrivener Publishing LLC

flexural waves propagation along the nanotube surface and of flexural vibrations of nanotube as the entire one-dimensional object.

Keywords: Quasi-low dimensional structures, electron and phonon spectra, localized and quasi-localized states, graphite, graphene, carbonic nanofilms and nanotubes, local and expanded defects, green functions, local densities of states

12.1 Introduction

It is well known [1, 2], that graphene is a zero-gap semiconductor. Moreover, its effective electronic mass vanishes near Fermi-level with appearance of V-like (Dirac) singularity in electron spectrum. Eventually, electron spectrum of graphene becomes highly sensitive to some sorts of distortions. Recent interest to different properties of graphene and related nano-arrangements is sufficiently aimed at controlled variation of electron density of states within energy range in close vicinity of ε_F. In particular, the search for possibilities to create either a finite semiconductor gap, or, in contrast, drastic increase of Fermi-level occupation in electron spectra of graphene and its nanoderivatives is in progress, as well as for possibilities of superconducting transition in such structures. Therefore, it is promising to look for the solution of tuning the electron spectrum of graphene near ε_F by a controlled production of both local and extended defects in carbon nanostructures [3–8].

Pronounced hybridization of atomic orbitals in a monolayer of graphene enhances its Debye temperature up to 2500 K, giving rise to the "low-temperature" features in vibrational thermodynamic properties of graphene and its derivatives, nanotubes and nanoribbons in particular, at temperatures above ambient. In the BCS consideration, strongly evidenced for a broad variety of superconductors, the unconventional ones, in particular (see, e.g., Ref. [9] and references therein), high temperature of superconducting transition T_c should be expected then in line with a mean frequency of phonons [10].

Within this approach, the absence of superconductivity in conventional graphene materials is explained by a low density of charge carriers in the vicinity of Fermi energy (ε_F), together with a lack of phonons for basic contribution to the electron–phonon coupling constant. To enhance the latter, it is necessary to increase the number of conducting electrons and saturate the fraction of phonon spectrum responsible for Cooper pairing. In fact, graphite with intercalated metal layers has manifested a superconducting transition at temperatures increased by a growing number of quasiflexural phonon modes with frequencies close to the K-point of the first Brillouin zone [3, 5, 7, 11] in addition to increment of charge carriers number. In the work [7], it was demonstrated that a similar increase of charge carriers and corresponding phonon numbers takes place in thin graphene nanofilm, i.e., bi- and trigraphene with a "step-edge" boundary.

In spite of numerous experiments and theoretical investigations nor in graphene and in carbon nanofilms [13–15], the nature of electron–phonon interaction is not clear. Therefore, a detailed analysis of phonon and electron properties in the carbon nanostructures and influence of different defects are actual problems.

12.1.1 Electron Spectra of Graphene

The 2D crystal structure of graphene is a complex lattice comprised of two close-packed 2D triangular sublattices oriented in such a way that (○ and ●), the atoms of one sublattice, occupy the centers of mass of the triangles of the other lattice.

Due to specific 2D-crystal structure of graphene, the spectral branches coincide in a single point of reciprocal space (K-point of 2D first Brillouin zone).

The other directions, outgoing from Γ point, are characterized by a finite gap between spectral branches, yielding a linear dispersion of electrons at such point, as $\varepsilon(\mathrm{K}) = \varepsilon_\mathrm{F}$ in graphene, and V-singularity of DOS at ε_F (curve 1 in Figure 12.1b).

In zero magnetic field, the electron spectrum of graphene is reasonably described within tight-binding approximation (see, e.g., Refs. [16, 17]) and corresponding Hamiltonian, read as

$$H = \sum_i \varepsilon_i |i\rangle\langle i| - \sum_{i,j} J_{ij} |i\rangle\langle j|. \tag{12.1}$$

In the case (see, e.g., Ref. [16]) of in-plane electron hopping restricted to the nearest neighbors $\forall J_{ij} = J \approx 2.8$ eV, and $\forall \varepsilon_i = \varepsilon_F = 3J$ (indices i and j label sites), the Hamiltonian (Equation 12.1) yields the following dispersion law:

$$\varepsilon_0(k) = \pm J\sqrt{1 + 4\cos\left(k\cdot\frac{a_1 - a_2}{2}\right)\left[\cos\left(k\cdot\frac{a_1 + a_2}{2}\right) + \cos\left(k\cdot\frac{a_1 - a_2}{2}\right)\right]}. \tag{12.2}$$

For high-symmetry directions in **k**-space, it is illustrated by Figure 12.1a with Fermi energy as the reference one.

The behavior of a real part of Green's function Re$G(\varepsilon)$ near the Fermi level, i.e., $\varepsilon(\mathrm{K}) = 0$ (curve 2 in Figure 12.1b) demonstrates that various sorts of defects should localize elementary excitations near this level [5, 6].

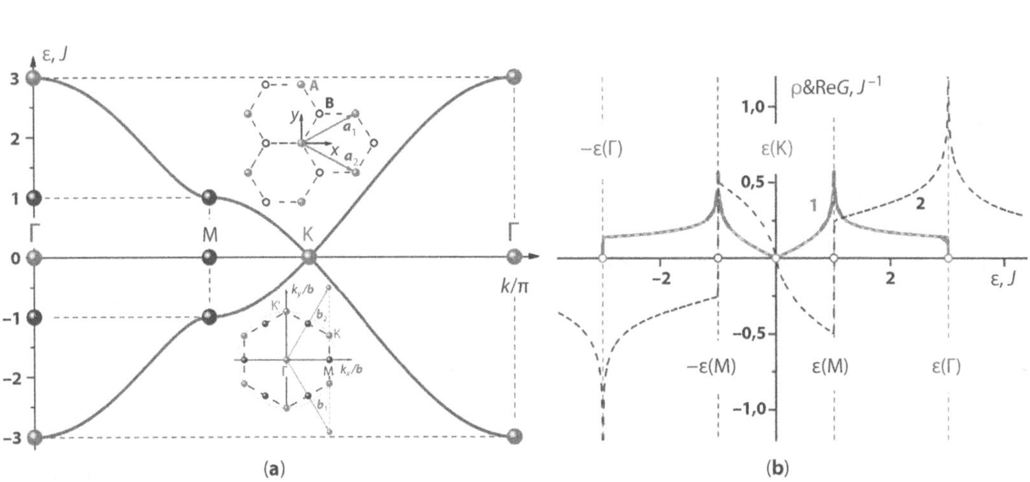

Figure 12.1 (Fragment a): dispersion curves of graphene along high-symmetry directions. Inset: choice of elementary cell and construction of first Brillouin zone of graphene. (Fragment b): DOS of graphene and real part of its Green's function (curves 1 and 2, respectively). Fermi energy is a reference one.

Indeed, the energies of quasilocal states are determined by the Lifshitz equation [18] for quasilocal states (see more detail, for example, in Ref. [19]). This equation is rewritten in the following:

$$\operatorname{Re} G(\varepsilon) = S(\varepsilon, \hat{\Lambda}) \tag{12.3}$$

where operator $\hat{\Lambda}$ describes perturbation caused by the effect and function $S(\varepsilon, \hat{\Lambda})$ describes its influence on quasiparticle spectra. So, enlarging of the Fermi level is needed for the transition of the graphene nanostructures in the superconducting state indeed may be enrich thanking appearing in this nanoobjects some defect structures.

Behavior of dependence ReG(ε) near εF (on interval $\varepsilon \in [\varepsilon(-M), \varepsilon(M)]$) testifies to the existence of solutions (Equation 12.3) for a very wide class of functions $S(\varepsilon, \hat{\Lambda})$, that is, for a very wide class of excitations. Thus, an increase in the population of the Fermi level necessary for the transition of graphene nanoformations to the superconducting state can indeed be achieved by the creation of certain defective structures in these nanoobjects.

12.1.2 The Phonon Spectrum of Graphene: General Provisions

The phonon spectra and vibrational characteristics of graphene and carbon nanofilms consisting of several graphene monolayers are interesting and important for practical applications no less than their much more actively studied electron spectra. First, the phonon spectra of graphite and graphene nanofilms have a number of interesting distinctive features: an unusually wide band of the phonon spectrum (so graphite Debye temperature is about 2500 K), and exceptionally strong anisotropy of the interatomic interaction (in graphite, the ratio of elastic module $C_{11}/C_{44} \sim 300$). Because of this, in a very wide frequency range, the phonon modes polarized along the layers and in the direction normal to the layers (quasiflexural modes) are practically independent, as well as the non-sound dispersion law of the quasiflexural modes in the long-wavelength region, which causes the "non-Debay" behavior of the low-temperature thermodynamic characteristics of these compounds. Secondly, and no less importantly, the analysis of phonon spectra and the calculation of the mean square amplitudes of atomic displacements of these structures make it possible to determine their dynamic stability, which is necessary both in their synthesis and in determining the operating conditions of the devices created on their basis.

Among a large number of both fundamental and applied studies of the various physical properties belonging to nanomaterials, the study of graphene nanofilms and graphene nanotubes undoubtedly occupies a special place. Although, at first glance from a practical point of view, the phonon spectra and vibrational characteristics of these structures do not seem as relevant as the spectra of some other quasiparticle excitations and the properties thereof (for example, electron and magnetic), this is clearly not the case. First of all, the phonon spectrum determines the stability of the structure, which is especially important to the formation of nanostructures. Second, many properties that are manifested by nanoobjects, either those that are already applicable and put into use at present, or those that are still sought after (such as superconductivity of graphene structures), occur with substantial and even defining participation from phonons.

Table 12.1 Graphite elastic modulus.

	C_{11}	C_{66}	C_{33}	C_{44}	C_{13}
Refs.	**TPa**				
[20]	106 ± 2	44 ± 2	3.65 ± 0.1	0.4 ± 0.04	---
[21]	106	44	3.7	0.37 ± 0.02	1.5

Another important reason is the analysis of dynamic stability of carbon nanostructures that is necessary for the "fine-tuning" of the process of their synthesis and for determining the working conditions of devices constructed on their basis.

In addition, the analysis of the mean-square amplitudes of atomic displacements makes it possible to clarify the limits of the applicability of the harmonic approximation when considering the vibrations of graphite and carbon nanofilms, that is, the use of the phonon and phonon modes to describe the vibrational characteristics of the very concepts, which is not *a priori* obvious. Indeed, the smallness of the elastic module C_{33} and C_{44} (see Table 12.1), at first glance, suggests that the contribution of anharmonisms to the vibrational characteristics of these compounds is essential.

Therefore (as in a free condition, and adsorbed on substrates), it is necessary to begin research of phonon spectra and vibrational characteristics of graphene and carbon nanofilms with the study of a phonon spectrum of graphite. Note that the vibrational characteristics of graphite are very good when studied experimentally.

The atoms of carbon are connected with a substrate by the same van der Waals interaction as graphene monolayers in a graphite crystal. This interaction involves only the lowest frequency region of the graphene phonon spectrum (~2%). Its change practically does not influence the interatomic interactions in the layer plane.

12.1.2.1 *Graphite Crystal Structure and Character of Force Constants between Its Atoms*

A strongly anisotropic layered graphite crystal (Figure 12.2) consists of graphene monolayers whose atoms form regular hexagons. Such a two-dimensional lattice is a complex lattice comprised of two close-packed 2D triangular sublattices (○ and ●), the atoms of one sublattice occupying the centers of mass of the triangles of the other lattice. The Bravais vectors, lying in the basal plane, can be chosen as (Figure 12.1a) $r_1 = \left(a_0\sqrt{3}/2;\; a_0/2;\; 0\right)$ and $r_2 = \left(a_0\sqrt{3}/2;\; -a_0/2;\; 0\right)$, where the parameter $a_0 \approx 2.45$ Å. The period of the lattice along the c axis equal to twice the interlayer distance is $r_3 = (0; 0; c_0)$, where the parameter $c_0 \approx 6.75$ Å. Thus, the unit cell of graphite contains four atoms.

We note that the atoms belonging to different sublattices in the *2D* lattice of graphene, containing two atoms per unit cell, are physically equivalent.

The local Green's functions characterizing the contributions of each atom to the phonon density of states (DOS) and the vibrational characteristics are the same for the atoms of different sublattices ($G^{\circ}(\omega) = G^{\bullet}(\omega)$). This equivalence breaks down in the graphite lattice because the interlayer interactions are different for each of the sublattices. Indeed, as clearly

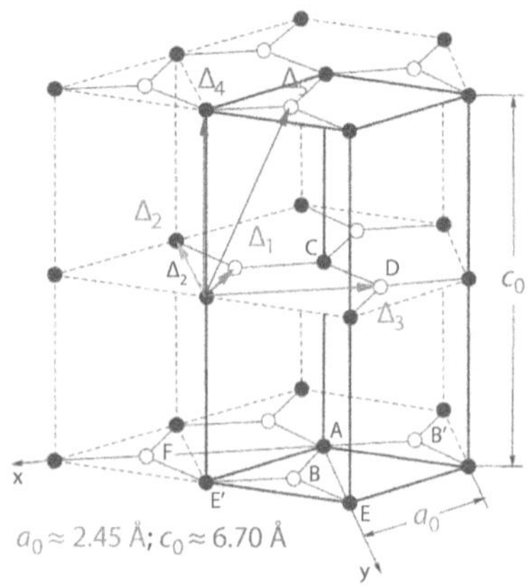

Figure 12.2 The structure of graphite with the designation of radius vectors for neighbors whose interaction between themselves is taken into account.

seen in Figure 12.1, the atoms of different sublattices from the basal plane are arranged differently with respect to the atoms of neighboring planes and therefore interact differently with them. Each atom of the sublattice (●) has in its nearest layers two neighboring atoms from the same sublattice that are located at distance $\Delta_4 = c_0/2 \approx 3.35$ Å and six neighboring atoms from the sublattice (○) that are located at the $\Delta_5 = \sqrt{\Delta_4^2 + \Delta_1^2} \approx 3.64$ Å; each atom of the sublattice (○) has 12 neighboring atoms that are located at the distance r_5 (six from the sublattice (●) and six from the sublattice (○)). That is, the atoms belonging to different sublattices of one graphene monolayer present in the graphite lattice are actually nonequivalent: the local Green's functions corresponding to these atoms and the vibrational characteristics determined by them, for example, the root-mean-square displacements of these atoms along different crystallographic directions, will be different.

The strong anisotropy of the interatomic interaction and other properties that are characteristic for graphite are due to the facts that the difference of the distances between the nearest neighbors in a layer and in neighboring planes is considerable and that the bonding forces in different crystallographic directions are of different types. For example, the interaction is of the covalent type between the nearest neighbors in the basal plane, which lie at distance $\Delta_1 = a_0/\sqrt{3} \approx 1.415$ Å, and of the van der Waals type between atoms located at the distances $\Delta_2 = a_0$ and $\Delta_3 = 2a_0/\sqrt{3} \approx 2.83$ Å (second and third neighbors in the basal plane) as well as between the atoms lying in neighboring layers at distances Δ_4 and Δ_5 from one another. In addition, graphite possesses metallic conductivity, which alters the interatomic interaction somewhat, primarily between the nearest neighbors. Since the coordinate z and the coordinates x, y of the basal plane in the crystal lattice of graphite transform according to different irreducible representations of the point symmetry group D_{6h} of the crystal, the force-constant matrix can be represented in the form

$$\Phi_{ik}(r,r') = \Phi_{ik}(r - r') \equiv \Phi_{ik}(\Delta)$$

$$= -(1-\delta_{iz}\delta_{kz})\cdot\left[\alpha(\Delta)\cdot\frac{\Delta_i\Delta_k}{\Delta^2} - \beta_x(\Delta)\cdot\delta_{ik}\right] - \beta_z(\Delta)\cdot\delta_{iz}\delta_{kz} \quad (12.4)$$

To describe the weak interlayer interaction, it is natural to take into account the interaction of atoms located in nearest-neighbor layers. For the nearest neighbors in the basal plane ($\Delta = \Delta_1$), the interaction between that is determined by a superposition of the covalent and metallic bonds, the matrices (Equation 12.4) will be characterized by three parameters. The interatomic interaction between the more distant neighbors ($\Delta = \Delta_2$; $\Delta = \Delta_3$; $\Delta = \Delta_4$, and $\Delta = \Delta_5$) can be assumed to be a van der Waals interaction and can be described by an isotropic pair potential $\varphi(\mathbf{\Delta}) = \varphi(\Delta)$. The corresponding force-constant matrices can be represented in the form (Equation 12.4) with $\beta_x(\Delta) = \beta_z(\Delta) = \beta(\Delta) \equiv \varphi'(\Delta)/\Delta$, and $\alpha(\Delta) \equiv \varphi''(\Delta) - \beta(\Delta)$.

For describing weak interlayer interaction, it is impossible to be restricted to interacting only between the atoms that are apart Δ_4 as in this case, counteraction to shear of layer as a whole will be ensured with exclusively noncentral forces that nature of origination thus remains indeterminate. The interacting account between the atoms that are apart Δ_5 (exceeding Δ_4 approximately on 8%) will allow not only to explain the nature of origination of noncentral forces, but will also allow to feature a relaxation of interlayer distances and interlayer interacting at formation of graphite nanofilms, and also to including on in exposition of interlayer noncentral forces. Thus, naturally, at exposition of force interactions between atoms of one layer to consider interacting first—the third neighbors, that is interacting between all atoms that have been had to each other is closer, than Δ_4. Further, we will see that only the account in-layer interactions between neighbors is not closer than the third will allow to feature the flexural stiffness of graphene layers. Existence of flexural stiffness plays a key role in providing stability of a crystal structure of graphite and stipulating, mainly, feature behaviors of its vibrational characteristics.

Thus, in the proposed model of dynamic loudspeaker of a lattice of graphite, it is featured by means of force constants: featuring five central and six noncentral interactions between atoms.

12.1.2.2 Force Constants and Flexural Rigidity of the Layers

We can receive the five equations for force constants using known experimentally data about elasticity module connected with structure of crystal and its matrices of force constants (see, for example, Ref. [19]):

$$\begin{cases} c_{iklm} = \dfrac{1}{V_0}(b_{imkl} + b_{kmil} - b_{lmki}\}; \\ b_{iklm} = -\dfrac{1}{2}\sum\limits_{\Delta}\Phi_{ik}(\Delta)\cdot\Delta_l\cdot\Delta_m \end{cases} \tag{12.5}$$

The condition of symmetry of the tensor of elastic moduli with respect to permutation of pairs of indices or symmetry Voigt matrix $C_{ik} = C_{ki}$ is a transition condition in the long-wavelength limit in the equations of the theory of elasticity. Since the coordinates in the basal plane and along the *c* axis transform according to different irreducible representations of the point symmetry group of the graphite, the relation $C_{13} = C_{31}$ is not satisfied identically, and provides an additional equation for the force constants.

$$\beta_{1z} + 6\beta_2 + 4\beta_3 = \frac{2}{3\Delta_1^2}\left[\Delta_4^2\beta_4 - 9\left(\Delta_1^2 - \Delta_4^2\right)\beta_5\right] \tag{12.6}$$

Four of the five elastic modules (C_{11}, C_{22}, C_{33}, and C_{44}) are reliably obtained from determining the sound velocities along the high-symmetry crystallographic directions, both acoustic (for example, Ref. [20]) and neutron [21] experiments. The experimentally determined values of the elastic module of graphite are presented in Table 12.1.

The elastic modulus C_{13} is determined by more complicated and less precise methods (for example, from measurements of Young's modulus and Poisson's ratio). This, considering the smallness of C_{13}, leads to differences in the definition of its values of the order of the values themselves, making it nearly impossible to use for determining the force constants.

Using the relation (12.6), the following equations are obtained for the elastic module of graphite from Equation 12.5:

$$C_{11} = \frac{\sqrt{3}}{12\Delta_4}\left[3(\alpha_1 + 6\alpha_2 + 4\alpha_3) + 4(\beta_{1x} + 6\beta_2 + 4\beta_3) + 9\frac{\Delta_1^2}{\Delta_5^2}\alpha_5 + 12\beta_5\right]; \tag{12.7}$$

$$C_{66} = \frac{\sqrt{3}}{12\Delta_4}\left[\alpha_1 + 6\alpha_2 + 4\alpha_3 + 4(\beta_{1x} + 6\beta_2 + 4\beta_3) + +3\frac{\Delta_1^2}{\Delta_5^2}\alpha_5 + 12\beta_5\right]; \tag{12.8}$$

$$C_{33} = \frac{2r_4\sqrt{3}}{9\Delta_1^2}\left(\alpha_4 + 9\frac{\Delta_4^2}{\Delta_5^2}\alpha_5 + \beta_4 + 9\beta_5\right); \tag{12.9}$$

$$C_{44} = \Delta_4\sqrt{3}\left[\frac{\alpha_5}{\Delta_5^2} + \frac{2(\beta_4 + 9\beta_5)}{9\Delta_1^2}\right] \tag{12.10}$$

The missing equations can be obtained from neutron diffraction data [21], Raman scattering data [22], and inelastic x-ray scattering data [23].

The following are obtained in Ref. [21]: $\omega_{TO}(\Gamma)/2\pi \approx 1.44$ THz and $\omega_{LO}(\Gamma)/2\pi \approx 3.76$ THz, which corresponds to Raman scattering data obtained in [24]. Thus, the following expressions are valid for the frequencies $\omega_{TO}(\Gamma)/2\pi \approx 1{,}44$ and $\omega_{LO}(\Gamma)/2\pi \approx 3{,}76$ (see Figure 12.3) in the model proposed:

$$m\omega_{TO}^2(\Gamma) = Q + 2\beta_4 + 2T - \sqrt{(Q - 2\beta_4)^2 + (T - 2\beta_4)^2}; \tag{12.11}$$

$$m\omega_{LO}^2(\Gamma) = G + 2F + R - \sqrt{(G - F)^2 + (F - R)^2} \tag{12.12}$$

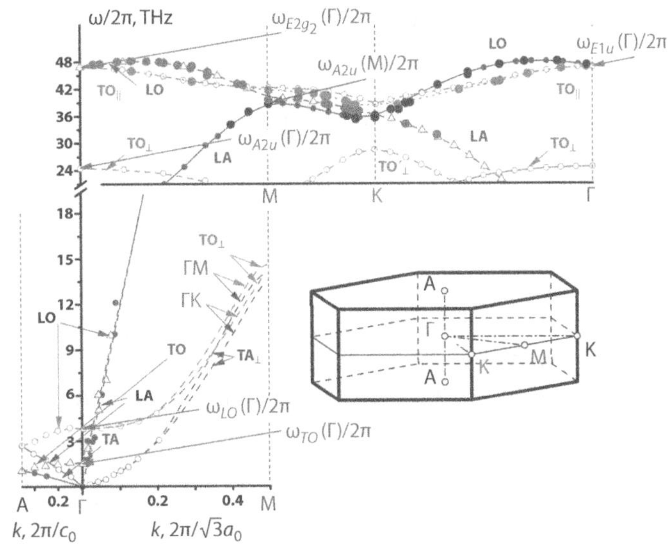

Figure 12.3 Dispersion curves of graphite along high-symmetrical directions in reciprocal space obtained experimentally, i.e., by inelastic neutron scattering [21] and by inelastic scattering of soft x-ray radiation [23]. Also shown the 1st Brillouin zone of graphite with the names of the corresponded high-symmetrical directions.

where m is the mass of a carbon atom. The following notation is introduced:

$$F\equiv 3\left(\frac{\Delta_4^2}{\Delta_5}\alpha_5+2\beta_5\right);\ G\equiv 2(\alpha_4+\beta_4);\ Q\equiv 3\left(\frac{\alpha_1+\alpha_3}{2}+\beta_{1x}+\beta_3\right);$$
$$T\equiv 3\left(\frac{\Delta_1^2}{\Delta_5}\alpha_5+2\beta_5\right);\ R\equiv -6\beta_2-12\beta_3+\frac{2\Delta_4^2}{\Delta_1^2}\beta_4+9\cdot\frac{2\Delta_4^2-\Delta_1^2}{\Delta_1^2}\beta_5. \tag{12.13}$$

The relations for the Raman frequency $\omega_{E2g2}/2\pi \approx 47.64$ THz and the frequency $\omega_{E1u}(\Gamma) \approx$ 22.6 THz and $\omega_{A2u}/2\pi \approx 26.04$ THz [22, 24] manifested in the infrared emission have the form

$$m\omega_{E2g2}^2(\Gamma)=Q+2\beta_4+2T+\sqrt{(Q-2\beta_4)^2+(T-2\beta_4)^2}; \tag{12.14}$$

$$m\omega_{E1u}^2(\Gamma)=Q+T; \tag{12.15}$$

$$m\omega_{A2u}^2(\Gamma)=2\left(R+G\right). \tag{12.16}$$

Force constants α_1, α_2, and α_3 are a linear sum $\alpha_1 + 6\alpha_2 + 4\alpha_3$ in Equations 12.7 and 12.8, and $\alpha_1 + \alpha_3$ in Equations 12.11–12.13. This indeterminacy can be eliminated by using, for example, the data of Refs. [21, 24] for the frequencies of in-plane polarized acoustic vibrations at the points K and M of the Brillouin zone of graphite. So, the frequencies $\omega_{TA||}$

(M)/2π ≈ (22.35 ± 5%) THz and $\omega_{LA_{\|}}(M)/2\pi \approx (39.7 \pm 5\%)$ THz. In the frame of the considered model, we have

$$m\omega_{TA_{\|}}(M) = 4(\beta_{1z} + 2\beta_2) + o(\alpha_4, \beta_4); \quad (12.17)$$

$$m\omega_{LA_{\|}}(M) = 2\alpha_1 + 6\alpha_2 + 3\alpha_3 + 2\beta_{1x} + 8\beta_2 + 6\beta_3 + o(\alpha_4, \beta_4). \quad (12.18)$$

Thus, the values of all force constants that characterize interaction between atoms of graphite from Equations 12.7 to 12.15 can be unequivocally received and checked up. These values [25] are resulted in Table 12.2. Further, we will use designations $\alpha_i \equiv \alpha(\Delta_i)$ and $\beta_i \equiv \beta(\Delta_i)$ (i = 1 ÷ 5).

Let's notice that noncentral interaction of the nearest neighbors in a layer plane on two orders exceeds the central interlayer interaction. It means that the role of noncentral interlayer interactions in formation of the returning force operating on atom that is displaced from position of balance in a direction along an axis c is dominating. Thus, β_{1x} and β_{1z} are positive that corresponds to an attraction of the nearest neighbors; β_2 and β_3 are negative that corresponds to pushing away of more remote atoms from each other.

Given, at first sight paradoxical, the fact speaks various types of interactions between the nearest neighbors (very high, but very short-range covalent bonds) and more remote atoms interaction with which is van der Waals. Thus, the equilibrium value of the potential describing it van der Waals interaction $r_0 \in (\Delta_4, \Delta_5)$, that is, it essentially exceeds Δ_2 and Δ_3. Therefore, in graphene layers, interaction of the nearest neighbors is covalent attraction, and interaction with the second and the third neighbors is van der Waals pushing away. The given circumstance plays a key role in formation of flexural rigidity of graphene monolayers, which in turn causes stability of crystal structures of graphite and carbon nanofilms and, thanks to the square-dependence of a dispersion $\omega(k)$ of flexural fluctuations (see, for example, the bottom fragment of Figure 12.3), causes non-Debye behavior of the low-temperature vibrational thermodynamic characteristics of considered connections.

Elastic modules C_{33} and C_{44}, connected with displacement along an axis c and defining speeds of a sound that extends or it is polarized along the given direction, in 30 ÷ 300 time less than elastic modules C_{11} and C_{66} and, defining speeds of the sound extending and polarized in basic planes [20, 21]. Therefore, if the vibrations polarized along c, would have at small frequencies sound, instead of quasiflexural character, root-mean-square displacements of atoms in the given direction already at low temperatures would get the values

Table 12.2 Graphite force constants.

	Δ				
	Δ_1	Δ_2	Δ_3	Δ_4	Δ_5
α, N/m	337.882	50.4759	19.647	2.5811	0.37061
β, N/m	β_x = 170.864				
	β_z = 96.3753	−10.1490	−8.661	−0.06537	0.035259

corresponding to melting of a crystal. That is already a fact of existence that film graphite at room temperatures testifies that the given vibrations are essentially defined by noncentral interatomic interaction and about presence of elastic pressure in the graphene layers forming a crystal lattice of graphite.

After the limit transition from the equations of dynamics of a lattice to the equations of the theory of elasticity, we obtain the expression for the acoustic mode polarized along an axis c and extending in a plane of layers plane of layers (that is a quasi-flexural mode),

$$\omega_3^2\left(k_x,k_y,0\right)\approx\frac{C_{44}}{\rho}\cdot k^2+\kappa^2k^4 \tag{12.19}$$

where the elastic module C_{44} is defined by expression (12.7) and ρ is graphite density, $k^2=k_x^2+k_y^2$. Having substituted in Equation 12.19 the values of parameters β_i from Table 12.2, for flexural rigidity of graphene layers κ, we will receive

$$\kappa=\frac{a^2}{4}\cdot\sqrt{\frac{\beta_{1z}-2\beta_3}{6m}}\approx 4{,}06\cdot10^{-7}\,\mathrm{m}^2/\mathrm{c} \tag{12.20}$$

Thus, the formula (12.19) will describe a quasiflexural phonon mode on the top fragment of Figure 12.3.

We will make the analysis of the phonon spectrum and vibrational characteristics of graphite in the following section, comparing them to the phonon spectrum and vibrational characteristics of graphene nanofilms.

12.2 Electron and Phonon Spectra of Ultrathin Graphene Nanofilms

12.2.1 Electron Spectra of Non-Defect Bigraphene

Let us consider the electron spectra of an ideal bigraphene. Similar to graphene monolayer, the honeycomb structure of basal layers in bigraphene is composed of the carbon close-packed triangles (Figure 12.4). These two graphene monolayers are coupled, similar to the bulk graphite, by van der Waals interaction. The interlayer spacing or film thickness is $h\sim 3.5$ Å. In contrast to the unit cell of graphite with two physically equivalent atoms, i.e., equal local Green's functions and local densities of states, the unit cell of bigraphene consists of four atoms, with different interaction of atoms pertained to different sublattices in one layer with those pertained to another layer. Their physical equivalence is, therefore, destroyed, while the atoms of different layers are, naturally, equivalent [12].

The electron spectrum of bigraphene, similar to that of graphene, can be described within a tight-binding approximation, that is, with the aid of the Hamiltonian (1). In assumption that, similar to graphene, the in-layer electron hopping is possible only for nearest neighbors $\forall J_{ij}=J\approx 2.8$ eV (see, e.g., Ref. [16]). The interlayer electron hopping is also assumed possible only between the nearest neighbors from different layers, i.e., those separated by distance h. The relevant hopping integral is deignated J'. It should be noted that such neighbors exist for half of bigraphene atoms, pertained to sublattices AI and AII (see Figure 12.4).

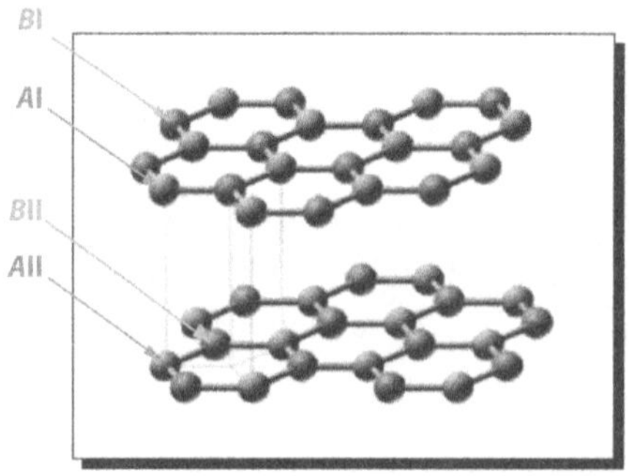

Figure 12.4 Structure of bigraphene.

The atoms of sublattices *BI* and *BII* miss such the neighbors because their nearest neighbors from adjacent layers are set at the distance $\sqrt{h^2+a^2}$. Here, $a \approx 1.415$ Å is the in-plane nearest-neighbor separation. Though it exceeds h for less than 10%, the interaction with atoms of sublattices *BI* and *BII* is omitted, as it does not qualitatively affect the behavior of spectral characteristics near ε_F (see e.g., Refs. [27, 28]).

Four atoms in the elementary cell of bigraphene yield dispersion relations with four branches of its electron spectrum, which are written as

$$\begin{aligned}\varepsilon_{1,2}(\boldsymbol{k}) &= \pm\sqrt{\varepsilon_0^2(\boldsymbol{k})+\frac{J'^2}{2}-J'\sqrt{\varepsilon_0^2(\boldsymbol{k})+\frac{J'^2}{4}}};\\ \varepsilon_{3,4}(\boldsymbol{k}) &= \pm\sqrt{\varepsilon_0^2(\boldsymbol{k})+\frac{J'^2}{2}+J'\sqrt{\varepsilon_0^2(\boldsymbol{k})+\frac{J'^2}{4}}}.\end{aligned} \tag{12.21}$$

where the $\varepsilon_0(\boldsymbol{k})$ function is given by Equation 12.2. In the K-points of first BZ, the function $\varepsilon_{1,2}(\boldsymbol{k})$ vanishes, in contrast to function $\varepsilon_{3,4}(\boldsymbol{k})$, which means that in these modes, Fermi level takes place within the energy gap.

The dispersion curves of bigraphene along the high-symmetry directions ΓK, ΓM, and KM for the case of $J' = 0.1J$ are illustrated by Figure 12.5. In the figure, the energy values in high-symmetry points Γ, M, and K of the first BZ are pointed out (for their location, see bottom part of Figure 12.1a).

$$\begin{aligned}&\varepsilon_{1,2}(\mathrm{K})=0; \quad \varepsilon_{3,4}(\mathrm{K})=\pm J';\\ &\varepsilon_{1,2}(\mathrm{M})=\pm\sqrt{J^2+J'^2/2-J'\sqrt{J^2+J'^2/4}};\\ &\varepsilon_{3,4}(\mathrm{M})=\pm\sqrt{J^2+J'^2/2+J'\sqrt{J^2+J'^2/4}};\\ &\varepsilon_{1,2}(\Gamma)=\pm\sqrt{9J^2+J'^2/2-J'\sqrt{9J^2+J'^2/4}};\\ &\varepsilon_{3,4}(\Gamma)=\pm\sqrt{9J^2+J'^2/2+J'\sqrt{9J^2+J'^2/4}};\end{aligned} \tag{12.22}$$

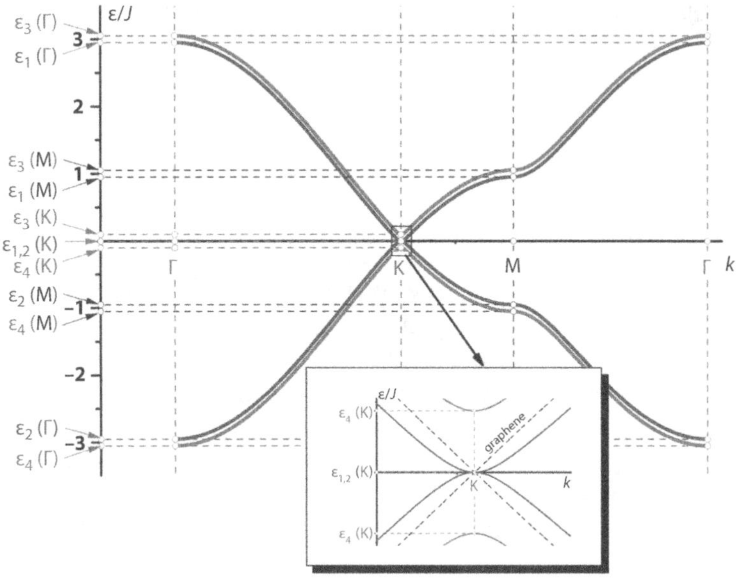

Figure 12.5 Dispersion of bigraphene along the high-symmetry directions. Inset: magnified area near the K-point of the first BZ; the dashed line is the graphene.

In the inset, the K-point area is shown at magnified scale. The same inset demonstrates the dispersion relations of graphene (Equation 12.2). The quasi relativistic character of electron spectrum of graphene is clearly seen, as well as the «usual» square-law run of the dispersion curve $\varepsilon_{1,2}(\boldsymbol{k})$ close to K-point. The spectral branches $\varepsilon_{3,4}(\boldsymbol{k}) \notin (-J', J')$.

In fact, for $\boldsymbol{k} \in \Gamma\mathrm{K}$, $\varepsilon_0(k) = \pm\left(1 + 2\cos\frac{ak\sqrt{3}}{2}\right)$ and, if k = K + κ (κ<<1), then $\varepsilon_0(\mathrm{K}+\kappa) \approx \mp\frac{3a\kappa}{2}$, is obtained that follows the linear (relativistic) dispersion law. The electron modes of bigraphene $\varepsilon_{1,2}(\boldsymbol{k})$ near the K-point with $\varepsilon_0(k) << J'$, are written as

$$\varepsilon_{1,2}^2(\mathrm{K}+\kappa) \approx \frac{J_1^2}{2} + \varepsilon_0^2(\mathrm{K}+\kappa) - \sqrt{\left(\frac{J_1^2}{2} + \varepsilon_0^2(\mathrm{K}+\kappa)\right)^2 - \varepsilon_0^4(\mathrm{K}+\kappa)} \approx \frac{J_1^2}{2} + \varepsilon_0^2(\mathrm{K}+\kappa) -$$

$$-\left[\frac{J_1^2}{2} + \varepsilon_0^2(\mathrm{K}+\kappa)\right] \cdot \left\{1 - \frac{\varepsilon_0^4(\mathrm{K}+\kappa)}{2\left[J_1^2/2 + \varepsilon_0^2(\mathrm{K}+\kappa)\right]^2}\right\} = \frac{\varepsilon_0^4(\mathrm{K}+\kappa)}{J_1^2 + 2\varepsilon_0^2(\mathrm{K}+\kappa)} \approx \frac{81 J^4 a^4 \kappa^4}{16 J_1^2}, \tag{12.23}$$

The trivial square law of dispersion is thus obtained:

$$\varepsilon_{1,2}(\mathrm{K}+\kappa) \approx \pm\left[\frac{9J^2 a^2 \kappa^2}{4J_1} - O(\kappa^2)\right]. \tag{12.24}$$

The effective electron mass, derived from equation $\varepsilon = (\hbar\kappa)^2/2m^*$ for considered branches, appears to be equal to

$$m^* = \frac{2\hbar^2 J'}{9J^2a^2}. \tag{12.25}$$

For the considered case of $J' = 0.1J$, the effective electron mass is $m^* \approx 2.75\cdot10^{-32}$ *kg*; at $J' \to J$, it approaches the value close to the order of magnitude of the free electron mass ($m_e \approx 9\cdot11^{-31}$kg). As $m^* \sim J'$, the interlayer hopping integral variation, allows one to change effective mass of charge carriers for the orders of magnitude.

It should be noted then that notwithstanding $J' \to 0$ at such the boundary transition from bigraphene to two noninteracting graphene monolayers and the effective mass $m^*\to0$, Equations 12.24 and 12.25 become invalid here as obtained in assumption of $\varepsilon_0(k)<<J'$.

Electron density of states (DOS) at energies close to ε_F is determined by the branches ε_1 and ε_2 only (electron modes ε_3 and ε_4 reveal a gap in this range), and Equation 12.3 yields $g_1(\varepsilon) = g_2(-\varepsilon)$. Then

$$g(\varepsilon) = \frac{\Sigma_0}{(2\pi)^2} \oint\limits_{\varepsilon(\kappa)=\varepsilon} \frac{dl_{1,2}}{\left|\partial\varepsilon_{1,2}/\partial\kappa\right|} \tag{12.26}$$

where $\Sigma_0 = 3a^2\sqrt{3}/2$ is the cross-square of Bravais lattice. Integration is performed over the closed isoenergetic line $\varepsilon(\kappa) = \varepsilon$. At $\varepsilon = 0$ (Fermi level), this line smears to point and near ε_F the circumferential contour of integration. Using Equation 12.26, we can write

$$g_{1,2}(\varepsilon_F) = \lim_{\varepsilon\to\varepsilon_F} \frac{\Sigma_0}{(2\pi)^2} \int\limits_0^{2\pi} \frac{\kappa d\varphi}{\left|\partial\varepsilon_{1,2}/\partial\kappa\right|} = \frac{J'}{2\pi J^2\sqrt{3}} = const \tag{12.27}$$

It means that at $\varepsilon = 0$, DOS has a finite constant value. Moreover, according to Equation 12.27, near the Fermi energy, DOS is analytical function with minimum at $\varepsilon = 0$ approaching $g(\varepsilon) \sim \varepsilon^2$ at ε_F.

The integral electron DOS can be described by the arithmetic mean of two LDOS $\rho_s(\varepsilon)$, apt to atoms from sublattices A and B (which follows from the above-mentioned physical equivalency of atoms pertained to different bigraphene layers) $\rho_{AI}(\varepsilon) = \rho_{AII}(\varepsilon) \equiv \rho_A(\varepsilon)$; $\rho_{BI}(\varepsilon) = \rho_{BII}(\varepsilon) \equiv \rho_B(\varepsilon)$; $g(\varepsilon) = [\rho_A(\varepsilon) + \rho_B(\varepsilon)]/2$. The LDOS for each of ideal sublattices can be represented as

$$\rho_s(\varepsilon_F) \approx \frac{\Sigma_0}{(2\pi)^2} \sum_{\alpha=1}^{q} \oint\limits_{\varepsilon(k)=\varepsilon} \frac{\left|\psi_s(\alpha,k)\right|^2}{\left|\nabla_k\varepsilon_\alpha(k)\right|} dl_\alpha. \tag{12.28}$$

In Equation 12.28, subscripts s and α refer to sublattices and branches, respectively; $\psi_s(\alpha,\boldsymbol{k})$ are mean eigenfunctions of atoms from each sublattice. The LDOSs, calculated by Jacobi matrix technique [29–31] for each of the bigraphene sublattices, are illustrated by Figure 12.6.

The Jacobi matrix technique was used in computations here for its efficiency in finding these characteristics, and because it does not use translational lattice symmetry explicitly, which is crucial for spectral calculations, when such symmetry is broken.

The two-dimensional van Hove singularities at energy values corresponding to Γ- and M-points of the first BZ are clearly seen on Figure 12.6. The inset presents the same dependences at magnified scale for energy range in the vicinity of ε_F. It shows, that near Fermi level, LDOS and integral DOS are analytical and their dependence on energy is sufficiently nonlinear (in contrast to DOS of graphene presented for a comparison in the inset). Moreover, the LDOS $\rho_A(\varepsilon)$, contrary to $\rho_B(\varepsilon)$ and integral DOS, vanishes at $\varepsilon \to 0$. It is worth noting that near $\varepsilon = 0$, the dependence $\rho_A(\varepsilon)$ is very smooth, much smoother than $\rho_B(\varepsilon)$, and the energy spectrum behavior for atoms of sublattice is a gap-like one.

Indeed, the substitution of zero eigenvalue in the equation for eigenfunctions of Hamiltonian (Equation 12.1), yields $\psi_{AI} \sim \psi_{AII} = O(\kappa^2)$, and $\psi_{BI} \sim \psi_{BII} \sim 1$. Then, at the Fermi level, $\rho_A(\varepsilon) \sim \varepsilon^2\rho_B(\varepsilon)$. The features in both of the LDOSs at $\varepsilon \approx \pm\, 0.1J = \pm\, J'$ are attributed to contributions of the modes $\varepsilon_{3,4}$. For the latter, the interval $\varepsilon\in[-J',J']$ is the gap, and the features at $\varepsilon \approx \pm\, 0.05J$ are caused by anisotropy of the isofrequency lines, which becomes noticeable starting just at these energy values.

We note that the complete agreement of the analytical results were obtained from an analysis of the behavior of the dispersion laws and eigenfunctions in the reciprocal space with the results shown in Figure 12.6, which were numerically obtained by the method of

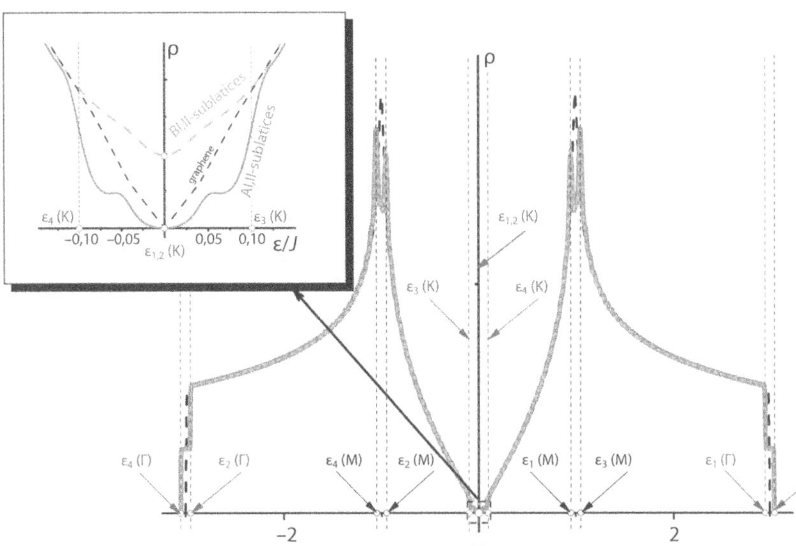

Figure 12.6 (Color online) LDOS of atoms pertained to different lattices of bigraphene; the dashed line is the DOS graphene (for comparison).

Jacobi matrices [29, 31], that is, without finding the dispersion laws and eigenfunctions in the inverse space.

12.2.2 Phonon Spectrum and Vibrational Characteristics of Graphene Nanofilms

12.2.2.1 Reconstruction and Relaxation at Nanofilms Formation

When considering the atomic dynamics of superthin films of graphite, we note that the flat form of a free graphene monolayer is not stable since even at $T = 0$, the root-mean-square displacement of atoms in the direction, perpendicular to the layer, logarithmically diverges. Therefore, in this section, the phonon spectrum and the root-mean-square amplitudes of vibrations in films consisting of two and three graphene monolayers are analyzed.

It has been shown in the previous section that in the graphite consisting of weakly bonded graphene monolayers, the interlayer interaction involves both central and non-central forces. This distinguishes graphite from other layered crystals formed by weakly bonded fragments, containing few monolayers (for example, from transition metal dichalcogenides [32]). Therefore, in graphite, the surface formation cannot be described within the limits of the Lifshits–Rozentsveig model [33], but is characterized by the surface reconstruction and surface relaxation. It is also quite natural to assume that the breakage of interlayer bonds, that is, weak van der Waals interactions, should not change both the distance between atoms in graphene layers and the force constants characterizing intralayer interaction. The surface reconstruction and relaxation will be reduced only to changes of interlayer distances and force constants α_4 and β_4, and α_5 and β_5, characterizing the interlayer interaction.

The fulfillment of the condition $\sigma_{iz}n_z = 0$ leads to the flat form of layers and to the same relation between force constants and lattice parameters as the condition $C_{13} = C_{31}$. For the case of thin films consisting of N monolayers, the mentioned condition can be written as

$$\beta_{1z} + 6\beta_2 + 4\beta_3 = \frac{N}{N-1} \cdot \frac{2}{3\Delta_2^2}\left[\Delta_4^2\beta_4 - 9(\Delta_1^2 - \Delta_4^2)\beta_5\right] \tag{12.29}$$

Distances Δ_4 and Δ_5 in the crystal lattice of graphite differ by less than 8%. Hence, neither one can be an equilibrium distance for a pair potential describing the interlayer van der Waals interaction. The equilibrium distance r_0 for this potential lies between these values: $\Delta_4 < r_0 < \Delta_5$. The small, of the order of the amplitudes of the harmonic atomic vibrations along the c axis, difference of the distances Δ_4 and Δ_4 from r_0 makes it possible to describe the interlayer van der Waals interaction by the Lennard–Jones potential (see, e.g., Ref. [19]):

$$\varphi_\perp(r) = \varphi_{L-J}(r) = 4\varepsilon\left[\left(\frac{\sigma}{r}\right)^{12} - \left(\frac{\sigma}{r}\right)^{6}\right]. \tag{12.30}$$

The parameters of this potential can be determined from the force constants α_4, β_4, α_5, and β_5 obtained in Section 12.1: $\sigma \equiv r_0 / \sqrt[6]{2} \approx 3.092$ Å; $\varepsilon \approx 152.3$ K.

Starting with Equations 12.29 and 12.30, for considered thin films of graphite, it is easy to find both interlayer distance $\tilde{\Delta}_4$, and force constants $\tilde{\alpha}_4$, $\tilde{\beta}_4$, $\tilde{\alpha}_5$, and $\tilde{\beta}_5$ that describe interlayer interaction in such objects.

For a two-layer film (bigraphene) $\tilde{\Delta}_4 \approx 3.636$ Å, $\tilde{\alpha}_4 \approx 372.82 \times 10^{-3}\,N/m$; $\tilde{\beta}_4 \approx 35.10 \times 10^{-3}\,N/m$; $\tilde{\Delta}_5 \equiv \sqrt{\tilde{\Delta}_4^2 + \Delta_1^2} \approx 3.902$ Å; $\tilde{\alpha}_5 \approx -87.44 \times 10^{-3}\,N/m$ $\tilde{\beta}_5 \approx 41.43 \times 10^{-3}\,N/m$.

For a three-layer film (trigraphene): $\tilde{\Delta}_4 \approx 3.453$ Å; $\tilde{\alpha}_4 \approx 1585.10 \times 10^{-3}\,N/m$; $\tilde{\beta}_4 \approx -15.34 \times 10^{-3}\,N/m$; $\tilde{\Delta}_5 \approx 3.713$ Å; $\tilde{\alpha}_5 \approx 162.60 \times 10^{-3}\,N/m$; $\tilde{\beta}_5 \approx 40.66 \times 10^{-3}\,N/m$.

12.2.2.2 Spectral Densities and Mean-Square Amplitudes of Atomic Displacements

The phonon density of states $g(\omega)$ and the spectral densities $\rho_i^{(s)}(\omega)$, corresponding to the displacements of the surface atoms of the sublattice *s* along the crystallographic direction *i* were calculated, for the model proposed in the preceding section for the crystal lattice of graphite, by means of Jacobi's matrices [29–31].

As shown in Ref. [34], in strongly anisotropic layered crystals, the interaction of the vibration modes polarized along the directions of the strong and weak bonds is proportional to the squared ratio of the weak interlayer to the strong intralayer interaction. In graphite, this ratio is $(C_{33}/C_{11})^2 \sim 10^{-3}$. Consequently, for frequencies $\omega > \omega_{TO}(\Gamma)$, when the isofrequency surfaces of the vibrational branches polarized in the plane of the layer become open along the *c* axis, the phonon spectrum acquires a practically two-dimensional character, and the functions $\rho_{ab}(\omega)$ and $\rho_c(\omega)$ are the phonon densities of states of graphene monolayers for independent atomic vibrations in the plane of the layer and in the direction perpendicular to it.

Figure 12.7 represents the calculated results for the bulk graphite phonon density of states g(ω)—top inset, and show partial contributions to g(ω) from atomic displacements in basal planes,

$$\rho_{ab}(\omega) = \frac{1}{6}\left[\rho_a^{(\circ)}(\omega) + \rho_a^{(\bullet)}(\omega) + \rho_b^{(\circ)}(\omega) + \rho_b^{(\bullet)}(\omega)\right] \tag{12.31}$$

and from displacements along the *c* axis.

$$\rho_c(\omega) = \frac{1}{6}\left[\rho_c^{(\circ)}(\omega) + \rho_c^{(\bullet)}(\omega)\right] \tag{12.32}$$

The spectral densities $\rho_i^{(s)}(\omega)$ are normalized to unit and $g(\omega) = \rho_{ab}(\omega) + \rho_c(\omega)$.

In an ideal crystal lattice, the spectral densities $\rho_i^{(s)}(\omega)$ normalized per unit fulfill the relation:

$$\rho_i^{(s)}(\omega) = \frac{V_0}{(2\pi)^3} \sum_{\sigma=1}^{3q} \oint_{\omega_\sigma(\mathbf{k})=\omega} \frac{\left|e_i^{(s)}(\mathbf{k},\sigma)\right|^2}{\left|\nabla_{\mathbf{k}}\omega_\sigma(\mathbf{k})\right|} dS_{\mathbf{k},\sigma}, \tag{12.33}$$

where V_0 and q are the unit-cell volume and the number of atoms per unit cell, index σ enumerates the vibrational modes, $e^{(s)}(\mathbf{k}, \sigma)$ are polarization vectors, and the integration extends over the isofrequency surfaces in reciprocal space. The phonon density of states $g(\omega)$ is given by

$$g(\omega)=\frac{V_0}{(2\pi)^3}\sum_{\sigma=1}^{3q}\oint_{\omega_\sigma(\mathbf{k})=\omega}\frac{dS_{k,\sigma}}{\left|\nabla_{\mathbf{k}}\omega_\sigma(\mathbf{k})\right|}=\frac{1}{3q}\sum_{i=1}^{3}\sum_{s=1}^{q}\rho_i^{(s)}(\omega) \tag{12.34}$$

Function $\rho_{ab}(\omega)$ (see Figure 12.7b and c) contains a kink at $\omega = \omega_{TO}(\Gamma)$, i.e., a singularity, analogous to the three-dimensional van Hove singularity corresponding to the transition from the quadratic dependence of density of states (DOS) at low frequencies, which is characteristic for crystal lattices, to a linear dependence characteristic for two-dimensional lattices. Other van Hove singularities in this function have logarithmic form that is characteristic for two-dimensional structures. The isofrequency surfaces with $\omega > \omega_{TO}(\Gamma)$ are cylindrical and can be regarded as isofrequency lines in the *2D* reciprocal space. The logarithmic van Hove singularities correspond to the rates of change of the topology of these isofrequency lines.

The function $\rho_c(\omega)$ (see Figure 12.7d and e) acquires two-dimensional character for $\omega > \omega_{LO}(\Gamma)$. Its form corresponds to the DOS of the two-dimensional scalar model, i.e., in terms of the density, it is analogous to the electronic DOS of graphene (see, for example, Refs. [35, 36]). So, the function $\rho_c(\omega)$ contains a singularity that is analogous to the so-called Dirac singularity in the electron density of states of graphene. This singularity likewise corresponds to the *K* point of the Brillouin zone.

Curves 2 in Figure 12.7 display the phonon densities of states for bigraphene (in the Figure 12.7a) as well as the contributions to them from atomic displacements along the layers (in Figure 12.7b and c) and in the direction perpendicular to them (in Figure 12.7d and e).

It is clearly evident that in a wide frequency range $\omega > \omega_{LO}(\Gamma)$, the densities of states of a film and a bulk sample are practically identical. In the frequency interval where the phonon spectrum of graphite exhibits three-dimensional behavior and the interaction between the vibrational modes polarized in the plane of the layers and in a direction perpendicular to the layers is quite strong, an appreciable difference is observed in the behavior of the corresponding spectral densities of bigraphene and the bulk sample.

In bigraphene, instead of the transverse phonon modes TA and TO, propagating along the *c* axis, there will be two discrete levels corresponding to the in-phase and antiphase displacements of layers in basal plane and along the *c* axis. The frequencies of these levels are denoted in Figure 12.7c and e as $\omega_a^{(-)}$ and $\omega_a^{(+)}$, respectively. The frequencies $\omega_a^{(+)}$ and $\omega_c^{(+)}$ correspond to the same atomic displacements as the frequencies $\omega_{TO}(\Gamma)$ and $\omega_{LO}(\Gamma)$ in the case of a bulk sample. The decrease of values $\omega_a^{(+)}$ and $\omega_c^{(+)}$ as compared to the frequencies $\omega_{TO}(\Gamma)$ and $\omega_{LO}(\Gamma)$ is due to the surface relaxation. The spectral density $\rho_{ab}(\omega)$ of bigraphene, as seen in Figure 12.7c, exhibits kinks at $\omega = \omega_a^{(-)}$ and $\omega = \omega_a^{(+)}$. For $\omega > \omega_a^{(+)}$ the spectral density acquires a two-dimensional form ($\rho_{ab}(\omega) \sim \omega$). The spectral density $\rho_c(\omega)$ of bigraphene, as seen in Figure 12.7e, exhibits kinks at $\omega = \omega_c^{(-)}$ and $\omega = \omega_c^{(+)}$. For $\omega \geq \omega_c^{(+)}$ this spectral density takes the form corresponding to that of the flexural waves propagating in the plane ($\rho_c(\omega) \to \text{const}$).

In stacks with more than one graphene layer, the relative position of two neighboring layers allows for two different orientations of the third layer. If we label the positions of the

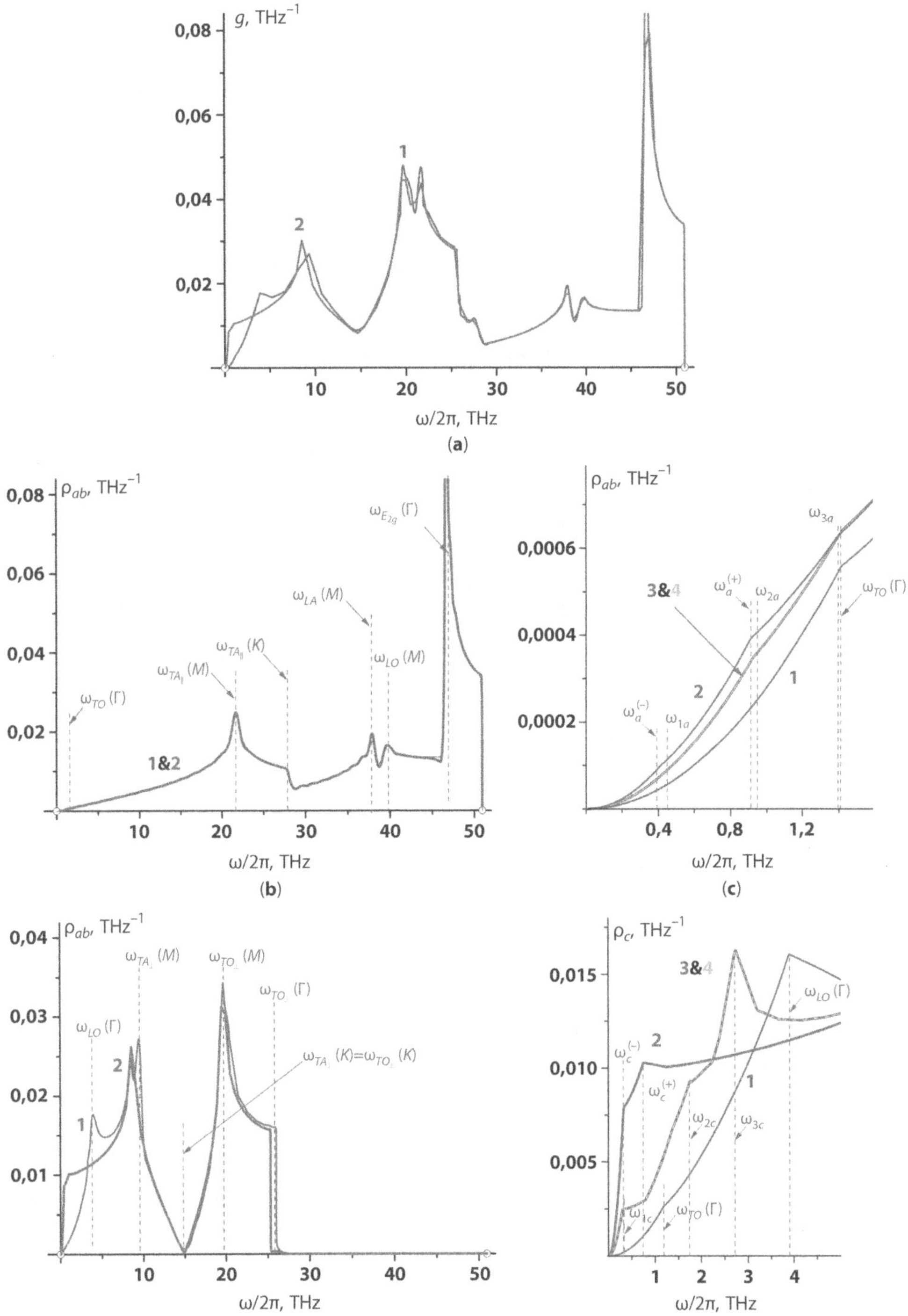

Figure 12.7 The phonon density of states of bulk graphite and bilayer graphene (part a, curves 1 and 2, respectively), as well as the partial contributions to them from the atomic displacements along the basal plane (parts b and c) and along axis *c* (parts d and e). In parts c and e, the corresponding dependence for trigraphene (curves 3 for ABA stacking, curves 4 for ABC stacking) is shown.

first two atoms as A and B, the third layer can be of type A, leading to the sequence ABA, or it can fill a third position different from A and B, that is, C. There are no more nonequivalent positions where a new layer can be placed, so that thicker stacks can be described in terms of these orientations. Regions with the stacking ABC (rhombohedral stacking) have also been observed in different types of graphite.

The ABC stacking has not four (as ABA) but eight atoms in its unit cell and, accordingly, its phonon spectrum contains twice more branches. This circumstance suggests an appreciable difference of behavior of phonon densities in the low-frequency region. Naturally, this difference can be manifested only at frequencies below the frequency of the first van Hove peculiarity, corresponding to the transition from the closed isofrequency surfaces to open ones along the *c* axis. This is manifested also in the temperature dependences of vibrational thermodynamic performances at low temperatures. The study of the influence of the structure on phonon spectra and vibrational performances of carbon nanofilms is of particular interest.

The spectral densities corresponding to atomic displacements along different crystallographic directions for a trigraphene of both ABA and ABC stacking are shown in Figure 12.7c and e (curve 3 for ABA, curve 4 for ABC).

It is clearly visible that

a. the frequency of the transition into the quasi two-dimensional behavior of spectral densities (for bulk graphite, it is the frequency of the first van Hove peculiarity) decreases with decreasing number of layers for $\rho_c(\omega)$, but it practically does not change for $\rho_{ab}(\omega)$;
b. for spectral densities $\rho_{ab}(\omega)$, the quasi two-dimensional behavior leads to $\rho_{ab}(\omega) \sim \omega$, that is, they show two-dimensional Debye character. In the long-wave limit, the spectral density $\rho_c(\omega)$ is comparable to the densities of states in a quasiflexural mode with the quadratic dispersion relation and it approaches to a constant value, which is inversely proportional to flexural rigidity of layers;
c. in spectral densities of nanofilms at frequencies below the frequency of the transition to the quasi-two-dimensional behavior, additional singularities appear, caused by the transformation of phonon branches with the wave vector along the *c* axis into discrete levels;
d. there is no essential difference between the spectral densities $\rho_{ab}(\omega)$ and $\rho_c(\omega)$ for trigraphenes with ABA stacking and ABC stacking, (see Figure 12.7c and e).

Let's note that if with the frequency approaching to zero the spectral density does not converge to zero, the mean-square displacements of atoms will diverge even at zero temperature. This circumstance causes the instability of linear chains and flat monolayers. It is apparent from Figure 12.7 that $\rho_c(\omega)$ of graphene nanofilms (especially of bigraphene) weakly depends on ω even at very low frequencies ($\omega \geq \omega_c^{(+)}$), which could lead to large mean-square displacements of atoms in the direction perpendicular to layers. It, at first sight, calls into question the stability of the lattice of a bigraphene and the applicability of the harmonic approach for the description of the lattice vibrations. In Figure 12.9, the temperature dependences of the mean-square amplitudes of atomic displacements of bigraphene, trigraphene, and bulk graphite along layers and in the perpendicular direction to them are presented.

The mean-square amplitudes atomic displacements of atoms s along crystallographic direction i are expressed in terms of the spectral density $\rho_i^{(s)}(\omega)$ as [19, 26]

$$\langle |u_i^{(s)}| \rangle_T \equiv \sqrt{\langle (u_i^{(s)})^2 \rangle_T}; \quad \langle (u_i^{(s)})^2 \rangle_T = \frac{\hbar}{2m_s} \int_D \frac{d\omega}{\omega} \cdot cth\left(\frac{\hbar\omega}{2kT}\right) \cdot \rho_i^{(s)}(\omega). \quad (12.35)$$

We note that in graphite, the mean-square amplitudes of atomic vibrations along the weak coupling direction (see Figure 12.8) at room temperature, calculated by means of spectral densities $\rho_{ab}(\omega)$, are approximately 0,12 Å, which is about 3% of the corresponding interatomic distance. Therefore, it is quite justified to calculate the vibrational performances of graphite by using harmonic approach.

It is evident from Figure 12.9 that the amplitudes of the atomic displacements along the graphene layers are practically independent of the thickness of the sample. The mean-square amplitudes of atomic displacements along the c axis increase strongly with the decreasing film thickness. Thus, the room-temperature amplitude of the transverse vibrations of an atom in the central layer in trigraphene (curve $3c'$) is twice the corresponding value of the bulk sample (curve ∞c). Curves $3c$ (atom in the surface layer in trigraphene) and $2c$ (atom in bigraphene) are shifted to even higher values.

It is evident from Figure 12.9 that the amplitudes of the atomic displacements along the graphene layers are practically independent of the thickness of the sample. The mean-square amplitudes of atomic displacements along the c axis increase strongly with the decreasing film thickness. Thus, the room-temperature amplitude of the transverse vibrations of an atom in the central layer in trigraphene (curve $3c'$) is twice the corresponding value of the bulk sample (curve ∞c). Curves $3c$ (atom in the surface layer in trigraphene) and $2c$ (atom in bigraphene) are shifted to even higher values. In Figure 12.9, the horizontal line marks the mean-square amplitudes of atomic vibrations along the c axis in bulk graphite at T = 3000 K. This temperature is approximately 1000 K lower than the melting temperature of graphite, $T_{me} \approx 3800 \pm 50$ K. Therefore, at T = 3000 K, the crystal lattice of graphite possesses a sufficient margin of stability. Since at room temperatures the value of the mean-square amplitudes of atomic vibrations of bigraphene and trigraphene lie appreciable below the dashed line, bigraphene and trigraphene possess an adequate margin of stability at room temperature.

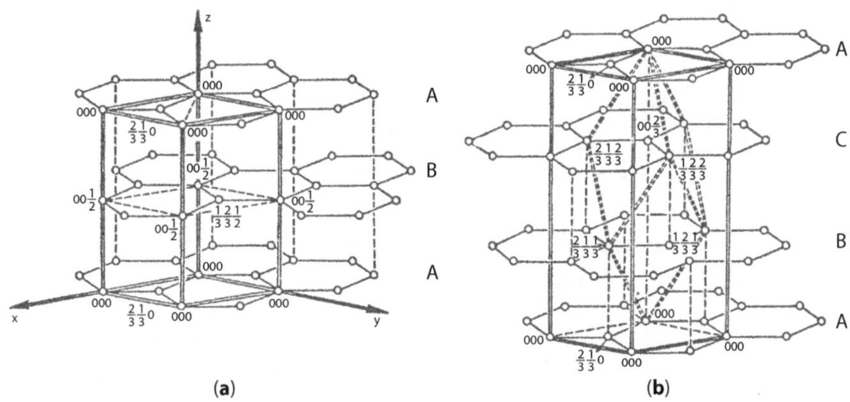

Figure 12.8 The crystal lattice of graphite: (a) ABAB stacking; (b) ABC stacking [22].

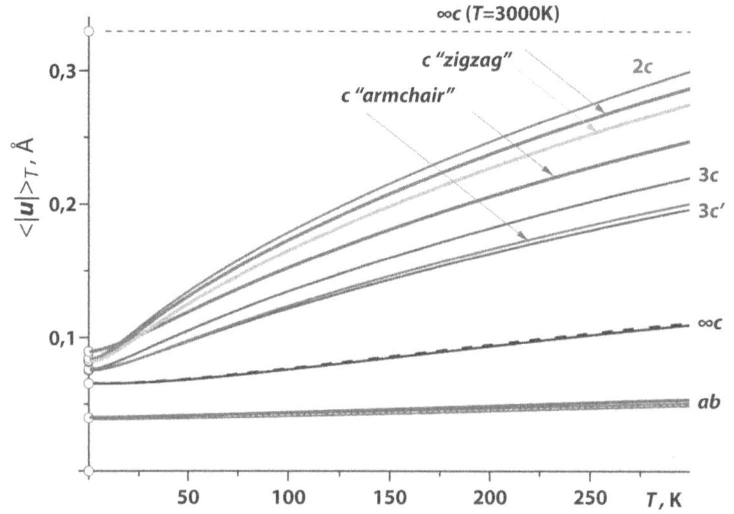

Figure 12.9 The temperature dependence of mean square amplitudes of atomic displacements along different crystallographic directions in bigraphene, trigraphene, and different configurations of a bigraphene–trigraphene interface. *"zigzag"* and *"armchair."*

In the same figure, the temperature dependences of the mean-square amplitudes of atoms within the defects of the "step-on-surface" type are shown, which will be discussed in detail in the next section. The dependences $\langle |u_i^{(s)}| \rangle$ given in Figure 12.9 for two basic configurations of such a defect (*zigzag* and *armchair*) show that such defects are stable up to 500 K.

It is necessary to note that graphene monolayers cannot exist in a free state as flat two-dimensional structures, since in that case there is a full splitting of the bulk phonon modes polarized in the layer plane and flexural modes. Therefore, the spectral density of the flexural mode with the dispersion law $\omega = \kappa k^2$ (formula (12.19) for $C_{44} = 0$) leads to the divergence of mean-square displacements even at $T = 0$. So, the graphene monolayers can exist in the flat shape only when adsorbed on some substrate. For studying the electronic spectra of graphene monolayers, dielectric substrates are usually used, in which bonds between carbon atoms and substrate atoms have van der Waals character. The contribution of the substrate to the phonon spectrum of graphene is manifested in intertwining of longitudinal modes with quasiflexural ones, which then will take their usual form (Equation 12.19). So, the phonon spectrum of the given heterostructure practically corresponds to the phonon spectrum of a graphene monolayer and will not differ essentially from $\omega_{TO}(\Gamma)$ for longitudinal mode and from $\omega_{LO}(\Gamma)$ for quasiflexural mode. The influence of substrate on the phonon spectrum of a graphene monolayer practically disappears at the frequencies exceeding ω_{TO} and ω_{LO}.

12.2.2.3 *Phonon Heat Capacity of Graphite and Graphene Nanofilms: Its "Non-Debye" Behavior*

Experimental study of the low-temperature heat capacity is an important and reliable source of information on the quasi-particle excitations and, in particular, on the phonon spectra. Calorimeter experiments are, as a rule, very accurate and, unlike the majority of optical and ultrasonic experiments, do not require good-quality single crystals.

It is evident from both the experimental data [21, 23], and the calculations given above that the phonon spectrum band for graphite and carbon nanofilms is very wide. Calculated Debye temperature for graphite is about 2500 K, and therefore, the phonon heat capacity increases with temperature much more slowly than with the majority of solids. At room temperatures, the phonon heat capacity is still far from saturation and the electronic contribution to the total heat capacity cannot be considered as negligibly small (when compared to the phonon contribution). Therefore, the analysis of the temperature dependences of the graphite and carbon nanofilm phonon heat capacity, based on microscopic calculations, is very much needed for correct interpretation of calorimeter measurements.

The temperature dependence of the molar isochoric phonon heat capacity $C_V(T)$ of a system whose all atoms have three degrees of freedom is described through its phonon density $g(\omega)$ as follows (see, for example, Ref. [19]):

$$C_V\left(T\right)=3R\cdot\int\limits_D\left(\frac{\hbar\omega}{2k_BT}\right)^2\cdot\sinh^{-2}\left(\frac{\hbar\omega}{2k_BT}\right)g\left(\omega\right)d\omega, \tag{12.36}$$

where k_B is the Boltzmann constant and $R\equiv N_Ak_B$ is the gas constant (N_A is the Avogadro number).

Usually, when explaining the temperature dependence of the phonon heat capacity at low temperatures, it is assumed that the main contribution to the phonon heat capacity comes from the long-wave low-frequency phonons with the sound dispersion relation $\omega(\boldsymbol{k}) = sk$, where s is the sound velocity depending on the direction of the sound propagation and polarization. For such phonons, the density of states has a so-called Debye form:

$$g^{(q)}\left(\omega\right)=\frac{q\omega^{q-1}}{\left[\omega_D^{(q)}\right]^q} \tag{12.37}$$

where q is the dimensionality of the lattice (number of degrees of freedom) and ω_D is the Debye frequency averaged through all directions and polarizations. Then the phonon heat capacity is given by the expression

$$C_D(T)=qR\left\{D_q\left(\frac{\Theta_D^{(q)}}{T}\right)-\frac{\Theta_D^{(q)}}{T}D_q'\left(\frac{\Theta_D^{(q)}}{T}\right)\right\}, \tag{12.38}$$

where $\Theta_D^{(q)}\equiv\hbar\omega_D^{(q)}/k_B$, $D_q\left(\frac{\Theta_D^{(q)}}{T}\right)$ is the so-called Debye function and

$$D_q(x)\equiv\frac{q}{x^3}\int\limits_0^x\frac{z^q\,dz}{e^z-1} \tag{12.39}$$

In the literature, we usually find these expressions for $q = 3$, see, for example Ref. [37].

Figure 12.10 shows the temperature dependences $C_V(T)$ of graphite, trigraphene, and bigraphene, which were calculated by formula (12.36) using the spectral densities shown in Figure 12.7. We use the same designation (numbering of curves, colors) as in Figure 12.7. In addition to the total heat capacity, we also see the contribution to the heat capacity from the atomic displacement along the graphene layers $C_V^{(ab)}(T)$ and in the direction normal to the layers $C_V^{(c)}(T)$. In graphite and bigraphene, the contribution of each of the layers is equal, and in trigraphene, the contributions from the inner and outer layers differ. Figure 12.10 also shows that a decrease in the number of layers of phonon heat capacity increases the basic growth due to atomic vibrations normally to the layers and any noticeable difference in the behavior of the specific heat phonon trigraphene with different structure was detected.

At low temperatures ($T << \Theta_D^{(q)}$), the heat capacity is proportional to T^q. In Figure 12.10, the contribution to the heat capacity from displacements of atoms along the basal plane at low temperatures is really close to the quadratic one, which is in a good agreement with the quasi-two-dimensional type of the spectral densities $\rho_{ab}(\omega)$ (Figure 12.7). We note that the presence at of the symmetry axis of the sixth order in graphene layers makes them elastic isotropic, which consequently leads to the agreement of $C_V^{(ab)}(T)$ with the Debye model. However, the main contribution to the low-temperature heat capacity of graphite and, especially, of graphene nanofilms comes, as is apparent from Figure 12.10, atomic vibrations along the c axis. These displacements propagate in the plane of layers as quasiflexural waves with non-acoustic dispersion relation (12.19). These displacements propagate in the plane of layers as quasiflexural waves with non-acoustic dispersion relation [38]. for one- and two-dimensional structures, if they are viewed in a three-dimensional space, when each atom has three degrees of freedom the dispersion laws of long-wave phonons, polarized normal to the layers of their chains, have a shape that is typical for flexural waves in elastic plates or rods (i.e., $\omega \sim k^2$, for example [39]). dispersion law of longwave phonons, polarized along the layers or chains, retains its usual acoustic nature (the frequency ω is proportional to the value of the quasi-wave vector k). At low temperatures the

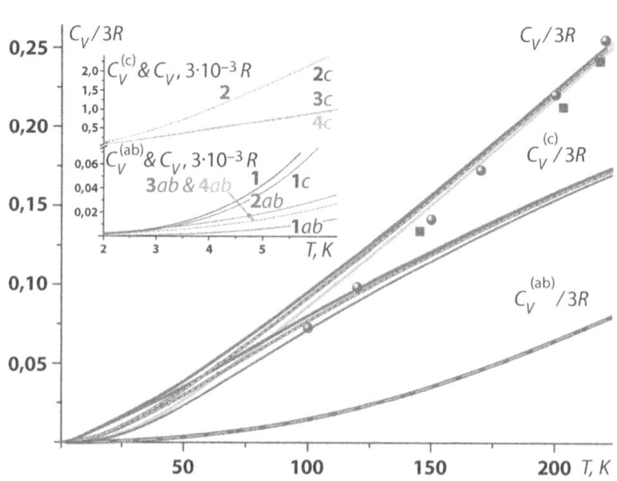

Figure 12.10 The temperature dependences of heat capacity and the contributions to it from different atomic displacements. Symbols represent the experimental data: ● –[10], ▪ – [11]. The colors and numbering of the curves on the insert correspond to Figure 12.7.

contribution of flexural modes to the vibrational heat capacity is proportional to the temperature T for two-dimensional structures and $\sqrt{T}$ for those that are one-dimensional.

Isolated one- and two-dimensional structures that are sufficiently large (i.e., having dimensions at which we can speak about a long-wave limit) cannot exist. The stability of real layered and chain crystals is determined by the weak interaction between the atoms of various chains or layers. The impact of this weak interaction is considered in Ref. [38], where in the long-wave approximation, it was demonstrated that the low-temperature heat capacity of the layered crystals is

$$C_V^{(2D)}(T) \sim \begin{cases} T^3; & T \ll \dfrac{C_{44}}{C_{11}} \cdot \Theta_{2D} \\ T^2; & \dfrac{C_{44}}{C_{11}} \cdot \Theta_{2D} \ll T \ll \sqrt{\dfrac{C_{33}}{C_{11}}} \cdot \Theta_{2D} \\ T; & \sqrt{\dfrac{C_{33}}{C_{11}}} \cdot \Theta_{2D} \ll T \ll \Theta_{2D} \end{cases} \tag{12.40}$$

where C_{ik} represents the elements of the elastic modulus tensor in Voigt notation (the z axis is normal to the layers), and the value Θ_{2D} is the Debye temperature of the isolated two-dimensional layer.

The extensive rectilinear portions of the temperature dependences of the total heat capacity for all three considered compounds are immediately striking. They stretch from temperatures of 50 ÷ 70 K to those that are above room temperature. For graphite, the calculated curve is in good agreement with the experiment [10] (•) and [11] (▪) (especially if we take into account that the heat capacity was calculated at constant volume, while it is usually measured at constant pressure).

Obviously, this rectilinear course of the heat capacity temperature dependence has nothing to do with the heat capacity's linear region in Equation 12.40 that is caused by the quadratic dispersion law of the quasiflexural vibrations. Indeed, the quasiflexural vibrations contribute only to the heat capacity components $C_V^{(c)}(T)$ (Curves 1*c*, 2*c*, 3*c*), whereas at $T > 70$ K, the given dependences have a noticeable negative curvature. However, since at temperatures above 50 K, it is impossible to neglect the heat capacity contribution from atomic vibrations in the layer plane $C_V^{(ab)}(T)$ (Curves 1*ab*, 2*ab*, 3*ab*), and these dependences have a positive curvature in this temperature interval, this leads to an almost rectilinear temperature dependence of the total heat capacity.

It should be emphasized that the rectilinear course of the temperature dependences $C_V(T)$ at $T \geq 70$ K is not a linear dependence in the same sense as how it is understood in Equation 12.40. Extrapolating these rectilinear sections of the $C_V(T)$ dependence to the x axis cuts off segments of about ~35 K on this axis, for all three considered compounds.

It is only for bigraphene that the temperature dependences $C_V^{(c)}(T)$ and $C_V(T)$ are proportional to the temperature along the interval $5\text{ K} \leq T \leq 70$ K and $5\text{ K} \leq T \leq 30$ K, respectively. At $T \leq 30$ K, the contribution to the heat capacity from the component $C_V^{(ab)}(T)$

becomes noticeable. At $T \leq 5$ K (which is about $0.002\Theta_D$!), there is a deviation of the spectral density $\rho_c(\omega)$ from the two-dimensional form at frequencies of $\omega < \omega_c^{(+)}$ (see Figure 12.7).

Note that at $T \to 0$, the temperature dependence $C_V^{(c)}(T)$ for graphite and graphene is practically quadratic. This can be seen on both the insets on Figure 12.11, which shows the temperature dependences of the quantities $\frac{\partial}{\partial T} C_V(T)$ and the contributions to this quantity that come from atomic displacements along the layers and perpendicular to them. Curves 1 (Figure 12.11) show the total value, Curves 2 and 3 show the dependences $\frac{\partial}{\partial T} C_V^{(ab)}(T)$ and $\frac{\partial}{\partial T} C_V^{(c)}(T)$, respectively.

The numbering of the curves on Figure 12.11 is similar to the numbering on Figure 12.10: Curves 1 (Figure 12.11) show the total value, Curves 2 and 3 show the dependences $\frac{\partial}{\partial T} C_V^{(ab)}(T)$ and $\frac{\partial}{\partial T} C_V^{(c)}(T)$, respectively.

On Figure 12.11a, the solid lines show the given characteristics for trigraphene, whereas on Figure 12.11b, we see these characteristics for bigraphene. For comparison, broken lines show similar dependences for graphite. The upper parts of Figure 12.11a and b show the dependences on the 0–300 K temperature scale. As was shown in [32], the flat shape of bigraphene and trigraphene is stable to temperatures of ~ 400÷500 K and the harmonic approximation is applicable until about 300 K. The bottom of Figure 12.11a and b shows the low-temperature region (to 70 K). It can be seen that the contributions from displacements along the layers (Curves 2) coincide for all three compounds, and up to 70 K, the contributions to the heat capacity from these displacements can be considered quadratic to a very high degree of accuracy. The contribution to the heat capacity from vibrations along the c axis are almost quadratic for graphite up to 70 K; for trigraphene, they are almost quadratic until 20 K, and in the interval of 5 to 10 K, the heat capacity of trigraphene is nearly linear with temperature. For bigraphene, the region along the $C_V^{(c)}(T)$ dependence that is linear with temperature is much broader, ranging between 5 and 70 K. This dependence points to the decisive role the quasiflexural mode $TA_\perp$ that has a quadratic dispersion law, plays in the contribution to that value (see Figure 12.1). Along the total heat capacity of bigraphene the linear temperature plot extends from 5 to about 30÷40 K. At lower temperatures the heat capacity of bigraphene grows more rapidly than T. In this temperature range, the heat capacity is defined as the "acoustic portion" of the quasiflexural mode $TA_\perp$ (see Equation 12.19, the elastic modulus C_{44} for bigraphene is about an order of magnitude smaller than for graphite, but still not zero), and as the exponential contribution to the heat capacity of discrete levels $\omega_i^{(\pm)}$.

Therefore, the phonon modes with a quadratic dispersion law as predicted by Lifshitz really exist in graphites and thin graphene nanofilms. The linear course of the heat capacity temperature dependence conditioned by these modes is actually observable only in bigraphene, since the phonon spectrum of bigraphene is the only one that has a sufficiently long identifiable frequency interval with $\omega_c^{(+)} \approx 0.7$ THz, starting from which its phonon spectrum becomes quasi-two-dimensional, to a frequency of about 5 THz, at which point the phonon mode $TA_\perp$ has an almost quadratic dispersion. In the cases of trigraphene and graphite, the phonon spectra acquire a two-dimensional nature starting from $\omega_{3c} \approx$ 2.8 THz and $\omega_{LO}(\Gamma) \approx 4$ THz, as a result of which the linear plot of the temperature dependences of heat capacity, caused by the quadratic dispersion of the quasiflexural mode, is not visible. Note that the frequencies ω_{3c} and $\omega_{LO}(\Gamma)$ are much less than the frequency

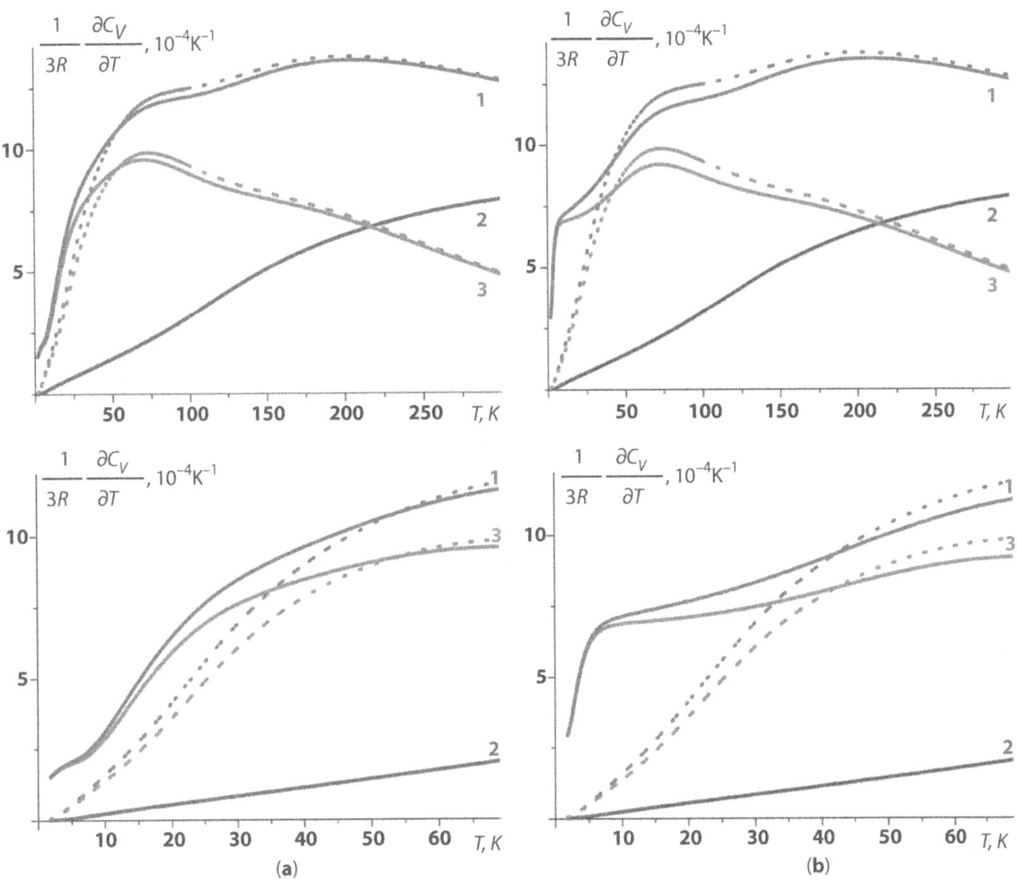

Figure 12.11 The derivatives with respect to temperature of the heat capacity and the contributions thereto from atomic displacement along the layers and normal to them. Trigraphene (a); bigraphene (b). On both fragments, the broken lines show the corresponding dependences for graphite, and the numbering on these curves is the same as on Figure 12.10.

$\omega_D\sqrt{C_{33}/C_{11}} \approx 9.5$ THz with which the quasi-two-dimensional behavior of the phonon spectrums starts in the long-wave approximation.

12.2.3 Phonon Spectra and Vibrational Heat Capacity of Graphene Nanotubes

Single-wall graphene nanotubes are really graphene sheets folded along a certain axis. The propagation of waves can only be discussed in terms of the direction of this axis, and it is only in terms of this axis that the quasi-wave vector $\boldsymbol{k}$ can be introduced. These tubes, the lengths of which are much greater than the diameter, can be regarded as quasi-one-dimensional structures and their quasi-particle spectra and the physical characteristics of these spectra can be expected to exhibit the distinguishing singularities typical for one-dimensional systems. In particular, the phonon spectra and the vibrational specific heat are expected to exhibit the manifestation of the contributions that come from flexural vibrations in quasi-one-dimensional structures [38]:

$$C_v^{(1D)}(T) \sim \sqrt{T}; \text{ at } \sqrt{\frac{C_{33}}{C_{11}}} \cdot \Theta_{1D} << T << \Theta_{1D}$$

Furthermore, the expression of one-dimensional singularities along the vibrational characteristics of graphene nanotubes must be markedly different from what is described in Ref. [38] for structures that are composed of weakly interacting one-dimensional chains. First of all, a tube still has a finite thickness and the separation of the variables in such a system takes place in cylindrical coordinates. Therefore, from two transverse phonon branches tending toward zero at $k \to 0$, only the one that is polarized along the cylinder radius $\boldsymbol{r}$ will be the flexural mode with a quadratic dispersion law, whereas the other, polarized along the tangent to the cylinder surface that is perpendicular to the axis, is a twisting mode with a normal acoustic dispersion law (see Ref. [39] for example). Second, the circumferential length of the graphene nanotubes is significantly greater than the distance between the carbon atoms, and the closure of the tube is expressed in the behavior of the local spectral densities of its individual atoms, starting from a sufficiently high torque. As such, for a "zigzag" tube, i.e., a tube that is folded along the side of the graphene hexagon having an n number of teeth per slice, this closure starts from the torque $2n$. Therefore, the spectral densities of the quasi-particle excitations in the nanotubes have a quasi-two-dimensional shape of the corresponding graphene spectra that are "modulated" by the characteristic features associated with the size quantization in such systems (see Ref. [40] for example).

Figure 12.12 shows the contributions to the nanotube phonon density of states normalized to one, from the displacements of its atoms along the different crystallographic directions: along the axis of the tube $\rho_l(\omega)$—Figure 12.12a; tangentially to the surface of the tube, perpendicular to its axis $\rho_\tau(\omega)$—Figure 12.12b; and along the normal to the surface of the tube $\rho_r(\omega)$—Figure 12.12c.

The given dependences are calculated using Jacobi matrices [29–31] for the "zigzag" nanotube having 14 teeth per slice, i.e., with a diameter of $d \approx 10.9$ Å. Such tubes are actively studied throughout experiments.

Note that all atoms of the nanotube are physically equivalent (their local densities of states coincide). On Figure 12.12a and b, thin solid lines (Curves 2) show the spectral densities $\rho_{ab}(\omega)/2$ of bigraphene, and on Figure 12.12c—the spectral density $\rho_c(\omega)$ (Figure 12.7).

A comparison of the corresponding spectral densities of the nanotubes and bigraphene allows us to suggest a certain analogy for the behavior of their contributions to the vibrational heat capacity. Indeed, as can be seen on Figure 12.12, the behavior of the temperature dependences of the heat capacity of the nanotubes (Figure 12.12a) and its derivative with respect to temperature (Figure 12.12b) coincide with the corresponding characteristics of bigraphene that are shown in Figures 12.10 and 12.11, at $2\text{ K} \leq T \leq 10\text{ K}$. The quasi-one-dimensional behavior starts to manifest at lower temperatures, when the main contribution to the heat capacity starts to come from phonons that have a wavelength that is significantly longer than the circumference of the tube cross section. Such vibrations can be described using the spectral densities generated by the same type of displacements of all $2n$ atoms at the edge of a normal tube slice.

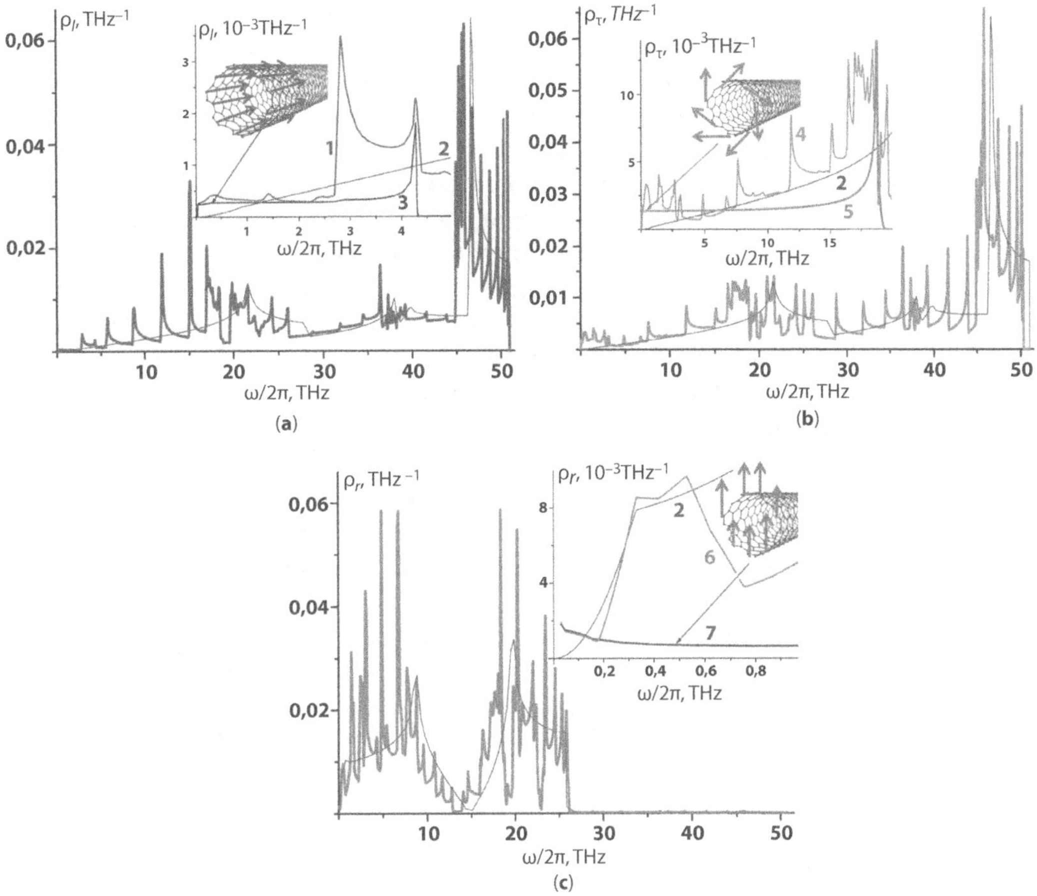

Figure 12.12 The contributions to the graphene nanotube phonon density of states from atomic displacements along the different crystallographic directions, the notification of fragments see in text.

Figure 12.12a–c shows the insets that present the low-frequency parts of the corresponding spectral density on an enlarged scale (Curves 1, 4, and 6, respectively), in comparison with the contributions to the phonon density of the tube that is generated by the displacement of the same type of all $2n$ atoms along the edge of a normal slice of the tube: along the tube axis (Curves 3); along the tangent to the surface of the tube, perpendicular to its axis (Curve 5), and along some single direction in the plane, normal to the tube axis (Curve 7).

On Figure 12.12b, Curve 5 is almost identical to the curve representing the density of states of the nanotube torsional mode. It has the typical shape of the density of states of a one-dimensional linear chain. Note that its behavior is not affected by any of the "modulation" associated with size quantization [40]. The contribution of this one-dimensional mode to the low-temperature heat capacity is ~ T, since its dispersion law in the long-wave region has the typical acoustic shape $\omega_{rot} \sim k$ [39]. As can be seen on Figure 12.13b (Curve 7), the linear temperature dependence of this contribution is preserved to temperatures above 100 K.

Curve 3 on Figure 12.12b in the low-frequency range determines the contribution to the tube's phonon density of states from its longitudinal vibrations as a single one-dimensional object. The dispersion law of these vibrations has a typical acoustic character, and therefore at $\omega \to 0$ this curve tends toward some constant value, and the contribution of the corresponding vibrations to the low-temperature heat capacity is also linear with the temperature (Curve 6, Figure 12.12b).

Curve 7 on Figure 12.12c in the low-frequency range determines the contribution to the tube's phonon density of states from its longitudinal vibrations as a single one-dimensional object. This contribution at $\omega \to 0$ must be proportional to $1/\sqrt{\omega}$, which is the density of states of the quasi-particles having a frequency $\omega_{fl} \sim k^2$ in the one-dimensional structure. At $T \to 0$ their contribution to the heat capacity must be proportional to $\sqrt{T}$. The growth with decreasing frequency of Curves 6 and 7 on Figure 12.13c at frequencies less than 2 THz, and of Curve 8 on Figure 12.13b at $T \leq 6$ K leads to the conclusion that in the long-wave limit ($\omega \leq 0.1$ THz) the contribution of flexural vibrations to the tube's phonon density of states becomes decisive and therefore, at $T \leq 1$ K, the temperature dependence of the nanotube heat capacity will actually be close to the root.

At temperatures ranging from 2 to 7 K, the heat capacity of the nanotube is determined by the quasiflexural vibrations of the graphene layer from which it is folded, i.e., by the quasiflexural wave that is propagating along the surface of the tube. Curve 1 on Figure 12.13b coincides with Curve 5 (it is also the same as Curve 3 on Figure 12.11b), which is the temperature dependence of the derivative with respect to temperature of the contribution to the bigraphene heat capacity from displacements normal to the plane of its layers. The similarity to bigraphene but not graphene is not accidental, since due to the deformability of the layer the quasiflexural mode will have a dispersion law that looks like Equation 12.19 with a "non-bending" or "acoustic" first term. Quasiflexural vibrations of the deformed graphene layer at this temperature interval define the behavior of the contribution to the heat capacity such as the vibrations along the tangent to the surface

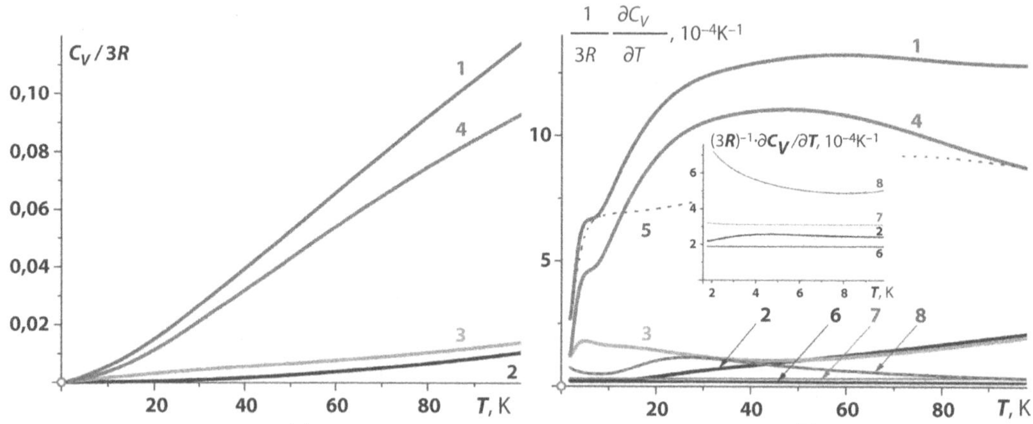

Figure 12.13 The temperature dependences of heat capacity (a) and the derivative of the heat capacity with respect to temperature (b). Along with the total values (Curves 1), the contributions thereto from different atomic displacements are also presented; Curves 2 shows the input from displacements along the axis of the tube; Curves 3 shows the displacement along the tangent to the surface of the tube, normal to its axis; Curves 4 shows the displacement along the normal to the surface of the tube. In addition, on (b), Curve 5 is the derivative with respect to temperature of the contribution to the heat capacity of bigraphene from displacements that are normal to the plane of its layers; Curve 6 shows the contribution of longitudinal vibrations of the tube as a whole; Curve 7 is the contribution of the torsional mode of the nanotube, Curve 8 is the contribution of the flexural vibrations of the tube as a whole.

of the tube, and normal to its axis (Curve 3) and the vibrations along the normal to the surface of the tube (Curve 4). The nonmonotonic behavior of Curve 3 on Figure 12.13b is caused specifically by the excess of the spectral density $\rho_\tau(\omega)$ (Curve 4) over Curve 5 at frequencies up to 2 THz, due to the contribution from the quasiflexural mode.

At higher temperatures, the heat capacity of the tube is mainly determined by the atomic vibrations along the normal to its surface, and those that are of the lowest frequency. Already at a temperature of $T \approx 40$ K (Figure 12.13a and b) the temperature dependence of the corresponding contribution (Curve 4) has a point of inflection. Starting from this temperature, the course of the total heat capacity (Curve 1) is close to being a straight line due to the "compensation" of the negative curvature of Curve 4 by positive curvature of Curves 2 and 3. As is in the cases considered in the previous sections, the given rectilinear course of the heat capacity temperature dependence has no relation to the flexural modes. In this temperature interval, our results are in good agreement with the results from Ref. [41], obtained for carbon nanobundles.

Thus, due to the bending vibrations of the graphene nanotube as a whole quasi-one-dimensional object, the root temperature dependence of the low-temperature heat capacity can exist at very low temperatures $T \leq 1$ K. At somewhat higher temperatures, the heat capacity of the nanotube is determined by the quasiflexural wave propagating along its surface, and in the interval 3–7 K, it is proportional to the temperature, as the contribution of the flexural mode in a quasi-two-dimensional system. The quasiflexural vibrations do not make any decisive contributions to the behavior of the graphene nanotube heat capacity at higher temperatures.

12.2.4 Negative Thermal Expansion in Graphene Nanostructures

Anisotropic interatomic interactions lead to a number of interesting features in linear thermal expansion, including nonmonotonicity of the temperature dependences of the coefficient of linear thermal expansion $\alpha_i(T) \equiv a_i^{-1}(T) \cdot \frac{\partial a_i(T)}{\partial T}$ (a_i is the crystal lattice constant in crystallographic direction i) and the possibility of negative values of this quantity in certain directions. Negative thermal expansion of graphite in the direction along the graphene layers that form it has very recently been observed experimentally (see, for example, [42]). In 1952, Lifshitz [38] predicted a negative coefficient of linear thermal expansion in directions along strong bonds in layers or chains for highly anisotropic layered structures in which the atomic vibrations polarized along a direction of weak bonding have a distinct quasi-bending character. In the long-wavelength region, the dispersion relation for the quasi-bending phonon mode has the form (Equation 12.19).

A dispersion relation similar to Equation 12.19 was, in fact, predicted for the transverse acoustic phonon mode of graphite polarized along the c axis. Twenty years later, this dispersion curve was detected in an inelastic neutron scattering experiment [21].

Negative coefficients of linear thermal expansion have also been observed in the direction along the layers in a number of layered crystals (see Ref. [43] for example) in whose phonon spectra these modes are entirely lacking or in which the quasi-bending distortion shows up very weakly in the corresponding dispersion relation (see Refs. [44, 45] for example). The negative coefficients of linear thermal expansion that occur when quasi-bending modes with a quadratic dispersion relation are present in the phonon spectrum and when these are absent can be explained by a microscopic analysis. Since the thermal expansion of solids is caused by anharmonic vibrations of the atoms in the lattice, it is extremely difficult to describe it on a microscopic level both because of the complexity of the relevant nonlinear equations and, mainly, because of the almost complete absence of information on the anharmonic force constants. For this reason, thermal expansion is usually described in terms of the so-called quasiharmonic approximation.

On expanding the lattice potential energy U in powers of the components of the small displacements of the atoms from their equilibrium positions u_i to third order, and taking the translational symmetry of the lattice for the temperature dependence of the principal values of the linear thermal expansion tensor $\alpha_{ii}(T)$ into account, we obtain

$$D_i\alpha_{ii}\left(T\right)=\frac{\partial u_{ii}\left(T\right)}{\partial T}=-\frac{1}{2}\sum_{r,r',k,l}\Phi_{ikl}\left(r,r\right)\frac{\partial}{\partial T}\left\langle u_k\left(r\right)u_l\left(r'\right)\right\rangle_T, \tag{12.41}$$

where u_{ik} is the strain (deformation) tensor; the coefficients $D_i=\sum_{r,k}\Phi_{ik}\left(r\right)\left|x_k\right|$; x_i are the components of the radius vectors of the equilibrium positions of the atoms r $\Phi_{ik}\left(r\right)\equiv\Phi_{ik}\left(r-r'\right)=\dfrac{\partial^2 U}{\partial u_i\left(r\right)\partial u_k\left(r'\right)}$ are the elements of the force matrix; and $\Phi_{ikl}\left(r,r'\right)\equiv\Phi_{ik}\left(r-r'',r'-r''\right)=\dfrac{\partial^3 U}{\partial u_i\left(r\right)\partial u_k\left(r'\right)\partial u_i\left(r''\right)}$ are the third-order force constants.

When constructing an elementary nonlinear theory of crystal lattice vibrations (see Ref. [19] for example), the anharmonicity only has to be taken into account in the terms related to the largest interatomic interaction forces, while the potential energy of the small interlayer (or interchain) interactions and noncentral forces can be treated in a harmonic approximation. Thus, in Equation 12.41, the terms containing correlators of the displacements of the atoms from different layers or correlators of the form $\left\langle u_x\left(r\right)u_z\left(0\right)\right\rangle_T$ can be neglected. For atomic displacements along the direction of the strong coupling, the correlators increase with temperature no more rapidly than the corresponding mean-square displacements and both of these temperature dependences are similar. Thus, the temperature dependence of the coefficient of linear thermal expansion along the strong coupling direction of a layered crystal can be described by the fairly simple formula [46, 47]

$$\alpha_{\parallel}\left(T\right)=A\frac{\partial}{\partial T}\left\langle u_{\parallel}^2\right\rangle_T\left[\delta-\Delta\left(T\right)\right];\ \Delta\left(T\right)\equiv\frac{\partial}{\partial T}\left\langle u_{\perp}^2\right\rangle_T\Big/\frac{\partial}{\partial T}\left\langle u_{\parallel}^2\right\rangle_T. \tag{12.42}$$

Since the temperature derivatives of the mean-square displacements are positive, Equation 12.42 implies that the atomic vibrations along the strong coupling direction cause expansion of the crystal in that direction, while the much higher amplitude vibrations along the weak coupling direction cause the crystal to contract (see Figure 12.14a). It is clear that $\Delta(T)$ has a maximum at the temperature corresponding to the transition of the temperature dependence of the mean-square displacement along the weak coupling direction to the classical limit (Figure 12.14b). Thus, near this temperature, the coefficient of linear thermal expansion along the strong coupling direction can take on negative values. Therefore, the compression of the structure along the strong coupling direction is caused by its expansion along the weak coupling direction. This effect has come to be known as the "membrane effect" for layered crystals [38].

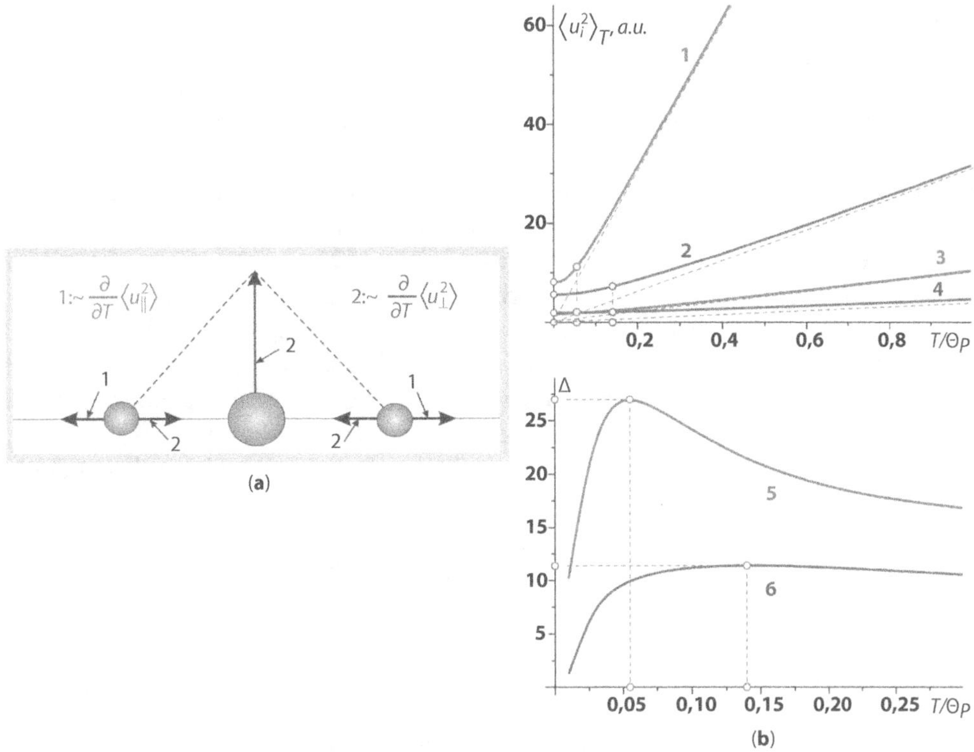

Figure 12.14 Illustrating the mechanism for the force that compresses the layers in highly anisotropic layered crystals owing to large-amplitude atomic vibrations transverse to the layers (a); the appearance of a maximum in the D(T) curve owing to anisotropy of the mean-square displacements of the atoms (b). In (b), curves 1 and 2 are the temperature dependences of the mean-square displacements in the direction normal to the layers and curves 3 and 4, along the layers; curves 5 and 6 show $\Delta(T)$.

We note that as the bending rigidity of the layers or chains increases, there are increases in both the mean-square displacement of the vibrations normal to the layers (chains) and the temperature derivative of these mean-square displacements.

Curves 1, 3, and 5 of Figure 12.14b correspond to a layered crystal with a low bending rigidity of the layers, and curves 2, 4, and 6, to a high bending rigidity. It is clear that, as the bending rigidity increases, the peak in the $\Delta(T)$ curve decreases in magnitude and becomes flatter. In this figure, $\Theta_P \equiv \hbar\omega_{max}/k$, where ω_{max} is the upper limit of the quasicontinuous phonon spectrum, and $\hbar$ and k, respectively, are the Planck and Boltzmann constants.

The parameters A and δ in Equation 12.42 can be expressed in obvious ways in terms of the anharmonic constants $\Phi_{ikl}(\boldsymbol{r}, \boldsymbol{r}')$, and $\langle u_\parallel^2 \rangle_T$ and $\langle u_\perp^2 \rangle_T$ are given by

for a layered crystal $\quad \langle u_\parallel^2 \rangle_T = \langle u_a^2 \rangle_T + \langle u_b^2 \rangle_T; \; \langle u_\perp^2 \rangle_T = \langle u_c^2 \rangle_T;$

(with the c axis directed normal to the layers)

for a quasi-chain structure $\quad \langle u_\parallel^2 \rangle_T = \langle u_a^2 \rangle_T; \; \langle u_\perp^2 \rangle_T = \langle u_b^2 \rangle_T + \langle u_c^2 \rangle_T$

(with the a axis directed along the chains)

It has been shown [47] that for most layered and chain structures, the membrane mechanism described by Equation 12.42 determines the negative thermal expansion. Equation 12.41 also includes the possibility of compression of the crystal with temperature owing to displacements or turns of the layers as a whole (so-called Poisson compression and liberation, respectively). These mechanisms do not play a significant role for most layered or chain structures, but for some polymers [48], ice [49], etc., their contribution can be dominant.

Further, we show the results of experimental and theoretical studies of manifestations of the membrane mechanism for the formation of negative coefficients of linear thermal expansion in layered crystals consisting both of weakly coupled monolayers (e.g., graphite and thin carbon nanofilms) and of trilayered "sandwiches" (e.g., niobium diselenide). Has been studied the nonmonotonicity of the coefficient of linear thermal expansion along different crystallographic directions in multilayer Eu-Ba-Cu-O high-temperature superconductor (HTSC) crystals. Type 1-2-3 HTSC are characterized by a strong local anisotropy of the interatomic interactions that is not preserved in the long-range order and does not lead to strong anisotropy of the elastic moduli [50, 51].

The mean-square displacements $\left\langle \left[u_i^{(s)}\right]^2 \right\rangle$ of the atoms in the s sublattice along the crystalline direction i are related to the phonon spectrum of the crystal by Equation 12.35. Figure 12.15 shows the temperature dependences of the mean-square displacements averaged over all the sublattices for different crystallographic directions: $\left\langle u_i^2 \right\rangle_T = q^{-1} \sum_{s=1}^{q} \left\langle \left[u_i^{(s)}\right]^2 \right\rangle_T$ (Figure 12.15a), as well as of the derivatives of these mean-square displacements with respect to temperature (Figure 12.15b) and of the ratio of these derivatives: $(T) = \frac{\partial}{\partial T}\left\langle u_c^2 \right\rangle_T \Big/ \frac{\partial}{\partial T}\left\langle u_{ab}^2 \right\rangle_T$ (Figure 12.15c).

These curves were calculated using the spectral densities shown in Figure 12.7. As in Figure 12.7, in Figure 12.15a–c, the curves 1 correspond to graphite and curves 2 and 3, to bi- and trigraphene, respectively. For bi- and trigraphene, the mean-square displacements of the atomic vibrations in the plane of the layers differ little from the corresponding dependences for bulk graphite. The mean-square displacement in the direction along the sixth-order axis, i.e., along the weak-bonding direction, is considerably higher and increases substantially as the layer thickness is reduced.

The derivatives $\frac{\partial}{\partial T}\left\langle u_c^2 \right\rangle_T$ with respect to temperature also increase rapidly in similar mode, so that near the temperature at which the classical limit of $\left\langle u_c^2 \right\rangle_T$ is approached, a maximum appears in the functions $\Delta(T)$. Because of the large bending rigidity of the layers, these maxima are rather flat; the shape of the $\Delta(T)$ curves in Figure 12.15c is similar to curve 6 in Figure 12.14b. Figure 12.15c also shows experimental measurements of the coefficient of linear thermal expansion α_{ab} of graphite [42, 52].

The high bending rigidity of the graphene layers is responsible for the large width of the temperature interval over which the coefficient of linear thermal expansion in the plane of the layers is negative and for the flat variation in the $\alpha_{ab}(T)$ curve within this interval. Given the extremely flat variation in these curves near the extrema, the agreement between the temperatures of the maximum in curve 1 of frame c and of the minimum in the $\alpha_{ab}(T)$

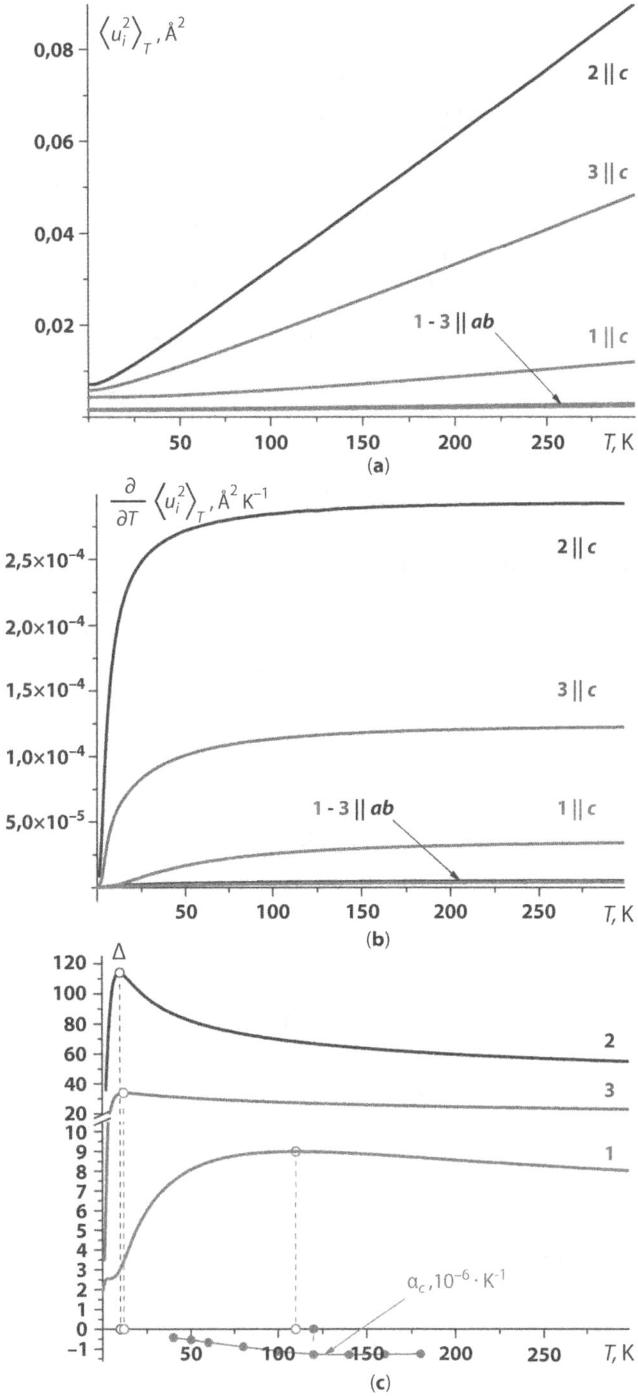

Figure 12.15 Temperature dependences of: the mean-square displacement (a); the derivatives of the mean-square displacements with respect to temperature (b); the ratios $\Delta(T)=\frac{\partial}{\partial T}\langle u_c^2\rangle_T \Big/ \frac{\partial}{\partial T}\langle u_{ab}^2\rangle_T$ of the derivatives of the mean-square displacements and of the experimental values of the coefficient of linear thermal expansion α_{ab} of graphite (c). In (a)–(c), the calculated curves 1 correspond to graphite, curves 2, to bigraphene, and curves 3, to trigraphene.

curve is fully satisfactory, and the thermal expansion of graphite is well explained by the microscopic model proposed in the previous section.

The anisotropy of the mean-square displacements in bi- and trigraphene is still greater than in graphite. Based on the shape of curves 2 and 3 in Figure 12.15c, we may conclude that the coefficient of linear thermal expansion $\alpha_{ab}(T)$ of these structures will be negative over the entire range where they exist (i.e., to temperatures of 400–500 K [32]), but the temperature of the minimum in this curve will be substantially lower than for graphite.

Figure 12.16 shows the temperature dependences of the mean-square displacements of the atoms in the nanotube in different directions (Figure 12.16a), the temperature derivatives of these displacements (Figure 12.16b), and the ratio $(T) = \frac{\partial}{\partial T}\left\langle u_\tau^2 \right\rangle_T \Big/ \frac{\partial}{\partial T}\left\langle u_n^2 \right\rangle_T$ of the derivatives (Figure 12.16c). This frame also shows the coefficient of linear thermal expansion $\alpha_R(T)$ [53] measured with a low-temperature capacitive dilatometer with a sensitivity of $2 \cdot 10^{-9}$ см.

The temperature variation $\Delta(T)$ shown in Figure 12.16 differs from these dependences in the preceding sections. Up to $T = 0.5$ K, i.e., roughly to $2 \cdot 10^{-4}\Theta_D$, a maximum does not appear. Similarly, a minimum is not observed in the $\alpha_R(T)$ curve up to a temperature of 2 K, which, given the high Debye temperature of graphene and graphene structures, can be regarded as extremely low for this system. The absence of extrema in these characteristics can be explained by the fact that the quasi-one-dimensional behavior of the phonon spectra shows up more strongly for tangential than for normal displacements. The experiments did not deal with isolated nanotubes, but with bundles of them with the surface normal vectors of the tubes in the same direction. The temperature of the minimum in the $\alpha_R(T)$ curve for such a bundle of tubes will be determined by the interactions between the nanotubes of which it consists.

The compression of the nanotubes in the radial direction is, therefore, caused by the rapid temperature rise in the amplitude of the rotational motion of their atoms. This kind of compression, as opposed to the effects discussed in the earlier sections of this chapter, may also lead to negative values of the volume expansion. In many ways, this is analogous to the behavior of the thermal expansion of fullerites [45]. This effect has been explained [46] rigorously in terms of quantum mechanics, but it was shown that it is also caused by rotational motions of the fullerene molecule. As opposed to fullerene, the torsional vibrations of carbon nanotubes can be treated quasiclassically by lattice dynamics methods. It is clear from Figure 12.16a that at temperatures below 10 K, the mean-square amplitudes of the displacements (both normal and tangential) do not exceed 0.15 Å, i.e., 0.1 times the distance between nearest neighbors in a tube. Thus, in order to find the phonon spectra, as well as the mean-square displacements and their derivatives, a harmonic approximation can be used, at least for the low temperatures where negative coefficients of linear thermal expansion have been observed.

12.3 Effect of Defects to Electron and Phonon Spectra

As noted in Section 12.1, a strong covalent bond between the atoms of a graphene monolayer leads to enormous values of its Debye temperature ($\Theta_D \sim 2500$ K). This determines the "low-temperature" oscillating behavior of the thermodynamic characteristics of graphene and graphite and other carbon-based structures of graphene (e.g., nanotubes) to a temperature exceeding the room, and also promotes to a high superconducting transition temperatures in these compounds.

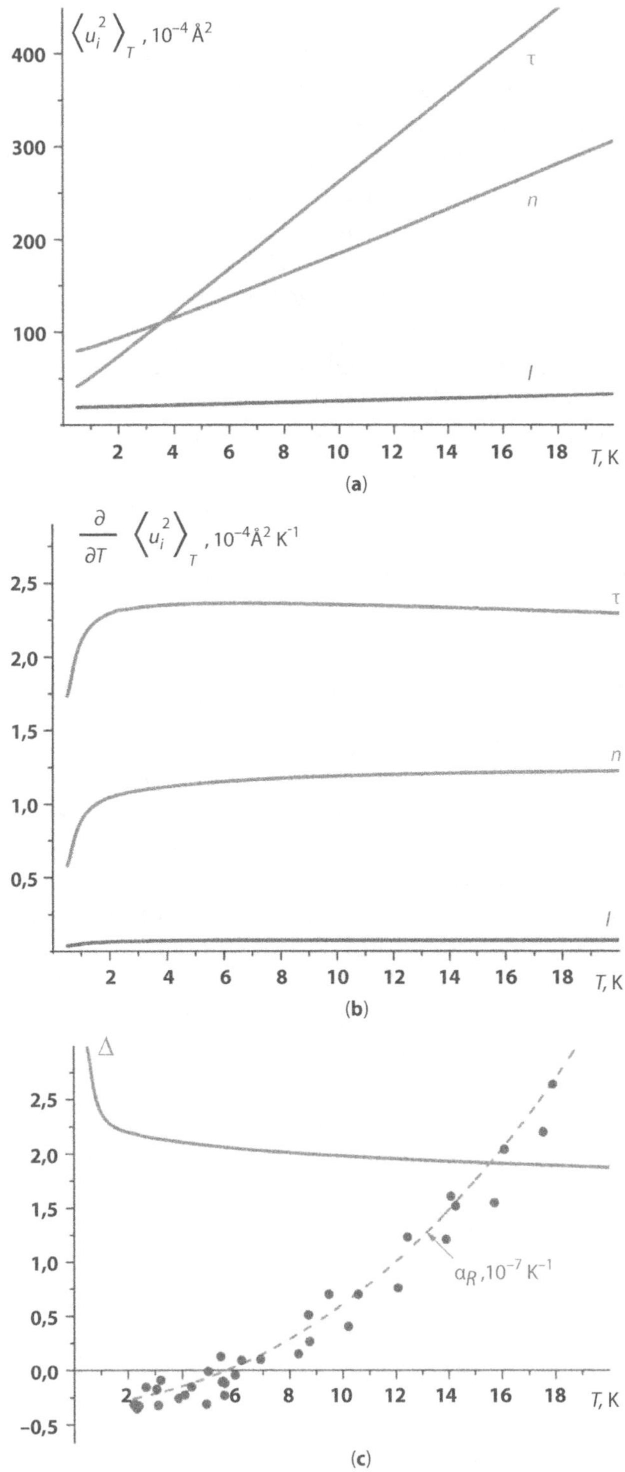

Figure 12.16 Mean-square displacements of atoms in a single-wall carbon nanotube along different directions and their derivatives with respect to temperature (frames a and b, respectively). Frame (c) compares the experimental dependence of the coefficient of linear thermal expansion $\alpha_R(T)$ with $\Delta(T)=\frac{\partial}{\partial T}\langle u_\tau^2\rangle_T \Big/ \frac{\partial}{\partial T}\langle u_n^2\rangle_T$.

Indeed, in the BCS theory, the superconducting transition temperature is proportional to the mean frequency of phonons [10] (in graphene, this value ~1500 K). And for the present time, there are many compelling evidences that the electron–phonon interaction (the BCS mechanism) is the basic mechanism responsible for the transition to the superconducting state, including, for superconductors with a high superconducting transition temperature (see, e.g., Ref. [9]).

The reason for the fact that the superconducting transition, both in graphite and in carbon nanotubes and nanofilms, has not been reliably detected to date is, in our opinion, a small number of charge carriers with energies close to Fermi energy (ε_F) and a small number of phonons, which should give the basic contribution to the electron–phonon coupling constant.

Consequently, for the effective coupling of electrons with phonons, it is desirable to increase the number of charge carriers and substantially enrich that part of the phonon spectrum whose contribution to Cooper pairing is determinative through the high values of the electron–phonon coupling constant [9, 10]. Already conducted in Section 12.1, the analysis of the behavior of the real part of the Green's function indicates that this problem can be solved by the controlled creation of certain defective structures in graphene nanoformations. An analysis of the variation of the electronic and phonon spectra of graphene nanoobjects due to the presence of sample boundaries, vacancies, and vacancy groups, as well as intercalation by metals, is considered in this part.

12.3.1 Electron Spectrum of Graphene with ZigZag-Bounder

Note above, that in perfect structure of graphene, the atoms of different sublattices are physically equivalent (Figure 12.1a), which means that their local Green's functions, and eventually, local densities of states (LDOSs) are coincident. The introduction to one of the sublattices of graphene defects, of course, violates this equivalence and leads to a significant distinction of the electron spectra of the atoms of different sublattices.

The example of defect, which contains only atoms in one of sublattices, is zigzag bounder. In this case, as is clearly demonstrated by Figure 12.17a, similar to vacancy, the atomic bonds are broken in one sublattice only, (see the sublattice **A**, Figure 12.17a), at the same time the atoms of the other sublattice have no dangling bonds. In contrast to vacancy, which is the point defect and is breaking translational symmetry along all directions in the crystal. In this direction, the one-dimensional vector of reciprocal lattice $\boldsymbol{\beta} = (0,\beta)$ and quasi-wave vector $\kappa = (0,\kappa)$ (Figure 12.17c) can be introduced. Then Hamiltonian (Equation 12.1) in a form of Jacobi matrix [54, 55] is written as

$$\hat{H} = \begin{pmatrix} \varepsilon_F - \Lambda(\kappa) & -\alpha(\kappa) & 0 & 0 & \dots & 0 & \dots \\ -\alpha(\kappa) & \varepsilon_F & -\beta(\kappa) & 0 & \dots & 0 & \dots \\ 0 & -\beta(\kappa) & \varepsilon_F & -\alpha(\kappa) & \dots & 0 & \dots \\ 0 & 0 & -\alpha(\kappa) & \varepsilon_F & \dots & 0 & \dots \\ 0 & 0 & 0 & -\beta(\kappa) & \dots & 0 & \dots \\ \dots & \dots & \dots & \dots & \dots & \dots & \dots \\ \dots & \dots & \dots & \dots & \dots & \dots & \dots \end{pmatrix} \quad (12.43)$$

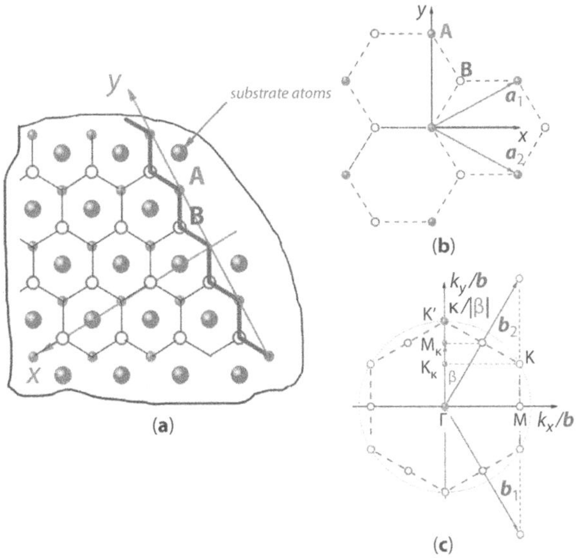

Figure 12.17 Geometry of the problem, choice of elementary cell (fragment a), and construction of first Brillouin zone of graphene (fragment b) and 1D Brillouin zone of graphene with a zigzag boundary (fragment c).

where $\alpha(\kappa)=2J\cos\frac{\kappa a}{2}$; $\beta(\kappa)\equiv\beta=J$. The variable $\Lambda(\kappa)$ describes the disturbance, introduced in the Hamiltonian (Equation 12.1) the formation of the boundary, (we assume that the disturbance captures only those atoms in which there is a failure of one connection). Construction of 2D first Brillouin zone for perfect graphene and 1D first Brillouin zone for considered problem is presented in Figure 12.17c.

The local Green's function $G^{(0)}(\varepsilon,\kappa,\Lambda)$, corresponding to the row of atoms of sublattice **A**, which lie on zigzag boundary (taken as the reference row – $n = 0$), can be written in a form of infinite continuous fraction [30, 31]:

$$G^{(0)}(\varepsilon,\kappa,\Lambda)=G_{00}(\varepsilon,\kappa,\Lambda)=\left(\varepsilon\hat{I}-\hat{H}\right)^{-1}_{00}$$
$$=\cfrac{1}{\varepsilon+\Lambda-\cfrac{\alpha^2(\kappa)}{\varepsilon-\cfrac{\beta^2}{\varepsilon-\cfrac{\alpha^2(\kappa)}{\cdots}}}}=\frac{1}{\varepsilon+\Lambda-\alpha^2K_\infty^{(\beta)}(\varepsilon,\kappa)} \tag{12.44}$$

where function $K_\infty^{(\beta)}(\varepsilon,\kappa)$ is continued fraction, corresponding to the Jacobi matrix, all diagonal matrix elements are equal ε, and on nondiagonal positions alternate values β and $\alpha(\kappa)$.

$$K_{\infty}^{(\beta)} = \begin{pmatrix} \varepsilon & \beta & 0 & 0 & \ldots & 0 & \ldots \\ \beta & \varepsilon & \alpha(\kappa) & 0 & \ldots & 0 & \ldots \\ 0 & \alpha(\kappa) & \varepsilon & \beta & \ldots & 0 & \ldots \\ 0 & 0 & \beta & \varepsilon & \ldots & 0 & \ldots \\ 0 & 0 & 0 & \alpha(\kappa) & \ldots & 0 & \ldots \\ \ldots & \ldots & \ldots & \ldots & \ldots & \ldots & \ldots \\ \ldots & \ldots & \ldots & \ldots & \ldots & \ldots & \ldots \end{pmatrix}_{00}^{-1}$$

$$= \frac{\varepsilon^2 - \beta^2 + \alpha^2(\kappa) + Z(\varepsilon)\sqrt{\left|(\varepsilon^2 - \varepsilon_0^2)(\varepsilon^2 - \varepsilon'^2)\right|}}{2\alpha^2(\kappa)\varepsilon}, \tag{12.45}$$

with the introduced variables: $\varepsilon_0^2 \equiv \left[\alpha(\kappa)+\beta\right]^2$ и $\varepsilon'^2 \equiv [\alpha(\kappa) - \beta]^2$.

Since $\beta = J > 0$, and in the range 1D first Brillouin zone $\kappa a \in [-\pi, \pi]$, (i.e., $\kappa \in [-M_\kappa, M_\kappa]$), where $\alpha(\kappa) = 2J\cos\frac{\kappa a}{2} \geq 0$, then the value $\varepsilon_0 > 0$ and quantity ε' become zero at the points K_κ. The dependencies of $\varepsilon_0(\kappa)$ и $\varepsilon'(\kappa)$ shown at Figure 12.18. The band of the quasicontinuous spectrum $D = \left[-\varepsilon_0(\kappa), -\left|\varepsilon'(\kappa)\right|\right] \cup \left[\left|\varepsilon'(\kappa)\right|, \varepsilon_0(\kappa)\right]$ is doubly-connected at all points, except the point K_κ ($\kappa = 2\pi/3a$), and, corresponding, the gap $[-|\varepsilon'(\kappa)|, |\varepsilon'(\kappa)|]$ is zero only at this point.

The function entering into the relation (12.45)

$$\begin{aligned} Z(\varepsilon) \equiv\; & \Theta(-\varepsilon_0 - \varepsilon) + i\cdot\Theta(\varepsilon+\varepsilon_0)\Theta(-|\varepsilon'| - \varepsilon) + \Theta(\varepsilon + |\varepsilon'|)\Theta(|\varepsilon'| - \varepsilon) + \\ & + i\cdot\Theta(\varepsilon - |\varepsilon'|)\Theta(\varepsilon_0 - \varepsilon) - \Theta(\varepsilon - \varepsilon_0) \end{aligned} \tag{12.46}$$

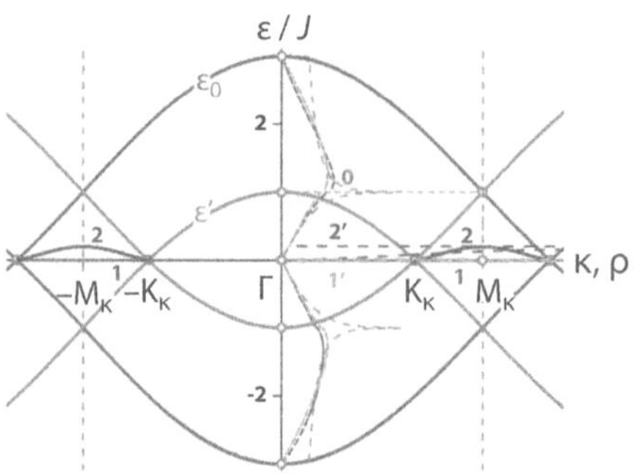

Figure 12.18 Dispersion curves splitting modes $\varepsilon_g(\kappa)$ at $\Lambda = 0$ (the line segment of curve 1) and at $\Lambda = 0.1$ (curve 2), and also spectral densities atoms, lying on the boundary (curves 1′ and 2′, respectively). Curve 0—the spectral density of ideal graphene.

is determined by the form of the Green's function both in the band of the quasicontinuous spectrum $\mathscr{D}$, and outside it. At $\varepsilon \in \mathscr{D}$ function $G^{(0)}(\varepsilon,\kappa,\Lambda)$ has the imaginary part and the energy distribution inside the band of the quasicontinuous spectrum for the atoms of the sublattice **A**, lying on zigzag boundary line defined as

$$\rho^{(0)}(\varepsilon,\kappa,\Lambda) = \pi^{-1}\,\mathrm{Im}\,G^{(0)}(\varepsilon,\kappa,\Lambda) \tag{12.47}$$

For each value of κ the integral

$$\int_{-\infty}^{\infty} \rho^{(0)}(\varepsilon,\kappa,\Lambda)\,d\varepsilon = 1.$$

When $\varepsilon \notin \mathscr{D}$ function $G^{(0)}(\varepsilon,\kappa,\Lambda)$ is purely real. Consider first the case $\Lambda = 0$, that is, we will assume, that the effect on the electronic spectrum of relaxation and reconstruction changes, which occur during the formation of the zigzag boundary, can be neglected. Function $G^{(0)}(\varepsilon,\kappa,\Lambda)$ in this case takes the form

$$G^{(0)}(\varepsilon,\kappa,0) = K_{\infty}^{(\alpha)}(\varepsilon,\kappa) = \frac{\varepsilon^2 + (\beta^2 - \alpha^2) + Z(\varepsilon)\cdot\sqrt{\left|(\varepsilon^2 - \varepsilon_0^2)(\varepsilon^2 - \varepsilon'^2)\right|}}{2\beta^2\,\varepsilon}. \tag{12.48}$$

It has pole $\varepsilon_g = 0$, which lies in the energy gap $[-|\varepsilon'(\kappa)|,|\varepsilon'(\kappa)|]$ and thus determines the non-dispersive localized energy level. The intensity of this level is defined as

$$\mu_g^{(0)}(\kappa,\Lambda) = 1 - \int_{D} \rho^{(0)}(\varepsilon,\kappa,\Lambda)\,d\varepsilon = \underset{\varepsilon=0}{\mathrm{re's}}\; G^{(0)}(\varepsilon,\kappa,\Lambda) \tag{12.49}$$

where the integral is taken over the entire doubly-connected band of the quasicontinuous spectrum $\mathscr{D}$. When $\Lambda = 0$ this residue is nonzero for $\alpha(\kappa) < \beta$, i.e., at $\kappa \in [-M_\kappa, -K_\kappa] \cup [K_\kappa, M_\kappa]$ (the line segments of curve 1 on Figure 12.18) and equal

$$\mu_g^{(0)}(\kappa,0) = \frac{\beta^2 - \alpha^2}{\beta^2}\cdot\Theta(\beta - \alpha) = |1 + 2\cos a\kappa|\cdot\Theta(-1 - 2\cos a\kappa) \tag{12.50}$$

The intensity of the discrete level ε_g at lines $n > 0$, formed by atoms, which are located under line $n = 0$, as follows from [56], equal

$$\mu_g^{(n)} = \mu_g^{(0)}(\kappa,\Lambda)\cdot P_n^2(\varepsilon_g) \tag{12.51}$$

where $P_n(\varepsilon)$ - degree of polynomial n, generated by Jacobi matrix (Equation 12.43). They satisfy the recurrence relation $\mathscr{H}_{n,n+1}P_{n+1}(\varepsilon) = (\varepsilon - \mathscr{H}_{n,n})P_n(\varepsilon) - \mathscr{H}_{n,n-1}P_{n-1}(\varepsilon)$ with initial conditions $P_{-1}(\varepsilon) = 0$; $P_0(\varepsilon) = 1$. It is simple to show and prove (for example, by the method of mathematical induction), that when $\varepsilon = \varepsilon_g = 0$ these polynomials are zero for odd n. Odd values n correspond to the lines of sublattice atoms **B** (see Figure 12.17a). Even-degree polynomials, which correspond to the lines of the atoms of the sublattice **A**, are non-zero and form a geometric progression

$$P_{2m-1}(0)=0;\quad P_{2m}(0)=(-1)^m\left(\frac{\alpha}{\beta}\right)^m \tag{12.52}$$

Thus, the discrete gap level $\varepsilon_g = 0$ available only in the electronic spectra of atoms sublattice **A**. Away from the boundary (with increasing parameter n) the intensity of the discrete level decreases exponentially (that is, it forms an infinitely decreasing geometric progression):

$$\mu_g^{(n)}(\kappa)=\mu_g^{(0)}(\kappa)\cdot q^n;\; q=P_n^2(0)=\frac{\alpha^2(\kappa)}{\beta^2}=2(1+\cos a\kappa), \tag{12.53}$$

the sum of which is one.

Figure 12.19 presented the functions $\rho_n(\varepsilon)$ - LDOS atoms, lying on lines with numbers n from 0 to 5 (the corresponding atom is shown on the inset of each of the fragments). Those functions are calculated using Jacobi matrices method, Hamiltonian $\hat{\mathscr{H}}$ (Equation 12.1) presented in the basis, obtained by orthonormalization of sequence $\left\{\hat{\mathscr{H}}^p\Psi_0\right\}_{p=0}^{\infty}$, where for generating function Ψ0, we have chosen a single excitation of atoms, labeled in inset of each fragment. All fragments are given the electron density of defect-free graphene for comparison. On the LDOS, atoms of the sublattice A ($n = 0, 2, 4$) have sharp (practically delta-functional) peaks when $\varepsilon = \varepsilon_g = 0$ (i.e., at the Fermi level). The area under the curves increases n with increasing, tending to one, at $n \to \infty$. Let's note that near the Fermi level, each of the LDOS data practically coincides with DOS of the non-defect graphene, that is, it preserves a V-shaped Dirac.

On the LDOS, atoms of the sublattice **B** ($n = 1, 3, 5$) peaks at the Fermi level are absent, and the areas under all curves are equal to one. Let's note that on these LDOSs, there are no characteristic V-figurative Dirac singularities, that is, near the Fermi level, the effective mass of the electrons is different from zero.

Let us now consider the possible relaxation processes at the boundary, i.e., we assume that the value of $\Lambda(\kappa)$ is nonzero. Then the Green's function (Equation 12.44) takes the form

$$G^{(0)}(\varepsilon,\kappa,\Lambda)=\frac{1}{2}\cdot\frac{\varepsilon^2+2\Lambda(\kappa)+\beta^2-\alpha^2(\kappa)+Z(\varepsilon)\cdot\sqrt{\left|(\varepsilon^2-\varepsilon_0^2)(\varepsilon^2-\varepsilon'^2)\right|}}{\Lambda(\kappa)\varepsilon^2+\left[\Lambda(\kappa)+\beta^2\right]\varepsilon+\Lambda(\kappa)\left[\beta^2-\alpha(\kappa)\right]} \tag{12.54}$$

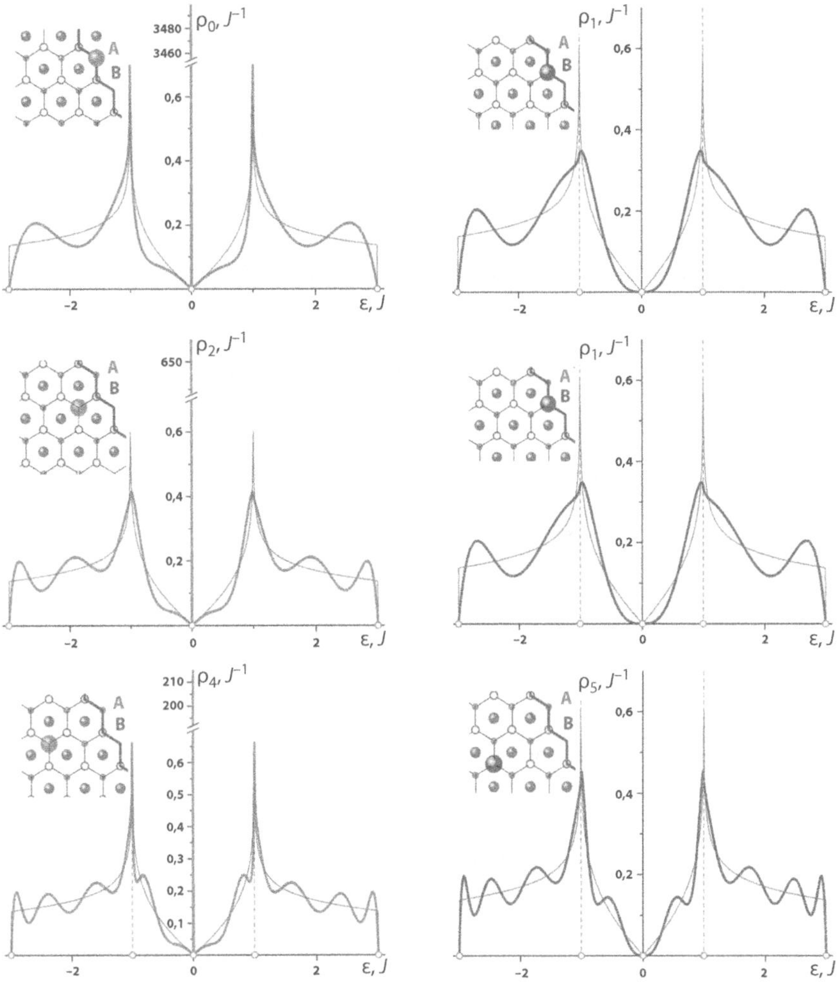

Figure 12.19 The evolution of electronic LDOS as they move away from the boundary (the corresponding atoms are shown on the inset of each fragment) at $\Lambda = 0$, when a nondispersive gap level with energy appears in the system, equal to the Fermi energy.

As we have assumed, the disturbance affects only the line $n = 0$, and along the direction of the boundary translational invariance is preserved, the disturbance $\Lambda(\kappa)$ can be represented in the form:

$$\Lambda(\kappa) = 2\tilde{J}\cos\kappa a \qquad (12.55)$$

and treat, for example, the occurrence due to the formation of one free-bond interaction between neighboring atoms, lying on this line ($\tilde{J}$-the corresponding hopping integral). In the graphene lattice, these atoms are the second neighbors; we do not consider the interaction between them on other lines.

Among two roots of denominator (Equation 12.54), we should then select the one vanishing at $\Lambda \to 0$, that is, the dependence

$$\varepsilon_g(\kappa,\Lambda)=\frac{\sqrt{\left[\beta^2-\Lambda^2(\kappa)\right]^2+4\alpha^2(\kappa)\Lambda^2(\kappa)}-\beta^2-\Lambda^2(\kappa)}{2\Lambda(\kappa)} \tag{12.56}$$

determines a dispersion law of the "border" wave, split of the band of quasi-continuous spectrum.

The residue of function (Equation 12.54) in this pole determines the intensity of corresponding discrete level, and, eventually, the portion of quasi-particles moved to the split-off wave. It equals to

$$\mu_g^{(0)}(\kappa,\Lambda)=-\frac{\beta^2-\Lambda^2(\kappa)}{\Lambda(\kappa)}\cdot\frac{\varepsilon_g(\kappa,\Lambda)}{\sqrt{\left[\beta^2-\Lambda^2(\kappa)\right]^2+4\alpha^2(\kappa)\Lambda^2(\kappa)}}\cdot\Theta\left(-\frac{\beta^2-\Lambda^2(\kappa)}{\Lambda(\kappa)}\cdot\varepsilon_g(\kappa,\Lambda)\right) \tag{12.57}$$

The condition of existence of the wave is that of positive meaning of the argument of Heaviside theta function

$$\frac{\beta^2-\Lambda^2(\kappa)}{\Lambda(\kappa)}\cdot\varepsilon_g(\kappa,\Lambda)<0 \tag{12.58}$$

and coincides with a condition of existence of nondispersive gap level for the case $\Lambda = 0$, i.e., $\kappa\in[-M_\kappa, -K_\kappa]\cup[K_\kappa, M_\kappa]$. Dispersion relations (Equation 12.56) at the intervals of existence of the split-off wave are presented by curve 2 at Figure 12.18. The splitting of the wave occurs from the Fermi level at $\kappa = \pm K_\kappa$. A dispersion law near Fermi level is linear (relativistic).

$$\varepsilon_g(K_\kappa + x,\Lambda) \approx \hbar v_F\cdot x, \tag{12.59}$$

where Fermi velocity v_F equals to group velocity in the point K_κ. The group velocity, corresponding to dispersion (Equation 12.56) as a function of quasi-wave vector κ, can be written in the following form:

$$v_{gr}(\kappa)=\frac{\partial}{\partial\kappa}\frac{\varepsilon(\kappa)}{\hbar}=-\frac{\Lambda'(\kappa)}{\hbar}\cdot\mu_g^{(0)}(\kappa,\Lambda)$$
$$-\frac{aJ^2\sin a\kappa}{\hbar}\cdot\frac{\Lambda(\kappa)}{\sqrt{\left[\Lambda^2(\kappa)-J^2\right]^2+16J^2\Lambda^2(\kappa)\cos^2\frac{\kappa a}{2}}}. \tag{12.60}$$

In Equation 12.60, the value $\mu_g^{(0)}(\kappa,\Lambda)$ is determined by the relationship (Equation 12.57), and the prime denotes differentiation with respect to κ. Then

$$v_F^{(g)} = v_{gr}\left(K_\kappa\right) = \frac{a}{\hbar}\frac{\sqrt{3}}{2}\frac{J^2\Lambda'\left(K_\kappa\right)}{J^2+\Lambda^2\left(K_\kappa\right)}, \qquad (12.61)$$

or, taking into account (Equation 12.55),

$$v_F^{(g)} = \frac{a}{\hbar}\frac{\sqrt{3}}{2}\frac{J^2\tilde{J}}{J^2+\tilde{J}^2} = \frac{v_F^{(0)}}{\sqrt{3}}\frac{J\tilde{J}}{J^2+\tilde{J}^2}, \qquad (12.62)$$

where $v_F^{(0)} = 3aJ/2\hbar$ - is the Fermi level of non-defect graphene with dispersion (Equation 12.2).

When Equation 12.55 is satisfied, the dispersion relation (Equation 12.56) corresponds to the density of states $g_g(\varepsilon) = \frac{a}{\pi\hbar}\cdot v_{gr}^{-1}(\varepsilon)$, где $v_{gr}(\varepsilon)$ is the group velocity, represented as energy function.

From Equations 12.55 and 12.56, we find that

$$\varsigma \equiv \cos\kappa a = -\frac{\tilde{J}\left(J^2-\varepsilon^2\right)+\sqrt{\tilde{J}^2\left(J^2-\varepsilon^2\right)^2+4\varepsilon J^2\tilde{J}\left(J^2-\varepsilon\tilde{J}\right)}}{4\tilde{J}\left(J^2-\varepsilon\tilde{J}\right)} \qquad (12.63)$$

(a sign before the radical is chosen so that the value $\varepsilon = 0$ corresponds to $\kappa = K_\kappa$). After that, the dependence $g_g(\varepsilon)$ is obtained with the help of Equations 12.60 and 12.63 by simple, but cumbersome calculations, the final result of which is due to the cumbersome nature we do not write.

It is easy to see from expressions (12.55), (12.56), and (12.63) that $\varepsilon_g \in \left[0, 2\tilde{J}\right]$. The dependence $g_g(\varepsilon) \to const \sim \left[v_F^{(g)}\right]^{-1}$, when $\varepsilon \to +0$, and it has a root divergence at $\varepsilon \to 2\tilde{J}$, which is typical for one-dimensional systems.

Since the gap wave (Equation 12.16) is attenuated as the distance from the boundary, its contributions to LDOS of different atoms will also be different and equal $\rho_g^{(n)}(\varepsilon) = g_g(\varepsilon)\cdot\mu_g^{(n)}(\varepsilon)$. The attenuation of the intensity of the split wave is described by the relation (12.51). In this case,

$$\begin{aligned} P_1\left(\varepsilon_g\right) &= \frac{\varepsilon_g+\Lambda}{\alpha} = \frac{\sqrt{\left(\beta^2-\Lambda^2\right)^2+4\alpha^2\Lambda^2}-\left(\beta^2-\Lambda^2\right)}{2\Lambda\alpha} \underset{\Lambda\to 0}{\longrightarrow} 0; \\ P_2\left(\varepsilon_g\right) &= \left[-\frac{\beta}{\Lambda}\cdot P_1\left(\varepsilon_g\right)\right] \underset{\Lambda\to 0}{\longrightarrow} \left(-\frac{\alpha}{\beta}\right). \end{aligned} \qquad (12.64)$$

The method of mathematical induction proved that

$$P_{2m+1}\left(\varepsilon_g\right)=\left(-\frac{\beta}{\Lambda}\right)^m\left[\cdot P_1\left(\varepsilon_g\right)\right]^{m+1}\underset{\Lambda\to 0}{\to}0;$$
$$P_{2m}\left(\varepsilon_g\right)=\left[-\frac{\beta}{\Lambda}\cdot P_1\left(\varepsilon_g\right)\right]^m\underset{\Lambda\to 0}{\to}\left(-\frac{\alpha}{\beta}\right)^m. \tag{12.65}$$

That is, on the atoms of the sublattice **B**, the split-off wave has, smaller than the atoms of the sublattice **A**, but they are different from zero (and tending to zero at $\Lambda\to 0$)

$$\mu_g^{(2m+1)}=\mu_g^{(0)}\cdot\left(-\frac{\beta}{\Lambda}\right)^{2m}\left[\cdot P_1\left(\varepsilon_g\right)\right]^{2(m+1)}\underset{\Lambda\to 0}{\to}0;$$
$$\mu_g^{(2m)}=\mu_g^{(0)}\cdot\left[-\frac{\beta}{\Lambda}\cdot P_1\left(\varepsilon_g\right)\right]^{2m}\underset{\Lambda\to 0}{\to}\left(-\frac{\alpha}{\beta}\right)^{2m}. \tag{12.66}$$

The relative fraction of the energy of the split-off wave, which is localized on the atoms of the sublattice **B**.

$$\sum_{m=0}^{\infty}\mu_g^{(2m+1)}=\frac{2\Lambda^2\alpha^2-\left(\beta^2-\Lambda^2\right)\left[\sqrt{\left(\beta^2-\Lambda^2\right)^2+4\alpha^2\Lambda^2}-\left(\beta^2-\Lambda^2\right)\right]}{4\Lambda^2\alpha^2-\left(\beta^2-\Lambda^2\right)\left[\sqrt{\left(\beta^2-\Lambda^2\right)^2+4\alpha^2\Lambda^2}-\left(\beta^2-\Lambda^2\right)\right]}\underset{\Lambda\to 0}{\to}0 \tag{12.67}$$

and the atoms of the sublattice **A**.

$$\sum_{m=0}^{\infty}\mu_g^{(2m)}=\frac{2\Lambda^2\alpha^2}{4\Lambda^2\alpha^2-\left(\beta^2-\Lambda^2\right)\left[\sqrt{\left(\beta^2-\Lambda^2\right)^2+4\alpha^2\Lambda^2}-\left(\beta^2-\Lambda^2\right)\right]}\underset{\Lambda\to 0}{\to}1 \tag{12.68}$$

Naturally, that $\sum_{m=0}^{\infty}\mu_g^{(2m+1)}+\sum_{m=0}^{\infty}\mu_g^{(2m)}=1.$

Figure 12.20 represented the evolution LDOS same atoms, that and on Figure 12.19, when relaxation perturbation $\Lambda(\kappa)$, which can be described by the relation (12.55) at $\tilde{J}=0.1J$. On LDOS atoms in sublattice **A** near Fermi level ($\varepsilon=0$) reveals a small peak of the width $2\tilde{J}$, which coincides completely with spectral density $\rho_g^{(2m)}(\varepsilon)=g_g(\varepsilon)\cdot\mu_g^{(2m)}(\varepsilon)$, calculated using formulas (12.63)–(12.68) for a gap wave with dispersion law (Equation 12.56). The functions

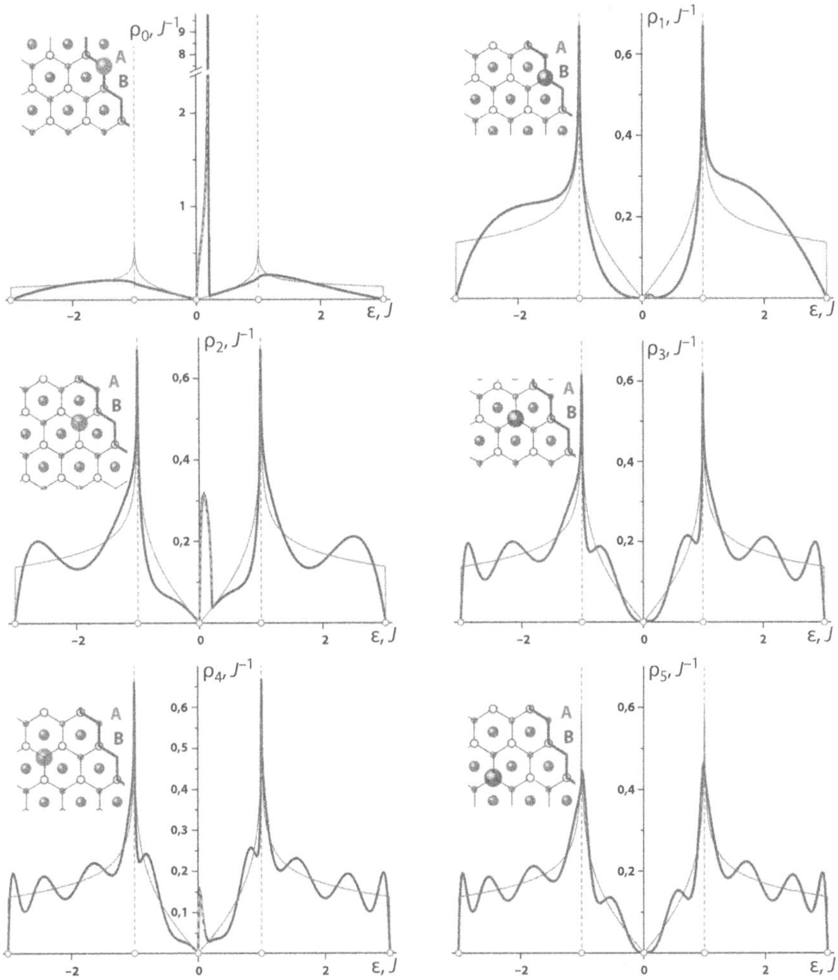

Figure 12.20 Evolution of electron LDOS at $\tilde{J}=0.1J$, as the distance from the boundary, when in the system occur the gap wave with dispersion (Equation 12.56).

$\rho_g^{(2m)}(\varepsilon)$ presented on fragments at Figure 12.20 by dashed lines, which are completely superimposed on the curves LDOS. It should be noted that at the same time, LDOSs keep their relativistic (Dirac) behavior at $\varepsilon \to -0$ is the Fermi velocity of non-defect graphene $v_F^{(0)}=3aJ/2\hbar$, and at $\varepsilon \to +0$ the Fermi velocity (Equation 12.62) in one-dimensional system, i.e., the angular coefficient of linear dependence $\rho_g^{(2m)}(\varepsilon \to +0)$ is related to the similar one $\rho_g^{(2m)}(\varepsilon \to -0)$ как $\sim\left[v_F^{(0)}/v_F^{(g)}\right]^2$.

For atoms in sublattice **B**, only LDOS of the atom, lying directly on the boundary, reveals a small peak of the width $2\tilde{J}$. It coincides completely with spectral density $\rho_g^{(1)}(\varepsilon)$. With a distance from the boundary the peaks $\rho_g^{(2m+1)}(\varepsilon)=\nu_g(\varepsilon)\cdot\mu_g^{(2m+1)}(\varepsilon)$ become vanishingly small. Near Fermi level, the behavior of LDOS of the atoms from sublattice **B**, as in the case of $\Lambda = 0$ is nonrelativistic and correspondent to the quasi-particles of finite mass.

12.3.2 The Deformation of the Phonon Spectrum of Graphene in the Formation of the ZigZag Boundary

Calculating the phonon spectrum and vibrational characteristics of graphene can actually exist either in the form of free microscopic flakes, always corrugated, or in the form of a flat monolayer, applied to the substrate. The existence of free graphene monolayers is impossible [57]. If in the study of the electron spectrum, it is possible to consider the substrate to be dielectric and completely eliminate it from consideration, then in calculating and analyzing the phonon spectrum, it is necessary to consider both the interaction of carbon atoms with the atoms of the substrate and the motion of the atoms of the substrate itself. Furthermore, the low-temperature thermodynamic vibrational characteristics of the system «graphene + substrate» will be determined precisely by the vibrations of the substrate atoms both because of its macroscopic thickness, and because of the high Debye temperature of graphene. Therefore, these characteristics, which are usually an important source of information on the phonon spectra, in this case, such information is practically not carried.

At the same time, on the basis of the currently available data on the theoretical and experimental study of the phonon spectra of graphite and graphene thin nanofilms [3–6, 11], one can draw an unambiguous conclusion that at frequencies exceeding the frequency, associated with weak van der Waals interaction interlayer, the phonon spectra of these objects match the phonon spectrum of a graphene monolayer. Since the connection of graphene to the substrate is also a van der Waals, then also, starting with a certain frequency ω^*, the phonon spectrum «graphene on a substrate» ceases to sense the presence of the substrate.

Figure 12.21 presented contributions to the phonon density of states being on a substrate of unlimited sample graphene atomic displacements along both directions in the plane of the graphene layer $\rho_a^{(\infty)}(\omega)+\rho_b^{(\infty)}(\omega)$ (fragment a, curve 1) and in the direction normal to the layer $\rho_c^{(\infty)}(\omega)$ (curve 2 on the same fragment). A symbol (∞) denotes the distance from the boundary, sufficient to completely neglect its influence. Curve 1 does not differ from the corresponding dependence for bi- or trigraphene [32], i.e., vibrations in plane of

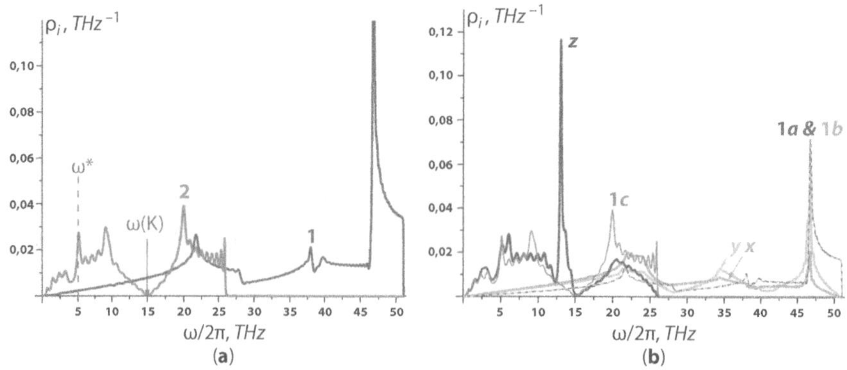

Figure 12.21 The (fragment a)—contributions to the phonon density of states graphene on a substrate from atomic displacements along the graphene layer and normal to it (respectively, the curves 1 and 2). The (fragment b)—the phonon spectral densities, corresponding to displacements along different crystallographic directions of the atom non-defect graphene (curves with number 1) and atom « peak » zigzag boundary (*x*—in the layer plane, normal to the boundary; *y*—along the boundary; *z*—normal to the plane of the layer).

the layer almost do not feel the influence of the substrate. On curve 2, the contribution of the substrate is more noticeable and concentrated in the frequency range $\omega \le \omega^* \sim 5THz$. At $\omega > \omega^*$, the function $\rho_c^{(\infty)}(\omega)$ practically coincides with the corresponding spectral density of a defect-free and unbounded graphene monolayer. We note that it is the characteristic contribution of the 3D substrate to the vibrational spectrum; it provides stability (i.e., « planar stability ») of the system under consideration.

In the frequency range $\omega \in (10, 20)$ *THz*, the spectral density $\rho_c^{(\infty)}(\omega)$ has minimum, analogous to the V-shaped singularity on the electron density. This is due to the splitting of the vibrational modes, which takes place in our system at $\omega > \omega^*$ (similarly to the splitting of phonon modes in graphite at frequencies exceeding the van Hove singularity frequency, which corresponds to the transition from closed to open isofrequency surfaces along the axis *c* [25]). Moreover, a scalar model can describe the modes, polarized along the axis *c*, on a hexagonal honeycomb lattice of graphene (similar to electrons in the strong coupling approximation). As in the electron spectrum, the frequency of this minimum corresponds to K-point first Brillouin zone, where the optical and acoustic modes are in contact, polarized along the axis *c* (see Ref. [23]).

As shown in Ref. [5] near the frequency $\omega(K) \approx 15THz$ under the influence of different types of defects can be formed quasilocalized states, enriching the phonon spectrum in a given frequency range. On the fragment, Figure 12.21b shows phonon spectral densities, corresponding to «peak» displacements located at zigzag boundary of the atom «peak» (i.e., sublattice **A**) along different crystallographic directions. Curves ***x*** and ***y*** correspond to spectral densities $\rho_x^{(0)}(\omega)$ and $\rho_y^{(0)}(\omega)$, generated by the displacement of a given atom in the plane of the graphene layer is normal to the boundary and along it, respectively (see Figure 12.17a); curve *z* corresponds to spectral densities $\rho_z^{(0)}(\omega)$, generated by the displacement perpendicular to the plane of the layer. On this spectral density is a pronounced peak near the frequency $\omega(K)$, which is analogous to peaks in electron LDOS, shown at Figures 12.19 and 12.20. With great certainty, we can assume that this peak is also due to the gap mode split off from quasiflex modes by the graphene phonon spectrum, analogous to the mode (Equation 12.56) in the electron spectrum.

This analogy is also confirmed by the character of the evolution with the distance from the boundary of the spectral densities $\rho_z^{(n)}(\omega)$ near the frequency $\omega(K)$, shown at Figure 12.22. As in the case of the electron spectrum, the spectral densities of the atoms of the sublattice **A** near $\omega(K)$ have maximum $(\rho_z^{(n)}(\omega) > \rho_z^{(\infty)}(\omega)$ $n=0,2,4)$, and for sublattice atoms **B** at given frequencies $\rho_z^{(n)}(\omega) < \rho_z^{(\infty)}(\omega)$, $n=0,2,4$.

Thus, the formation of the zigzag boundary substantially enriches the phonon spectrum in the frequency range near the frequency of contact between the acoustic and optical branches polarized normally to the plane of the graphene layer.

Thereby, forming a boundary zigzag chirality of a graphene monolayer on a dielectric substrate leads to a significant change in its electron spectrum near the Fermi level as well as the spectrum of its quasiflexural phonons in the frequency range near the value of the frequency, which corresponds to the point K its first Brillouin zone.

In quasiparticle spectra, waves are formed, which are split off from the bands of the quasicontinuous spectrum and propagate along the boundary, attenuating with distance from it. Moreover, these waves propagate only atoms of the sublattice, to which those atoms belong, in which the formation of the boundary there is breaks off one of the bonds.

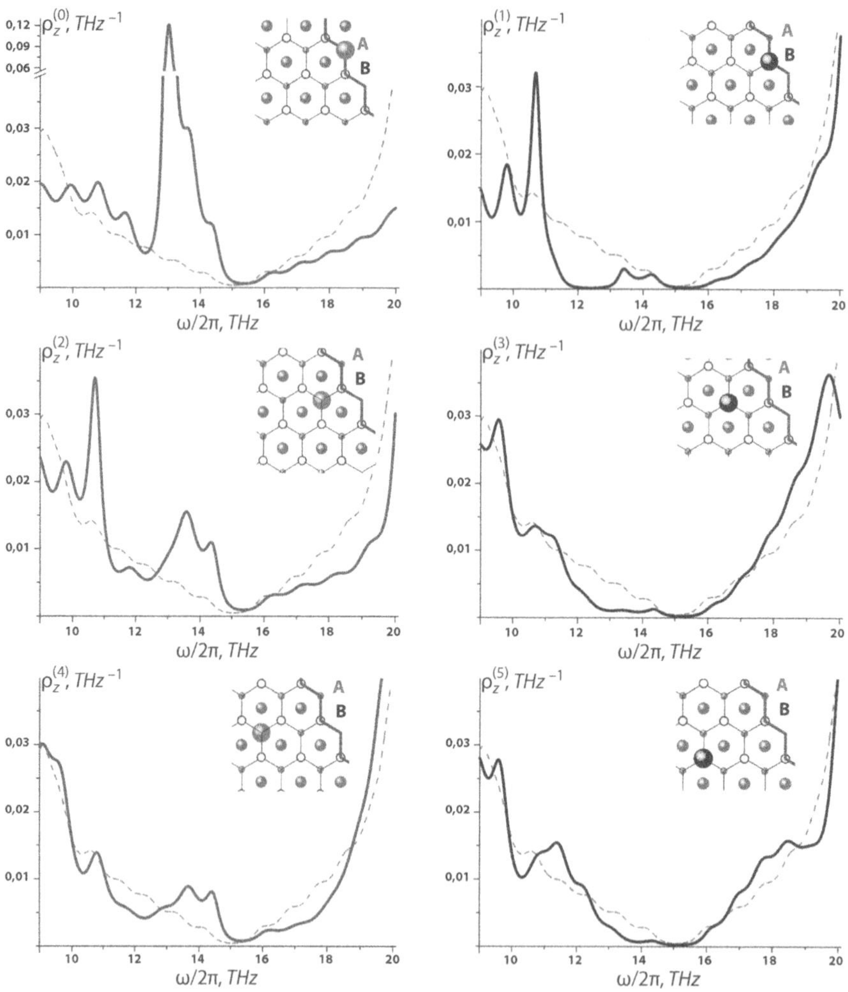

Figure 12.22 The evolution of the distance from the zigzag boundary phonon spectral densities generated atomic displacements normal to the plane of the graphene layer.

The dispersion of these waves is determined by the nature of the relaxation processes in the formation of the boundary. In the electron spectrum, the dispersion is relativistic, but corresponds to much lower values of the group velocity, than in an infinite graphene monolayer.

The split gap waves form sharp peaks at local densities of states, which substantially enrich both the electron spectrum near the Fermi level and the phonon spectrum near the point of intersection of the acoustic and optical branches, polarized perpendicular to the plane of the graphene monolayer. The special role of quasiflexural phonons in the electron–phonon interaction in layered compounds, in particular in graphene structures, is mentioned, for example, in Refs. [9] and [58]. These phonons in the given frequency domain practically do not interact with the phonons of other polarizations, and have high group velocities, which can determine their determining contribution to the electron–phonon interaction. All these factors may contribute to the formation in the graphene samples the Cooper pairs and transition of the graphene with zigzag boundary in the superconducting state.

12.3.3 The Electron Spectrum of Graphene with Point Defects

In Section 12.1, it was noted that the behavior near the Fermi level of the real part of the Green's function Re$G(\varepsilon)$ shows the high probability that under the influence of various kinds of defects, elementary excitations will be localized near this level. The singularity on the density of states at $\varepsilon = \varepsilon(K) = \varepsilon_F$ determines the behavior of the real part of the Green's function near ε_F, which for a wide class of perturbations, caused by defects, ensures the existence on the interval [$-\varepsilon(M)$, $\varepsilon(M)$] (in the model under consideration $\varepsilon(M) = J$) solutions of the Lifshitz equation (Equation 12.3), corresponding quasi-localized states, and an increase in the population of the Fermi level, necessary for the transition of the graphene nanostructures to the superconducting state. It can actually be achieved by the creation of these nanoobjects of certain defect structures.

As an illustration on Figure 12.23a, the graphical solution of the Lifshitz equation is given for the case when there is an isolated substitutional impurity in a graphene nitrogen atom. The local spectral density of such an impurity atom is calculated in Ref. [59]. For an isolated substitutional impurity differing from the atom of the main lattice by the energy values in the impurity site i = 0 ($\varepsilon_0 = \tilde{\varepsilon}_0$) and the overlap integral $J_{i0} = (1 + \eta)J$ function $S(\varepsilon, \tilde{\varepsilon}_0, \eta)$ has the form

$$S(\varepsilon, \tilde{\varepsilon}_0, \eta) = \frac{(1+\eta)^2}{\tilde{\varepsilon}_0 + \varepsilon\eta(2+\eta)}. \tag{12.69}$$

The curve $S(\varepsilon)$ for nitrogen impurity represented on Figure 12.23a according to [59] (in this case $\tilde{\varepsilon}_0 \approx \varepsilon_F - 0.525J$; $\eta \approx -0.5$). Equation 12.3, as can be seen from the figure, has a solution both on the interval [$-\varepsilon(M)$, $\varepsilon(K)$] - point $\varepsilon_{ql}^{(1)}$, and on the interval - point $\varepsilon_{ql}^{(2)}$. Calculated in [59] the local density of states (LDOS) nitrogen impurities (Figure 12.23b) have quasilocal maxima at both these intervals.

Although, because of the difference in these intervals between the imaginary part of the Green's function from zero, the locations of the quasilocal maxima are different from $\varepsilon_{ql}^{(1)}$ и $\varepsilon_{ql}^{(2)}$, the presence or absence of solutions of the Lifshitz equation on the interval [$-\varepsilon(M)$,

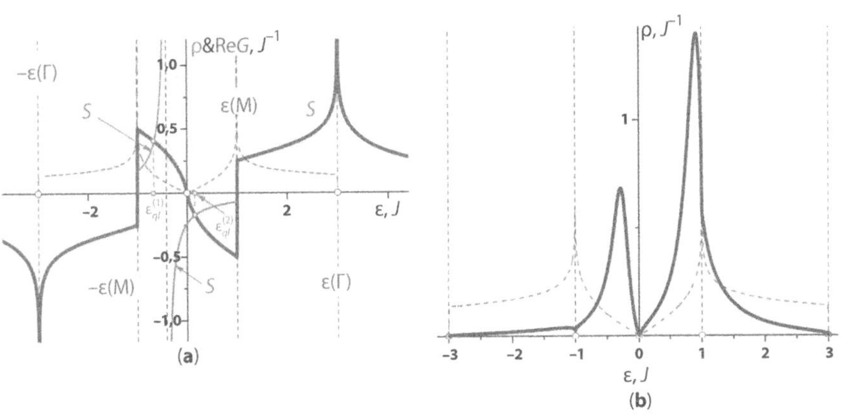

Figure 12.23 The solution of Equation 12.3 for a carbon substitution impurity with nitrogen (a) and (fragment a) and LDOS (fragment b).

ε(M)] at given defect parameters determines the presence or absence on this interval of quasilocalized states. So, also considered in Ref. [59], boron impurity ($\tilde{\varepsilon}_0 \approx \varepsilon_F + 0.525J$; $\eta \approx 0.5$) on the interval [–ε(M), ε(M)], Equation 12.3 has no solutions [5]. There are no quasilocalized states in this interval. In this case, as was shown in Ref. [60], an electron spectrum, along with the quasicontinuous part, will contain two symmetric discrete levels $\varepsilon_l^{(\pm)} \approx \pm 3.0698J$ (the band of the quasicontinuous spectrum in the case under the consideration –3*J* < ε < 3*J* <). The quantities $\varepsilon_l^{(\pm)}$ are poles of the local Green's function. The deductions at these poles (so-called the intensity of the discrete levels) $\mu_l^{(\pm)} \approx 0.139$. The area under the curve ρ(ε) n this case is less than one by the sum of the intensities of the discrete levels.

Thus, the behavior of the real part of the Green's function of graphene at –ε(M) < ε < ε(M) indeed indicates a high sensitivity of the density of states at given values of the energy to various kinds of perturbations, caused by defects and other changes in the crystal structure. In particular, the possibility of the formation of localized excitations with energies close to the Fermi energy (see, for example, Refs. [36, 59–62]). In this case, generally speaking, it is not essential whether this perturbation is degenerate, when the function $S(\varepsilon, \hat{\Lambda})$ can be written in explicit form, or not—quasi-localized states will arise (occur) by enriching the electron spectrum near the Fermi level.

Very interesting properties show the electron spectrum of graphene-containing vacancies that are an example of such a nondegenerate perturbation.

For example, interesting features of the local density of states near the Dirac point (that is, the Fermi level) on the neighbors of a single vacancy [5], and the behavior of the local density of states depends on the sublattice to which the selected node belongs.

We note at once that the possibility of practical realization and observation in a real experiment of the features of the electron spectrum of graphene with vacancies noted in these studies is far from obvious, in particular, the strong inhomogeneity of the local densities of states (LDOSs) noted in Ref. [5]. To create the vacancy in graphene (i.e., to extract a single atom), the energy is needed ~ 18–20 eV. [64], and this energy can be achieved when bombarded by ions or electrons in the plasma with energy > 86 keV. It is very likely that not one isolated vacancy is formed, but some vacancy groups.

In the strong-coupling approximation with the Hamiltonian (Equation 12.1), we assume that electron jumps within the layer are possible both between nearest neighbors $J_{ij}(a) \equiv J \approx$ 2.8 eV, and between the second neighbors $J_{ij}\left(a\sqrt{3}\right) \equiv J' \leq 0.1J$ (there, $a \approx$ 1.415 Å—the distance between the nearest neighbors in a graphene layer). The Fermi energy corresponds to the energy at the K-point of the first Brillouin zone, and for the dispersion, law we can write

$$\varepsilon(k) - \varepsilon_F = \varepsilon_0(k) \cdot \left[1 \mp J' \left|\varepsilon_0(k)\right| / J^2\right] \tag{12.70}$$

where $\varepsilon_0(\boldsymbol{k})$ is the law of the dispersion of electrons in graphene with allowance for the interaction between only the nearest neighbors (2) ($\Delta_v \equiv -\varepsilon_v(\Gamma) = 3J(1 + 3J'/J)$) and narrows the conduction band ($\Delta_c \equiv \varepsilon_c(\Gamma) = 3J \cdot (1 - 3J'/J)$).

On Figure 12.24, the electron density of states (DOS) of the ideal graphene as considering the interaction of only nearest neighbors (curve 1), and taking into account the interaction with the second neighbors (curve 3—a value *J'* is assumed to be 0.1*J*). The same dependencies, for comparison, are presented in all subsequent figures (Figures 12.25–12.28) as dashed lines.

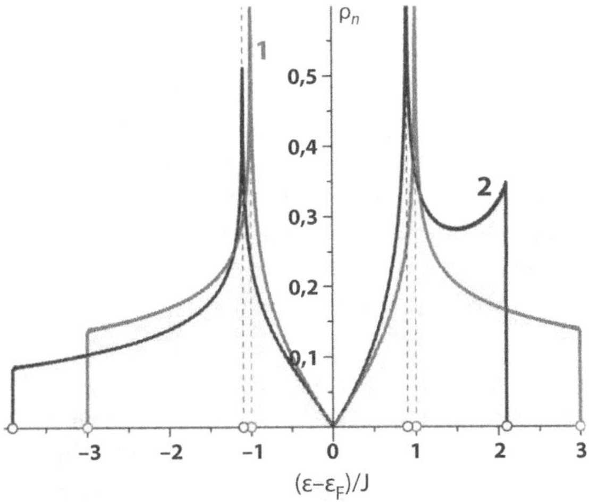

Figure 12.24 The electron densities of graphene states, for different values of interaction with the second neighbors: curve 1 corresponds to the value $J' = 0$, and curve 2 $J' = 0.1J$.

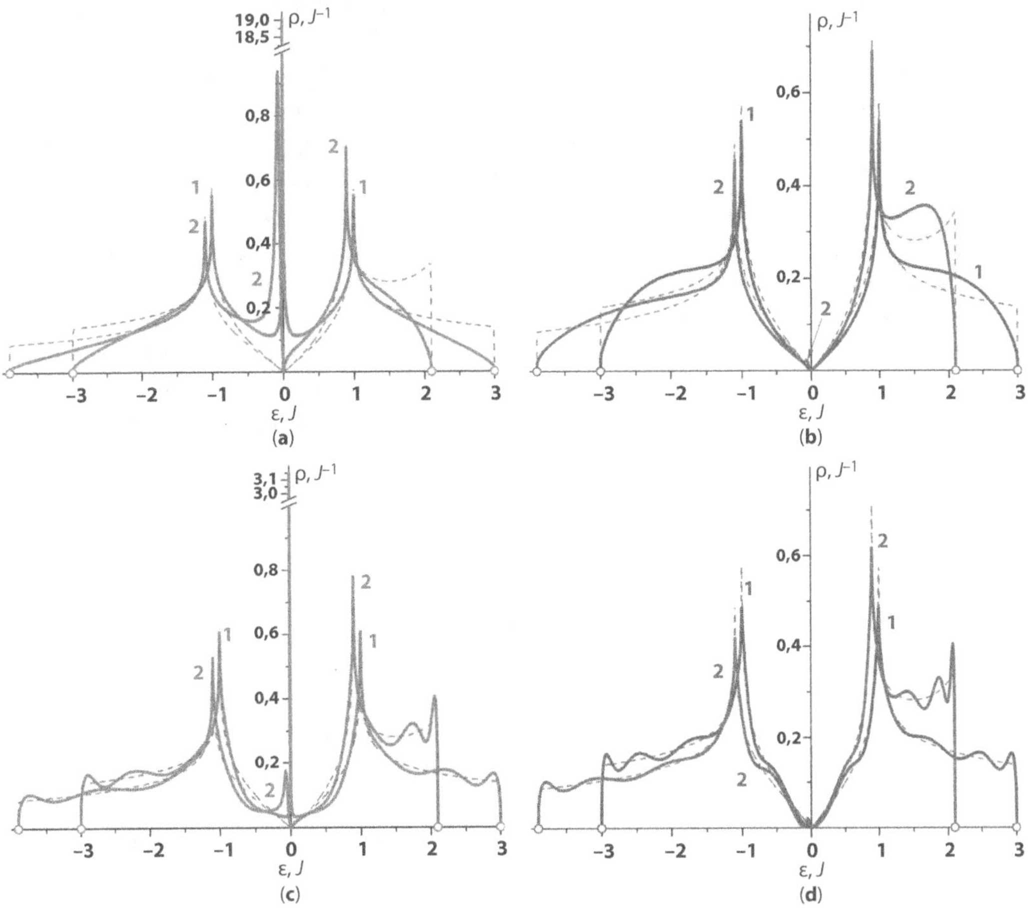

Figure 12.25 LDOS of the first, second, seventh, and tenth neighbors of an isolated vacancy (fragments a, b, c, and d, respectively).

Both DOS have V-shaped Dirac features at $\varepsilon = \varepsilon(K) = \varepsilon_F$, the slope angles of both lines (and hence the Fermi velocities) coincide. These DOS demonstrate behavior typical of two-dimensional structures: steps on the edges of the spectrum (at $\varepsilon = \varepsilon(\Gamma) = \varepsilon_F \pm 3J\,(1 \mp 3\,J'/J))$ and the logarithmic singularities at $\varepsilon = \varepsilon(M) = \varepsilon_F \pm J\cdot(1 \mp J'/J)$. If curve 1 mirror-symmetrical with respect to the line $\varepsilon = \varepsilon_F$, than curve 2 is shifted relative to this line to the low-energy region and its "center of gravity" is located in the valence band.

For a perfect graphene LDOS, each atom coincides with the total density of states. The formation in a graphene lattice of an isolated vacancy leads to the fact that LDOS of the atoms, located near the vacancy, will differ from each other. Figure 12.25 submitted to LDOS of the first, second, seventh, and tenth neighbors of an isolated vacancy.

In Refs. [62, 65, 66], the case of the interaction of only the nearest neighbors was shown, that when $\varepsilon = \varepsilon_F$ influenced by the vacancies on the total electron DOS of graphene formed sharp peak. As can be seen on Figure 12.25 (curves 1), LDOSs have sharp peaks at a given energy value only for atoms, which belong to the sublattice containing no vacancies (sublattice **A**). For atoms of the sublattice **B** in which there is the vacancy, LDOS at $\varepsilon = \varepsilon_F$ equals to zero. Moreover, the Dirac singularity inherent in an ideal system is preserved for the second vacancy neighbors, and for slightly more distant atoms near the Fermi level is formed some microgap. With further distance from the vacancy, of course, the LDOSs of all atoms tend to the DOS of an ideal graphene with a V-shaped Dirac singularity at $\varepsilon = \varepsilon_F$. Section 12.3.1 noted similar behavior of LDOS atoms near the zigzag boundary of the graphene, and it can be proven similarly (see formulas (12.51)–(12.53) and (12.64)–(12.68)).

Thus, the allowance for the interaction with the second neighbors does not eliminate, but only slightly modifies the strong inhomogeneity of the electron spectra of atoms of different sublattices, which is due to the formation of an isolated vacancy in graphene.

The presence in the system of several closely spaced vacancies can significantly affect both the form of the electron LDOS of the neighboring atoms, and the inhomogeneity of these characteristics, in particular, the population of the nearest neighborhood of the Fermi level, which is generated by a single isolated vacancy. This section contains LDOS number of atoms, adjacent to bivacancies, which educated, as two nearby vacancies (Figure 12.26), and two vacancies, which are the second neighbors of each other (Figure 12.27).

In the first case, vacancies are in both sublattices of graphene and any atom is in the sublattice containing the vacancy. This leads to the fact that, in contrast to the case of one isolated vacancy considered in the previous section, both at the total DOS and at all LDOS near $\varepsilon = \varepsilon_F$ resonance peaks are not formed (see Figure 12.26). This result is consistent with the data [19].

The inhomogeneity of the behavior of the electron local density of states is not of a qualitative nature, although the alternation of atoms, on whose LDOS near the Fermi level there is a distinctly expressed V-shaped Dirac singularity (Figure 12.26a and c) with atoms, which LDOS are close to the local densities of electronic states of ordinary semiconductors with a very small gap and the usual nonrelativistic quadratic dispersion law (Figure 12.26b and d).

Taking into account the interaction of the second neighbors does not have a significant effect on the behavior of the electron density of states near the Fermi level, but leads only to a small asymmetry of the corresponding curves.

In the case of a bivacancy formed by two vacancies, which are the second neighbors of each other, that is, the behavior of the local densities of states is in the same sublattice,

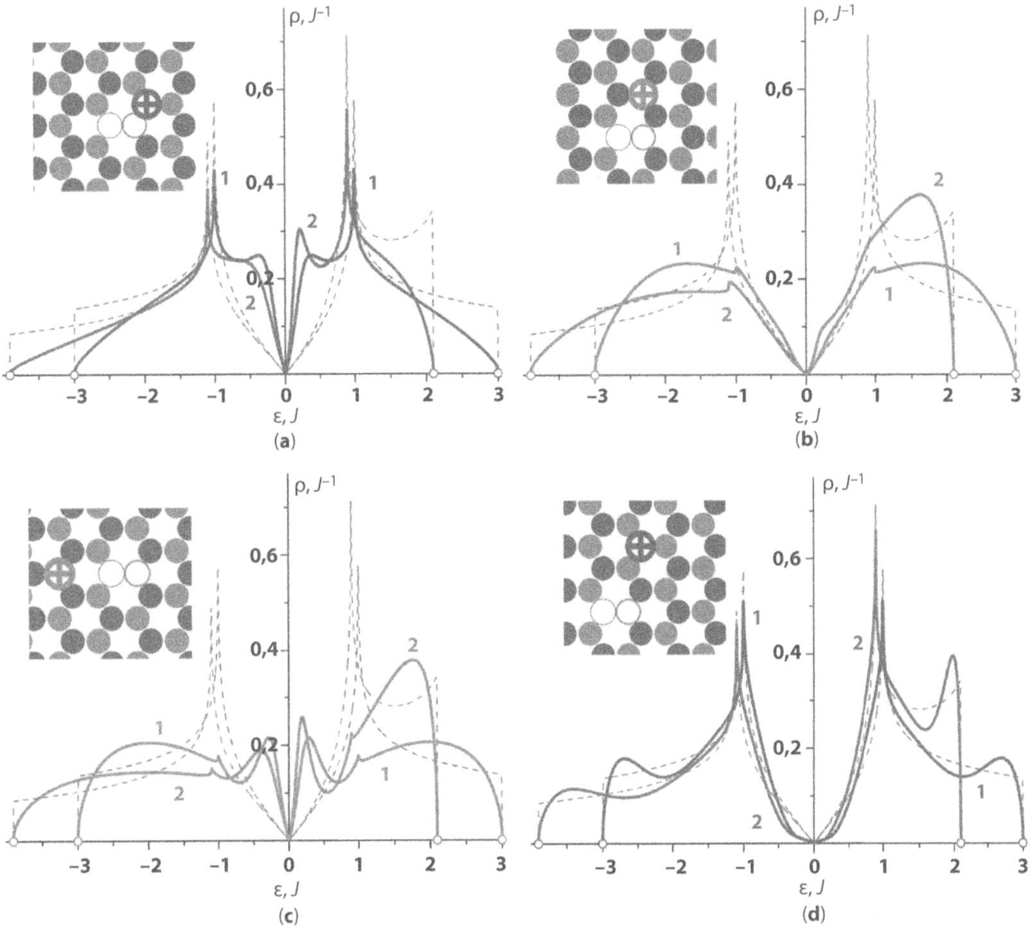

Figure 12.26 LDOS neighbors of the bivacancy, formed by two nearby vacancies. The insets of each fragment show the location of the corresponding atom. Curves 1 correspond to the value $J' = 0$, and curves 2 $J' = 0.1J$.

completely analogous to the case of one isolated vacancy, only expressed much more strongly due to increased capacity of the defect (Figure 12.27a and c). On LDOS of the sublattice atoms, which does not contains vacancies near $\varepsilon = \varepsilon_F$, formed sharp resonance peaks. Their heights exceed the heights of similar peaks on Figure 12.25 by more than two times.

As in the case of an isolated vacancy (see Figure 12.25), taking into account the interaction of the second neighbors leads to a smearing of this peak and its displacement into the energy range of the valence band, and the population of the Fermi level proper decreases significantly. Moreover, in the case of this bivacancy, it is more clearly seen that, despite the low population of the Fermi level, the corresponding LDOS demonstrates the behavior characteristic of metals, with the nonrelativistic quadratic dispersion law.

Atoms of a single sublattice with vacancies, as in the case of one isolated vacancy, do not have such a peak near their own LDOS near $\varepsilon = \varepsilon_F$ (Figure 12.5b and d). In the case $J' = 0$, the proof of this is identical to that given in the previous section. Taking into account the interaction of the second neighbors, blurring this peak and shifting it from the Fermi level, leads to the formation at the corresponding local densities of states of a small burst. In the

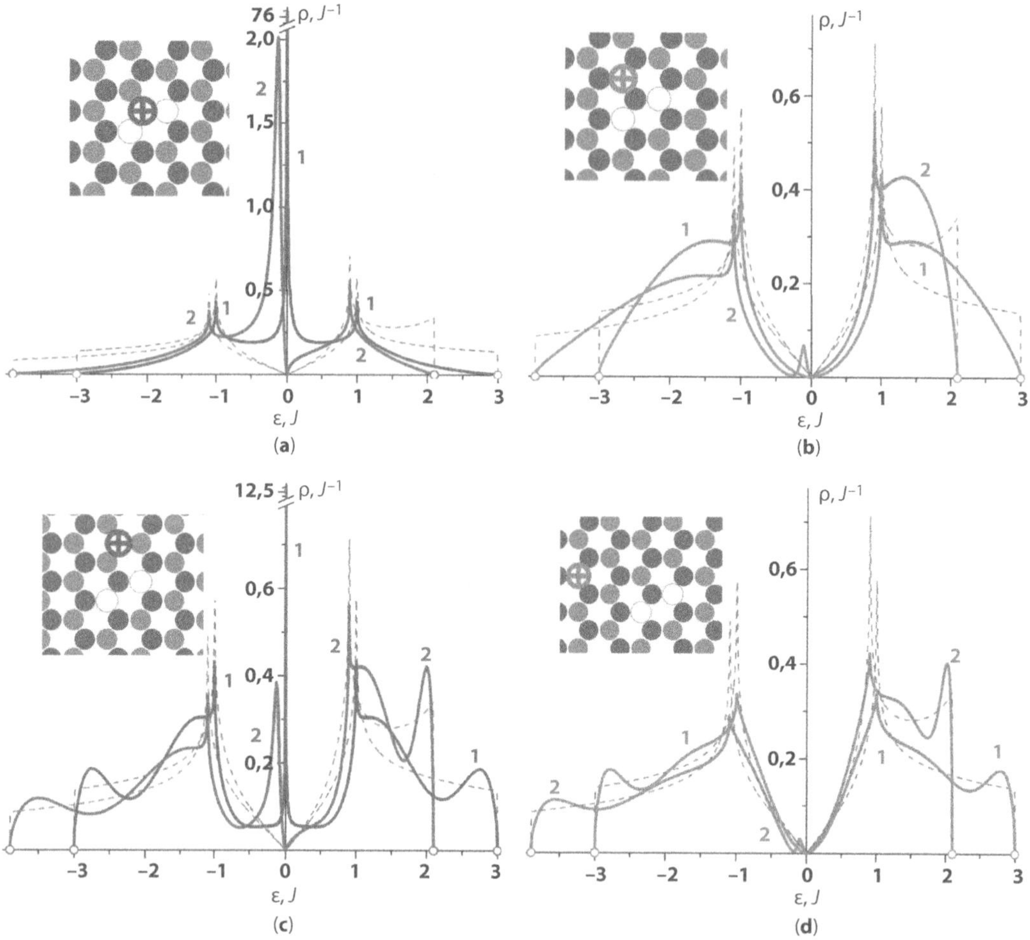

Figure 12.27 LDOS of the neighbors of the bivacancy, formed by two vacancies, which are the second neighbors of each other. All notations are similar to the notation on Figure 12.26.

case of this bivacancy, it is more noticeable than in the case of an isolated vacancy, but it is two orders of magnitude smaller than on LDOS atoms of another sublattice, which are approximately at the same distances from the defect.

Note that for this bivacancy, it is more noticeable, that when $J' = 0$, and taking into account the interaction of the second neighbors, the LDOS behavior of the sublattice atoms, which contains vacancies near the Fermi level, is characteristic of semiconductors in the quadratic nonrelativistic electron dispersion law.

Now consider the defect, which is a group of four vacancies. Some "central" atom is knocked out along with all three of its closest neighbors. The behavior of the LDOS of the different atoms (Figure 12.6) is qualitatively similar to the case of a vacancy, formed by two vacancies in one sublattice. For atoms of one sublattice with a "central," i.e., the nearest neighbors of the defect, on these characteristics at $J' = 0$, sharp resonance peaks are formed near the Fermi level at which with increasing J' blurred and shifted into the valence. The behavior of the LDOS of these atoms is characteristic for metals with a low carrier concentration and a quadratic dispersion law for electrons (Figure 12.28a and c).

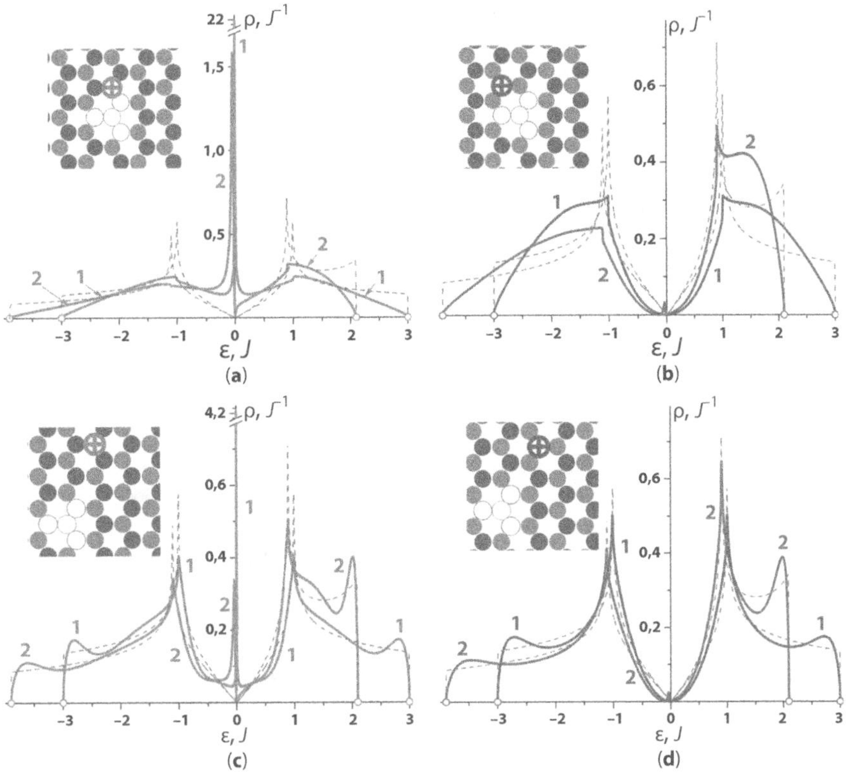

Figure 12.28 LDOS of the neighbors of the vacancy group formed by four vacancies—some "central" atom and three of its nearest neighbors are knocked out. All notations are similar to the notation on Figure 12.4.

The local density of states of the atoms of the sublattice containing the three extreme vacancies of this group demonstrates the behavior characteristic of semiconductors with a very narrow gap. Moreover, the interaction with the second neighbors leads to the formation in the valence band near the Fermi level of a small peak (burst), whose magnitude is several tens of times smaller than on LDOS atoms of another sublattice.

We also present (Figure 12.29) the evolution of the LDOS of the vacancy neighbors calculated by us, as we move away from it, taking into account the relaxation of the interaction of the nearest vacancy neighbors. When considering the interaction between all the other atoms, we confined ourselves to taking into account the nearest neighbors, but we took into account the interaction between the nearest neighbors of the vacancy, in which one of the bonds breaks off (these atoms are at each other's distance from the second neighbors). When we calculated, we considered this interaction equal to $0.1J$.

All LDOSs calculated using Jacobi matrices, the Hamiltonian (Equation 12.1) in the form of which it is represented in the basis, are obtained by orthonormalization of the sequence $\left\{\hat{\mathscr{H}}^p\Psi_0\right\}_{p=0}^{\infty}$, whereas the generating function Ψ_0 a single excitation of the atom is designated on the inset of each fragment.

It is clearly seen that on LDOS atoms of a sublattice that does not contain a vacancy, there are peaks analogous to peaks on Figure 12.2. At $\varepsilon \to -0$, these functions have a pronounced

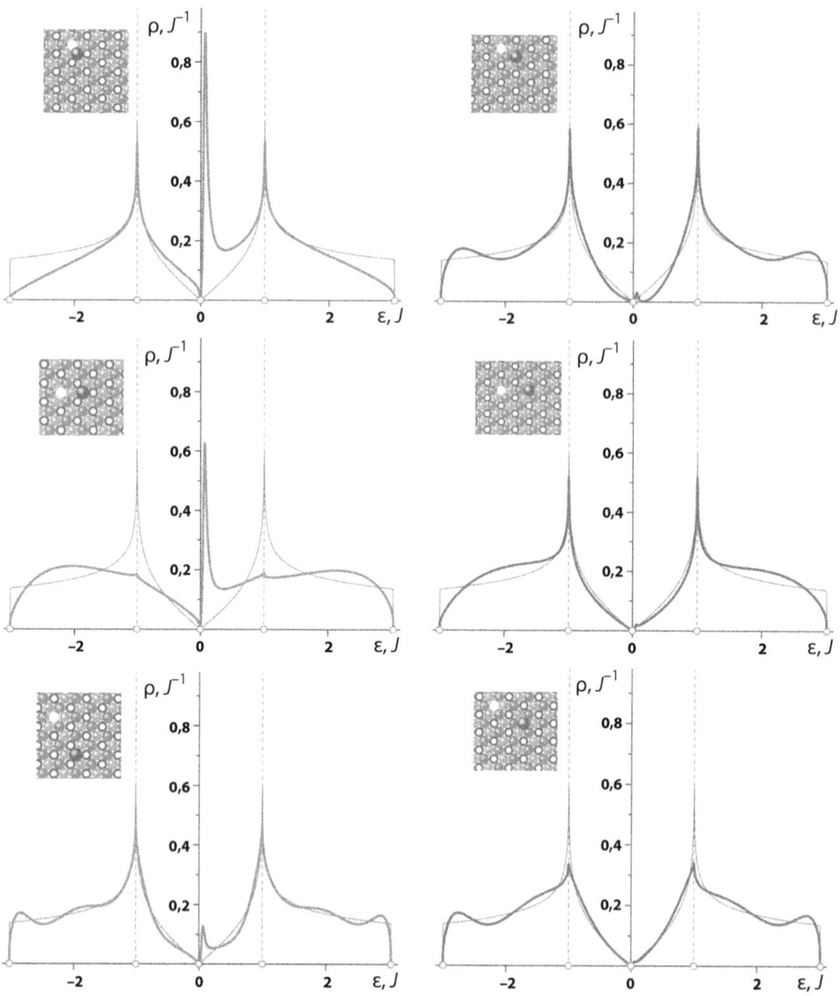

Figure 12.29 The evolution of electron LDOS neighbors search by moving away from it.

relativistic behavior. Atoms of another sublattice have a small peak at the LDOS of the second vacancy neighbors, rapidly decaying as they move away from it. This attenuation is accompanied by a "restoration" of the relativistic behavior of these dependences.

The calculation results obtained for simple models based on the strong-coupling approximation (Hamiltonian) correlate well with the results of calculations from the first principles for both single vacancies and their complexes [63].

12.3.3.1 The Defect of the "Step-Edge on the Surface" Type of a Graphene Nanofilm and Its Effect on the Electron and Phonon Spectra

In the description of the electron spectrum of bilayer graphene in the approximation of strong coupling with electron hopping only to the nearest neighboring atoms as vacancies (see, for example, Refs. [9, 10]), the boundaries of the sample are characterized by the breakage of one or more bonds. Therefore, perturbations introduced by such defects into the electron spectrum should be in many respects similar. In this section, for the same bigraphene model (when describing the interlayer electron hopping, only hops between the nearest atoms of the sublattices AI and AII into account) will be analyzed LDOS atoms, located near the step-edge on the surface of the bigraphene, that is, the interface «bigraphene»-«graphene». When calculating both the electron and phonon spectra, in order to avoid the appearance of additional parameters, we neglect the finite dimensions of the whole sample and consider the system, where half-planes filled with bigraphene and graphene are considered. Before considering the electron properties of a given structure, one should find out the very possibility of its existence and determine the temperature intervals for the stability of such a system.

12.3.3.2 Phonon Spectrum and Root-Mean-Square Amplitudes (rms) of Vibrations of Step Atoms on the Surface of Carbon Nanofilms

Unlike a bigraphene, the bigraphene–graphene structure in the absence of a substrate cannot exist as a flat structure of sufficiently large sizes. Rather the out-of-plane phonon displacements should be responsible for observing the phonon spectrum and vibrational characteristics of this system. Though it was demonstrated before [67] that bigraphene and graphene keep plane geometry at temperatures much above the ambient one, the stability limits of step-edge atoms with dangling bonds remains unexplored. For straightforward estimation of the stability limits for step-edge atoms, the structure of the type "trigraphene step-edge" can be chosen, when another graphene monolayer represents the substrate. It is absolutely clear that real substrates of macroscopic thickness would provide even less values of mean-square amplitudes of atomic displacements and, eventually, the broader temperature ranges of structure stability.

Phonon spectrum and vibrational properties of the boundary are strongly dependent on its configuration. Those of two types, "*armchair*" (Figure 12.1c) и "*zigzag*" (Figure 12.1 d–e)), are considered. For the latter, "*zigzag*" on two cases are studied, either with a single dangling bond on the atom pertained to lattice A(●) (Figure 12.1d), or that for sublattice B(○) (see Figure 12.30).

The phonon densities of state, originated from displacements in c-axis direction of the atoms pertained to both sublattices of bulk graphite, and moreover, those of carbon nanofilms are featured by a deep minimum at $\omega = \omega(K)$ (see Figure 12.31, curves 1 and 2, respectively). This minimum resembles the Dirac V-singularity in electron density of graphene and, similarly, is responsible for imperfection-induced quasi-local states with frequencies close to $\omega(K)$ [5]. Note that, namely, the polarized normal to graphene layers (quasiflexural) vibrations [9] should play a crucial role in, say, the Cooper pairing at superconducting transition.

In Figure 12.31, curve 3 is the phonon spectral density, set by normal to the layer displacement of the boundary atom of step-edge "*zigzag*" A. The quasi-local maximum is clearly

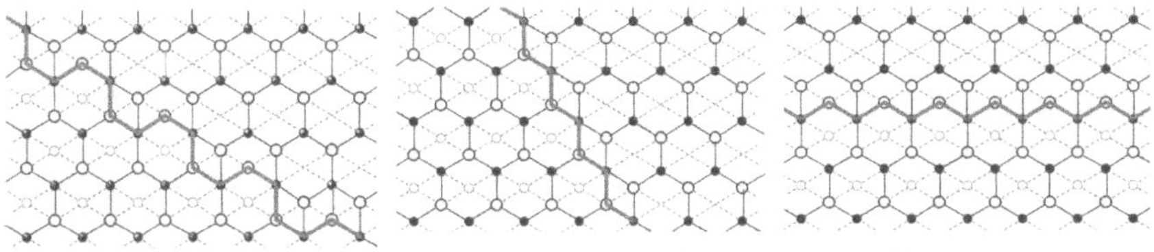

Figure 12.30 The step-edge configurations under study.

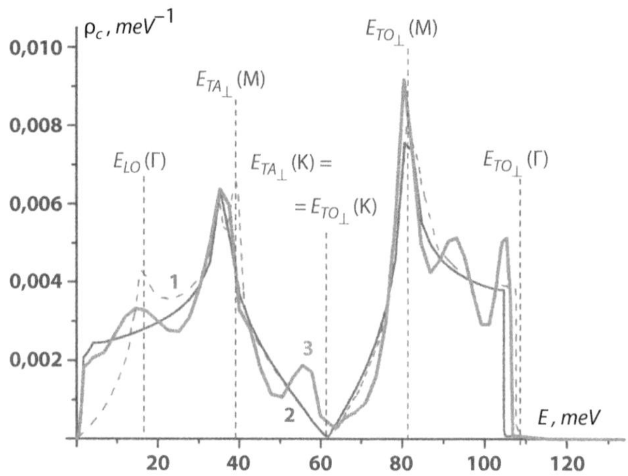

Figure 12.31 Spectral densities generated by displacement along the crystallographic axis *c* for the atoms of the graphite—curve 1, clean bigraphene—curve 2, and the atom in step-edge *"zigzag"A*—curve 3.

seen in a frequency range slightly below $E(\mathrm{K}) \equiv \frac{\hbar\omega(\mathrm{K})}{e}$ (*e* is the electron charge), as well as a pronounced growth of the number of phonon states $E = E(\mathrm{K})$.

The temperature dependence of *rmd* displacements of atom *s* in a crystallographic direction *i* relates to corresponding spectral density (Equation 12.35). Calculations of root-mean-square displacement (*rmd*) amplitudes of atomic displacements for atoms of free-standing epitaxial bigraphene, the trigraphene here, and their interfaces of various configuration (see Figure 12.30) shown on Figure 12.9. It can be seen that for all the cases considered, the root-mean-square amplitudes of atomic displacements (*rmd*) in the plane of the layers *ab* are changed very slightly and remain essentially the same as for the bulk graphite. At the same time, *rmd* along the *c* axis shows a pronounced growth with decreasing number of layers (in Figure 12.2*c* for bigraphene; 3*c* и 3*c*′ to the extreme and middle layers of the trigraphene, respectively; ∞c for massive graphite). The *rmd* of the step-edge atoms lie between the curves, as can be seen from the figure, the shifts of the step atoms lie between the curves 3*c* and 2*c* and even at the temperatures above the ambient one do not reach the value $\infty c(T = 3000K)$ values of *rmd* of the bulk graphite at temperature 3000 K (the melting temperature of graphite is ~4000 K).

Therefore, the step-edge in epitaxial bigraphene with the most rigid condition of the trigraphene-surface configuration meets the stability requirements for further investigation of electron spectrum.

12.3.3.3 Local Electron Density of States of Atoms Located Near the Step-Edge

It was shown [5] that imperfections favor the occurrence of quasilocal states in electron spectra of carbon nanofilm. The presence of vacancies in graphite causes the appearance of sharp resonance in electron density of states near $\varepsilon = \varepsilon_F$, and in Ref. [67], the similar effect was predicted for bigraphene. Similar to vacancy, the considered imperfection is due to breaking the atomic bonds, and the similar effect on electron spectrum can be expected here, for at least particular configurations.

Figure 12.32 presents electron LDOS for the atoms, located near bigraphene step-edge of the *"armchair"* type a. It is seen that the most pronounced LDOS transformations occur

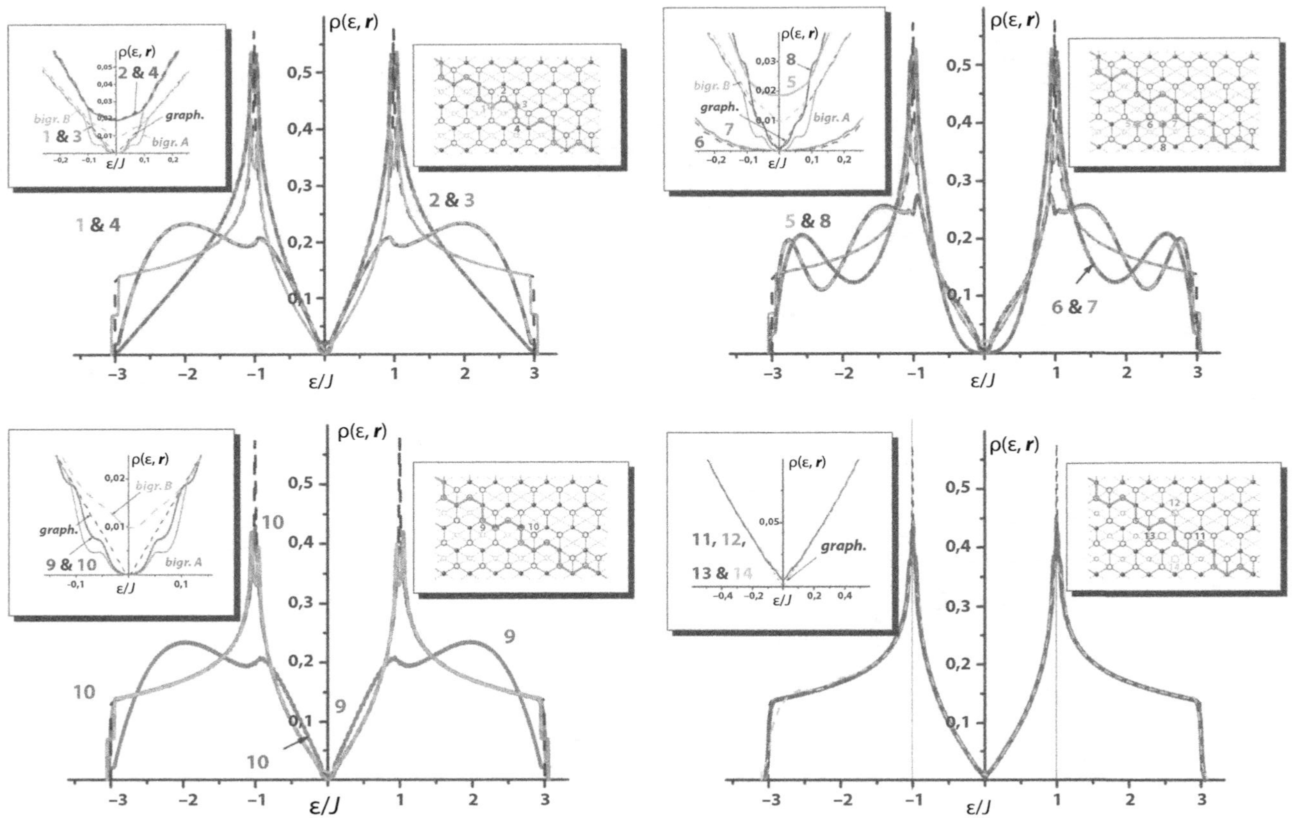

Figure 12.32 Atomic LDOS for the *"armchair"* step at the surface of bigraphene. The upper-right part of each panel demonstrates the corresponding atoms distribution nearby «step-edge» (labels and color of the atoms are the same for their LDOS); at the left hand the magnified portions of corresponding LDOS are presented near $\varepsilon = \varepsilon_F$.

for the atoms located in the plane of the edge (layer 1 on the top plane step). The LDOS of the atoms on the bottom plane approaches rather rapidly that of graphene or bigraphene, from the either occupied side of the step. In the case of *"armchair"* none of the LDOSs reveal resonances near $\varepsilon = \varepsilon_F$.

The LDOS behavior near $\varepsilon = \varepsilon_F$ for atoms of sublattice *B*, located at the boundary (upper left fragment, brown and blue curves), indicates higher, than for clean bigraphene (orange curve in inset), value of effective electron mass.

An even more intriguing behavior is observed for LDOS of atoms, located near the step-edge *"zigzag"A*, presented by Figure 12.33. The LDOS of atoms from sublattices *A*I (with a single dangling bond due to step formation), of those from *B*II reveal sharp resonances at $\varepsilon = \varepsilon_F$ (the heights of atoms *A*I, for several orders of magnitude exceed those of *B*II). The maximum heights are decreased quite slowly with going away from the boundary inside bigraphene, and disappear immediately at the «graphene» side of step-edge (layer II).

At the "bigraphene" side of step the Fermi-level of atoms belonging to sublattices *A*II and *B*I appears in the gap of width ~*J*′. At the «graphene» side of step-edge, the LDOS corresponds to those of ideal bigraphene, but their tendency toward LDOS of graphene is manifested immediately apart from the step-edge.

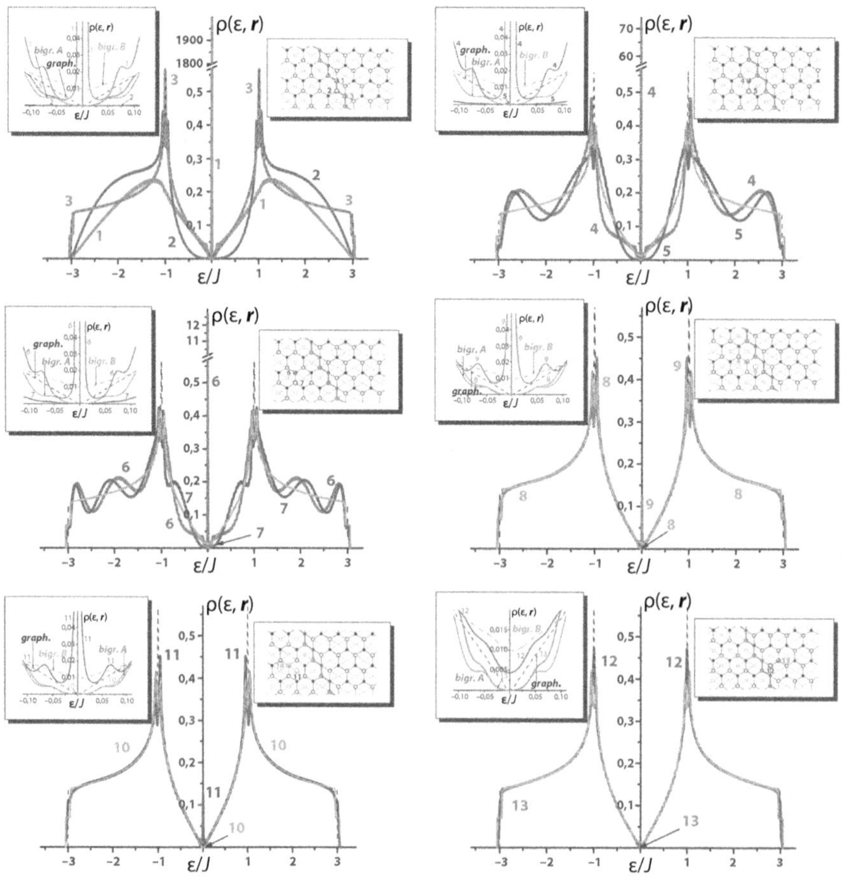

Figure 12.33 Atomic LDOS for the *"zig-zag"* step at the surface of bigraphene *A* (the same designations as in Figure 12.32).

The same behavior is manifested by LDOS in the case of step-edge *"zigzag"-B*. Similar to the previous case, the somewhat weaker resonance should be found in LDOS $\rho_{AI}(\varepsilon)$ and $\rho_{BII}(\varepsilon)$, though in this case, the bonds are broken on atoms *BI*. Apparently the evolution of LDOS for the given orientation of the boundary is most strongly influenced by the choice of the model.

Thus, the step-edge on the surface, which is located on some dielectric substrate of bigraphene, is a stable nanoformation with a number of interesting properties. So, for the step-edge *"zigzag"-A* the local densities of electronic states corresponding to sublattice atoms *AI* and *BII* have sharp resonant maxima at energies near the Fermi level. The corresponding heights of the maxima decay slowly enough from the boundary toward the bigraphene and immediately disappear on the graphene monolayer (after the step-edge).

12.3.3.4 The Phonon Spectrum of Graphite Intercalated by Metal

The graphite structures intercalated by metals are interesting in that the temperature of the superconducting transition T_c for such compounds essentially depends on the type of intercalant. For intercalated graphite by Yb – C_6Yb, they transition into the superconducting state at the temperatures 6,5 K, and for C_6Ca – 11,5 K [68, 69].

In the formation of a superconducting state, the main role is played by the electron–phonon interaction and, therefore, changes T_c in these compounds should be determined mainly by the features of their phonon spectrum. Electron spectrum C_6Ca and C_6Yb should not differ qualitatively, although the structure of these compounds is different from each other. In C_6Yb, metals form an HCP structure and in the unit cell of this compound contains 14 atoms, while in C_6Ca, metals form an FCC lattice and an elementary cell consists of seven atoms. However, the density of states in HCP and FCC of the lattices of these densely packed differ slightly from one another (at identical values of force constants or overlap integrals, the difference is manifested, starting with the fourth moment). The overlap integrals of metal–metal and metal–carbon in this case should differ little (and calcium and ytterbium belong to the same group in the Periodic Table of Elements), and therefore a rather significant difference in the electronic spectra for the compounds under consideration seems unlikely.

The presence in a phonon spectrum of quasiflexural oscillations of a graphite singularity, analogous to the Dirac singularity in the electronic spectrum of graphene, can cause the formation at maxima close to the frequency of this singularity of maxima on the phonon density of states analogous to quasilocal oscillations, often arising in the low-frequency region of the quasicontinuous spectrum of various lattices under the influence of heavy or weakly bound impurities, and has by now been well studied (see, for example, Ref. [19]).

Data on acoustic, optical, and other properties of C_6Ca and C_6Yb, which would allow us to reconstruct, as was done in the first section of this paper, the parameters of interatomic interaction, are not available to date. This forces to make some assumptions, which, however, should not be qualitatively and quantitatively even appreciably affect the behavior of the considered spectral characteristics.

These structures consist of graphene monolayers, between which lie 2D triangular metal lattices with period $a\sqrt{3}$. The atoms of both sublattices of the graphene monolayers are situated one under the other. The lattice period in a direction perpendicular to the layers in both compounds is $c' \approx 4{,}5$ Å, see, for example, Refs. [68, 69].

We neglect the interaction of carbon atoms and the metal in different layers, and the interaction of metallic atoms located in one layer is considered to be purely central, that is, the interaction of the matrix of force constants describing this interaction has the form (2.2) at $\beta_z(\Delta) = \beta_x(\Delta) = 0$, where $\Delta = a\sqrt{3}$. Value $\alpha(a\sqrt{3})$ can be determined from data on the Young's modulus of polycrystals Ca and Yb, from which it is known that $E_{Ca} \approx 2{,}6{\cdot}10^{10}$ Pa и $E_{Yb} \approx 1{,}815{\cdot}10^{10}$ Pa. Value is $\alpha(a\sqrt{3})$ in monolayers of calcium $\approx 4.0N/m$, and in ytterbium monolayers $\approx 2.75\ N/m$.

The distance between the metal and carbon atoms closest to each other is $r_{\text{C-Me}} \equiv \sqrt{\left(\frac{c'}{2}\right)^2 + \frac{a^2}{3}} \approx 2{,}66$ Å, i.e., greater than between the second ($a \approx 2{,}45$ Å), but less than between the third ($2a\sqrt{3} \approx 2{,}83$ Å) neighbors in graphene monolayers. Consequently, the potential describing this interaction can likewise naturally be assumed to be an isotropic pair potential, i.e., $\beta_z(r_{\text{C-Me}}) = \beta_x(r_{\text{C-Me}}) = \beta(r_{\text{C-Me}})$.

Since intercalation does not change an interatomic distances in the graphene monolayers, the force constants describing the corresponding interatomic interactions likewise do not change, and $\beta(r_{\text{C-Me}})$ can be found from the condition $C_{13} = C_{31}$, and in the present case assumes the form

$$\beta_z\left(\frac{a}{\sqrt{3}}\right) + 6\beta(a) + 4\beta\left(\frac{2a}{\sqrt{3}}\right) = 2\left[\left(\frac{c'}{2a}\right)^2 - 13\right]\beta\left(r_{\text{C-Me}}\right), \qquad (12.71)$$

whence, like for the interaction of carbon with calcium and for the interaction of carbon with ytterbium $\beta(r_{\text{C-Me}}) \approx 3.096\ N/m$.

Experimental data to determine the force constant characterizing the interaction between the central metal atoms, and carbon does not exist. Therefore, it is assumed on the basis of the distance between the carbon atoms and the metal intercalated into graphite that the value of this quantity will lie in the range from $\alpha(r_{\text{C-Me}}) \approx 19.647\ N/m$ to $\alpha(a) \approx 50.4759\ N/m$. In the current section, three variants of the values of this quantity: $\alpha(r_{\text{C-Me}}) =$ 30, 40, and 50 N/m.

Figure 12.34 displays the frequency dependences of the partial contributions to the phonon density of states from the displacements of the atoms of carbon and metal, intercalated in the graphite, in a direction perpendicular to the layers. The area under the curves corresponding to the partial contribution of the intercalated metal atoms is marked by hatching. The left fragments of this figure correspond to the compound C_6Ca, right – C_6Yb, and the fragments from top to bottom correspond to different values of the quantity $\alpha(r_{\text{C-Me}})\sim$ 30, 40 and 50 N/m, respectively. In the compound C_6Ca in the partial contributions of the intercalating metal and carbon seen, the interaction between carbon and the intercalant increases these peaks shift to the center of the frequency interval $[\omega_{TA\perp}(\text{M}), \omega_{\text{TO}\perp}(\text{M})]$ on the corresponding characteristic of pure graphite (see Figure 12.7). This displacement enriches with phonons the frequency interval near the K point of the first Brillouin zone through which the Fermi level of the electrons in graphite passes.

In the compound C_6Yb (ytterbium is more than four times heavier than calcium), these resonance peaks have lower frequencies, and one can talk about an appreciable increase of the number of phonons in the interval near the K point of the first Brillouin zone only for $\alpha(r_{\text{C-Me}}) = 30$, 40, and 50 N/m. The foregoing correlates well with the difference in the

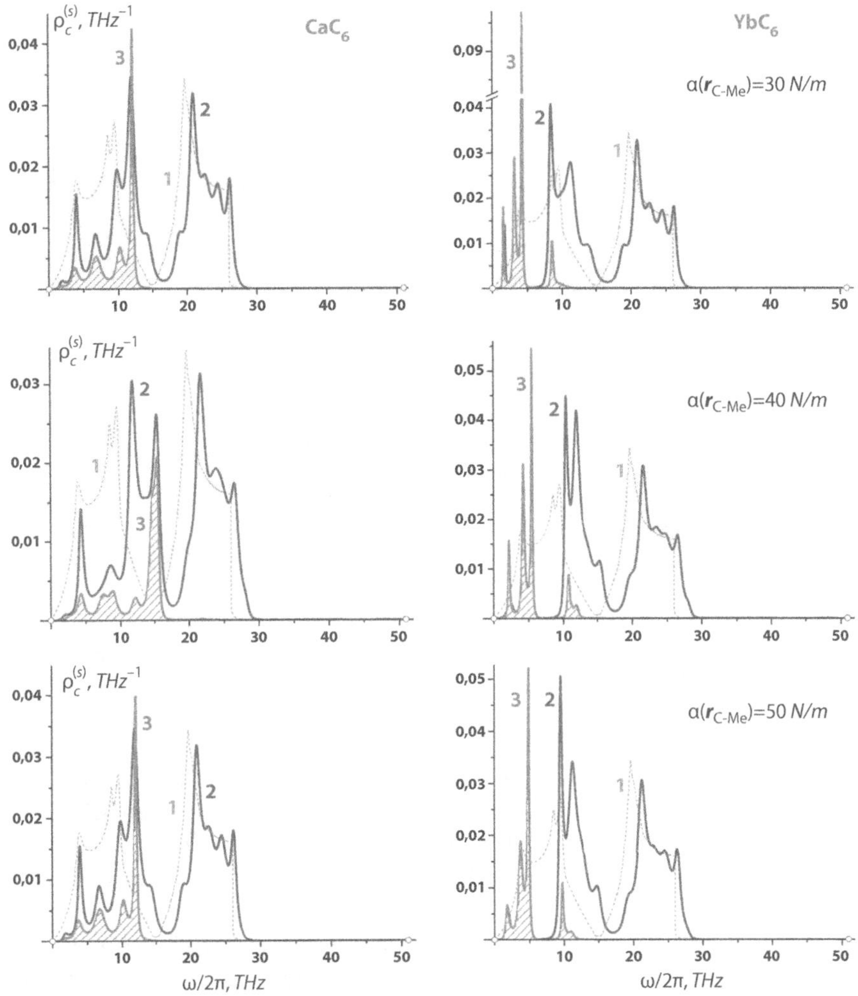

Figure 12.34 Partial contributions to the phonon density of states of intercalated graphite from displacements along the axis c of carbon (curve 2) and metal atoms (curve 3, areas under the curves are hatched) The dashed lines in all panels (curves 1) correspond to pure graphite.

superconducting transition temperatures for C_6Ca and C_6Yb. The temperature of the superconducting transition in C_6Ca is almost 1.8 times higher than in C_6Yb.

Figure 12.35 for intercalated compounds C_6Ca (left) and C_6Yb (right) shows the total phonon density of states and partial contributions to these quantities from the intercalated metal atoms (upper fragments, areas under the curve $\rho^{(Me)}(\omega)$ are hatched).

The lower fragments of this figure represent the partial contributions to the phonon state densities from all displacements of the carbon atoms $\rho^{(C)}(\omega)$ and from the displacements of carbon atoms along the axis c $\rho_c^{(C)}(\omega)$. For comparison on all fragments on Figure 12.35, the dashed line shows the phonon density of states of pure graphite. It can be seen that the metallic intercalation appreciably reconstructs the phonon spectrum not only in the low-frequency region corresponding to the bands of the quasi-continuous spectrum of intercalating metals, but also at a much higher frequency interval $[\omega_{TA\perp}(M), \omega_{TO\perp}(M)]$.

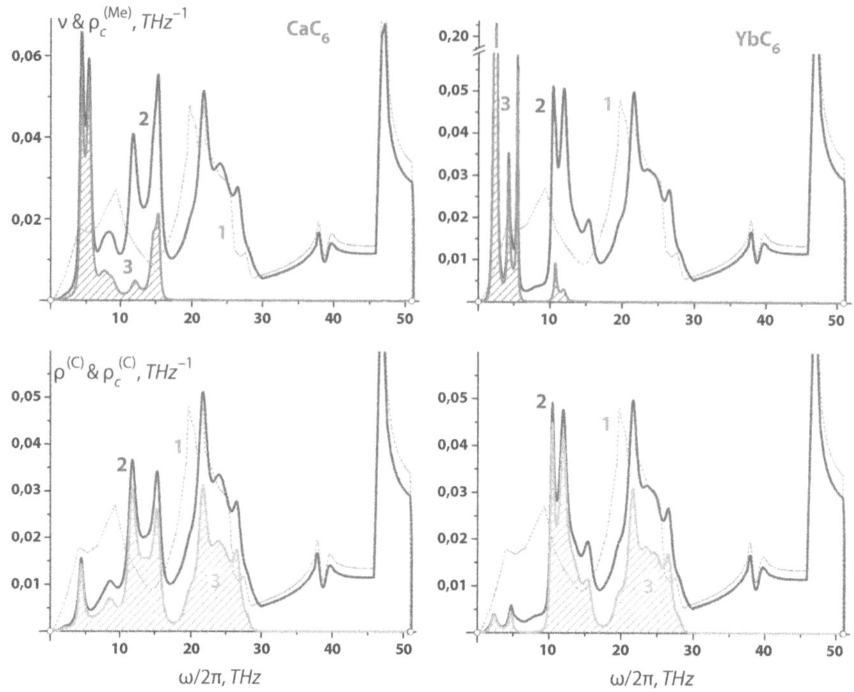

Figure 12.35 The upper fragments are the phonon density of states of intercalated graphite and the contribution to them from the displacements of the metal atoms (the areas under the curves are hatched). The lower fragments are the partial contributions to the phonon state densities from all displacements of the carbon atoms and from the displacements of these atoms to the axis *c* (the areas under the curves are hatched). Parameter $\alpha(r_{C\text{-}Me}) = 50\ N/m$. The dashed lines on all fragments are the phonon density of states of pure graphite.

Moreover, the presence of an intercalant significantly affects the vibrational spectrum of carbon atoms, mainly on vibrations polarized along the axis *c* (in a direction perpendicular to the graphene layers). In the case of intercalation by lighter calcium atoms, this effect is much more pronounced.

We note that the features of the phonon spectra and the vibrational characteristics of graphene systems are an outgrowth of the manifestations of the strong anisotropy of interatomic interactions inherent in such structures. Such a strong anisotropy is inherent in many compounds and causes a whole series of interesting and important features in their electronic (in particular, superconducting), magnetic, and other characteristics, and also affects the nature of phase transformations and critical phenomena in many objects [70]. The unusual behavior of the phonon subsystem in graphite in combination with metallic intercalated layers can significantly affect the electronic properties, in particular, the superconducting transition temperature T_c. Some qualitative considerations in this regard are given in Ref. [5].

It can be stated unequivocally that in order to increase T_c, it is necessary to have high phonon frequencies, large values of the interaction constant of electrons with these phonons, and a significant density of electron states on the Fermi surface. Such properties are in particular of compounds of metals with light elements: hydrides, borides, and also carbides and nitrides, since they have high frequencies associated with vibrations of light atoms H,

B, C, N. For example, according to [71–73], in MgB_2, electrons strongly interact only with two flexural modes.

This assumption can be confirmed by the fact that the transition to the superconducting state is observed in a number of compounds obtained by intercalation of graphite with various metals, which leads to analogous changes in the phonon spectrum [3, 4, 74, 75]. So, in Ref. [74], on the phonon density of the graphite intercalant CaC_6 neutron diffraction revealed a significant increase (compared with pure graphite) of the number of phonons with energies near E(K) and a good agreement of the experimental data with theoretical calculations from the first principles is shown [76, 77]. Similar results were obtained in Ref. [5] by us within the framework of classical lattice dynamics. On Figure 12.36, the results of works are compared in Refs. [74] and [5]. The agreement of the presented results should be considered quite satisfactory, especially if we consider the intercalant as an ideal crystal structure, which inevitably makes the peaks at spectral densities more abrupt.

In favor of an important role in the formation of a superconducting state is the fact that, as shown in [5] (and agrees with experiment [75]), intercalation of graphite with ytterbium considerably less enriches the region of the phonon spectrum near E(K). The superconducting transition temperature in CaC_6 is approximately equal to 11.5 K, and in YbC_6—6.5 K.

We also note that when intercalating graphite with lithium [78], whose mass is more, less than half the mass of the carbon atom, the superconducting transition temperature of the whole 1.9 K. The vibrations of such a light atom slightly rearrange the phonon spectrum of graphite for energies near $E = E(\mathrm{K})$, and concentrated in higher frequencies $E \geq E^{\Gamma}_{TO\perp}$.

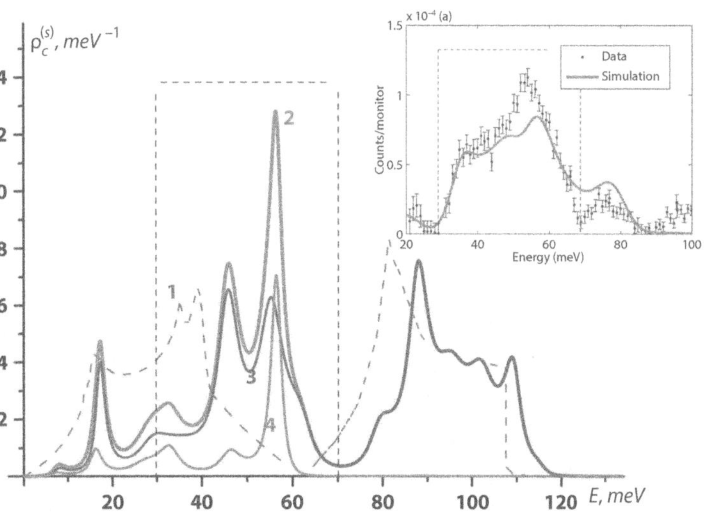

Figure 12.36 Partial contributions of vibrations along the axis c in phonon DOS pure (curve 1) and calcium-intercalated graphite (curve 2), and also the contributions to this dependence on the vibrations of the carbon atoms and the intercalant (curves 3 and 4, respectively). Inset: the symbols—the results of neutron diffraction measurements of the phonon density CaC_6 [74], the solid line made in this work the simulated spectrum reproduces.

12.4 Conclusion

Electron spectra of graphene, as well as the carbon nanofilms synthesized on its basis, are featured by sharp minimum at the energy value, which corresponds to K-point of first Brillouin zone and equal to Fermi energy in our case. In contrast to electron spectrum of graphene, where this value corresponds to Dirac singularity, in graphite and nanofilms composed of several, at least two, graphene layers, this minimum is analytical and effective mass is finite.

The similar minimum at $\omega = \omega(K)$ is also typical of a spectrum of atomic vibrations, polarized at the normal to graphene single layers. It is caused by two-dimensional behavior at frequencies above the van Hove frequencies, which correspond for each polarization to a transition from closed to open isofrequency surfaces along the axis normal to graphene layers. In this frequency range, the vibrations polarized in and perpendicular to the plane of the layers do not, for all practical purposes, interact with one another and can be studied independently.

The quasiflexural modes with a quadratic dependence of the frequency on the quasi-wave vector, which were predicted by Lifshitz in 1952 for the phonon spectra of layered and chain structures and detected in 1972 in experiments on inelastic scattering of neutrons on graphite, are manifested in the behavior of the temperature dependences of low-temperature heat capacity of bigraphene and graphene nanotubes.

The rectilinear temperature dependence of heat capacity, inherent in many layered compounds including graphite, is not caused by the quadratic dispersion of quasiflexural modes over the broad temperature range (from about 40 K and above). In order for the quadratic dispersion of flexural phonons to manifest in the heat capacity, there must be a sufficiently extended range in the phonon spectrum of the layered or chain crystal, along which the spectrum has a low-dimensional character and the dispersion law of the transverse phonon mode, polarized along the direction of weak coupling, can still be considered quadratic.

Graphene nanotubes cannot be described by the model built by Lifshitz for chain structures; however, their low temperature heat capacity exhibits both flexural vibrations of the tube as a whole one-dimensional formation (at $T \leq 1$ K) and quasiflexural waves (at $3\text{ K} \leq T \leq 7\text{K}$), propagating along the surface of the tube.

It has been shown that negative thermal expansion (or, more generally, nonmonotonicity in the temperature dependence of the coefficient of linear thermal expansion) is intrinsic to structures with a strong anisotropy in their interatomic interactions. On a microscopic level, the methods of crystal lattice dynamics have been used to explain the nature of a force that compresses a crystal with rising temperature as it expands rapidly in an orthogonal direction. The relationship of this force to the temperature derivatives of the mean square atomic displacements in different directions has been found. It has been found that the reason for the negative expansion of graphene nanotubes is the high amplitudes of their torsional vibrations, in which the quasi-one-dimensionality of the crystal structure of these systems shows up most strongly.

The V-shaped feature, which inheres in the electronic density of states of graphene as well as the electronic and phonon spectra results, in the presence of a wide class of defects, in the formation of sharp resonance peaks whose energies will lie near the minimum of this feature, which corresponds to the K-point in the first Brillouin zone. The fact that the energies of these peaks lie at the center of the quasicontinuous spectrum and not near its edges, just as in the case of "ordinary" quasilocal states, can result in oscillations of the dependences of their values on the distance to the defect.

In electron spectra of graphene with zigzag boundary, there appear the waves, split of the bands of quasi-continuous spectrum, which propagate along the boundary and decay with a distance from it. They, moreover, propagate only via the atoms of sublattice, which contains atoms with dangling bond, appeared with a formation of boundary. Dispersion of these waves is determined by character of relaxation processes during formation of the boundary. Dispersion in electron spectrum is relativistic, but corresponds to the significantly less values of group velocity, compared to infinite monolayer of graphene. The similar behavior is demonstrated by electron spectra of graphene with isolated vacancies and vacancy groups. The split gap waves lead to a formation on local densities of state of the sharp resonances, which enrich significantly electron spectrum near Fermi level, as well as phonon spectrum near the point of intersection of acoustic and optical branches, polarized normally to the plane of graphene monolayer.

The "step-edge on the surface" defect is stable even for the "trigraphene–bigraphene" system and leads to an increase in the number of both electronic and phonon states (for quasiflex phonons) near the energy corresponding to the K-point of the first Brillouin zone. In this case, the formation of quasilocalized states is possible in this energy region. A similar increase in the number of electronic and phonon states occurs in graphite intercalated with metals, where a superconducting transition is observed and contributes to an increase in the temperature of this transition.

The quasiflexural phonons within a considered frequency range do not practically interact with differently polarized phonons, and also possess high group velocities, which there dominant contribution to electron–phonon coupling. The presented results demonstrate a possibility to facilitate superconductivity in a graphene matter by a controlled creation of defects like vacancy or zigzag boundary, which distort the atomic bonding in a particular sublattice of graphene monolayer.

References

1. Wallace, P.R., The band theory of graphite. *Phys. Rev.*, 71, 622, 1947.
2. Novoselov, K. S. Nobel lecture. Graphene: Materials in the flatland. *Rev. Mod. Phys.*, 83, 837–849, 2011.
3. Dean, M.P.M., Howard, C.A., Saxena, S.S., Ellerby, M., Nonadiabatic phonons within the doped graphene layers of XC_6 compounds. *Phys. Rev. B*, 81, 045405, 2010.
4. Upton, M.Y., Forrest, T.R., Walters, A.C., Howard, C.A., Ellerby, M., Said, A.H., McMorrow, D.F., Phonons and superconductivity in YbC_6 and related compounds. *Phys. Rev. B*, 82, 134515, 2007.
5. Eremenko, V.V., Sirenko, V.A., Gospodarev, I.A., Syrkin, E.S., Feodosyev, S.B., Bondar, I.S., Minakova, K.A., Feher, A., *J. Phys: Con. Ser.*, 969 012021.
6. Feher, A., Feodosyev, S., Gospodarev, I., Kotlyar, O., Kravchenko, K., Manzhelii, E., Syrkin, E., Impurity levels in the electron spectra of graphene. *Superlattices Microstruct.*, 53, 55, 2013.
7. Eremenko, V.V., Sirenko, V.A., Gospodarev, I.A., Syrkin, E.S., Feodosyev, S.B., Bondar, I.S., Saxena, S., Feher, A., Minakova, K.A., Electron spectra of graphene with local and extended defects. *Low Temp. Phys.*, 42, 99, 2016.

8. Eremenko, V.V., Sirenko, V.A., Gospodarev, I.A., Syrkin, E.S., Feodosyev, S.B., Bondar, I.S., Minakova, K.A., Anisotropic behavior and inhomogeneity of atomic local densities of states in graphene with vacancy groups. *J. Sci.: Adv. Mater. Devices*, 1, 167, 2016.
9. Maksimov, E.G., Room-temperature superconductivity - Myth or Reality? *Phys. Usp.*, 51, 167, 2008 (in Russian).
10. Eliashberg, G.M., Interactions between electrons and lattice vibrations in a superconductor. *Sov. Phys. JETP*, 11, 966; 12, 1437, 1960.
11. Eremenko, V.V., Sirenko, V.A., Gospodarev, I.A., Syrkin, E.S., Feodosyev, S.B., Bondar, I.S., Feher, A., Minakova, K.A., *Low Temp. Phys.*, 43, 1657, 2017.
12. Pisana, S., Lazzeri, M., Casiraghi, C., Novoselov, K.S., Geim, A.K., Ferrari, A.C., Mauri, F., Breakdown of the adiabatic Born–Oppenheimer approximation in graphene. *Nat. Mater.*, 6, 198, 2007.
13. Das, A., Pisana, S., Chakraborti, B., Piscane, S., Sahai, S.K., Waghmare, U.V., Novoselov, K.S., Krishnamurthy, H.R., Geim, A.K., Ferrari, A.C., Sood, A.K., Monitoring dopants by Raman scattering in an electrochemically top–gated graphene transistor. *Nat. Nanotechnol.*, 3, 210, 2008.
14. Lazzeri, M. and Mauri, F., Nonadiabatic Kohn Anomaly in a Doped Graphene Monolayer. *Phys. Rev. Lett.*, 97, 266407, 2006.
15. Tsang, J.C., Freiting, M., Perebeinos, V., Liu, J., Avouris, P., Doping and phonon renormalization in carbon nanotubes. *Nat. Nanotechnol.*, 2, 725, 2007.
16. Novoselov, K.S., Graphene: materials in the Flatland. *Phys. Usp.*, 181, 1299, 2011 (in Russian).
17. Castro Neto, A.H., Guinea, F., Peres, N.M.R., Novoselov, K.S., Geim, A.K., The electronic properties of graphene. *Rev. Mod. Phys.*, 81, 109, 2009.
18. Lifshits, I.M., Some problems of the electron theory of metals. *Rep. AS USSR*, 48, 83, 1945 (in Russian).
19. Kossevich, A.M., The crystal lattice (phonons solitons dislocations), *Wiley VCH Verlag Berlin GmBH, Berlin*, 326, 1999.
20. Blakslee, O.L., Proctor, D.G., Spence, G.B., Elastic constants of compression-annealed pyrolytic graphite. *J. Appl. Phys.*, 41, 3373, 1970.
21. Nicklow, R., Wakabayashi, N., Smith, Y.G., Lattice dynamics of pyrolytic graphite. *Phys. Rev. B*, 5, 4951, 1972.
22. Dresselhaus, M.S. and Dresselhaus, G., Intercalation compounds of graphite. *Adv. Phys.*, 30, 139, 1981.
23. Maultzsch, J., Reich, S., Thomsen, C., Requardt, H., Ordejyn, P., Phonon dispersion in graphite. *Phys. Rev. Lett.*, 92, 075501, 2004.
24. Reich, S. and Thomsen, C., Raman spectroscopy of graphite. *Phil. Trans. R. Soc. Lond.*, 362, 2271, 2004.
25. Gospodarev, I.A., Kravchenko, K.V., Syrkin, E.S., Feodosyev, S.B., Quasi–two–dimensional features in the phonon spectrum of graphite. *Low Temp. Phys.*, 35, 589, 2009.
26. Eremenko V.V., Sirenko, V.A., Gospodarev, I.A., Syrkin, E.S., Feodosyev, S.B., Dolbin, A.V., Minakova, K.A., Role of acoustic phonons in the negative thermal expansion of layered structures and nanotubes based on them. *Low Temp. Phys.*, 42, 401, 2016.
27. Castro Neto, A.H., Guinea, F., Peres, N.M.R., Novoselov, K.S., Geim, A.K., The electronic properties of graphene. *Rev. Mod. Phys.*, 81, 109, 2009.
28. Castro, E.V., Lopez-Sanhco, M.P., Vozmediano, M.A.H., New type of vacancy – induced localized states in multilayer graphene. *Phys. Rev. Lett.*, 104, 036802, 2010.
29. Peresada, V.I., New computational method in the theory of the crystal lattice. *Condensed Matter Physics* (in Russian), p. 172, FTINT AN UkrSSR, Kharkov, 1968.
30. Peresada, V.I., Afanasyev, V.N., Borovikov, V.S., On the calculation of the distribution function of one-magnon excitations in ferromagnets. *Sov. Low Temp. Phys.*, 1, 227, 1975.

31. Haydock R., Heine V., Kelly M.J., Electronic structure based jn the local atomic environment for toght-binding bands. *Journ. of Phys. C*, 20, 2845, 1972.
32. Galetich, I.K., Gospodarev, I.A., Grishaev, V.I., Eremenko, A.V., Kravchenko, K.V., Sirenko, V.A., Feodosyev, S.B., Vibrational characteristics of the niobium dichalcogenide. Bulk samples and nano-films. *Superlattices Microstruct.*, 45, 564, 2009.
33. Lifshits, I.M. and Rozentsveig, L.N., The dynamics of the crystal lattice filling the semi-space. *JETP*, 18, 1012, 1948 (in Russian).
34. Kosevich, A.M., Syrkin, E.S., Feodosyev, S.B., Peculiar features of phonon spectra of low-dimensional crystals. *Phys. Low-Dim. Struct.*, 3, 47, 1994.
35. Novoselov, K.S., Gein, A.K., Morozov, S.V., Jiang, D., Katsnelson, M.I., Grigorieva, I.V., Dubonos, S.V., Firsov, A.A., Two-dimensional gas of massless Dirac fermions in graphene. *Nature*, 438, 197, 2005.
36. Skrypnyk, Yu.V. and Loktev, V.M., Spectral function of graphene with short-range impurity centers. *Low Temp. Phys.*, 34, 818, 2008.
37. Landau, L.D. and Lifshits, E.M., Stastistical Physics (Vol. 5 of A Course of Theoretical Physics). *Pergamon Press*, 1969.
38. Lifshits, I.M., About thermal properties of chain and layered structures at low temperatures. *JETP*, 22, 475, 1952 (in Russian).
39. Landau, L.D. and Lifshits, E.M., Theory of Elasticity (Vol. 7 of A Course of Theoretical Physics), *Pergamon Press*, 1970.
40. Dresserhaus, M.S. and Eklund, P.C., Phonons in carbon nanotubes. *Adv. Phys.*, 49, 705, 2000.
41. Bagatskii, M.I., Barabashko, M.S., Dolbin, A.V., Sumarokov, V.V., Sundqvist, B., The specific heat and the radial thermal expansion of bundles of single-walled carbon nanotubes. *Low Temp. Phys.*, 38, 523, 2012.
42. Riley, D.P., The thermal expansion of graphite: Part II. Theoretical. *Proc. Phys. Soc. London*, 57, 486, 1945.
43. Belen'kii, G.L., Suleimanov, R.A., Abdullaev, N.A., Shteinshraiber, V.Ya., Thermal-expansion of layered crystals-lifshits models. *Sov. Phys. Solid State*, 26, 12, 2142, 1984.
44. Wakabayashi, N., Smith, H.G., Shanks, R., Two dimensional Kohn anomally in $NbSe_2$. *Phys. Lett. A*, 50, 367, 1974.
45. Abdullaev, N.A., Mamedov, T.G., Suleimanov, R.A., Thermal expansion of single crystals of the layered compounds $TlGaSe_2$ and $TlInS_2$. *Low Temp. Phys.*, 27, 676, 2001.
46. Feodosyev, S.B., Gospodarev, I.A., Syrkin, E.S., Anomalies of Linear Expansion Coefficient in Highly Anisotropic Crystals at Low Temperature. *Phys. Status Solidi B*, 150, K 19, 1988.
47. Eremenko, V., Dolbin, A., Minakova, K., Sirenko, V., Syrkin, E., Feodosyev, S., Gospodarev, I., The Phonon Mediated Anomalies of Thermal Expansion in Transition-Metal Componds and Emergent Nanostructures. *Solid State Phenomena*, 257, 81, 2017.
48. Van Smaalen de Boer, J.L., Haas, C., Kommadeur, J., Anisotropic thermal expansion in crystals with stacks of planar molecules, such as tetracyanoquinodimethanide (TCNQ) salts. *Phys. Rev. B*, 31, 3496, 1985.
49. Katrusiak, A., Rigid O., Molecule model of anomalous thermal expansion of ices. *Phys. Rev. Lett.*, 77, 4366, 1996.
50. Gospodarev, I.A., Eremenko, A.V., Kravchenko, K.V., Sirenko, A.F., Sirenko, V.A., Syrkin, E.S., Feodosyev, S.B., Shabakaeva, Yu.A., Distinctive features of thermal expansion of niobium diselenide. *Phys. Solid State*, 55, 898, 2013.
51. Eremenko, V.V., Gospodarev, I.A., Ibulaev, V.V., Sirenko, V.A., Feodosyev, S.B., Shvedun, M.Yu., Anisotropy of temperature dependences of lattice parameters $Eu_{1+x}(Ba_{1-y}Ry)_{2-x}Cu3O_{7-d}$ in the quasi-harmonic limit. *Low Temp. Phys.*, 32, 12, 1189, 2006.
52. Bailey, A.C. and Yates, B.J., Anisotropic thermal expansion of pyrolytic graphite at low temperatures. *J. Appl. Phys.*, 41, 5088, 1970.

53. Dolbin, A.V., Esel`son, V.B., Gavrilko, V.G., Manzhelii, V.G., Vinnikov, N.A., Popov, S.N., Sundqvist, B., Radial thermal expansion of single-walled carbon nanotube bundles at low temperatures. *Low Temp. Phys.*, 34, 678, 2008.
54. Peresada, V.I. and Syrkin, E.S., *Sov. Low Temp. Phys.*, 3, 113, 1977.
55. Maradudin, A.A., Surface waves, in: *Modern Problems of Surface Physics*, pp. 11–400, I.J. Lalov (Ed.), ISCMP, 1980.
56. Kotlyar, O.V., Feodosyev, S.B., Local oscillations in crystal lattices with a simply connected region of a quasicontinuous phonon spectrum. *Low Temp. Phys.*, 32, 256, 2006.
57. Landau, L.D., To the theory of the phase transition. *JETP*, 7, 627, 1937 (in Russian).
58. Ochoa H., Castro E.V., Katsnelson M.I., Guinea F., Scattering by flexural phonons in suspended graphene under back gate induced strain. *Physica E*, 44, 963, 2012.
59. Bena, C. and Kivelson, S.A., Quasiparticle scattering and local density of states in graphite. *Phys. Rev. B*, 72, 125432, 2005.
60. Feher, A., Feodosyev, S., Gospodarev, I., Kotlyar, O., Kravchenko, K., Manzhelii, E., Syrkin, E., Impurity levels in the electron spectra of graphene. *Superlattices Microstruct.*, 53, 55, 2013.
61. Skrypnyk Yu.V. and Loktev V.M., Impurity effects in a two-dimensional system with the Dirac spectrum. *Phys. Rev. B*, 73, 241402, 2006.
62. Pereira, V.M., Guinea, F., Lopes dos Santos, J.M.B., Peres, N.M.R., Castro Neto, A.H., Electron Waves in chemistry substituted graphene. *Phys. Rev. Lett.*, 96, 036801, 2006.
63. Pokropivny, A.V., Ni, Y., Chalopin, Y., Solonin, Y.M., Volz, S., Tailoring properties of graphene with vacancies. *Phys. Status Solidi B*, 251, 555, 2014.
64. Smith, B.W. and Luzzi, D.E., Electron irradiation effects in single wall carbon nanotubes. *J. Appl. Phys.*, 90, 3509, 2001.
65. Wu, S., Jing, L., Li, Q., Shi, Q.W., Chen, J., Wang, X., Average density of States in disordered graphene systems. *Condensed Matter*, 7, 208, 2007.
66. Kang, J., Bang, J., Ryu, B., Chang, K.J., Effect of atomic-scale defects on the low-energy electronic structure of graphene: Perturbation theory and local-density-functional calculations. *Phys. Rev. B*, 77, 115453, 2008.
67. Gospodarev, I.A., Eremenko, V.V., Kravchenko, K.V., Sirenko, V.A., Syrkin, E.S., Feodosyev, S.B., Vibrational characteristics of niobium diselenide and graphite nanofilms. *Low Temp. Phys.*, 36, 344, 2010.
68. Weller, T.E., Ellerby, M., Saxena, S.S., Smith, R.P., Skipper, N.T., Superconductivity in the intercalated graphite compounds C_6Yb and C_6Ca. *Nat. Phys.*, 1, 39, 2005.
69. Emery, N., Herold, C., d'Astuto, M., Garcia, V., Bellina, Ch., Mareche, J.F., Lagrange, P., Loupias, G., Superconductivity of Bulk CaC_6. *Phys. Rev. Lett.*, 95, 087003, 2005.
70. Sirenko, V.A., Critical phenomena in superconductors and uniaxial antiferromagnets (Review Article). *Low Temp. Phys.*, 38, 799, 2012.
71. Nagamatsu, J., Nakagawa, N., Muranaka, T., Zenitani, Y., Akimitsu, J., Superconductivity at 39K in magnesium diboride. *Nature*, 410, 63, 2001.
72. Maksimov, E.G., Magnitskaya, M.V., Ebert, S.V., Savrasov, S.Yu., Accounts in the first principle of the critical temperature superconducting transition in NbC and her dependence from pressure. *JETP Lett.*, 80, 548, 2004.
73. Pickett, W.E., Design for a room-temperature superconductor. *J. Supercon. Novel Magn.*, 19, 291, 2006.
74. Dean, M.P.M., Walters, A.C., Howard, C.A., Weller, T.E., Calandra, M., Mauri, F., Ellerby, M., Saxena, S.S., Ivanov, A., McMorrow, D.F., Non-adiabatic phonons within the doped graphene layers of XC_6 compounds. *Phys. Rev. B*, 82, 014533, 2010.

75. Upton, M.H., Walters, A.C., Howard, C.A., Rahnejat, K.C., Ellerby, M., Hill, J.P., McMorrow, D.F., Alatas, A., Leu, B.M., Ku, W., Phonons in superconducting CaC_6 studied via inelastic X-ray scattering. *Phys. Rev. B*, 76, 220501, 2007.
76. Calandra, M. and Mauri, F., Possibility of superconductivity in graphite intercalated with alkaline earths investigated with density functional theory. *Phys. Rev. Lett.*, 95, 237002, 2005.
77. Kim, J.S., Boeri, L., Kremer, R.K., Razavi, F.S., Effect of pressure on superconducting Ca-intercalated graphite CaC_6. *Phys. Rev. B*, 74, 214513, 2006.
78. Belash, I.T., Bronnikov, A.D., Zharikov, O.V., Pal'nichenko, A.V., Superconductivity of graphite intercalation compound with lithium C_2Li. *Solid State Commun.*, 69, 921, 1989.

13

Complex Refractive Index (RI) of Graphene

Sosan Cheon[1]* and Kenneth David Kihm[2†]

[1]*School of Mechanical and Aerospace Engineering, Seoul National University, Seoul, Republic of Korea*
[2]*Mechanical, Aerospace, and Biomedical Engineering, University of Tennessee, Knoxville, Tennessee, U. S. A.*

Abstract

Graphene is known as the first realized 2-D material of single atomic layer, and its unique electrical, thermal, and mechanical properties have attracted the vast interest from science and engineering in the past decade. Optical transparency is one of the most important features of graphene in both theoretical physics and engineering applications. While prediction of graphene's "optical conductivity" has been the main interest of theoretical works, its "complex refractive index (RI)" is more intuitive and practically useful to fully understand optical responses of graphene. We present a comprehensive review of delineating progresses of graphene's complex RI both in theoretical predictions and experimental measurements: (1) analytical derivation, (2) numerical calculation by *ab-initio* method, (3) reflection spectroscopy, (4) ellipsometry, (5) picometrology, (6) simultaneous reflection and transmission measurements, (7) surface plasmon resonance (SPR), (8) attenuated total internal reflection (ATR), and (9) tandem use of SPR and ATR.

Keywords: Graphene, complex refractive index, optical conductivity, optical properties, calculations, measurements

13.1 Introduction

Graphene, one of the most actively studied materials for the last decade [1], shows extreme mechanical [2], electrical [3, 4], and thermal [5] properties, due to its inherent 2-D nature of the honeycomb carbon atom structure. These superior physical properties, combined with the optical transparency (Figure 13.1a) [6], make graphene a promising candidate for flexible and transparent electronics (Figure 13.1b) [7]. The refractive index (RI) of graphene is likely the most important optical properties in its diverse applications. A comprehensive review of the recent progress on complex RI of pristine graphene is reported, covering both theoretically predicted and experimentally measured complex RI database for graphene.

In Section 13.2, the theoretical framework for the optical property of graphene is introduced, focusing on the conversion of optical conductivity or dielectric constant into complex RI. Afterwards, both the analytical derivations and the numerical calculation results are presented for the graphene's optical properties. Section 13.3.1 presents discussions on various far-field detection methods to measure the complex RI of graphene, including

**Corresponding author*: cjsthtks@gmail.com
†*Corresponding author*: kkihm@utk.edu

Tobias Stauber (ed.) Handbook of Graphene: Volume 2, (389–412) © 2019 Scrivener Publishing LLC

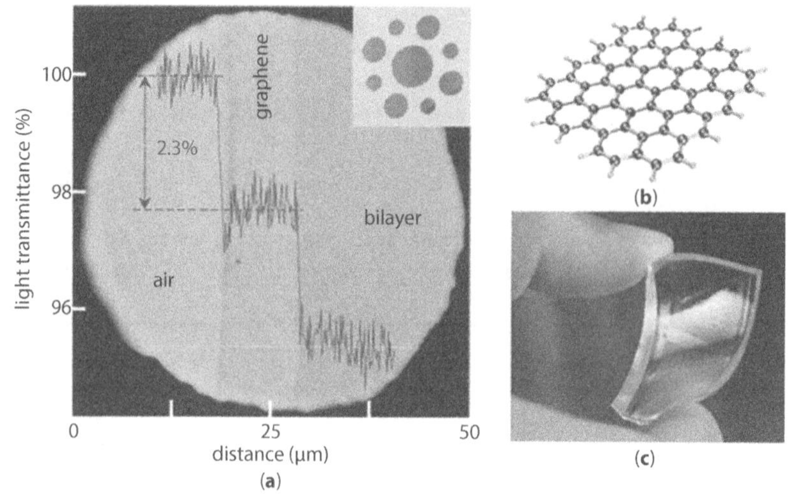

Figure 13.1 (a) Graphene suspended on a 50 μm aperture hole, absorbing 2.3% of white light per layer (From Ref [6], by permission of the American Association for the Advancement of Science), (b) atomic structure of graphene [8], and (c) bending electric circuits made by graphene electrodes and a flexible substrate (From Ref. [7], by permission of the Springer Nature).

reflection spectroscopy (including Fabry–Perot etalon), ellipsometry, picometrology, and simultaneous reflection and transmission measurements. Section 13.3.2 presents discussions on near-field optical characterization methods that enable more sensitive detection of the complex RI of graphene. These near-field techniques include surface plasmon resonance (SPR), attenuated total internal reflection (ATR), and the tandem use of SPR and ATR to ensure the unique determination of optical properties of graphene with quantitative uncertainty ranges defined. In Section 13.4, all of the presently reviewed complex RI data are summarized in a single graph and also listed in a table.

Note that this review focuses on the inherent RI properties of pristine graphene that is free from the optical property variations caused by the external factors, such like the effects of substrate and contact of graphene with it, electrochemical or atomic doping, defects and disorders in graphene, the interaction of multiple graphene layers, or imposed mechanical strain.

13.2 Theoretical Predictions of Complex RI of Graphene

This chapter presents discussions on theoretical determination of the complex RI of graphene. Since theoretical studies traditionally focused on evaluating the optical conductivity and dielectric permittivity, conversion of these quantities to the more convenient complex RI is presented in Section 13.2.1. Then, both the purely analytical methods (Section 13.2.2) and the *ab-initio* numerical calculation method using density functional theory (DFT) (Section 13.2.3) are briefly discussed. A notable limitation of most theoretical studies is that their outcomes provide the complex RI of graphene in terms of often complicated formulas with parameters that need to be further elaborated and/or experimentally verified.

13.2.1 Conversion of Optical Conductivity and Dielectric Constant to Complex RI

Despite the fact that the theoretical studies focus mostly on the optical conductivity and dielectric constant of graphene, its complex RI is more intuitive for experimental examinations and practical applications. The real part of RI accounts for the refraction of light due to the change of speed of light in graphene, while the imaginary part describes the absorption or attenuation of light in a material. The conventional definition of RI is concerning bulk materials as a collective response of abundant electrons to the external electromagnetic waves [9]. Thus, one may think that this classical concept of RI may not be applicable to "one-atom-thick" graphene. However, provided a proper conversion of the relevant physical quantities (Equation 13.5), graphene can also be described as a complex RI with a nominal thickness. This has been proven by comparing the model based on optical conductivity with zero thickness and that based on complex RI with finite thickness [10, 11]. They yielded the same calculation results for reflection, transmission, and absorption of graphene in negligible error.

Absorbing materials such as graphene carry real and imaginary parts of their optical properties. There are three different ways of presenting the optical properties: complex RI ($\tilde{\boldsymbol{n}}$), complex dielectric constant (dielectric permittivity divided by vacuum permittivity, $\tilde{\varepsilon} \equiv \frac{\varepsilon}{\varepsilon_0} = (\tilde{n})^2$ for graphene, being nonmagnetic), and complex optical conductivity ($\tilde{\sigma}$), where tilde means a complex-valued quantity. These three quantities, $\tilde{\boldsymbol{n}}$, $\tilde{\varepsilon}$, and $\tilde{\sigma}$ are given by the Maxwell equation [10–13]:

$$\nabla \times \mathbf{H} = \mathbf{J} + \frac{\partial \mathbf{D}}{\partial t} \tag{13.1}$$

$$= (\sigma' - i\omega\varepsilon_0\varepsilon')\mathbf{E} \tag{13.2}$$

$$= (\sigma' - i\sigma'')\mathbf{E} \tag{13.3}$$

$$= -i\omega\varepsilon_0(\varepsilon' + i\varepsilon'')\mathbf{E} \tag{13.4}$$

$$\therefore \tilde{\varepsilon} = \frac{\tilde{\sigma}}{i\omega\varepsilon_0} = \tilde{n}^2 \tag{13.5}$$

where **H**, **J**, **D**, **E**, and ε_0 are the magnetic field, electric current density, dielectric displacement, electric field, and vacuum permittivity, respectively, and ′ and ″ denote the real and imaginary parts of the complex quantities. Note that the signs in front of the imaginary quantities conform the electromagnetic waves described by $e^{i(\mathrm{k\,x\text{-}w}t)}$, so that the fields decays (not amplified) in materials ($\tilde{\mathbf{n}} = n + ik$, $\tilde{\sigma} = \sigma' - i\sigma''$, and $\tilde{\varepsilon} = \varepsilon' + i\varepsilon''$).

Equation 13.5 allows conversion of either $\tilde{\sigma}$ or $\tilde{\varepsilon}$ into complex RI ($\tilde{n}$). However, care should be taken as the input quantities should also be in complex. For example, the real-valued optical conductivity $\sigma = e^2/4$ [6] cannot be converted to $\tilde{n}$ without additional information on the imaginary part of $\tilde{\sigma}$ (considering $\sigma = e^2/4$ as a real-valued quantity and relative permittivity of graphene as vacuum's (ε_0), one can obtain the complex RI [14], but the values are much different from the experimentally reported values).

Another issue is that graphene has optical properties that are anisotropic, i.e., the responses to the in-plane and out-of-plane electromagnetic fields are completely different. However, in most cases, the response to the in-plane component is dominant, which is easily envisaged from the π-electron orbitals (Figure 13.2a) in the restricted 2-D graphene layer of a nominal thickness of mere 0.335 nm [15]. Therefore, only the in-plane RI values are discussed in this review, although the out-of-plane quantities can also be derived from the basically identical theory.

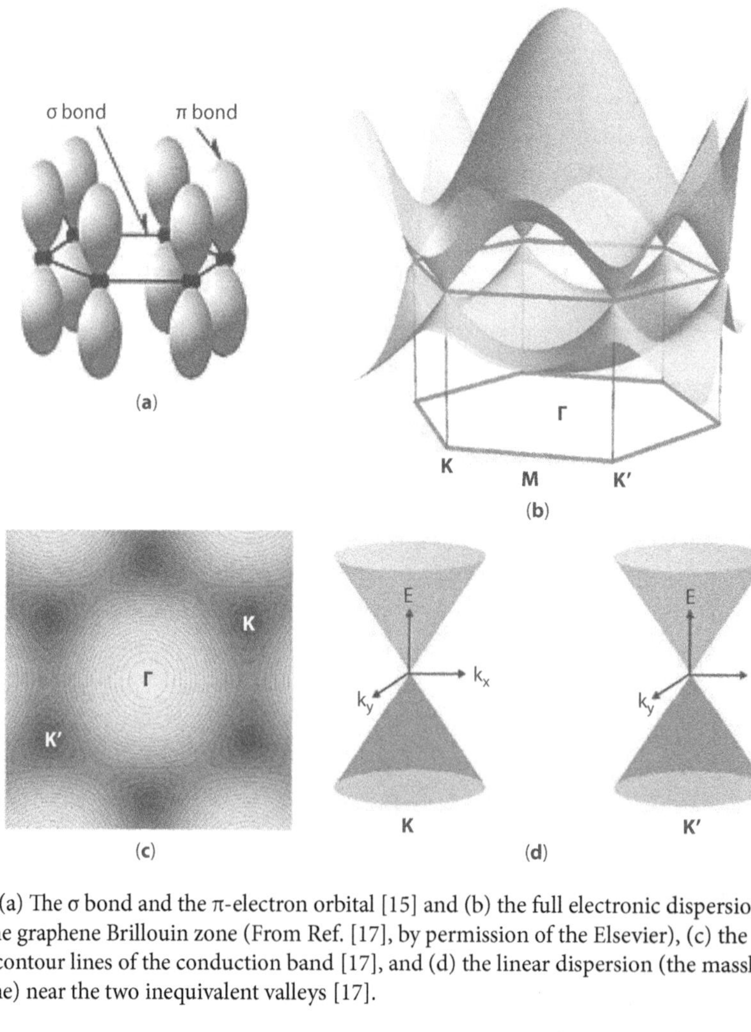

Figure 13.2 (a) The σ bond and the π-electron orbital [15] and (b) the full electronic dispersion of the π-bands in the graphene Brillouin zone (From Ref. [17], by permission of the Elsevier), (c) the corresponding equi-energy contour lines of the conduction band [17], and (d) the linear dispersion (the massless spectrum of the Dirac cone) near the two inequivalent valleys [17].

13.2.2 Analytical Determination of Graphene's Complex RI

The analytical approaches have been made to determine the optical conductivity of graphene, which can be readily converted to complex RI using Equation 13.5. The real part of the optical conductivity of graphene can be determined from consideration of the electronic transitions in graphene, while the imaginary part can be determined from Kramers–Krönig relation [16]. Figure 13.2b schematically shows graphene's full-dispersion electronic π–band structure for allowed electronic states in false-colored surfaces (abscissa: momentum, ordinate: energy) near the Fermi energy where the upper and the lower cones meet. The equal-energy contour lines of the conduction band (the upper half of the band structure in Figure 13.2b) are shown in Figure 13.2c. Figure 13.2d describes the linear electronic dispersion approximation [18] for the band structure near the Fermi energy for a single graphitic layer by massless Dirac Hamiltonian that can quantum mechanically describe the relativistic electrons in graphene.

This linear electronic band structure approximation, $E = \pm\, v_F|\mathbf{k}|$ (E: energy level, v_F: Fermi velocity, $\mathbf{k}$: momentum, i.e., horizontal axes of Figure 13.2b and d), provides the universal real-part optical conductivity of graphene as

$$\sigma' = \sigma_{univ} = e^2/4\hbar \tag{13.6}$$

where the few-percent of the next-nearest-neighbor hopping effect [19] has been disregarded. Strictly speaking, this is valid only for pristine graphene at absolute zero temperature for relatively low light energy ranging from infrared to long wave visible or near IR range [20, 21]. Figure 13.3a shows the complex RI of graphene for the simplest case of the universal conductivity and under assumption for the real part of the dielectric constant being identical as in vacuum [14]. For the wavelength span from 200 nm to 1000 nm, these crude values of the real and imaginary parts of RI fall approximately in the same order of magnitude with the measured RI values as shown in Section 13.3. However, these values do not predict the correct dispersion (wavelength dependence) of complex RI of graphene.

When the light energy is higher for shorter wavelengths in the visible or near UV range, however, these higher energy photons can excite electrons into the higher-energy nonlinear bands further away from the intersection points and the aforementioned linear band approximation will become invalid [17]. In addition, the excitonic effects near the saddle-point singularity (M point in Figure 13.2b) need to be further examined by Fano resonance of electronic state and the band continuum [22]. Furthermore, graphene is almost always subjected to a certain level of doping when exposed to a non-vacuum environment, which in turn alters the Fermi energy from the original k_x-k_y plane for pristine graphene (Figure 13.2c). In this case, the Pauli exclusion principle blocks the transition of electrons that are at the energy levels lower than $|E_F|$, thus the absorption of low energy photons (long wavelength) is suppressed and the resulting optical conductivity should be modified accordingly [21].

The imaginary part of the optical conductivity is relatively easily given by the Kramers–Krönig relation [16] in association with Kubo formula [23] that accounts for the response of the electrons in graphene to the time-dependent perturbation of EM wave, i.e., the incident light:

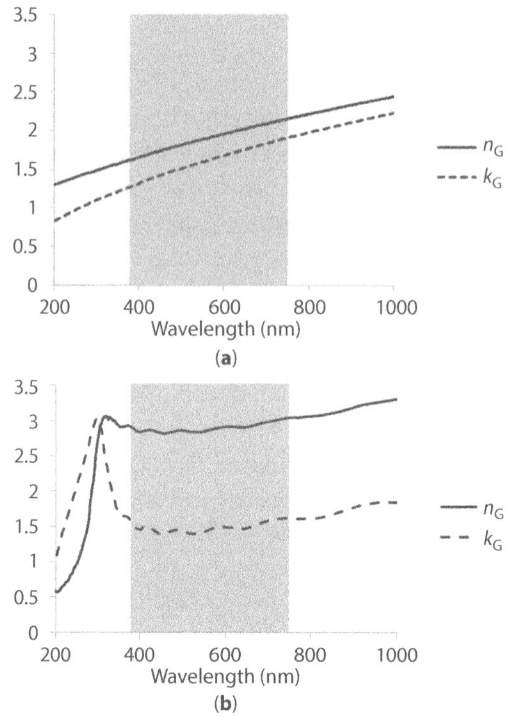

Figure 13.3 (a) Complex RI of graphene derived from the universal optical conductivity and the vacuum permittivity [24], and (b) the complex RI determined from the *ab-initio* DFT calculations. The authors converted the dielectric function data by DFT [25] into complex RI using Equation 13.5 and generated this graph showing its wavelength dependence.

$$\sigma''(\omega) = -\frac{1}{\pi} P \int_{-\infty}^{\infty} \frac{\sigma'(\omega')}{\omega' - \omega} d\omega' \tag{13.2.7}$$

where P denotes the Cauchy principal value and ω is the angular frequency of incident light. In principle, substituting the above real (not shown here, but see Equations 7–14 in the reference [19]) and imaginary (Equation 13.2.7) parts of the optical conductivity into Equation 13.5 can determine the complex RI of graphene in principle. In reality, however, numerical determination of complex RI of graphene from these will be very cumbersome because unknown parameters of the above analytically derived expressions must be assumed or experimentally determined.

13.2.3 Numerical Determination of Graphene's Complex RI

The more detailed prediction of the complex RI of graphene is given by density functional theory (DFT), which numerically calculates the band structures and the electronic transitions upon the incident light. The aforementioned assumptions and approximations that were essential for the analytical determination of the complex RI in Section 13.2.2 can now be released to the most extent by adopting this *ab-initio* numerical method. The accurate atomic structure of graphene is calculated from the relaxation of carbon atoms, followed by the electronic band structure calculations. Finally, the electrons' optical transition determines the imaginary part of dielectric constant of graphene as [25]

$$\varepsilon''_{ij}(\omega) = \frac{4\pi^2 e^2}{Vm^2\omega^2}\sum_{\mathrm{k}nn'\sigma}\langle \mathrm{k}n\sigma | p_i | \mathrm{k}n'\sigma\rangle\langle \mathrm{k}n'\sigma | p_j | \mathrm{k}n\sigma\rangle \times f_{\mathrm{k}n}(1-f_{\mathrm{k}n'})\delta(e_{\mathrm{k}n'}-e_{\mathrm{k}n}-\hbar\omega) \tag{13.8}$$

where e is the electron charge, m is its mass, V is the volume (or the area for 2-D materials) of the system, f_{kn} is the Fermi distribution, $|\mathbf{k}n\sigma$ is the crystal wave function corresponding to the n-th eigenvalue with crystal momentum $\mathbf{k}$ and spin σ, p is the momentum, and i and $j = x, y$, or z. The real part of the complex dielectric constant is given by the Kramers–Krönig relation:

$$\varepsilon'(\omega) = \frac{1}{\pi}P\int_{-\infty}^{\infty}\frac{\varepsilon''(\omega')}{\omega'-\omega}d\omega' \tag{13.9}$$

The original dielectric function data [25] have been converted to complex RI by using Equation 13.5, and the results are shown in Figure 13.3b. The DFT calculations require predetermination of multiple variables, including the atomic potential, k-point grid and the integration process, exchange–correlation functional, the smearing, and the cutoff energy. Despite the fact that these parameters are set to closely predict the relatively well-known material properties, e.g., the dielectric function of bulk graphite, uncertainties are inevitable when using these variables in predicting complex RI values of single-layer graphene. Nevertheless, the calculated values of the real part of RI (n_G) of about 3.0 for the visible range falls within the close proximity of measured data ranging from 2 to 3 for single-layer graphene.

In summary, analytical methods to predict the complex RI of graphene use the linear electronic band approximation near the Fermi energy, but employ various assumptions and corrections to improve the prediction accuracy. However, these analytically derived results are generally incapable of providing the complex RI in numbers because of multiple unknown variables. Numerical methods based on DFT further improve the prediction accuracy using the *ab-initio* calculations. While the DFT results readily provide numerical data for complex RI values as functions of wavelength (Figure 13.3b), the DFT calculation processes often render uncertainties originated from the various parameters involved in the theory. Therefore, the more realistic and comprehensive values of the complex RI should be ultimately determined by the experimental measurements of graphene layers, which will be covered in the Sections 13.3 and 13.4.

13.3 Measurements of Complex RI of Graphene

Graphene's complex RI measurements have two fundamental difficulties: (1) there are two unknowns to be determined, i.e., the real and imaginary parts of RI, and (2) graphene is much thinner than the wavelength of visible light, leading to insufficient contrast for optical detections. The first difficulty should be resolved either by spectroscopic measurements with data fitting using a proper optical model, or conducting two independent measurements that can simultaneously determine the real and imaginary parts of graphene's RI. The second difficulty is more inherent for graphene and may be alleviated by using various optical techniques that can enhance the optical contrast or using very accurate and precise measurement elaborations.

13.3.1 RI Measurements of Far-Field Response

Far-field response refers to the optical phenomena where the incident light interacts directly with the graphene but the resulting signals, the reflectance and/or transmittance, are captured by a detector that is far away from the sample. In this configuration, the light-graphene interaction is weak because the optical path length in graphene is mere an order of 1 nm (the real part of graphene RI of (2 ~ 3) × the thickness of 0.335 nm), which is in the order of hundredth of the incident light wavelength. To overcome this limitation of the far-field methods and extract sufficient signal-to-noise ratio, multiple efforts have been explored including the use of special substrates to enhance the reflection intensity (reflection spectroscopy) or implementation of advanced measurements (ellipsometry, picometry, and simultaneous reflection and transmission measurements). Section 13.3.1.1 presents improved reflection spectroscopy, Section 13.3.1.2 presents ellipsometry, Section 13.3.1.3 discusses picometrology measurements, and Section 13.3.1.4 presents simultaneous reflection/transmission measurements.

13.3.1.1 Reflection Spectroscopy

Complex RI measurements of graphene require more sensitive detection of the optical signals than ordinary imaging to identify graphene layers. The principle of reflection microscopy is based on the fact that optical reflection signals from graphene can easily be enhanced by *optimizing* the substrate type and thickness underneath the graphene, and thus, the reflection method adopts composite substrate of SiO_2/Si, where SiO_2 thickness is optimized for maximum reflection.

The reflection contrast from graphene on an arbitrarily selected substrate may not be intense enough even to identify the graphene layers (Figures 13.4a–c). In contrast, the optical reflection from graphene can be as high as 0.15, depending on the thickness of the oxide layer on the Si wafer, based on Fresnel's predictions [26] (Figure 13.4d). In reflection spectroscopy, the reflection signal is measured as a function of the incident light wavelength. This data can provide the graphene's complex RI from the following steps: (1) Assume an optical dispersion model that describes the wavelength dependence of $\widetilde{n}_G$, with one or two experimentally determined coefficients, (2) calculate the reflectance spectra by Fresnel's equations and multilayer interference considerations while changing the experimentally

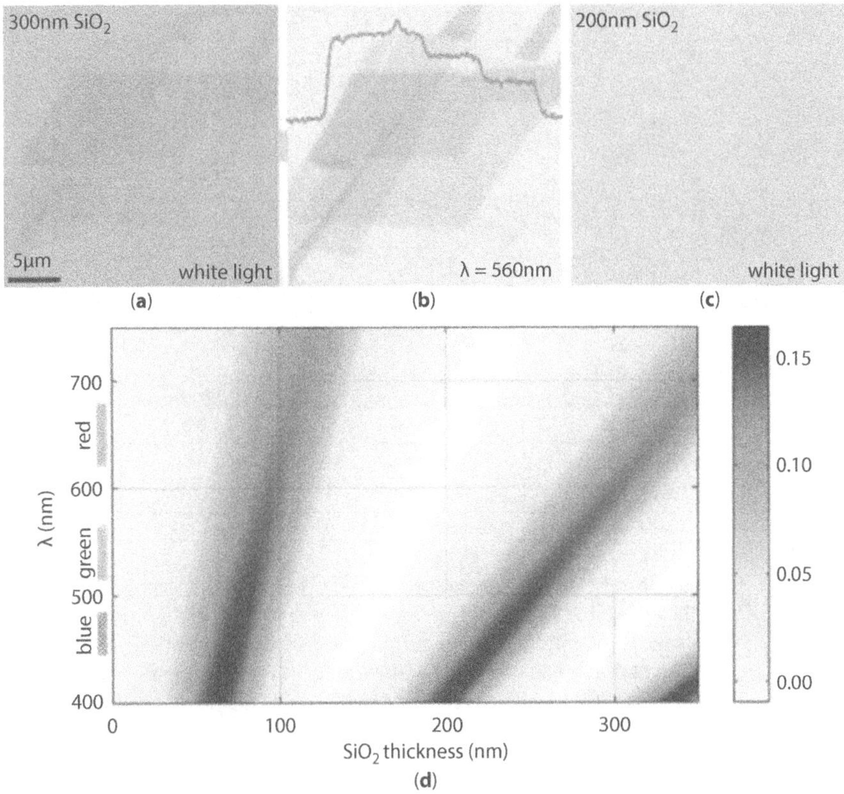

Figure 13.4 Graphene on Si wafer with two different oxide thicknesses, 300 nm for both (a) and (b) and 200 nm for (c). The 300-nm oxide layer is known optimal for graphene detection with 560 nm wavelength. The 2D map (d) shows the reflection contrast as a function of wavelength and SiO_2 thickness. (From Ref. [26], by permission of AIP.)

determined coefficient(s), and (3) find out the best coefficient(s) to finally get the best fit and the corresponding complex RI of graphene as a function of wavelength.

The simplest dispersion model bluntly assumes that $\widetilde{n_G}$ is independent of wavelength where the two experimentally determined coefficients are n_G and k_G, such as in [24] and [27]. The raw data and the fitting results are given in Figure 13.5a and c, but the value obtained in [27] is quite different from those reported later, which is attributed to the crude assumption of wavelength-independent $\widetilde{n_G}$, which has been proven to be not always accurate (see the following sections).

Another dispersion model was suggested [28] based on the earlier observation that the light absorption of graphene is universal in the visible range and that graphite has the sharp resonance of 5.1 eV far away from the visible spectrum (1.65–3.26 eV) [29]. Consequently, the complex RI of graphene was set, $\widetilde{n_G} = n + iC_1\lambda/3$, where n is a constant and C_1 is an experimentally determined coefficient. Their measurement shows the best fitting for $n = 3.0$ and $C_1 = 5.446$ μm^{-1} as an example (Figure 13.5c), which turns out to be fairly acceptable for graphene's RI [22].

Note that if the reflection contrast is measured for two or more substrates with different responses to the graphene's RI, it can uniquely determine graphene's RI without the need

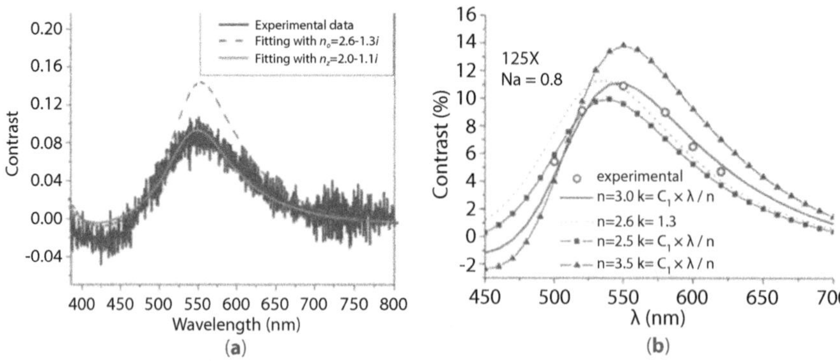

Graphene	n	k
Monolayer	2.69±0.02	1.52±0.02
Bilayer	2.38±0.02	1.66±0.02
Trilayer	2.27±0.02	1.60±0.02

(c)

Figure 13.5 (a) Reflection contrast spectrum of graphene on Si/SiO2 (285 nm) (From Ref. [27], by permission of AIP). From the fitting, the RI of graphene, which is assumed wavelength-independent, turns out to be 2.0–1.1*i*. 2.6–1.3*i* is the RI of graphite (different from the different convention). (b) Contrast of a graphene on a 291 nm thick SiO2 film on Si, with various theoretical curves by different RI models (From Ref. [28], by permission of the AIP). (c) Measured refractive indices of 1- to 3-layer graphene using reflection spectroscopy and assuming wavelength-independent RI [24].

for spectroscopic measurements or optical modeling. This has been realized for another atomic 2-D layer of MoS_2 [30], which can be applied to the graphene measurements as well.

13.3.1.2 Ellipsometry

As a most widely used far-field method for graphene RI measurements, ellipsometry measures the polarization state change $\rho = \frac{r_p}{r_s} = \tan\Psi e^{i}$ of light beam after it is reflected from graphene with oblique incidence. Here, r_p and *rs* are the reflection coefficients for the *p*- and *s*-polarized light, both of which are functions of complex RI, and Ψ and Δ are the amplitude ratio and phase difference of the two polarizations. Usually, the data is acquired for a light beam of tunable wavelength to get the spectroscopic data. The conventional ellipsometry requires a fitting process for the inversion of the raw data into $\widetilde{n_G}$, involving the optical dispersion relation [31–34]. The number of fitting parameters normally exceeds 4 and the reliability of the best fit result can depend on the optical model for graphene and the uncertainties in the measurements (Figure 13.6).

Because the ellipsometry measures two observables, namely, the intensity and the phase change of incident polarized light (or the complex-valued polarization state), the unique determination of graphene's RI is possible without the optical dispersion modeling in principle, while the fitting calculations are highly time consuming. Recent articles [35–37] successfully demonstrated this, either by very precise measurements to obtain smooth spectral curves for the "point-by-point fitting" result for graphene's complex RI [35] or complex optical conductivity [36], or by B-spline fitting method for graphene's

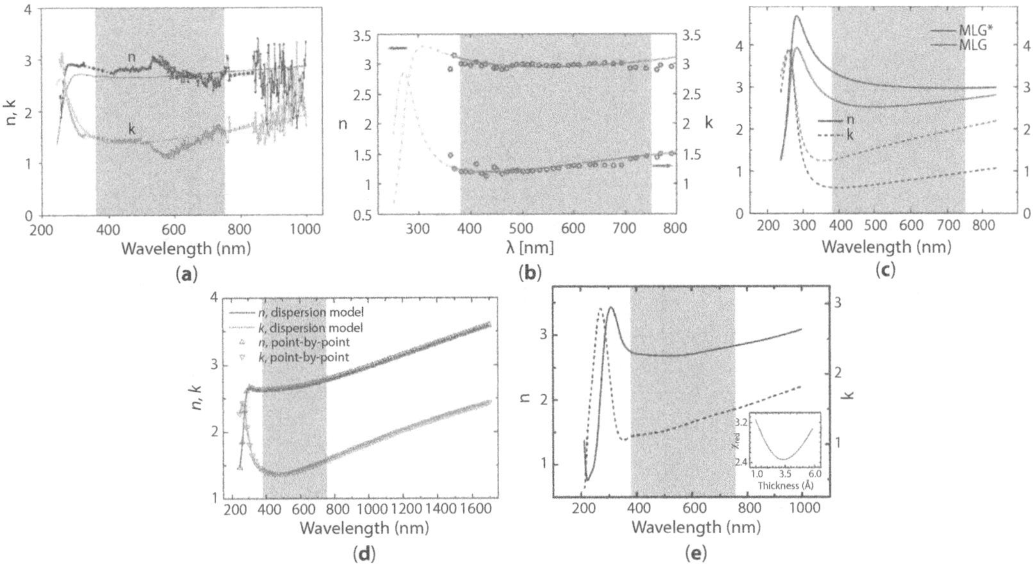

Figure 13.6 Graphene's RI as functions of wavelength measured by spectroscopic ellipsometry. Green shades are inserted to denote the visible range. (a) Graphene modeled by an anisotropic material with an arbitrary horizontal and a Cauchy vertical component (From Ref. [31], by permission of the APS.) (b) Fano line shape (From Ref. [32], by permission of the AIP.) (c) Lorentz–Drude model [34], (d–e) point-by-point fit (From Refs. [35, 37], by permission of the AIP). In (e), the MLG* (blue) used the nominal structure (0.335 nm graphene on substrate), while the MLG (red) includes adsorbed water layer and interlayer effects in practice. The results represent different research laboratories as well as different data fitting methods.

complex RI [37]. The resultant $\widetilde{n_G}(\lambda)$ values from these non-spectroscopic ellipsometry [35, 37] follow the similar trends, except for the peaks in ultraviolet incidence (Figure 13.6d and e).

13.3.1.3 *Picometrology*

As a far-field non-spectroscopic method, which does not require a dispersion modeling, picometrology [38] uses both the amplitude and phase change as two observables when light traverses the graphene edge area (Figure 13.7a and b). The two observables are merged into the complex-valued reflection coefficient $\tilde{r}$ and the existence of graphene with thickness d introduces the change, $\tilde{r}' = \tilde{r} + (1+\tilde{r})^2(1-\widetilde{n_g}^2)\frac{\pi i}{\lambda}d$, determination of $\widetilde{n_g}$ possible from the comparison of the two measured observables. Intensity images by picometrology, its phase-contrast images, and two-channel scans are shown in Figure 13.7c and d, and the resultant complex RI values of graphene are given in Figure 13.7e.

In principle, this approach can determine the complex RI of graphene uniquely for each incident wavelength, because it uses the two different observables for the single wavelength. However, the scheme bears potential error sources affiliated with imperfect Gaussian profile of the focused spot and/or slight defocus and local variations of the SiO_2 thickness. For the inversion of the raw data to $\widetilde{n_G}$, for example, the Fourier transform and integration of convolutions of Gaussian, Dawson, and Heaviside step functions are adopted.

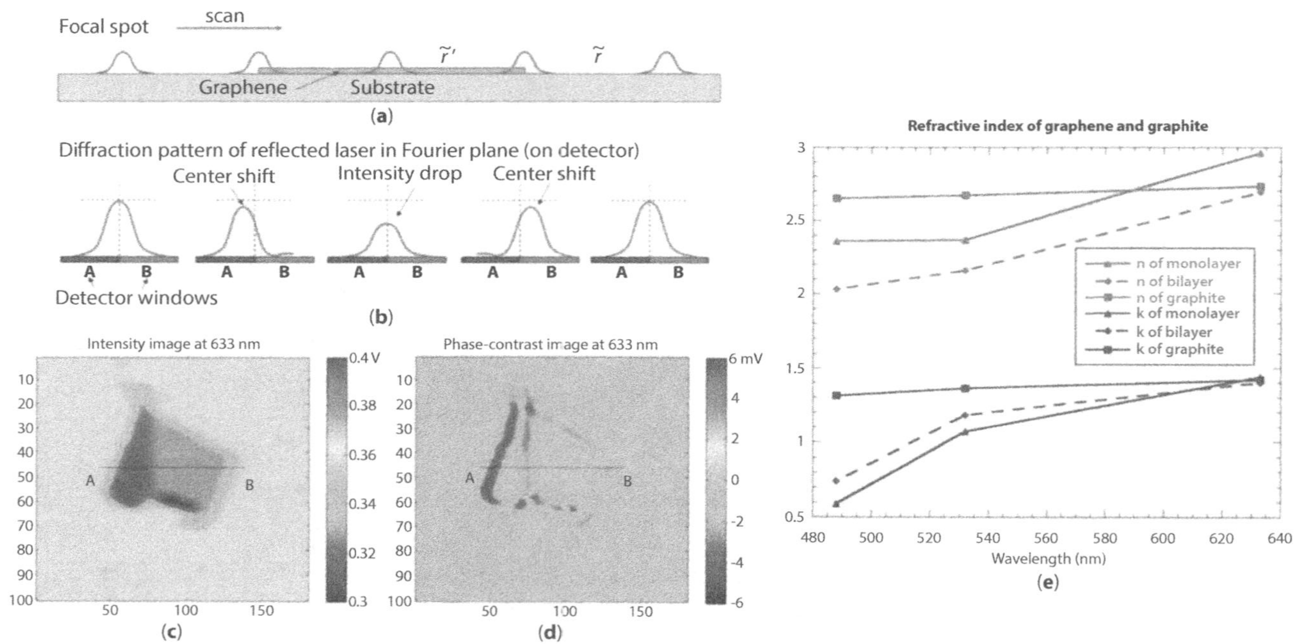

Figure 13.7 (a) (Principles of picometrology) A focused Gaussian beam scans across a graphene sample. (b) Across the edge of the graphene, the center of the diffraction pattern is shifted from the original position due to the phase difference between $\tilde{r}$ and $\tilde{r}'$, and the corresponding intensity (c) and phase-contrast (d) images. (e) Resulting $\widetilde{n_G}$ values from the inversion of the intensity and phase-contrast data for graphene, with the reference values for graphite (From Ref. [38], by permission of the OSA).

13.3.1.4 Simultaneous Reflection and Transmission Measurements

More recently, simultaneous measurements of the differential reflection $\delta R = \frac{R - R_0}{R_0}$ and differential transmission $\delta T = \frac{T - T_0}{T_0}$ for a 2-D material on a glass slide allow derivation of its complex RI as a function of wavelength (Figure 13.8) [10]. In principle, the unique determination of the complex RI of graphene is possible with the two observables above. Although the reported results were for MoS_2, $MoSe_2$, and WSe_2, it can be applied for other 2-D materials including graphene layer.

Because this method doesn't involve the optical modeling and dispersion modeling of complex RI, it is a non-spectroscopic method, which requires the two measurement observables to be independent of each other. However, the reflection and the transmission are rather strongly coupled variables, i.e., one follows another as clearly presented in Figure 13.8b. Nevertheless, they are not completely entangled each other, i.e., one cannot be readily calculated or converted from another, so the unique determination of the best solution for $\widetilde{n_G}$ is still possible

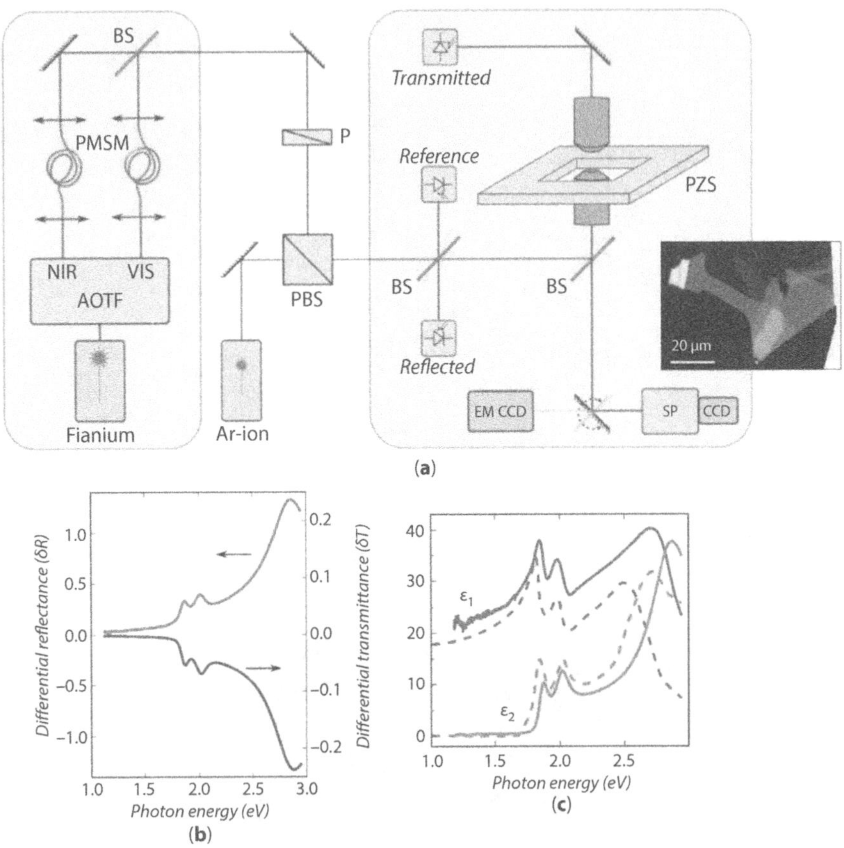

Figure 13.8 (a) Simplified schematic of the experimental setup for the simultaneous reflection and transmission measurements and the raster-scanned image of graphene layers (From Ref. [10], by permission of the AIP). (M: mirror, BS: beam splitter, PBS: polarizing beam splitter, DCM: dichroic mirror, P: polarizer, PMSM: polarization-maintaining single-mode fiber, PD: photodetector, SP: spectrometer, Fianium: supercontinuum laser, AOTF: acousto-optical tunable fiber, Ar-ion: 514 nm Ar+ laser, PZS: piezo stage.) (b) Differential reflectance and transmittance spectra and (c) extracted dielectric function of monolayer MoS2.

with high precision measurements. For this reason, the intensity values should be measured with high accuracy and precision to get reliable data to avoid excessive uncertainties.

13.3.2 RI Measurements of Near-Field Response

The word "near-field" refers to a thin region in the vicinity of the interface between two (or more) media, where the electromagnetic wave field strength changes drastically. Examples are the water/gold/glass or water/graphene/gold/glass for SPR (Figure 13.9a) and

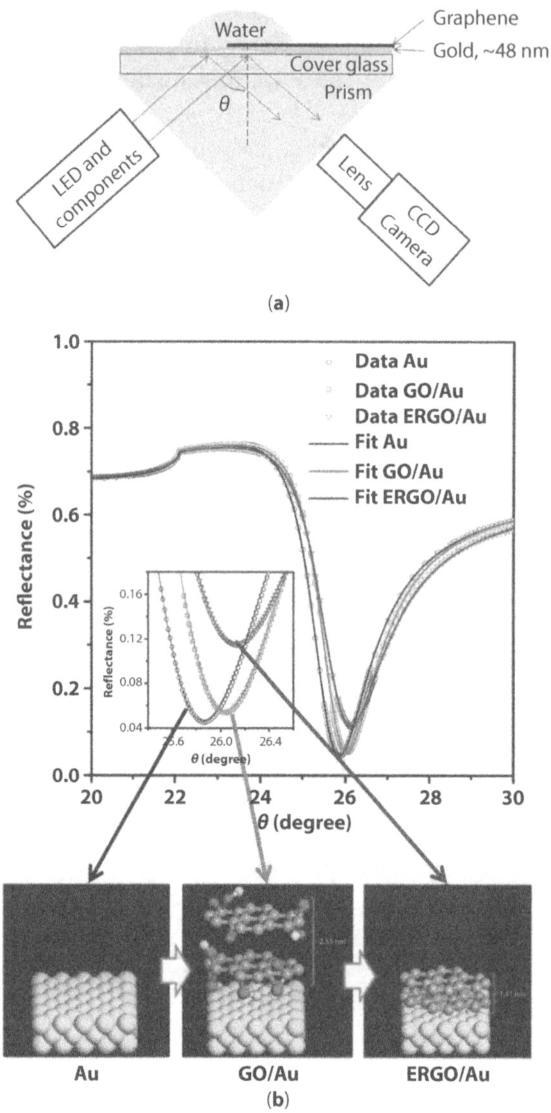

Figure 13.9 (a) The SPR reflectance imaging layout, where the far-field reflected light is captured by a CCD camera after a near-field light–graphene interaction. The water is used for the sensitivity enhancement. (From Ref. [42], by permission of the OSA.) (b) SPR reflectance as functions of incident angle for the bare Au film, graphene oxide (GO) on Au, and the electrochemically reduced graphene oxide (ERGO) on Au. (From Ref. [43], by permission of the ACS.)

substrate/graphene/glass for ATR (Figure 13.10a). There is no strict or universal definition on the thickness of the near-field region but in this review, it refers to either the surface plasmon polariton penetration depth (for SPR) or the evanescent wave field depth (for ATR). Note that the two depths are in the same order of a few hundred nanometers, and change with incident angle of light, wavelength, and optical configurations [39].

The biggest merit of the near-field methods is that the interaction between the electromagnetic fields and the sample in the near-field is much stronger than the far-field light–sample interactions that are conventional in optical microscopy. This leads to the higher signal-to-noise ratio for the near-field methods, especially when the sample is so thin that it is confined in the near-field regions, like graphene or other extremely thin materials [39]. The stronger signals are due to the enhanced electromagnetic fields approaching the interface as well as the van der Waals force that keeps graphene in contact with that interface.

The near-field methods covered in this chapter do not directly detect the electromagnetic waves in near-field regions by using near-field scanning probes. Instead, they measure the

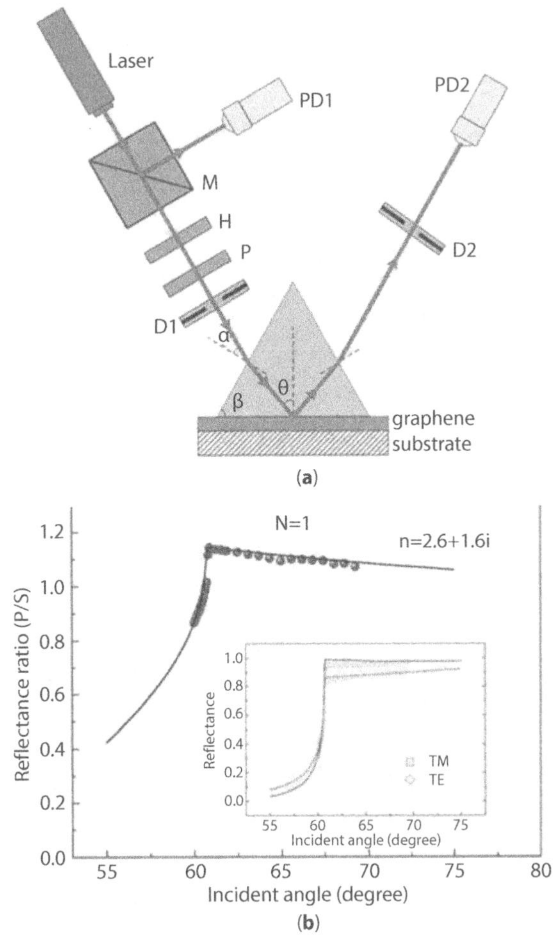

Figure 13.10 (a) The ATR reflectance measurement layout. (M: beam splitter, H: half-wave plate: P: polarizer, PD: detector, D: aperture diaphragm.) (b) Experimental and calculated angular dependence of optical reflectance ratio of p- and s-polarized light. (From Ref. [45], by permission of the AIP.)

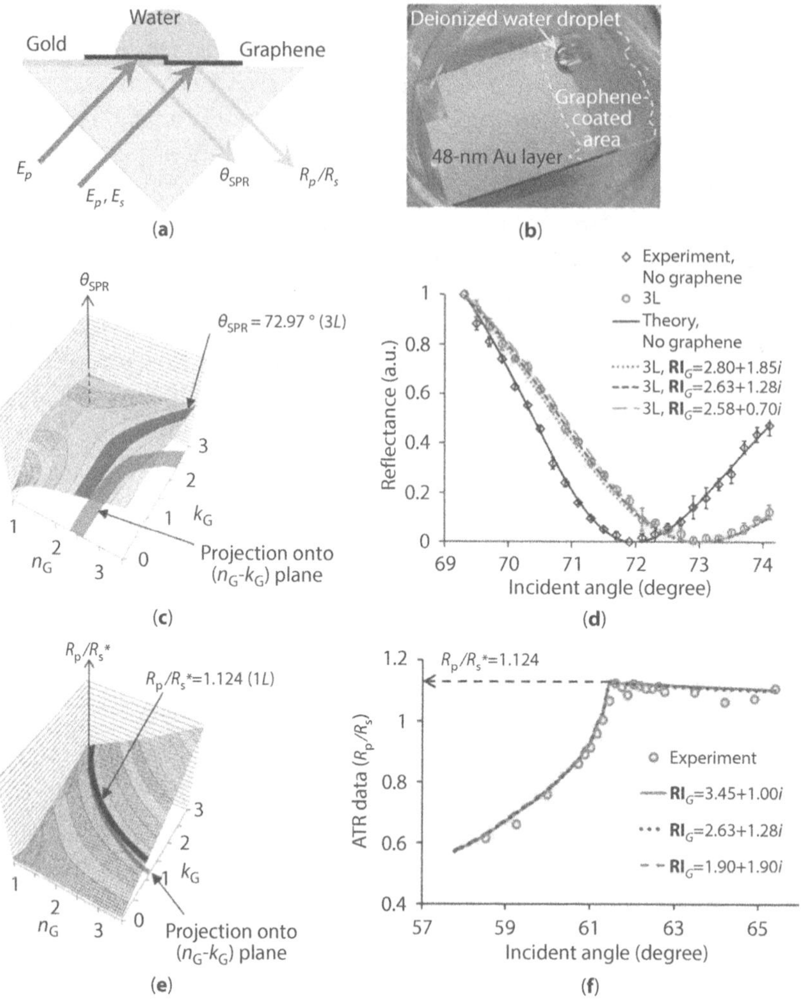

Figure 13.11 (a) A graphene sample laid on a BK7 glass slide coated with 48-nm thick Au film, and (b) a schematic of the tandem measurements for the SPR angle (θ_{SPR}) and the reflectance ratio by ATR (R_p/R_s*). The deionized water environment is imposed for enhanced measurement sensitivity. Uncertainties associated with the single constraints of SPR (c, d) and ATR (e, f). (c) 3-D plot of the RI sensitivity calculation of SPR angle, with the measured value of θ_{SPR} = 72.97°, (d) experimental results for SPR data curves and the data fitting with three different $\widetilde{n_G}$ (RIG in the marking), (e) RI sensitivity calculations of ATR reflectance ratio, with the measured value of R_p/R_s* = 1.124, and (f) measured ATR data and its fitting with three arbitrarily selected $\widetilde{n_G}$. (From Ref. [44], by permission of the Springer Nature.)

far-field reflection light away from the samples that contain the near-field information generated by graphene on the interfaces (Figures 13.9a, 13.10a, and 13.11a).

13.3.2.1 Surface Plasmon Resonance (SPR) and Attenuated Total Internal Reflection (ATR)

Surface plasmon refers to the surface electromagnetic wave that can be excited on a metal–dielectric interface [40] by *p*-polarized light for a special configuration suggested by Krechtmann (Figure 13.9a). (Note that it is different from the localized SPR excited by metal nanoparticles [41].) At a specific incident angle, the excitation of the surface plasmons is maximized and consequently the reflection is minimized for a specific incident angle, which is called surface plasmon resonance (SPR). The SPR is widely used in biomolecular sensing, thanks to the very high sensitivity to the RI change, and this merit enables the reliable detection of sub-nm-thin graphene layers by SPR. Various information such as the layer number [42] and the optical properties of graphene [43, 44] can be obtained when the SPR reflectance is measured with respect to the incident angle.

In Figure 13.9b, typical reflectance vs. incident angle data curves are presented for three different samples, i.e., water, graphene oxide (GO), and electrochemically reduced graphene oxide (ERGO). The measured SPR curves are fitted against the Fresnel's calculations to find out the RI values of GO and ERGO that match the experimental data the best (Figure 139b). The exact formulas for the SPR reflection curves are constructed by using the transfer matrix methods ([11, 44]). However, this method resorts to only one observable (SPR reflectance) to measure the two components of $\widetilde{n_G}$. Therefore, very accurate measurements are required for the unique determination of $\widetilde{n_G}$, but practically speaking, it will be very difficult to circumvent the ambiguities associated with multiple solution pairs that are all well inside the uncertainty range [11]. Note that the reported RI values in [43] are not for pure graphene, but for GO and ERGO, where the Raman spectra of both show different trends from the typical pure graphene samples.

Another method that shares a similar limitation is ATR [45], where the optical configuration is very similar to that of SPR (Figure 13.10a), except for using a clean prism with no Au layer laid and inclusion of both polarities, while SPR is usually measured for the p-polarization only. The measurement and the data fitting schemes are also similar to those explained for SPR in the previous paragraph, but the shape of the data curve is largely different (Figure 13.10b). For the same reason as described above for SPR, the ATR shares the fundamental limitation that the unique determination of $\widetilde{n_G}$ is difficult. It has been presented that the unique determination of the solution pair of $\widetilde{n_G}$ solely by ATR measurements results in large uncertainty [44].

13.3.2.2 Tandem Use of SPR and ATR

As explained in the previous section, when trying to fit a single measurement observable (e.g., reflectance) as a function of an experimental parameter (e.g., incident angle) to get both the real and imaginary parts of $\widetilde{n_G}$, it is inevitable to accommodate the fundamental limitation on the RI sensitivity and uncertainty. Although the methods in Section 13.4.1 use near-field phenomena that are much more sensitive to $\widetilde{n_G}$ than the most far-field methods, the aforementioned limitation is not completely resolved because of the fact that they use a single observable against two unknowns.

A tandem use of SPR and ATR was realized as a solution to this fundamental limitation [44], where the configuration for the simultaneous measurements and the actual graphene sample are shown in Figures 13.11a and 13.11b, respectively. The graphene layer on the Au-coated region is for the θ_{SPR} measurements and the graphene on the uncoated part is for the ATR R_p/R_s measurements, while both regions are covered with water for enhanced measurement sensitivity.

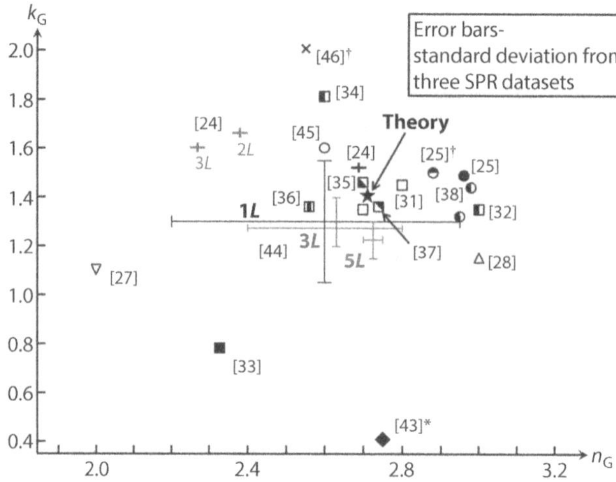

Figure 13.12 Reported complex RI values by various methods: reflection spectroscopy [24, 27–28], DFT [25], [25]†, spectroscopic ellipsometry [31–37], [46]†, picometrology [38], SPR [43]*, tandem use of SPR and ATR [44], and ATR [45]. * and † denote reduced graphene oxide and HOPG (highly oriented pyrolytic graphite), respectively. (From Ref. [44], by permission of the Springer Nature.)

Measurements results of SPR (Figure 13.11d) and ATR (Figure 13.11f) show that the experimental results can be reproduced within the measurement uncertainties for all the combination of (n_G, k_G), which are shown as the gray bands on the projection planes in Figure 13.11c or e. This again demonstrates that either the SPR (θ_{SPR}) or the ATR (R_p/R_s) observable alone cannot uniquely determine the complex RI of graphene. However, the two projections in Figures 13.11c and 13.11e cross at a single point, which means that the two observables are independent of each other. More importantly, the pair of (n_G, k_G) falling on the crossing point satisfies both the SPR and the ATR measurement constraints and uniquely determines $\widetilde{n_G}$.

In Figure 13.12, the SPR/ATR measured RI results are shown for 1-, 3-, and 5-layer graphene samples. The averaged RI value for the three samples is 2.71+1.41*i*, while the deviations due to the different layer numbers remain within ±3%. As the number of the graphene layer increases, the uncertainty becomes smaller due to the higher sensitivity from the amplified near-field interaction of light with graphene. Most major published complex RI results to date are also shown by various far-field as well as near-field methods and the largest deviation of [2] may be attributed to the aforementioned oversimplified assumption of no dispersion (Section 13.3.1) [27, 44]. Note that the symbols * and † are corresponding to RI data measured for the electrochemically reduced graphene oxide (ERGO) and graphite, respectively.

13.4 Summary

In this review, the optical properties of graphene have been reviewed comprehensively, ranging from the definition of basic properties (Section 13.2), theoretical works (Section 13.2), and experimental measurements by both far-field (Section 13.3) and near-field (Section 13.4) methods. All the reported values of complex RI of graphene in the text are in Figure 13.12 and Table 13.1, which shows the widespread data due to the difficulty of the accurate measurement.

Table 13.1 Measured and calculated complex refractive index (RI) values of graphene for $\lambda = 633$ nm.

References	Methods	Descriptions	RI (n_G, k_G)	Note	Sample
44	SPR and ATR measurements	Measurements of θ_{SPR} and R_p/R_s and simultaneous matching with calculation results	(2.58±0.38, 1.3±0.25)		CVD, 1L
			(2.63±0.20, 1.28±0.10)		CVD, 3L
			(2.63±0.20, 1.28±0.10)		CVD, 5L
	DFT	Same as in Ref. [25]	(2.71, 1.41)	Oscillatory behavior similar to that in Ref. [25]	Graphene (2 carbon atoms in a unit cell)
27	Reflection spectroscopy	Reflected light from a graphene sample by a spectrometer (illumination: white)	(2.0, 1.2)	Assumed that n_G and k_G are λ-independent.	Exfoliation, 1L
24			(2.69±0.02, 1.52±0.02)		CVD, 1L
			(2.38±0.02, 1.66±0.02)		CVD, 2L
			(2.27±0.02, 1.60±0.02)		CVD, 3L
28		Same as above, but the experimental data are from Ref. [6] and [29]	(3.0, 1.15)	Modeling: $K_G \quad 5.446 \quad m^{-1} \dfrac{\lambda}{n_G}$, and fitted only for n_G.	Exfoliation, 1L-2L (Single result)

(*Continued*)

Table 13.1 Measured and calculated complex refractive index (RI) values of graphene for $\lambda = 633$ nm. (*Continued*)

References	Methods	Descriptions	RI (n_G, k_G)	Note	Sample
31-34	Spectroscopic ellipsometry	Polarization state change of light beam of tunable wavelength, when reflected at the graphene with oblique incidence	(2.8, 1.45) (2.7, 1.35) (3.0, 1.35) (2.3, 0.8) (2.6, 1.81)	Needs a dispersion modeling with 4 or more fitting parameters.	Exfoliation, 1*L*
35		Point-to-point fitting with no need of dispersion modeling	(2.71, 1.46)	Solution optimization is required.	CVD, 1*L*
36			(2.56, 1.36) -1*L* data	Optical conductivity is obtained.	CVD, 1*L*-2*L*
37		B-spline method with no need of dispersion modeling	(2.74, 1.36)	Fitting elaboration is required.	Exfoliation, 1*L*
45	Polarization-dependent ATR	Attenuated total internal reflection of monochromatic, p- and s-polarized light at air/graphene/prism with changing incident angle	(2.6, 1.6)	Determination of unique n_G and k_G will be challenging.	CVD, 1*L*
			(2.6, 1.6)		CVD, 2*L*
			(2.6, 2.1)		CVD, 4*L*
38	Picometrology	Amplitude and phase change of reflected light when a focused monochromatic light traverses a graphene boundary	(2.95, 1.32) (2.98, 1.44)	Beam profile other than Gaussian and slight defocusing should be noted.	Exfoliation, 1*L*

(*Continued*)

Table 13.1 Measured and calculated complex refractive index (RI) values of graphene for $\lambda = 633$ nm. (*Continued*)

References	Methods	Descriptions	RI (n_G, k_G)	Note	Sample
25	Density functional theory (DFT)	First principles calculation by full potential linear muffin-tin orbital (FP-LMTO) within local density approximation (LDA) and generalized gradient approximation (GGA)	(2.96, 1.49)	Oscillatory behavior of the dielectric functions for the visible wavelengths	Graphene (2 carbon atoms in a unit cell)
	DFT	Same as above	(2.88, 1.50)		Graphite
46	Spectroscopic ellipsometry	Same as above	(2.56, 2.03)	Measurements with several wavelengths	HOPG (highly oriented pyrolytic graphite), ZYA grade after tape cleavage
43	SPR	Reflectance of monochromatic and *p*-polarized light from water/graphene/Au/Cr/glass with changing incident angle	(2.75, 0.41)	Used graphene oxide samples with 6 fitting parameters to be determined	Reduced graphene oxide laid on an Au/Cr/glass substrate

The theoretical works (Sections 13.2.2 and 13.2.3) paved a way to understand the optical response of graphene in a wide range of wavelengths and suggested various optical applications. The experimental methods to measure the RI of graphene, evolved for more sensitive and robust techniques that can overcome the fundamental obstacles. As a result, reliable measurements of $\widetilde{n_G}$ are now possible, despite of graphene's mere 0.335 nm thickness. These techniques are expected to be applied for development of various next-generation 2-D materials [47, 48] as well, for future applications in transparent and flexible electronics based on 2-D materials.

References

1. Allen, M.J., Tung, V.C., Kaner, R.B., Honeycomb carbon: A review of graphene. *Chem. Rev.*, 110, 132, 2010.
2. Lee, C., Wei, X., Kysar, J.W., Hone, J., Measurement of the elastic properties and intrinsic strength of monolayer graphene. *Science*, 321, 5887, 385, 2008.
3. Bolotin, K.I., Sikes, K.J., Jiang, Z., Klima, M., Fudenberg, G., Hone, J., Kim, P., Stormer, H.L., Ultrahigh electron mobility in suspended graphene. *Solid State Commun.*, 146, 9-10, 351, 2008.
4. Novoselov, K.S., Geim, A.K., Morozov, S.V., Jiang, D., Zhang, Y., Dubonos, S.V., Grigorieva, I.V., Firsov, A.A., Electric field effect in atomically thin carbon films. *Science*, 306, 5696, 666, 2004.
5. Cai, W., Moore, A.L., Zhu, Y., Li, X., Chen, S., Shi, L., Ruoff, R.S., Thermal transport in suspended and supported monolayer graphene grown by chemical vapor deposition. *Nano Lett.*, 10, 5, 1645, 2010.
6. Nair, R.R., Blake, P., Grigorenko, A.N., Novoselov, K.S., Booth, T.J., Stauber, T., Peres, N.M.R., Geim, A.K., Fine structure constant defines visual transparency of graphene. *Science*, 320, 5881, 1308, 2008.
7. Sanderson, K., Graphene electrode promises stretchy circuits. *Nature*, doi: 10.1038/news .2009.28. 2009.
8. Graphene, http://archive.cnx.org/contents/790bacf3-6512-4957-bbed-ac887a4fca7c@4/graphene
9. Jackson, J.D., *Classical electrodynamics*, 3rd edition, John Wiley & Sons, Inc., Hoboken, NJ, 1999.
10. Morozov, Y.V. and Kuno, M., Optical constants and dynamic conductivities for single layer MoS2, MoSe2, and WSe2. *Appl. Phys. Lett.*, 107, 083103, 2015.
11. Cheon, S., *Refractive index characterization of CVD graphene using near-field optical techniques*, Doctorate thesis, Seoul National University, 2015.
12. Balanis, C.A., *Advanced engineering electromagnetics*, John Wiley & Sons, Inc., Section 2.8.1, 1989.
13. Kirby, B.J., *Micro- and nanoscale fluid mechanics: Transport in microfluidic devices*, Cambridge University Press, Section 5.3, 2010.
14. Skulason, H.S., Gaskell, P.E., Szkopek, T., Optical reflection and transmission properties of exfoliated graphite from a graphene monolayer to several hundred graphene layers. *Nanotechnology*, 21, 295709, 2010.
15. Lami, E.B-., Faucheu, J., Noël, A., Latex routes to graphene-based nanocomposites. *Polym. Chem.*, 6, 5323, 2015.
16. David Tong: Lectures on Kinetic Theory, 4. Linear Response, http://www.damtp.cam.ac.uk/user/tong/kintheory/four.pdf
17. Mak, K.F., Ju, L., Wang, F., Heinz, T.F., Optical spectroscopy of graphene: From the far infrared to the ultraviolet. *Solid State Commun.*, 152, 1341, 2012.

18. Semenoff, G.W., Condensed-matter simulation of a three-dimensional anomaly. *Phys. Rev. Lett.*, 53, 2449, 1984.
19. Stauber, T., Peres, N.M.R., Geim, A.K., Optical conductivity of graphene in the visible region of the spectrum. *Phys. Rev. B*, 78, 085432, 2008.
20. Ando, T., Zheng, Y., Suzuura, H., Dynamical conductivity and zero-mode anomaly in honeycomb lattices. *J. Phys. Soc. Jpn.*, 71, 1318, 2002.
21. Gusynin, V.P., Sharapov, S.G., Carbotte, J.P., Unusual microwave response of Dirac quasiparticles in graphene. *Phys. Rev. Lett.*, 96, 256802, 2006.
22. Mak, K.F., Shan, J., Heinz, T.F., Seeing many-body effects in single- and few-layer graphene: Observation of two-dimensional saddle-point excitons. *Phys. Rev. Lett.*, 106, 046401, 2011.
23. Gusynin, V.P. and Sharapov, S.G., Transport of Dirac quasiparticles in graphene: Hall and optical conductivities. *Phys. Rev. B*, 73, 245411, 2006.
24. Ghamsari, B.G., Tosado, J., Yamamoto, M., Fuhrer, M.S., Anlage, S.M., Measuring the complex optical conductivity of graphene by Fabry-Pérot reflectance spectroscopy. *Sci. Rep.*, 6, 34166, 2016.
25. Klintenberg, M., Lèbegue, S., Ortiz, C., Sanyal, B., Fransson, J., Eriksson, O., Evolving properties of two-dimensional materials: From graphene to graphite. *J. Phys.: Condens. Matter*, 21, 335502, 2009.
26. Blake, P. and Hill, E.W., Making graphene visible. *Appl. Phys. Lett.*, 91, 063124, 2007.
27. Ni, Z.H., Wang, H.M., Kasim, J., Fan, H.M., Yu, T., Wu, Y.H., Feng, Y.P., Shen, Z.X., Graphene thickness determination using reflection and contrast spectroscopy. *Nano Lett.*, 7, 9, 2758, 2007.
28. Bruna, M. and Borini, S., Optical constants of graphene layers in the visible range. *Appl. Phys. Lett.*, 94, 031301, 2009.
29. Bassani, F. and Pastori Parravicini, G., Band structure and optical properties of graphite and of the layer compounds GaS and GaSe. *Nuovo Cimento*, B 50, 95, 1967.
30. Zhang, H., Ma, Y., Wan, Y., Rong, X., Xie, Z., Wang, W., Dai, L., Measuring the refractive index of highly crystalline monolayer MoS_2 with high confidence. *Sci. Rep.*, 5, 8440, 2015.
31. Kravets, V.G., Grigorenko, A.N., Nair, R.R., Blake, P., Anissimova, S., Novoselov, K.S., Geim, A.K., Spectroscopic ellipsometry of graphene and an exciton-shifted van Hove peak in absorption. *Phys. Rev. B*, 81, 155413, 2010.
32. Matković, A., Beltaos, A., Milićević, M., Ralević, U., Vasić, B., Javanović, D., Gajić, R., Spectroscopic imaging ellipsometry and Fano resonance modeling of graphene. *J. Appl. Phys.*, 112, 123523, 2012.
33. Wurstbauer, U., Röling, C., Wurstbauer, U., Wegscheider, W., Vaupel, M., Thiesen, P.H., Weiss, D., Imaging ellipsomtry of graphene. *Appl. Phys. Lett.*, 97, 231901, 2010.
34. Martínez, E.O., Gabás, M., Barrutia, L., Pesquera, A., Centeno, A., Palanco, S., Zurutuza, A., Algora, C., Determination of a refractive index and an extinction coefficient of standard production of CVD-graphene. *Nanoscale*, 7, 1491, 2015.
35. Nelson, F.J., Kamineni, V.K., Zhang, T., Comfort, E.S., Lee, J.U., Diebold, A.C., Optical properties of large-area polycrystalline chemical vapor deposited graphene by spectroscopic ellipsometry. *Appl. Phys. Lett.*, 97, 253110, 2010.
36. Chang, Y.-C., Liu, C.-H., Liu, C.-H., Zhong, Z., Norris, T.B., Extracting the complex optical conductivity of mono- and bilayer graphene by ellipsometry. *Appl. Phys. Lett.*, 104, 261909, 2014.
37. Weber, J.W., Calado, V.E., van de Sanden, M.C.M., Optical constants of graphene measured by spectroscopic ellipsometry. *Appl. Phys. Lett.*, 97, 091904, 2010.
38. Wang, X., Chen, Y.P., Nolte, D.D., Strong anomalous optical dispersion of graphene: Complex refractive index measured by picometrology. *Opt. Express*, 16, 26, 22105, 2008.

39. Kihm, K.D., *Near-field characterization of micro/nano-scaled fluid flows*, Springer-Verlag Berlin Heidelberg, 2011.
40. Schasfoort, R.B.M. and Tudos, A.J., *Handbook of surface plasmon resonance*, RSC Publishing, 2008.
41. Willets, K.A. and Van Duyne, R.P., Localized surface plasmon resonance spectroscopy and sensing. *Annu. Rev. Phys. Chem.*, 58, 267, 2007.
42. Cheon, S., Kihm, K.D., Park, J.S., Lee, J.S., Lee, B.J., Kim, H., Hong, B.H., How to optically count graphene layers. *Opt. Lett.*, 37, 18, 3765, 2012.
43. Xue, T., Cui, X., Chen, J., Chen, J., Liu, C., Wang, Q., Wang, H., Zheng, W., A switch of the oxidation state of graphene oxide on a surface plasmon resonance chip. *ACS Appl. Mater. Interfaces*, 5, 2096, 2013.
44. Cheon, S., Kihm, K.D., Kim, H.G., Li, G., Park, J.S., Lee, J.S., How to reliably determine the complex refractive index (RI) of graphene by using two independent measurement constraints. *Sci. Rep.*, 4, 6364, 2014.
45. Ye, Q., Wang, J., Liu, Z., Deng, Z.-C., Kong, X.-T., Xing, F., Chen, X.-D., Zhou, W.-Y., Zhang, C.-P., Tian, J.-G., Polarization-dependent optical absorption of graphene under total internal reflection. *Appl. Phys. Lett.*, 102, 021912, 2013.
46. Jellison, G.E., Jr., Hunn, J.D., Lee, H.N., Measurement of optical functions of highly oriented pyrolytic graphite in the visible. *Phys. Rev. B*, 76, 085125, 2007.
47. Radisavljevic, B., Radenovic, A., Brivio, J., Giacometti, V., Kis, A., Single-layer MoS_2 transistors. *Nat. Nanotechnol.*, 6, 147, 2011.
48. Li, L., Yu, Y., Ye, G.J., Ge, Q., Ou, X., Wu, H., Feng, D., Chen, X.H., Zhang, Y., Black phosphorus field-effect transistors. *Nat. Nanotechnol.*, 9, 372, 2014.

14

Fractional Quantum Hall Effect in Graphene, a Topological Approach

Janusz E. Jacak

Department of Quantum Technologies, Faculty of Fundamental Problems of Technology, Wrocław University of Science and Technology, Wrocław, Poland

Abstract

Recent experimental progress in Hall measurements in graphene allowed for observation of fractional quantum Hall effect (FQHE) in first six subbands (with $n = 0$ and $n = 1$ Landau levels) in monolayer graphene and in eight subbands (with $n = 0$, $n = 1$, and $n = 2$ Landau levels) in bilayer graphene. The observed new features, especially in bilayer system, do not agree with a conventional model of composite fermions revealing phenomena previously not encountered in conventional GaAs Hall systems. This sheds a new light on the foundations of correlated multiparticle states responsible for FQHE manifestation in terms of topology different in the bilayer system in comparison to the monolayer one and going beyond local models of band structure. We have developed the general nonlocal topological approach to FQHE that allowed for an identification of topology-type factor responsible for the oddness of Hall physics in bilayer systems. The presented theory includes composite fermion (CF) model as its special case and explains insufficiency of CF construction for bilayer graphene. A consistence of the developed topological approach with current experimental data in monolayer and bilayer graphene is demonstrated.

Keywords: Monolayer graphene, bilayer graphene, fractional quantum Hall effect, hierarchy of filling fractions, correlated state, braid group commensurability

14.1 Introduction

Fractional quantum Hall effect (FQHE) is still refractory to its complete understanding despite intensive experimental and theoretical studies. Specific long-range correlations induced by the interaction in the case of a quenched kinetic energy competition in Landau levels (LLs) are responsible for FQHE organization exceeding a conventional framework of local quantum mechanics and cannot be described upon the standard broken symmetry scheme of phase transitions in condensed matter. In formation of nonlocal specific correlations in FQHE, the 2D topology plays the more important role than particularities of system organization. FQHE is essentially the same regardless of materials, their band structure, and even absence of external magnetic field in fractional Chern insulators. Quantum Hall effects are absent in 3D systems.

Email: janusz.jacak@pwr.edu.pl

Tobias Stauber (ed.) Handbook of Graphene: Volume 2, (413–454) © 2019 Scrivener Publishing LLC

FQHE has been discovered in 1982 [1] in 2DEG of GaAs heterostructure. Since this first observation, the effect has been widely studied both experimentally and theoretically [2–4] mostly in conventional GaAs systems. An interest in FQHE has been rapidly renewed after observation of this phenomenon in graphene [5–8]. An experimental examination of FQHE in graphene is especially challenging because in this system, one can control LL filling fraction independently of magnetic field by electric field application. This is due to a specific band structure in graphene, where a relatively small lateral voltage (ca. 10–60 V) can shift the Fermi level along the ladder of LLs independently of the magnetic field value. Graphene is perfectly two dimensional and, moreover, due to the exceptional gapless band structure, this semiconductor exhibits a modified LLs quantization in comparison to GaAs 2DEG. The specific for graphene band structure is dominated by the presence of so-called Dirac cones in corners of the hexagonal Brillouin zone. In the apexes of these cones, the valence band meets with the conduction band resulting in linear in momentum local Hamiltonian [9]. Substitution of ordinary parabolic semiconductor bands with linear crossing band cones perturbs ordinary Landau quantization and leads to nonequidistant LL structure. The pseudo-relativistic Dirac dynamics causes LL energy proportional to $\sqrt{n}$ (n is number of the LL) instead of linear $\sim n$ LL energy dependence in conventional 2DEG. Despite this essential difference in LL quantization, both integer quantum Hall effect (IQHE) and FQHE are observed in graphene [5–9] quite similar as in conventional 2DEG. The pseudo-relativistic band structure close to the Dirac points causes, however, some specific modifications in both quantum Hall effects. These modifications resolve themselves mostly to subband reorganization in graphene in comparison to conventional 2DEG, which convincingly proves that the 2D topology supports organization both of IQHE and FQHE independently of material particularities. The topology constraints imposed on trajectories crucial for IQHE and FQHE in terms of the Feynman path quantization are apparently beyond the band structure even in its limiting pseudo-relativistic form as encountered in graphene. This is understandable if to remind that the crystal field is actually induced by local electric-type interaction and cannot change the topological features common to all 2D charged systems exposed to strong magnetic field. Hence, the IQHE and FQHE in graphene should repeat the general topology scheme.

An impressive development in techniques of manufacturing monolayer and bilayer graphene samples induced recently a rapid boom in Hall experiment in this material. Within last years there were reported several new observations of FQHE in monolayer and bilayer graphene with an unprecedented accuracy and range, exceeding previous studies in conventional 2DEG materials. This is caused by larger stability of Hall correlated states in Dirac-like material, and observation of IQHE in graphene is possible even up to room temperature. FQHE is also noticeable in graphene in higher temperatures than in GaAs 2DEG even up to 10–20 K [7, 8]. In this situation, the reduction of temperature to millikelvin scale allows for revealing new features of fragile correlated Hall states with exceptional accuracy and resolution. In monolayer graphene, FQHE features have been observed in six consecutive subbands of the LL structure [10–13], with filling factor $-6<\nu=\frac{N}{N_0}<6$ (the negative mirror filling rates correspond to hole states in valence band symmetrically located with respect to electrons in conduction band in graphene). In bilayer graphene, the range of observation is even larger and reaches eight consecutive subbands [14–18]. Interestingly, the observations of FQHE in bilayer graphene reveal some distinctions in comparison to FQHE manifestation in monolayer graphene and to conventional semiconductor 2DEG [10–13]. This fact is, in particular, in conflict with a conventional imagination that FQHE can be explained in terms of hypothetical composite fermions (CFs) [19] proposed as electrons equipped with localized on each particle even number of flux quanta of some auxiliary magnetic field. According to CF theory, such effective dressed quasiparticles should be present both in monolayer and bilayer Hall systems, which, however, does not agree with experimental observations in bilayer graphene [14–18].

In the present chapter, we identify the topological features common for all 2D Hall systems and identify the specific topological effect that may explain the oddness of correlated multiparticle states in the bilayer system in consistence with experimental observations. We summarize the experimental observations of FQHE in graphene and compare them with theoretical predictions both for monolayer and bilayer graphene.

The progress in Hall experiment in graphene has been achieved by successful mastering of a new technique for sampling of suspended graphene scrapings (both monolayer and bilayer) [12–18] and of graphene sheets supported by hexagonal boron nitride (hBN) crystal substrate [10, 11]. The absence of a substrate in the case of suspended samples or avoiding of by-substrate-induced lattice mismatch for hBN substrate commensurate with hexagonal graphene structure, favor in both cases the formation of fragile correlations of FQHE induced by interaction of 2D electrons. It must be, however, emphasized that in both configurations, graphene samples of an ultrahigh cleanliness are required with electron mobility of order of 200 000 $cm^2/(Vs)$, which evidenced the triggering role in the FQHE correlation formation of a long free path exceeding the sample size.

Two carbon atoms in the Bravais cell in graphene result in two equivalent crystalline planar sublattices, which together with vanishing of the forbidden semiconductor gap in Dirac points in two nonequivalent corners of the Brillouin zone lead to the four-fold spin-valley degeneracy of LLs [9, 20]. The presence of Dirac points in graphene causes chirality of carriers leading to the specific Berry phase-like shift in LL energy spectrum. In the monolayer graphene case, this Berry phase is equal π and causes IQHE plateaus in monolayer graphene to occur at half-filling of consecutive LLs, according to the formula for IQHE filling rates $\nu=4(n+\frac{1}{2})$ (n enumerates LLs) [9].

In bilayer graphene, the LL organization is different than in monolayer case [9, 21]. Due to hopping of electrons between two sheets of bilayer system, an extra degeneracy of $n = 0$ and $n = 1$ LLs state occurs [21, 22]. This property is visible in the form of effective local Hamiltonian upon the tight binding approximation for graphene, which, after inclusion of the interlayer hopping in bilayer structure, gains the term $\pi\pi$ of lowering oscillatory operator π, vanishing both $n = 0$ and $n = 1$ states—cf. Appendix 14.6. The Barry phase twice larger in bilayer graphene in comparison to the monolayer case shifts IQHE plateau positions to edges of LLs. LLs in bilayer graphene are also four-fold spin-valley degenerated except for the eight-fold degenerated lowest LL (LLL) (due to the extra $n = 0$ and 1 degeneracy) [9, 20–23].

In the case of monolayer graphene, the filling fractions for FQHE are observed in six first subbands of LLs with $n = 0$ and $n = 1$ [10–13], which reproduce a hierarchy similar to the one in conventional semiconductor 2DEG. In bilayer graphene, observation of FQHE features reaches even subbands with $n = 2$ [11, 15]. Especially interesting are, however, observations of unusual even denominator fillings for FQHE in bilayer graphene in the LLL including the most pronounced feature at $\nu=-\frac{1}{2}$ [14], which does not find any counterpart in monolayer system. In particular, this state cannot be explained upon the CF approach, as for CFs the Hall metal state is predicted at $\pm\frac{1}{2}$ filling ratio [19]. Modification of FQHE in bilayer graphene evidences changes in topology distinguishing a two 2D sheet system in comparison to single sheet one.

In the present chapter, we propose explanation of FQHE hierarchy in monolayer and bilayer graphene by the topological commensurability approach [24–26] going beyond the CF model. We identify and describe the topological reason for differences between monolayer and bilayer systems leading to distinct Hall state organization in both situations. We explain the oddness of FQHE hierarchy in bilayer graphene including the exotic even denominator fractions for FQHE in the LLL and the unconventional states in subbands with $n = 2$ [15]. We also propose how to explain the observed rearrangement of FQHE series in the LLL of bilayer graphene in suspended samples in comparison to samples supported by hBN substrate in terms of various schemes of SU(4) symmetry lifting [14, 15]. In topology terms, we also interpret the experimentally studied phase transition in Hall states in bilayer graphene in response to application of vertical electric field tuning or even blocking the interlayer hopping [17]. Within the topological nonlocal braid group approach, we explain the structure of FQHE filling hierarchy and explain the reason of the failure of CF model in the bilayer graphene case.

14.2 CF Model of FQHE in Topology Terms

The correlated incompressible Hall states are attributed to interaction of electrons and may be identified numerically by exact diagonalization of the Coulomb interaction in small model systems. Taking into account the single-particle LL wave functions for a particular material band structure (like in graphene modeled upon standard tight binding approximation), one can find fractions corresponding to correlated states by numerical minimization of Coulomb energy in the basis of these functions (as, e.g., for bilayer graphene in Ref. [27]). To clarify the states exhibiting energy minimization, various phenomenological trial wave functions are proposed upon different schemes and ideas about the nature of correlations, including CFs, paired states of Pfaffian type, charge density waves, spin correlated states, Halperin multicomponent generalization of Laughlin function, and others. The illuminating prediction of the ground state in 2DEG at filling fraction $\nu=\frac{1}{q}$ by the Laughlin function [2] focused attention on unconventional symmetry of this function expressed by Jastrow polynomials. Initially this symmetry has been implemented by superfermions [28] renamed next to CFs and acquiring the Laughlin phase by conceptual attachment of quantized local flux of some auxiliary magnetic field to each electron, which would produce Aharonov–Bohm phase shift in agreement with Laughlin symmetry. Moreover, the mapping of FQHE onto IQHE in resultant magnetic field decreased by averaged field of CF local fluxes gave rise to model trial wave functions in the LLL at fractional fillings by wave functions of higher LLs completely filled and projected onto the LLL according to some intuitive schemes yielding a holomorphic function as required in the LLL [19]. An agreement with exact diagonalization supported the CF model despite an unclear and artificial character of the pinned flux quanta concept. Nevertheless, many fractions in the LLL experimentally observed as FQHE in conventional 2DEG are out of reach for standard CF model (e.g., $\frac{3}{8},\frac{3}{10},\frac{4}{11},\frac{5}{13},\frac{5}{17},\frac{6}{17},\frac{4}{13},\frac{7}{11},\ldots$) [4].

On the other hand, a different approach to trial wave functions for FQHE has been proposed by Halperin [29] in the form of a multicomponent generalization of Laughlin wave function. Some of multicomponent trial functions also occur very close to ground states yielded by exact diagonalization. The closeness in energy of various candidates to true ground states taken from completely different approaches with apparently distinct forms of trial wave functions evidences that different approaches are convergent to the true picture of FQHE correlations.

To enhance an insight into this situation, the topological arguments can be helpful and they can shed a light on both CF and multicomponent Laughlin–Halperin trial wave functions and their closeness. The topology approach of braid group commensurability allows for tractable discrimination of filling rates and for systematic definition of trial wave functions for FQHE states using the unitary representations of subgroups of the braid group accommodated to incompressible correlated states at particular filling rates. The method and resulting trial wave functions elucidate simultaneously the CF construction revealing its constraints and applicability range and show the linkage to multicomponent Halperin approach. The coincidence of FQHE hierarchy in various materials like conventional semiconductor GaAs 2DEG, graphene with pseudo-relativistic LLs, and even fractional Chern insulators evidences an existence of a common factor in organization of correlated incompressible states despite the sharply different single-particle band structure in various materials (and even at absence of external magnetic field in fractional Chern insulators). The unifying property of all these systems is the 2D topology and planar dynamics quantization in Feynman path integral terms, which can be systematically handled upon the braid group approach.

The main line of FQHE hierarchy, $\nu=\frac{1}{q}$, q - *odd*, is described by Laughlin multiparticle functions [2] with exponent q of the Jastrow polynomials. A heuristic concept of electrons dressed with magnetic field flux quanta, i.e., CF concept [30], was helpful in identification of filling fractions $\nu=\frac{n}{n(q-1)\pm 1}$ (q - *odd*, n - 1,2,...). The CF theory is an effective single-particle model of a strongly correlated multiparticle system [19]. The rigorous definition of quasiparticles in 2D Fermi liquid upon the quantized magnetic field is, however, precluded by discontinuity of the mass operator induced by the Coulomb interaction [31], and this specific to 2D property is called as quantization of electron separation [2, 32]. The concept of CF thus goes beyond an idea of quasiparticle locally dressed with interaction and the auxiliary field quanta pinned to CFs must display an essentially nonlocal effect, topological in nature and specific to 2D-charged systems exposed to strong perpendicular magnetic field. This basic topological effect is, however, not defined explicitly in the CF model. Moreover, as mentioned above, a long series of filling ratios in the LLL is out of reach in the conventional CF hierarchy. Hence, one can suspect that CFs describe phenomenologically a more fundamental property caused by the specific topology of 2D charged system in magnetic field and of long-range correlations triggered by electron repulsion at sufficiently large electron mobility (of order of 10^6 cm^2/(Vs) as required to observe FQHE).

In order to decipher the hidden origin and nature of CFs and related correlations of FQHE, one can employ the algebraic-topology braid group-based approach [33]. The method utilizes the Feynman path integral quantization formalism appropriately lifted to describe multiparticle systems [24, 28, 34]. Within this approach, the trajectories of particle position interchanging (braids) in the configuration space of N-particle system are considered including quantum indistinguishability of identical particles. These braid loops (mixing initial and final enumeration of particles, unified due to their indistinguishability) can be attached in any point to the open multiparticle trajectory connecting point $z_1,\ldots,z_n$ (the initial point at time instant t) with point $z'_1,\ldots,z'_N$ (the final point at time instant t') in the configuration space. Because braid loops are mutually nonhomotopic, the resulted open trajectories with attached distinct braid loops are also topologically inequivalent (cannot be transformed one into another one in a continuous way—as illustrated in Figure 14.1). All trajectories thus fall to disjoint classes of nonhomotopic trajectories (which cannot be unified by continuous deformation). These classes form the domain of the Feynman path integral [24, 34],

$$I(z_1,\ldots,z_N,t;z'_1,\ldots,z'_N,t') = \sum_{l\in\pi_1(\Omega)} e^{i\alpha_l}\int d\lambda_l e^{iS[\lambda_l(z_1,\ldots,z_N,t;z'_1,\ldots,z'_N,t')]/\hbar}, \tag{14.1}$$

where $I(z_1,\ldots,z_N,t;z'_1,\ldots,z'_N,t')$ is the propagator, i.e., the matrix element of the evolution operator of the total system in position representation that determines the probability of quantum transition from the point, $z_1,\ldots,z_N$, in time instant t to another point in the configuration space, $z'_1,\ldots,z'_N$, in time instant t', $d\lambda_l$ is the measure in the path space sector enumerated by braid group element $l \in\pi_1(\Omega)$, $\pi_1(\Omega)$ is the first homotopy group of the configuration space Ω (it is just called the braid group), $\Omega = (M^N - \Delta)/S_N$, M is 2D plane here, M^N is N-fold normal product, Δ is the collection of diagonal points in the normal product (when at least two coordinates z_i coincide) subtracted in order to assure particle number conservation, the quotient structure by the permutation group S_N accounts for quantum indistinguishability of particles, and $S[\lambda_l(z_1,\ldots,z_N,t;z'_1,\ldots,z'_N,t')]$ is the classical action for the trajectory λ_l joining selected points in the configuration space Ω between time instances t, t' and lying in lth sector of the trajectory space. The whole space of trajectories is decomposed into disjoint sectors enumerated by braid group element discrete index l. The discontinuous decomposition of the domain of the path integral into disjoint sectors (topologically inequivalent) precludes a definition of the path measure $d\lambda$ uniformly on the whole space of paths, and for each sector, the measure $d\lambda_l$ must be defined separately and finally the contributions of all sectors must be summed with unitary factors $e^{i\alpha_l}$ (unitarity is caused by the causality). It has been proved [34] that these unitary factors establish a one-dimensional unitary representation (1DUR) of the braid group. Distinct unitary weights in the path integral (i.e., distinct 1DURs of the braid group) determine different sorts of quantum

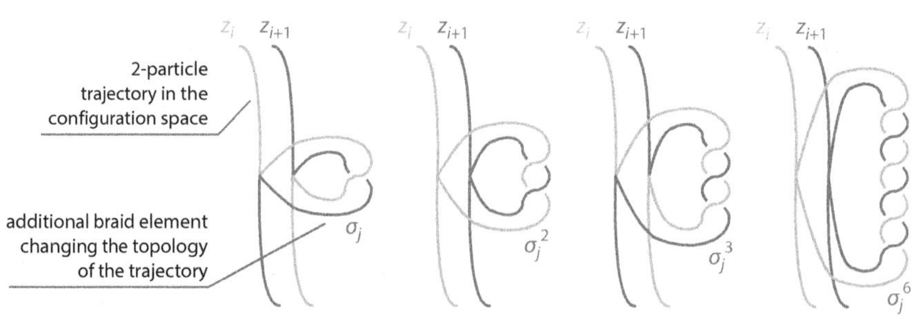

Figure 14.1 Example of nonhomotopic trajectories obtained by addition of various braids to 2-particle trajectory.

particles corresponding to the same classical ones. Braids describe particle exchanges, thus their 1DURs assign quantum statistics. Equivalently, the 1DUR of a particular braid defines a phase shift of the multiparticle wave function $\Psi(z_1,\ldots,z_N)$ when its arguments $z_1,\ldots,z_N$ (classical coordinates of particles on the plane) mutually exchange according to this braid (let us emphasize that in 2D, these exchanges are not permutations [35]).

All quantum multiparticle correlated states (including correlated states of FQHE) must be thus characterized unavoidably by a certain 1DUR of the braid group for a particular system. In 3D, the braid group of the N particle systems are always the N-element permutation groups (regardless of a charge, interaction, or magnetic field presence). There exist only two 1DURs for an arbitrary permutation group: $\sigma_j \rightarrow \begin{cases} e^{i0} = 1 \\ e^{i\pi} = -1 \end{cases}$, where σ_j, $j = 1,\ldots, N - 1$, are braids (multiparticle classical trajectories) for exchanges of positions of jth particle with $(j + 1)$th particle, when other particles remain at rest (in 3D, σ_j are simply permutation of jth and $(j + 1)$th positions; in 2D, trajectory of an exchange is also important). 1DUR=1 defines bosons and 1DUR=-1 defines fermions in 3D multiparticle systems. For 2D multiparticle systems, the braid groups are different than the permutation group [24, 33, 35] and their 1DURs are different as well, $\sigma_j \rightarrow e^{i\alpha}$, $\alpha \in (-\pi, \pi]$. Various 1DURs define 2D anyons (including 2D fermions for $\alpha = \pi$ and bosons for $\alpha = 0$) at the absence of quantizing magnetic field.

14.2.1 Braid Groups for 2D Electrons at Magnetic Field Presence

In 2D, for charged repulsing electrons in the presence of strong perpendicular magnetic field, the braid group approach does not resolve itself to anyons only! Strong magnetic field perpendicular to the basal plane considerably changes the braid group structure, which appears to be that topological factor that conditions FQHE manifestation according to the same scheme even in completely different systems with different single-particle properties and allows for CF construction in some specific situation.

Namely, for magnetic field strong enough, the planar cyclotron orbits may be too short to match neighboring particles uniformly distributed on the plane (with classical positions fixed by the repulsion of electrons—the classical distribution of 2D-charged particles at $T = 0$ K is the static triangular Wigner lattice), which precludes the existence of the braid group generators σ_j, i.e., exchanges of neighboring particles. The braids σ_j, which for charged 2D particles at magnetic field presence must be built from pieces of classical cyclotron orbits, cannot be defined in this case. Too short σ_j braids, which cannot be implemented, must be thus rejected from the braid group. Nevertheless, it has been proved [25, 33] that remaining in braid group, other braids forming a subgroup of the original group are large enough to match neighboring particles. This subgroup is called the cyclotron braid subgroup. The generators of the cyclotron subgroups are multiloop braids—such braids have in 2D larger size than single-loop braids [25]. The cyclotron subgroups allow for the definition of quantum statistics at presence of strong magnetic field via their 1DURs. In particular, 1DURs of the cyclotron braid subgroups generated by fermionic 1DUR of initial full braid group define CFs and allow for construction of related multiparticle wave functions for FQHE using symmetry constraints precisely defined by the form of cyclotron braid subgroup generators. These wave functions (without a need of any projection onto the LLL, contrary to Jain's idea of CFs) pretty well agree with exact diagonalization on small models—cf. Appendix 14.8. CFs are thus not equipped with auxiliary flux quanta but acquire the needed Laughlin phase shift according to 1DUR of the cyclotron subgroup generated by multiloop braids. The trial wave functions do not need to be projected from higher LLs (as in the CF model) but are uniquely defined according to symmetry imposed by appropriate 1DUR and by the particular form of cyclotron braid group generators. These functions in the LLL generalize the Laughlin function in a similar manner as the Halperin functions do, but are systematically and uniquely defined by the structure and symmetry of cyclotron braid generators for specific filling ratio—cf. Appendix 14.8.

From this point of view, the FQHE filling hierarchy is determined rather by the specific structure of cyclotron braid subgroups and is independent of single-particle properties of a particular Hall system, like the pseudo-relativistic LL structure of monolayer or bilayer graphene. Single-particle band properties are unimportant for FQHE hierarchy unless they can change the topology of braid trajectories. This topology is immune to dynamics particularities. The Coulomb interaction defines, however, the initial Wigner crystal distribution of classical particles—the obligatory prerequisite for commensurability constraints imposed by cyclotron braids—this is the essential role of the interaction. Hence, the FQHE hierarchy is repeated in various systems in similar form despite single-particle band structure differences. Similarly CFs can be utilized but only when this phenomenological effective picture agrees with the braid group commensurability approach—cf. Appendix 14.7 and Appendix 14.8. CFs prove to agree with the simplest braid commensurability instance, whereas the more complicated braid commensurability situations (as encountered in bilayer graphene or in higher LLs of monolayer systems) do not admit CF model, however.

The construction of appropriate cyclotron braids is possible only at some specific filling rates of the LLL when the commensurability constraints imposed on cyclotron orbit size versus particle separation are fulfilled. The discrimination of filling rates by this commensurability condition results in filling hierarchy in full consistence with experimental observations of FQHE hierarchy. In this way, there are also reproduced fractions experimentally observed in conventional semiconductor 2DEG that are out of the CF series—cf. Appendix 14.7 and Appendix 14.8. The standard CF model agrees with braid group approach for the simplest case of the commensurability only (the '$x = 1$' case precisely defined in Appendix 14.7). The short summary of the topological braid group approach to FQHE including graphene is given below.

14.2.2 FQHE, Cyclotron Braids, and Commensurability Condition

To clarify the structure of the cyclotron braid subgroups, let us note that in the presence of strong magnetic field, the 2D braids for charged particles must be built from pieces of cyclotron orbits and these orbits, of definite size in each LL, must precisely fit to interparticle spacing fixed by Coulomb interaction. Otherwise, an implementation (definition) of braids is precluded at magnetic field presence in charged 2D system. Hence, the commensurability of planar braids with particle spacing is an unavoidable condition to define the braid group and to establish the quantum statistics (by 1DURs of the defined braid group). The quantum statistics is a necessary prerequisite to any multiparticle correlated state. Therefore, the commensurability condition selects the magnetic field strength (or equivalently, Landau-level filling rate) at which a correlated state can be formed. Various types of the braid commensurability thus define all possible filling fractions for correlated states in a Hall system.

The archetype of the braid commensurability in 2D N electron system in strong magnetic field is the accurate fitting of cyclotron orbits of interacting 2D electrons in the LLL to interelectron spacing, which happens at the completely filled LLL, i.e., at filling factor $\nu = \frac{N}{N_0} = 1$, where $N_0 = \frac{B_0 Se}{hc}$ is the LL degeneracy ($\frac{hc}{e}$ is the magnetic field flux quantum). In other words, the condition $\nu = 1$ is equivalent with the commensurability condition,

$$\frac{S}{N} = \frac{S}{N_0} = \frac{hc}{eB_0}, \tag{14.2}$$

i.e., the size of the cyclotron orbit in the LLL at $\nu = 1$, $\frac{hc}{eB_0}$, fits to interparticle spacing, $\frac{S}{N}$ (S is the 2D plane size of the system; in the thermodynamic limit, $\frac{S}{N}$ is constant even if S and N tend to infinity). For $\nu = 1$, the braid group can thus be established and the corresponding correlated state manifests itself as the IQHE.

For stronger magnetic fields, $B > B_0$, $\nu < 1$ and the commensurability (Equation 14.2) fails. It means that ordinary cyclotron orbits $\frac{hc}{eB}$ are too short and the corresponding braids, σ_j, must be rejected from the braid group as unavailable ones. However, among the remaining braid group elements, there are still present braids σ_j^q, which for q odd integer, similarly as σ_j define exchanges of jth and $(j + 1)$th particles. In distinction to σ_j, the braids σ_j^q realize exchange with additional $\frac{q-1}{2}$ loops [25, 33]. These additional loops "take away" $\frac{q-1}{2}$ flux quanta, when the external field is passing through the planar multiloop orbit. This is the origin of the auxiliary flux tubes pinned to CFs. In 2D, the external field flux per particle, $\frac{BS}{N}$, is thus reduced by $\frac{q-1}{2}\frac{hc}{e}$ similarly as in the CF concept. Braids σ_j^q define exchanges along multiloop cyclotron 2D orbits. Exclusively in 2D, the multiloop cyclotron orbits share the same external field flux per particle as the single-loop cyclotron orbit, thus per single loop of multiloop orbit must fall only some fraction of $\frac{BS}{N}$, just flux quantum. Noticeably, this is in contrary to 3D case, when each scroll of spiral in 3D adds a new surface pierced by the same field B, but not in 2D. A flux is the product of a surface and a field, hence its smaller value can be achieved at lower field conserving the surface. The division of the external $\frac{BS}{N}$ flux into pieces per each loop is equivalent to the reduction of the effective field for a single loop to the value B_0 at which flux quantum has the size $\frac{S}{N}$. Hence, in 2D, the multiloop cyclotron orbits related to σ_j^q have a larger size as accommodated to reduced flux portion per each loop and eventually σ_j^q can reach particles out of reach for single-loop braids σ_j. The braid cyclotron subgroup generated by $\sigma_j^q, j = 1,\ldots,N-1$ is the proper braid group for $\nu < 1$ provided that the new commensurability condition holds:

$$\frac{S}{N} = \frac{qBSe}{hc}, \tag{14.3}$$

where the r.h.s. of this condition expresses the q times larger range of multiloop cyclotron orbit in 2D. From this commensurability condition, it follows the relation: $\nu = \frac{N}{N_0} = \frac{1}{q}$ displaying the main line of LLL fillings for FQHE (described by the Laughlin function with qth order of the Jastrow polynomial). The braid group commensurability is thus the kernel of the Laughlin function derivation—the multiparticle wave function of N-correlated 2D charged interacting particles (electrons) at magnetic field must transform according to 1DUR of the related cyclotron braid subgroup [24, 34, 36]. This feature, together with the requirement that the multiparticle function in the LLL must be a holomorphic function of $z_1,\ldots,z_N$ ($z_j = x_j + iy_j$—complex coordinate on the plane of jth particle), i.e., without poles of polynomial form, uniquely results in the form of the Jastrow polynomial $\prod_{i>j}^{N}(z_i - z_j)^q$ (multiplied by a factor independent of particle interchanges—in the Laughlin function, $e^{-\sum_{i=1}^{N}|z_i|^2/4l_B^2}$, l_B is the magnetic length). The latter term is common for all states from the LLL, thus the derivation of the Laughlin function resolves itself to the determination of the uniform with respect to all particles polynomial factor, which can be deduced according to the 1DUR and the form of a generator of related cyclotron braid subgroup as described above.

The FQHE phenomena in monolayer graphene and especially in bilayer graphene supply an opportunity to verify the cyclotron braid group commensurability approach because the "relativistic" LL structure is different in comparison to conventional GaAs 2DEG and it is challenging to verify the independence of topology-induced essence of FQHE with respect to single-particle "relativistic" dynamics. On the other hand, the topology of double sheet in bilayer graphene is different than its counterpart in monolayer case that should cause different commensurability conditions in bilayer system in comparison to the monolayer one. This will result in specific hierarchy of filling rates for FQHE in bilayer graphene not observed in monolayer graphene or in conventional semiconductor 2D systems.

It must be emphasized that for FQHE formation, the interaction of electrons is essential similarly as for any other correlated state. The strong Coulomb repulsion of electrons determines the steady uniform distribution of 2D electrons—in the form of triangle Wigner crystal lattice as the classical lowest energy state at $T = 0$ K. Such a classical distribution of electrons rigidly fixed by interaction is the start point for quantization in terms of Feynman path integral including summation of contributions assigned by the braid group 1DUR [33]. This explains the central role of 1DURs of braid groups in definition of quantum statistics of multiparticle systems. Each different 1DUR defines a different sort of quantum particles corresponding to the same classical ones [24, 34]. Hence, the determination of the braid group for the multiparticle system (in 2D, not the permutation group) is an unavoidable prerequisite for any quantum correlated state.

The braid group is a topological object, π_1 homotopy group of N-particle configuration space, $\pi_1((M^N - \Delta)/S_N)$, collecting all classes of nonhomotopic trajectory loops in the configuration space, where points that differ only by enumeration of particles are unified (due to indistinguishability of particles). The braid group does not reflect the dynamics details of the system but rather identifies only the topology restrictions imposed on interparticle trajectories conditioned by the geometry-type global features, like the dimension of the manifold M or the presence of the quantizing magnetic field. This states behind the similarity of FQHE manifestation in various materials, despite local dynamics differences.

As mentioned above, the Coulomb repulsion of electrons on the plane is a central premise for the braid group definition at magnetic field presence, because the interparticle separation rigidly fixed by the Coulomb repulsion (in the classical equilibrium state) must interfere with the planar cyclotron orbits, discriminating in this way possible correlation types, by the braid commensurability condition. Exclusively in 2D, the multiloop cyclotron orbits

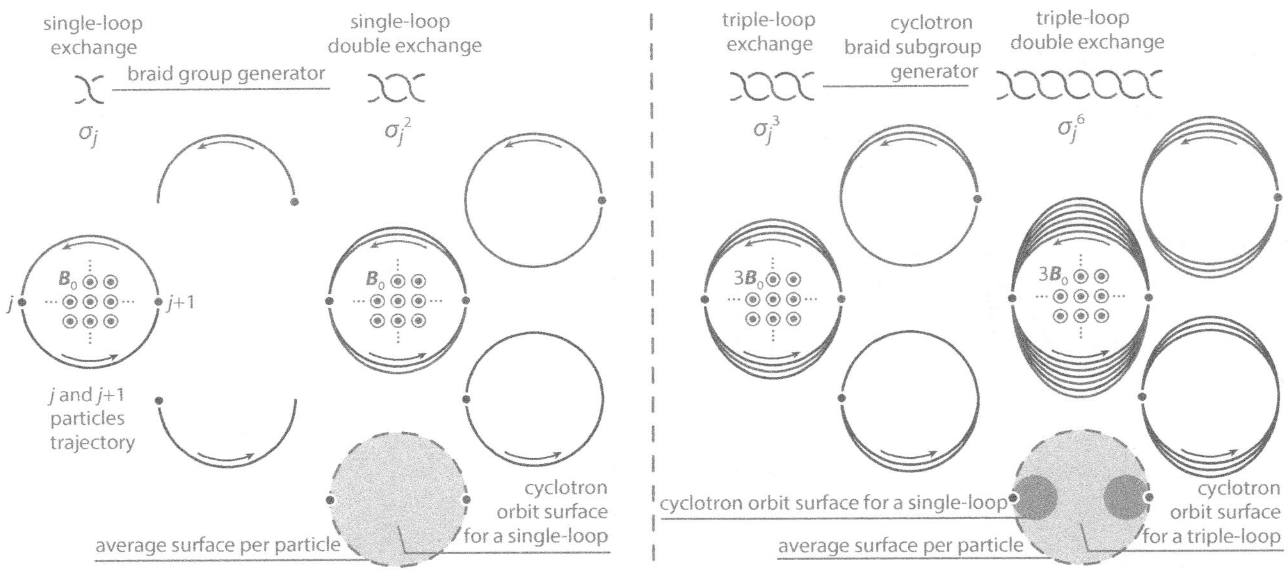

Figure 14.2 The geometrical presentation of the braid generator σ_j (top-left panel)—it corresponds to the ordinary single-loop exchange of neighboring particles, jth and $(j + 1)$th (left panel). The cyclotron semi-orbits of both particles may realize together this braid at $\nu = 1$. It is visible that the closed cyclotron full-orbits correspond to the double exchange of matched particles (top-right panel). The three-loop braid generator σ_i^3 (right panel) must be considered at $\nu = \frac{1}{3}$ in order to match neighboring particles at three times stronger magnetic field $3B$. The additional loops are needed in this case to enhance an effective cyclotron radius. Schematic illustration of cyclotron orbit enhancement in 2D due to multiloop trajectory structure (bottom-right panel). Single-loop orbit fits to particle separation at $\nu = 1$ (bottom-left panel). For multiloop orbit in 2D, an external magnetic field flux must be divided among all loops. For each loop falls only fraction of the external field flux that leads to an enhancement of its size.

have a larger size that allows the braids σ_j^q (q odd positive integer) to fit with interparticle separation too long for single-loop exchanges. This opportunity happens, however, only at some "magic" fractional fillings of the LLL—the same ones at which FQHE is observed. The surface spanned by 2D cyclotron orbit is the same regardless of its multiloop character; hence, the portion of the flux per each loop is smaller for multiloop orbit in comparison to a single-loop orbit, which subsequently results in multiloop orbit size growth. This is illustrated in Figure 14.2, right panel.

In Figure 14.2 (bottom-left panel), the scheme of the cyclotron orbit at magnetic field B_0 is shown as accommodated to the quantum of magnetic field flux, i.e., $B_0A=\frac{hc}{e}$. This is the definition of the single-loop cyclotron orbit size A at B_0 in the LLL. A fits with the interparticle separation in the case of the completely filled LLL, $\frac{S}{N}$, S—the sample area, N—the number of particles. If only single-loop orbits are considered, then at larger field, e.g., $q = 3$-times larger, $3B_0$, the cyclotron orbit accommodated again to the flux quantum is too small in comparison to the interparticle separation $\frac{S}{N}$—as is sketched in the bottom-right panel of Figure 14.2 (dark gray shape). However, in the case when tree-loop orbits are considered, then, exclusively in 2D space, the external flux $\frac{3B_0S}{N}=3B_0A$ passing this orbit must be divided among three loops. In the case of the uniform division, per each loop it falls B_0A fraction of the total $3B_0A$ flux. Therefore, each loop in this situation accommodated to the flux quantum $\frac{hc}{e}=B_0A$ has the size A, the same one as for the single-loop orbit at three times weaker magnetic field. Three loops contribute together to the total flux $3B_0A$ per particle as needed—cf. Figure 14.2 (bottom-right panel). It means that the three-loop orbits surprisingly fit to the interparticle separation $\frac{S}{N}=A$.

The braid group generator must be defined by the half of the cyclotron orbit (cf. Figure 14.2), thus the braid with one additional loop corresponds to the cyclotron orbit with three loops—such a braid generator has the form σ_j^3. The group generated by σ_j^3, $j = 1,\ldots,N-1$ (new elementary braid exchanges) is obviously the subgroup of the original braid group because its generators σ_j^3 are built from original group generators σ_j. This subgroup is called as the cyclotron braid subgroup. It is clear that 1DURs of this subgroup define statistics of 2D charged particles at sufficiently strong magnetic fields, i.e., at fields corresponding to the fractional fillings of the LLL provided that the braid commensurability condition is fulfilled (in the presented example, for $\nu=\frac{1}{3}$). The generalization to more loops attached to the braid generator one by one results in the double increase of the number of loops in multiloop cyclotron orbits and thereby in filling fractions $\nu=\frac{1}{q}$, q - odd integer.

This approach can be generalized to higher LLs also in exact agreement with the experimental observations [26]. The generalization to higher LLs resolves itself to the observation that the braid commensurability condition can be written here as follows:

$$x\frac{S}{N-\beta N_0}=\frac{(2n+1)hc}{eB}, \tag{14.4}$$

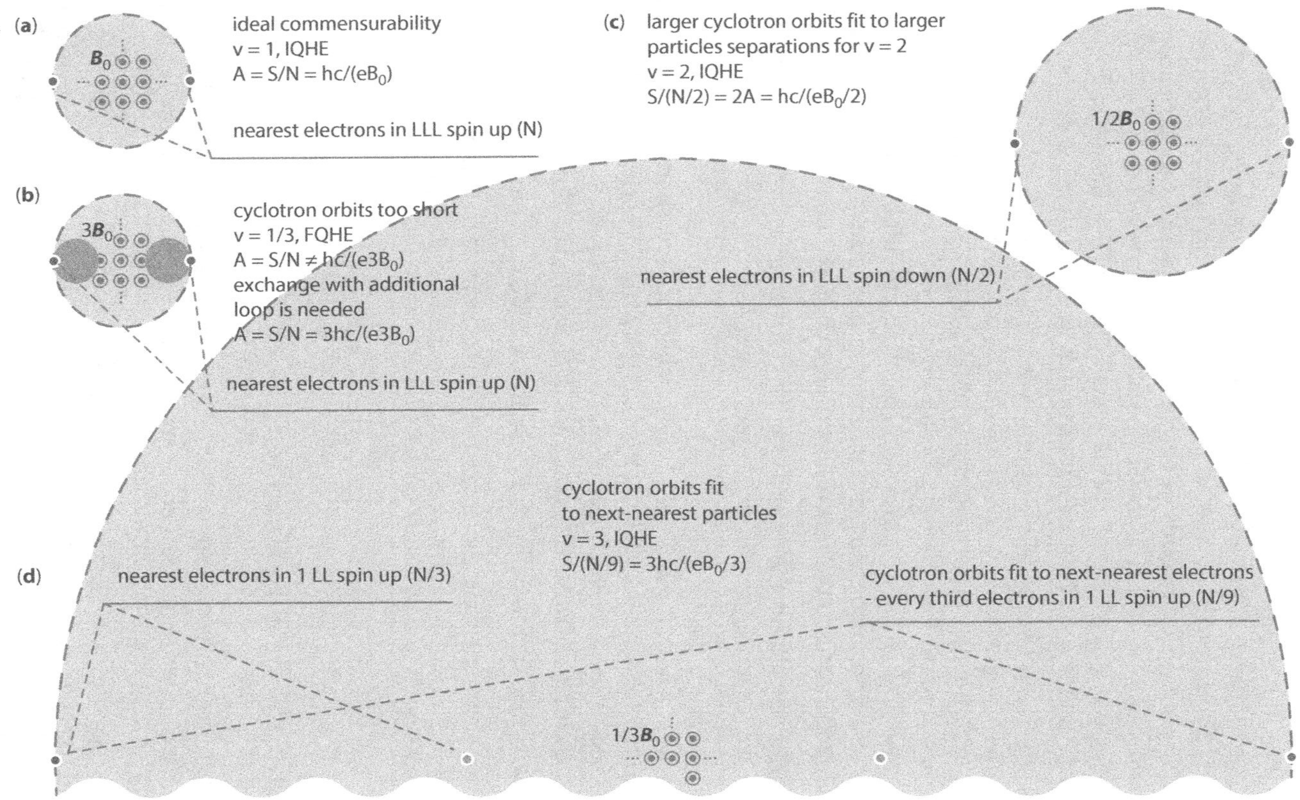

Figure 14.3 Illustration of the commensurability between the cyclotron orbits (schematically drawn as circular ones) and the interparticle spacing. At $\nu = 1$, nearest particles may exchange positions along perfectly accommodated single-loop cyclotron orbits (a). For stronger fields $\nu < 1$, single-loop cyclotron orbits are too short to match neighboring particles (b). For $\nu = 2$, orbits again fit to interparticle spacing in the spin-down subband of the LLL (c). For $\nu = 3$, i.e., for complete filling of spin-up subband of the first LL, cyclotron orbits fit accurately to every third particle separation in this subband (d). In the cases (a), (c), (d), we deal with the IQHE, whereas in (b) at $\nu = \frac{1}{3}$ with the FQHE due to larger three-loop orbits.

with x – positive integer and $\beta = \begin{cases} 2n \text{ for spin} \uparrow \\ 2n+1 \text{ for spin} \downarrow \end{cases}$ (Figure 14.3). The cyclotron orbits in higher LLs are larger than those in the LLL due to higher energy and in nth LL are of size $\frac{(2n+1)hc}{eB}$. These larger orbits may fit to equidistantly separated particles, though not to any ones but rather to every x-th particle (next-nearest neighbors of the xth order, Figure 14.3 bottom). Thus, the commensurability (Equation 14.4) also allows for the definition of the generators σ_j' in the form of ordinary single-loop braids linking every xth particle. This happens in the completely filled higher LLs ($n \geq 1$). The resulting statistics (expressed by 1DUR of these braid groups with longer braids for $n > 1$) is the same as for IQHE (the longer braids are single-loop as for IQHE at $\nu = 1$). Noticeably, such an opportunity for commensurability may be encountered in higher LLs not only for complete filled these levels but also at some fractional their fillings, which has been described in Ref. [26] in satisfactory good correspondence with the experimental observations available now up to the third LL for 2DEG in conventional semiconductor Hall systems [37–40]. In LLs with n ≥ 1 (i.e., at weaker magnetic field in comparison to B_0 for $\nu = 1$), the cyclotron orbits are sufficiently large to match every second, every third, and so on particles. This is because the size of the cyclotron orbits in n-th LL grows proportionally to the factor $2n + 1$ present in the Landau kinetic energy, i.e., the cyclotron orbit size attains the value $\frac{(2n+1)hc}{eB}$ for particles with kinetic energy $(2n+1)\frac{eB}{2mc}$ in nth LL. However, for $n \geq 1$, too-short cyclotron orbits may also sometimes occur but not always as it was in the LLL ($n = 0$). In higher LLs, too-short cyclotron orbits may happen close to the subband edges, i.e., for sufficiently small density of particles and thus for their larger separation exceeding large cyclotron orbits at $n \geq 1$.

The quantization of the transverse resistance R_{xy} at a particular filling rate ν (also in higher LLs) is always similar as for ordinary FQHE in the LLL, i.e., equals to $\frac{h}{e^2\nu}$, regardless of the correlations expressed by the exponent in the Jastrow polynomial displaying single-loop, $q = 1$, or multiloop, $a > 1$, braid exchanges, both accessible for FQHE in higher LLs in contrary to the LLL.

14.3 Hierarchy of FQHE in Graphene

Due to the specific band structure in graphene with Dirac points at corners of the hexagonal Brillouin zone [6], the LL spectrum is not equidistant as for ordinary 2DEG but is proportional to $\sqrt{n}$ instead of n [9]. This pseudo-relativistic form of LL energy emerges from the linearity in momentum of the local Hamiltonian in the vicinity of Dirac points. The massive degeneracy of each LL subband in graphene is, however, the same one as in the conventional 2DEG and equals to $\frac{BS}{hc/e}$ (B is the external magnetic field, S is the sample surface, $\frac{hc}{e}$ is the magnetic field flux quantum). Nevertheless, the number of subbands per each LL in graphene is different than in a conventional semiconductor case and equals 4 in graphene, which corresponds to the Zeeman spin splitting and to the valley pseudo-spin splitting (absent in conventional semiconductors) due to two inequivalent Dirac points mixed with two sublattices in crystal lattice of the graphene sheet [9]. The Zeeman splitting in graphene is small [6], and the valley splitting is small as well [9], thus the four-fold approximate spin-valley additional degeneracy may be assumed (referred to SU(4) symmetry). The LLL subbands are divided between particles and holes from the conduction and valence bands [9]. Hence, the bottom of the LLL is shifted by 2 upward in terms of the filling factor (in monolayer case). Conventionally there are assigned filling rates for holes from the valence band as negative numbers, mirror reflection of those positive numbers denoting filling rates for electrons in the conduction band. An additional opportunity in graphene, beyond the ability of conventional 2DEG, is a possibility to control a transition between particles and holes in graphene by the shift of the Fermi level passing the Dirac point. Experimentally it is realized by application of a lateral relatively small voltage.

For the bilayer graphene, the situation slightly differs [9, 21]. Due to an interlayer electron hopping, the Hamiltonian for bilayer graphene attains back the quadratic form with respect to the momentum. Hence, the LL spectrum in bilayer graphene, $\sqrt{n}\sqrt{n+1}\hbar\omega_c \simeq n\hbar\omega_c$, resembles the one of the ordinary 2DEG for higher n, but with four-fold spin-valley degeneracy for each LL level except for the LLL which has eight-fold degeneracy due to an extra degeneracy of states $n = 0$ and $n = 1$ [9, 21]. As usual in graphene, the division of the LLL subbands equally between particles and holes causes that the bottom for uniformly charged carriers (electrons or holes) is placed in the center of the eight-fold degenerated LLL. The extra degeneracy of the LLL is caused by inclusion of states with $n = 0$ and $n = 1$ to the LLL in the bilayer graphene in opposition to the monolayer one. In the bilayer graphene, the Berry phase shift for chiral particles is also different (twice larger) in comparison to the monolayer one, and equals to 2π [9]. Hence, the consecutive plateaus of IQHE are located in bilayer graphene at integer filling rates whereas in monolayer graphene were located at half-fillings of LLs [9, 21].

14.3.1 FQHE Hierarchy in Monolayer Graphene

For Fermi level shifted (by a lateral voltage) to the conduction band and the magnetic field strong enough that $\nu \in (0,1)$, we deal with fractionally filled first conduction subband of the particle LLL marked as n = 0,2↑ (in this notation, 2 marks the valley pseudospin component and the arrow ↑ marks the orientation of the ordinary spin along the magnetic field). For $N < N_0$, the filling rate, $\nu = N/N_0$, is fractional (the degeneracy N_0 of each subband is $N_0 = \frac{BS}{hc/e}$).

To decipher FQHE hierarchy in this subband in graphene, we apply the braid group topological approach developed for ordinary 2DEG system [25, 33]. In order to implement braid group generators, the cyclotron orbit must commensurate with interparticle separation. An archetype of the commensurability is, $\frac{S}{N} = \frac{hc}{eB_0} = \frac{S}{N_0}$ (where S is the sample surface, N is number of electrons, N_0 is LL degeneracy), as for $\nu = \frac{N}{N_0} = 1s$ and IQHE. Various more complicated patterns of the commensurability define filling fractions for FQHE [25, 33].

In the case of graphene, an important property follows from the fact that cyclotron orbits in graphene are defined by the bare kinetic energy $T = \hbar\omega_c(n + \frac{1}{2})$ with $\omega_c = \frac{eB}{mc}$, similarly to the conventional semiconductor 2DEG (as in noninteracting 2D gas), despite the different pseudo-relativistic version of Landau-level energy. This is because the "relativistic" oddness is caused by a peculiar crystal field (electric interaction of ions and electrons) that does not change the bare kinetic part of Landau energy. Hence, the dimension of braid cyclotron orbits for graphene repeats the corresponding orbit size from the noninteracting gas. The difference between the conventional 2DEG system and graphene will be thus related with different number of LL subbands in graphene in comparison to the conventional 2DEG. The uniform shift in filling factors will also be caused by the Barry phase shift for chiral valley pseudospin in graphene [9].

Therefore, the cyclotron orbit size in the subband $n = 0,2\uparrow$ is equal to $\frac{hc/e}{B} = \frac{S}{N_0}$. Because this orbit size is lower than the interparticle spacing expressed by $\frac{S}{N}$ (as $N < N_0$), the multiloop braids with enhanced size are needed to match neighboring particles [25, 33]. The commensurability condition reads here as, $q\frac{S}{N_0} = \frac{S}{N}$, which gives $\nu = \frac{N}{N_0} = \frac{1}{q}$, ($q$ odd integer [25]). For the local band holes in this subband, the particle-hole symmetric filling rates $\nu = 1 - \frac{1}{q}$ are expected.

The next possible commensurability occurs when the last loop of the multiloop cyclotron orbit is commensurate with the every lth particle separation (as in lth LL), whereas the $q - 1$ former loops take away an integer number of flux quanta (which are commensurate with nearest-neighboring particles). For $l = 2,3,\ldots$ the last loop reaches every l-th particle (next neighbors). In this manner, we obtain the hierarchy of fillings for FQHE in this LLL subband in the following form (the same as for CF model): $\nu = \frac{l}{l(q-1) \pm 1}$, $\nu = 1 - \frac{l}{l(q-1) \pm 1}$, where $l = 1,2,\ldots$ and minus in the denominator corresponds to the possibility of the reverse eight-figure orientation of the last loop with respect to the antecedent loop in the multiloop orbit. Additionally we notice that filling rates for the Hall metal states can be achieved in the limit $l \to \infty$ in the above formula, which corresponds to the situation when the residual flux passing the last loop tends to zero. This means that the last loop can reach in such a case the infinitely distant particles as for fermions at the absence of a magnetic field, which is referred to the case of the Hall metal archetype for $\nu = \frac{1}{2}$ in the conventional 2DEG. In the limit $l \to \infty$, we thus arrive with the hierarchy for the Hall metal states in the form: $\nu = \frac{1}{q-1}$, $\nu = 1 - \frac{1}{q-1}$.

One can also observe that other variants of commensurability may concern multiloop orbits. Namely, each loop of the multiloop structure may be in principle accommodated to particle separation in a different and mutually independent manner matching nearest or next-nearest neighbors in various schemes. One of such possibilities may correspond with the situation when in q-loop orbit $q - 1$ loops are accommodated to every x-th particle ($x = 1,2,3,\ldots$), whereas the last one fits to every l-th particle separation. This commensurability scheme is observed in ordinary 2DEG Hall systems within the LLL for exotic fractions, e.g., $\nu = \frac{4}{11}, \frac{5}{13}, \frac{3}{8}, \frac{3}{10}, \ldots$ (beyond the CF hierarchy corresponding to $x = 1$). It is noticeable, however, that this series of exotic FQHE filling fractions are not observed in the LLL in graphene as of yet, though they are observed in the first LL in monolayer graphene (as will be identified below).

At lowering the magnetic field, one can achieve the completely filled subband $n = 0,2\uparrow$, which corresponds to the IQHE state. For lower magnetic fields, the next subband, the last one in the LLL, $n = 0,2\downarrow$, is gradually filled with electrons. In this subband, the cyclotron orbit size $\frac{S}{N_0}$ is still lower than the interparticle separation $\frac{S}{N - N_0}$ (because $N - N_0 < N_0$) similarly as in the antecedent subband. Therefore, the multilooped structure of orbits must be repeated here from the previous subband only shifted ahead by 1. After complete filling of this subband, the LLL is completely filled as well. This gives the IQHE according to its main-line hierarchy $\nu = 4(n+\frac{1}{2})$ at $n = 0$.

In an analogous way, one can consider the fillings of the following LLs. The nearest one corresponds to $n = 1$. This level has four subbands of electron type. The bare kinetic energy in this LL (in all its subbands) is equal to $\frac{3\hbar\omega_c}{2}$. In this subband, the cyclotron orbits are of size $\frac{3S}{N_0}$ (larger in comparison to the LLL) and they must be accommodated with interparticle separation between electrons in this subband, $\frac{S}{N-2N_0}$. For small amount of electrons in the subband (close to the subband edge), we may thus deal with the multilooped orbits if single-loop orbits are too short, $\frac{3S}{N_0} < \frac{S}{N-2N_0}$. The q-loop orbits satisfy the commensurability condition $q\frac{3S}{N_0} = \frac{S}{N-2N_0}$, q odd integer, which defines the main series for FQHE(multiloop) in this subband, $\nu = 2+\frac{1}{3q}$, shifted toward subband edge. This main line can be supplemented to the complete related hierarchy, $\nu = 2+\frac{l}{l3(q-1)\pm 1}$, $\nu = 3-\frac{l}{l3(q-1)\pm 1}$ $l = i/3$, i = 1,2,..., with the Hall metal hierarchy in the limit $l \to \infty$, similarly as described in the case of the LLL. These series of filling rates are located closer to the subband edges in comparison to the FQHE rates in the LLL, due to larger size of cyclotron orbits in $n = 1$ subband. Simultaneously, in the central part of this higher LL subband, the new type of commensurability is possible, not accessible in the LLL. This new commensurability occurs when $\frac{3S}{N_0} = \frac{xS}{N-2N_0}$ and x = 1,2,3, i.e., when the cyclotron orbit size exceeds particle separation. Then the single-loop orbit (large enough in this subband) can fit with every x-th particle (x order next-nearest neighbors). From this new commensurability opportunity, one finds fractions $\nu = \frac{7}{3}, \frac{8}{3}, 3$ corresponding to single-loop cyclotron orbits (similar as for IQHE). Thus, for $\nu = \frac{7}{3}, \frac{8}{3}$ we deal with the FQHE(single-loop). This is a new Hall feature manifesting itself only in higher LLs, where cyclotron orbits may be larger than the interparticle separation and single-loop orbits can reach next-nearest neighbors.

Let us note that the special case of the commensurability, $\frac{3S}{N_0} = \frac{1.5S}{N-2N_0}$, one can identify at $\nu = \frac{5}{2}$. This commensurability concerns rather the paired particles and not the single ones. The pairing does not change the cyclotron radius (being invariant upon doubling of mass and charge) but reduces twice the carrier number $\frac{N-2N_0}{2}$, which gives the above commensurability for pairs at $\nu = \frac{5}{2}$. Hence, at this filling rate, one can expect a manifestation of IQHE-type correlation but for paired electrons (the considered correlation corresponds to p-like pairing due to the spin polarization in this subband).

The similar scheme of commensurability may be applied to the following subband with $n = 1$.

An interesting new possibility for commensurability happens for q-loop orbits commensurate with next-nearest neighbors. As mentioned previously, the size of particular loops in the multiloop structure may be in general accommodated to the interparticle spacing in an

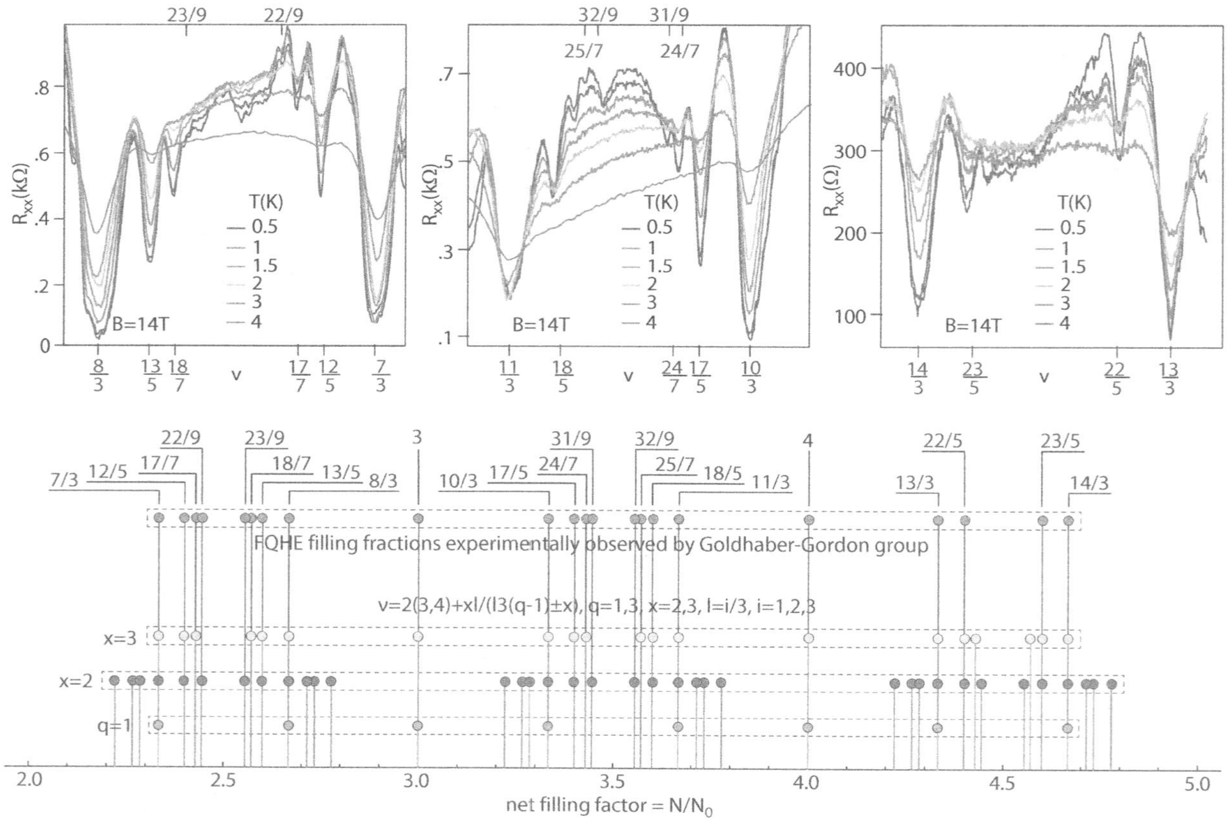

Figure 14.4 Fitting of cyclotron braid hierarchy for FQHE in monolayer graphene in three first subbands in the first LL ($n = 1$), $\nu \in [2,5]$. Upper panel—R_{xx} after experiment [11], lower panel—the corresponding theoretical hierarchy according to cyclotron braid approach. The larger residual longitudinal resistance (in upper panel for $\frac{12}{5}, \frac{17}{7}, \frac{22}{9}$ and for other fractions with denominators 5,7,9), corresponds to correlated states of next-nearest electrons, of every second ($x = 2$) or of every third ($x = 3$) particles, according to the commensurability series $\nu = 2(3,4) + \frac{xl}{l3(q-1) \pm 1}$ with $q = 3$, $x = 2,3$, $l = \frac{i}{3}$, $i = 1,2,3$ (lower panel) ($x = 1$ corresponds to CF-like commensurability)—uncorrelated electrons enhance the resistance.

independent way resulting in abundance of possible new filling rates. In particular, it results in the additional hierarchy in all subbands of the first LL, $\nu=2(3,4,5)+\frac{xl}{l3(q-1)\pm 1}$, $\nu=3(4,5,6)-\frac{xl}{l3(q-1)\pm 1}$ which for $q = 3$, $x = 2,3$, $l = i/3$, $i = 1,2,3$ reproduces $\nu=\frac{7}{3},\frac{8}{3},\frac{12}{5},\frac{13}{5},\frac{17}{7},\frac{18}{7},\frac{22}{9},\frac{23}{9},\frac{10}{3},\frac{11}{3},\frac{17}{5},\frac{18}{5},\frac{24}{7},\frac{25}{7},\frac{13}{3},\frac{14}{3},\frac{22}{5},\frac{23}{5}$. This opportunity for FQHE pretty well agrees with the recent observations of FQHE in three first subbands of the $n = 1$ LL in monolayer graphene at ultra-low temperatures [11]—cf. Figure 14.4.

14.3.2 FQHE Hierarchy in Bilayer Graphene

In bilayer graphene, the topology of braid trajectories considerably changes in comparison to the monolayer system. The bilayer graphene is not strictly two dimensional and this opens a new possibility for topology of trajectories as visualized in Figure 14.5.

Two sheets of the bilayer graphene lie in a close distance, and electrons can hop between them. Multiloop cyclotron orbits (and related braids) may thus reside in both layers simultaneously, i.e., loops may be distributed among both sheets. This makes a difference in comparison to the monolayer case because each sheet contributes to the total flux of external magnetic field independently with own surface that strongly affects cyclotron orbit size and braid commensurability condition (cf. Figure 14.5).

The simplest commensurability instance in the LLL (subband $n = 0,2\uparrow$) with three-loop cyclotron orbit located in both sheets of bilayer graphene (as illustrated in Figure 14.5) results in filling fraction $\nu=\frac{1}{2}$, not $\frac{1}{3}$ as in monolayer case. This exceptional fraction has been observed in the experiment (actually for holes at $\nu=-\frac{1}{2}$) [14] and cannot be explained by CF model (CF model predicts a Hall metal state at $\nu=\pm\frac{1}{2}$).

The fact that the second loop of any pair of loops may be located in the opposite sheet of bilayer graphene with respect to the first loop—cf. Figure 14.6—is the source of the oddness of FQHE hierarchy in bilayer graphene.

In general, in bilayer system, the loops of a multiloop orbit may be located partly in both 2D sheets. To account for this effect in topological terms adjusted to braid trajectories and commensurability requirements, one must neglect the contribution of a single loop in the multiloop structure when the total flux of the external field is divided into fractions per each loop. This single loop captures its own flux, whereas the remaining loops must share the same flux similarly as passing through any cyclotron orbit in monolayer case. Removing of the single loop must be done independently of how loops are distributed among two sheets. When we consider a selected loop located in the opposite sheet with respect to the antecedent loop, the next loops must fill both sheets of bilayer structure as additional ones regardless of their specific distribution—thus all these loops take part in dividing of the external field flux exactly in the same manner as in monolayer case, provided that the selected loop is omitted together with the flux passing this loop. This trick reduces the bilayer system to monolayer one in braid loop topology sense. Thus, we can write out the commensurability condition in the bilayer graphene in the case of too-short single-loop cyclotron orbits in the following form (as an example, for the subband $n = 0,2\uparrow$ of the LLL):

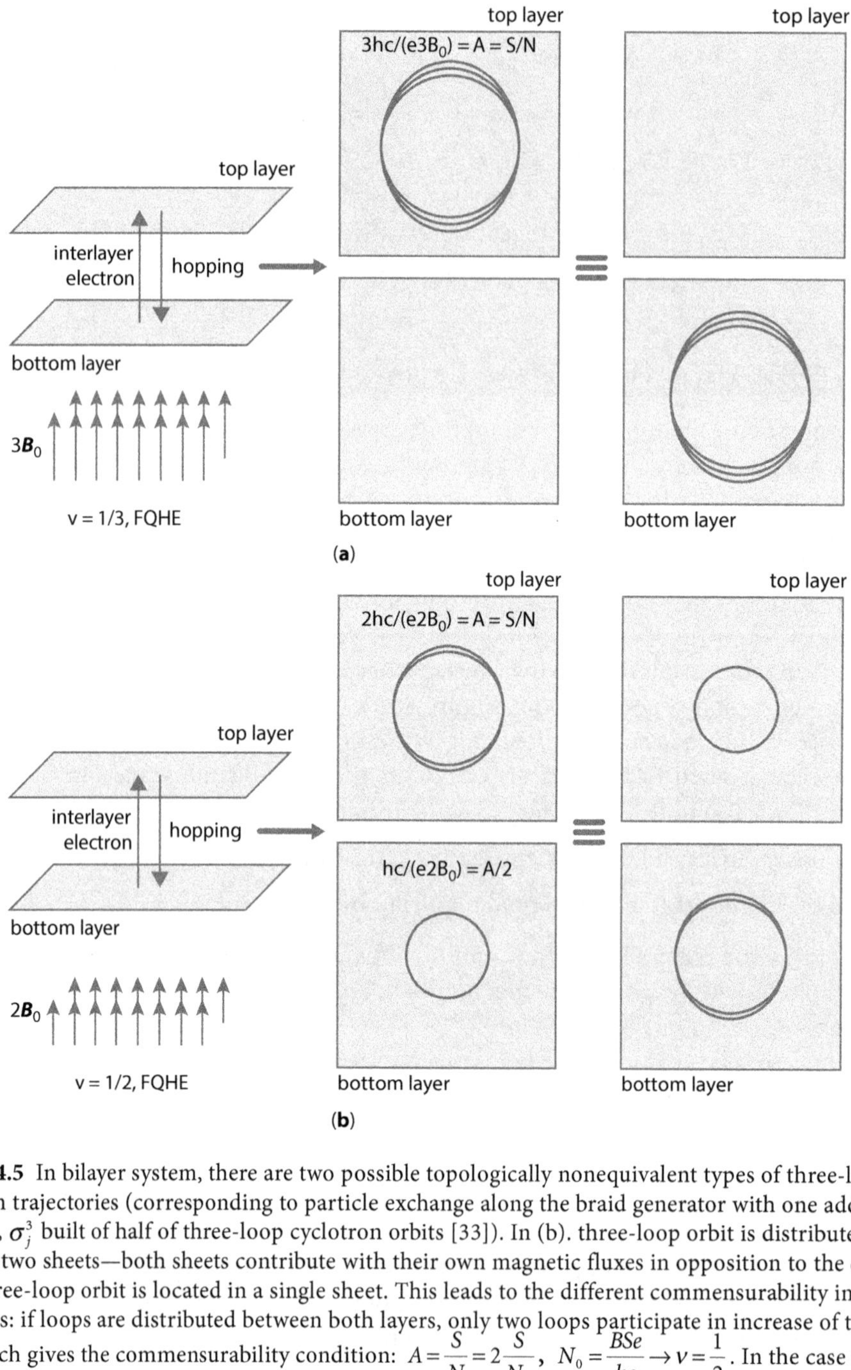

Figure 14.5 In bilayer system, there are two possible topologically nonequivalent types of three-loop cyclotron trajectories (corresponding to particle exchange along the braid generator with one additional loop, i.e., σ_j^3 built of half of three-loop cyclotron orbits [33]). In (b). three-loop orbit is distributed between two sheets—both sheets contribute with their own magnetic fluxes in opposition to the case when three-loop orbit is located in a single sheet. This leads to the different commensurability in both situations: if loops are distributed between both layers, only two loops participate in increase of the orbit size, which gives the commensurability condition: $A = \frac{S}{N} = 2\frac{S}{N_0}$, $N_0 = \frac{BSe}{hc} \rightarrow \nu = \frac{1}{2}$. In the case when all three loops are placed in a single layer (a), the commensurability repeats the one from the monolayer case, $\nu = \frac{1}{3}$.

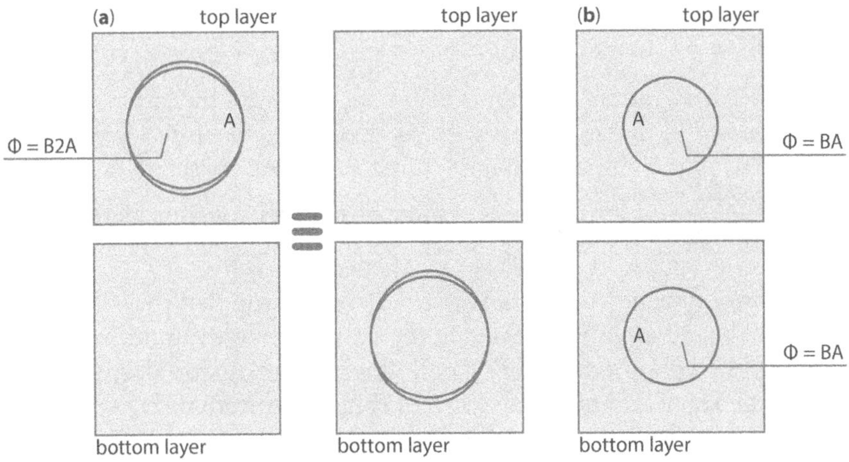

Figure 14.6 If two-loop orbit is divided among two layers (b), then the size of both loops is the same as for a single-loop (in figure $A = S/N$ as for $\nu = 1$ at B_0), but if both loops are placed in same layer, then the size of the double-loop orbit is twice larger (a). This causes different commensurability in both situations at the same magnetic field.

$$
\begin{gathered}
(q-1)\frac{hc}{eB}=\frac{S}{N}, \\
\nu=\frac{N}{N_0}=\frac{1}{q-1}=\frac{1}{2},\frac{1}{4},\frac{1}{6},\ldots,
\end{gathered}
\tag{14.5}
$$

where N is the number of electrons in each sheet, $N_0=\frac{BSe}{hc}$ is the degeneracy of any subband, S is the surface of the sample (the surface of the single sheet), and q is an odd integer (it must be odd in order to assure that the half of the cyclotron orbit defines the braid, similarly as in monolayer case). After avoiding a single loop, the next loops must duplicate the former ones, no matter in which way loops are distributed among both sheets. Thus, only $q - 1$ loops take part in the enhancement of the effective q-loop cyclotron orbit in bilayer graphene taking into account the same instances of commensurability as in the monolayer case.

It must be emphasized that for multiloop orbits in bilayer graphene, the total number of loops still is q (despite avoiding one loop in the commensurability condition (Equation 14.5))—therefore the generators of the corresponding cyclotron subgroup are of the form σ_j^q, which results in the standard Laughlin correlations with the Jastrow polynomial with the exponent q. Due to the commensurability (Equation 14.5), the resulting main line of filling fractions is $\nu=\frac{1}{q-1}$ (p-odd) in the first particle-type subband of the LLL, i.e., in the subband $n = 0,2\uparrow$. The even denominators in this main series for the FQHE hierarchy for bilayer graphene coincide well with the experimental observations [14].

For holes in this subband (holes corresponding to empty states in the almost-filled subband of particle type) we can write, $\nu=1-\frac{1}{q-1}$. The generalization to the full hierarchy of

FQHE in this subband thus attains the form, $\nu = \frac{l}{l(q-2) \pm 1}$, $\nu = 1 - \frac{l}{l(q-2) \pm 1}$, where $l > 1$ describes lth order next-nearest neighbors commensurate with the last loop of $q - 1$ loops, similarly as discussed in the monolayer case (as previously, the limit $l \to \infty$ defines hierarchy for Hall metal). For commensurability of first $q - 2$ loops with xth order next-nearest neighbors, $\nu = \frac{xl}{l(q-2) \pm x}$, but similarly to monolayer graphene, this hierarchy line is also not observed as of yet in the LLL of bilayer graphene.

In the following subbands of the LLL, $n = 0,2\uparrow$ (assuming that this subband succeeds the former one), the hierarchy is repeated in the same form, only uniformly shifted ahead by one (because the commensurability conditions are similar for all subbands with the same n due to the same size of the cyclotron orbits). A novelty occurs, however, in the next two subbands of the LLL, $n = 1,2\uparrow$ and $n = 1,2\downarrow$. Because of the larger size of cyclotron orbits for $n = 1$, the FQHE main series in the first of these subbands of the LLL, $n = 1,2\uparrow$, attains the form,

$$\begin{gathered} \frac{3hc}{eB} = \frac{3S}{N_0} < \frac{S}{N - 2N_0}, \\ (p-1)3\frac{hc}{eB} = (p-1)\frac{3S}{N_0} = \frac{S}{N - 2N_0}, \\ \nu = \frac{N}{N_0} = 2 + \frac{1}{3(q-1)} = 2 + \frac{1}{6}, 2 + \frac{1}{12}, 2 + \frac{1}{18}, \ldots, \end{gathered} \tag{14.6}$$

The generalization of this main series for holes in the subband and to the full FQHE hierarchy in this subband looks as follows: for subband holes, $\nu = 3 - \frac{1}{3(q-1)}$ and for the full FQHE hierarchy in this subband, $\nu = 2 + \frac{l}{l3(q-2) \pm 1}$, $\nu = 3 - \frac{l}{l3(q-2) \pm 1}$, $l = \frac{i}{3}$, $i =$ 1,2,3,... (the Hall metal hierarchy may be obtained in the limit $l \to \infty$).

In the subband $n = 1,2\uparrow$, the cyclotron orbit may be larger than the particle separation (similarly as for $n = 1$ in monolayer graphene), which allows single-loop commensurability with next-nearest neighbors. For $\frac{3}{N_0} = \frac{x}{N - 2N_0}$ for $x = 1,2,3$, one obtains the filling rates, $\nu = \frac{7}{3}, \frac{8}{3}, 3$. These rates are related with single-loop correlations similar as for IQHE (though the first two are for not integer filling rates) and are referred to as FQHE(single-loop). Similarly as in the monolayer case, one can consider paired state for $x = 1.5$ in the above formula, which corresponds to the perfect commensurability of cyclotron orbits of electron pairs with the separation of these pairs at electron filling rate $\nu = \frac{5}{2}$. Filling of the last subband $n = 1,2\downarrow$ in the LLL in bilayer graphene undergoes the similar constraints; as for all subbands with n = 1, the cyclotron orbits have the same size and the FQHE hierarchy is only shifted by 1 from the antecedent subband.

The situation significantly changes, however, in the next LL (the first one beyond the LLL). The cyclotron orbits are determined here by the bare kinetic energy with $n = 2$, which gives the cyclotron orbit size: $\frac{5hc}{eB}=\frac{5S}{N_0}$. These orbits are large, thus multiloop orbits may be needed only at small density of electrons close to subband edges. In the subband $n=2,1\uparrow$, the main multiloop series and the full hierarchy for FQHE(multiloop) is shifted toward subband edges: $\nu=4+\frac{1}{5(p-1)}$, $\nu=4+\frac{l}{l5(p-2)\pm 1}$, $l=\frac{i}{5}$, $i>1$, respectively (for subband holes, 5– substitutes 4+ in both above formulae). As previously, the limit $l\to\infty$ determines the Hall metal hierarchy. Because the orbit size for n = 2 may be larger than particle separation (especially in central part of the subband), the commensurability of this orbit with next-nearest neighbors ought to be taken into account. Similarly as in monolayer subband with n = 2, one can expect the presence of four ($2n$) satellite FQHE(single-loop) states symmetrically located around the central paired state. In the subband n = 2,1,↑, these satellite states occur at $\nu=\frac{21}{5},\frac{22}{5},\frac{23}{5},\frac{24}{5}$ and the central paired state at $\nu=\frac{9}{2}$. Such states are visible in experiment in conventional 2DEG for subbands with n = 2—cf. Figure 14.10, whereas in bilayer graphene, the experimental picture is different [15]. This oddness is again caused by specific topology of double-sheet structure.

In order to solve this puzzle, let us note that in bilayer system, it may occur distinct topological realizations of single-loop orbits impossible in monolayer system. This new opportunity is visualized in Figure 14.7—when a part of a single loop is located in one sheet whereas the rest of this loop in the opposite one in such a way that particles interchange along the braid built from cyclotron orbit pieces located in opposite sheets. Such a topology of a single loop can be realized due to hopping of electrons. Because both interchanging electrons may have their individual trajectories located simultaneously in different sheets, the mutual distance between electrons may not be conserved in opposition to the monolayer case (left panel in Figure 14.7). Hence, the braid built from pieces of the orbits as in Figure 14.7 (central panel) defines exchange of electrons laying *closer* than the orbit size $\frac{5hac}{eB}$ (right panel in Figure 14.7). This corresponds to effective reduction of the cyclotron orbit size, or in other words, to a formal leakage of flux passing the cyclotron orbit. The resulting commensurability can thus be associated to smaller effective cyclotron orbits despite its nominal larger value for n = 2. Orbits can change only by integer number of flux quanta, thus for initial nominal flux for n = 2, $\frac{5hc}{e}$, we get the final reduced single-loop

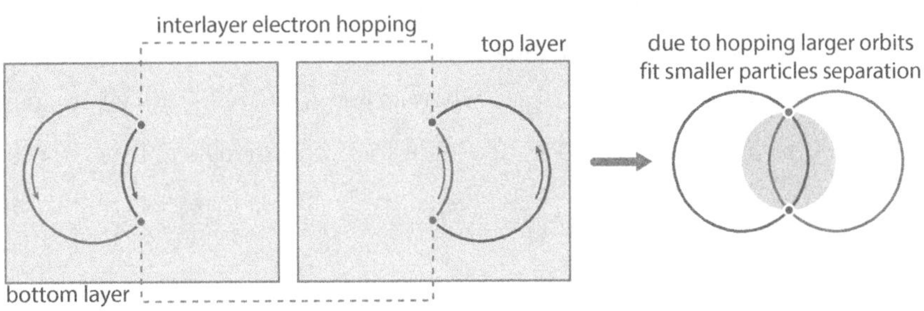

Figure 14.7 When electrons can hop between two sheets of bilayer graphene, the topology of single-loop braid may change. Both interchanging particles can hop between sheets and may be in opposite layers when traversing own orbits—in this case, they do not conserve their mutual distance. This results in leakage of flux of the cyclotron orbit and such smaller orbit can match particles laying closer (right).

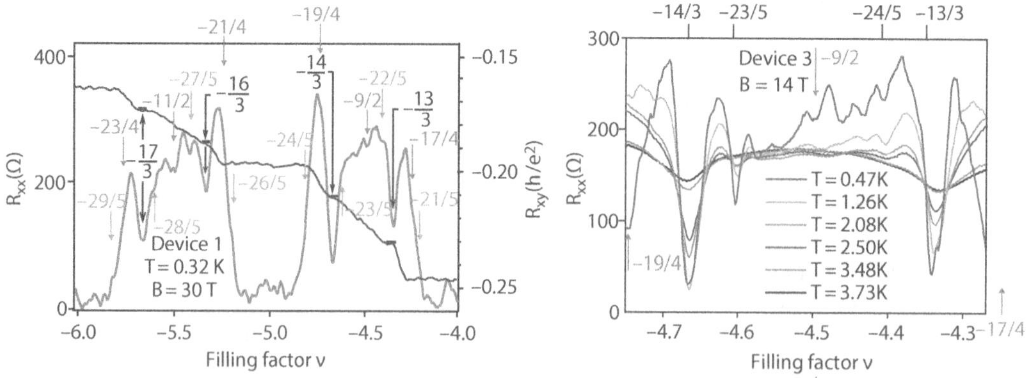

Figure 14.8 Longitudinal resistivity R_{xx} measured in bilayer graphene (encapsulated in hBN with open face) for $n = 2$ subbands (first LL)—experiment [15]. The series of fractions with denominators 3 agrees with single-loop braid commensurability at leakage of flux to opposite sheet in bilayer structure; the same for fractions with denominators 2 and 4. Fractions with denominators 5 correspond to single-loop braid commensurability for $n = 2$ (experiment [15] is repeated with different samples).

flux possibilities $\frac{hc}{e}$, $\frac{2hc}{e}$, $\frac{3hc}{e}$ and $\frac{4hc}{e}$. These effective orbits give the following new fractions for FQHE(single-loop) due to commensurability with nearest and next-nearest neighbors: $\nu = 4+\frac{1}{3}$ and $4+\frac{2}{3}$ for commensurability of orbit $3\frac{hc}{e}$ with nearest and every second neighbors, respectively. These fractions are visible in experiment in three first subbands with $n = 2$ in bilayer graphene [15]—cf. Figures 14.8 and 14.9. The corresponding states are more stable as associated with single-loop correlations similarly to IQHE. These states have nothing in common with CFs as the related correlations are described by single-loop braids. It must be emphasized that the pairs of states $4(5,6,7)+\frac{1}{3}$ and $4(5,6,7)+\frac{2}{3}$ are not the particle and hole partners (as particle $\frac{1}{3}$ and hole $\frac{2}{3}$ multiloop states in the LLL)—these pairs with denominator 3 in subbands of $n = 2$ LL correspond to single-loop braid commensurabilities of nearest and next-nearest (every second) neighbors, respectively (the small asymmetry experimentally observed in corresponding local minima of R_{xx} for these pairs is thus consistent with the difference in the correlations—for $4(5,6,7)+\frac{2}{3}$, there are correlated every second electrons, whereas for $4(5,6,7)+\frac{1}{3}$ all electrons).

Orbits reduced to $\frac{hc}{e}$ are too short for single-loop commensurability, whereas orbits $\frac{2hc}{e}$ and $\frac{4hc}{e}$ give $\nu = \frac{1}{2}$ and $\nu = \frac{1}{4}, \frac{1}{2}, \frac{3}{4}$, respectively. These features are also noticeable in the experiment—cf. Figure 14.8.

14.3.3 Specific to Bilayer Graphene FQHE Hierarchy Change Caused by the Type of the LLL Degeneracy Lifting

Bilayer graphene has a different subband structure in comparison with monolayer graphene and with conventional semiconductor 2DEG, as illustrated in Table 14.1.

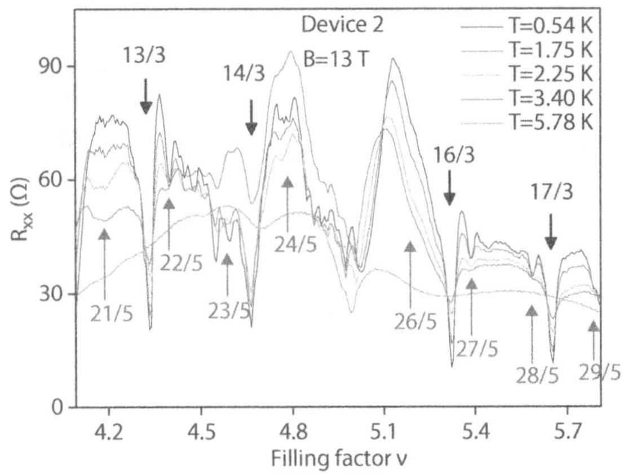

Figure 14.9 Resistivity R_{xx} for bilayer graphene experiment [15] for two first subbands with $n = 2$ from first LL ($\nu \in (4,6)$) (a third sample). The pronounced FQHE features for fractions with denominators 3 for single-loop commensurability (due to leakage of flux between two sheets of bilayer structure) and fractions with denominators 5 also for single-loop braid commensurability are marked.

In bilayer graphene, the degeneracy of $n = 0$ and $n = 1$ states results in eight-fold degeneration of the LLL, doubling four-fold spin-valley degeneracy of the LLL in comparison to the monolayer case. The degeneracy is not exact, and for enhancing magnetic field amplitude, both the Zeeman splitting and the valley splitting grow. Stress, deformation, and structure imperfections also cause lifting of the valley degeneracy. Moreover, the Coulomb interaction

Table 14.1 Comparison of subband arrangement in bilayer graphene, monolayer graphene, and in GaAs 2DEG and the corresponding filling rate $\nu = \frac{N}{N_0}$ range (the nominal size of the cyclotron orbit corresponding to $\boldsymbol{n}$ is $(2n+1)\frac{hc}{eB}$; in the bilayer system, the orbit size can, however, be reduced by flux leakage to opposite sheet).

Type of system	Subbands of the LLL	Subbands of the first LL	Subbands of the second LL
Bilayer graphene	$\nu \in (0,4]$ $n = 0,2,\uparrow$ $n = 0,2,\downarrow$ $n = 1,2,\uparrow$ $n = 1,2,\downarrow$ cond. band	$\nu \in (4,8]$ $n = 2,1,\uparrow$ $n = 2,1,\downarrow$ $n = 2,2,\uparrow$ $n = 2,2,\downarrow$	$\nu \in (8,12]$ $n = 3,1,\uparrow$ $n = 3,1,\downarrow$ $n = 3,2,\uparrow$ $n = 3,2,\downarrow$
Monolayer graphene	$\nu \in (0,2]$ $n = 0,2,\uparrow$ $n = 0,2,\downarrow$ cond. band	$\nu \in (2,6]$ $n = 1,1,\uparrow$ $n = 1,1,\downarrow$ $n = 1,2,\uparrow$ $n = 1,2,\downarrow$	$\nu \in (6,10]$ $n = 2,1,\uparrow$ $n = 2,1,\downarrow$ $n = 2,2,\uparrow$ $n = 2,2,\downarrow$
GaAs 2DEG	$\nu \in (0,2]$ $n = 0,\uparrow$ $n = 0,\downarrow$	$\nu \in (2,4]$ $n = 1,\uparrow$ $n = 1,\downarrow$	$\nu \in (4,6]$ $n = 2,\uparrow$ $n = 2,\downarrow$

causes mixing of $n = 0{,}1$ states that lifts their degeneracy within the LLL. Of particular importance, however, is an order of subbands after the degeneracy lifting allowing for mutually inverted ordering of LLL subbands with distinct $n = 0{,}1$. The orders $n = 0{,}1$ or $n = 1{,}0$ lead to distinct FQHE filling hierarchy. In the case when the LLL subbands with $n = 1$ is filled earlier than the $n = 0$ subband, one arrives with the following hierarchy for the first subband $n = 1,2\uparrow$: multilooped orbits for $\nu = \frac{l}{l3(p-2)\pm 1}$, $\nu = 1 - \frac{l}{l3(p-2)\pm 1}$, single-looped orbits for $\nu = \frac{1}{3}, \frac{2}{3}$, and a paired state for $\nu = \frac{1}{2}$. For the next subband, $n = 0,2\uparrow$, we get the hierarchy in the form: multilooped orbits for $\nu = 1 + \frac{l}{l(p-2)\pm 1}$, $\nu = 2 - \frac{l}{l(p-2)\pm 1}$ and no single-loop orbits. The comparison of reverted orderings of two first LLL subbands is summarized in Table 14.2. One can notice that the state at $\nu = \frac{1}{2}$ corresponds to FQHE only when the subband with $n = 0$ is filled earlier than the subband with $n = 1$.

One can also consider the situation in the LLL of bilayer graphene, when the degeneracy of $n = 0{,}1$ states is lifted in such a way that both levels cross at certain filling factor $\nu^* < 1$ (cf. Ref. [27], where mixing between $n = 0{,}1$ states has been analyzed numerically on small models on torus or sphere). Let us assume for an example, that the $n = 1$ subband ($n = 1,2\uparrow$)

Table 14.2 Comparison of filling hierarchy in the LLL level in bilayer graphene for two mutually inverted successions of two lowest subbands: $n = 0, 2\uparrow$, $n = 0, 2\uparrow$ (1 and 2 lines), and $n = 0, 2\uparrow$, $n = 0, 2\uparrow$ (3 and 4 lines): FQHE at $\nu = \frac{1}{2}$ exists for upper order of subbands and it disappears for lower subband order.

LL subb.	FQHE(single-loop), paired–no FQHE, IQHE	FQHE(multiloop) (q - odd, $l = \frac{i}{2n+1}$, i = 1,2,3,...)	Hall metal
1) $n = 0, 2\uparrow$	1	$\frac{1}{(q-1)}$ (red $\frac{1}{2}$, ...), $1-\frac{1}{(q-1)}$, $\frac{l}{l(q-2)\pm 1}$, $1-\frac{l}{l(q-2)\pm 1}$	$\frac{1}{q-2}$, $1-\frac{1}{q-2}$
2) $n = 0, 2\uparrow$	$\frac{4}{3}, \frac{5}{3}, \left(\frac{3}{2}\ paired\right), 1, 2$	$1+\frac{1}{3(q-1)}$, $2-\frac{1}{3(q-1)}$, $1+\frac{l}{3l(q-2)\pm 1}$, $2-\frac{l}{3l(q-2)\pm 1}$	$1+\frac{1}{3(q-2)}$, $2-\frac{1}{3(q-2)}$
1) $n = 0, 2\uparrow$	$\frac{1}{3}, \frac{2}{3}, \left(\frac{1}{2}\ paired\right), 1$	$\frac{1}{3(q-1)}$ ($\frac{}{6}$,...), $1-\frac{1}{3(q-1)}$, $\frac{l}{3l(q-2)\pm 1}$, $1-\frac{l}{3l(q-2)\pm 1}$	$\frac{1}{3(q-2)}$, $1-\frac{1}{3(q-2)}$
2) $n = 0, 2\uparrow$	1,2	$1+\frac{1}{q-1}$, $1+\frac{l}{l(q-1)\pm 1}$, $2-\frac{1}{q-1}$, $2-\frac{l}{l(q-2)\pm 1}$	$1+\frac{1}{q-2}$, $2-\frac{1}{q-2}$

is energetically favorable up to some filing fraction ν^*. At this filling rate, the subband $n = 1,2\uparrow$ crosses with the subband $n = 0,2\uparrow$ and the latter starts to be of lower energy for $1 + \nu^* > \nu > \nu^*$. The hierarchy of fractional fillings corresponding to such a situation looks like for ordinary filling of the subband $n = 1,2\uparrow$, though with an insertion of $n = 0,2\uparrow$ subband hierarchy. Depending on the value of ν^*, various patterns are possible by a combination of hierarchy patterns as illustrated in Table 14.2.

14.4 Comparison with Experiment

Recent progress in experiments with monolayer graphene on BN substrate [10, 11] and with suspended small graphene sheets [12, 13] allowed for observation of more and more Hall features referred to FQHE in subsequent subbands of two first LLs. While the sequence of fillings in the lowest subband of monolayer LLL fits well to CF predictions, an explanation of the FQHE filling structure in next subbands strongly deviates from this simple picture. CF theory fails in all subbands of the first LL in monolayer graphene [11–13]. Various scenarios of breaking of the approximate SU(4) spin-valley symmetry in graphene do not solve this problematic situation despite many theoretical attempts, which evidences insufficiency of the CF model in this case.

More efficient in understanding of FQHE in graphene is the braid group-based commensurability approach. The hierarchy for FQHE predicted in this way is consistent with all experimental data known as of yet—the corresponding filling fractions can be perfectly reproduced by the topology cyclotron braid group method.

Effectiveness of the CF model in the LLL in monolayer system is linked with the fact that exclusively in the LLL, cyclotron orbits are always shorter than the interparticle spacing and additional loops are necessary to exchange neighboring particles along braids. These additional loops can be simulated by auxiliary fictitious field flux quanta attached to CFs. In the case, however, when the more complicated commensurability conditions support particular FQHE states in the LLL (known as out of CF hierarchy, e.g., $\nu = \frac{5}{13}, \frac{4}{11}, \frac{3}{10}, \ldots$) or in higher LLs, then the CF model fails. The braid group approach reproduces all features described correctly by CF model and moreover explains hierarchy details inaccessible for CF approach. The usefulness of the CF model is especially limited in higher LLs because in these levels, the simple multiloop commensurability is needed only close to the subband edges, whereas the central regions of all subbands in higher LLs correspond to cyclotron orbits larger than particle separation. Hence, in this case, the single-loop-type commensurability with next-nearest neighbors is involved beyond the CF concept. For example, in the first LL in monolayer graphene, the following doublets of fillings are experimentally observed: $\left(\frac{7}{3}, \frac{8}{3}\right), \left(\frac{10}{3}, \frac{11}{3}\right), \left(\frac{13}{3}, \frac{14}{3}\right), \left(\frac{16}{3}, \frac{17}{3}\right)$, corresponding to single-loop commensurability for nearest and next-nearest (every second) neighbors. These doublets in monolayer graphene are visible in experiments [10–13]. The number of centrally located filling rates for FQHE(single-loop) grows next with the LL number as $2n$ (it is observed experimentally in conventional 2DEG at $n = 2$ four of fillings with denominators 5 [37], as illustrated in Figure 14.10). In monolayer graphene, the repeating doublets of filling ratios (with denominators 3) for $n = 1$ have been observed in very accurate measurements in suspended samples [12, 13] besides of

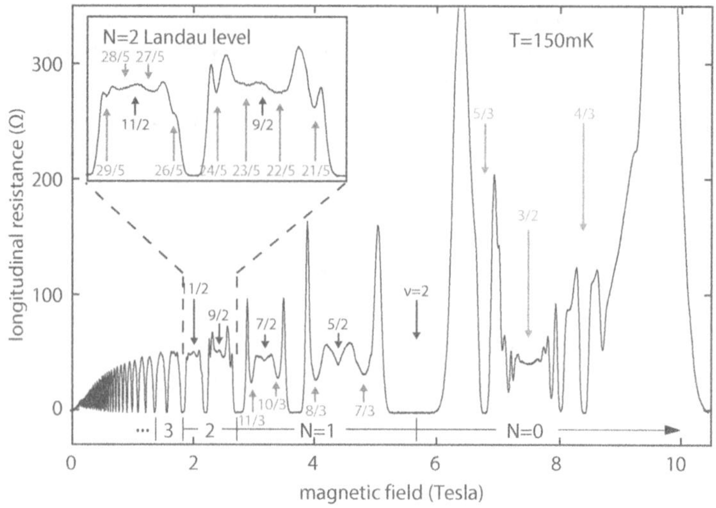

Figure 14.10 For comparison—measurement of resistivity R_{xx} in conventional 2DEG for a wide range of magnetic fields corresponding to n = 1,2 in the high-mobility GaAs/AlGaAs heterostructure (after Ref. [37]). In red color, there are indicated fractions for the FQHE(single-loop)—doublets with denominators 3 in subbands with n = 1 and four with denominators 5 for n = 2, in consistence with braid commensurability predictions. The pair with denominators 3 for n = 0 (blue, 5/3, 4/3) corresponds to 3-loop orbits. At 11/2, 9/2, 7/2, 5/2 the braid group approach predicts paired states but for 3/2 and 1/2 Hall metal. A similar structure of FQHE is predicted for monolayer graphene, though the data for n = 2 in monolayer graphene are not available as of yet.

those on the BN substrate [10, 11]. Worth noting is the observation [11] that stability of corresponding FQHE(single-loop) states is of similar strength as of IQHE states and higher in comparison to FQHE(multiloop) states as is visible in Figure 14.11. This evidences that stronger correlations related to single-loop braids similar to those in IQHE states are present.

The commensurability braid group approach reproduces successfully all positions of observed features in two lowest LLs of the monolayer graphene. The elongate plateaus at edges of subbands with IQHE in center also embrace minima related to near subbband edges located FQHE(multiloop) states being in this way out of the experimental resolution. In higher LL, the new features were also observed besides doublets mentioned above but associated with not vanished longitudinal resistivity opposite to other FQHE states. This property suggests that not all electrons participate in corresponding correlated states of every second or every third particles. Such features in the first LL of monolayer graphene recently reported [11] at $\nu=\frac{7}{3},\frac{8}{3},\frac{12}{5},\frac{13}{5},\frac{17}{7},\frac{18}{7},\frac{22}{9},\frac{23}{9},\frac{10}{3},\frac{11}{3},\frac{17}{5},\frac{18}{5},\frac{24}{7},\frac{25}{7},\frac{13}{3},\frac{14}{3},\frac{22}{5},\frac{23}{5}$ are reproduced one-to-one by the commensurability series $\nu=2(3,4)+\frac{xl}{l3(q-1)\pm 1}$ with q = 3, x = 2, 3, $l=\frac{i}{3}$, i = 1,2,3 as shown in Figure 14.4. One can notice, however, that the FQHE filling rates $\frac{7}{3},\frac{8}{3},\frac{10}{3},\frac{11}{3},\frac{13}{3},\frac{14}{3}$ are independently single-loop states that are more stable than the multiloop ones, which is consistent with the experimental data presented in Figure 14.11 and in the upper panel of Figure 14.4.

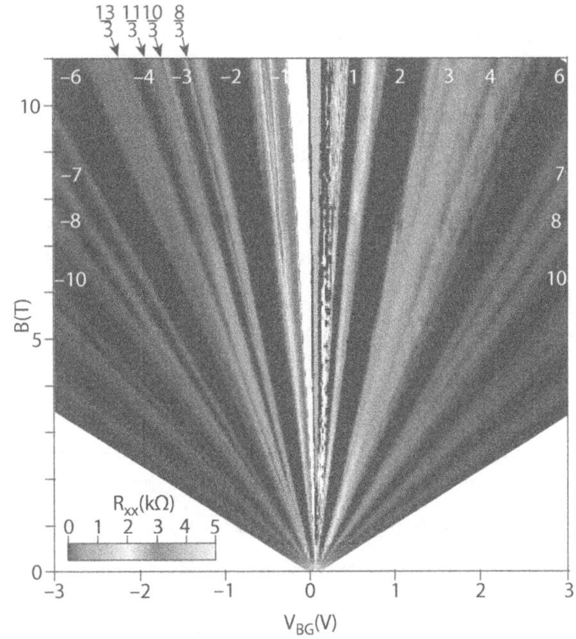

Figure 14.11 Fan diagram for $R_{xx}(\nu, B)$ in monolayer graphene up to 11 T from experiment [11]. The remarkable property is the closeness in value of R_{xx} of FQHE features for fractions with denominators 3 for $n = 1$ with those for IQHE, which supports the FQHE single-loop braid correlations in corresponding states similar as in the case of IQHE.

In bilayer graphene, the manifestation of FQHE deviates from the CF picture also in the LLL due to double-layer topology oddness in this system. As was presented in Section 14.3.2, in the lowest subband of the LLL in bilayer graphene, the even denominator filling fractions for FQHE appear [14]. The commensurability braid group approach for bilayer graphene reproduces all the observed experimental FQHE hierarchies, including the pronounced state at $\nu=-\frac{1}{2}$—cf. Figure 14.12, the illustration in Figure 14.5 and in Table 14.2.

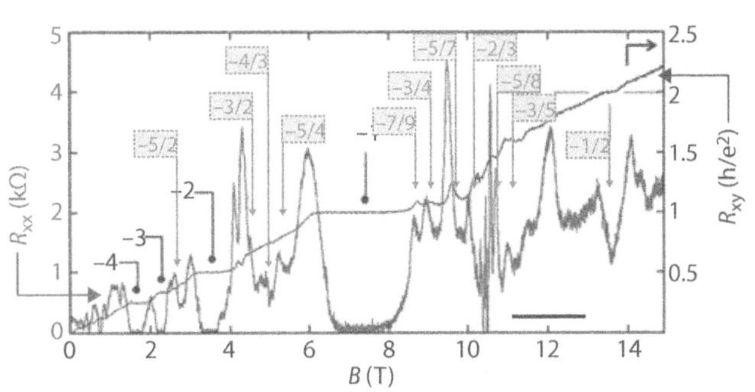

Figure 14.12 Observation of FQHE at T = 0.25 K in bilayer suspended graphene, magnetoresistance R_{xx} (blue curve), and R_{xy} (black curve) at the lateral voltage −27 V, after experiment [14]. In red is added fitting with the cyclotron braid group hierarchy (as in Table 14.2, 1 and 2 lines) for mirror valence-band FQHE states including $\nu=-\frac{1}{2}$ (the mirror to $\nu=\frac{1}{2}$).

Let us emphasize that the FQHE state at $\nu=\frac{1}{2}$ has been discovered earlier in bilayer structure of conventional 2DEG [41, 42], which is also consistent with the commensurability braid group predictions and evidences that this not-CF fraction is caused by the double-layer topology and not by specific material properties or band structure of bilayer systems.

The interesting observation in bilayer graphene is reported by comparison of FQHE measurement in bilayer graphene in suspended samples and in samples on the BN substrate [14–18]. Surprisingly, the FQHE state at $\nu=-\frac{1}{2}$ observed in bilayer structure in the suspended sample disappears in bilayer graphene on BN substrate. We propose to explain this behavior by our commensurability braid group approach (as detailed in Section 14.3.3), via noting that the occurrence of the FQHE state depends on the order of the LLL subbands $n = 0,1$ when their degeneracy is lifting, as shown in Table 14.2. We thus suppose that the external conditions related to the presence of the BN substrate cause the ordering 1,0 of the LLL subbands resulting in disappearance of the FQHE state at $\nu=-\frac{1}{2}$ (as illustrated in Table 14.2), which is, however, no case for the suspended sample with opposite, 0,1, lowest subband ordering.

The most spectacular observations of FQHE in bilayer graphene were reported recently [15] in the first LL beyond the LLL, i.e., for $n = 2$ in bilayer graphene (in four subbands with $n = 2$, i.e., for $\nu \in (4,8]$). The unprecedented accuracy of Hall measurements of bilayer graphene encapsulated in hBN but for samples with so-called "open face" [15] revealed pronounced FQHE features in the subbands with $n = 2$ at filling rates with denominators 3. The fractions with denominators 5 also are noticeable but weaker in comparison to those with denominators 3 (actually, 2/5 and 3/5 are clearly visible, whereas 1/5 and 4/5 could be identified as only small local bends in longitudinal resistivity curves, and cannot be attributed to FQHE). We have explained this astonishing behavior by specific double-layer topology of bilayer graphene and by an effective leakage of flux due to electron hopping as described quantitatively in Section 14.3.2. The resulted commensurability hierarchy for FQHE in the first LL with $n = 2$ in bilayer system (as derived in Section 14.3.2) is perfectly consistent with the experimental data [15]. We noticed the agreement with topological predictions not only for fractions with denominators 3 and 5, but also with denominators 2 and 4 (at low temperature, ~0.5 K—cf. Figure 14.8). The related correlations for all these features (including fractions with denominators 3, 5, 4, 2 for $n = 2$) have nothing in common with CFs because all of them correspond to single-loop commensurability instances.

In view of topological effects in bilayer graphene resulted from hopping of electrons between graphene sheets and linked to a specific cyclotron braid commensurability presented above, one can expect the phase transition visible in FQHE hierarchy and caused by blocking the interlayer hopping of electrons. The hopping of electrons can be tuned by a transverse electric field applied perpendicularly to the basal plane. The applied voltage can open a band gap at the charge neutrality point and may change the topology of multiloop trajectories in bilayer case reducing them to only instances available in the monolayer case. Such an experiment has been performed [17] for fully encapsulated bilayer graphene between two hBN layers with perpendicular electric field applied (the displacement field $D \in (-100, 100$ mV/nm). The experiment demonstrates significant rearrangement of FQHE hierarchy in $n = 0$ and

$n = 1$ LLL subbands. The authors of Ref. [17] argue that the reason of the observed phase transitions is linked with different ordering of LLL valley subbands induced by the voltage as the observed transitions concern fractional states $\nu = \frac{2}{3}, \frac{5}{3}$, respectively. An inspection presented in Ref. [17] data reveals the transition also at $\nu = \frac{1}{2}$, which agrees with conditioning of this Hall fraction by interlayer hopping. Its blocking should result in disappearance of the FQHE feature at $\frac{1}{2}$. The experiment [17] did not reach $n = 2$ subbands of the first LL, but in these subbands, the expected phase transitions due to reducing of interlayer hopping might be even more explicit. One can expect that by blocking interlayer hopping, the pronounced features with denominators 3 in $n = 2$ subbands should be canceled.

14.5 Conclusion

The commensurability of cyclotron braids with interparticle spacing in homogeneous 2D charged systems in magnetic field is utilized in order to verify possibility of arrangement of correlated multiparticle Hall states and to decipher the "magic" hierarchy of filling ratios for FQHE. In this way, we have derived the FQHE hierarchy in perfect consistence with available experimental observations in monolayer graphene. By specific topology instances for braids in bilayer graphene, we have also successfully explained the recent experimental FQHE observations in the bilayer system, both in the LLL and in first LL with $n = 2$. The oddness of FQHE hierarchy evidenced experimentally in the bilayer graphene in comparison to monolayer graphene and conventional 2DEG in eight LLs subbands has been clarified.

The new opportunities for commensurability in the LLL and in higher LLs in graphene bilayer were identified in double sheet topology with electron interlayer hopping leading to different than FQHE in monolayer Hall systems. The even denominator main line of the fractional filling hierarchy in the bilayer graphene is derived in consistence with experimental observations. The unconventional hierarchy of FQHE observed recently in $n = 2$ spin-valley subbands in bilayer graphene is explained by topology arguments specific to two-sheet system with interlayer tunneling. An experimentally noticed oddness of FQHE hierarchy in the subbands of first LL in bilayer graphene with $n = 2$ (different than in monolayer systems for $n = 2$ subbands) has been successfully explained in terms of specific to bilayer system topology. The presented topology-induced hierarchy for the monolayer and bilayer graphene found the confirmation in all up-to-date available experimental observations in graphene on BN substrate as well as for suspended samples including monolayer graphene up to sixth spin-valley subband and bilayer graphene up to eighth spin-valley subband. The topology foundations of FQHE also allow for explanation of observed hierarchy phase transitions in vertical electric field in graphene bilayer.

The specific to graphene in its monolayer and bilayer manifestation of FQHE evidences that the particularities of single electron band structure do not influence this effect, which appears to be determined by the topological factors independent of local dynamics in various systems.

Appendix 14.6 Degeneracy of LLL in Tight Binding Approximation for Bilayer Graphene

The electron band structure for graphene is commonly modeled upon simple tight binding approach neglecting interaction, cf. e.g., [9, 21, 22]. The low-energy tight binding Hamiltonian in bilayer graphene close to K point in Brillouin zone attains the form

$$H_K = \frac{1}{2m}\begin{bmatrix} 0 & \pi^2 \\ \pi^{+2} & 0 \end{bmatrix}, \tag{14.7}$$

where $\pi = \tilde{p}_x + i\tilde{p}_y$, $\tilde{\mathbf{p}} = \mathbf{p} + \frac{e}{c}\mathbf{A}$, $m = \frac{\gamma_1}{2v^2}$, γ_1 is the hopping amplitude (for interlayer transition indicated in Figure 14.13), and v is the velocity of Dirac-type excitations in monolayer graphene. For the second nonequivalent valley point K', $H_{K'} = H_K^*$. The zero energy state (ground state) for Hamiltonian (Equation 14.7) is double degenerated since for lowering oscillatory operator π, one has, $\pi\,|\,0> = 0$ and $\pi^2\,|\,10> = 0$, where $|\,0>$, $|\,1>$ are oscillatory-type states of nonrelativistic Landau spectrum. The similar property also holds for the valley point K', which, together with Zeeman degeneracy and above $n = 0{,}1$ degeneracy, results in eight-fold degeneracy of the LLL in bilayer graphene. Higher LLs in bilayer graphene do not repeat $n = 0{,}1$ degeneracy and are four-fold spin-valley degenerated similarly as in the case of monolayer graphene.

For monolayer graphene, the local Hamiltonian was of the following form:

$$H_K = \xi v\begin{bmatrix} 0 & \pi \\ \pi^+ & 0 \end{bmatrix}, \tag{14.8}$$

where $v = \frac{\sqrt{3}}{2}\frac{a\gamma_0}{\hbar}$, $\xi = \pm 1$ enumerates valleys, a is the lattice constant, $\gamma_0 >> \gamma_1$ is in-plane hopping between nearest neighbors (Figure 14.13). Hamiltonian (Equation 14.8) produces the linear low-energy gapless spectrum $|\,e\,| = vp$ in monolayer graphene, whereas in bilayer Hamiltonian (Equation 14.7), it contributes to the dispersion $|e| = \frac{p^2}{2m}$. Opening of the gap in Dirac valley points K and K' for parabolic dispersion is induced by off-diagonal operator in Equation 14.7 driven by tunneling (hopping) of electrons between two sheets of bilayer structure (assuming the direct hopping γ_1 strongly prevails over next-neighbor hopping γ_3, γ_4—cf. Figure 14.13 [22]).

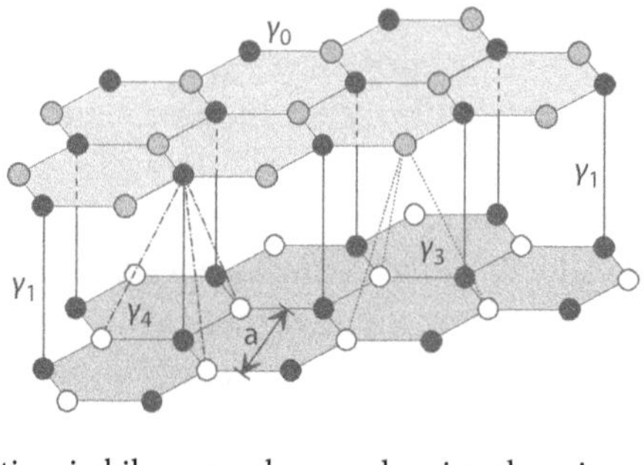

Figure 14.13 Position of sublattices in bilayer graphene and various hopping ways for electrons indicated.

The structure of tight binding local Hamiltonian (Equations 14.7 and 14.8) gives at magnetic field presence the "relativistic" Landau-level quantization [9, 21, 22], $\epsilon_n = \pm\hbar\omega_c\sqrt{n(n-1)}$, $\epsilon_n = \pm\hbar\omega_c\sqrt{n}$ for bilayer and monolayer graphene, respectively.

Appendix 14.7 Cyclotron Braid Commensurability for FQHE States in the LLL in Conventional 2DEG

One can identify the correlated states at fractional fillings generalizing the genuine pattern of the correlation of IQHE, $\frac{S}{N}=\frac{S}{N_0}$ when cyclotron orbit size $\frac{S}{N_0}$ ($N_0=\frac{eBS}{hc}$ is the LL degeneracy) fits to electron separation $\frac{S}{N}$. At fractional fillings of the LLL, the cyclotron orbits $\frac{hc}{eB}$ are smaller than $\frac{S}{N}$ and cyclotron orbits cannot match neighboring particles. For establishing of any correlated state, the particle exchanges are, however, necessary to define statistics of quantum particles. Exclusively in 2D, multiloop cyclotron orbits have larger size in comparison to single-loop ones at the same magnetic fields [43, 44]. It follows from the distribution of the external field B flux per particle among all loops of the multi-loop cyclotron orbit all located in the same plane. The condition for commensurability thus attains the more general form:

$$\frac{BS}{N}=(q-1)\frac{hc}{ex}\pm\frac{hc}{ey}, \tag{14.9}$$

where q is the number of loops of single cyclotron orbit (q must be odd integer in order to ensure the corresponding braid to describe particle exchange—the braid generator with n additional loops corresponds to $2n + 1 = q$-loop cyclotron orbit [33, 43]). At magnetic fields in 2D, the braids are built from half-pieces of cyclotron orbits provided that these orbits accurately fit to neighboring particle separation at the uniform particle distribution caused by the electric repulsion. In condition (14.9), $x \geq 1$ (integer) indicates the commensurability q-1 single loops from q-loop cyclotron orbit to every xth particle on the plane; $y \geq x$ (also integer) indicates the commensurability of the last loop of the q-loop orbit) with every yth particle; $\pm$ indicates the same or opposite (of eight-figure-shape) orientation of the last, i.e., qth loop. From Equation 14.9, we obtain the following conditions:

$$\begin{aligned} \nu &= \frac{N}{N_0}=\frac{xy}{(q-1)y\pm x}, \textit{ for band electrons,} \\ \nu &= 1-\frac{xy}{(q-1)y\pm x}, \textit{ for band holes,} \end{aligned} \tag{14.10}$$

for the general hierarchy of correlated states in the LLL describing the FQHE hierarchy. For $x = 1$, the hierarchy (Equation 14.10) reproduces the CF hierarchy. For $x > 1$, the hierarchy (Equation 14.10) is beyond the ability of the CF model and displays filling ratios for FQHE in the LLL including those outside CF hierarchy observed in experiment in conventional 2DEG [4]. The comparison with the experimental data is summarized in Figure 14.14.

The CF model agrees with the simplest commensurability case ($x = 1$) and breaks down in more complicated commensurability instances as given by Equation 14.10 for $x > 1$. In Figure 14.14, in red color are indicated filling rates out of main CF hierarchy, but visible in experiments and successfully reproduced by the hierarchy (Equation 14.10).

The limit $y \to \infty$ displays the hierarchy of the Hall metal exactly in the same manner as for the archetype of the Hall metal at $\nu = 1/2$ (the last orbit is then infinite and fits to infinitely distant particles as in the normal Fermi liquid without any magnetic field [45]). The general Hall metal hierarchy in the LLL thus has the form

$$\begin{aligned} \nu &= \frac{x}{q-1}, \textit{ for electrons}, \\ \nu &= 1 - \frac{x}{q-1}, \textit{ for holes}. \end{aligned} \tag{14.11}$$

Note that the Hall metal correlation can manifest itself at fractions not necessarily with even denominators (for x even, beyond the CF concept), similarly as the hierarchy (Equation 14.10) displays fractions both with odd and even denominators in compliance with the experimental observations [4]. Some fractions are repeated in various lines of the general

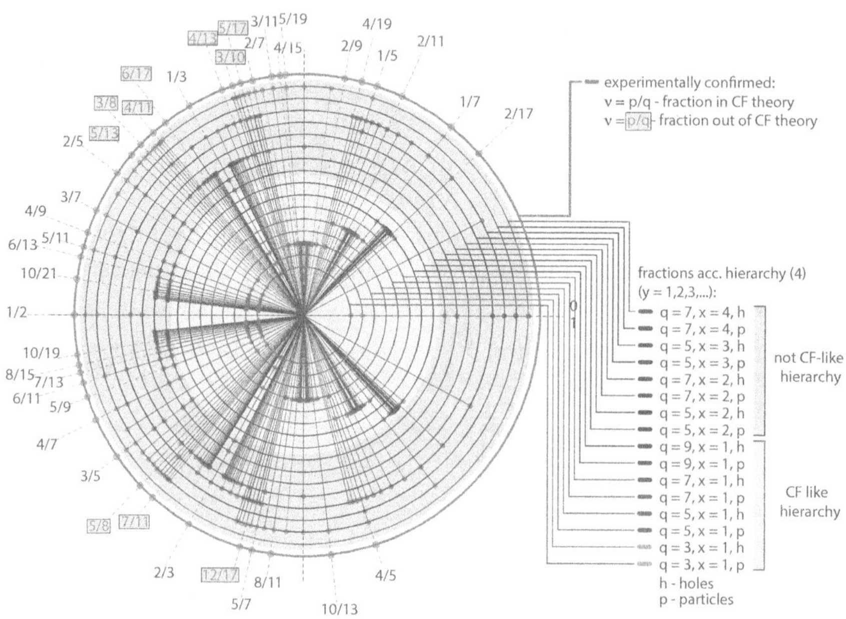

Figure 14.14 Comparison of the hierarchy (Equation 14.10) with all measured fractional filling rates for FQHE features in the LLL (spin polarized). The hierarchy series according (Equation 14.10) for several y each are displayed; filling rates beyond the CF hierarchy are shown in red (Hall metal state fraction 1/2 is marked).

hierarchy (Equation 14.10). This fact reveals the possibility of various types of commensurability of multiloop cyclotron orbits with interparticle spacing $\frac{S}{N}$. The advantage of one commensurability over the others (alternative ones at the same filling ratio) is related with energy minimization, i.e., with the minimization of the Coulomb interaction.

Appendix 14.8 Trial Wave Functions for FQHE States in the LLL in Conventional 2DEG

For the simplest line of the hierarchy (Equation 14.10) with $x = y = 1$, i.e., $\nu = \frac{1}{q}$, q - *odd*, the corresponding wave functions have been given by Laughlin in the form [2],

$$\Psi_q(z_1, z_2, \ldots, z_N) = A \prod_{i,j,i>j}^{N,N} (z_i - z_j)^q e^{-\sum_i^N \frac{|z_i|^2}{4l^2}}, \tag{14.12}$$

where $z_i = x_i + iy_i$ is ith particle classical position on the complex plane (the argument of the quantum multiparticle wave function), $l = \sqrt{\frac{hc}{eB}}$ is the magnetic length, the product $\prod_{i,j,i>j}^{N,N} (z_i - z_j)^q$ is the Jastrow polynomial, and A is an appropriate normalization constant. The defining characteristic of the Laughlin function is that the q-fold zero at each particle keeps particles apart, and thus diminishes the Coulomb interaction energy. The function (Equation 14.12) must transform itself according to the one unitary representation (1DUR) of the cyclotron braid subgroup with generators σ_i^q. And indeed, for the 1DUR of the full braid group given by $\sigma_i \rightarrow e^{i\alpha}$ with $\alpha = \pi$ (fermionic), one gets $e^{iq\pi}$ as the 1DUR of σ_i^q, which coincides with the Laughlin phase.

For the hierarchy (Equation 14.10), the generators (describing elementary exchanges) of the appropriate more complicated cyclotron braid subgroups are defined as follows (for ± in Equation 14.10):

$$\begin{aligned}
&b_i^{q,x,y,+} = \\
&(\sigma_i \cdot \sigma_{i+1} \cdot \ldots \cdot \sigma_{i+x-2} \cdot \sigma_{i+x-1} \cdot \sigma_{i+x-2}^{-1} \cdot \ldots \cdot \sigma_{i+1}^{-1} \cdot \sigma_i^{-1})^{q-1} \\
&\cdot \sigma_i \cdot \sigma_{i+1} \cdot \ldots \cdot \sigma_{i+y-2} \cdot \sigma_{i+y-1} \cdot \sigma_{i+y-2}^{-1} \cdot \ldots \cdot \sigma_{i+1}^{-1} \cdot \sigma_i^{-1}, \\
&\textit{and} \\
&b_i^{q,x,y,-} = \\
&(\sigma_i \cdot \sigma_{i+1} \cdot \ldots \cdot \sigma_{i+x-2} \cdot \sigma_{i+x-1} \cdot \sigma_{i+x-2}^{-1} \cdot \ldots \cdot \sigma_{i+1}^{-1} \cdot \sigma_i^{-1})^{q-1} \\
&\cdot (\sigma_i \cdot \sigma_{i+1} \cdot \ldots \cdot \sigma_{i+y-2} \cdot \sigma_{i+y-1} \cdot \sigma_{i+y-2}^{-1} \cdot \ldots \cdot \sigma_{i+1}^{-1} \cdot \sigma_i^{-1})^{-1},
\end{aligned} \tag{14.13}$$

with 1DURs (for $\alpha = \pi$) $e^{iq\pi}$ (for +) and $e^{i(q-2)\pi}$ (for -) (with supplement of the above notation for $x(y) = 1$, $\sigma_i \cdot \sigma_{i+1} \cdot \ldots \cdot \sigma_{i+x-2} \cdot \sigma_{i+x-1} \cdot \sigma_{i+x-2}^{-1} \cdot \ldots \cdot \sigma_{i+1}^{-1} \cdot \sigma_i^{-1} = \sigma_i$). Examples of these braid generators are depicted in Figure 14.15.

The related modification of the Jastrow polynomial in the Laughlin function (Equation 14.12) must thus be as follows (in the LLL, the true wave function must be holomorphic function uniquely defined by its nodes):

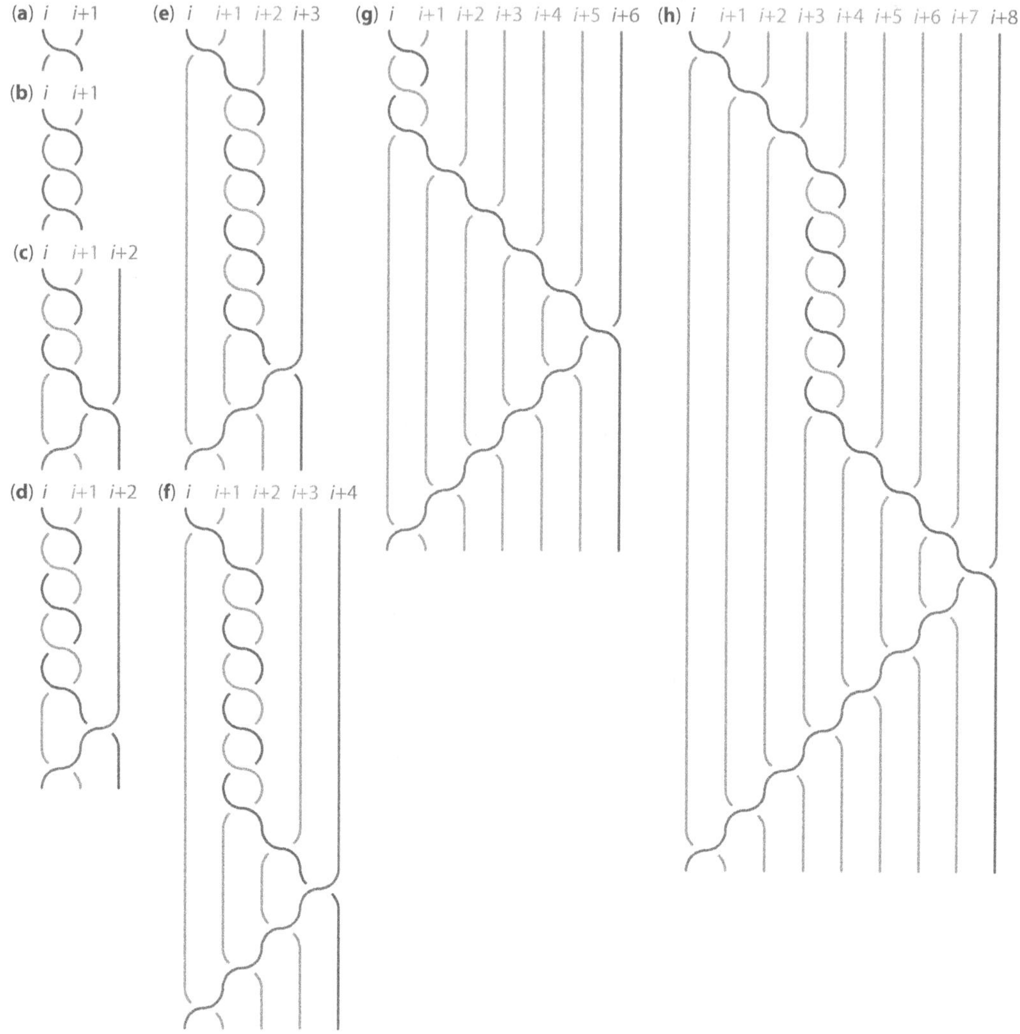

Figure 14.15 The braid cyclotron subgroups generators for several selected filling fractions (e, f, h—examples of generators for filling fractions that cannot be derived using standard CF theory; a, b, c, d, g—examples of fractions from Jain-like hierarchy). Generators are as follows: a) $\nu = 1 \rightarrow \sigma_i$, b) $\nu = 1/3 \rightarrow b_i^{3,1,1,+} = \sigma_i^3$, c) $\nu = 2/5 \rightarrow b_i^{3,1,2,+} = \sigma_i^3 \sigma_{i+1} \sigma_i^{-1}$ d) $\nu = 2/7 \rightarrow b_i^{5,1,2,-} = \sigma_i^5 \sigma_{i+1}^{-1} \sigma_i^{-1}$, e) $\nu = 3/8 \rightarrow b_i^{7,2,3,-} = \sigma_i \sigma_{i+1}^7 \sigma_{i+2}^{-1} \sigma_{i+1}^{-1} \sigma_i^{-1}$, f) $\nu = 4/11 \rightarrow b_i^{7,2,4,-} = \sigma_i \sigma_{i+1}^7 \sigma_{i+2} \sigma_{i+3}^{-1} \sigma_{i+2}^{-1} \sigma_{i+1}^{-1} \sigma_i^{-1}$, g) $\nu = 6/13 \rightarrow b_i^{3,1,6,+} = \sigma_i^3 \sigma_{i+1} \cdots \sigma_{i+5} \sigma_{i+4}^{-1} \cdots \sigma_{i+1}^{-1} \sigma_i^{-1}$, h) $\nu = 5/13 \rightarrow b_i^{7,4,8,+} = \sigma_i \sigma_{i+1} \sigma_{i+2} \sigma_{i+3}^7 \sigma_{i+4} \cdots \sigma_{i+7} \sigma_{i+6}^{-1} \cdots \sigma_i^{-1}$.

$$
\begin{aligned}
&\Psi_q^{x,y,+}(z_1,z_2,\ldots,z_N)=\\
&\quad A\prod_{i,j=1;i<i\,mod\,x+(j-1)x}^{N,N/x}(z_i-z_{i\,mod\,x+(j-1)x})^{q-1}\\
&\quad\times\prod_{i,j=1;i<i\,mod\,y+(j-1)y}^{N,N/y}(z_i-z_{i\,mod\,y+(j-1)y})e^{-\sum_i^N\frac{|z_i|^2}{4l^2}},\\
&\Psi_q^{x,y,-}(z_1,z_2,\ldots,z_N)=\\
&\quad A\prod_{i,j=1;i<i\,mod\,x+(j-1)x}^{N,N/x}(z_i-z_{i\,mod\,x+(j-1)x})^{q-1}\\
&\quad\times\prod_{i,j=1;i<i\,mod\,y+(j-1)y}^{N,N/y}(z_{i\,mod\,y+(j-1)y}-z_i)e^{-\sum_i^N\frac{|z_i|^2}{4l^2}}.
\end{aligned}
\tag{14.14}
$$

The above functions for CF-like hierarchy ($x = 1$) attain the form (they define in a unique manner the unclear projection onto LLL in the CF theory),

$$
\begin{aligned}
&\Psi_q^{x=1,y,+}(z_1,z_2,\ldots,z_N)=\\
&\quad A\prod_{i,j=1,i<j}^{N,N}(z_i-z_j)^{q-1}\\
&\quad\times\prod_{i,j=1;i<i\,mod\,y+(j-1)y}^{N,N/y}(z_i-z_{i\,mod\,y+(j-1)y})e^{-\sum_i^N\frac{|z_i|^2}{4l^2}},\\
&\Psi_q^{x=1,y,-}(z_1,z_2,\ldots,z_N)=\\
&\quad A\prod_{i,j=1,i<j}^{N,N}(z_i-z_j)^{q-1}\\
&\quad\times\prod_{i,j=1;i<i\,mod\,y+(j-1)y}^{N,N/y}(z_{i\,mod\,y+(j-1)y}-z_i)e^{-\sum_i^N\frac{|z_i|^2}{4l^2}}.
\end{aligned}
\tag{14.15}
$$

The functions (Equation 14.14) are proposed as the trial wave functions for correlated states for filling rates (Equation 14.10) for which elementary exchanges of particles are defined by braids (Equation 14.13) and generalize the Laughlin function (Equation 14.12) for the case when $x, y > 1$ with some analogy to Halperin multicomponent functions [29].

The energy gain in the Laughlin state is due to lowering of the Coulomb repulsion energy $<\Psi|\sum_{i.j,i>j}^{N,N}\frac{e^2}{|z_i-z_j|}|\Psi>$. It is clear that the energy reducing for the function (Equation 14.14) is the weaker the higher x is (for the same q and y). It follows from the dilution of correlated particles for $x > 1$ (the correlation concerns every xth electron only) as expressed in modified Laughlin-type function (Equation 14.14) by reducing of the domain of the product. This leads to the diminishing of the repulsion energy gain due to the averaging of the Coulomb energy, $\sum_{i.j,i>j}^{N,N}\frac{e^2}{|z_i-z_j|}$, with the wave function (Equation 14.14 instead of 14.15 or 14.12) because q-1 fold zero in these functions prevents approaching not all electrons in the case of function (Equation 14.14) but only its $1/x$ fraction (opposite to the case of function (Equation 14.12 or 14.15) for which $x = 1$). Therefore, states with lower x are more stable. Thus, states with $x = 1$ energetically prevail over states with $x > 1$ and are more stable.

Table 14.3 Comparison of energy values obtained by exact diagonalization and by Monte Carlo simulation for some exemplary filling fractions for FQHE (Monte Carlo Metropolis simulation for proposed topology-based wave functions, for 200 particles).

q	x	y	**Hierarchy fraction, $\nu = N/N_0$**	**Energy from Monte Carlo simulation for functions, according to Equations 14.14 and 14.15**	**Energy from exact diagonalization [46]**
3	1	2	$\frac{2\cdot 1}{(3-1)\cdot 2+1}=\frac{2}{5}$	−0.432677	−0,432804
3	1	3	$\frac{3\cdot 1}{(3-1)\cdot 3+1}=\frac{3}{7}$	−0.441974	−0,442281
3	1	4	$\frac{4\cdot 1}{(3-1)\cdot 4+1}=\frac{4}{9}$	−0.446474	−0,447442
3	1	5	$\frac{5\cdot 1}{(3-1)\cdot 5+1}=\frac{5}{11}$	−0.451056	−0,450797
5	1	2	$\frac{2\cdot 1}{(5-1)\cdot 2+1}=\frac{2}{9}$	−0.342379	−0,342742
5	1	3	$\frac{3\cdot 1}{(5-1)\cdot 3+1}=\frac{3}{13}$	−0.348134	−0,348349
5	1	4	$\frac{4\cdot 1}{(5-1)\cdot 4+1}=\frac{4}{17}$	−0.351857	−0,351189

To confront the energy values obtained from exact diagonalization for different FQHE fillings [46], the numerical estimation of energy for newly proposed functions (Equations 1.14 and 1.15) were performed according to Monte Carlo Metropolis scheme [47–49]. Some exemplary results revealing very good overlap with the exact diagonalization are presented in Table 14.3.

Nevertheless, it should be commented that from the point of view of the commensurability condition governed the form of the cyclotron braid generator corresponding to multiloop cyclotron orbits, none of the loops cannot be featured, thus each loop can be accommodated to the particle separation independently. Thus, for q-looped orbit, one would deal with the ordered series $x_1 \leq x_2 \leq \cdots \leq x_q$ simplified in Equation 14.10 to $x_1 = \cdots = x_{q-1} = x$, $x_q = y$. Apparently, the Coulomb repulsion minimization prefers $x_1 = \cdots = x_{q-1}$ for which the minimization domain restriction (resulting in weaker interaction energy reducing) is more convenient than for distinct distributions of x_i. This explains the choice of the uniform behavior of $q - 1$ loops (i.e., $x_1 = \cdots = x_{q-1} = x$), but this is not a rule and for many fractions, various energetically competitive commensurability opportunities might be considered.

Another observation related to various types of correlation identified by the braid commensurability criterion agrees with experimental data for the longitudinal resistivity R_{xx} [4], which is zero for states with all correlated particles (i.e., with $x = 1$), whereas the residual its value grows with $x > 1$ probably due to scattering on portion of noncorrelated electrons.

The trial wave functions (Equations 14.14 and 14.15) may be applied to model FQHE states in the LLL of monolayer graphene.

Acknowledgments

This chapter is supported by NCN projects P.2011/02/A/ST3/00116 and P.2016/21/D/ST3/00958.

References

1. Tsui, D.C., Störmer, H.L., Gossard, A.C., Two-dimensional magnetotransport in the extreme quantum limit. *Phys. Rev. Lett.*, 48, 1559, 1982.
2. Laughlin, R.B., Anomalous quantum Hall effect: An incompressible quantum fluid with fractionally charged excitations. *Phys. Rev. Lett.*, 50, 1395, 1983.
3. Prange, R.E. and Girvin, S.M., *The quantum Hall effect*, Springer Verlag, New York, 1990.
4. Pan, W., Störmer, H.L., Tsui, D.C., Pfeiffer, L.N., Baldwin, K.W., West, K.W., Fractional quantum Hall effect of composite fermions. *Phys. Rev. Lett.*, 90, 016801, 2003.
5. Wallace, P.R., The band theory of graphite. *Phys. Rev.*, 71, 622, 1947.
6. Castro Neto, A.H., Guinea, F., Peres, N.M.R., Novoselov, K.S., Geim, A.K., The electronic properties of graphene. *Rev. Mod. Phys.*, 81, 109, 2009.
7. Geim, A.K. and MacDonald, A.H., Graphene: Exploring carbon flatland. *Phys. Today*, 60, 35, 2007.
8. Novoselov, K.S., Geim, A.K., Morozov, S.V., Jiang, D., Katsnelson, M.I., Grigorieva, I.V., Dubonos, S.V., Firsov, A.A., Two-dimensional gas of massless Dirac fermions in grapheme. *Nature*, 438, 197, 2005.

9. Goerbig, M.O., Electronic properties of graphene in a strong magnetic field. *Rev. Mod. Phys.*, 83, 1193, 2011.
10. Dean, C.R., Young, A.F., Cadden-Zimansky, P., Wang, L., Ren, H., Watanabe, K., Taniguchi, T., Kim, P., Hone, J., Shepard, K.L., Multicomponent fractional quantum Hall effect in graphene. *Nat. Phys.*, 7, 693, 2011.
11. Amet, F., Bestwick, A.J., Williams, J.R., Balicas, L., Watanabe, K., Taniguchi, T., Goldhaber-Gordon, D., Composite fermions and broken symmetries in graphene. *Nat. Commun.*, 6, 6838, 2015.
12. Feldman, B.E., Krauss, B., Smet, J.H., Yacoby, A., Unconventional sequence of fractional quantum Hall states in suspended graphene. *Science*, 337, 1196, 2012.
13. Feldman, B.E., Levin, A.J., Krauss, B., Abanin, D.A., Halperin, B.I., Smet, J.H., Yacoby, A., Fractional quantum Hall phase transitions and four-flux states in graphene. *Phys. Rev. Lett.*, 111, 076802, 2013.
14. Ki, D.K., Falko, V.I., Abanin, D.A., Morpurgo, A., Observation of even denominator fractional quantum Hall effect in suspended bilayer graphene. *Nano Lett.*, 14, 2135, 2014.
15. Diankov, G., Liang, C.-T., Amet, F., Gallagher, P., Lee, M., Bestwick, A.J., Tharratt, K., Coniglio, W., Jaroszynski, J., Watanabe, K., Taniguchi, T., Goldhaber-Gordon, D., Robust fractional quantum Hall effect in the N = 2 Landau level in bilayer graphene. *Nat. Commun.*, 7, 13908, 2016.
16. Kou, A., Feldman, B.E., Levin, A.J., Halperin, B.I., Watanabe, K., Taniguchi, T., Yacoby, A., Electron-hole asymmetric integer and fractional quantum Hall effect in bilayer graphene. *Science*, 345, 55, 2014.
17. Maher, P., Wang, L., Gao, Y., Forsythe, C., Taniguchi, T., Watanabe, K., Abanin, D., Papić, Z., Cadden-Zimansky, P., Hone, J., Kim, P., Dean, C.R., Tunable fractional quantum Hall phases in bilayer graphene. *Science*, 345, 61, 2014.
18. Kim, Y., Lee, D.S., Jung, S., Skákalová, V., Taniguchi, T., Watanabe, K., Kim, J.S., Smet, J.H., Fractional quantum Hall states in bilayer graphene probed by transconductance fluctuations. *Nano Lett.*, 15, 7445, 2015.
19. Jain, J.K., *Composite fermions*, Cambridge UP, Cambridge, 2007.
20. Zhang, Y., Jiang, Z., Small, J.P., Purewal, M.S., Tan, Y.-W., Fazlollahi, M., Chudov, J.D., Jaszczak, J.A., Störmer, H.L., Kim, P., Landau-level splitting in graphene in high magnetic fields. *Phys. Rev. Lett.*, 96, 136806, 2006.
21. McCann, E. and Fal'ko, V.I., Landau level degeneracy and quantum Hall effect in a graphite bilayer. *Phys. Rev. Lett.*, 96, 086805, 2006.
22. McCann, E. and Koshino, M., The electronic properties of bilayer graphene. *Rep. Prog. Phys.*, 76, 056503, 2013.
23. Greiter, M., Microscopic formulation of the HH hierachy of quantized Hall states. *Phys. Lett. B*, 336, 48, 1994.
24. Wu, Y.S., General theory for quantum statistics in two dimensions. *Phys. Rev. Lett.*, 52, 2103, 1984.
25. Jacak, J. and Jacak, L., Recovery of Laughlin correlations with cyclotron braids. *Europhys. Lett.*, 92, 60002, 2010.
26. Jacak, J. and Jacak, L., The commensurability condition and fractional quantum Hall effect hierarchy in higher Landau levels. *JETP Lett.*, 102, 19, 2015.
27. Papic, Z. and Abanin, D.A., Topological phases in the zeroth Landau level of bilayer graphene. *Phys. Rev. Lett.*, 112, 046602, 2014.
28. Wilczek, F., *Fractional statistics and anyon superconductivity*, World Sc., Singapore, 1990.
29. Halperin, B.I., Theory of the quantized Hall conductance. *Helv. Phys. Acta*, 56, 75, 1983.
30. Jain, J.K., Composite-fermion approach for the fractional quantum Hall effect. *Phys. Rev. Lett.*, 63, 199, 1989.

31. Abrikosov, A.A., Gorkov, L.P., Dzialoshinskii, I.E., *Methods of quantum field theory in statistical physics*, Dover Publ. Inc., Dover, 1975.
32. Haldane, F.D.M., Fractional quantization of the Hall effect: A hierarchy of incompressible quantum fluid states. *Phys. Rev. Lett.*, 51, 605, 1983.
33. Jacak, J., Gonczarek, R., Jacak, L., Jóźwiak, I., *Application of braid groups in 2D Hall system physics: Composite fermion structure*, World Scientific, 2012.
34. Laidlaw, M.G. and DeWitt, C.M., Feynman functional integrals for systems of indistinguishable particles. *Phys. Rev. D*, 3, 1375, 1971.
35. Birman, J.S., *Braids, links and mapping class groups*, Princeton UP, Princeton, 1974.
36. Imbo, T.D., Imbo, C.S., Sudarshan, C.S., Identical particles, exotic statistics and braid groups. *Phys. Lett. B*, 234, 103, 1990.
37. Eisenstein, J.P., Lilly, M.P., Cooper, K.B., Pfeiffer, L.N., West, K.W., New physics in high Landau levels. *Phys. E*, 6, 29, 2000.
38. Dolev, M., Gross, Y., Sabo, R., Gurman, I., Heiblum, M., Umansky, V., Mahalu, D., Characterizing neutral modes of fractional states in the second Landau level. *Phys. Rev. Lett.*, 107, 036805, 2011.
39. Willett, R.L., The quantum Hall effect at 5/2 filling factor. *Rep. Prog. Phys.*, 76, 076501, 2013.
40. Knothe A., Jolicoeur T., Phase diagram of a graphene bilayer in the zero-energy Landau level. *Phys. Rev. B*, 94, 235149, 2016.
41. Suen, Y.W., Engel, L.W., Santos, M.B., Shayegan, M., Tsui, D.C., Observation of a $\nu = 1/2$ fractional quantum Hall state in a double-layer electron system. *Phys. Rev. Lett.*, 68, 1379, 1992.
42. Eisenstein, J.P., Boebinger, G.S., Pfeiffer, L.N., West, K.W., He, S., New fractional quantum Hall state in double-layer two-dimensional electron systems. *Phys. Rev. Lett.*, 68, 1383, 1992.
43. Jacak, J., Jóźwiak, I., Jacak, L., New implementation of composite fermions in terms of subgroups of a braid group. *Phys. Lett. A*, 374, 346, 2009.
44. Jacak, J., Jóźwiak, I., Jacak, L., Wieczorek, K., Cyclotron braid group structure for composite fermions. *J. Phys.: Condens. Matter*, 22, 355602, 2010.
45. Störmer, H.L., Du, R.R., Kang, W., Tsui, D.C., Pfeiffer, L.N., Baldwin, K.W., West, K.W., The fractional quantum Hall effect in a new light. *Semicond. Sci. Technol.*, 9, 1853, 1994.
46. Balram, A.C., Tőke, C., Wójs, A., Jain, J.K., Fractional quantum Hall effect in graphene: Quantitative comparison between theory and experiment. *Phys. Rev. B*, 92, 075410, 2015.
47. Ciftja, O. and Wexler, C., Monte Carlo simulation method for Laughlin-like states in a disk geometry. *Phys. Rev. B*, 67, 075304, 2003.
48. Morf, R. and Halperin, B.I., Monte Carlo evaluation of trial wavefunctions for the fractional quantized Hall effect: Spherical geometry. *Z. Phys. B Condens. Matter*, 68, 391–406, 1987.
49. Metropolis, N., Rosenbluth, A.W., Rosenbluth, M.N., Teller, A.M., Teller, E., Equation of state calculations by fast computing machines. *J. Chem. Phys.*, 21, 1087, 1953.

15

Graphene Plasmonic: Switching Applications

Ali Farmani

Department of Electronic Engineering, Khoram abbad, Iran

Abstract
This chapter focuses on the graphene plasmonic-based optical switching devices operating in the infrared to terahertz spectrum. Graphene plasmonic due to high strong light–matter interaction and nanoscale size has received significant attention in recent years, as they have found an important role in many areas of optical science and engineering by offering schemes to control electromagnetic fields. Another area of science that has been under the spotlight for the last few years relates to graphene plasmonic metasurface, which is formed of graphene layer packed into a structure. This material exhibits exceptional electronic and optical properties, intriguing many research groups across the world including us. However, our interest is specifically in studying the interaction between electromagnetic waves with surface modes of graphene in multilayer configuration (such as Otto configuration) and switching applications that might follow. In this chapter, relying on theoretical models and numerical simulations, we show that by designing and manipulating optical and electrical properties of graphene such as chemical potential, temperature, scattering rate, as well as incident beam characteristics such as beam width, one can have the structures based on graphene as a good platform for not only infrared optical switching devices but also terahertz switching devices.

Keywords: Graphene plasmonic, optical switches, metasurface, Otto configuration, terahertz, graphene conductivity

15.1 Graphene Plasmonic

Graphene plasmonic is one of the hot topics due to the extraordinary features of graphene, such as the ability to control the dynamic conductivity via doping or external voltage. Graphene-based surface wave supporting structures, on the other hand, provide long-range propagating surface waves and fabrication of these structures is relatively simple and straightforward [1]. Because of its extraordinary properties such as transparency in the white light spectrum, high electronic mobility, low ohmic loss, high surface area, and adjustable band gap, graphene has recently attracted great attention from a researcher in both theoretical and experimental aspects. As mentioned above, low ohmic loss of graphene facilitates the formation of long-range propagating surface waves, which can be used to achieve tunable switching devices [2]. Moreover, optical properties of the graphene and therefore optical properties of the mentioned surface waves can be controlled by applying

Email: Farmani.a@lu.ac.ir

Tobias Stauber (ed.) Handbook of Graphene: Volume 2, (455–506) © 2019 Scrivener Publishing LLC

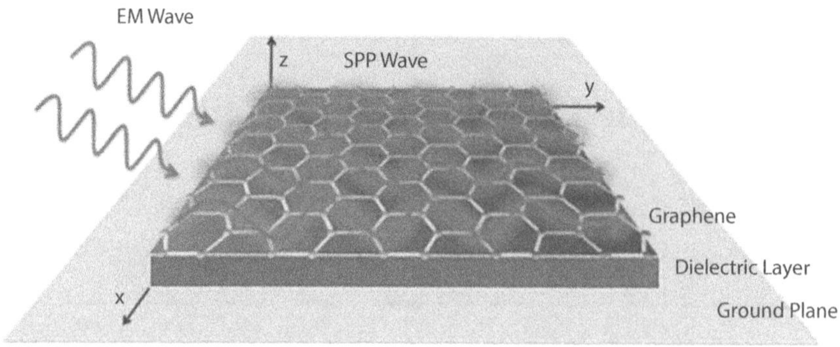

Figure 15.1 The free space graphene plasmonic structure.

an external electric voltage. Hence, by coupling the incident beam to a graphene-containing structure, a large lateral and transverse shift can be realized whose magnitude can be controlled by an applied voltage. This way, the combination of graphene-based surface waves tunability of optical and electrical properties of graphene makes it possible to produce an ultra-broadband graphene-based electro-optical switch [3, 4]. This graphene-based electro-optical switch has excellent characteristics such as tunable power switching, ultrafast switching signal propagation, ultralow-power consumption, low insertion loss, and high scalability. For example, a graphene plasmonic-based structure is shown in Figure 15.1. As can be seen, the incident beam easily coupled to the surface waves of the structure and the surface waves experience high confinement on the surface of graphene. Hence, by tuning coupling condition through modifying the optical properties of graphene as well as incident light beam, it can be used for controlling the propagation of surface wave on graphene layer, which in turn leads to obtaining the switching mechanism [5, 6]. To access the optimum properties for switch, in the following we first study the categories of switches, then introduce the main parameters of switches, and finally review the switching mechanism [7].

15.2 Category of Switching Devices

Classification of switching devices in a unit group is a difficult task. Typical classification schemes group the switching devices by the applications such as wide area network (WAN), and metropolitan area network (MAN), the input stimuli such as optical beam, electrical signals, and so on, the switching mechanisms such as the manipulation of refractive index or changing the direction of incident light, the materials used such as three-dimensional or two-dimensional material and plasmonic material, the various wavelength range such as infrared, and terahertz region, technologies such as total internal reflection, photonic crystal, liquid crystal, electro and acoustic–optic systems, and micro-electro-mechanical systems (MEMS), or the switches properties such as tunability [8–14]. Each of these categories may be helpful in different switching development stages. Possibly the best way to start is grouping the switches by switching mechanism and switching materials.

Typical switching devices can be divided into two categories: free space switches such as holographic switches and MEMS, and waveguide switches such as Mach-Zehnder.

Usually in free space switches, by modifying one of the incident beam properties such as propagation speed, polarization, and the direction, the switching condition is met. In waveguide switches, which are usually polarization dependent, the switching condition is met by increasing the attenuation for one polarization [15–24].

On the other hand, switching devices can be categorized based on their material including semiconductor, noble metal, and two-dimensional material that have been developed for switching the incident free space beam [25–29].

15.2.1 Switching Devices Characteristics

In order to investigate the switching devices performance, a set of parameters is used. In the free space switches, such as the structure shown in Figure 15.2 and waveguide switches, some main common parameters are introduced that will be briefly studied below.

The most important parameters and their brief definitions are listed below:

- Insertion loss: is the loss of signal power emanating from the insertion of a device in a transmission line or optical fiber and is usually expressed in decibels (dB) and expressed with the ratio of output power to input power ($IL(dB)=10\ Log(P_{in}/P_t)$) as shown in Figure 15.3.
- Extinction ratio: is the ratio of two optical power levels of a digital signal (one level and zero level). For true detection of the level of signals, a high ratio of these signals is needed.
- Polarization extinction ratio: is similar to extinction ratio but expresses the ratio of the power of two perpendicular polarizations, i.e., transverse electric (TE) and transverse magnetic (TM).
- Switching speed: A measure of the rate at which the optical logic device is capable of altering the logic state of its output in response to actuation. It is a function of the delay encountered by the device, which in turn is a function of the device technology.
- Working temperature: is usually the temperature that corresponds to maximum sensitivity.
- Power consumption: refers to the applied energy over the time for altering the logic state.

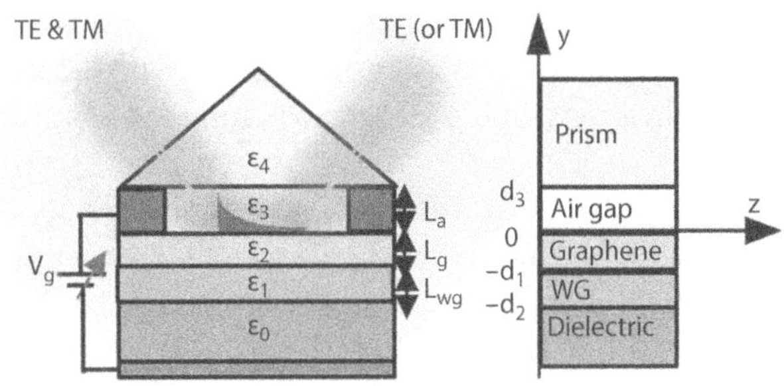

Figure 15.2 The schematic of an optical graphene plasmonic switch.

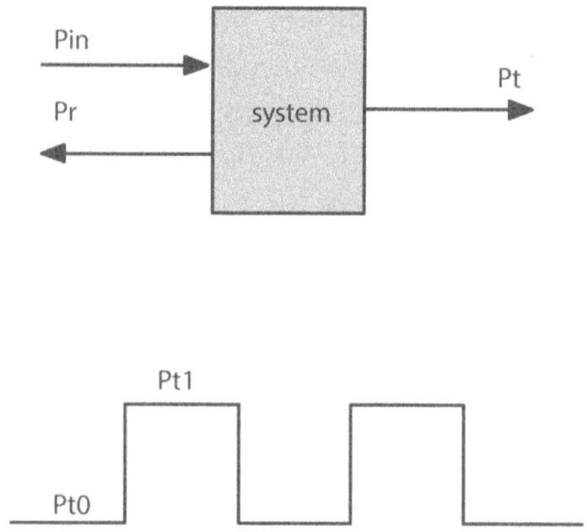

Figure 15.3 Upper panel: The insertion loss that is expressed with the ratio of output power to input power, i.e., Pt/Pin (dB). Lower panel: The extinction ratio that is the ratio of two optical power levels, i.e., Pt1/Pt0 (dB).

- Interference: is usually expressed with intersymbol interference (ISI) and is a form of distortion of a signal in which one symbol interferes with subsequent symbols.
- Footprint: is a parameter that refers to integration capability of the switch with other devices.
- Scalability: is the capability of switches to handle a growing amount of input and output.
- Fabrication process: the rapid prototype is considered.
- Integration capability: refers to CMOS compatibility of the switches for integrating with another component.
- Cost: refers to the type of technology such as MEMS, and liquid crystal, and the scale of the switch.

All of the abovementioned parameters are used to characterize the properties of a particular switching device. An ideal switch would possess high tunability, low power consumption, high switching speed, low insertion loss, and small response time. Investigators usually make efforts to approach only some of these ideal parameters, disregarding the others. On the other hand, real applications usually do not require switches with all perfect characteristics at once. For example, if a switch being developed shows lower switching speed than previously studied ones, it does not mean that it is not worthwhile to continue the research since it might show characteristics other than low switching speed that may be advantageous in some applications where a high switching speed is not the main requirement [30–35].

It is worth to mention that the performance of a switching device also depends on the environmental conditions. Under this group of parameters, external magnetic and electric fields, incident beam waist, temperature, ohmic contact, and so on can be included. These parameters must be considered in order to prevent degradation of the switching response due to the environment parameters.

15.2.2 Switching Mechanism

Switching devices mainly are divided into two categories based on their switching mechanism, including (a) modulating of incident beam, in which by considering different condition and parameters incident beam can be passed or blocked by the switches; and (b) modifying one of incident beam properties such as propagation speed, polarization, and propagation direction [36–42].

The first systems block or pass the signal in different directions and commonly present in various architectures such as Crossbar, Spanke, and Benes. The second category mainly works with the total internal reflection (TIR) and attenuated total reflection (ATR) conceptions, and works based on MEMS and hologram plane. In the next section, we briefly discuss two phenomena (Goos–Hänchen and Imbert–Fedorov shifts), which are based on TIR, for switching application in free space switches [43].

15.2.3 Goos–Hänchen Shift

As mentioned in an earlier section, typical switching devices are divided into two categories: free space switches and waveguide switches. The aim of this work is to study the free space switches based on the lateral (Goos–Hänchen shift) and transverse (Imbert–Fedorov shift) shift of incident beam [44–50].

When an incident light beam impinges at the boundary of the two different mediums under total internal reflection (TIR), a part of electromagnetic field penetrates into the second medium and forms an evanescent wave whose amplitude decays exponentially in a direction perpendicular to the boundary of two mediums, as shown in Figures 15.4 and 15.5. Although this evanescent wave is not traveling in a direction normal to the boundary, it carries active electromagnetic power in a direction along the interface; hence, after reentering into the incident medium, the reflected energy is laterally shifted with respect to the nonspecular reflection. This lateral shift is known as Goos–Hänchen (GH) shift in honor of F. Goos and H. Hänchen, who in 1947 were the first to study this phenomenon experimentally, as shown in Figure 15.4.

Moreover, the reflected beam in longitudinal direction (i.e., in the plane of the incident) experiences both spatial displacements and angular deflections. The spatial displacement, which is referred to as spatial Goos–Hänchen (SGH), and, in the same way, the angular deflection can occur in the plane of the incident, which is referred to as angular

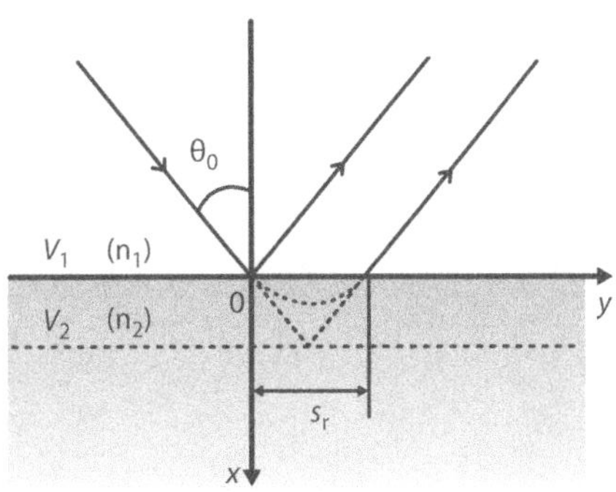

Figure 15.4 The conception of Goos–Hänchen shift.

Figure 15.5 The lateral shift through Goos–Hänchen shift.

Goos–Hänchen (AGH). The GH shift was then theoretically explained by Artmann who proposed an analytical formula to describe the results achieved in the GH experiment.

Thereafter, multiple theories have been proposed in the literature to explain this phenomenon. These include the stationary-phase approach by Fragstein, the time-average Poynting vector by Renard, the time delay scattering by Chiu *et al.*, and the angular spectrum method by Brekhovskikh, and later on by McGuirk *et al.* Moreover, to increase the lateral shift, the plasmonic structure is proposed, as shown in Figure 15.5.

15.2.4 Imbert–Fedorov Shift

In total internal reflection (TIR) of an optical beam with finite width from the interface of two different homogeneous media, the reflected beam experiences both spatial displacements and angular deflections. The spatial displacement can occur in the transversal direction (i.e., normal to the plane of the incident), which is referred to as spatial Imbert–Fedorov (SIF) [51].

In addition, the angular deflection can occur in normal to the plane of the incident, which is referred to as angular Imbert–Fedorov (AIF), as depicted in Figure 15.6.

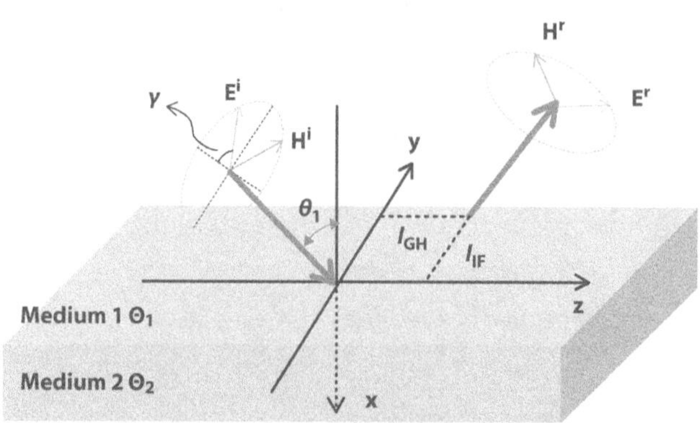

Figure 15.6 The Imbert–Fedorov shift. L_{GH} is attributed to longitudinal shifts and l_{IF} is referred to the transverse shifts.

Fedorov and Imbert, respectively, experimentally and theoretically showed perpendicular displacement of the center of gravity of totally reflected light beams from the interface of two homogeneous media [51].

15.3 Graphene Properties

Our aim in this subsection is to review graphene and its properties including optical, electrical, and thermal properties, to point out the good tunability of this material for switching applications. Then, the free space switching devices are reviewed in detail.

15.3.1 Graphene

Graphene is a 2D one-atom thickness layer with carbon (C) atoms organized in a honeycomb lattice, as shown in Figure 15.7. It is the building block of the carbon allotrope structures, such as zero-dimensional (0D) fullerenes, one-dimensional (1D) nanotubes, and three-dimensional (3D) graphite. Given its importance, it is quite surprising that graphene is the last one being studied extensively among all known carbon allotropes, which is mainly due to the instability problem of 2D crystals. It came as a huge surprise when Professors Andre Geim and Konstantin Novoselov's group from Manchester University in the UK in 2004 experimentally isolated a single layer of graphene via mechanical exfoliation of graphite flakes using Scotch tape, which eventually led to their 2010 Nobel Prize in Physics. Since then, the study of this new material has become one of the hottest topics for material scientists due to its exotic thermal, optical, and electrical properties arising from the two-dimensional crystal structure. These unique properties make graphene a suitable material for a vast variety of applications, to name a few, as energy conversion and storage, sensors, optical switches, electronic devices, transparent electrodes in solar cells, flexible displays in optoelectronic devices, field emission sources, and as highly tunable devices. In the following, the optical, electrical, and thermal properties of graphene will be briefly presented [52].

Figure 15.7 The structure of a single graphene sheet.

15.3.2 Graphene Optical Properties

The measured white light absorbance of pristine single-layer graphene is 2.3% with a slight reflectance (<0.1%), and this absorbance increases linearly by increasing the graphene layers. The abovementioned values and the observed linearity are in accordance with the theoretical calculation results obtained with a model of non-interacting massless Dirac fermions. The transparency of graphene only depends on the fine structure constant $\sigma=2\pi e^2/hc$, (c is the speed of light, h is Planck's constant), which describes the coupling between the incident light and relativistic electrons. The absorbance of n-layer graphene can thus be simply expressed as $n\pi\sigma$. However, deviation from this behavior is found with the incident photons with energy less than 0.5 eV (or wavelength larger than ~2480 nm), which is attributed to the effects of finite-temperature, doping, and intraband transitions.

The ultrafast optical pump–probe spectroscopy has been applied to study the carrier dynamics and relative relaxation timescales of graphene layers grown on SiC. An initial fast relaxation transient (70–120 fs) followed by a slower relaxation process (0.4–1.7 ps) has been identified, which is associated with the carrier–carrier intraband and carrier–phonon interband scattering process, respectively.

With the help of infrared spectroscopy, it is found that the interband transitions and optical transitions of monolayer and double-layer graphene are layer-dependent and can be modulated through the electrical gating, which holds promise for the infrared optics and optoelectronics switching devices. Moreover, the Fermi level of graphene can be changed with slight changing in chemical potential through external voltage; hence, the devices with low power consumption can be realized [53].

15.3.3 Graphene Electrical Properties

Figure 15.8 shows the graphical hexagonal lattice of graphene. The primitive cell of graphene is a rhombus with a basis of two non-equivalent carbon atoms A and B situated at the two sublattices. In real space, the cell can be described by the two basic vectors a1 and a2 intersecting at an angle of 60° and with equal length of 2.46 Å.

Figure 15.9 shows the band structure of graphene along the high symmetry lines in the first irreducible Brillion zone (IBZ). A peculiar feature of the graphene band structure is that the conduction band and the valence band touch each other symmetrically at the charge neutrality point K (K') with the linear energy-momentum dispersion around the

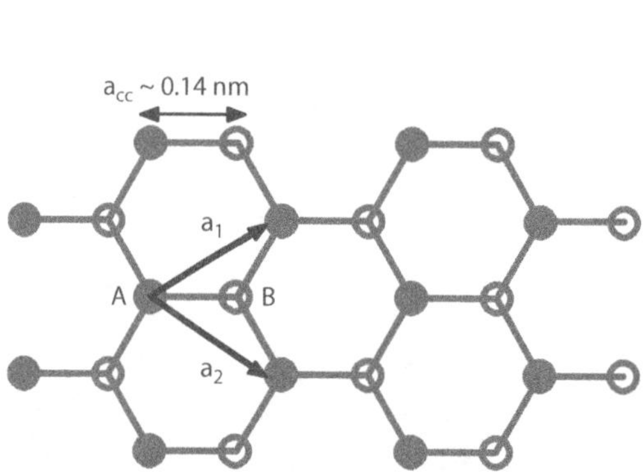

Figure 15.8 The monolayer graphene structure.

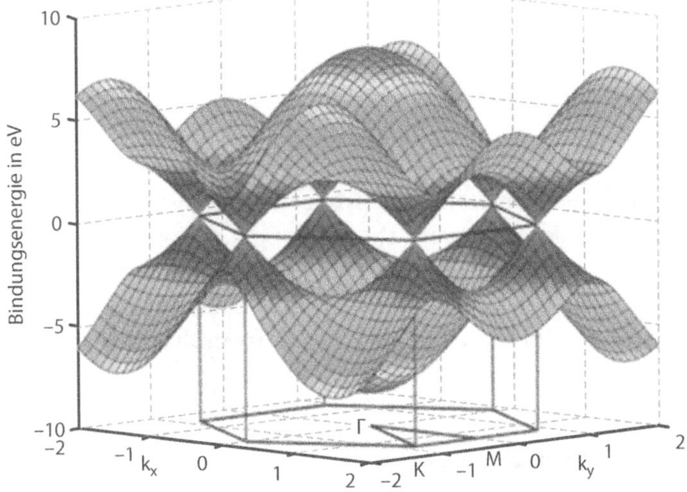

Figure 15.9 Electronic dispersion in the honeycomb lattice and zoom of the energy bands close to one of the K Dirac points.

Fermi level. Therefore, graphene is usually presented as metal for zero band gap and semiconductor for nonzero band gap.

The band structure of graphene clearly suggests that its electronic properties are mainly determined by the states around the Fermi level (namely, at the charge neutrality point). This could also be reflected from the point of view of orbital hybridization. The single C atom has four valence electrons occupying the $2s^2$, $2p_x^1$, and $2p_y^1$ orbitals. Upon the sp^2 hybridization, one of the electrons in the 2s orbital is promoted into the $2p_z$ orbital, and the threefold symmetric hybridized orbitals are generated by the superposition of the 2s, $2p_x$, and $2p_y$ orbitals, leaving the $2p_z$ orbital intact. In the honeycomb lattice, each C atom forms three bonds with the neighboring C atoms via the planar sp^2 hybridized orbitals, while the overlap of the remaining free $2p_z$ orbitals presents symmetry orientation. The planar σ bonds have a much higher bond strength than π bonds perpendicular to the carbon plane.

Thus, π (bonding) and π^* (antibonding) bands are located closer to the Fermi level than σ (bonding) and σ* (antibonding) bands. In other words, the π and σ bonds are primarily responsible for the electronic and mechanical properties of graphene, respectively. The Dirac equation is adopted to depict the linear dispersion in terms of massless relativistic particles. Therefore, the electrons in graphene are called Dirac fermions and the charge neutrality point is referred to as Dirac point. Based on the Dirac equation, one could determine that the Dirac fermions in graphene possess an effective Fermi velocity of around 300 times smaller than the speed of light. This makes graphene a reliable system for physicists now to study the quantum electrodynamics phenomena in the lab.

Over the years, many unusual properties of graphene have been experimentally unveiled. It has been demonstrated that the suspended graphene shows low-temperature mobility approaching 200,000 cm^2/V s for carrier densities below 5×10^9 cm^{-2}, which is not observable in semiconductors or nonsuspended graphene sheets. It also exhibits a strong ambipolar electric field effect with the concentration of charge carriers up to 10^{13} cm^{-2} and room-temperature mobilities of ~10,000 cm^2/V s, when the gate voltage is applied. Moreover, an unusual half-integer quantum Hall effect (QHE) for both electron and hole carriers in graphene has been observed by tuning the chemical potential with the use of the electric field effect. Such QHE can be shown at room temperature as well, and the fractional QHE was achieved when pristine graphene devices were probed, which allows for the isolation of the sample from substrate-induced perturbations.

Although graphene has been proposed as a replacement for Si in digital logic circuits due to its exceptional electron transport properties, the gapless nature of the massless Dirac fermions, however, severely limits the pace of graphene in the semiconductor world, in which Si is still dominant in digital logic circuits. Although that pure graphene-based transistor with a cut-off frequency as high as 300 GHz has been achieved, the poor on/off ratio of ~1,000 at room temperature is still far away from that (>10,000) of silicon transistors [53].

15.3.4 Graphene Thermal Properties

Due to tightly packed carbon atoms, graphene could also play an important role in heat dissipation in future nano-optics and electronics. According to Moore's law, semiconductor-based devices have been gradually decreasing in size and are now in the nanometer regime. Further reductions in the scale of the devices, however, are limited by the ability to dissipate heat from the device, because of the significantly enhanced heat generation rate per unit surface area and the decreasing thermal conductivity of device components. For example, the heat generation rate for nanoscale devices can reach as high as that of a rocket nozzle (1,000 W cm^{-2}) and the thermal conductivity reduces significantly from 150 W m^{-1} K^{-1} for bulk Si to 10 W m^{-1} K^{-1} for Si nanowires. Therefore, heat removal is a major bottleneck to the further miniaturization of logic circuit chips. Graphene has been shown to exhibit superior room temperature thermal conductivity, which varies from 3,000 to 5,000 W m^{-1} K^{-1} depending on the size of the measured graphene sheet (for a pristine monolayer graphene), but reduces to 600 W m^{-1} K^{-1} when placed on a SiO_2/Si substrate. This reduction in thermal conductivity is attributed to the leaking of phonons across the graphene–silica interface and strong interface scattering. Nevertheless, this value is still about 2 times and 50 times higher than copper and silicon, respectively, which are used in electronics today [53].

15.3.5 Graphene-Based Switches

Owing to the two-dimensional nature of single layer graphene, graphene has a maximum 100% surface-to-volume ratio and is highly suitable for switching applications based on interactions of surface plasmons of graphene with incident beam, such as free space optical switches. Every carbon atom in graphene acts as an active site for the incident beam present in the surroundings to interact with. Moreover, electrical properties of graphene such as small chemical potential, high charge carrier mobility at room temperature, metallic conductivity, and low Johnson noise also contribute to the rise of graphene as one of the best suitable plasmonic materials for switching devices.

In the following, the free space switching based on GH and IF shifts and application of these shifts in the optical component will be briefly reviewed.

Since the first observation of GH and IF shifts, these effects have been vastly investigated both theoretically and experimentally. The profound attention to these effects is mainly because of their application in the realization of different optical switching devices such as polarizer switches. Moreover, due to noteworthy features of lateral shift effect, it is used in designing many other optical devices such as sensors, lasers, absorbers, and filters.

GH and IF shifts have been studied in various wavelength ranges including visible, infrared, terahertz, and microwave.

Historically, this shift was mostly studied in structures containing two homogeneous materials with different optical characteristics, e.g., at the interface of two different dielectric medias, non-absorptive and absorptive medias, and dielectrics and metals, and at the interface of a homogeneous medium and one- or two-dimensional periodic structures, and also at the interface of different homogeneous media, including linear lossy and lossless dielectric, chiral media, metallic, and nonlinear media.

In all these cases, the magnitude of GH shift is usually smaller or comparable to the wavelength (or wavelength range) of the incident beam. Until now, this amplified GH shift has been studied in different structures such as multilayer structures with metasurfaces (like graphene-containing structures), metamaterial, hyperbolic metamaterial, nonlinear metamaterial, and photonic crystal. Except for the metasurfaces, fabrication of all structures requires relatively complicated lithography steps, which increase the cost and complexity of the related devices. Although vast research on large tunable lateral shift has been done, the GH shift is too small (on the order of wavelength) [54].

15.3.6 Experimental and Theoretical Improvement Switching Tunability

The enhancement of GH and IF shifts can be achieved by introducing a resonant coupling between the incident light beam and the reflecting structure, when the reflecting structure supports propagating optical modes (in a direction parallel to the interface). The resonance coupling between the incident beam and the propagating modes of the reflecting structure happens when the phase matching condition between these two waves is met. Thus far, the enhanced GH and IF shifts have been studied in total reflection of light beams from different structures such as photonic crystals supporting surface or waveguide mode, metallic structures supporting plasmonic surface waves, and graphene-containing structures supporting surface waves [55–59].

To increase the lateral shift, three configurations are introduced in which will be reviewed here:

- Surface plasmon resonance structure
- Symmetrical metal-cladding waveguide
- Prism–waveguide coupling system

15.3.6.1 Surface Plasmon Resonance Structure

When using incident light to excite surface waves, there are two configurations that are well known: Otto and Kretschmann configurations.

Andreas Otto was the first to propose a configuration for optically exciting nonradiative surface plasmons. Recognizing the difficulty with matching the wave vector of the incident light to that of the surface plasmons, Otto realized that the wave vector of the incident-free space light could be increased by propagation within a higher index dielectric material.

The field from the incident photon in a high-index dielectric could then be coupled evanescently through a thin dielectric film with a low refractive index to excite surface plasmons at the interface between the metal and the low-index dielectric. The typical Otto configuration uses a prism to do this as illustrated in the right panel of Figure 15.10. The surface plasmons propagate in the x direction, which is also the direction of surface charge oscillation. The incident beam must have a component of p or TM polarization to couple to this surface charge oscillation. The electric field for TE polarization lies along the y-direction, which is orthogonal to the surface charge oscillation, so this polarization cannot excite surface plasmons.

In this configuration, the light is incident upon the prism and refracted toward its bottom surface. The prism is spaced by a small distance from the surface of the metal. The refractive index of the dielectric spacer is less than that of the prism and the angle of incident light upon the bottom surface of the prism is sufficiently high that if there were no metal layer underneath the spacer, the light in the prism would undergo total internal reflection and an evanescent field would exist within the low-index dielectric below the prism. In the presence of the metal, however, this evanescent field can excite the surface plasmons if the coupling conditions are satisfied. Because energy from the evanescent field is removed from the reflected beam when the surface plasmons are excited, there is no longer total internal reflection. The reflectivity drops, sometimes nearly to zero, and thus this configuration is a form of "attenuated total reflection." It is worth to mention that the strong dependence of the surface plasmons coupling efficiency on the thickness of the dielectric spacer is the important reason that this configuration has not become popular by considering the material with the fixed optical properties.

Shortly after Otto described his optical configuration for exciting surface plasmons, Kretschmann and Raether proposed another prism-based configuration, which has since

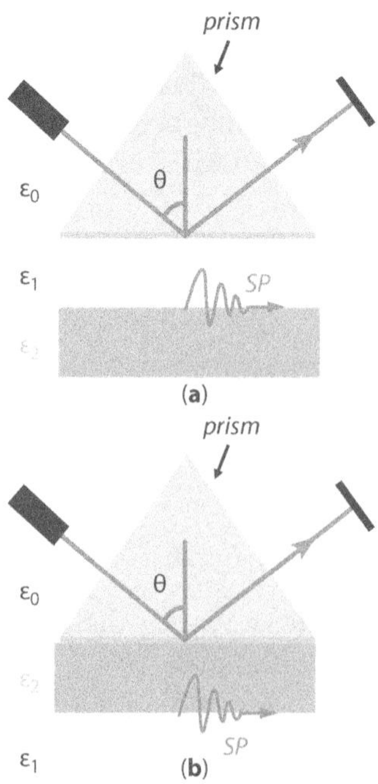

Figure 15.10 The Kretschmann (lower panel) and Otto (upper panel) configurations.

become the most popular configuration for surface plasmons excitation. It is typically referred to as the "Kretschmann configuration" or the "Kretschmann–Raether configuration" [60–68].

The light is again incident on a prism in this configuration, but there is a thin metal film on the bottom surface of the prism instead of a dielectric film. At the appropriate angle of incident light, the energy and momentum of the incident light are efficiently transferred to the surface plasmon, and there is again a substantial reduction in the reflected light intensity. The condition for resonance is identical to that in the Otto configuration. In contrast to the Otto configuration, the surface plasmons propagate along the bottom surface of the metal film where it is easily accessible for measurements and interactions.

Similar to the Otto configuration, the thickness of the metal film is critical for adjusting the loading and obtaining efficient coupling to the surface plasmons. Typical optimum metal film thicknesses are usually 40–60 nm depending on the specific wavelength and metal, although for aluminum the film must be much thinner than 15 nm.

In the Kretschmann configuration, the metal film is evaporated onto the glass block. The light again illuminates the glass block, and an evanescent wave penetrates through the metal film. This configuration is depicted in the left panel of Figure 15.10.

For example, to obtain a larger GH shift, Yin *et al.* described a surface plasmon resonance configuration, with which a GH shift of greater than 50 wavelengths was observed, since much more light energy was coupled into the medium under the metal film.

As a TM-polarized light beam incidents upon the interface between prism and metal and the resonance condition is satisfied, the surface plasmon wave will be excited. In the experiments, the position of the reflected beam is detected by a position-sensitive detector (PSD). By periodically modulating the incident polarization, the difference of the lateral displacement between TE light and TM light is measured. However, since the TE-polarized incidence cannot excite any surface plasmon resonance, there is no enhanced GH shift. Therefore, it serves as a perfect reference beam, and the measured relative beam shift between TM and TE excitation indeed indicates an absolute beam displacement for a TM wave at the surface plasmon resonance region. As shown in the following figures (i.e., Figures 15.11 and 15.12), gold and silver are the best materials for implementation of surface plasmonic waveguides to be used for achieving large GH shift.

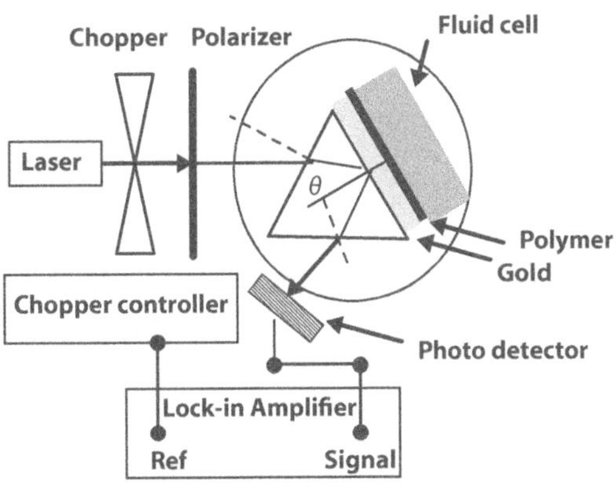

Figure 15.11 The plasmonic structure based on noble metal (i.e., gold) for lateral shift.

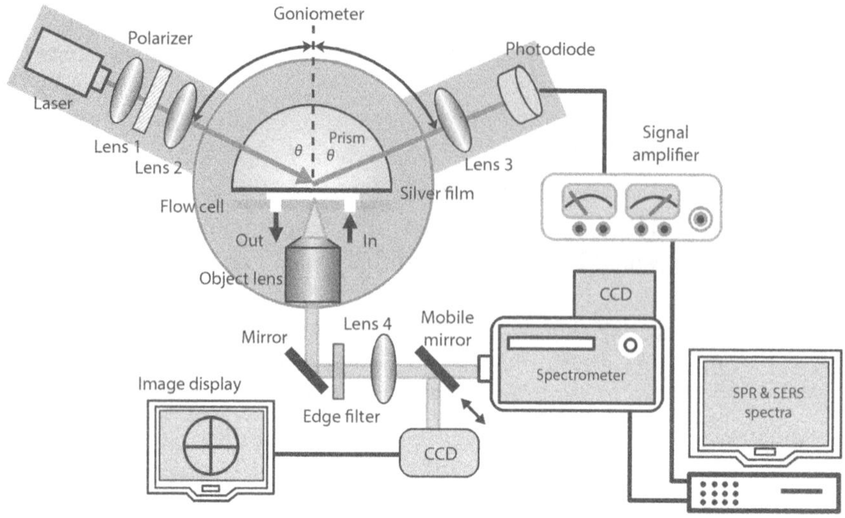

Figure 15.12 The Otto plasmonic setup-based silver for lateral shift.

15.3.6.2 Symmetrical Metal-Cladding Waveguide

The schematic diagram of the symmetrical metal-cladding waveguide (SMCW) is shown in Figure 15.13, which includes three layers: an upper metal film serving as the upper cladding as well as the coupling layer, a glass slab with millimeter thickness acting as the guiding layer, and a relatively thick metal film deposited upon the other side of the glass slab working as the substrate.

The incident angle is selected to be located at the maximum reflectivity of one certain reflectivity dip, where the GH shift is not remarkable. Since the magnitude of the GH shift is strongly dependent on the energy coupling between the incident light and the guided mode, it is reasonable to take this position of the reflected beam as the reference of the GH shift. After removing the photodiode (PD) out of its position without changing the position of the incident beam and the structure, we put on the position-sensitive detector (PSD) and let the reflected beam impinge onto the PSD at the center perpendicularly.

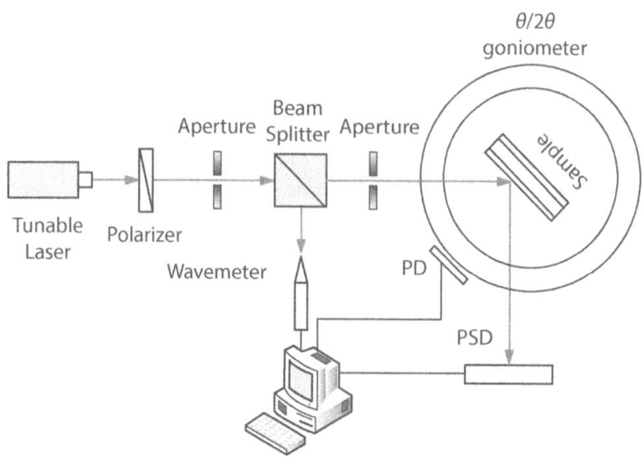

Figure 15.13 The experimental setup for lateral shift based on metal clad structure.

For instance, Wang *et al.* proposed a scheme to observe a large GH shift on a metal-cladding waveguide. Analogously, Chen *et al.* theoretically and experimentally reported a scheme to study the large GH shift in symmetrical metal-cladding waveguides. Furthermore, the tunable GH shift by applying an external voltage to the metal-cladding structure has also been studied by Yu *et al.*

Moreover, Salasnich reported the theoretical values of 80 λ_0 and 2 λ_0 for the SGH and SIF shifts, respectively.

However, in metallic plasmonic structures, the maximum achievable GH and IF shift values are limited by the short propagation length of optical surface modes, which is stemmed from the large ohmic losses of the metals. Furthermore, the GH and IF shift values in this structure cannot be tuned by applying an actuating signal in a straightforward manner. The ability to tune the shift values is an essential requirement for realization of GH and IF shift-based optical active devices. Although the propagation length of the surface modes of metal plasmonic structures can be increased by selective patterning of the metallic layer and introducing a metamaterial structure, still the lack of the tunability of the shift value remains an issue to be solved. On the other hand, the fabrication of metamaterial structures for application in terahertz and higher frequency ranges is challenging.

Moreover, these structures are not usually CMOS compatible and therefore cannot be integrated with silicon-based photonic integrated circuits.

15.3.6.3 Prism–Waveguide Coupling System

The prism–waveguide is a good candidate for obtaining a highly tunable lateral displacement of the reflected light. The prism–waveguide coupling system is shown in Figure 15.14. The permittivity of the substrate, the guiding layer, the air gap, and the prism are depicted in this figure. The thickness of the guiding layer and the air gap are used for final optimum lateral shift as illustrated in Figure 15.15. Since the radiative damping mainly depends on the thickness of the air gap, there is also a critical thickness in the prism–waveguide coupling system. Moreover, the experimental setup of the prism–waveguide is shown in Figure 15.16.

In comparison to the above-mentioned structures, graphene-based surface wave supporting structures provide long-range propagating surface waves, and fabrication of these structures is relatively simple and straightforward. As mentioned earlier, low ohmic loss of graphene facilitates the formation of long-range propagating surface waves, which can be used to achieve a large GH shift. Moreover, optical properties of the graphene and therefore the optical properties of the mentioned surface waves can be controlled by applying an external electric voltage. Hence, by coupling the incident beam to a graphene-containing

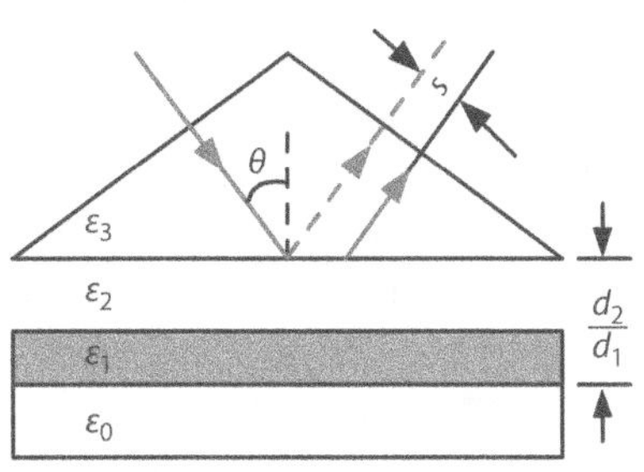

Figure 15.14 The prism waveguide for enhancing GH shift.

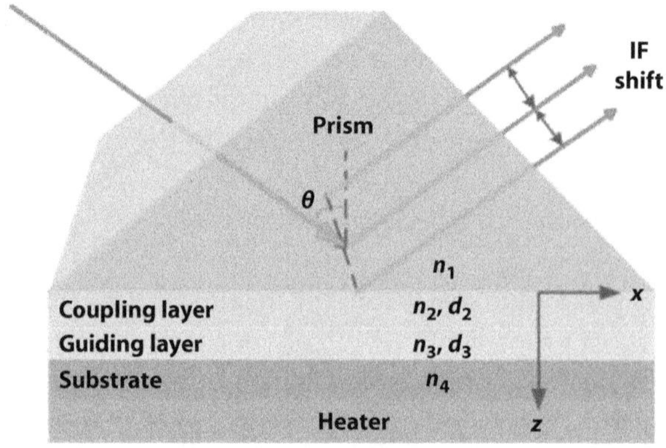

Figure 15.15 The prism plasmonic waveguide for increasing lateral shift.

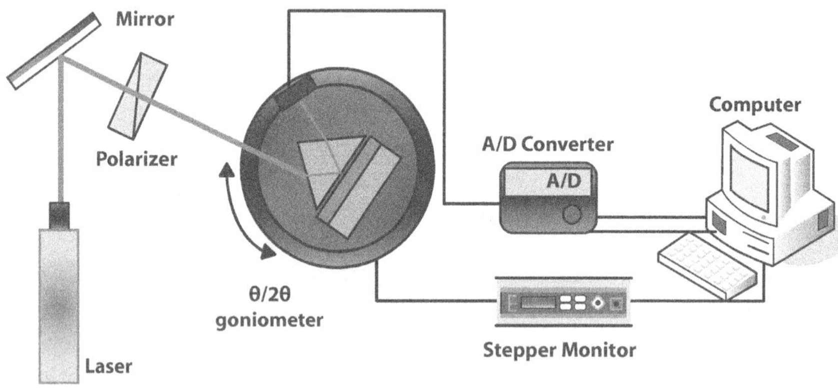

Figure 15.16 The experimental setup for prism waveguide structure.

structure, a large GH shift can be realized whose magnitude can be controlled by an applied voltage. This technique, the graphene-based surface waves with tunable optical and electrical mechanism of graphene, makes it possible to produce an ultra-broadband graphene switch. This graphene switch has brilliant features such as tunable power switching, ultrafast switching signal propagation, and ultralow power consumption.

On the other hand, because of anisotropy in optical properties of graphene, reflection response of the graphene-containing structures depends on the polarization of the incident light, as shown in Figure 15.17. For example, it can be shown (by proposing a new measurement method) that the difference between the GH shift for TM and TE polarized incident waves can be as high as 31.16 mm. Although the tunable GH shift in graphene-based waveguides has already been reported in the literature both theoretically and experimentally, because of lack of attention to the problem of proper coupling between the incident beam and surface waves of the structure, the value of the GH shift is either small (in the order 2 λ_0 to 110 λ_0) or requires a large tuning electrical voltage (in the range of 30 V). Therefore, the goal of achieving a large and tunable GH shift in graphene-containing structures in response to slight variations of the applied voltage still remains a problem to be considered.

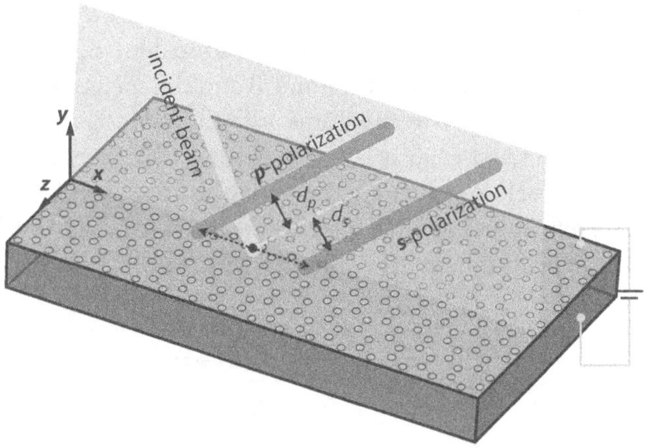

Figure 15.17 The lateral shift in a graphene plasmonic structure.

15.4 Research Methods

In this section, we introduce the nonspecular reflection effects and in particular theory of the Goos–Hänchen (GH) and Imbert-Fedorov (IF) shifts as well as the calculation of surface conductivity of graphene. We first briefly review the theory of GH and IF shifts. We describe the three theoretical models to obtaining the GH shift. Then we calculate the reflection coefficient, which is used for obtaining a displacement of the incident beam. Finally, we give a unified theory based on Kubo formula for calculating the surface conductivity of both graphene and graphene plasmonic metasurface (GPM) [68, 69].

15.4.1 Goos–Hänchen Shift

It is well known that the reflection point will experience a lateral displacement from its incident counterpart when an incident light beam is totally reflected through an interface between two materials with different optical properties (see Figure 15.18) [69].

This displacement is called the GH shift, and its magnitude is given as

$$\Delta z = \frac{2}{k_1}\frac{\partial \varnothing}{\partial \beta} \tag{15.1}$$

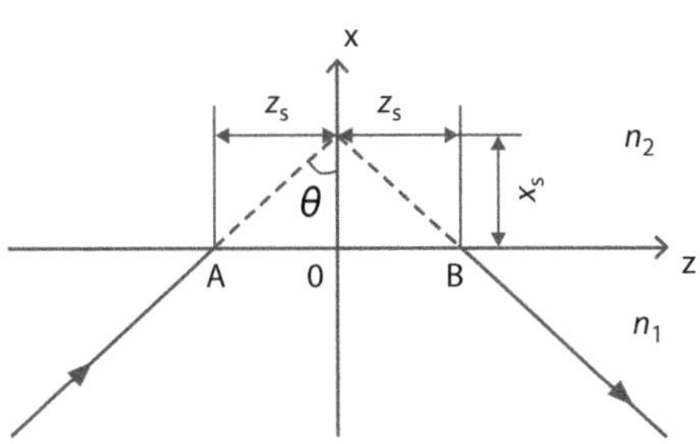

Figure 15.18 The lateral shift.

where $\kappa 1= k_0 n_1 cos\theta$ is the vertical component of the wave vector, $k_0 = 2\pi/\lambda$ is the wave number with light wavelength λ in the free space, θ is the incident angle, and -2ϕ is the total reflection phase shift between the reflection point B and the incident point A.

Consider the interface between two mediums with different optical properties as shown in Figure 15.18, when the incident angle is larger than the critical angle; $\theta > \theta c = arcsin\ (n_2/n_1)$, where n_1 and n_2 are the refractive indexes of the two mediums, and the incident light will be totally reflected and its reflection coefficient is $r = exp\ (i\varphi) = exp\ (-i2\phi)$, where the reflection phase shift for the TM and the TE polarizations are given as

$$\varnothing_{TE} = \arctan[\sqrt{\frac{n_1^2 sin^2\theta - n_2^2}{n_1 cos\theta}}] \tag{15.2}$$

$$\varnothing_{TM} = \arctan[(\frac{n_1}{n_2})^2\sqrt{\frac{n_1^2 sin^2\theta - n_2^2}{n_1 cos\theta}}] \tag{15.3}$$

To define the magnitude of the lateral shift of $2z_s$, we assume a simple wave consisting of two plane waves with slightly different incident angles. Assuming that the z-component of the wave vector is $\beta_z \pm \Delta\beta_z$, the amplitude of the incident wave at the interface of two different mediums can be expressed as

$$\begin{aligned} &A(z) \\ &= [\exp(i\Delta\beta_z) + \exp(-i\Delta\beta_z)]\exp(i\Delta\beta_z) \\ &= 2\cos(\Delta\beta_z))\exp(i\Delta\beta_z) \end{aligned} \tag{15.4}$$

where $\Delta\beta$ is a small quantity, and the total reflection phase shift can be expanded by the differential formula

$$\varnothing(\beta \pm \Delta\beta) = \varnothing(\beta) \pm \frac{d\varnothing}{d\beta}\Delta\beta \tag{15.5}$$

Hence, the complex amplitude of the reflected beam at the interface is

$$\begin{aligned} B(z) = \{&[\exp[(i\Delta\beta_z - 2\Delta\varnothing)] \\ &+ [\exp[(-i\Delta\beta_z \\ &- 2\Delta\varnothing)]\}\exp[(i\beta_z \\ &- 2\varnothing) \\ &= 2\cos[\Delta\beta(z \\ &- 2z_S)]\exp[i(\beta_z - 2\varnothing)] \end{aligned} \tag{15.6}$$

where

$$z_s = \frac{d\varnothing}{d\beta} \tag{15.7}$$

is the simple form for the lateral shift of the beam. The vertical distance between the reflected light beam and the predicted reflected light beam by the geometrical optics is

$$S = 2z_s cos\theta = \frac{2d\varnothing}{d\beta} cos\theta = \frac{2}{k_0 n_1}\frac{d\varnothing}{d\beta} \tag{15.8}$$

The obtained expression of the GH shift is identical to Equation (15.8), and its derivation procession is so-called as the stationary-phase approach, which is proposed by Artmann. Replacing Equations (15.5) and (15.6) into Equation (15.8), the GH shift at the total reflection between two semi-infinite mediums is obtained as

$$S_{TE} = \frac{2sin\theta}{k_0\sqrt{n_1^2 sin^2\theta - n_2^2}} \tag{15.9}$$

$$S_{TM} = \frac{S_{TE}}{[\left(\frac{n_1}{n_2)^2} + 1\right] sin^2\theta - 1} \tag{15.10}$$

15.4.2 Gaussian Beam Model

As shown in Figure 15.19, an interface is illuminated by a Gaussian light, whose field at the interface z = 0 is expressed as

$$\varphi_i(x, z=0) = \exp(-\frac{x^2}{2w_x^2} + i\beta_0 x) \tag{15.11}$$

This Gaussian beam can be represented as

$$\varphi_i(x, z=0) = \frac{1}{\sqrt{2\pi}}\int A(\beta)\exp(i\beta x)d\beta \tag{15.12}$$

where $w_x = w_0\, sec\theta$ with w_0 as the width of the waist, and β is the x-component of the wave vector with the incident angle θ. The angular spectrum distribution of the incident beam is

$$A(\beta) = w_x \exp[-(w_x^2/2)(\beta - \beta_0)^2] \tag{15.13}$$

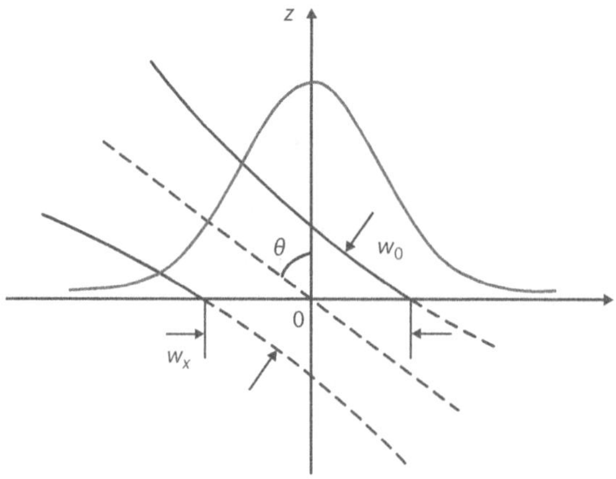

Figure 15.19 The Gaussian beam spectrum.

Upon contacting with the interface, the incident beam is reflected and the reflection coefficient of each frequency component is non-uniform. The profile of the reflected beam is given by

$$\varphi_r(x,z=0) = \frac{1}{\sqrt{2\pi}}\int r(\beta)A(\beta)\exp(i\beta x)d\beta \tag{15.14}$$

The corresponding GH shift can be calculated from the above integral by finding the x-component position of max $|\psi\ (x,\ z\ =\ 0)|$. In addition, the range of the profile of the reflected beam is $(-k, k)$, where k is the wave vector.

Actually, the Gaussian light beam is similar to the plane wave and can be observed as an ideal plane wave when its width of the waist is increased to be infinite. Here, a detailed investigation on the relationship between the stationary-phase approach and the Gaussian beam model is set. Supposing that the width of the incident beam is large enough, namely the half-width in β space is extremely small, the reflection phase shift in the reflection coefficient $r(\beta)=|r|\ exp\ (i\varphi)$ can be regarded as a linear function of β. On performing the Taylor expansion of φ around $\beta = \beta_0$, we obtain Equation (15.15). With the ignorance of the higher-order infinitesimal, the reflection coefficient is approximately written as

$$\varphi(\beta)=\varphi(\beta_0)+\left.\frac{d\varphi}{d\beta}\right|_{\beta=\beta_0}(\beta-\beta_0) + \sigma(\beta-\beta_0) \tag{15.15}$$

$$r(\beta) \approx |r|\exp[i\varphi'(\beta_0)]\exp(i\left.\frac{d\varphi}{d\beta}\right|_{\beta=\beta_0}\beta) \tag{15.16}$$

where

$$\varphi'(\beta_0)=\varphi'(\beta_0)-i\frac{d\varphi}{d\beta}\bigg|_{\beta=\beta_0}\beta_0 \tag{15.17}$$

By substituting Equation (15.16) into Equation (15.14), the profile of the reflected beam is expressed as

$$\begin{aligned}\varphi_r(x,z=0)=&\frac{1}{\sqrt{2\pi}}|r|\exp[i\varphi'(\beta_0)]\\&\times\int A(\beta)\exp\{i\beta[x\\&-(-i\frac{d\varphi}{d\beta}\bigg|_{\beta=\beta_0})]\}d\beta\end{aligned} \tag{15.18}$$

Comparing Equation (15.14) with Equation (15.11), it can be seen that besides an added constant term $|r|$ and an added phase factor exp $[i\varphi'(\beta_0)]$, the centric position of the reflected light distribution shifts from that of the incident light distribution x = 0 to the position $x = -(d\varphi/d\beta)|_{\beta=\beta 0}$, which is identical to the GH shift defined by Equation (15.8).

Therefore, the stationary-phase approach is the first-order approximation of the Gaussian beam model [69].

15.4.3 Imbert–Fedorov (IF) Shift Concept

It is well known that the interaction of a plane wave with an interface is described by Snell's law and the Fresnel formula. For a real optical beam that has a finite width (i.e., a distributed plane wave spectrum), besides the GH shift, there are two nonspecular effects deviating from the geometrical optics. As shown in Figure 15.20, other nonspecular shifts are the Imbert–Fedorov (IF) shift (i.e., a spatial shift perpendicular to the plane of incidence) and the angular IF shift, respectively.

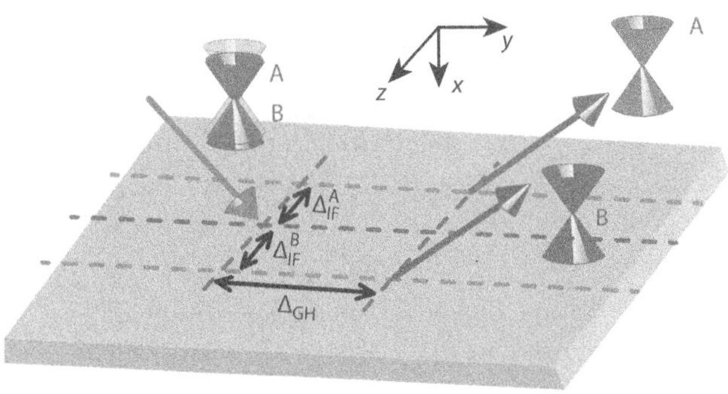

Figure 15.20 The IF shift of incident light beam.

Consider a system consisting of two homogeneous isotropic mediums of dielectric constants ε_1 and ε filling the half spaces $z < 0$ and $z \geq 0$, respectively.

A monochromatic beam of light of wavelength λ and waist w propagates along the central wave vector k in the region $z < 0$ before impinging upon the plane interface of equation $z = 0$ that separates medium 1 from medium 2. The expressions for these nonspecular effects can be written as

$$S = \frac{1}{K} Im[\rho] \tag{15.19}$$

$$A = -\frac{\theta_0^2}{2} Re[\rho] \tag{15.20}$$

where S and A refer to spatial and angular for IF shifts, respectively, $\theta_0 = 2/k\omega_0$, and $k = 2\pi/\lambda_0$. The parameter ρ can be expressed for IF shifts as

$$\rho = \begin{cases} 2i(r_{TE} cos\theta), mode: TM \\ 2i(r_{TM} cos\theta), mode: TE \end{cases} \tag{15.21}$$

where r_{TE} and r_{TM} are the reflection coefficients, which can be obtained through Maxwell equation and transfer matrix method, as mentioned in the next section.

15.4.4 Reflection Calculation

The reflection coefficient is used in ray optics and extremely useful to analyze the reflection characteristics of electromagnetic waves in a planar multilayer structure. This section aims to introduce the Maxwell equation and transfer matrix method to calculate the reflection.

Considering the TM surface waves propagating along the z-direction at the interface of graphene and coupling structure, as shown in Figure 15.21, the magnetic field can be expressed as

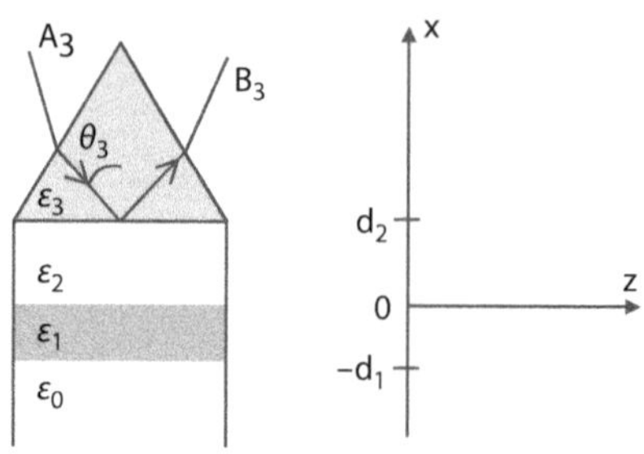

Figure 15.21 The schematic of multilayer structure.

$$H_y(x) = \begin{cases} A_3 e^{\alpha_3(x-d_2)} + B_3 e^{-\alpha_3(x-d_2)} & x > d_2 \\ A_2 e^{\alpha_3(x-d_2)} + B_2 e^{-\alpha_3(x-d_2)} & 0 < x < d \\ A_1 e^{\alpha_3(x-d_2)} + B_1 e^{-\alpha_3(x-d_2)} & -d_1 < x < \\ A_0 e^{\alpha_3(x-d_2)} & x < -d_1 \end{cases} \tag{15.22}$$

where

$$H_y(x) = \begin{cases} \alpha_j = (\beta^2 - k_0^2 \varepsilon_j)^{1/2} & j = 0,1,2,3 \\ \beta = k_0 \sqrt{\varepsilon_3} sin\theta_3 \end{cases} \tag{15.23}$$

The oscillating field in the prism results in an imaginary parameter α_3 in that region. α_1 is complex due to the complex permittivity of graphene, and both parameters of α_0 and α_2 are real. Using the boundary condition, the expression of reflectivity is

$$r = \frac{B_3}{A_3} = (\frac{\gamma_{23} + \gamma_{012} e^{-2\alpha_2 d_2}}{1 + \gamma_{23}\gamma_{012} e^{-2\alpha_2 d_2}}) \tag{15.24}$$

where

$$\gamma_{012} = (\frac{\gamma_{12} + \gamma_{01} e^{-2\alpha_1 d_1}}{1 + \gamma_{12}\gamma_{01} e^{-2\alpha_1 d_1}}) \tag{15.25}$$

and

$$\gamma_{23} = (\frac{\varepsilon_2 \alpha_3 - \varepsilon_3 \alpha_2}{\varepsilon_2 \alpha_3 + \varepsilon_3 \alpha_2}) \tag{15.26}$$

and

$$\gamma_{01} = (\frac{\varepsilon_0 \alpha_1 - \varepsilon_1 \alpha_0}{\varepsilon_0 \alpha_1 + \varepsilon_1 \alpha_0}) \tag{15.27}$$

and

$$\gamma_{12} = (\frac{\varepsilon_1 \alpha_2 - \varepsilon_2 \alpha_1}{\varepsilon_1 \alpha_2 + \varepsilon_2 \alpha_1}) \tag{15.28}$$

To obtain reflection in a multilayer structure, transfer matrix method is introduced as another technique [69].

To calculate the reflection phase, we use a distributed circuit model for the proposed multilayer structure in which each layer is modeled by a transmission line section with its corresponding characteristic impedance, *Zi*, and propagation constant, β_i. Impedance and propagation constant of each layer depend on both polarization of the incident wave and its incident angle. For TM waves propagating in a medium with permittivity of ε_i, and permeability of μ_i, with the angle θ_i with respect to the normal direction, as it is shown in Figure 15.22, the characteristic impedance is defined as $Z_i=\sqrt{\mu_i/\varepsilon_i}$ and the propagation constant is $\beta_i=\omega\sqrt{\mu_i{}^*\varepsilon_i}$. Figure 15.22 shows the distributed circuit model of the multilayer structure in which every layer of thickness *d*, is modeled by a transmission line section of the same length with its corresponding characteristic impedance and propagation constant. Having the transmission line model of the multilayer structure, the reflection coefficient for an incident TM wave with arbitrary incident angle can be found by using the transfer matrix method. In this method, the tangential components of electromagnetic fields at two ends of the medium of length *di* (which we modeled with a transmission line section) related to each other via a 2×2 transfer matrix. Since the boundary condition at the interface of two different mediums is applied to the tangential field components, the transfer matrix can be used to relate the incident and reflected waves at the interface of two mediums (corresponding to two transmission line sections). Furthermore, the transfer matrix of a multilayer structure can be constructed by multiplying the transfer matrix of its successive layers.

Therefore, the transfer matrix for the graphene-containing structure introduced in the previous subsection can be calculated as

$$\begin{pmatrix} A_{11} & A_{12} \\ A_{21} & A_{22} \end{pmatrix} = M_1 M_2 M_3 M_4 \tag{15.29}$$

in which M_1, M_2, M_3, and M_4 represent the transfer matrix of silicon, SiO_2, graphene, and air gap regions, respectively, and are given in the following equations:

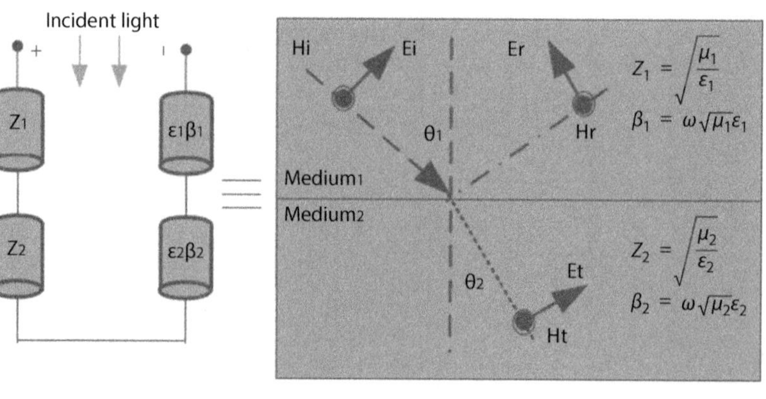

Figure 15.22 The transfer matrix model of multilayer structure.

$$M_4 = \begin{pmatrix} \cos(\beta_0 d_1) & jZ_0 \sin(\beta_0 d_1) \\ j\sin(\beta_0 d_1)/Z_0 & \cos(\beta_0 d_1) \end{pmatrix} \tag{15.30}$$

$$M_3 = \begin{pmatrix} 1 & 0 \\ \sigma_g & 1 \end{pmatrix} \tag{15.31}$$

$$M_2 = \begin{pmatrix} \cos(\beta_2 d_2) & jZ_2 \sin(\beta_2 d_2) \\ j\sin(\beta_2 d_2)/Z_2 & \cos(\beta_2 d_2) \end{pmatrix} \tag{15.32}$$

$$M_1 = \begin{pmatrix} \cos(\beta_1 d_1) & jZ_1 \sin(\beta_1 d_1) \\ j\sin(\beta_1 d_1)/Z_1 & \cos(\beta_1 d_1) \end{pmatrix} \tag{15.33}$$

In the above formulas, β's are the propagation constants of TM waves in each region and can be calculated as

$$\beta_4 = d_4 \sqrt{\varepsilon_4 - (n_1^2 sin^2\theta)} \tag{15.34}$$

$$\beta_2 = d_2 \sqrt{\varepsilon_2 - (n_1^2 sin^2\theta)} \tag{15.35}$$

$$\beta_1 = d_1 \sqrt{\varepsilon_1 - (n_1^2 sin^2\theta)} \tag{15.36}$$

Having the overall transfer matrix of the multilayer structure, the reflection coefficient can be represented in terms of matrix elements as

$$R = \frac{\kappa \times (A_{22} - A_{11}) - (\kappa^2 \times A_{12} - A_{21})}{\kappa \times (A_{22} + A_{11}) - (\kappa^2 \times A_{12} + A_{21})} \tag{15.37}$$

where $\kappa = \beta_0/\kappa_0$ is the free space wave number, and the reflection coefficient can be decomposed into its magnitude and phase as in the following equation:

$$R(\theta,\omega) = |R(\theta,\omega)| e^{j\phi(\theta,\omega)} \tag{15.38}$$

15.5 Graphene Surface Conductivity Calculation

The purpose of this section is to present approaches for calculating the graphene surface conductivity. The surface conductivity of graphene is attributed to intraband and interband transitions. Intraband transition can be calculated via semiclassical Boltzmann's equation, and interband transition can be calculated using time-dependent perturbation theory in quantum mechanics. In the following, we focus our attention on the calculation of only intraband transition, which is known to be of interest for optical devices [69].

15.5.1 Kubo Formula

We begin the calculation of surface conductivity by deriving Boltzmann's equation with the aid of transport properties in solids. To this end, first, we ignore the possibility of collisions taking place between *t-dt* and *t*. If no collisions occurred, then the *r* and *k* coordinates of every electron would evolve according to the semiclassical equations of motion.

Since *dt* is infinitesimal, we can find the explicit solution to these equations to linear order in *dt*, an electron at *r* and *k* at times *t* must have been at *r-v(k)dt*, *k-F dt/ħ*, at time *t-dt*. In the absence of collisions, this is the only point electrons at *r* and *k* can have come from, and every electron at this point will reach the point *a* and *k*. Consequently,

$$\begin{cases} p = \hbar k \\ \dfrac{dp}{dt} = \hbar \dfrac{dk}{dt} \\ \dfrac{dr}{dt} = v(k) \\ \hbar \dfrac{dk}{dt} = -e(E + \dfrac{1}{c}\, v \times H) = F(r,k) \end{cases} \tag{15.39}$$

To take collision into account, we must add two correction terms of the following formula:

$$f(r,k,t) = f(r - v(k)dt, k - F\ dt/d\hbar, t - dt) \tag{15.40}$$

The right-hand side is wrong because it assumes that all electrons get from *r-v dt*, *k-Fdt/ħ* to *r*, *k* in the time *dt*, ignoring the fact that some are deflected by collisions. It is also wrong because it fails to count those electrons found at *r*, *k* at time *t* not as a result of their unimpeded semiclassical motion since time *t-dt*, but as a result of a collision between *t-dt* and *t*. Adding these corrections, we find to leading order in *dt*:

$$\begin{cases} f(r,k,t) = f\left(r - v(k)dt, k - F\dfrac{dt}{d\hbar}, t - dt\right) \\ \qquad +(\partial f(r,k,t)/\partial t)|_{out}\, dt \\ \qquad +(\partial f(r,k,t)/\partial t)|_{in}\, dt \end{cases} \tag{15.41}$$

If we expand the left side to linear order in *dt*, then in the limit as *dt* → *0*, the above expression can be simplified as

$$\frac{\partial f}{\partial t} + \nabla k\ f.\frac{\partial k}{\partial t} + \nabla f.\frac{\partial r}{\partial t} = (\frac{\partial f}{\partial t})_{collision} \tag{15.42}$$

$$\frac{\partial f_1}{\partial t} + \frac{1}{\hbar}\nabla kF \cdot f_0(k) + v(k)\cdot\nabla f_0(k) = \frac{f_1(k)}{\tau(k)} \tag{15.43}$$

$$\frac{1}{\hbar}\nabla kF \cdot f_0(k) = -\frac{f_1(k)}{\tau(k)} \tag{15.44}$$

This is the celebrated Boltzmann equation. The terms on the left side are often referred to as the drift terms, and the terms on the right side as the collision terms.

To further simplify Equation 15.42, it is customary to apply the relaxation time approximation. Based on this approximation, the distribution function deviates from the equilibrium distribution function f_0, and it is expected that it will decay exponentially in time to the equilibrium value. It may be written as

$$(\frac{\partial f}{\partial t})_{collision} \to -(\frac{f(k) - f_0(k)}{(\tau(k)})_{collision} \tag{15.45}$$

where τ (k) is the relaxation time, and $f_0(k)$ is the equilibrium distribution function. Using this relaxation time approximation, the Boltzmann equation is

$$\frac{\partial f}{\partial t} + \nabla k\ f.\frac{\partial k}{\partial t} + \nabla f.\frac{\partial r}{\partial t} = \to -(\frac{f(k) - f_0(k)}{(\tau(k)})_{collision} \tag{15.46}$$

where

$$\frac{\partial k}{\partial t} = \frac{1}{\hbar} F \tag{15.47}$$

And by considering the below formula

$$F(k) = f_0(k) + f_1(k) \tag{15.48}$$

Frequently, we will be interested in applying fields to the system and seeking only the linear response, i.e., we can write the distribution function as $f = f0 + f1$, where $f1$ is the deviation from the equilibrium distribution function $f0$. This may be substituted in the Boltzmann equation in the collision approximation and only first-order terms in the applied fields retained. The result is the linearized Boltzmann equation.

For the sake of simplicity, we neglect the variations of temperature and hence $v = 0$. The relativistic energy of a particle in motion can be described as

$$\mathrm{E} = \sqrt{}\ ((mc\hat{}2)\hat{}2 + (pc)\hat{}2) \tag{15.49}$$

Benefited from linear energy dispersion in graphene, the carriers have zero effective mass, simplifying Equation 15.49 as $E = c\,|\,p\,|$. Substituting c and P with Fermi velocity and $\hbar^*k$, respectively, the energy of carriers can be expressed as

$$\mathrm{E} = \pm vf\,\hbar|k| \tag{15.50}$$

where positive and negative signs represent the upper (conduction band) and lower (valence band) of the Dirac cone. By assuming a uniform electric field $\varepsilon = \varepsilon\,\hat{x}$ and above equation can be rewritten as

$$\frac{1}{\hbar} q\varepsilon \frac{\partial f_0(k)}{\partial k_x} = \frac{q\mathcal{E}}{\hbar}\frac{\partial f_0(k)}{\partial E} v_x(k)\hbar = qv_x(k)\varepsilon\frac{\partial f_0(k)}{\partial E} \tag{15.51}$$

where

$$\begin{cases} qv_x(k)\varepsilon\dfrac{\partial f_0(k)}{\partial E} = -\dfrac{f_1(k)}{\tau(k)} \\ or \\ f_1(k) = -\ qv_x(k)\varepsilon\ \tau(k)\ \dfrac{\partial f_0(k)}{\partial E} \end{cases} \tag{15.52}$$

The current density in this case is

$$\begin{aligned} j_x &= \rho v_x(k) = \frac{nq}{A} v_x(k) \\ &= \frac{4q}{A} \sum_k f_1(k) v_x(k) \end{aligned} \tag{15.53}$$

By substituting $f_1(k)$ into the above equation, the current density can be expressed as

$$j_x = \frac{-4q^2\varepsilon}{A} \sum_k v_x(k)^2 \tau(k) \frac{\partial f_0(k)}{\partial E} \tag{15.54}$$

The factor of 4 is the result of degeneracy in graphene (spin up and down). We consider the surface conductivities of graphene in the x and y directions are the same, resulting in substituting $v_x(k)^2$ with $(v_f/2)^2$. It should be noted that $df_0(k)/dE$ is minimal except near the Fermi level. According to the above description and the fact that $j = \sigma E$, Equation (15.54) can be stated as

$$\sigma_{mono} = \frac{-4q^2\tau}{A} \frac{v_f^2}{2} \sum_k \frac{\partial f_0(k)}{\partial E} \tag{15.55}$$

where σ_{mono} is the surface conductivity of monolayer graphene.

By considering the following formula of Schrödinger equation:

$$\Delta k = k_x k_y = \left(\frac{2\pi}{l}\right)\left(\frac{2\pi}{l}\right) = \frac{(2\pi)^2}{A} \tag{15.56}$$

Hence, the surface conductivity of graphene can be expressed as

$$\sigma_{mono} = \frac{-4q^2\tau}{A} \frac{v_f^2}{2\Delta k} \sum_k \frac{\Delta k\, \partial f_0(k)}{\partial E} \tag{15.57}$$

Assuming large-area graphene (i.e., $\Delta k \to 0$), Equation (15.54) can be rewritten as

$$\sigma_{mono} = \frac{-4q^2\tau v_f^2}{2(2\pi)^2} \int_{\phi=0}^{2\pi} \int_{k=0}^{\infty} k\,dk\,d\phi \frac{\partial f_0(k)}{\partial E} \tag{15.58}$$

the solution of Equation (15.57) is

$$\sigma_{mono} = \frac{-q^2\tau v_f^2}{\pi}\int_0^{\infty}\frac{\pm E}{\hbar v_f}\frac{\pm dE}{\hbar v_f}\frac{\partial f_0(E)}{\partial E}$$
$$= \frac{-q^2\tau}{\pi\hbar^2}\left\{\int_0^{\infty} E\,dE\frac{\partial f_0(E)}{\partial E} + \int_0^{-\infty} E\,dE\frac{\partial f_0(E)}{\partial E}\right\} \tag{15.59}$$
$$= \frac{-q^2\tau}{\pi\hbar^2}\int_{-\infty}^{+\infty}|E|\,dE\frac{\partial f_0(E)}{\partial E} = \frac{e^2\tau}{\pi\hbar^2}2K_BT\;Ln\left(2cosh(\frac{E_f}{2K_BT}\right)$$

The surface conductivity of monolayer graphene is obtained (Equation (15.59)) under an external static electric field. To obtain the surface conductivity in the frequency domain, a time-harmonic electric field $(\varepsilon=\varepsilon e^{i\omega t}\hat{x})$ is applied onto the graphene. By considering the equation of motion, we have

$$m\frac{dv}{dt} = -e\varepsilon - m\frac{v}{\tau} \tag{15.60}$$

Substituting $d/dt=i\omega$, the velocity as a function of time-harmonic field is

$$v = -\frac{e\varepsilon}{i\omega m + \frac{m}{\tau}} \tag{15.61}$$

Regarding $j=-nev$, $j=\sigma\varepsilon$ the surface conductivity as a function of frequency can be expressed as

$$\sigma_{graphene} = \frac{\sigma_{mono}}{i\omega + \frac{1}{\tau}} \tag{15.62}$$

In this case, the surface conductivity of graphene is derived from Kubo formula in a complex term consisting of interband and intraband transitions. It is worth mentioning that graphene is often modelled by an ultrathin layer with negligible thickness ($d \rightarrow 0$), and its electromagnetic properties are characterized by a surface conductivity tensor. In addition, we considered the effect of the magnetic field to be zero. According to this formulation, the surface conductivity of graphene is composed of two terms: interband and intraband, which are expressed in the following formula [69]:

$$\sigma_{graphene} = \sigma_{interband} = \sigma_{intraband} \tag{15.63}$$

where

$$\sigma_{interband} = -J\frac{e^2}{4\pi\hbar}\ln\left(\frac{\left|2\mu_c + (\omega - j\tau^{-1})\hbar\right|}{\left|2\mu_c - (\omega - j\tau^{-1})\hbar\right|}\right) \tag{15.64}$$

$$\sigma_{intraband} = -J\frac{e^2\kappa_B T}{\pi\hbar^2(\omega - j\tau^{-1})}\left(\frac{\mu_c}{\kappa_B T} + 2\ln(e^{-\mu_c/\kappa_B T} + 1)\right) \tag{15.65}$$

where ω is the radian frequency, τ^{-1} is the scattering rate representing the loss mechanism, e is the charge of the electron, $B= 1.38*10^{-23}$ J/k is the Boltzmann constant, $\hbar$ is the reduced Planck constant, and μ_C is the chemical potential. The chemical potential of the graphene and therefore the complex conductivity of the graphene can be tuned by applying an external voltage. As a result, the graphene can be treated as either a metal or a dielectric.

On the other hand, graphene can be modelled as an ultrathin nonlocal anisotropic surface, characterized by a tensor conductivity:

$$\overleftrightarrow{\sigma}(\omega, \mu_c(\boldsymbol{E}_0), \Gamma, \mathrm{T}, \boldsymbol{B}_0) = \hat{\boldsymbol{x}}\hat{\boldsymbol{x}}\sigma_{xx} + \hat{\boldsymbol{x}}\hat{\boldsymbol{y}}\sigma_{xy} + \hat{\boldsymbol{y}}\hat{\boldsymbol{x}}\sigma_{yx} + \hat{\boldsymbol{y}}\hat{\boldsymbol{y}}\sigma_{yy} = \begin{pmatrix} \sigma_{xx} & \sigma_{xy} \\ \sigma_{yx} & \sigma_{yy} \end{pmatrix} \tag{15.66}$$

where ω is the angular frequency, m_c is the chemical potential, depending on electric biasing **E**, Γ is the phenomenological scattering rate, defined as $\Gamma = 1/2\tau$ (τ is the relaxation time), T is temperature, and $\mathbf{B}_0$ is the applied magnetic biasing. It is worth mentioning that $\sigma_{xx}(\sigma_{yy})$ and $\sigma_{xy}(\sigma_{yx})$ are longitudinal (diagonal) and transverse (off-diagonal) conductivities, respectively, and can be obtained using Kubo formula:

$$\sigma_{xx}(\mu_c(E_0), B_0) = \frac{e^2 v_f^2 |eB_0|(\omega - j2\Gamma)\hbar}{-j\pi} \times \sum_{n=0}^{\infty}\left\{\frac{f_d(M_n) - f_d(M_{n+1}) + f_d(-M_{n+1}) - f_d(M_n)}{(M_{n+1} - M_n)^2 - (\omega - j2\Gamma)^2\hbar^2} \times \left(1 - \frac{\Delta^2}{M_n M_{n+1}}\right) \times \frac{1}{M_{n+1} + M_n} + \frac{f_d(M_n) - f_d(M_{n+1}) + f_d(-M_{n+1}) - f_d(M_n)}{(M_{n+1} - M_n)^2 - (\omega - j2\Gamma)^2\hbar^2} \times \left(1 + \frac{\Delta^2}{M_n M_{n+1}}\right)\frac{1}{M_{n+1} + M_n}\right\} \tag{15.67}$$

$$\sigma_{yx}(\mu_c(E_0), B_0) = -\frac{e^2 v_f^2 e B_0}{\pi} \times \sum_{n=0}^{\infty} \{f_d M_n l) - f_d(M_{n+1}) - f_d(-M_{n+1}) + f_d(-M_n)\} \times \left\{ \left(-\frac{\Delta^2}{M_n M_{n+1}} \right) \frac{1}{(M_{n+1} - M_n)^2 - (\omega - j2\Gamma)^2 \hbar^2} + \left(1 + \frac{\Delta^2}{M_n M_{n+1}} \right) \frac{1}{(M_{n+1} + M_n)^2 - (\omega - j2\Gamma)^2 \hbar^2} \right. \tag{15.68}$$

where

$$M_n = \sqrt{\Delta^2 + 2nv_f^2 |eB_0| \hbar} \qquad f_d(M_n) = \frac{1}{1 + e^{(M_n - \mu_c)/k_B T}} \tag{15.69}$$

$\hbar$ is the reduced Plank constant, -e is the electron charge, n_f is the Fermi velocity, and Δ is an excitonic energy gap associated to the electron interactions in the presence of the magnetic biasing. To show the excitation of guided mode resonance (GMRs), the dispersion relation is studied. To ensure the propagation of the GMRs supported by the graphene sheet, the propagation constant of the GMRs must be substantially larger than the propagation constant of the air, i.e., $k\rho >> k_0$. Regarding this condition, the dispersion relation of the hybrid TM-TE GMR when graphene is sandwiched between two media (h-BN and air) can be expressed as

$$k_\rho \approx \frac{jk_0}{2} \left[\frac{\varepsilon_{r1} + \varepsilon_{r2}}{\eta_0 \sigma_L} + \frac{\eta_0}{2\sigma_L} (\sigma_L^2 + \sigma_H^2) \right] + \frac{jk_0}{2} \sqrt{ \left[\frac{\varepsilon_{r1} + \varepsilon_{r2}}{\eta_0 \sigma_L} + \frac{\eta_0}{2\sigma_L} (\sigma_L^2 + \sigma_H^2) \right]^2 - 2(\varepsilon_{r1} + \varepsilon_{r2}) } \tag{15.70}$$

where

$$\sigma_L(\omega) = \sigma_0 \frac{1 + j\omega\tau}{(1 + j\omega\tau)^2 + (\tau\omega_c)^2} \qquad \sigma_H(\omega) = \sigma_0 \frac{\tau\omega_c}{(1 + j\omega\tau)^2 + (\tau\omega_c)^2} \tag{15.71}$$

where η_0=377 and ε_0=1 are impedance and dielectric constant of the free space, respectively. Note that the surrounding media of graphene are taken to be nonmagnetic [69].

15.5.2 Graphene Conductivity Calculation *via* Kubo Formula

In Figure 15.23, we have plotted band structure of graphene with no external voltage and with an external voltage. Moreover, we have shown in Figure 15.24 the variation of surface conductivity of the graphene as the chemical potential alters from below the threshold value (i.e., $\mu_C<\hbar\omega/2$) to above the threshold value (i.e., $\mu_C>\hbar\omega/2$). As shown in Figure 15.24, when $\mu_C=\hbar\omega/2$, the imaginary part of surface conductivity is positive; in this case, the intraband conductivity term becomes the dominant term in the surface conductivity of graphene, and the graphene will behave as metal ultrathin metal supporting TM surface waves (distinguished by the green region in Figure 15.24).

Moreover, the dependence of chemical potential on external voltage is shown in Figure 15.25.

On the other hand, when $\mu_C<\hbar\omega/2$, the imaginary part of surface conductivity is negative. In this case, the interband conductivity is the dominant term and graphene will behave as a semiconductor and the structure shown in Figure 15.24 will support TE surface waves (distinguished by the red region in Figure 15.24). The ability to control the propagation of the surface waves results from tunability of the complex conductivity

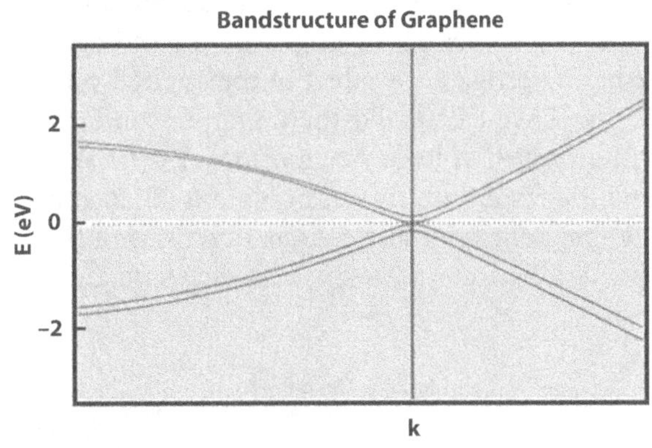

Figure 15.23 The band structure of monolayer graphene before and after applied external voltage.

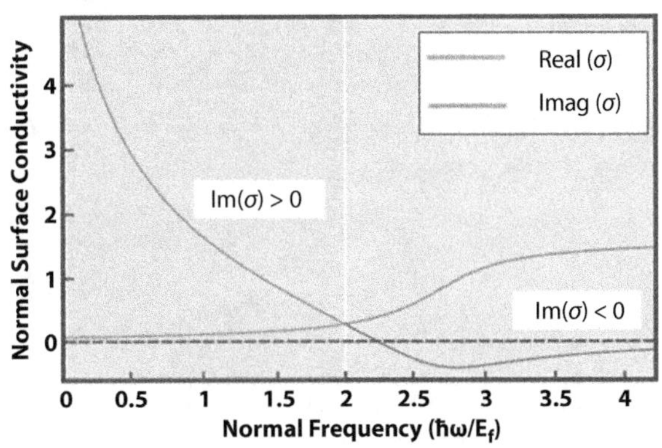

Figure 15.24 The normal surface conductivity of graphene.

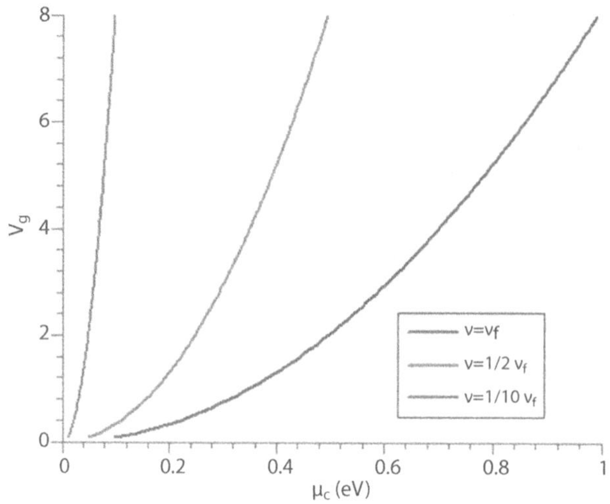

Figure 15.25 The chemical potential of graphene as a function of external voltage.

σ_{gr}, which according to the abovementioned Kubo formulation depends on the electrically controllable chemical potential of graphene μ_c, and also the wavelength of the surface waves. To further investigate the effect of the applied voltage on the surface conductivity of graphene, we have calculated the complex conductivity at room temperature ($T = 300\ K$) for the applied voltage range of $0 < V < 10$ V, in the wavelength range $0.8\ \mu m < \lambda < 1.8\ \mu m$ and the results are shown in Figures 15.26 and 15.27 where the imaginary and real parts of graphene are illustrated as functions of the applied voltage V and wavelength λ in Figures 15.26 and 15.27, respectively. As it can be seen, both real and

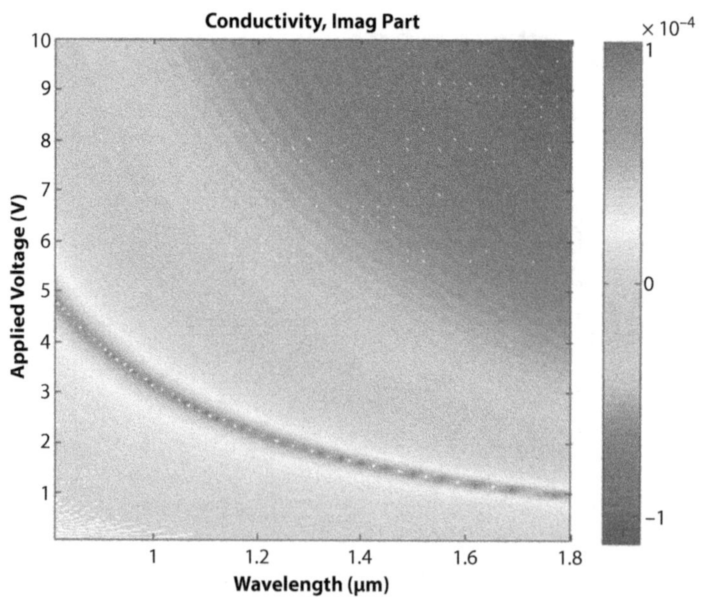

Figure 15.26 The imaginary part of surface conductivity of graphene.

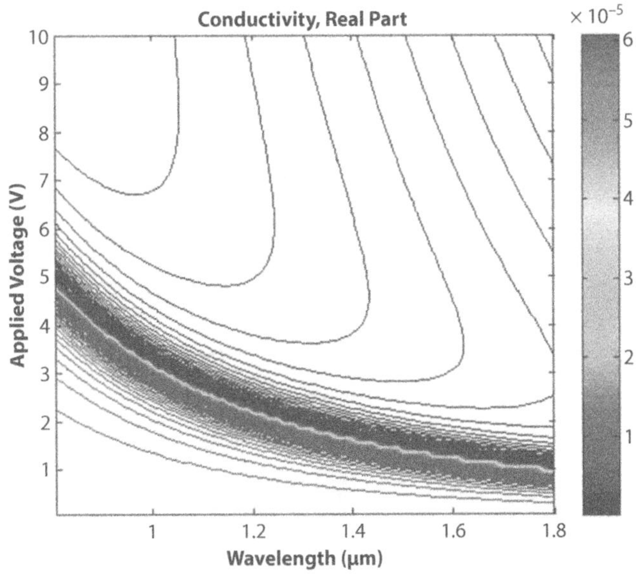

Figure 15.27 The real part of surface conductivity of graphene.

imaginary parts of surface conductivity of graphene are divided into two parts at the threshold value: $\mu_C = \hbar\omega/2$. In Figure 15.27, the real part of the conductivity of graphene refers to the ohmic loss. Unlike the traditional plasmonic materials, the value of ohmic loss of graphene is small; the surface wave can be greatly propagated without reduction along the surface of graphene. From Figure 15.27, a positive peak value of the imaginary part of surface conductivity appears at the threshold value. In both Figures 15.26 and 15.27, by altering external voltage, the threshold values shift to lower or higher wavelength; the complex conductivity of a graphene can be tuned by adjusting the chemical potential via an external voltage shifting the Fermi level above or below the threshold value; hence, the propagation of surface wave on graphene can be tuned.

15.5.3 Graphene Conductivity of Metasurface Structure

The chemical potential of the graphene (and GPM) and therefore its surface conductivity can be tuned by applying an external voltage. Figure 15.28 illustrates the graphene surface conductivity as a function of external electric and magnetic fields. Figure 15.28a shows the real and imaginary parts of the surface conductivity of the graphene for three different values of the chemical potential at the temperature of T=3 K and for a scattering rate of 2 ps. By considering the structure in the terahertz region, chemical potential values of μ_C = 0.01 eV, 0.02 eV, and 0.03 eV are, respectively, analogous to the applied voltage of V = 34 mV, 49 mV, and 60 mV. According to Figure 15.28a, the threshold wavelength, at which a sharp transition happens in the surface conductivity, shifts to lower values by increasing external voltage.

Moreover, Figure 15.29 shows the real and imaginary parts of the surface conductivity of the graphene for chemical potential 0.01 eV, and B_0= 0.3T.

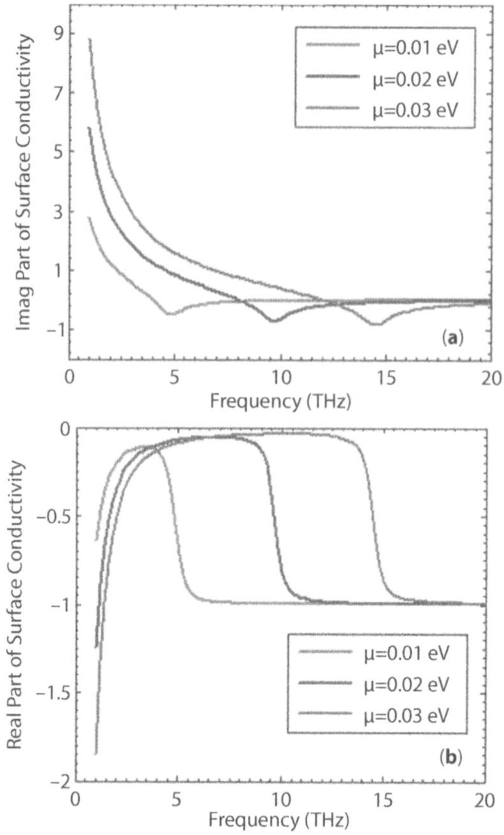

Figure 15.28 The real and imaginary part of GPM as a function of chemical potential.

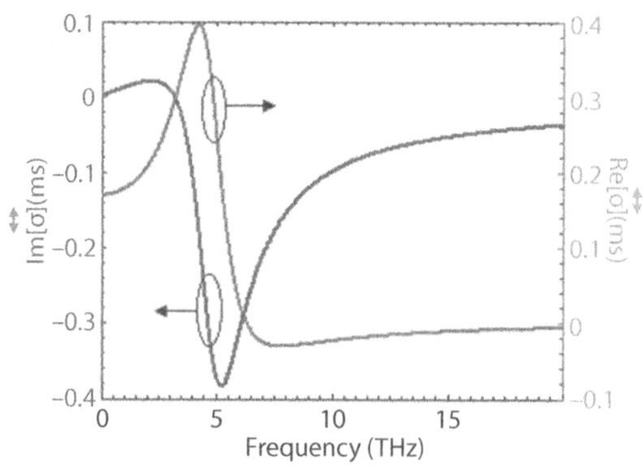

Figure 15.29 The real and imaginary part of graphene surface conductivity as a function of chemical potential, for B=0.3T and μ_c=0.01 eV.

15.6 Graphene-Based Switching Devices

In recent years, a plethora of research has been reported on GH shifts of the reflected light in the surface plasmon resonant systems. However, very little attention has been given on the free space switching devices based on GH shift. Also either proposed devices cannot be tuned or their lateral shift is in order of incident wavelength.

Thus, the main purpose of this section is as follows:

i. We propose a new free space switching device with graphene-containing structure based on GH shift.
ii. The distributed circuit modeling based on an analytical method, which is used for the analysis of the abovementioned structure.
iii. It is shown that the proposed structure is able to tune the surface plasmons within a small external voltage, which leads to the low power consumption tunable switching devices.

15.6.1 Circuit Model Properties

In this section, we propose a distributed circuit model to analytically study the tunable enhanced lateral displacement of electromagnetic waves in total reflection of light beams from a graphene-containing structure. The graphene-containing structure considered here supports transverse magnetic surface modes whose dispersion properties can be controlled by applying an appropriate electrical voltage to the graphene. Using this property of the structure, coupling of the incident wave to the surface modes of the structure is used to enhance the lateral displacement of the totally reflected wave, while it is also shown that this large lateral shift can be controlled by adjusting the dispersion properties of the surface modes by applying an electrical voltage. Using the proposed circuit model, phase of the reflected plane wave is calculated to obtain the GH shift under stationary phase approximation. This approximation is then modified by considering the finite spatial width of the incident beam.

15.6.2 Structure Properties

According to Figure 15.30, the structure supporting plasmonic surface wave is formed by placing a graphene sheet on top of a SiO_2 covered silicon substrate. We have considered the thickness of SiO_2 to be 300 nm. The use of SiO_2 increases the propagation length of surface waves by significantly reducing the propagation loss. Because of their large propagation constant, the surface waves cannot be directly excited by incident light from free space. Therefore, the prism coupling scheme is used for excitation of these waves where a dielectric layer with a large enough refractive index is placed on top of the structure supporting the surface wave with an air-gap spacing. Coupling between the incident light beam and the surface waves occurs when the central propagation constant of the incident beam in the prism region is close to that of the surface waves. The strength of coupling can be tuned by controlling the thickness of the air-gap region, d, which is on the order of incident wavelength $\approx \lambda$.

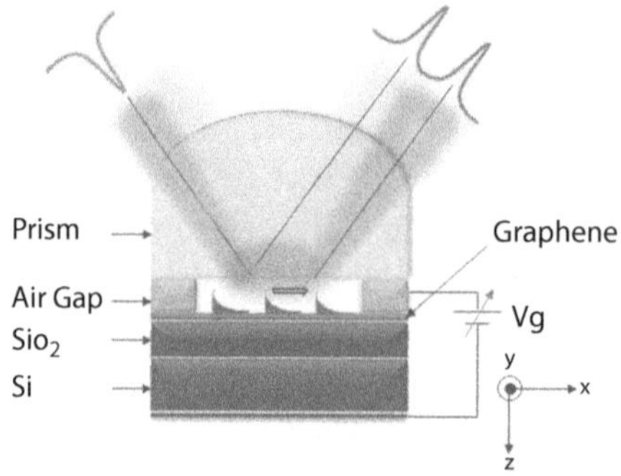

Figure 15.30 The proposed graphene-containing structure for huge GH shift of TM incident waves.

In this structure, the five layers of the proposed structure are considered to be nonmagnetic, and the prism has the refractive index of 1.5. We assume that the chemical potential of the graphene layer is tuned in a way that $\mu_C=\hbar\omega>2$, and therefore, only TM surface mode can be propagated on the graphene interface. In this case, the intraband conductivity-dominant chemical potential is nonzero ($\mu_c \neq 0$) and the surface conductivity of graphene is obtained by Kubo formula with τ =2 ps, μ_c =0.2 eV (corresponding) to the applied voltage of v =0.5 V. In addition, we assume that incident light has the wavelength of λ =1.55 μm.

In this structure (Figure 15.30), excitation of the surface waves on the graphene surface can be achieved by adjusting the chemical potential of graphene by applying an external voltage. By applying an external voltage, the Fermi level can be shifted above or below the threshold value of $\mu_C=\hbar\omega/2$. When the Fermi level is shifted above the threshold value, this structure supports TM modes (i.e., modes with no magnetic field component in the plane of interface). On the other hand, when the Fermi level is shifted below the threshold value, this structure supports TE modes (i.e., modes with no electric field component in the plane of interface). Incident light can be coupled to these surface modes through a prism region as shown in Figure 15.30. As mentioned in the earlier section, we use the phase of the reflected wave from the graphene containing structure to calculate the GH shift according to the stationary phase approximation.

15.6.3 Computational Method

To calculate the reflection phase, we use a distributed circuit model based on transfer matrix method (see Chapter 3 for more detail) for the proposed multilayer structure in which each layer is modeled by a transmission line section with its corresponding characteristic impedance, as shown in Figure 15.31.

As was noted in earlier and shown in Figure 15.31, because the thickness of the graphene layer (0.3 nm) is much smaller than the wavelength in the range of interest (i.e., infrared range), graphene is modeled with its surface conductivity. The surface conductivity itself can be modeled by a distributed or lumped circuit.

Since here we only consider TM surface waves that occur only when the chemical potential of graphene is above the threshold value, the surface conductivity is mostly caused by intraband transitions and can be written as

$$\sigma_{graphene} l = \frac{\sigma_{mono}}{i\omega\tau + 1} \tag{15.72}$$

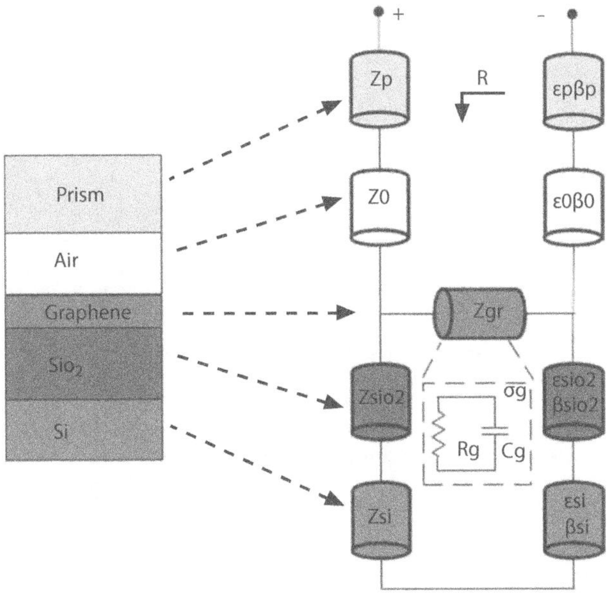

Figure 15.31 The distributed circuit model of the graphene-containing structure for TM polarization.

The abovementioned surface conductivity can be modeled by the impedance of a parallel RC circuit ($Z=R/J\omega RC+1$). According to this model, whenever the chemical potential increases, the equivalent impedance of the graphene layer decreases.

15.6.4 Results

Using the distributed circuit model and transfer matrix formulation (see Chapter 3) given in the previous section, the phase of reflected wave for incident wavelength of λ =1.55 μm and incident angle of 30° < θ < 90°, when chemical potential of graphene is μ_c = 0.2 eV, μ_c = 0.4 eV, and μ_c = 0.6 eV, is extracted and plotted in Figure 15.32. The chemical potential of graphene can be set to μ_c = 0.2 eV, μ_c = 0.4 eV, and μ_c = 0.6 eV, with the applied voltage of v = 0.5 V, v = 1.2 V, and v = 3 V, respectively. As can be seen in Figure 15.32, in the presence of graphene with the chemical potential of μ_c = 0.2 eV, μ_c = 0.4 eV, and μ_c = 0.6 eV, the reflection phase undergoes a sharp transition at the incident angle of θ = 55.88°, θ = 57.8°, and θ = 59.5°, respectively. The occurrence of a sharp transition in the phase of totally reflected plane waves from a multilayer structure is a sign of resonance between the incident wave and lateral modes of the multilayer structure. In other words, at the wavelength of λ = 1.55 μm and incident angle of θ = 55.88°, θ = 57.8°, and θ = 59.5°, the incident TM polarized wave excites the surface wave of the graphene-containing structure, when the voltage applied to the graphene is v = 0.5 V, v = 1.2 V, and v = 3 V, respectively.

According to Arman's formula (see Chapter 3), the reflection phase can be used to calculate the GH shift under stationary phase approximation. Using the derivative of the reflection phase in Figure 15.32, the normalized GH shift (normalized to the incident wavelength λ =1.55 μm) is calculated and shown in Figure 15.33. On the other hand, when actual graphene-containing structure is considered, the value of GH shift significantly increases at the incident angles where the reflection phase experiences sharp changes, i.e., when the

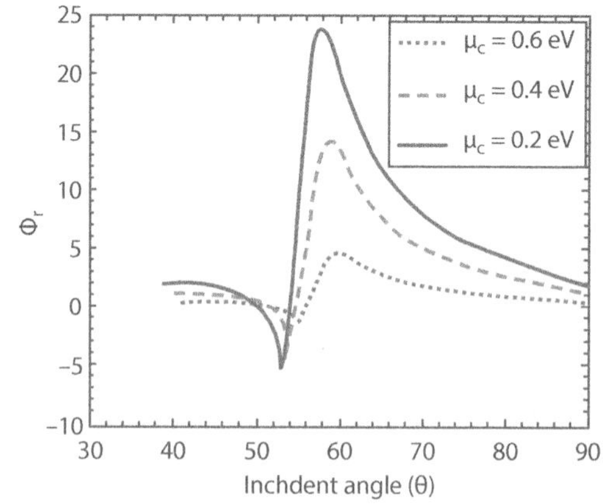

Figure 15.32 Calculated phase of reflected beam as a function of incident angle with various mc.

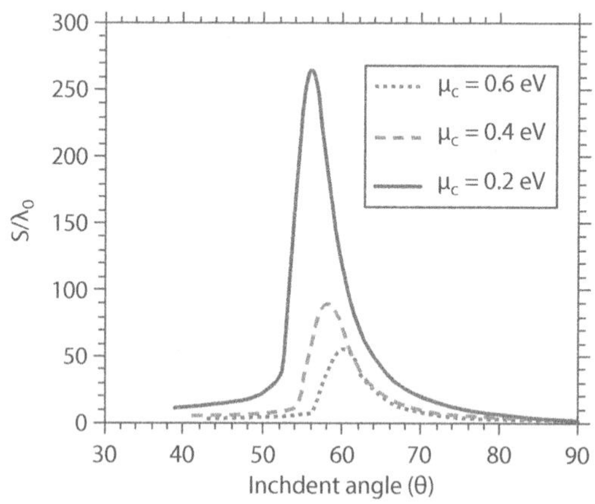

Figure 15.33 The calculated GH shift as a function of incident angle with various μ_c.

incident wave excites the surface wave of the graphene-containing structure. According to our calculations, when the chemical potential of graphene layer is $\mu_c = 0.2$ eV, and $\mu_c = 0.6$ eV, the resonance of GH shift has the extremely large value of 270 λ, 90 λ, and 55 λ, respectively. The reduction in value of the GH shift for higher values of the chemical potential can be attributed to the increase in the imaginary part of surface conductivity of graphene, which leads to a higher propagation loss of surface modes of the structure. Surface waves with higher propagation loss have smaller propagation lengths. Therefore, the power coupled to these modes (from the incident light beam) propagates a smaller length in the lateral dimension and the reflected wave experiences a smaller lateral displacement.

15.6.5 Conclusion

Here, a giant tunable GH shift in total reflection of light beams from a graphene-containing structure was analytically investigated here by proposing a distributed circuit model for switching devices. The considered graphene-containing structure supports TM surface modes with adjustable dispersion characteristics in response to an applied electrical voltage.

This property of the structure was then exploited to achieve an enhanced GH shift in total reflection of light beams. Furthermore, it was shown that the value of the giant GH shift can be varied by altering the dispersion properties of the surface modes by using a variable applied electrical voltage. The results were then modified by considering the finite spatial width of the incident beam in a more accurate formulation. According to the results presented here, by coupling of incident wave to the surface modes of the structure, giant GH shift values as high as 270 λ are achievable and this giant shift can be varied more than 215 λ by 2.5 V.

These results suggest that the giant GH shift in total reflection of light beams from graphene-containing structures can be used in designing integrated optical devices such as optical switches.

15.7 Graphene Plasmonic Metasurface-Based Switching Structures

In this section, we propose a graphene plasmonic metasurface (GPM) structure with highly tunable enhanced lateral displacements in the center of gravity of totally reflected light beam. Multiple reflections of the incident beam, and the resonance coupling between the incident beam and the surface modes of the graphene metasurface in each reflection, are employed to enhance the GH and IF shifts in the proposed structure. Then, the effects of different parameters including the incident beam waist, temperature, the scattering time, and the chemical potential of the graphene on the shift values are studied. Because of the strong light confinement in the surface modes of the graphene metasurface, the dispersion properties of these modes, and, therefore, the coupling strength between the incident beam and these modes, are highly sensitive to the parameters of the reflecting structure and the incident beam itself. The high sensitivity of the coupling strength between the incident beam and the surface modes is then exploited to tune the shift values.

In designing active optical devices, it is desirable to realize the shift variations through inducing changes in physical properties of the incident beam or the GPM structure rather than the geometrical parameters of the structure. To investigate these interesting features, here we propose a GPM structure that supports TM surface modes, and the incident beams illuminating the structure will experience multiple reflections. Under certain conditions (i.e., for certain frequencies and incident angles), the reflected beams will be resonantly coupled to the surface waves of the structure. Multiple resonance couplings between the incident beam and the surface modes of the structure result in substantial enhancement of the GH and IF shifts. The strong confinement of the surface modes provides the ability to tune the shift values with small adjustments in the geometrical and physical properties of the structure.

15.7.1 Structural Properties

The GPM is formed by introducing a periodic perturbation in the graphene layer of a graphene-containing structure. Such a structure is shown in the inset of Figure 15.34. Compared to a planar graphene layer, surface modes of a GPM structure are more confined in a direction perpendicular to the propagation direction and have a smaller group velocity. Therefore, introducing small adjustments in the physical properties of the graphene layer can result in abrupt variations in the coupling strength between the incident beam and the surface modes of the structure. As a result, the GH and IF shift values will experience large

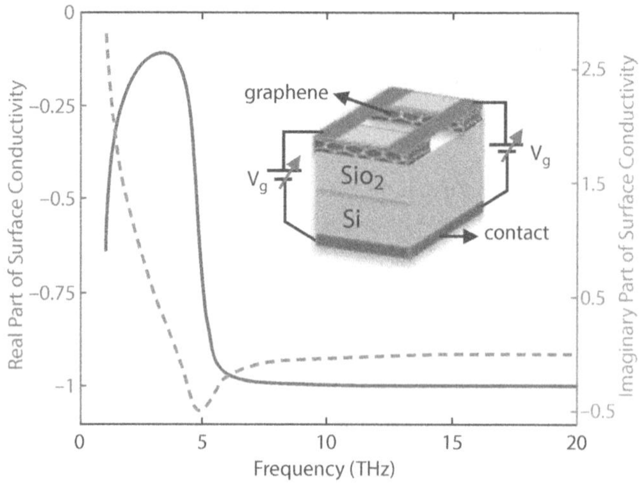

Figure 15.34 The real and imaginary parts of surface conductivity as a function of frequency. The inset shows the perspective view of a GPM structure.

changes. The graphene layers in a GPM structure are usually modeled with ultrathin layers whose optical properties are characterized by a surface conductivity tensor.

Here, we will use Kubo formulation for calculation of the surface conductivity of the graphene. Results in Figure 15.34 are calculated for $m_c = 0.01$ eV, scattering time of $\tau = 2$ ps, and temperature of T = 3 K. For this value of chemical potential, the condition $\mu_C = \hbar\omega/2$ holds at the threshold frequency of $f_{th} = 5$ THz, and the GPM structure supports TM and TE surface waves for frequencies lower and higher than f_{th}, respectively.

The diagram and normalized group velocity v_g/C_0 (C_0: speed of light in free space) for a simple graphene-containing structure with no perturbation and a GPM structure are shown in Figure 15.35. Geometrical parameters for the GPM considered in these calculations are provided in the inset of Figure 15.35. According to this figure, compared to the original graphene-containing structure, the group velocity of the GPM structure is reduced by a factor of 3. As mentioned above, the reduction of the group velocity of the surface waves results in improvement of the GH and IF shift values.

The cross-sectional view of the proposed structure and the schematic process of the multiple reflections of the incident beam from this structure is depicted in Figure 15.36. As can be seen, the proposed structure is composed of multiple reflecting surfaces, aligned in a way that the reflected beam of each surface illuminates the successive surfaces. Each reflecting GPM surface supports an optical surface mode, and the coupling of the incident beam to these modes results in large GH and IF shifts. Overall shift of the output beam is simply the sum of all shift values in the reflection of the incident beam from each surface. Every reflecting surface is composed of a SiO_2/Si substrate with a SiO_2 thickness of t_{SiO2}=300 nm, a GPM with the geometrical parameters given in Figure 15.35, a prism layer of the refractive index of np = 3.42 (which is needed for phase matching between incident beam and surface modes of the GPM), an air gap between the prism and the GPM, and the metallic contacts needed for applying the tuning voltage to the GPM layer. The length of each graphene layer in GPM is considered to be 363 λ (in accordance with the maximum lateral shift accessible in a single reflecting structure), which is a typical value in terahertz applications of the graphene.

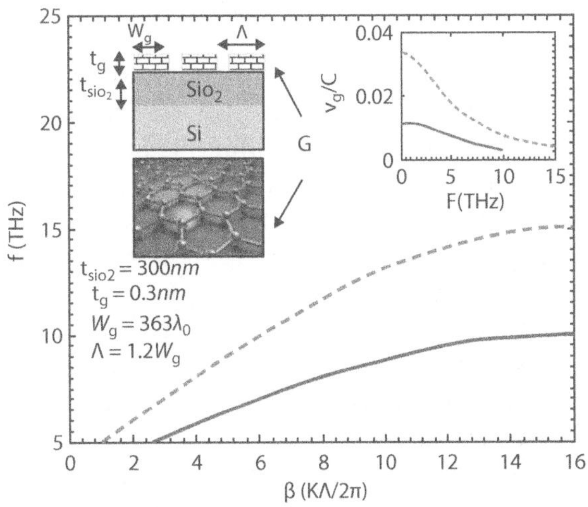

Figure 15.35 Dispersion diagram of the TM surface modes of a simple graphene-containing structure, respectively, on Si (red) and a GPM (blue). The inset depicts the normalized group velocity vg/C0. For the graphene layer, Λ = Wg, and for the GPM, Λ = 1.2 Wg.

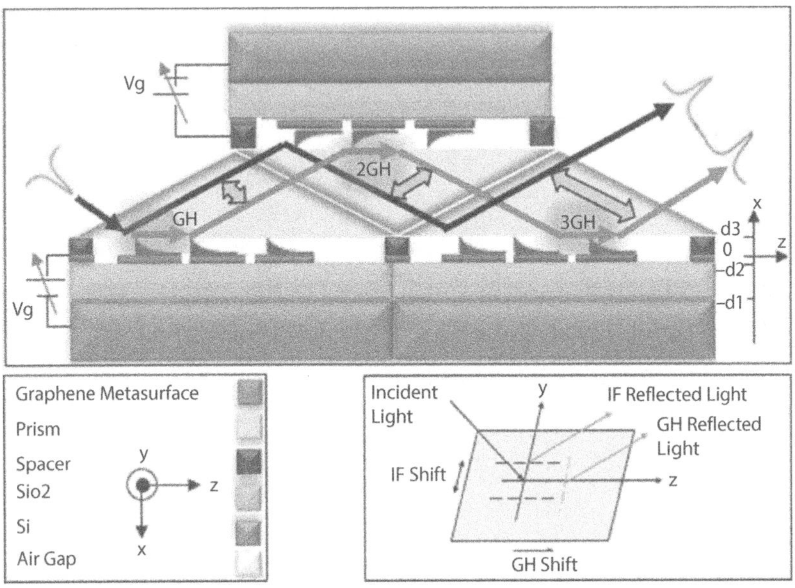

Figure 15.36 The proposed GPM structure for achieving GH and IF shifts. Inset shows the coordinate system of the GH and IF shifts.

15.7.2 Computational Method

When the waist of the incident beam is large enough compared to the incident wavelength, the SPA can be used to calculate the GH and IF shifts (see Chapter 3 for more details). According to the IM approach, the reflected light can be expressed as two waves: a directly reflected wave (exactly at the position of incident light) and a shifted wave due to the GH or IF effect.

Using the IM approach, the spatial and angular GH and IF shifts of the proposed structure can be calculated.

15.7.3 Results

Using the IM approach, the spatial and angular GH and IF shifts of the proposed structure with the parameters of $\mu_C = 0.01\ eV$, $\tau = 2\ ps$, $T = 3\ K$, and the incident wavelength of $\lambda = 60\ mm$ are calculated and plotted in Figures 15.37 and 15.38.

It should be noted that, in these calculations, the incident beam waist is assumed to be much larger than the incident wavelength so that the SPA can be used. Figure 15.37 shows the SGH shift for different incident angles. According to this figure, the phase matching condition and the resonance coupling between these two waves occur at an incident angle of $\theta = 48.19°$. The GH shift reaches its maximum value at this incident angle, which is calculated to be SGH = 1089 λ. The AGH shift is also plotted in the inset of Figure 15.37, which also shows that its maximum value occurs at the same incident angle, and is equal to AGH = 128.5 λ. Here calculated values of the SGH and AGH shifts are substantially larger than previously reported results for graphene-based structures.

In the same manner, Figure 15.38 represents the values of the SIF and AIF for different incident angles. By considering phase matching between the incident wave and the surface

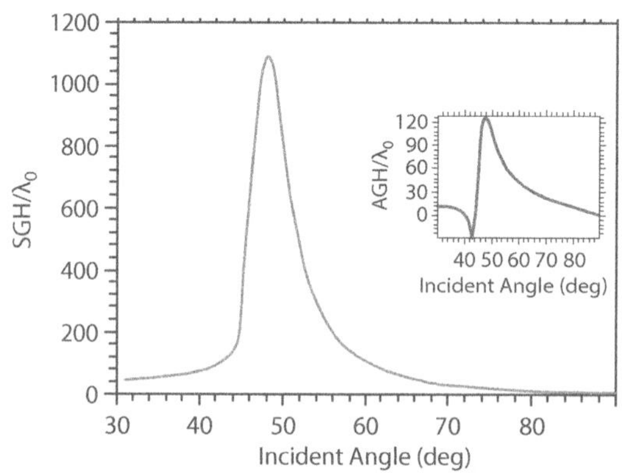

Figure 15.37 The spatial GH shift of the proposed structure for the parameter sets of $\mu_C = 0.01$ eV, $\tau = 2$ ps, and T = 3 K. The inset shows the angular GH shift for the same structure.

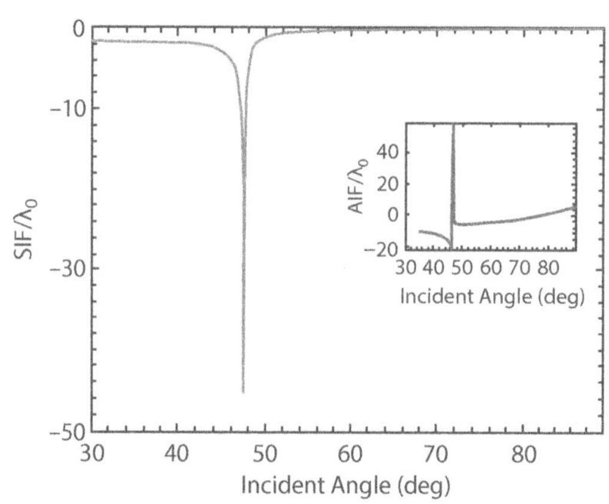

Figure 15.38 The spatial IF shift of the proposed structure for the parameter sets of $\mu_c = 0.01\ eV$, $\tau = 2\ ps$, and $T = 3\ K$. The inset shows the angular IF shift for the same structure.

modes of the GPM structure at resonance angle of θ = 48.19. The SIF is calculated to be SIF = −44.66 λ, and the AIF is calculated to be AIF = 60 λ. As in the case of GH shift, the calculated values of the SIF and AIF shifts for the proposed structure are substantially larger than previously reported results for graphene-based structures.

According to the above results, very large GH and IF shift values are achievable in the proposed structure. However, for practical applications such as designing optical devices, the effect of variations in different parameters such as temperature, scattering rate, and chemical potential of the graphene, and incident beam waist on the shift values should be considered. In this section, the effect of the chemical potential variations on the GH and IF shifts is studied first, and then the effects of the other parameters mentioned above are studied.

Based on the results of these studies, the optimum structure, in which the highest shift variations can be attained, is introduced. In the following calculations, we consider the incident wavelength to be 60 μm, which corresponds to the frequency of f = 5 THz. This frequency is chosen based on the possibility of the application of the proposed structure in the realization of terahertz devices such as switches. To investigate the effect of the chemical potential variations on the GH and IF shift values, we have calculated these shifts using the methods described in the previous section. The results of the calculations of GH and IF shifts are depicted in Figures 15.39 and 15.40, respectively.

Considering the optical modes of the structure, using a substrate with higher refractive indices increases the overall refractive index of the structure and results in optical

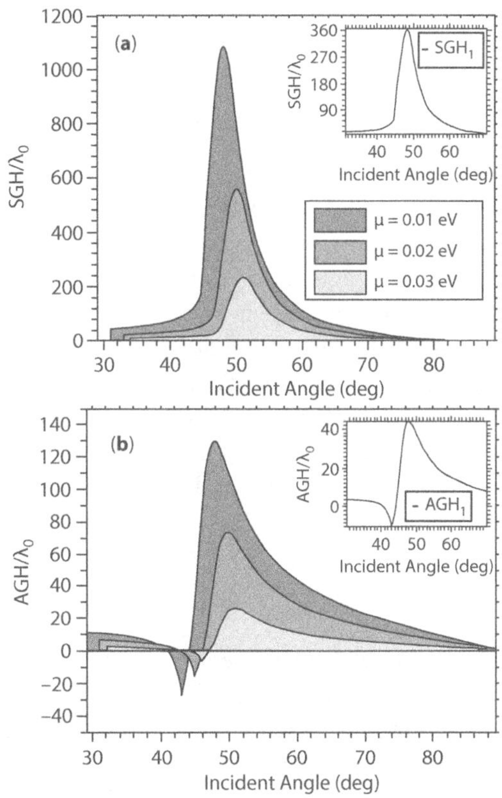

Figure 15.39 (a) The spatial GH shift and (b) the angular GH shift.

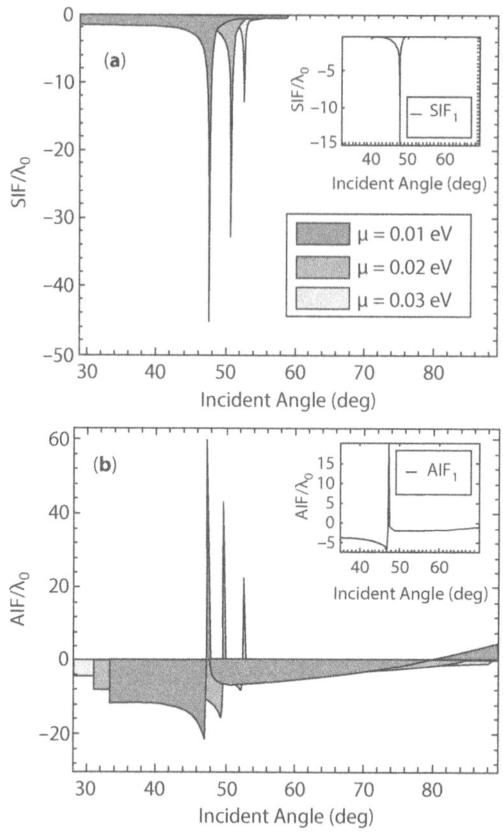

Figure 15.40 (a) The spatial IF shift and (b) the angular IF shift.

modes with smaller group velocities. For example, the dispersion diagram of a simple graphene-containing structure with silicon, sapphire, and SiO_2 substrates is plotted in Figure 15.41. As can be seen in this figure, dispersion diagram of the graphene-containing structure with silicon and sapphire substrate is located below the dispersion diagram with the SiO_2 substrate. According to the discussion above, the GH and IF shift values can be increased by decreasing the group velocity of the surface modes of the reflecting structure. Therefore, based on the dispersion diagrams in Figure 15.41, one expects that using the silicon or sapphire substrates can result in higher shift values. But, as will be discussed below, using these substrates increases the propagation loss of the surface modes and therefore reduces the achievable shift values.

15.7.4 Conclusion

Here highly tunable enhanced lateral displacements of the totally reflected light beams from a GPM structure were investigated. A structure was proposed in which the multiple reflections of the incident beam and the resonance coupling between the incident beam and the surface modes of the proposed structure in each reflection were employed to enhance the GH and IF shifts. The wide range of the lateral shift variations, along with the relatively

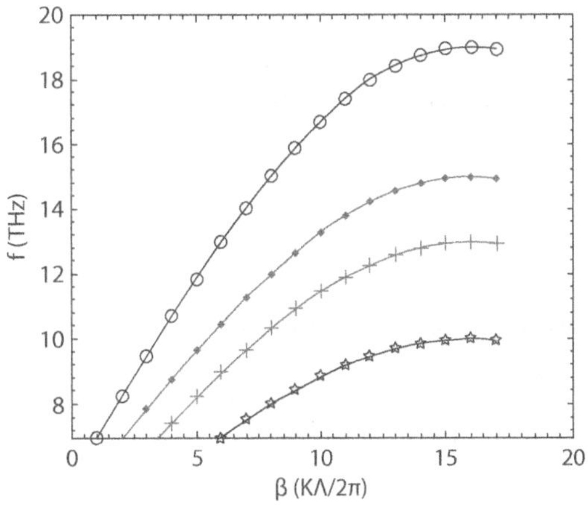

Figure 15.41 The dispersion diagram of the TM surface modes of a simple graphene-containing structure with Si (black), sapphire (green), graphene (red), and the SiO_2 substrate.

small required actuation power, recommends the application of the proposed structure in the realization of optical devices such as switches.

15.8 Future Roadmap

The field of graphene plasmonic and graphene plasmonic metasurfaces is in its infancy. In this thesis, we talked about several examples of switching devices with graphene-containing structure; however, there is a vast research based on graphene-containing structure with unlike mechanism such as sensors.

Hence, there are various roadmaps to follow of the topics we presented in this thesis; it is indeed impossible to summarize all of these directions here, but we refer to a subset of them that could be followed in the future.

- Optical sensor using graphene plasmonic metasurfaces
- Terahertz modulator based on graphene-containing structure
- Optical memory and logic gates [70]
- Efficient coupling of incident beam to graphene plasmonic metasurface for solar cell applications
- Two-dimensional optical logic gates

15.9 Closing Thoughts

Graphene plasmonic is undeniably interesting. It originates from its one-atom-thickness, tunable surface conductivity, strong interaction with incident beam, and low losses in broadband wavelength ranges; one feature that makes it particularly satisfactory is its tunability and the design flexibility it offers. Hence, graphene plasmonic metasurfaces have been broadly studied by condensed matter physicists and engineers. This work and several others suggest that graphene is also a good candidate for plasmonic and photonics applications.

Finally, graphene-containing structure can be introduced for the next generation of integrated plasmonics devices.

References

1. Fei, Z. *et al.*, Gate-tuning of graphene plasmons revealed by infrared nano-imaging. *Nature*, 487, 7405, 82, 2012.
2. Farmani, A. *et al.*, Design of a tunable graphene plasmonic-on-white graphene switch at infrared range. *Superlattices Microstruct.*, 112, 404–4145, 2017.
3. Farmani, A., Mir, A., Sharifpour, Z., Broadly tunable and bidirectional terahertz graphene plasmonic switch based on enhanced Goos-Hänchen effect. *Applied Surface Science*, 453, 358–364, 2018.
4. Baqir, M.A. *et al.*, Nanoscale, tunable, and highly sensitive biosensor utilizing hyperbolic metamaterials in the near-infrared range. *Applied Optics*, 57.31, 9447–9454, 2018.
5. Grigorenko, A.N., Polini, M., Novoselov, K.S., Graphene plasmonics. *Nature Photonics*, 6, 11, 749, 2012.
6. Alihosseini, F., Ahmadi, V., Mir, A., Design and analysis of a tunable liquid crystal switch/filter with metallic nano-slits. *Liq. Cryst.*, 42, 11, 1638–1642, 2015.
7. Farmani, A., Farhang, M., Sheikhi, M.H., High performance polarization-independent quantum dot semiconductor optical amplifier with 22 dB fiber to fiber gain using mode propagation tuning without additional polarization controller. *Opt. Laser Technol.*, 93, 127–132, 2017.
8. Cheng, M. *et al.*, Giant and tunable Goos–Hanchen shifts for attenuated total reflection structure containing graphene. *JOSA B*, 31, 10, 2325–2329, 2014.
9. Cheng, M. *et al.*, Spatial and angular shifts of terahertz wave for the graphene metamaterial structure. *J. Phys. D: Appl. Phys.*, 48, 28, 285105, 2015.
10. Cheng, M. *et al.*, Gate-voltage control of angular and spatial shifts for a dielectric slab containing graphene. *Eur. Phys. J. D*, 70, 7, 158, 2016.
11. Dadoenkova, Y.S. *et al.*, Goos-Hänchen shift at the reflection of light from the complex structures composed of superconducting and dielectric layers. *J. Appl. Phys.*, 118, 21, 213101, 2015.
12. Dadoenkova, Y.S. *et al.*, Effect of lateral shift of the light transmitted through a one-dimensional superconducting photonic crystal. *Photonics Nanostruct. Fundam. Appl.*, 11, 4, 345–352, 2013.
13. Aiello, A. and Woerdman, J.P., Role of spatial coherence in Goos–Hänchen and Imbert–Fedorov shifts. *Opt. Lett.*, 36, 16, 3151–3153, 2011.
14. Bao, Q. and Loh, K.P., Graphene photonics, plasmonics, and broadband optoelectronic devices. *ACS Nano*, 6, 5, 3677–3694, 2012.
15. Hao, J. *et al.*, 1.5 mm light beam shift arising from 14 pm variation of wavelength. *JOSA B*, 27, 6, 1305–1308, 2010.
16. He, J., Yi, J., He, S., Giant negative Goos–Hänchen shifts for a photonic crystal with a negative effective index. *Opt. Express*, 14, 7, 3024–3029, 2006.
17. Hermosa, N., Reflection beamshifts of visible light due to graphene. *J. Opt.*, 18, 2, 025612, 2016.
18. Dong, W.T., Gao, L., Qiu, C.-W., Goos–Hänchen shift at the surface of chiral negative refractive media. *Metamaterials, 2008 International Workshop on*, IEEE, 2008.
19. Emadi, R. *et al.*, Design of low loss waveguide switch using graphene strips at THz frequencies. *Millimeter-Wave and Terahertz Technologies (MMWaTT), 2016 Fourth International Conference on*, IEEE, 2016.

20. Fan, Y. *et al.*, Electrically tunable Goos–Hänchen effect with graphene in the terahertz regime. *Adv. Opt. Mater.*, 4, 11, 1824–1828, 2016.
21. Merano, M. *et al.*, Demonstration of a quasi-scalar angular Goos–Hänchen effect. *Opt. Lett.*, 35, 21, 3562–3564, 2010.
22. Miri, M. *et al.*, Approximate expressions for resonant shifts in the reflection of Gaussian wave packets from two-dimensional photonic crystal waveguides. *JOSA B*, 29, 4, 683–690, 2012.
23. Namdar, A., Talebzadeh, R., Jamshidi-Ghaleh, K., Surface wave-induced enhancement of the Goos–Hänchen shift in single negative one-dimensional photonic crystal. *Opt. Laser Technol.*, 49, 183–187, 2013.
24. Goswami, N., Kar, A., Saha, A., Long range surface plasmon resonance enhanced electro-optically tunable Goos–Hänchen shift and Imbert–Fedorov shift in ZnSe prism. *Opt. Commun.*, 330, 169–174, 2014.
25. Goswami, S. *et al.*, Optimized weak measurements of Goos–Hänchen and Imbert–Fedorov shifts in partial reflection. *Opt. Express*, 24, 6, 6041–6051, 2016.
26. Hamedi, H.R., Radmehr, A., Sahrai, M., Manipulation of Goos-Hänchen shifts in the atomic configuration of mercury via interacting dark-state resonances. *Phys. Rev. A*, 90, 5, 053836, 2014.
27. Madani, A. and Entezar, S.R., Tunable enhanced Goos–Hänchen shift in one-dimensional photonic crystals containing graphene monolayers. *Superlattices Microstruct.*, 86, 105–110, 2015.
28. Merano, M., Optical beam shifts in graphene and single-layer boron-nitride. *Opt. Lett.*, 41, 24, 5780–5783, 2016.
29. Merano, M. *et al.*, Observation of Goos-Hänchen shifts in metallic reflection. *Opt. Express*, 15, 24, 15928–15934, 2007.
30. Ju, L. *et al.*, Graphene plasmonics for tunable terahertz metamaterials. *Nat. Nanotechnol.*, 6, 10, 630, 2011.
31. Juzeliūnas, G. *et al.*, Effective magnetic fields in degenerate atomic gases induced by light beams with orbital angular momenta. *Phys. Rev. A*, 71, 5, 053614, 2005.
32. Li, C.-F., Negative lateral shift of a light beam transmitted through a dielectric slab and interaction of boundary effects. *Phys. Rev. Lett.*, 91, 13, 133903, 2003.
33. Li, J.-S., Wu, J.-F., Zouhdi, S., Thermal tunable terahertz wave Goos–Hänchen shift. *IEEE Photonics Technol. Lett.*, 26, 21, 2162–2165, 2014.
34. Lin, L.Y., Goldstein, E.L., Tkach, R.W., Free-space micromachined optical switches for optical networking. *IEEE J. Sel. Top. Quantum Electron.*, 5, 1, 4–9, 1999.
35. Liu, F. *et al.*, Goos–Hänchen and Imbert–Fedorov shifts at the interface of ordinary dielectric and topological insulator. *JOSA B*, 30, 5, 1167–1172, 2013.
36. Dadoenkova, Y.S. *et al.*, Transverse magneto-optic Kerr effect and Imbert–Fedorov shift upon light reflection from a magnetic/non-magnetic bilayer: Impact of misfit strain. *J. Opt.*, 19, 1, 015610, 2016.
37. Dadoenkova, Y.S. *et al.*, Influence of misfit strain on the Goos–Hänchen shift upon reflection from a magnetic film on a nonmagnetic substrate. *JOSA B*, 33, 3, 393–404, 2016.
38. Luo, L. and Tang, T., Goos-Hänchen effect in Kretschmann configuration with hyperbolic metamaterials. *Superlattices Microstruct.*, 94, 85–92, 2016.
39. Madani, A. and Entezar, S.R., Surface polaritons of one-dimensional photonic crystals containing graphene monolayers. *Superlattices Microstruct.*, 75, 692–700, 2014.
40. Nie, Y. *et al.*, Detection of chemical vapor with high sensitivity by using the symmetrical metal-cladding waveguide-enhanced Goos–Hänchen shift. *Opt. Express*, 22, 8, 8943–8948, 2014.
41. Low, T. and Avouris, P., Graphene plasmonics for terahertz to mid-infrared applications. *ACS Nano*, 8, 2, 1086–1101, 2014.
42. Luo, C. *et al.*, Electrically controlled Goos–Hänchen shift of a light beam reflected from the metal-insulator-semiconductor structure. *Opt. Express*, 21, 9, 10430–10439, 2013.

43. Yallapragada, V.J. *et al.*, Direct measurement of the Goos–Hänchen shift using a scanning quadrant detector and a polarization maintaining fiber. *Rev. Sci. Instrum.*, 87, 10, 103109, 2016.
44. Yin, X. and Hesselink, L., Goos–Hänchen shift surface plasmon resonance sensor. *Appl. Phys. Lett.*, 89, 26, 261108, 2006.
45. Li, X. *et al.*, Experimental observation of a giant Goos–Hänchen shift in graphene using a beam splitter scanning method. *Opt. Lett.*, 39, 19, 5574–5577, 2014.
46. Lin, I.-T., *Optical properties of graphene from the THz to the visible spectral region*, University of California, Los Angeles, 2012.
47. Liu, X. *et al.*, Physical origin of large positive and negative lateral optical beam shifts in prism–waveguide coupling system. *Opt. Commun.*, 283, 13, 2681–2685, 2010.
48. Chiu, K.W. and Quinn, J.J., On the Goos–Hänchen effect: A simple example of a time delay scattering process. *Am. J. Phys.*, 40, 12, 1847–1851, 1972.
49. Cowan, J.J. and Aničin, B., Longitudinal and transverse displacements of a bounded microwave beam at total internal reflection. *JOSA*, 67, 10, 1307–1314, 1977.
50. Zhang, L.-J., Chen, L., Liang, C.-H., Goos–Hänchen shift at the interface of nonlinear left-handed metamaterials. *J. Electromagnet. Waves Appl.*, 22, 7, 1031–1041, 2008.
51. Bliokh, K.Y. and Aiello, A., Goos–Hänchen and Imbert–Fedorov beam shifts: An overview. *J. Opt.*, 15, 1, 014001, 2013.
52. Farmani, A., Miri, M., Sheikhi, M.H., Tunable resonant Goos–Hänchen and Imbert–Fedorov shifts in total reflection of terahertz beams from graphene plasmonic metasurfaces. *JOSA B*, 34, 6, 1097–1106, 2017.
53. Farmani, A. *et al.*, Tunable graphene plasmonic Y-branch switch in the terahertz region using hexagonal boron nitride with electric and magnetic biasing. *Appl. Opt.*, 56, 32, 8931–8940, 2017.
54. Farmani, A., Miri, M., Sheikhi, M.H., Analytical modeling of highly tunable giant lateral shift in total reflection of light beams from a graphene containing structure. *Opt. Commun.*, 391, 68–76, 2017.
55. Abbas, M. and Qamar, S., Goos–Hänchen shift of partially coherent light fields in double quantum dots. *JOSA B*, 34, 2, 245–250, 2017.
56. Chuang, Y.-L., Qamar, S., Lee, R.-K., Goos–Hänchen shift of partially coherent light fields in epsilon-near-zero metamaterials. *Sci. Rep.*, 6, 26504, 2016.
57. Chen, C.-W. *et al.*, Optical temperature sensing based on the Goos–Hänchen effect. *Appl. Opt.*, 46, 22, 5347–5351, 2007.
58. Chen, L. *et al.*, Mechanism of giant Goos–Hänchen effect enhanced by long-range surface plasmon excitation. *J. Opt.*, 13, 3, 035002, 2011.
59. Farmani, A., Jafari, M., Miremadi, S.S., A high performance hardware implementation image encryption with AES algorithm. *Third International Conference on Digital Image Processing (ICDIP 2011)*. Vol. 8009. International Society for Optics and Photonics, 2011.
60. Chen, S. *et al.*, Observation of the Goos–Hänchen shift in graphene via weak measurements. *Appl. Phys. Lett.*, 110, 3, 031105, 2017.
61. Hu, X. and Wang, J., Ultrabroadband compact graphene–silicon TM-pass polarizer. *IEEE Photonics J.*, 9, 2, 1–10, 2017.
62. Huang, Y.Y. *et al.*, Large positive and negative lateral shifts near pseudo-Brewster dip on reflection from a chiral metamaterial slab. *Opt. Express*, 19, 2, 1310–1323, 2011.
63. Yin, X. *et al.*, Large positive and negative lateral optical beam displacements due to surface plasmon resonance. *Appl. Phys. Lett.*, 85, 3, 372–374, 2004.
64. Zahidi, Y., Redouani, I., Jellal, A., Goos–Hänchen shifts in AA-stacked bilayer graphene superlattices. *Phys. E*, 81, 259–267, 2016.

65. Farmani, A., Miri, M., Sheikhi, M.H., Design of a High Extinction Ratio Tunable Graphene on White Graphene Polarizer. *IEEE Photonics Technol. Lett.*, 30, 2, 153–156, 2017.
66. Zang, M. *et al.*, Temperature-dependent Goos–Hänchen shift in the terahertz range. *Opt. Commun.*, 370, 81–84, 2016.
67. Zeller, M.A., Cuevas, M., Depine, R.A., Critical coupling layer thickness for positive or negative Goos–Hänchen shifts near the excitation of backward surface polaritons in Otto-ATR systems. *J. Opt.*, 17, 5, 055102, 2015.
68. Farmani, A., Quantum-Dot Semiconductor Optical Amplifier: Performance and Application for Optical Logic Gates. *Majlesi J. Telecommun. Devices*, 6, 3, 2017.
69. Farmani, A. and Bahar, H.B., Hardware Implementation of 128-Bit AES Image Encryption with Low Power Techniques on FPGA to VHDL. *Majlesi J. Electr. Eng.*, 6, 4, 2012.
70. Pirzadi, M., Mir, A., Bodaghi, D., Realization of ultra-accurate and compact all-optical photonic crystal or logic gate. *IEEE Photonics Technol. Lett.*, 28, 21, 2387–2390, 2016.

16

Theoretical Study and Numerical Modeling of Graphene's Electromagnetic Response

Amanatiadis Stamatios* **and Kantartzis Nikolaos**

Department of Electrical and Computer Engineering, Aristotle University of Thessaloniki, Thessaloniki, Greece

Abstract

In this chapter, the electromagnetic properties of graphene are investigated analytically and its numerical modeling is thoroughly presented for advanced applications. Graphene is deemed an infinitesimally thin layer, addressed by its tensorial surface conductivity and extracted through Kubo formula. Initially, the properties of the reflected and transmitted plane waves, owing to an incident wave illumination of graphene, are calculated via the appropriate boundary conditions. Various interesting phenomena are observed, such as the gyrotropy and nonreciprocity at the microwave and millimeter-wave regime. Furthermore, the dyadic Green function of an infinite dimension graphene sheet is extracted along with the propagation properties of the highly confined surface plasmon polariton waves, supported onto graphene at the far-infrared spectrum. The theoretical study of the electromagnetic response is concluded by a parametric investigation that reveals the nature of graphene's exotic capabilities. However, the investigation of more demanding structures, encountered in modern applications, is not feasible analytically. To this end, advanced algorithms, based on the popular finite-difference time-domain method, are presented for the precise modeling of graphene. The latter is efficiently treated as a surface boundary condition, while validation is performed in terms of direct comparisons with the outcomes of the aforementioned theoretical study.

Keywords: Gyrotropy, surface wave, Green function, boundary conditions, electromagnetic response, numerical methods

16.1 Introduction

Carbon is one of the first and most significant elements discovered since ancient times, whose atoms can bond together in diverse ways, designated as allotropes. The most common of the latter are the amorphous carbon (atoms placed in irregular noncrystallic lattice), the graphite of hexagonal crystallic lattice, and the diamond of tetrahedral structured carbon atoms in cubic lattice. As anticipated, these differences on the formation of the allotropes result in a serious differentiation of their properties. As an example, the diamond is transparent and very hard, while the graphite is soft and opaque.

**Corresponding author*: samanati@auth.gr

Tobias Stauber (ed.) Handbook of Graphene: Volume 2, (507–548) © 2019 Scrivener Publishing LLC

Nonetheless, the preceding carbon properties are not the most impressive ones, because of graphene, the two-dimensional honeycomb-patterned carbon allotrope. Its emergence is attributed to A. Geim and K. Novoselov, who won the 2010 Nobel Prize in Physics "for groundbreaking experiments the two-dimensional material graphene." In their fundamental work [1], they isolated an one-atom-thick carbon film, assumed to be thermodynamically unstable. Throughout the years to follow, graphene earned the notable attention of the research community, due to its extraordinary features. In particular, it exceeds the diamond hardness and presents comparable mechanical, thermal, and optical traits with bulk materials, notwithstanding its infinitesimal thickness.

Furthermore, one of the most important properties of graphene is its capability to allow the propagation of strongly confined surface plasmon polariton (SPP) waves at the far-infrared spectrum [2–4]. Although these waves are, also, observed on other noble metals, such as gold and silver, the spectral band, where graphene SPP waves propagate, is unique. This fact can lead to the design of advanced devices at this practically virgin spectrum. Moreover, gyrotropic effects are revealed at lower frequencies [5], appointing graphene as a strong candidate for field detection among other conventional media.

As a consequence, it is indeed critical to study the electromagnetic response of graphene in order to reveal the nature of the abovementioned features. In this chapter, the analysis initiates via the demonstration of the finite surface conductivity of the two-dimensional material and all the interesting phenomena are theoretically interpreted, employing the appropriate boundary conditions on graphene. Additionally, numerical algorithms that model graphene accurately and effectively are presented for the design and response extraction of complex components as parts of advanced applications in the industrial and biomedical sector.

16.2 Graphene Surface Conductivity

The electromagnetic response of graphene and the consequent analysis is feasible through a macroscopic model of the two-dimensional material, such as the surface conductivity. However, the extraction of the material's properties initiates from the study of its lattice via quantum physics and Schrödinger equation. Specifically, the density of electrons on graphene surface is divided into the intrinsic, n_i, where any bias field is absent, and the extrinsic, n_e, one. The former depends on temperature T through the expression [6]

$$n_i \approx 9.5\times 10^5 T^2, \tag{16.1}$$

while the latter is influenced, mainly, by the applied electrostatic bias field E_0. Considering, a homogenous material of dielectric constant, ε_b, as a graphene substrate, where the bias field is applied, the flux displacement, D_n, is calculated as

$$D_n = \varepsilon_b E_0 = \frac{en_e}{2}. \tag{16.2}$$

Moreover, the extrinsic electron density is extracted via the specific honeycomb lattice geometry of graphene and given by [7]

$$n_e = \frac{2}{\pi(\hbar v_F)^2}\int_0^\infty E[f_d(E) - f_d(E+2\mu_c)]\mathrm{d}E, \tag{16.3}$$

where $\hbar$ is the reduced Planck constant, v_F is the Fermi velocity, corresponding to the velocity of the electrons with the highest kinetic energy (for graphene, it is equal to 9.71×10^5 m/s), $f_d(E)$ is the Fermi–Dirac distribution

$$f_d(E) = \frac{1}{1+\mathrm{e}^{(E-\mu_c)/k_B T}}, \tag{16.4}$$

and μ_c is the chemical potential. Although its exact definition is very complicated, it can be considered as the required energy to add the Nth electron at an $N - 1$ electron system. In the case of absence of any interaction between them, this energy coincides to the highest occupied energy level. The chemical potential value at the absolute zero temperature, namely $T = 0$ K, is defined as the Fermi energy and coincides to 0 eV for pure and not-biased graphene. However, the chemical potential dependency on the temperature is very weak, and thus this term is used instead of the Fermi energy in the following analysis. Also, the chemical potential can be controlled and deviate from its zero value either by chemical doping or electrostatic bias field E_0. The formula that relates the bias field to the chemical potential is described, by means of Equations (16.2) and (16.3),

$$\frac{2\varepsilon_b E_0}{e} = \frac{2}{\pi(\hbar v_F)^2}\int_0^\infty E\left[f_d(E) - f_d\left(E+2\mu_c\right)\right]\mathrm{d}E. \tag{16.5}$$

The connection of this microscopic model to the macroscopic one is achieved through the Kubo formula [8] that combines the time-independent Schrödinger equation with a time-dependent excitation in order to extract the linear response of the system. This linear response corresponds to the surface conductivity that connects the surface current distribution, $\vec{J}$, that is the macroscopic representation of the moving electrons, to the electric field, $\vec{E}$, that is the time-dependent excitation, via

$$\vec{J} = \sigma\vec{E}. \tag{16.6}$$

Note that the high-frequency excitation field, $\vec{E}$, must not be confused with the electrostatic bias one E_0. In addition, the surface conductivity of graphene depends on the electrostatic bias, due to the influence on the free electrons in contrast to the excitation field that its effect on the two-dimensional material's properties is negligible since its intensity is several orders of magnitude lower at realistic cases.

Furthermore, graphene electrons can be influenced by a magnetostatic bias field, B_0, as already mentioned. Considering graphene on the xz-plane, as depicted in Figure 16.1, the bias fields are perpendicular to its surface, namely parallel to the y-axis. For an excitation $\vec{E} = E_x\hat{x}$, graphene electrons accelerate toward the negative values of the x-axis, while

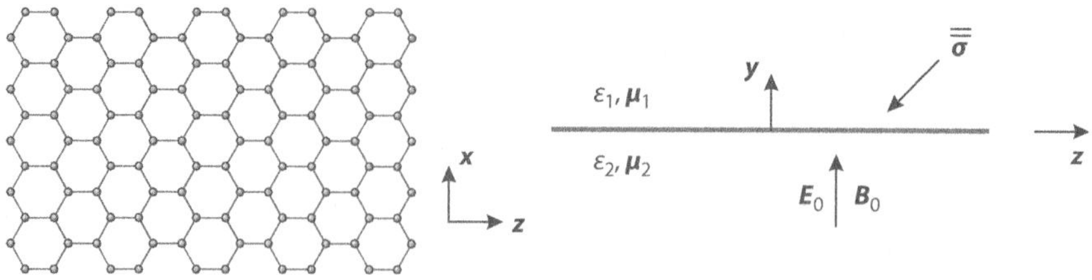

Figure 16.1 Graphene is placed on the ***xz***-plane and the bias fields $\boldsymbol{E}_0$ and $\boldsymbol{B}_0$ are perpendicular to its surface.

Lorentz forces are enabled, due to their motion and magnetostatic bias, toward the positive values of the z-axis. Hence, the resulting surface current consists of two components

$$\vec{J} = \sigma_d E_x \hat{\boldsymbol{x}} - \sigma_o E_x \hat{\boldsymbol{z}}, \tag{16.7}$$

where σ_d and σ_o are the parallel and perpendicular, to the excitation, surface conductivity terms, respectively. A similar procedure can be realized for an excitation that is parallel to z-axis, $\boldsymbol{E} = E_z \hat{\boldsymbol{z}}$, and the derived surface conductivity is

$$\vec{J} = \sigma_o E_z \hat{\boldsymbol{x}} + \sigma_d E_z \hat{\boldsymbol{z}}. \tag{16.8}$$

It is, now, obvious that the presence of both electrostatic and magnetostatic bias fields leads to a surface current distribution on graphene, estimated in term of

$$\vec{J} = \overline{\overline{\boldsymbol{\sigma}}} \vec{E}, \tag{16.9}$$

where

$$\overline{\overline{\boldsymbol{\sigma}}} = \sigma_d \left(\hat{\boldsymbol{x}}\hat{\boldsymbol{x}} + \hat{\boldsymbol{z}}\hat{\boldsymbol{z}} \right) + \sigma_o \left(\hat{\boldsymbol{x}}\hat{\boldsymbol{z}} - \hat{\boldsymbol{z}}\hat{\boldsymbol{x}} \right) \tag{16.10}$$

The surface conductivity terms, namely σ_d and σ_o, are extracted through the application of the Kubo formula [9], i.e.,

$$\sigma_d\left(\mu_c(E_0), B_0\right) = \frac{e^2 v_F^2 |B_0| (\omega - j2\Gamma)\hbar}{-j\pi}$$
$$\times \sum_{n=0}^{\infty} \left[\frac{f_d(M_n) - f_d(M_{n+1}) + f_d(-M_{n+1}) - f_d(-M_n)}{(M_{n+1} - M_n)^2 - (\omega - j2\Gamma)^2 \hbar^2} \left(1 - \frac{\Delta^2}{M_n M_{n+1}}\right) \frac{1}{M_{n+1} - M_n} \right.$$
$$\left. + \frac{f_d(-M_n) - f_d(M_{n+1}) + f_d(-M_{n+1}) - f_d(M_n)}{(M_{n+1} + M_n)^2 - (\omega - j2\Gamma)^2 \hbar^2} \left(1 + \frac{\Delta^2}{M_n M_{n+1}}\right) \frac{1}{M_{n+1} + M_n} \right] \tag{16.11}$$

$$\sigma_o(\mu_c(E_0), B_0) = \frac{e^2 v_F^2 e B_0}{\pi} \Sigma_{n=0}^{\infty}[f_d(M_n) - f_d(M_{n+1}) - f_d(-M_{n+1}) + f_d(-M_n)]$$
$$\times\left[\left(1 - \frac{\Delta^2}{M_n M_{n+1}}\right)\frac{1}{(M_{n+1} - M_n)^2(\omega - j2\Gamma)^2\hbar^2}\right. \tag{16.12}$$
$$\left. + \left(1 + \frac{\Delta^2}{M_n M_{n+1}}\right)\frac{1}{(M_{n+1} + M_n)^2 - (\omega - j2\Gamma)^2\hbar^2}\right]$$

where Γ is the scattering rate, *e* is the electron charge, Δ is the energy band gap due to the magnetostatic bias field, and

$$M_n = \sqrt{\Delta^2 + 2n v_F^2 \,|\, eB_0 \,|\, \hbar} \tag{16.13}$$

are the Landau levels. The latter correspond to the quantized energy levels of the electrons due to the application of the magnetostatic bias. Every single term at the summations corresponds to a transition of the electrons between Landau levels, as illustrated in Figure 16.2.

There are two basic transition mechanisms, the intraband and the interband, that correspond to the interactions between the same and different band (valence and conductance), respectively. The required total energy for any transition depends on the difference between the two interaction levels, while the transition possibility is proportional to the chance that the starting level is highly occupied by electrons, while the ending one is empty.

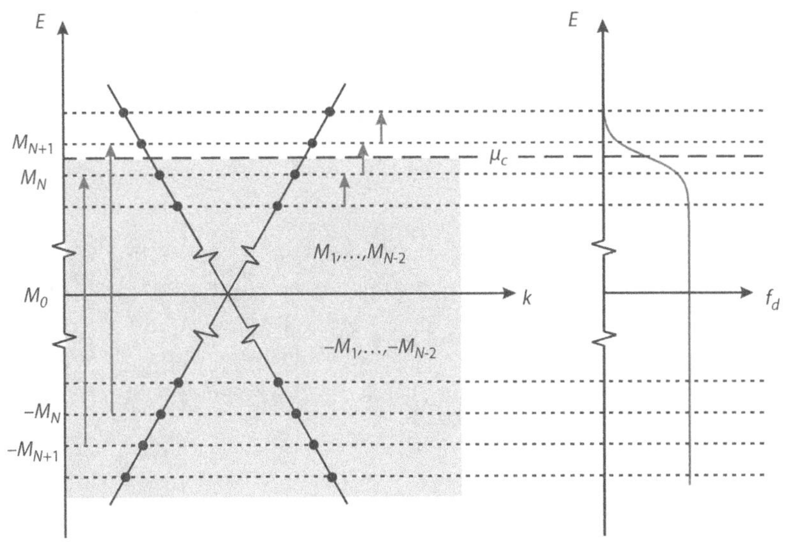

Figure 16.2 Energy dispersion diagram in magnetostatically biased graphene and representation of the Fermi–Dirac distribution. Arrows on the left and right indicate the interband and intraband electron transitions, correspondingly.

This possibility can be estimated through the Fermi–Dirac distribution, and it is maximized for transitions that cross the level of the chemical potential since $f_d(\mu_c) = 0.5$.

Let us, now, consider that the chemical potential is between levels M_n and M_{n+1}. The interband transitions of the lowest energy involve levels $-N$ and $N + 1$ or $-N - 1$ and N. However, the energy is related to the frequency through the normalized Planck constant; therefore, the interband transitions are observed at frequencies

$$\hbar\omega_{\text{inter}} \geq M_n + M_{n+1}. \tag{16.14}$$

The specific spectrum of these transitions can be clarified by assuming realistic values of graphene parameters.

First of all, the energy band gap, Δ, is significant at lower temperatures, namely below $T =$ 100 K. Nevertheless, at higher temperatures, such as the ones that are assumed at real applications, the increased density of free electrons reverses the influence of the magnetostatic field, at the lattice level, and the energy band gap is vanished. Therefore, at room temperature, the band-gap value is $\Delta = 0$. Moreover, the applied magnetic field values are relatively low, i.e., $B_0 \leq 1$ T, and the energy between two consecutive Landau levels is 0.036 eV. When the chemical potential is $\mu_c \gg 0.036$ eV, the Landau levels are approximately $M_n \approx M_{n+1} \approx \mu_c$, since $N \gg 1$, and $\hbar\omega_{\text{inter}} \geq 2\mu_c$. For real applications, μ_c is larger than 0.05 eV and the interband transitions are observable at frequencies higher than 10 THz. On the contrary, intraband transitions involve levels of the same band and they are dominant at the far-infrared regime, namely until approximately 10 THz.

Based on the prior analysis, at room temperature $T = 300$ K and a chemical potential that is significantly larger that the first Landau level $\mu_c \gg M_1$, the two surface conductivity terms are simplified to [7]

$$\sigma_d(\mu_c(E_0),B_0) = \frac{e^2(j\omega+2\Gamma)}{\pi\hbar^2}\left[\frac{1}{\omega_c^2+(j\omega+2\Gamma)^2}\int_0^{\infty} E\left(\frac{\partial f_d(-E)}{\partial E}-\frac{\partial f_d(E)}{\partial E}\right)\mathrm{d}E - \int_0^{\infty}\frac{f_d(E)-f_d(-E)}{\omega_c^2+(j\omega+2\Gamma)^2+4(E/\hbar)^2}\mathrm{d}E\right] \tag{16.15}$$

$$\sigma_o(\mu_c(E_0),B_0) = -\frac{e^2 v_F^2 e B_0}{\pi\hbar^2}\left[\frac{1}{\omega_c^2+(j\omega+2\Gamma)^2}\int_0^{\infty}\left(\frac{\partial f_d(E)}{\partial E}+\frac{\partial f_d(-E)}{\partial E}\right)\mathrm{d}E + \int_0^{\infty}\frac{1}{\omega_c^2+(j\omega+2\Gamma)^2+4(E/\hbar)^2}\mathrm{d}E\right] \tag{16.16}$$

where ω_c is the cyclotron frequency or gyrofrequency that is introduced due to the application of magnetostatic bias on graphene. This parameter describes the frequency of an electron moving perpendicular to the direction of a uniform magnetic bias field B_0 (constant magnitude and direction). The motion of the electron is circular owing to the magnetic Lorentz force and is calculated through

$$\omega_c = \frac{eB_0 v_F^2}{\mu_c}. \tag{16.17}$$

The two transition mechanisms are, now, more evident, since the first integral is attributed to the intraband and the second one to the interband transitions. Furthermore, at frequencies where the intraband contributions dominate, the surface conductivity can be further simplified. Specifically,

$$\sigma_{d,\text{intra}} = \frac{e^2 k_B T}{\pi\hbar^2}\left[\frac{\mu_c}{k_B T} + 2\ln\left(e^{-\frac{\mu_c}{k_B T}} + 1\right)\right]\frac{j\omega + 2\Gamma}{\omega_c^2 + (j\omega + 2\Gamma)^2} \tag{16.18}$$

$$\sigma_{o,\text{intra}} = \frac{e^2 k_B T}{\pi\hbar^2}\left[\frac{\mu_c}{k_B T} + 2\ln\left(e^{-\frac{\mu_c}{k_B T}} + 1\right)\right]\frac{\omega_c}{\omega_c^2 + (j\omega + 2\Gamma)^2} \tag{16.19}$$

These relations are very interesting, as there is a clear indication of the Drude model on graphene.

Throughout the previous analysis, the phenomenological scattering rate, Γ, has been introduced. It is presumed independent of energy and quantifies the quality of the graphene lattice, since it expresses the frequency that a free electron interacts with other electrons or carbon atoms. Explicitly, the higher values of Γ indicate many impurities of the lattice, as electrons encounter more obstacles and their scattering is more frequent, while lower values indicate high-quality lattices. The conversely proportional quantity is the relaxation time, τ, that can be computed via the electron density and the electron mobility, μ_s,

$$\tau = \frac{\mu_s \hbar\sqrt{(n_e + n_i)\pi}}{ev_F}. \tag{16.20}$$

The mobility depends on the electron density, too, since the larger the density, the lower the mobility. For pure graphene, typical values vary between 10^4 and 10^5 cm^2/Vs. Thus, the typical values of Γ range from 0.11 meV (excellent quality) to 10 meV (poor quality).

16.3 Electromagnetic Response on Electrically Biased Graphene

The extraction of the surface conductivity of graphene enables the deeper electromagnetic analysis of the two-dimensional material. Consequently, graphene is considered as an equivalent infinitesimally thin surface characterized via its surface conductivity at room temperature, i.e., T = 300 K, throughout the remaining chapter. Moreover, in this section, the magnetostatic bias field is considered B_0 = 0 T, in order to examine the scalar surface conductivity scenario.

16.3.1 Plane Wave Propagating through Graphene

Initially, the influence of graphene at a propagating plane wave that is incident normally on the two-dimensional material's surface is examined. The plane wave travels along the positive values of the y-axis, namely from medium 2 to 1, as illustrated in Figure 16.1. Graphene is located on the xz-plane at y = 0 and the electric field of the incident wave $\vec{E}^{inc}$ is

$$\vec{E}^{inc} = \vec{E}_0^{inc} e^{-jk_2 y}, \tag{16.21}$$

where $k_2 = \omega\sqrt{\varepsilon_2\mu_2}$ is the wavenumber of medium 2. The propagation is interrupted because of graphene and two additional plane waves are generated, the reflected $\vec{E}^{ref}$ and the transmitted $\vec{E}_0^{tran}$ ones, expressed as

$$\vec{E}^{ref} = \vec{E}_0^{ref} e^{jk_2 y}, \vec{E}^{tran} = \vec{E}_0^{tran} e^{-jk_1 y}, \tag{16.22}$$

respectively, where $k_1 = \omega\sqrt{\varepsilon_1\mu_1}$ is the wavenumber of medium 1. These three plane waves are connected through the application of the appropriate boundary conditions on the surface of graphene for both electric and magnetic field components

$$\hat{n}_0 \times (\vec{E}^{tran} - \vec{E}^{inc} - \vec{E}^{ref})\Big|_{y=0} = 0, \tag{16.23}$$

$$\hat{n}_0 \times (\vec{H}^{tran} - \vec{H}^{inc} - \vec{H}^{ref})\Big|_{y=0} = \sigma_d \vec{E}^{tran}\Big|_{y=0}, \tag{16.24}$$

where $\hat{n}_0 = \hat{y}$ is the normal vector to the material's surface. Apparently, graphene is introduced as an equivalent surface current, controlled via the surface conductivity, σ_d, while the magnetic field components are connecting to the electric ones through

$$\vec{E}^{inc} = \eta_2(\vec{H}^{inc} \times \hat{y}), \vec{E}^{ref} = -\eta_2(\vec{H}^{ref} \times \hat{y}), \vec{E}^{tran} = \eta_1(\vec{H}^{tran} \times \hat{y}), \tag{16.25}$$

where $\eta_1 = \sqrt{\mu_1/\varepsilon_1}$ and $\eta_2 = \sqrt{\mu_2/\varepsilon_2}$ are the wave impedances of the two media. It is worth to mention that the sign of the reflected wave is opposite to the others due to the different propagation direction. Inserting the electric components (16.26) in the magnetic field boundary conditions (16.25)

$$\left(-\frac{\vec{E}^{\text{tran}}}{\eta_1} + \frac{\vec{E}^{\text{inc}}}{\eta_2} - \frac{\vec{E}^{\text{ref}}}{\eta_2} \right)\Bigg|_{y=0} = \sigma_d \vec{E}^{\text{tran}}\Big|_{y=0}, \tag{16.26}$$

while at $y = 0$

$$\vec{E}_0^{\text{tran}} - \vec{E}_0^{\text{inc}} - \vec{E}_0^{\text{ref}} = 0, \tag{16.27}$$

$$-\frac{\vec{E}_0^{\text{tran}}}{\eta_1} + \frac{\vec{E}_0^{\text{inc}}}{\eta_2} - \frac{\vec{E}_0^{\text{ref}}}{\eta_2} = \sigma_d \vec{E}_0^{\text{tran}}. \tag{16.28}$$

Finally, the reflection and the transmission coefficients are extracted through the solution of Equations (16.27) and (16.28)

$$R = \left| \frac{\vec{E}_0^{\text{ref}}}{\vec{E}_0^{\text{inc}}} \right| = \frac{\eta_1 - \eta_2 - \sigma_d \eta_1 \eta_2}{\eta_1 + \eta_2 + \sigma_d \eta_1 \eta_2}, \tag{16.29}$$

$$T = \left| \frac{\vec{E}_0^{\text{tran}}}{\vec{E}_0^{\text{inc}}} \right| = \frac{2\eta_1}{\eta_1 + \eta_2 + \sigma_d \eta_1 \eta_2}. \tag{16.30}$$

The observation of these coefficients reveals that the subsequent reflected and transmitted waves present not only different amplitude, compared to the incident one, but their phase is altered, too, because of the complex nature of graphene's conductivity. Moreover, it is obvious that the coefficients are connected via $T = R + 1$.

16.3.2 Surface Plasmon Polariton Waves on Graphene

The theoretical investigation of graphene's capability to support SPP waves is realized through the determination of the electromagnetic fields due to a high-frequency excitation. To this aim, the extraction of the dyadic Green function is required. The electromagnetic fields, due to a volume current distribution, $\vec{J}_s(\vec{r}')$, near the surface of graphene on the xz-plane and $y = 0$, as shown in Figure 16.3, are estimated at any point $\vec{r}$ through the electric Hertzian potential

$$\vec{E}^{(n)}(\vec{r}) = (k_n^2 + \nabla\nabla)\vec{\boldsymbol{\pi}}^{(n)}(\vec{r}), \tag{16.31}$$

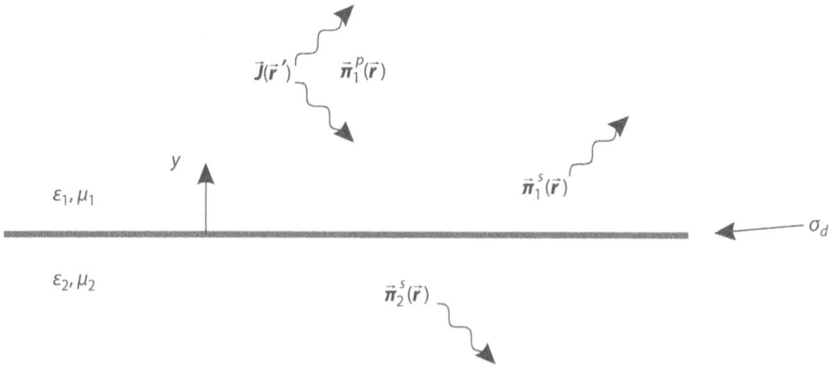

Figure 16.3 Calculation of the dyadic Green function for a graphene layer on the ***xz***-plane. The dielectric constant and the magnetic permeability of the upper space are ε_1 and μ_1, while those of the lower space are ε_2 and μ_2.

$$\vec{H}^{(n)}(\vec{r}) = j\omega\varepsilon_n \nabla \times \vec{\pi}^{(n)}(\vec{r}), \tag{16.32}$$

where k_n and $\vec{\pi}^{(n)}$ are the wavenumber and the Hertzian potential of medium n, respectively. Assuming that the current source is located in region 1, the potentials are defined as

$$\vec{\pi}^{(1)}(\vec{r}) = \vec{\pi}_1^{\,p}(\vec{r}) + \vec{\pi}_1^{\,s}(\vec{r}) = \int_{\Omega'} [\bar{g}_1^{\,p}(\vec{r},\vec{r}\,') + \bar{g}_1^{\,s}(\vec{r},\vec{r}\,')] \frac{\vec{J}s(\vec{r}\,')}{j\omega\varepsilon_1} \mathrm{d}\Omega', \tag{16.33}$$

$$\vec{\pi}^{(2)}(\vec{r}) = \vec{\pi}_2^{\,s}(\vec{r}) = \int_{\Omega'} [\bar{g}_2^{\,s}(\vec{r},\vec{r}\,')] \frac{\vec{J}s(\vec{r}\,')}{j\omega\varepsilon_1} \mathrm{d}\Omega', \tag{16.34}$$

With $\vec{\pi}_1^{\,p}(\vec{r})$, $\vec{\pi}_n^{\,s}(\vec{r})$ and $\bar{g}_1^{\,p}(\vec{r},\vec{r}\,')$, $\bar{g}_n^{\,s}(\vec{r},\vec{r}\,')$ the Hertzian potentials and the dyadic Green functions of source p (principal) and scattering waves s (scattered) on medium n, correspondingly, while Ω' is the support of the current. With y parallel to graphene interface normal, the principle dyadic Green function can be written as [10]

$$\bar{g}_1^{\,p}(\vec{r},\vec{r}\,') = \bar{I}\frac{e^{-jk_1R}}{4\pi R} = \bar{I}\frac{1}{2\pi}\int_{-\infty}^{\infty} e^{-p_1|y-y'|} \frac{\mathrm{H}_0^{(2)}(k_\rho \rho)}{4p_1} k_\rho \,\mathrm{d}k_\rho, \tag{16.35}$$

where $k_\rho = \sqrt{k_x^2 + k_z^2}$ is the radial wavenumber, $p_n^2 = k_\rho^2 - k_n^2, \rho = \sqrt{(x-x')^2 + (z-z')^2}$, $R = |\vec{r} - \vec{r}\,'| = \sqrt{(y-y')^2 + \rho^2}$, $\bar{I} = \hat{x}\hat{x} + \hat{y}\hat{y} + \hat{z}\hat{z}$ the unit dyadic and $\mathrm{H}_0^{(2)}(k_\rho \rho)$ the Hankel function of second kind and zero order. The Hankel function is utilized because of the stratified setup, under study, in order to facilitate the rest of the analysis.

The objective, now, is to determine the scattered Green functions and to this aim the appropriate boundary conditions are enforced on the surface of graphene. In detail,

$$\hat{\boldsymbol{n}}_0 \times (\vec{\boldsymbol{E}}^{(1)} - \vec{\boldsymbol{E}}^{(2)})|_{y=0} = 0, \tag{16.36}$$

$$\hat{\boldsymbol{n}}_0 \times (\vec{\boldsymbol{H}}^{(1)} - \vec{\boldsymbol{H}}^{(2)})|_{y=0} = \vec{\boldsymbol{J}}|_{y=0} = \sigma_d \vec{\boldsymbol{E}}_{\mathrm{t}}^{(1)}|_{y=0}. \tag{16.37}$$

The normal, to graphene surface, vector is $\hat{\boldsymbol{n}}_0 = \hat{\boldsymbol{y}}$, while subscript t indicates the tangential, to graphene, component of the electric field. Thus, the boundary conditions for all the components are summarized to

$$E_{1,\alpha}(y = 0^+) = E_{2,\alpha}(y = 0^-),\ \alpha = x, z, \tag{16.38}$$

$$H_{2,x}(y = 0^-) - H_{1,x}(y = 0^+) = \sigma_d E_{1,z}(y = 0^+), \tag{16.39}$$

$$H_{2,z}(y = 0^-) - H_{1,z}(y = 0^+) = -\sigma_d E_{1,x}(y = 0^+), \tag{16.40}$$

The boundary conditions on Hertzian potentials at $(x, y = 0, z)$ are easily calculated via Equations (16.31) and (16.32)

$$\pi_{1,\alpha} = N^2 M^2 \pi_{2,\alpha}, \quad \alpha = x, z, \tag{16.41}$$

$$\varepsilon_1 \pi_{1,y} - \varepsilon_2 \pi_{2,y} = \frac{\sigma_d}{j\omega} \nabla \pi_1, \tag{16.42}$$

$$\varepsilon_2 \frac{\partial \pi_{2,\alpha}}{\partial y} - \varepsilon_1 \frac{\partial \pi_{1,\alpha}}{\partial y} = \frac{\sigma_d}{j\omega} k_1^2 \pi_{1,\alpha}, \quad \alpha = x, z, \tag{16.43}$$

$$\frac{\partial \pi_{1,y}}{\partial y} - \frac{\partial \pi_{2,y}}{\partial y} = (1 - N^2 M^2)\left(\frac{\partial \pi_{2,x}}{\partial x} + \frac{\partial \pi_{2,z}}{\partial z}\right), \tag{16.44}$$

where $N^2 = \frac{\varepsilon_2}{\varepsilon_1}$ and $M^2 = \frac{\mu_2}{\mu_1}$. The appropriate representation of the Hertzian potentials is critical, since the following analysis can be simplified considerably. To this end, the scattered Hertzian potentials are similar to the principal one [11]

$$\pi_{1,\alpha}^p = \int_{-\infty}^{\infty} \frac{\mathrm{H}_0^{(2)}(k_\rho \rho)}{4 p_1} k_\rho \,\mathrm{d}k_\rho \int_{\Omega'} \mathrm{e}^{-p_1 |y - y'|} \frac{J_{s,\alpha}(\vec{\boldsymbol{r}}')}{j\omega\varepsilon_1} \mathrm{d}\Omega', \quad \alpha = x, y, z, \tag{16.45}$$

$$\pi_{1,\alpha}^s = A_\alpha(\vec{\boldsymbol{r}}') \int_{-\infty}^{\infty} \mathrm{e}^{-p_1 y} \frac{\mathrm{H}_0^{(2)}(k_\rho \rho)}{4 p_1} k_\rho \,\mathrm{d}k_\rho, \quad \alpha = x, y, x, \tag{16.46}$$

$$\pi^{s}_{2,\alpha} = B_\alpha(\vec{r}')\int_{-\infty}^{\infty} \mathrm{e}^{p_2 y}\frac{\mathrm{H}_0^{(2)}(k_\rho\rho)}{4p_1}k_\rho\,\mathrm{d}k_\rho,\quad \alpha = x,y,z. \tag{16.47}$$

The index of term p_n depends on the propagating medium, while its sign is determined through the propagation direction. Moreover, the unknown terms $A_\alpha(\vec{r}')$ and $B_\alpha(\vec{r}')$ can be expressed as

$$A_\alpha(\vec{r}') = R_\beta\int_{\Omega'}\frac{\mathrm{e}^{-p_1 y'}}{j\omega\varepsilon_1}J_{s,\alpha}(\vec{r}')\mathrm{d}\Omega' = R_\beta V_\alpha(\vec{r}'), \tag{16.48}$$

$$B_\alpha(\vec{r}') = T_\beta\int_{\Omega'}\frac{\mathrm{e}^{-p_1 y'}}{j\omega\varepsilon_1}J_{s,\alpha}(\vec{r}')\mathrm{d}\Omega' = T_\beta V_\alpha(\vec{r}'), \tag{16.49}$$

where the unknowns are, now, the scalar terms R_β and T_β. Index β depends on the direction of the source vector, relatively to graphene, and $\beta = n$ is for the normal component, $\beta = t$ for the tangential, and $\beta = c$ for the conjugation between the tangential and the normal. Specifically, in the case that the source vector is tangential to graphene, namely $\alpha = x, z$, the Hertzian potential presents not only a tangential component but a normal, too. Then, the boundary conditions (16.41)–(16.44) are applied and the unknown scalar terms are calculated

$$R_t = \frac{M^2 p_1 - p_2 - j\sigma_d\omega\mu_2}{M^2 p_1 + p_2 + j\sigma_d\omega\mu_2} = \frac{N^H(k_\rho,\omega)}{Z^H(k_\rho,\omega)}, \tag{16.50}$$

$$R_n = \frac{N^2 p_1 - p_2 + \dfrac{\sigma_d p_1 p_2}{j\omega\varepsilon_1}}{N^2 p_1 + p_2 + \dfrac{\sigma_d p_1 p_2}{j\omega\varepsilon_1}} = \frac{N^E(k_\rho,\omega)}{Z^E(k_\rho,\omega)}, \tag{16.51}$$

$$R_c = \frac{2p_1\left(N^2M^2 - 1 + \dfrac{\sigma_d p_2 M^2}{j\omega\varepsilon_1}\right)}{Z^E Z^H}, \tag{16.52}$$

$$T_t = \frac{1+R_t}{N^2M^2} = \frac{2p_1}{N^2 Z^H}, \tag{16.53}$$

$$T_n = \frac{p_1(1-R_n)}{p_2} = \frac{2p_1}{Z^E}, \tag{16.54}$$

$$T_c = \frac{2p_1\left(N^2M^2 - 1 + \dfrac{\sigma_d p_1}{j\omega\varepsilon_1}\right)}{N^2 Z^H Z^E}. \tag{16.55}$$

Finally, the dyadic Green functions of the scattered waves are

$$\overline{g}_1^s(\vec{r},\vec{r}') = \hat{y}\hat{y}g_{1,n}^s(\vec{r},\vec{r}') + (\hat{y}\hat{x}\frac{\partial}{\partial x} + \hat{y}\hat{z}\frac{\partial}{\partial z})g_{1,c}^s(\vec{r},\vec{r}') + (\hat{x}\hat{x} + \hat{z}\hat{z})g_{1,t}^s(\vec{r},\vec{r}'), \tag{16.56}$$

$$\overline{g}_2^s(\vec{r},\vec{r}') = \hat{y}\hat{y}g_{2,n}^s(\vec{r},\vec{r}') + (\hat{y}\hat{x}\frac{\partial}{\partial x} + \hat{y}\hat{z}\frac{\partial}{\partial z})g_{2,c}^s(\vec{r},\vec{r}') + (\hat{x}\hat{x} + \hat{z}\hat{z})g_{2,t}^s(\vec{r},\vec{r}'), \tag{16.57}$$

where the resulting Sommerfeld integrals are expressed

$$g_{1,\beta}^s(\vec{r},\vec{r}') = \frac{1}{2\pi}\int_{-\infty}^{\infty} R_\beta \frac{\mathrm{H}_0^{(2)}(k_\rho\rho)\mathrm{e}^{-p_1(y+y')}}{4p_1} k_\rho\,\mathrm{d}k_\rho, \quad \beta = t,n,c, \tag{16.58}$$

$$g_{2,\beta}^s(\vec{r},\vec{r}') = \frac{1}{2\pi}\int_{-\infty}^{\infty} T_\beta \frac{\mathrm{H}_0^{(2)}(k_\rho\rho)\mathrm{e}^{p_2 y}\,\mathrm{e}^{-p_1 y'}}{4p_1} k_\rho\,\mathrm{d}k_\rho \quad \beta = t,n,c. \tag{16.59}$$

As in the case of a simple dielectric interface, the denominators $Z^{H,E}(k_\rho,\omega)$, in Equations (16.50)–(16.55), implicate pole singularities in the spectral plane associated with surface wave phenomena. Furthermore, both wave parameters $p_n^2 = k_\rho^2 - k_n^2, n = 1,2$, lead to branch points at $k_\rho = \pm k_n$, and thus the k_ρ-plane is a four-sheeted Riemann surface. The standard hyperbolic branch cuts [12] that separate the one proper sheet (where $\mathrm{Re}\{p_n\} > 0$, such that the radiation condition as $y \to \infty$ is satisfied) and the three improper sheets (where $\mathrm{Re}\{p_n\} > 0$) are the same as in the absence of surface conductivity σ_d.

The Sommerfeld integrals (16.58) and (16.59) of the scattered fields, due to graphene's presence, implicate pole singularities, mainly via R_β and T_β. These singularities represent propagating surface waves on graphene surface, and there are two basic kinds, the transverse magnetic (TM) and the transverse electric (TE) ones. The dispersion equation of the surface waves of the first kind, which are also known as H-waves, is extracted via

$$Z^H(k_\rho,\omega) = M^2 p_1 + p_2 + j\sigma_d\omega\mu_2 = 0 \tag{16.60}$$

while for the TE waves, which are also characterized as E-waves, through

$$Z^E(k_\rho,\omega) = N^2 p_1 + p_2 + \frac{\sigma_d p_1 p_2}{j\omega\varepsilon_1} = 0. \tag{16.61}$$

The surface wave field can be obtained from the residue contribution of the Sommerfeld integrals [13]. As an example, considering a point source $\vec{J}_s = A_0\delta(x)\delta(y)\delta(z)\hat{y}$, where is the amplitude, the Hertzian potentials (16.33) and (16.34) are calculated

$$\vec{\pi}^{(1)}(\vec{r}) = -\frac{A_0 k_\rho^2 R_n'}{4\omega\varepsilon_1} e^{-p_1 y} \frac{H_0^{(2)}(k_\rho \rho_0)}{k_\rho\sqrt{k_1^2 - k_\rho^2}} \hat{y}, \tag{16.62}$$

$$\vec{\pi}^{(2)}(\vec{r}) = -\frac{A_0 k_\rho^2 T_n'}{4\omega\varepsilon_1} e^{p_2 y} \frac{H_0^{(2)}(k_\rho \rho_0)}{k_\rho\sqrt{k_1^2 - k_\rho^2}} \hat{y}, \tag{16.63}$$

where $\rho_0 = \sqrt{x^2 + z^2}$ and

$$R_n' = \frac{N^E}{\dfrac{\partial Z^E}{\partial k_\rho}} = \frac{N^2 p_1 - p_2 + \dfrac{\sigma_d p_1 p_2}{j\omega\varepsilon_1}}{\dfrac{N^2}{p_1} + \dfrac{1}{p_2} + \dfrac{\sigma_d}{j\omega\varepsilon_1}\left(\dfrac{p_1^2 + p_2^2}{p_1 p_2}\right)}, \tag{16.64}$$

$$T_n' = \frac{2p_1}{\partial Z^E/\partial k_\rho} = \frac{2p_1}{\dfrac{N^2}{p_1} + \dfrac{1}{p_2} + \dfrac{\sigma_d}{j\omega\varepsilon_1}\left(\dfrac{p_1^2 + p_2^2}{p_1 p_2}\right)}. \tag{16.65}$$

The electromagnetic fields are now evaluated via Equations (16.31) and (16.32)

$$\vec{E}^{(1)}(\vec{r}) = \frac{A_0 k_\rho^2 R_n'}{4\omega\varepsilon_1} e^{-p_1 y}\left[-j\left(\frac{x}{\rho_0}\hat{x} + \frac{z}{\rho_0}\hat{z}\right) H_1^{(2)}(k_\rho\rho_0) + \frac{(k_1^2 + p_1^2) H_0^{(2)}(k_\rho\rho_0)}{k_\rho\sqrt{k_1^2 - k_\rho^2}}\hat{y}\right], \tag{16.66}$$

$$\vec{E}^{(2)}\vec{r} = \frac{A_0 k_\rho^2 T_n'}{4\omega\varepsilon_1} e^{p_2 y}\left[-j\left(\frac{x}{\rho_0}\hat{x} + \frac{z}{\rho_0}\hat{z}\right) H_1^{(2)}(k_\rho\rho_0) + \frac{(k_2^2 + p_2^2) H_0^{(2)}(k_\rho\rho_0)}{k_\rho\sqrt{k_1^2 - k_\rho^2}}\hat{y}\right]. \tag{16.67}$$

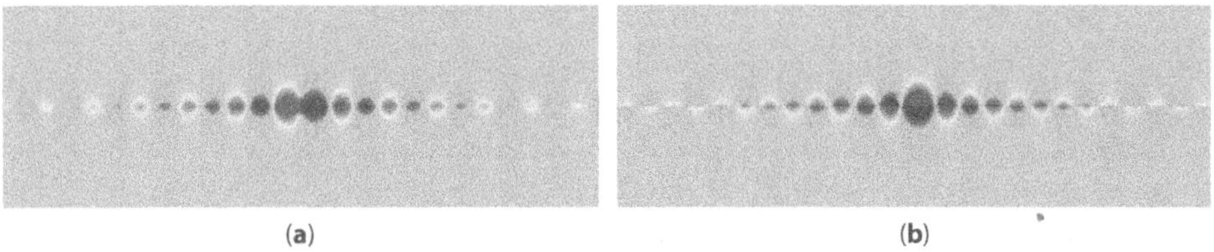

Figure 16.4 Distribution of (a) tangential and (b) normal, to graphene surface, electric field components for a TM surface wave propagation.

The exponential terms, as discussed previously, must satisfy the radiation condition $y \to \infty$ in order to observe physical phenomena, thus $\mathrm{Re}\{p_n\} > 0$, $n = 1,2$. The electric field distribution of a typical TM surface wave on graphene is presented in Figure 16.4, which reveals the strong confinement to the two-dimensional material's surface.

The propagating surface waves on graphene are characterized as surface plasmon polariton (SPP) modes, similarly to the noble metals, such as gold and silver, at the near-infrared and visible spectrum. These surface plasmons are the motion of the free electrons at the interface between the two dielectrics of different sign of dielectric constant [14]. This phenomenon is observed in nature, principally on the interface between a dielectric and a noble metal at the aforesaid frequencies, as the latter approaches its plasma resonance, since it is approximated through the Drude model. Graphene presents similar behavior and is capable to support surface plasmons, yet at lower frequencies, i.e., the far-infrared regime. When surface plasmons are coupled to an external excitation, the propagating electromagnetic fields are characterized as surface plasmon polariton waves or modes. Their basic properties are

- Their propagation is feasible on the interface between two materials.
- The resulting wave is strongly confined on the interface and the intensity of the electromagnetic field is particularly strong.
- The wavelength and the propagation velocity are significantly decreased compared to the free space propagation, and thus they are characterized as slow waves.

The features that totally characterize the SPP waves are the wavelength, λ_{spp}, the propagation length, L_{spp}, which is defined as the distance along the surface for the SPP intensity to decay by a factor of $1/e$, and the confinement, ζ, denoted as the distance from the surface that the intensity of the SPP decay by a factor of $1/e$. These features can be calculated through the complex wavenumber k_ρ

$$\lambda_{\mathrm{spp}} = \frac{2\pi}{\mathrm{Re}(k_\rho)}, \tag{16.68}$$

$$L_{\mathrm{spp}} = -\frac{1}{2\mathrm{Im}(k_\rho)}, \tag{16.69}$$

$$\zeta = \frac{1}{\mathrm{Re}\left(\sqrt{k_\rho^2 - k_n^2}\right)}, \tag{16.70}$$

where k_n is the wavenumber of medium n. Finally, the ratio of the complex wavenumber to the free space one is defined as the effective refractive index $n_{\mathrm{eff}} = k_\rho / k_0$. This parameter is dimensionless and provides an intuitive perception of the surface wave propagation. Therefore, it is critical to extract analytically the complex wavenumber.

Focusing on graphene, the two kinds of SPP waves are studied separately. Initially, resolving the dispersion equation of the transverse electric SPP waves (16.60), the complex wavenumber is

$$k_\rho = k_0 \sqrt{\mu_{r1}\varepsilon_{r1} - \frac{1}{(M^4-1)^2}\left(M^2\sigma_d\eta_0\mu_{r2} \mp \sqrt{(\sigma_d\eta_0\mu_{r2})^2 - (M^4-1)(\varepsilon_{r1}\mu_{r1} - \varepsilon_{r2}\mu_{r2})}\right)^2}, \tag{16.71}$$

where ε_{rn} and μ_{rn} are the relative dielectric constant and magnetic permeability, respectively, of medium n. This relation is simplified in the case that the two media present identical magnetic properties, namely $\mu_{r1} = \mu_{r2} = \mu_r$ and $M = 1$,

$$k_\rho = k_0 \sqrt{\mu_r\varepsilon_{r1} - \left(\frac{(\varepsilon_{r1} - \varepsilon_{r2})\mu_r + \sigma_d^2\eta_o^2\mu_r^2}{2\sigma_d\eta_o\mu_r}\right)^2}. \tag{16.72}$$

It is further simplified when the electrical properties are also identical, namely $\varepsilon_{r1} = \varepsilon_{r2} = \varepsilon_r$ and $N = 1$,

$$k_\rho = k_0 \sqrt{\mu_r\varepsilon_r - \left(\frac{\sigma_d\eta_o\mu_r}{2}\right)^2}. \tag{16.73}$$

while, for a suspended graphene layer (standing on the free space, where $\varepsilon_r = \mu_r = 1$), it becomes

$$k_\rho = k_0 \sqrt{1 - \left(\frac{\sigma_d\eta_o}{2}\right)^2}. \tag{16.74}$$

Many useful remarks can be extracted by analyzing the final expression, where $\sigma_d = \sigma_d' + j\sigma_d''$ ($\sigma_d' = \mathrm{Re}\{\sigma_d\}$ and $\sigma_d'' = \mathrm{Im}\{\sigma_d\}$ are the real and imaginary parts, accordingly). First of all, when a pure real conductivity exists (at lower temperatures and chemical potential values), the $p_0 = \sqrt{k_\rho^2 - k_0^2} = -j\sigma_d''\frac{\omega\mu_o}{2}$ term is imaginary and the condition $\mathrm{Re}\{p_n\}>0$ is not satisfied; consequently, there is not any propagating surface wave. On the other hand, when the surface conductivity term had a positive imaginary value $\sigma_d = j\sigma_d''$, where $\sigma_d''>0$, the complex wavenumber is larger than the free space ($p_0 = \sigma_d'' \frac{\omega\mu_o}{2} > 0$) and as a result a TE SPP wave can propagate. This condition appeared at frequencies larger than approximately 10 THz where the interband contribution dominates. However, at lower frequencies, where intraband term is significant, the conductivity is a negative imaginary number and $p_0 < 0$, violating the radiation condition. In the general case, where the surface conductivity is complex valued, $\sigma_d = \sigma_d' + j\sigma_d''$ results in

$$p_0 = (\sigma_d'' - j\sigma_d') \frac{\omega\mu_0}{1} \tag{16.75}$$

and the TE surface wave propagates at frequencies over the far-infrared region where. The confinement ζ^{TE} of this SPP mode in a homogeneous material is estimated

$$\zeta^{TE} = \frac{2}{\sigma_d'' \omega\mu}. \tag{16.76}$$

Additionally, the complex wavenumber of the transverse magnetic SPP waves are extracted by resolving Equation (16.61). Assuming, the general scenario of graphene inside media of different electric and magnetic properties, a quadrant equation appears

$$\frac{\varepsilon_{r1}}{\sqrt{\left(\frac{k_\rho}{k_0}\right)^2 - \varepsilon_{r1}\mu_{r1}}} + \frac{\varepsilon_{r2}}{\sqrt{\left(\frac{k_\rho}{k_0}\right)^2 - \varepsilon_{r2}\mu_{r2}}} = j\sigma_d\eta_0. \tag{16.77}$$

This equation is greatly simplified, since graphene is a homogeneous medium ($\varepsilon_{r1} = \varepsilon_{r2} = \varepsilon_r$ and $\mu_{r1} = \mu_{r2} = \mu_r$)

$$k_\rho = k_0 \sqrt{\varepsilon_r\mu_r - \left(\frac{2\varepsilon_r}{\sigma_d\eta_0}\right)^2}. \tag{16.78}$$

Finally, the special case of a suspended graphene layer

$$k_\rho = k_0 \sqrt{1 - \left(\frac{2}{\sigma_d\eta_0}\right)^2}, \tag{16.79}$$

while, presuming the complex valued surface conductivity $\sigma_d = \sigma_d' + j\sigma_d''$,

$$p_0 = \frac{-2\omega\varepsilon_0}{|\sigma_d|^2}\left(\sigma''_d + j\sigma'_d\right). \tag{16.80}$$

As in the TE SPP waves, when the conductivity is purely real, the surface wave is absent, since the condition Re$\{p_n\} > 0$ is not satisfied. Similarly, at frequencies beyond the far-infrared regime, where the interband contribution dominates, the radiation condition is violated and thus no TM surface waves appear. Nonetheless, at lower frequencies, the imaginary part of the surface conductivity is negative due to the intraband contribution, and a TM SPP wave appears presenting strong confinement

$$\zeta^{TM} = -\frac{|\sigma_d|^2}{2\omega\varepsilon\sigma''_d}. \tag{16.81}$$

Summarizing, transverse electric SPP waves are propagating beyond the far-infrared frequencies, while the propagation of transverse magnetic ones appears at lower spectrum. Moreover, at lower temperatures, where the conductivity is real, surface waves of any kind are absent.

16.3.3 Parametric Analysis of the Propagating Waves on Graphene

The analytical extraction of the reflection and transmission coefficients of a propagating plane wave toward graphene as well as the propagation properties of the SPP waves on graphene are crucial for their characterization and categorization. However, the true nature of these phenomena is perceived via the thorough investigation concerning graphene parameters, such as the chemical potential and the scattering rate, as well as the universal parameters, namely the frequency and the surrounding media electromagnetic properties.

16.3.3.1 Frequency Response

Initially, the parametric analysis focuses on graphene's frequency response. Typical and realistic parameters of the two-dimensional material are selected; specifically the carrier density is 10^{12} electrons/cm^2 and the mobility μ_s = 90000 cm^2/Vs resulting in a chemical potential μ_c = 0.1 eV, while a high-quality lattice (Γ = 0.33 meV) is employed. The surface conductivity is calculated through Equation (16.15), since the magnetic bias is absent, at an extremely wide spectrum, from the microwave to the visible regime. The frequency response of graphene's surface conductivity is sketched in Figure 16.5 that is separated in two subregions. The first one includes microwave, millimeter-wave, and far-infrared frequencies (Figure 16.5a), where the intraband contributions are dominant and the surface conductivity can be estimated through the popular Drude model. Additionally, the imaginary part is always negative. On the other hand, the second subregion covers the remaining infrared and the visible regime (Figure 16.5b), where interband transitions prevail, mainly over 50 THz. Here, the conductivity imaginary part is turned positive, while at even higher frequencies approximates the zero value, indicating that graphene acts as a low-loss

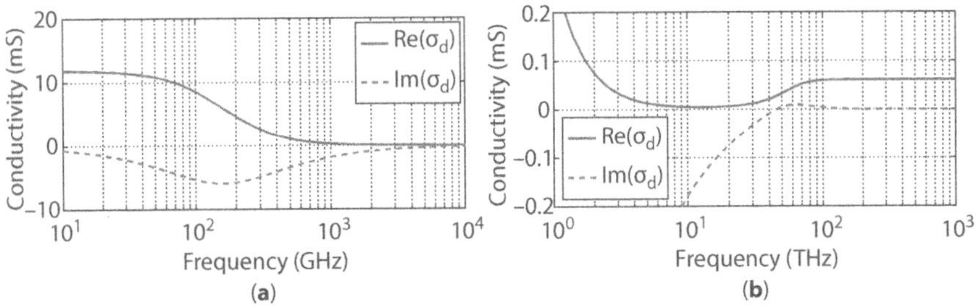

Figure 16.5 Frequency response of graphene's surface conductivity when (a) intraband and (b) interband electron transitions dominate.

conductor at the visible spectrum. Consequently, the remarks of Section 16.3.2, concerning the imaginary part behavior, are observed in this analysis, too.

Moreover, the imaginary part is proven to be critical for the characterization of the SPP waves on graphene. The frequency response of their propagation properties is extracted through Equations (16.68)–(16.70) and depicted in Figure 16.6. In the first region (Figure 16.6a), where the intraband contributions prevail, the imaginary part of the surface conductivity is negative, and the resulting surface wave is transverse magnetic; thus, the complex wavenumber is calculated via Equation (16.79). At lower frequencies, microwave and the majority of the millimeter-wave regime, the SPP wave is weakly confined and basically it approximates the free-space propagation. However, as the frequency approaches the far-infrared spectrum, the slow propagating TM SPP wave appears, while after 2 THz, the propagation is optimized, since the propagation length is larger than the wavelength.

This performance is extended in the second subregion (Figure 16.6b), until the point that the interband transitions are significant. At this point, the propagation length is severely decreased indicating high losses, while after 50 THz, the imaginary part of the conductivity is turned positive. Now, the complex wavenumber is estimated through Equation (16.74) and the resulting surface wave is transverse electric. However, the TE SPP wave on graphene is very weakly confined on the surface, while at even higher frequencies, namely the visible spectrum, it completely vanishes. Therefore, in contrast to the noble metals, graphene supports the propagation of highly confined SPP waves mainly at the far-infrared regime.

However, some more interesting features are observed in Figure 16.7, where the reflection and transmission coefficients of a normally incident plane wave are depicted. Focusing on

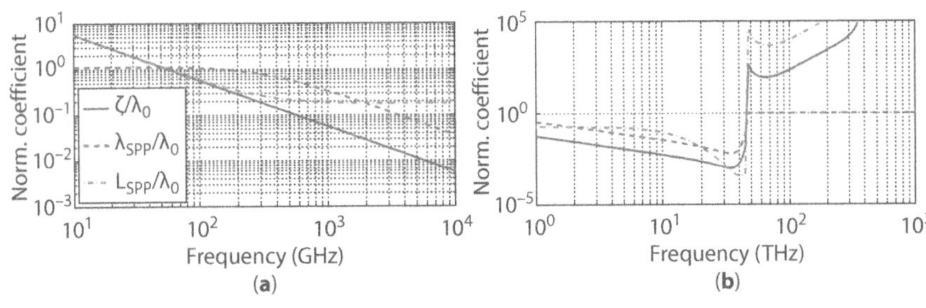

Figure 16.6 Normalized, to the free space wavelength, propagation properties of the SPP waves on graphene when (a) intraband and (b) interband electron transitions dominate.

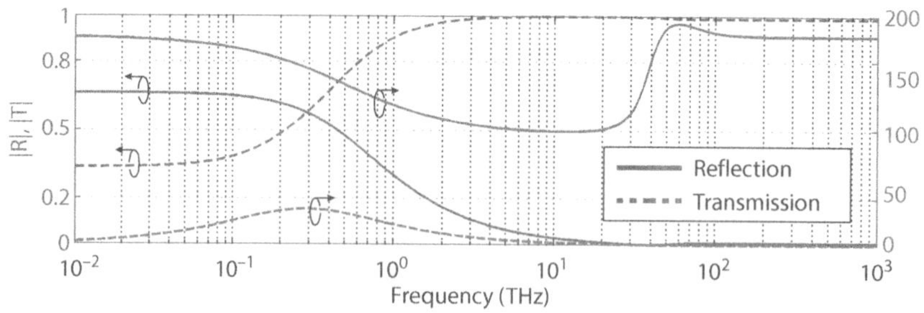

Figure 16.7 Frequency response of reflection and transmission coefficients of a normally incident plane wave on graphene.

the transmitted wave, it is noticed that its coefficient is increased at higher frequencies, while at the millimeter-wave region, it presents a phase difference of approximately 40°. On the other hand, the reflected wave has a more complicated behavior, excluding the reduction at higher frequencies. Specifically, the phase difference approximates a perfect conductor in the entire spectrum, apart from the far-infrared regime where it drops to 90°.

16.3.3.2 Chemical Potential Response

Furthermore, the electrostatic bias field is able to alter the carrier density on graphene, and thus the chemical potential is influenced. The response of the surface conductivity, due to the chemical potential alteration, is calculated via Equation (16.18) and illustrated in Figure 16.8, at 100 GHz and 30 THz for a scattering rate $\Gamma = 0.33$ meV. Observing Figure 16.8a, namely at 100 GHz, the interband contribution is several orders of magnitude even at the lower values of the chemical potential. Nevertheless, at the higher frequency, Figure 16.8b at 30 THz, the intraband transitions are dominant for larger absolute values of the chemical potential, while the interband are more significant for lower absolute values. Note that the dotted lines correspond in $2|\mu_c| = \hbar\omega$ that is near to the imaginary part change of sign.

This last observation is extremely useful for the characterization of the resulting SPP wave on graphene. The propagation properties are calculated via Equations (16.68)–(16.70) and sketched in Figure 16.9 for the aforementioned frequencies. Specifically, at the millimeter-wave spectrum, e.g., Figure 16.9a at 100 GHz, a TM SPP wave is propagating on

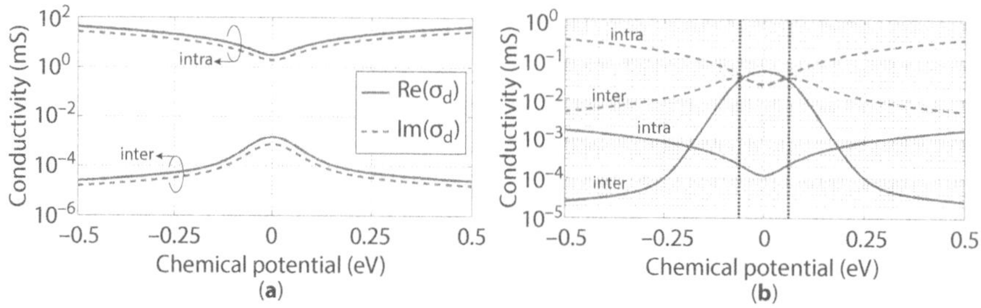

Figure 16.8 Graphene surface conductivity versus the chemical potential at (a) 100 GHz and (b) 30 THz. The indications *intra* and *inter* refer to intraband and interband transitions, respectively.

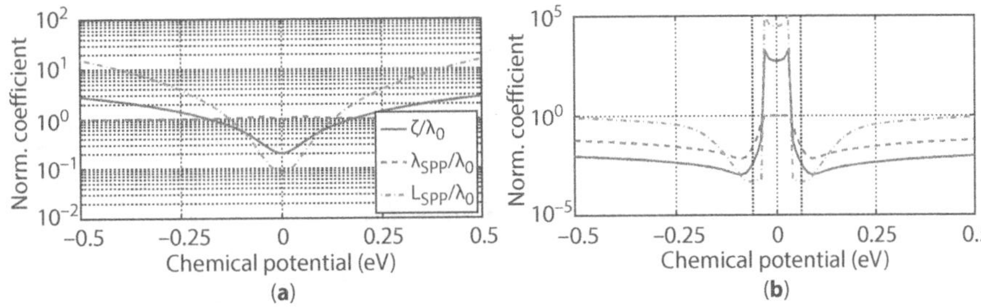

Figure 16.9 Normalized, to the free-space wavelength, propagation properties of the SPP waves on graphene versus the chemical potential at (a) 100 GHz and (b) 30 THz. The indications *intra* and *inter* refer to intraband and interband transitions, respectively.

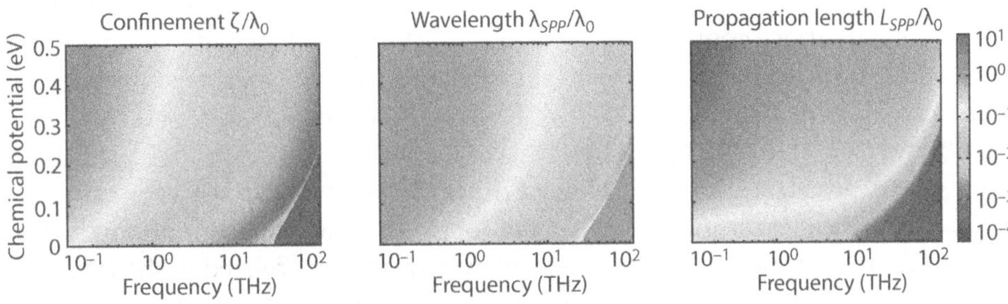

Figure 16.10 Normalized, to the free-space wavelength, propagation properties of the SPP waves on graphene versus the chemical potential and the frequency.

graphene that is more confined on its surface for the lowest absolute values of the chemical potential. On the other hand, in Figure 16.9b at 30 THz, the surface wave is retained as a strongly confined TM SPP wave at the larger absolute values of the chemical potential. However, at the region $2|\mu_c| = \hbar\omega$, the SPP wave is transformed to a weakly confined TE wave of similar propagation properties to a free-space propagating one.

The previous analysis revealed, also, two basic points. The first one refers to the even symmetry of the chemical potential, while the second one indicates the significant influence of the frequency and the chemical potential to graphene SPP propagation properties. Thus, the combined investigation is realized through the extraction of Figure 16.10, where the negative values of chemical potential are neglected due to the abovementioned even symmetry. Obviously, the slowest and strongest confined SPP waves on graphene are the TM ones and at higher frequencies and lower chemical potential values. Similar findings are extracted via the observation of the propagation length, that is, also, optimized at the larger chemical potential values. Additionally, at the lower right region at the diagrams, where the interband transitions are more evident, the propagation losses are severely increased, until the weakly confined TE SPP wave appears.

16.3.3.3 Scattering Rate Response

The scattering rate parametric investigation is, also, important, since it is related to graphene lattice quality. To this aim, the SPP propagation characteristics are depicted in Figure 16.11 for different values of chemical potential and at 2 THz, where the performance of the SPP

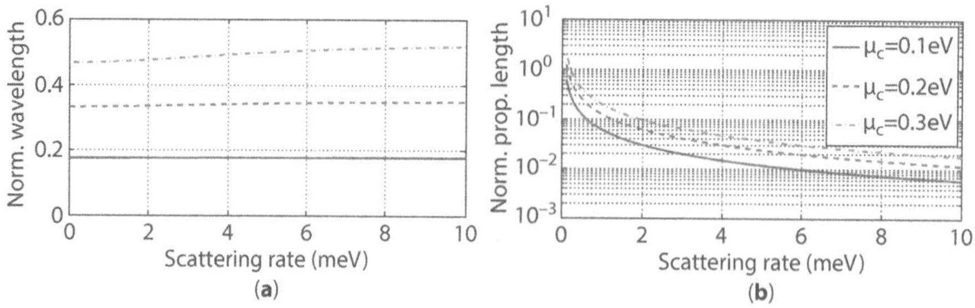

Figure 16.11 Normalized, to the free-space wavelength, (a) wavelength and (b) propagation length of graphene SPP wave versus the scattering rate at 2 THz for different chemical potential values.

waves is enhanced. Note that the confinement is absent, since it is not affected by the scattering rate, while the surface conductivity is neglected, although altered, due to the nonpractical conclusions. Initially, the influence on the SPP wavelength is increased for larger values of the chemical potential. Specifically, the alteration is almost 1% for $\mu_c = 0.1$ eV, while for $\mu_c = 0.2$ eV and $\mu_c = 0.3$ eV, this rate is increased to 5% and 10%, respectively. Additionally, focusing on the propagation length (Figure 16.11b), the influence of the scattering rate is apparent. Specifically, as the scattering rate increases, the propagation length degrades drastically. This observation is expected, since this parameter is related to the lattice quality and a poor one leads to increased impurities, i.e., propagation losses. Additionally, as the chemical potential increases, the propagation length is proportional, due to the inherent larger SPP wavelength, though. Summarizing, the scattering rate has a significant effect on the SPP propagation length that is drastically decreased at larger Γ values, while the SPP wavelength can be affected for larger chemical potential values.

16.3.3.4 *Effect of the Adjacent Media*

Finally, a parametric analysis is conducted for the surrounding, to graphene, media in terms of their electromagnetic properties. The previous analysis considered a suspended graphene layer, namely graphene standing in free space that is an unrealistic setup. To this end, the propagation characteristics of a propagating SPP wave on graphene surrounded by general electromagnetic media are depicted in Figure 16.12 (surface conductivity is neglected, as

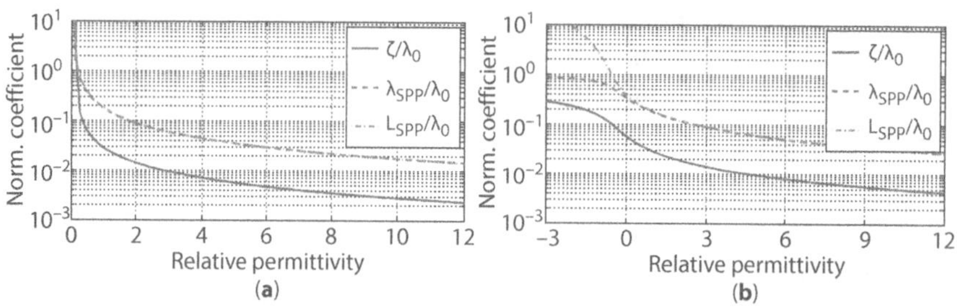

Figure 16.12 Normalized, to the free-space wavelength, propagation properties of the SPP waves on graphene of $\mu_c = 0.1$, $\Gamma = 0.33$ meV at 2 THz, propagating in (a) a homogeneous medium and (b) free-space and a dielectric substrate.

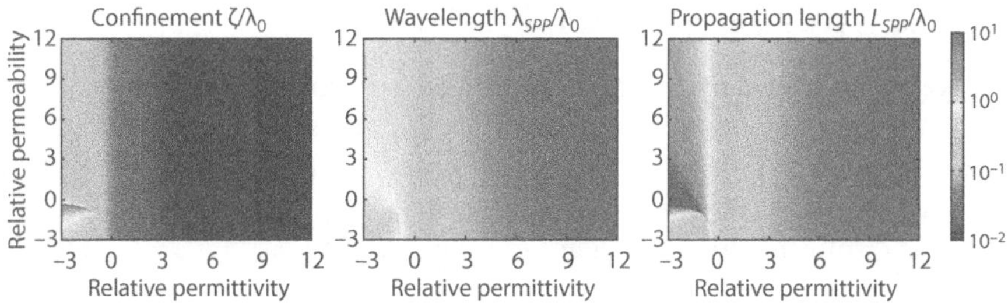

Figure 16.13 Normalized, to the free-space wavelength, propagation properties of the SPP waves on graphene versus the substrate electric and magnetic properties.

it is not affected straightforwardly) [15]. First of all, the case of a homogeneous medium is examined in Figure 16.12a, extracting the complex wavenumber through Equation (16.78) for selected graphene parameters $\mu_c = 0.1$ eV and $\Gamma = 0.33$ meV at 2 THz. Obviously, the surface wave is confined stronger on graphene presenting decreased wavelength and propagation length as the relative permittivity is increased. Note that the propagation characteristics are normalized to the free-space wavelength in order to be straightforwardly comparable.

Also, the realistic case of a propagating surface wave on graphene that has a dielectric substrate and air above is examined in Figure 16.12b. The SPP propagation properties are extracted via the correct branch of Equation (16.77). Although the behavior is analogous to the previous homogeneous case, the influence of the dielectric is weaker. However, materials with negative dielectric constant are, also, examined revealing that moderately confined SPP waves can propagate that present wavelength and propagation length that approach the free-space one. These media do not exist in nature, at the investigated far-infrared frequencies, but artificial media can be implemented in order to achieve the aforementioned interesting properties.

Until now, there is not a single reference to the magnetic properties of the surrounding media. For this reason, in Figure 16.13, the combined investigation of electric and magnetic properties of the substrate is performed. The influence of the magnetic properties is negligible when the dielectric constant is positive. However, at negative dielectric values, the increment of the magnetic permeability tends to decrease the wavelength and the propagation length and strengthen the confinement of the SPP wave on graphene. Finally, in the case of a double-negative substrate ($\varepsilon_r < -1$ and $\mu_r < -1$), the SPP wave is not able to propagate, since leaky waves are developed.

16.4 Electromagnetic Response on Magnetically Biased Graphene

Until this point, graphene is considered electrically biased, exclusively. However, a magnetostatic bias field can be applied having an effect on the surface conductivity properties of the two-dimensional material. Specifically, in this scenario the conductivity is a tensorial quantity in contrast to the scalar one of the previous section. Hence, the electromagnetic response differentiates significantly and, herein, the features of a magnetically biased graphene are thoroughly studied. Note that the temperature is retained to room temperature, namely $T = 300$ K.

16.4.1 Plane Wave Propagating through Graphene

The influence of magnetically biased graphene to a normally incident plane wave can be extracted through Equations (16.29) and (16.30), adjusting appropriately the surface conductivity. In particular, the scalar term is replaced with the tensorial one of Equation (16.10). Recall that the plane wave propagates toward the positive values of the y-axis and the reflection and transmission coefficients are computed by

$$\overline{\boldsymbol{R}} = \frac{[(\eta_1 - \eta_2 - \sigma_d \eta_1 \eta_2)(\eta_1 + \eta_2 + \sigma_d \eta_1 \eta_2) - (\sigma_o \eta_1 \eta_2)^2](\hat{x}\hat{x} + \hat{z}\hat{z}) + 2\sigma_o \eta_1^2 \eta_2 (\hat{z}\hat{x} - \hat{x}\hat{z})}{(\eta_1 + \eta_2 + \sigma_d \eta_1 \eta_2)^2 + (\sigma_o \eta_1 \eta_2)^2}, \tag{16.82}$$

$$\overline{\boldsymbol{T}} = 2\eta_1 \frac{(\eta_1 + \eta_2 + \sigma_d \eta_1 \eta_2)(\hat{x}\hat{x} + \hat{z}\hat{z}) + \sigma_o \eta_1 \eta_2 (\hat{z}\hat{x} - \hat{x}\hat{z})}{(\eta_1 + \eta_2 + \sigma_d \eta_1 \eta_2)^2 + (\sigma_o \eta_1 \eta_2)^2}, \tag{16.83}$$

respectively. As expected, these coefficients are, also, tensorial quantities and as a result a linearly polarized plane wave obtains an additional electric field component that is perpendicular to the initial one. So, the transmitted and reflected plane waves are rotated, concerning their polarization vector, or even convert to a different polarization type, such as an elliptical or a circular one. The investigation of the polarization alteration due to the magnetically biased graphene is realized after the determination of the basic parameters of a general elliptically polarized plane wave.

The considered plane wave propagates along the positive y-axis in order to maintain convention of the previous analysis. The electric field, for a harmonic time variation, is separated into two perpendicular electric field components

$$E_x = E_{x0} e^{j\phi_x}, \tag{16.84}$$

$$E_z = E_{z0} e^{j\phi_z}, \tag{16.85}$$

where $E_{x0} = |E_x|$ and $E_{z0} = |E_z|$ are the amplitudes, ϕ_x and ϕ_z the initial phase, and $\Delta\phi = \phi_z - \phi_x$ the phase difference of the components. The axial ratio of the plane wave is defined as the ratio between the half-axle of the ellipse, as illustrated in Figure 16.14a and computed [16] as

$$\mathrm{AR} = \frac{\mathrm{OA}}{\mathrm{OB}} = \frac{\sqrt{\frac{1}{2}\left\{E_{x0}^2 + E_{z0}^2 + \left[E_{x0}^4 + E_{z0}^4 + 2E_{x0}^2 E_{z0}^2 \cos(2\Delta\phi)\right]^{\frac{1}{2}}\right\}}}{\sqrt{\frac{1}{2}\left\{E_{x0}^2 + E_{z0}^2 - \left[E_{x0}^4 + E_{z0}^4 + 2E_{x0}^2 E_{z0}^2 \cos(2\Delta\phi)\right]^{\frac{1}{2}}\right\}}}. \tag{16.86}$$

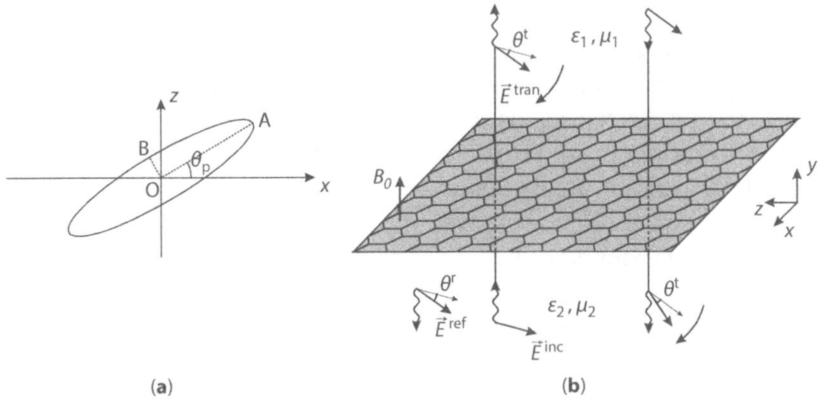

Figure 16.14 (a) Elliptically polarized plane wave propagating along y-axis and (b) incident plane wave on magnetically biased graphene.

Moreover, the tilt of the ellipse, regarding the x-axis is calculated as

$$\theta_p = \frac{\pi}{2} - \frac{1}{2}\tan^{-1}\left[\frac{2E_{x0}E_{z0}}{E_{x0}^2 - E_{z0}^2}\cos(\Delta\phi)\right]. \tag{16.87}$$

The polarization of the plane wave can be, now, evaluated through the values of the axial ratio and the tilt. Specifically, for an axial ratio that is larger than 100, the plane wave is considered as linearly polarized, while a circularly one has axial ratio 1, identically. For any other value, the plane wave is elliptically polarized.

Back to the reflection and transmission coefficients (16.82) and (16.83), the electric field components are transformed as the incident plane wave interacts with graphene and the polarization angle is rotated. Therefore, gyrotropy is observed. Moreover, the rotation is realized toward the same direction, with respect to the y-axis, regardless of the direction of the incident plane wave, as demonstrated in Figure 16.14b. The latter indicates that magnetically biased graphene generates nonreciprocal effects. In particular, let us consider a plane wave, propagating along the positive y-axis, with a polarization angle rotated for θ^t, which when returning back through the negative y-axis has again a rotation angle of θ^t toward the same direction. So, the polarization angle of the initial plane wave is rotated for $2\theta^t$, proving the nonreciprocity of the system.

The rotation angle as well as the possible polarization conversion depend on the two terms of the tensorial surface conductivity. These terms are influenced from the applied bias fields and the quality of the graphene lattice, as described in Equations (16.11) and (16.12). Thus, it is important to perform a parametric analysis with respect to the basic parameters of a magnetically biased graphene. Firstly, the scattering rate is studied, considering a chemical potential $\mu_c = 0.1$ and $B_0 = 1$T. The first Landau level of Equation (16.13) is calculated as $M_n = 0.035$ eV, which is much lower than the chemical potential, and hence the approximate expressions (16.18) and (16.19) of the surface conductivity can be safely and accurately used, as the analysis reaches up to the far-infrared regime, where the more interesting effects are revealed.

The frequency response in Figure 16.15 indicates that the real part of the parallel, to the excitation field, term σ_d presents a maximum value, while the imaginary part is positive at lower frequencies, until the maximum of its real counterpart. This differentiation of the surface conductivity response is attributed to the magnetostatic bias field and the frequency, while the maximum value of its real part almost coincides to the cyclotron frequency of Equation (16.17). Furthermore, the increment of the scattering rate (i.e., a poorer graphene lattice) leads to a reduction of this maximum value. Note that the rotation angle of the transmitted wave is, also, reduced, as observed in Figure 16.16, at lower frequencies and if the cyclotron frequency is surpassed, the rotation angle is reversed, until approximating the zero value. On the other hand, the axial ratio reveals that for a linearly polarized plane wave, the transmitted one retains the linear polarization at the entire range, excluding the region near the cyclotron, where it is converted to an elliptically polarized wave.

The surface conductivity presents a similar behavior, concerning the parametric investigation of the magnetostatic bias field and the chemical potential, in Figure 16.17 and Figure 16.18, for $\Gamma = 0.33$ meV. The main difference is the displacement of the real part's maximum and the sign switch of the σ_d imaginary part, as the cyclotron frequency is affected by these graphene parameters. Specifically, the cyclotron frequency (16.17) is proportional to the magnetic bias field and inversely proportional to the chemical potential. Moreover, the perpendicular term of the surface conductivity σ_o presents an interesting

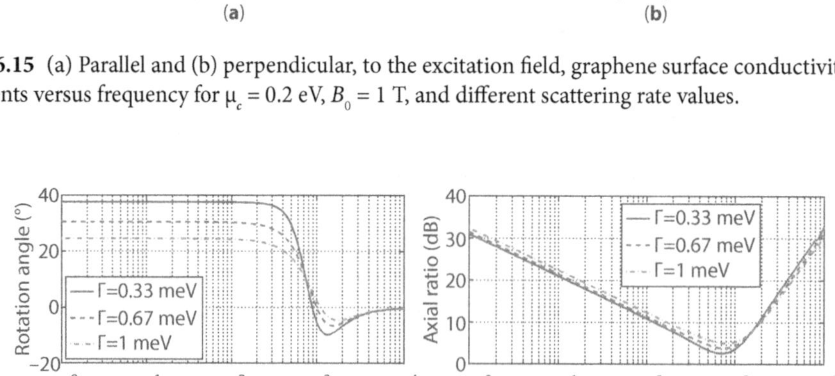

Figure 16.15 (a) Parallel and (b) perpendicular, to the excitation field, graphene surface conductivity components versus frequency for $\mu_c = 0.2$ eV, $B_0 = 1$ T, and different scattering rate values.

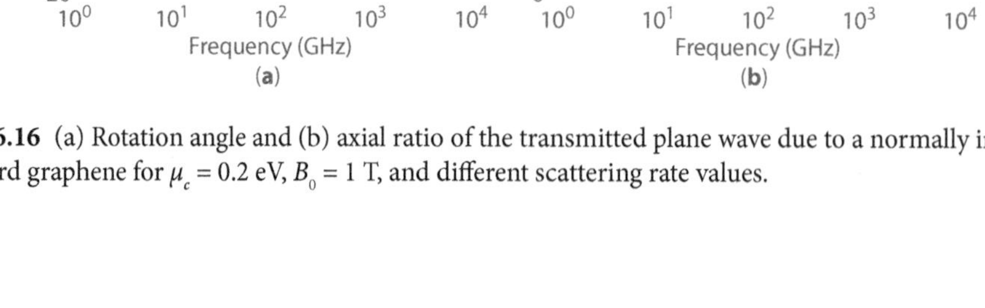

Figure 16.16 (a) Rotation angle and (b) axial ratio of the transmitted plane wave due to a normally incident one toward graphene for $\mu_c = 0.2$ eV, $B_0 = 1$ T, and different scattering rate values.

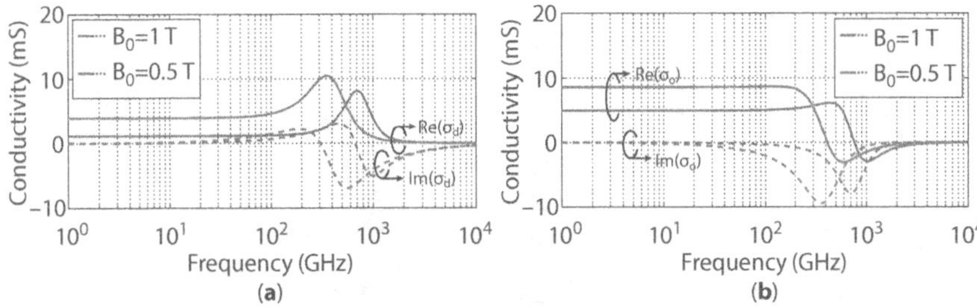

Figure 16.17 (a) Parallel and (b) perpendicular, to the excitation field, graphene surface conductivity components versus frequency for $\mu_c = 0.2$ eV, $\Gamma = 0.33$ meV, and different magnetic bias field values.

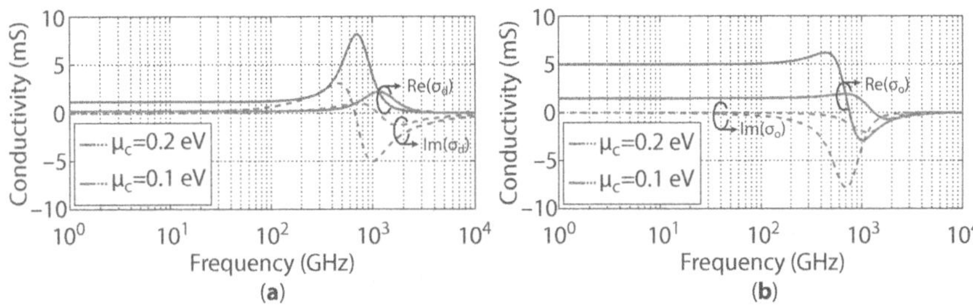

Figure 16.18 (a) Parallel and (b) perpendicular, to the excitation field, graphene surface conductivity components versus frequency for $B_0 = 1$ T, $\Gamma = 0.33$ meV, and different chemical potential values.

behavior, since the augmentation of the magnetostatic field increases, as expected, because of the more intense Lorentz forces. Finally, the observation of higher frequencies for $\mu_c =$ 0.1 eV unveils an oscillation on both surface conductivity terms. The explanation of this behavior is that μ_c is only three times larger than the first Landau level and the accuracy of the approximated expression degrades. Thus, the complete relations (16.11) and (16.12) are required.

Obviously, the magnetostatic bias field and the chemical potential influence graphene's surface conductivity (when combined), since the former determines the Landau energy levels and the latter their location. For this reason, the rotation angle and the axial ratio are examined for the combination of these two parameters, at 100 GHz and for the typical value of $\Gamma = 0.33$ meV, in Figure 16.19. The two bias fields (the chemical potential is connected to the electrostatic bias fields) have a negligible impact when any one of them is weak. Nevertheless, the rotation angle increases drastically for stronger bias fields, while an incident linearly polarized plane wave is converted to an elliptically polarized transmitted one. Observe that at the increment of the chemical potential, graphene surface conductivity increases and the transmitted wave is weaker.

16.4.2 Surface Plasmon Polariton Waves on Graphene

The electromagnetic response of a magnetically biased graphene, due to a volume current distribution, $\vec{J}_s(\vec{r}')$, near the surface of a two-dimensional material, is extracted similarly

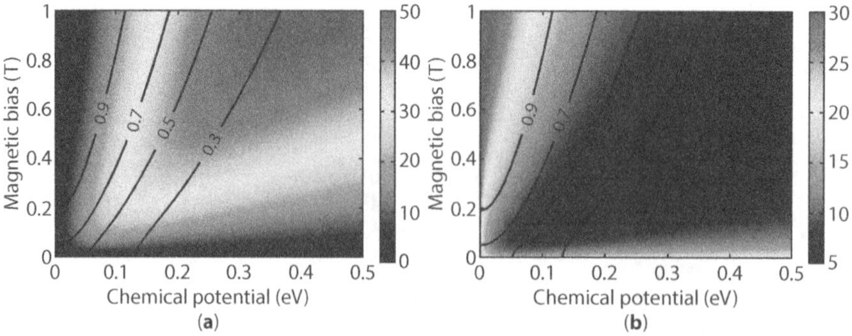

Figure 16.19 (a) Rotation angle (degrees) and (b) axial ratio (dB) of the transmitted plane wave, due to a normally incident illumination toward a graphene layer of Γ = 0.33 meV, versus chemical potential and magnetic bias field at 100 GHz. Black lines denote the transmission coefficient value.

to the analysis of Section 16.3.2 via the dyadic Green function. The Hertzian potentials (16.33)–(16.34) are again employed, considering the current source in region 1, as depicted in Figure 16.3. The differentiation from the formulation in Section 16.3.2 is the tensorial nature of graphene's surface conductivity, because of the magnetostatic bias. Explicitly, Lorentz forces on graphene electrons enable the σ_o term in Equation (16.10) and thus the radial complex wavenumber k_ρ has to be divided into two components, i.e., k_x and k_z, for a graphene layer at the xz-plane. So, the Hankel function does not simplify the analysis and the dyadic Green function of the source (16.35) is now implemented as

$$\bar{\bar{g}}_1^p\left(\vec{r}-\vec{r}'\right)=\bar{\bar{I}}\frac{e^{-jk_1R}}{4\pi R}=\bar{\bar{I}}\frac{1}{4\pi^2}\int_{-\infty}^{\infty}\int_{-\infty}^{\infty}\frac{e^{-p_1|y-y'|}}{2p_1}e^{-j\vec{k}_\rho(\vec{r}-\vec{r}')}\,dk_x\,dk_z. \tag{16.88}$$

Moreover, the boundary conditions of the Hertzian potentials (16.41)–(16.44) are, also, affected due to the σ_o term and are represented as

$$\pi_{1,x}=N^2M^2\pi_{2,x}, \tag{16.89}$$

$$\frac{\partial\pi_{1,y}}{\partial y}-\frac{\partial\pi_{2,y}}{\partial y}=-jk_x\left(1-M^2N^2\right)\pi_{2,x}, \tag{16.90}$$

$$\omega k_x\left(\varepsilon_2\pi_{2,y}-\varepsilon_1\pi_{1,y}\right)=-\left(\sigma_d\left(k_1^2-k_x^2\right)+\sigma_o k_x k_z\right)\pi_{1,x}+\left(\sigma_d jk_x-\sigma_o jk_z\right)\frac{\partial\pi_{1,y}}{\partial y}$$
$$-j\omega\left(\varepsilon_1\frac{\partial\pi_{1,x}}{\partial y}-\varepsilon_2\frac{\partial\pi_{2,x}}{\partial y}\right), \tag{16.91}$$

$$\omega k_z\left(\varepsilon_2\pi_{2,y}-\varepsilon_1\pi_{1,y}\right)=-\left(\sigma_o\left(k_1^2-k_x^2\right)-\sigma_d k_x k_z\right)\pi_{1,x}+\left(\sigma_o jk_x+\sigma_d jk_z\right)\frac{\partial\pi_{1,y}}{\partial y}. \tag{16.92}$$

Applying the appropriate boundary conditions (16.89)–(16.92) and following a similar procedure to that of Section 16.3.2, the scattered dyadic Green functions are extracted for both regions, i.e.,

$$\overline{\overline{g}}_1^s(\vec{r},\vec{r}')=\hat{x}\hat{x}g_{1,xx}^s(\vec{r},\vec{r}')+\hat{y}\hat{y}g_{1,yy}^s(\vec{r},\vec{r}')+\hat{z}\hat{z}g_{1,zz}^s(\vec{r},\vec{r}')+\hat{x}\hat{y}g_{1,xy}^s(\vec{r},\vec{r}')$$
$$+\hat{y}\hat{x}g_{1,yx}^s(\vec{r},\vec{r}')+\hat{z}\hat{y}g_{1,zy}^s(\vec{r},\vec{r}')+\hat{y}\hat{z}g_{1,yz}^s(\vec{r},\vec{r}'), \tag{16.93}$$

$$\overline{\overline{g}}_2^s(\vec{r},\vec{r}')=\hat{x}\hat{x}g_{2,xx}^s(\vec{r},\vec{r}')+\hat{y}\hat{y}g_{2,yy}^s(\vec{r},\vec{r}')+\hat{z}\hat{z}g_{2,zz}^s(\vec{r},\vec{r}')+\hat{x}\hat{y}g_{2,xy}^s(\vec{r},\vec{r}')+$$
$$\hat{y}\hat{x}g_{2,yx}^s(\vec{r},\vec{r}')+\hat{z}\hat{y}g_{2,zy}^s(\vec{r},\vec{r}')+\hat{y}\hat{z}g_{2,yz}^s(\vec{r},\vec{r}'), \tag{16.94}$$

where the inner functions are

$$g_{1,\alpha\beta}^s\left(\vec{r},\vec{r}'\right)=\frac{1}{4\pi^2}\int_{-\infty}^{\infty}\int_{-\infty}^{\infty}R_{\alpha\beta}\frac{\mathrm{e}^{-p_1\left(y+y'\right)}}{2p_1}\mathrm{e}^{-j\vec{k}\rho\left(\vec{r}-\vec{r}'\right)}\,\mathrm{d}k_x\,\mathrm{d}k_z,\alpha,\beta=x,y,z, \tag{16.95}$$

$$g_{2,\alpha\beta}^s\left(\vec{r},\vec{r}'\right)=\frac{1}{4\pi^2}\int_{-\infty}^{\infty}\int_{-\infty}^{\infty}T_{\alpha\beta}\frac{\mathrm{e}^{p_2 y}\mathrm{e}^{-p_1 y'}}{2p_1}\mathrm{e}^{-j\vec{k}_\rho\left(\vec{r}-\vec{r}'\right)}\,\mathrm{d}k_x\,\mathrm{d}k_z,\alpha,\beta=x,y,z, \tag{16.96}$$

while $R_{\alpha\beta}$ and $T_{\alpha\beta}$ are the scalar terms of the functions that depend on the surface conductivity and propagation properties of the electromagnetic waves on graphene. For a set of general electromagnetic properties of the surrounding materials ($\varepsilon_{r1}\neq\varepsilon_{r2}$ and $\mu_{r1}\neq\mu_{r2}$), the scalar terms are considerably complicated. Thus, the special case of a graphene sheet inside a homogeneous medium ($\varepsilon_{r1}=\varepsilon_{r2}=\varepsilon_b$ and $\mu_{r1}=\mu_{r2}=\mu_b$) is examined. Bearing in mind this assumption, one arrives at

$$R_{xx}=\frac{k_b^2k_zs_d-p_b^2k_xs_o-jp_bk_bk_z\left(s_d^2+s_o^2\right)}{k_zD\left(k_\rho\right)}, \tag{16.97}$$

$$T_{xx} = \frac{p_b^2\left(k_z s_d - k_x s_o\right) + jp_b k_b k_z}{k_z D\left(k_\rho\right)}, \tag{16.98}$$

$$R_{xy} = T_{xy} = \frac{-jp_b s_o k_\rho^2}{2k_z D\left(k_\rho\right)}, \tag{16.99}$$

$$R_{yy} = \frac{p_b k_x k_z\left[2p_b s_d + 2jk_b\left(s_d^2 + s_o^2\right)\right] + p_b^2 s_o\left(k_z^2 - k_x^2\right)}{2k_x k_z D\left(k_\rho\right)}, \tag{16.100}$$

$$T_{yy} = \frac{2k_b k_x k_z\left(jp_b - k_b s_d\right) + p_b^2 s_o\left(k_x^2 - k_z^2\right)}{2k_x k_z D\left(k_\rho\right)}, \tag{16.101}$$

$$R_{yx} = -T_{yx} = \frac{-jp_b\left[k_x k_z s_d + \left(k_b^2 - k_x^2\right)s_o\right]}{k_z D\left(k_\rho\right)}, \tag{16.102}$$

$$R_{yz} = -T_{yz} = \frac{-jp_b\left[k_x k_z s_d + \left(k_z^2 - k_b^2\right)s_o\right]}{k_x D\left(k_\rho\right)}, \tag{16.103}$$

$$R_{zz} = \frac{p_b^2 k_z s_o + k_b^2 k_x s_d - jp_b k_b k_x\left(s_d^2 + s_o^2\right)}{k_x D\left(k_\rho\right)}, \tag{16.104}$$

$$T_{zz} = \frac{p_b^2\left(k_x s_d + k_z s_o\right) + jp_b k_b k_x}{k_x D\left(k_\rho\right)}, \tag{16.105}$$

$$R_{zy} = T_{zy} = \frac{jp_b s_o k_\rho^2}{2k_x D\left(k_\rho\right)}, \tag{16.106}$$

where $D(k_\rho)$ is a function of the radial wavenumber and s_d, s_o are normalized terms of σ_d,σ_o, respectively, expressed as

$$D\left(k_\rho\right) = \left(p_b^2 - k_b^2\right)s_d + jp_b k_b\left(1 + s_d^2 + s_o^2\right), \tag{16.107}$$

$$s_d = \frac{\eta_b \sigma_d}{2}, s_o = \frac{\eta_b \sigma_o}{2}, \quad (16.108)$$

while k_b and η_b are the wavenumber and wave impedance of the homogeneous medium, respectively, with $p_b^2 = k_\rho^2 - k_b^2$. The existence of a surface wave depends on the Re$\{p_b\} > 0$ convention in order to satisfy the radiation condition at $y \to \pm\infty$, as thoroughly examined in Section 16.3.2.

The integrals of the scattered dyadic Green functions (16.95) and (16.96) are of the Sommerfeld kind and there are singularities at the zero points of $R_{\alpha\beta}$ and $T_{\alpha\beta}$. Except for the trivial points $k_x = k_z = 0$ and $k_\rho = k_b$, the singularities of the $D(k_\rho)$ function are related to surface wave propagation phenomena. Thus, it is important to resolve this function, then extract k_ρ, and explore the propagation properties of the magnetoplasmons. The dispersion equation of the complex wavenumber for the case of a graphene layer in a homogeneous medium ($\varepsilon_{r1} = \varepsilon_{r2} = \varepsilon_b$, $\mu_{r1} = \mu_{r2} = \mu_b$) is estimated as

$$k_\rho^\pm = k_b \sqrt{\frac{1}{4s_d^2}\left(-j\left(s_d^2 + s_o^2 + 1\right) \pm \sqrt{4s_d^2 - (s_d^2 + s_o^2 + 1)^2}\right)^2 + 1}, \quad (16.109)$$

where the positive values correspond to transverse magnetic SPP waves and the negative values to transverse electric SPP ones. Evidently, this separation depends on the sign of the parallel surface conductivity term imaginary part, namely $\sigma_d'' = \mathrm{Im}\{\sigma_d\}$. Commonly to the scenario that the magnetostatic bias is absent, TM SPP waves appear when $\sigma_d'' < 0$, while TE SPP waves emerge when $\sigma_d'' > 0$. Otherwise, the radiation condition Re$\{p_b\} > 0$ is not satisfied.

The surface conductivity of graphene is extracted via Equations (16.11) and (16.12) and Figure 16.17 is recalled for the selected values of $\mu_c = 0.2$ eV and $\Gamma = 0.33$ meV. The sign of σ_d'' is positive at the lower frequencies, but it becomes negative near the cyclotron frequency (16.17). It is, also, noticed that this frequency depends on the magnetostatic bias strength as well as the chemical potential, controlled through the electrostatic bias. Consequently, the applied static fields can be optimally adjusted to achieve the desired properties of graphene.

The absence of a magnetostatic bias leads to weakly confined TE SPP waves beyond the far-infrared regime, as examined in Section 16.3.2. However, these waves appear, also, at microwave and millimeter-wave frequencies, when the magnetostatic bias increases. The influence of the bias fields is perceived in Figure 16.20, where the propagation properties of

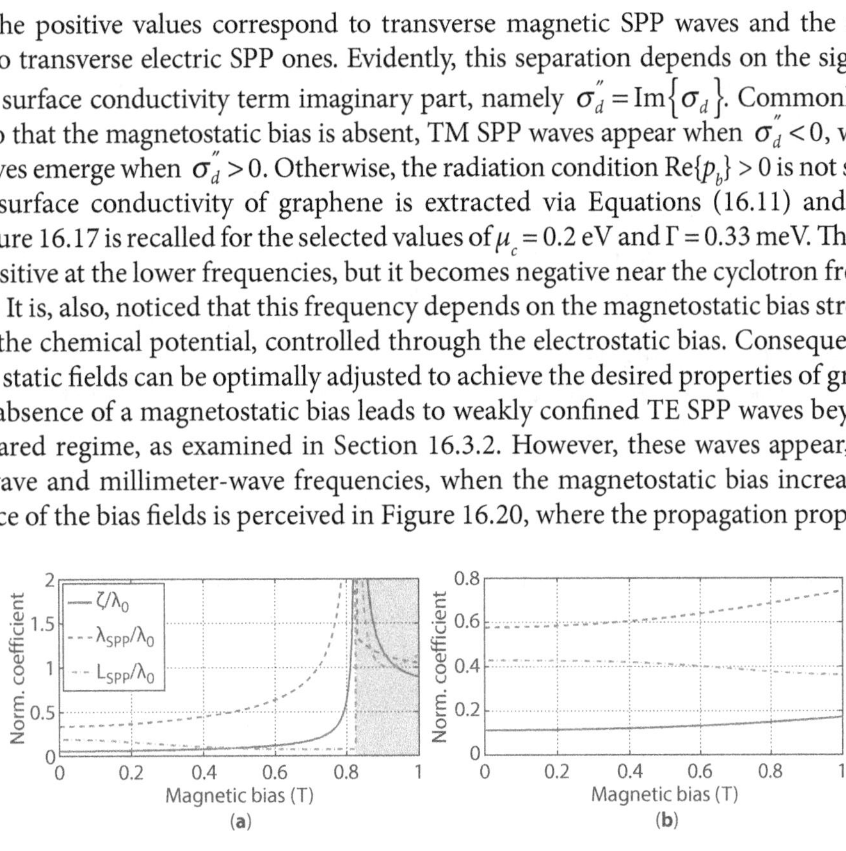

Figure 16.20 Normalized, to the free-space wavelength, propagation properties of magnetoplasmons at 1 THz for a magnetically biased graphene of $\Gamma = 0.33$ meV and a chemical potential of (a) 0.1 eV and (b) 0.2 eV. The shaded region indicates $\sigma_d'' < 0$.

the magnetoplasmons (extracted via the complex wavenumber (16.109)) is depicted. The increment of the magnetostatic bias field results in a proportional raise of the cyclotron frequency and at 1 THz for $\mu_c = 0.1$ eV (Figure 16.20a), the σ_d'' term becomes positive for $B_0 > 0.82$ T. Thus, a weakly confined TE SPP wave appears in contrast to the strongly confined TM SPP wave at lower magnetostatic values. Moreover, near the cyclotron frequency ($0.9\omega_c < 2\pi f < 1.1\omega_c$), any kind of SPP wave vanishes and a leaky wave is observed. Conversely, increasing the chemical potential to 0.2 eV (Figure 16.20b), the cyclotron frequency does not approach 1 THz and the magnetostatic bias has to be increased further to transform the TM SPP to a TE one.

16.5 Numerical Modeling of Graphene

The theoretical analyses of the preceding sections revealed the majority of the exotic electromagnetic phenomena in graphene, while the parametric investigation indicated its critical features. However, the considered structures do not approximate realistic devices that are significantly more complex. For this reason, advanced numerical algorithms are required in order to accurately and efficiently model graphene, thus paving the way for the design and electromagnetic response extraction of complicated devices. Particularly, the popular finite-difference time-domain (FDTD) method is utilized and graphene is introduced via appropriate modifications of the conventional algorithm. The main objectives are the representation of graphene as a two-dimensional material, maintaining its true nature as well as the reliable modeling of its frequency dispersion.

16.5.1 Graphene as an Equivalent Surface Current Density

The traditional FDTD algorithm solves the electromagnetic propagation via the approximation of Maxwell equations through a robust finite-differencing scheme [17]. The basic quantities are the electric and magnetic field, while the electric current density is utilized as the excitation. Recalling that graphene is represented as a surface conductivity, it is introduced to the algorithm as an excitation field, $\vec{J}_{\text{gr}}$, controlled via the conductivity and the instantaneous electric field

$$\vec{J}_{\text{gr}}\left(\vec{r}',\omega\right)=\sigma\left(\omega\right)\vec{E}\left(\vec{r}',\omega\right). \tag{16.110}$$

Therefore, graphene's contribution is launched as an equivalent surface current and its location in the original Yee cell [18] coincides to the electric field components, as depicted in Figure 16.21, due to the strong dependency on the electric field. At each time step of the algorithm, the prior graphene surface current is added in the calculation of the electric field via

$$E_x\Big|^{n+1}_{i,j+\frac{1}{2},k-\frac{1}{2}}=E_x\Big|^{n+1}_{i,j+\frac{1}{2},k-\frac{1}{2}}-\frac{2\Delta t}{\left(\varepsilon_1+\varepsilon_2\right)+\sigma\Delta t}\frac{1}{\Delta y}J_{x,\text{gr}}\Big|^{n}_{i,j+\frac{1}{2},k-\frac{1}{2}}, \tag{16.111}$$

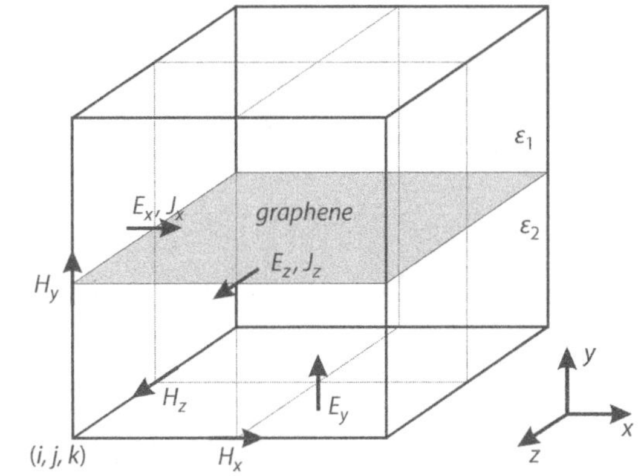

Figure 16.21 Graphene location in a Yee cell and its modeling via a surface current density.

$$E_z\Big|^{n+1}_{i+\frac{1}{2},j+\frac{1}{2},k} = E_z\Big|^{n+1}_{i+\frac{1}{2},j+\frac{1}{2},k} - \frac{2\Delta t}{(\varepsilon_1+\varepsilon_2)+\sigma\Delta t}\frac{1}{\Delta y}J_{z,\mathrm{gr}}\Big|^{n}_{i+\frac{1}{2},j+\frac{1}{2},k}, \tag{16.112}$$

where it is assumed that the material is placed on the xz-plane. Due to this fact, the perpendicular to graphene surface component, i.e., E_y, is calculated through the conventional algorithm. Moreover, the $\frac{1}{\Delta y}$ term is placed to normalize the current density with respect to the cell size, thus representing an equivalent surface current.

16.5.2 The Recursive Convolution Method

To this point, the surface conductivity of graphene is studied at the frequency domain. However, the FDTD algorithm is exclusively implemented in the time domain and the inverse Fourier transformation of Equation (16.110) has to be performed

$$\vec{J}_{\mathrm{gr}}(\vec{r}',t) = \sigma(t) * \vec{E}(\vec{r}',t), \tag{16.113}$$

where the * symbol denotes a convolution between the surface conductivity and the electric field. At a random time step $t = (n+1)\Delta t$, the discrete convolution for a causative conductivity function (such as any realistic) is implemented as

$$\vec{J}(\vec{r})\big|^{n+1} = \sum_{m=0}^{n+1}\left\{\sigma\left[(n+1-m)\Delta t\right]\vec{E}(\vec{r})\big|^{m}\Delta t\right\}. \tag{16.114}$$

One can easily notice that this expression is very demanding with regard to memory allocation and computational resources, since all the past electric field values have to be stored. However, there is a class of functions where the convolution is attained through recursive schemes [19]. These functions consist of the combination of an exponential function and a harmonic one as

$$\sigma(t) = Ae^{-\alpha t}\cos(\omega_0 t)u(t), \tag{16.115}$$

$$\sigma(t) = Ae^{-\alpha t}\sin(\omega_0 t)u(t), \tag{16.116}$$

where $u(t)$ is the Heaviside step function. Although it may seem that the available functions are limited, the most basic ones that describe the popular Debye, Lorentz, and Drude models are included in Equations (16.115) and (16.116).

Initially, the surface conductivity is expressed as a complex exponential function of $\gamma = \alpha - j\omega_0$, while Equations (16.115) and (16.116) can be recovered by means of

$$\sigma(t) = \text{Re}\{Ae^{-\gamma t}u(t)\}, \tag{16.117}$$

$$\sigma(t) = \text{Im}\{Ae^{-\gamma t}u(t)\}. \tag{16.118}$$

The exponential nature of the function is quite convenient, since Equation (16.114) can be calculated via

$$\vec{J}(\vec{r})|^{n+1} = \sum_{m=0}^{n+1}\left\{Ae^{-\gamma(n+1-m)\Delta t}\vec{E}(\vec{r})|^{m}\,\Delta t\right\}, \tag{16.119}$$

$$\vec{J}(\vec{r})|^{n+1} = e^{-\gamma\Delta t}\sum_{m=0}^{n}\left\{Ae^{-\gamma(n-m)\Delta t}\vec{E}(\vec{r})|^{m}\,\Delta t\right\} + A\Delta t\vec{E}(\vec{r})|^{n+1}. \tag{16.120}$$

It can be easily noticed, now, that the summation of the first term coincides to the surface current density at the previous time step

$$\vec{J}(\vec{r})|^{n} = \sum_{m=0}^{n}\left\{Ae^{-\gamma(n-m)\Delta t}\vec{E}(\vec{r})|^{m}\,\Delta t\right\}, \tag{16.121}$$

and Equation (16.120) is expressed as

$$\vec{J}(\vec{r})|^{n+1} = e^{-\gamma\Delta t}\vec{J}(\vec{r})|^{n} + A\Delta t\vec{E}(\vec{r})|^{n+1}. \tag{16.122}$$

Finally, the appropriate selection of the complex part leads to the accurate calculation of the surface current due to a surface conductivity of Equation (16.115) or (16.116). Moreover, the efficiency of the conventional FDTD method is maintained with respect to the computational resources and time. The described technique is a modification of the recursive convolution method (RCM) for the calculation of a current density by means of the conductivity and the electric field.

16.5.3 Graphene Modeling through the Recursive Convolution Method

16.5.3.1 *Electrically Biased Graphene*

The surface conductivity of graphene is computed through Equations (16.18) and (16.19) for the far-infrared regime, where all the interesting phenomena are observed. Additionally, the absence of the magnetostatic bias simplifies even more the expression, since the cyclotron frequency is zero. Thus,

$$\sigma_d(\omega) = \frac{e^2 k_B T}{\pi\hbar^2 (j\omega + 2\Gamma)} \left[\frac{\mu_c}{k_B T} + 2\ln\left(e^{-\frac{\mu_c}{k_B T}} + 1 \right) \right] = \frac{A_{\mu_c}}{j\omega + 2\Gamma}, \tag{16.123}$$

where the term A_{μ_c} depends on the chemical potential. Then, the inverse Fourier transform of Equation (16.123) is applied and one obtains

$$\sigma_d(t) = A_{\mu_c} e^{-2\Gamma t} u(t), \tag{16.124}$$

to enable the transition in the time domain as required by the FDTD algorithm. The last expression coincides to Equation (16.117), for $A = A_{\mu_c}$ and $\gamma = 2\Gamma$, indicating that the exponential function is real-valued, while the surface current at each time step is calculated through Equation (16.122)

$$\vec{J}_{\text{gr}}(\vec{r})|^{n+1} = e^{-2\Gamma\Delta t} \vec{J}_{\text{gr}}(\vec{r})|^{n} + A_{\mu_c} \Delta t \vec{E}(\vec{r})|^{n+1}. \tag{16.125}$$

Separating the two components of Equation (16.125) yields the final updating equations

$$J_{x,\text{gr}}|^{n+1}_{i,j+\frac{1}{2},k-\frac{1}{2}} = e^{-2\Gamma\Delta t} J_{x,\text{gr}}|^{n}_{i,j+\frac{1}{2},k-\frac{1}{2}} + A_{\mu_c} \Delta t E_x|^{n+1}_{i,j+\frac{1}{2},k-\frac{1}{2}}, \tag{16.126}$$

$$J_{z,\text{gr}}|^{n+1}_{i+\frac{1}{2},j+\frac{1}{2},k} = e^{-2\Gamma\Delta t} J_{z,\text{gr}}|^{n}_{i+\frac{1}{2},j+\frac{1}{2},k} + A_{\mu_c} \Delta t E_z|^{n+1}_{i+\frac{1}{2},j+\frac{1}{2},k}. \tag{16.127}$$

The developed numerical model of graphene is validated via thorough simulations. Specifically, the analysis concentrates on the surface wave propagation properties detection because of their significance in the design of advanced applications. Two different simulations are conducted to cover the far-infrared region. The computational space in the first one is discretized into $400 \times 240 \times 400$ cells of $\Delta x = \Delta y = \Delta z = 1.5$ μm, while the time step is set at 2.8 fs and the domain is terminated via a perfectly matched layer (PML) [20, 21]. Correspondingly, the computational space of the second problem is divided into $200 \times 100 \times 200$ cells of $\Delta x = \Delta y = \Delta z = 0.1$ μm, a time step of 0.19 fs, and a similar PML termination. Finally, the frequency region of the former simulation is 1–4 THz, while that of the latter is 4–10 THz.

First of all, the electric field distribution along the surface of a graphene of $\mu_c = 0.2$ eV and $\Gamma = 0.33$ meV is extracted at 2 THz and compared to the theoretical values computed via Equation (16.66). The remarkable matching of both electric components in Figure 16.22 constitutes a solid indication of the algorithm's overall accuracy.

Moreover, the numerical results coincide to the theoretical ones, concerning the propagation properties of graphene surface wave, as depicted in Figures 16.23, 16.24, and 16.25, independently to the chemical potential or the scattering rate choice. As observed, the influence of the latter on the confinement and the wavelength of the SPP wave is negligible, in contrast to the propagation length that degrades for increased scattering rate values.

However, the surface wave propagation on a suspended graphene, namely surrounded by air, is not a realistic scenario. For this reason, additional simulations are conducted, where the substrate is replaced by a theoretical one of $\varepsilon_r = 2$ and $\varepsilon_r = 4$. The comparison to the theoretical values in Figure 16.26 prove the validity of the developed algorithm at any scenario.

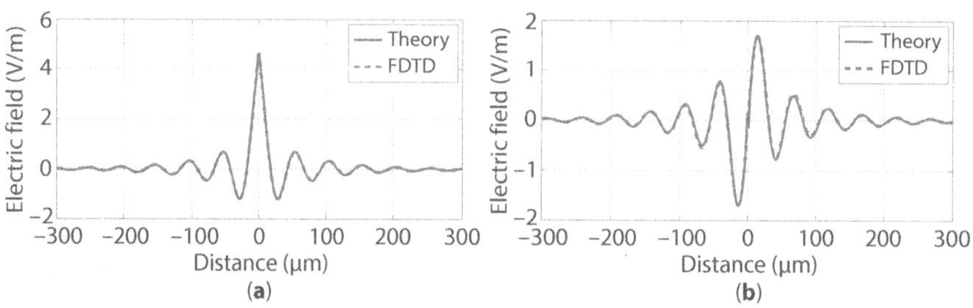

Figure 16.22 Electric field distribution on the surface of graphene with $\mu_c = 0.2$ eV and $\Gamma = 0.33$ meV at 2 THz, concerning (a) the normal and (b) the parallel component to graphene.

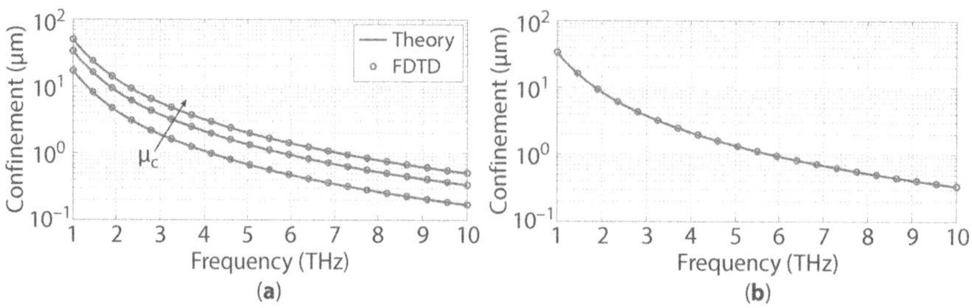

Figure 16.23 Confinement of the surface wave onto graphene for (a) $\mu_c = 0.1$, 0.2, and 0.3 eV, $\Gamma = 0.33$ meV and (b) $\mu_c = 0.2$ eV, $\Gamma = 0.33$, 3.3, and 33 meV.

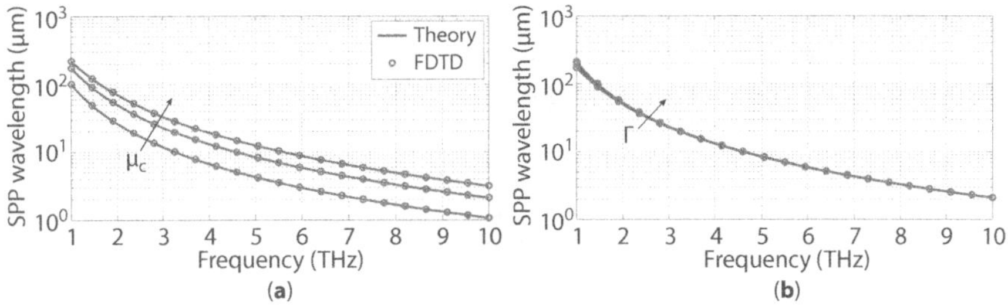

Figure 16.24 Wavelength of the surface wave onto graphene for (a) μ_c = 0.2, 0.2, and 0.3 eV, Γ = 0.33 meV and (b) μ_c = 0.2 eV, Γ = 0.33, 3.3, and 33 meV.

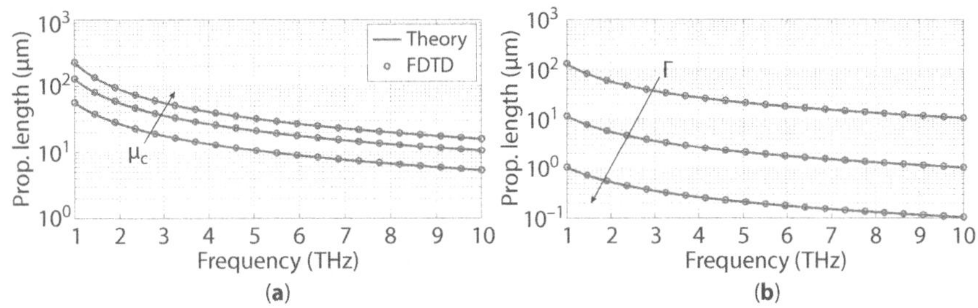

Figure 16.25 Propagation length of the surface wave onto graphene for (a) μ_c = 0.1, 0.2, and 0.3 eV, Γ = 0.33 meV and (b) μ_c = 0.2 eV, Γ = 0.33, 3.3, and 33 meV.

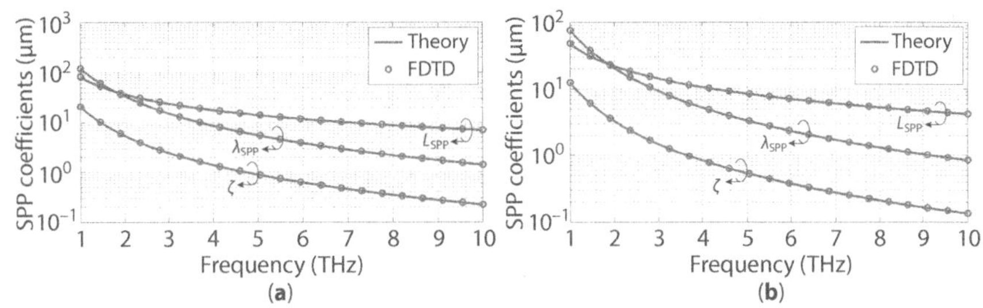

Figure 16.26 Propagation properties of a graphene surface wave with μ_c = 0.2 eV and Γ = 0.33 meV. The superstrate is air and the substrate has a dielectric of (a) ε_r = 2 and (b) ε_r = 4.

16.5.3.2 Magnetically Biased Graphene

The modeling process of graphene as a scalar surface conductivity is deemed inadequate when a magnetostatic bias field is applied additive to the electrostatic one. In this case, the conductivity is transformed into the tensorial quantity (16.10) and the equivalent surface current is calculated via Equation (16.9). Considering that graphene is located on the xz-plane, the surface current components at the frequency domain are extended as

$$J_{x,\text{gr}}(\omega) = \sigma_d(\omega)E_x(\omega) + \sigma_o(\omega)E_z(\omega), \tag{16.128}$$

$$J_{z,\text{gr}}(\omega) = \sigma_d(\omega)E_z(\omega) - \sigma_o(\omega)E_x(\omega). \tag{16.129}$$

The numerical analysis is, also, performed until the far-infrared region, while the magnetostatic field is supposed to be $B_0 \leq 1$T. Thus, the surface conductivity terms are evaluated through the simplified expressions (16.18) and (16.19) as

$$\sigma_d(\omega) = A_{\mu_c} \frac{j\omega + 2\Gamma}{\omega_c^2 + (j\omega + 2\Gamma)^2}, \tag{16.130}$$

$$\sigma_o(\omega) = A_{\mu_c} \frac{\omega_c}{\omega_c^2 + (j\omega + 2\Gamma)^2}, \tag{16.131}$$

where the A_{μ_c} term from Equation (16.123), depending on the chemical potential, is utilized. The inverse Fourier transformation is applied in the two surface conductivity terms to render

$$\sigma_d(t) = A_{\mu_c} \mathrm{e}^{-2\Gamma t} \cos(\omega_c t) u(t), \tag{16.132}$$

$$\sigma_o(t) = A_{\mu_c} \mathrm{e}^{-2\Gamma t} \sin(\omega_c t) u(t), \tag{16.133}$$

which belong to the class of functions that the RCM process can manipulate. Specifically, the functions match Equations (16.115) and (16.116) for $\gamma = 2\Gamma - j\omega_c$ as well as $A = A_{\mu_c}$, and they can be implemented as the complex exponential functions (16.117) and (16.118). Moreover, the main difference between Equations (16.132) and (16.133) is the selection of the appropriate part of the complex function. This feature is reclaimed by defining the complex surface currents J_{cx} and J_{cz} that exclusively depend on the corresponding electric component [22]

$$J_{cx}(\vec{r},t) = A_{\mu_c} \mathrm{e}^{-(2\Gamma - j\omega_c)t} u(t) * E_x(\vec{r},t), \tag{16.134}$$

$$J_{cz}(\vec{r},t) = A_{\mu_c} \mathrm{e}^{-(2\Gamma - j\omega_c)t} u(t) * E_z(\vec{r},t). \tag{16.135}$$

In addition, the location of the complex surface current components in the Yee cell coincides to electric ones and taking into account Equation (16.122) results in

$$J_{cx}\Big|_{i,j+\frac{1}{2},k-\frac{1}{2}}^{n+1} = \mathrm{e}^{-(2\Gamma - j\omega_c)\Delta t} J_{cx}\Big|_{i,j+\frac{1}{2},k-\frac{1}{2}}^{n} + A_{\mu_c} \Delta t E_x\Big|_{i,j+\frac{1}{2},k-\frac{1}{2}}^{n+1}, \tag{16.136}$$

$$J_{cz}\Big|_{i+\frac{1}{2},j+\frac{1}{2},k}^{n+1} = \mathrm{e}^{-(2\Gamma - j\omega_c)\Delta t} J_{cz}\Big|_{i+\frac{1}{2},j+\frac{1}{2},k}^{n} + A_{\mu_c} \Delta t E_z\Big|_{i+\frac{1}{2},j+\frac{1}{2},k}^{n+1}. \tag{16.137}$$

Finally, the real-valued surface currents are now calculated through Equations (16.128) and (16.129)

$$J_{x,\mathrm{gr}}\Big|_{i,j+\frac{1}{2},k-\frac{1}{2}}^{n+1}=\mathrm{Re}\left\{J_{cx}\Big|_{i,j+\frac{1}{2},k-\frac{1}{2}}^{n+1}\right\}+\mathrm{Im}\left\{J_{cz}\Big|_{i,j+\frac{1}{2},k-\frac{1}{2}}^{n+1}\right\}, \tag{16.138}$$

$$J_{z,\mathrm{gr}}\Big|_{i+\frac{1}{2},j+\frac{1}{2},k}^{n+1}=\mathrm{Re}\left\{J_{cz}\Big|_{i+\frac{1}{2},j+\frac{1}{2},k}^{n+1}\right\}-\mathrm{Im}\left\{J_{cx}\Big|_{i+\frac{1}{2},j+\frac{1}{2},k}^{n+1}\right\}. \tag{16.139}$$

However, the location of the real-valued surface currents is not identical to the complex ones, as demonstrated in Figure 16.27, and further treatment is required. Particularly, the pairs of components $J_{x,\mathrm{gr}}-J_{cz,\mathrm{gr}}$ and $J_{z,\mathrm{gr}}-J_{cx,\mathrm{gr}}$ are slightly shifted. For this reason, the complex currents are approximated through the neighboring values at these positions via

$$J_{cz}\Big|_{i,j+\frac{1}{2},k-\frac{1}{2}}^{n+1}=\frac{1}{4}\left(J_{cz}\Big|_{i+\frac{1}{2},j+\frac{1}{2},k}^{n+1}+J_{cz}\Big|_{i-\frac{1}{2},j+\frac{1}{2},k}^{n+1}+J_{cz}\Big|_{i+\frac{1}{2},j+\frac{1}{2},k-1}^{n+1}+J_{cz}\Big|_{i-\frac{1}{2},j+\frac{1}{2},k-1}^{n+1}\right), \tag{16.140}$$

$$J_{cx}\Big|_{i+\frac{1}{2},j+\frac{1}{2},k}^{n+1}=\frac{1}{4}\left(J_{cx}\Big|_{i+1,j+\frac{1}{2},k+\frac{1}{2}}^{n+1}+J_{cx}\Big|_{i+1,j+\frac{1}{2},k-\frac{1}{2}}^{n+1}+J_{cx}\Big|_{i,j+\frac{1}{2},k+\frac{1}{2}}^{n+1}+J_{cx}\Big|_{i,j+\frac{1}{2},k-\frac{1}{2}}^{n+1}\right). \tag{16.141}$$

The developed algorithm is validated via detailed simulations, similarly to the electrically biased graphene case. However, in this scenario, the propagation of a linearly polarized plane wave is, also, studied. The computational domain is divided into 10 × 50,000 × 10 cells of $\Delta x=\Delta y=\Delta z=2.5$ μm, while the time step is set at 5.7 fs. Herein, periodic boundaries and the total-field/scattered-field scheme [23] of the FDTD algorithm are utilized to ensure the propagation of the plane wave. Moreover, the frequency range of the simulation is extremely wide, starting from 10 GHz up to 10 THz. Comparing the numerical results to their theoretical counterparts proves, one more time, the superior accuracy of the method, as illustrated in Figures 16.28 and 16.29.

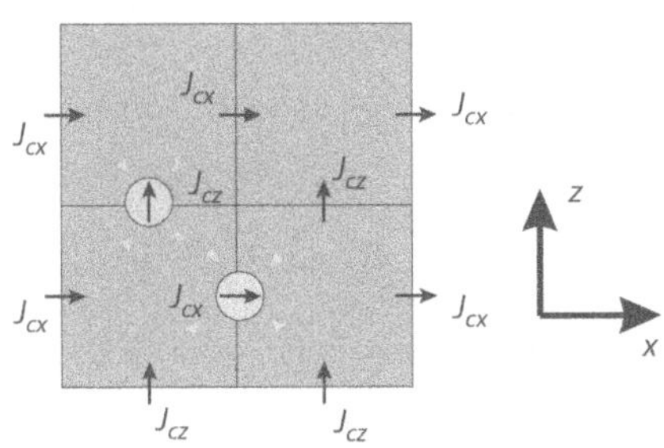

Figure 16.27 The location of the complex surface current components of graphene.

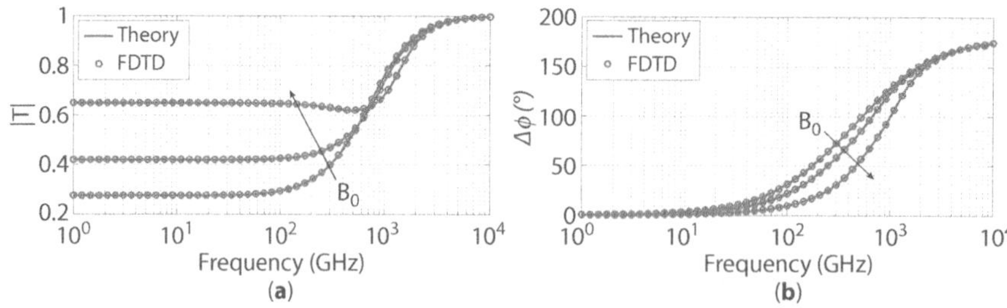

Figure 16.28 (a) Amplitude of the transmission coefficient and (b) phase difference between the electric components of the polarization for a normally incident plane wave on graphene with $\mu_c = 0.2$ eV, $\Gamma = 0.33$ meV, and $B_0 = 0.25$, 0.5, and 1 T.

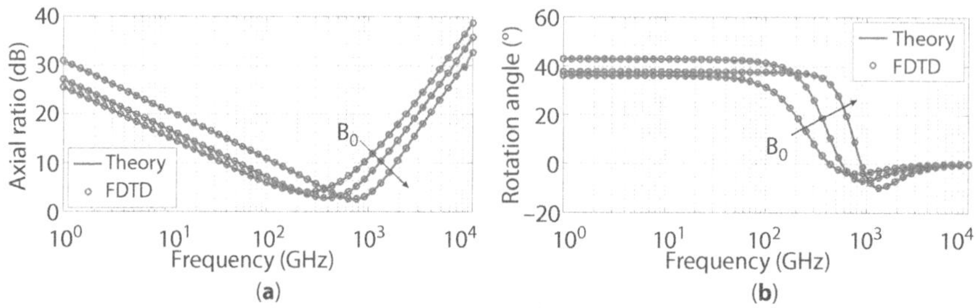

Figure 16.29 (a) Axial ratio and (b) polarization rotation angle of a normally incident plane wave on graphene with $\mu_c = 0.2$ eV, $\Gamma = 0.33$ meV, and $B_0 = 0.25$, 0.5, and 1 T.

Finally, the surface plasmon polariton waves on a magnetically biased graphene is investigated for two different magnetostatic bias fields, namely $B_0 = 0.5$ T and $B_0 = 1$ T. The simulations concern the far-infrared regime and the region is divided into two parts, identically to the electrically biased scenario. Numerical results are compared to the theoretical estimations, in Figure 16.30, indicating the matching and the accuracy of the modified FDTD algorithm. Consequently, the latter can be used for more complex devices to extract their electromagnetic response for advanced applications.

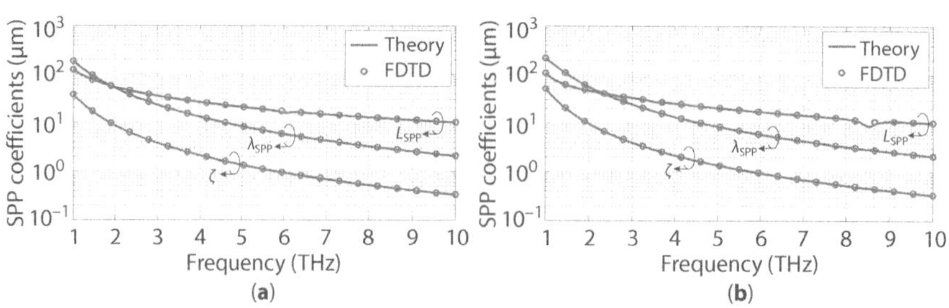

Figure 16.30 Propagation properties of surface waves on magnetically biased graphene of $\mu_c = 0.2$ eV, $\Gamma = 0.33$ meV, and (a) $B_0 = 0.5$ and (b) $B_0 = 1$ T.

16.6 Conclusion

In this chapter, the electromagnetic response of graphene has been thoroughly investigated analytically. Graphene has been considered as an infinitesimally thin layer characterized via its surface conductivity that depends on the quality of the material as well as the applied bias fields. The electromagnetic analysis revealed several exotic properties of graphene, such as the capability to support highly confined surface plasmon polariton waves at the far-infrared regime and the gyrotropy at the millimeter-wave spectrum. Moreover, various parametric analyses have been conducted to study the true nature of graphene's properties. Finally, a numerical implementation of graphene has been presented that treats the two-dimensional material as a surface. The developed algorithms are based on the powerful finite-difference time-domain method and numerical simulations proved the remarkable accuracy and efficiency in terms of the computational time and resources.

References

1. Novoselov, K.S., Geim, A.K., Morozov, S.V., Jiang, D., Zhang, Y., Dubonos, S.V., Grigorieva, I.V., Firsov, A.A., Electric field effect in atomically thin carbon films. *Science*, 306, 5696, 666–669, 2004.
2. Mikhailov, S. and Ziegler, K., New electromagnetic mode in graphene. *Phys. Rev. Lett.*, 99, 1, 016803, 2007.
3. Hanson, G.W., Dyadic Green's functions and guided surface waves for a surface conductivity model of graphene. *J. Appl. Phys.*, 103, 6, 064302, 2008.
4. Nikitin, A.Y., Garcia-Vidal, F.J., Martin-Moreno, L., Analytical expressions for the electromagnetic dyadic Green's function in graphene and thin layers. *IEEE J. Sel. Top. Quantum Electron.*, 19, 3, 4600611–4600611, 2013.
5. Sounas, D.L. and Caloz, C., Gyrotropy and non reciprocity of graphene for microwave applications. *IEEE Trans. Microwave Theory Tech.*, 60, 4, 901–914, 2012.
6. Philip Wong, H.-S. and Akinwande, D., *Carbon Nanotube and Graphene Device Physics*, Cambridge University Press, 2010.
7. Hanson, G.W., Dyadic Green's functions for an anisotropic, non-local model of biased graphene. *IEEE Trans. Antennas Propag.*, 56, 3, 747–757, 2008.
8. Kubo, R., Statistical-mechanical theory of irreversible processes. I. general theory and simple applications to magnetic and conduction problems. *J. Phys. Soc. Jpn.*, 12, 6, 570–586, 1957.
9. Gusynin, V., Sharapov, S., Carbotte, J., Magneto-optical conductivity in graphene. *J. Phys.: Condens. Matter*, 19, 2, 026222, 2006.
10. Chew, W.C., *Waves and Fields in Inhomogeneous Media*, vol. 522, IEEE Press, New York, 1995.
11. Bagby, J. and Nyquist, D., Dyadic green's functions for integrated electronic and optical circuits. *IEEE Trans. Microwave Theory Tech.*, 35, 2, 207–210, 1987.
12. Ishimaru, A., *Electromagnetic Wave Propagation, Radiation, and Scattering*, Prentice-Hall, 1991.
13. Sommerfeld, A., *Partial Differential Equations in Physics*, vol. 1, Academic Press, 1949.
14. Economou, E., Surface plasmons in thin films. *Phys. Rev.*, 182, 2, 539, 1969.
15. Amanatiads, S.A. and Kantartzis, N.V., Substrate controllable transverse magnetic surface waves onto a graphene layer at far-infrared frequencies. *7th International Congress on Advanced Electromagnetic Materials in Microwaves and Optics*, pp. 256–258, 2013.
16. Balanis, C.A., *Antenna Theory: Analysis and Design*, John Wiley & Sons, 2016.

17. Taflove, A. and Hagness, S.C., *Computational Electrodynamics*, Artech House Publishers, 2000.
18. Yee, K., Numerical solution of initial boundary value problems involving Maxwell's equations in isotropic media. *IEEE Trans. Antennas Propag.*, 14, 3, 302–307, 1966.
19. Kelley, D.F. and Luebbers, R.J., Piecewise linear recursive convolution for dispersive media using FDTD. *IEEE Trans. Antennas Propag.*, 44, 6, 792–797, 1996.
20. Berenger, J.P., A perfectly matched layer for the absorption of electromagnetic waves. *J. Comput. Phys.*, 114, 2, 185–200, 1994.
21. Amanatiads, S.A., Kantartzis, N.V., Tsiboukis, T.D., A loss-controllable absorbing boundary condition for surface plasmon polaritons propagating onto graphene. *IEEE Trans. Magn.*, 51, 3, 1–4, 2015.
22. Amanatiads, S.A., Kantartzis, N.V., Ohtani, T., Kanai, Y., Precise modeling of magnetically biased graphene through a recursive convolutional FDTD method. *IEEE Trans. Magn.*, 54, 3, 2017.
23. Taflove, A. and Umashankar, K., Radar cross-section of general three-dimensional scatterers. *IEEE Trans. Electromagn. Compat.*, 4, 433–440, 1983.

17

Graphene-Like $A_N B_{8-N}$ Compounds on Metals and Semiconductors

Sergei Yu. Davydov

Ioffe Institute, Russian Academy of Sciences, St. Petersburg, Russia

Abstract

Analytical expressions for densities of states of free graphene-like $A_N B_{8-N}$ compounds (GLC) and flat and buckled epitaxial monolayers on a metallic and semiconducting substrates have been obtained by the tight-binding theory using a low-energy approximation. Characteristic features of the densities of states as functions of the layer–substrate coupling constant and the buckling factor have been analyzed. Based on a proposed model of the electronic spectrum, adsorption theory is constructed, which allows determination of the role of the adatom level position, the adatom–substrate coupling constant, and the gap inherent to GLC in the free state with heteropolar bonds in the formation of the adatom electronic structure. The cases of free-standing and epitaxial GLCs on a metal surface are considered.

Keywords: Graphene-like compound, substrate, metal, semiconductor

17.1 Introduction

In recent years, interest in a theoretical description of various two-dimensional (2D) structures increased noticeably (e.g., Refs. [1–6] and references therein). In this case, significant attention is focused on graphene-like $A_N B_{8-N}$ compounds [7–14] and structures constructed on this base [15–20]. The point is that the $A_N B_{8-N}$ compounds (at $A \neq B$) are characterized by nonzero energy gap widths, unlike graphene, silicone, and germanene, which are semimetals or zero-gap semiconductors. It is the circumstance that makes $A_N B_{8-N}$ compounds promising elements of device structures. However, if we deal with real device structures, we should consider multilayer structures or, as a minimum, epitaxial layers formed on solid-state substrates, not free 2D sheets. In this case, a substrate is not only a basis for 2D-layers, but also it can favor their formation and stability [12, 13].

The works cited above (like the majority of other works performed in this field) present the results of numerical calculations performed using various variants of the density functional theory. Here, we used an approach based on the Green function method and the tight-binding theory [21]. The approach makes it possible to obtain analytical expressions for the electronic spectrum and the density of states (DOS) of epitaxial layers.

Email: Sergei_Davydov@mail.ru

Tobias Stauber (ed.) Handbook of Graphene: Volume 2, (549–592) © 2019 Scrivener Publishing LLC

17.2 Graphene-Like Compounds on Metals

17.2.1 General Consideration

To describe the electronic structure of an epitaxial layer, we used here the so-called adsorption approach [21]. This approximation consists in constructing a 2D lattice of the epitaxial not from A and B atoms but, taking into account their interaction with substrate, from A and B adatoms placed on a solid-state substrate, as schematically shown in Figure 17.1.

We have to find an epitaxial graphene-like compound (GLC) Green function

$$\mathbf{G} = \begin{pmatrix} G^{AA} & G^{AB} \\ G^{BA} & G^{BB} \end{pmatrix} \tag{17.1}$$

of a monolayer of interacting adatoms A and B. In accordance with the adsorption approach, we begin with the Green function of adsorbed adatoms A and B that do not interact to one another:

$$\mathbf{g} = \begin{pmatrix} g^{A} & 0 \\ 0 & g^{B} \end{pmatrix}, \quad g^{A(B)}(\omega) = (\Omega_{a(b)} + i\Gamma_{a(b)}(\omega))^{-1}, \tag{17.2}$$

where $\Omega_{a(b)} = \omega - \tilde{\varepsilon}_{a(b)}$, $\tilde{\varepsilon}_{a(b)} = \varepsilon_{a(b)} + \Lambda_{a(b)}(\omega)$, ω is the energy variable, $\varepsilon_{a(b)}$ is the energy of p orbital of atom $A(B)$,

$$\Gamma_{a(b)}(\omega) = \pi V_{a(b)}^{2} \rho_{sub}(\omega) \tag{17.3}$$

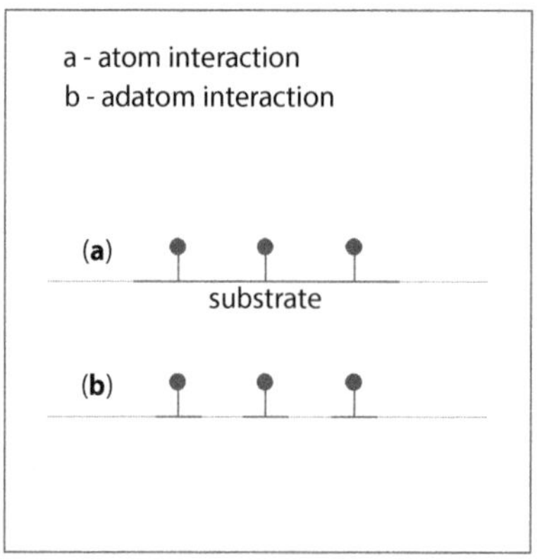

Figure 17.1 Schematic of the replacement of (a) atomic lattice adsorbed on a substrate by (b) a lattice of adatoms.

is the function of the half-width of a quasi-level of adatom $A(B)$, $V_{a(b)}$ is the matrix element of interaction between atom $A(B)$ and a substrate, $\rho_{sub}(\omega)$ is the density of states of the substrate, and

$$\Lambda_{a(b)}(\omega) = P\frac{1}{\pi}\int_{-\infty}^{\infty}\frac{\Gamma_{a(b)}(\omega')d\omega'}{\omega-\omega'} \tag{17.4}$$

is the function of the quasi-level shift of adatom $A(B)$ (P is the principal value symbol). Equations (17.2)–(17.4) are written, assuming that all adatoms A (and adatoms B too) occupy equivalent positions.

Figure 17.2 shows a cluster of the 2D structure that is necessary to derive expressions for the Green functions $G_{i,j}^{A(B)}(\omega, \mathbf{k})$, where i and j are numbers of the lattice sites. Coordinates (x, y) of atoms numerated in Figure 17.2, expressed in units of the distance between the nearest neighboring atoms a, are 0 – (0, 0), 1 – $(-\sqrt{3}/2, 1/2)$, 2 – $(\sqrt{3}/2, 1/2)$, 3 – (0, −1), 11 $(-\sqrt{3}, 0)$, 21 $(\sqrt{3}, 0)$, 12 – $(-\sqrt{3}/2, 3/2)$, 22 – $(\sqrt{3}/2, 3/2)$, 31 – $(-\sqrt{3}/2, -3/2)$, and 32 – $(\sqrt{3}/2, -3/2)$. Including interaction t (transition energy) between p_z orbitals of nearest atoms A and B and using the Dyson equations [21], we obtain the following relationships:

$$\begin{aligned} G_{0,0}^{AA} &= g_{0,0}^{A} + g_{0,0}^{AA}t(G_{1,0}^{BA} + G_{2,0}^{BA} + G_{3,0}^{BA}), \\ G_{1,0}^{BA} &= g_{11}^{B}t(G_{0,0}^{AA} + G_{11,0}^{AA} + G_{12,0}^{AA}), \\ G_{2,0}^{BA} &= g_{22}^{B}t(G_{0,0}^{AA} + G_{22,0}^{AA} + G_{21,0}^{AA}), \\ G_{3,0}^{BA} &= g_{33}^{B}t(G_{0,0}^{AA} + G_{31,0}^{AA} + G_{32,0}^{AA}), \end{aligned} \tag{17.5}$$

where $g_{ij}^{A(B)} = g^{A(B)}\delta_{ij}$ and δ_{ij} is the Kronecker symbol. With allowance for the transformation properties [21], the Green function takes the form

$$\begin{aligned} G^{AA(BB)}(\omega, \mathbf{k}) &= \frac{g^{A(B)}(\omega)}{1 - t^2 g^A(\omega) g^B(\omega) f^2(\mathbf{k})}, \\ f(\mathbf{k}) &= \sqrt{3 + 2\cos(k_x a\sqrt{3}) + 4\cos(k_x a\sqrt{3}/2)\cos(3k_y a/2)}, \end{aligned} \tag{17.6}$$

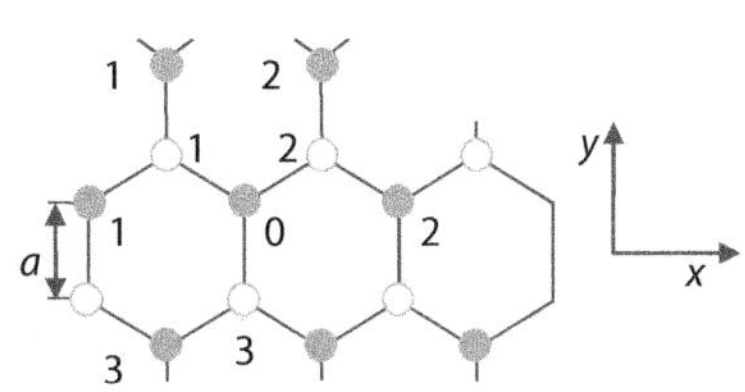

Figure 17.2 To the derivation of Equation (17.6): I is for atoms belonging to sublattice A, II is for atoms belonging to sublattice B; the numbers are ordinal numbers of the sites.

or

$$G^{AA(BB)}(\omega, \mathbf{k}) = \frac{\Omega_{b(a)} + i\Gamma_{b(a)}(\omega)}{(\Omega_a + i\Gamma_a(\omega))(\Omega_b + i\Gamma_b(\omega)) - t^2 f^2(\mathbf{k})}, \tag{17.7}$$

where $\mathbf{k} = (k_x, k_y)$ is the wave vector for electron motion in the sheet plane. The electronic spectrum of the system is determined from equation $\Omega_a\Omega_b = t^2f^2(\mathbf{k})$, which gives

$$\begin{aligned} E_{\pm}(\omega, \mathbf{k}) &= \varepsilon(\omega) \pm R(\omega, \mathbf{k}), \\ R(\omega, \mathbf{k}) &= \sqrt{\Delta^2(\omega) + t^2 f^2(\mathbf{k})}, \end{aligned} \tag{17.8}$$

where $\varepsilon(\omega) = (\tilde{\varepsilon}_a + \tilde{\varepsilon}_b)/2$, $\Delta(\omega) = (\tilde{\varepsilon}_a - \tilde{\varepsilon}_b)/2$; hence, $\Omega_{a(b)} = \omega - \varepsilon(\omega) \mp \Delta(\omega)$. Note that Equation (17.8) describes the valence π band for the sign "minus" and the π* conduction band for the sign "plus."

The density of states (DOS) of GLC calculated per one atom is

$$\begin{aligned} \rho_{AB}(\omega) &= \rho_A(\omega) + \rho_B(\omega), \\ \rho_{A(B)}(\omega) &= -\frac{1}{2\pi N}\sum_{\mathbf{k}} \operatorname{Im} G^{AA(BB)}(\omega, \mathbf{k}), \end{aligned} \tag{17.9}$$

where $\rho_{A(B)}(\omega)$ is the DOS on adatom $A(B)$, $N = N_A = N_B$ is the number of atoms in the A and B sublattices (the number of unit cells); the summation is performed over the first Brillouin band. The real and imaginary parts of the Green function (17.7) are given in Appendix 17.A, (1).

17.2.2 Free GLC Layers

In the case of free (unbounded with substrate) GLC, we have $\Gamma_a(\omega) = \Gamma_b(\omega) = 0$, $\Lambda_a(\omega) = \Lambda_b(\omega) = 0$, so that $E_{\pm}(\mathbf{k}) = \varepsilon \pm \sqrt{\Delta^2 + t^2 f^2(\mathbf{k})}$, where $\varepsilon = (\varepsilon_a + \varepsilon_b)/2$ and $\Delta = (\varepsilon_a - \varepsilon_b)/2$. Here and in what follows, we describe the electronic spectrum, using a low-energy approximation, taking $f(\mathbf{k}) \approx (3a/2)|\mathbf{q}|$, where $\mathbf{q} = \mathbf{K}-\mathbf{k}$, $\mathbf{K} = a^{-1}(2\pi/3\sqrt{3}, 2\pi/3)$ is the wave vector of the Dirac point [22]. Then, at $\mathbf{q} = 0$, the gap width in the spectrum of the A_NB_{8-N} is $2|D| = |\varepsilon_a - \varepsilon_b|$. The graphene, silicone, and germanene spectra do not have gaps. The DOS for free GLC is

$$\rho^0_{AB}(\omega, \mathbf{k}) = (\rho^0_A(\omega, \mathbf{k}) + \rho^0_B(\omega, \mathbf{k}) = \delta(\Omega - R(\mathbf{k})) + \delta(\Omega + R(\mathbf{k}), \tag{17.10}$$

where $\Omega = \omega - \varepsilon$. Using a low-energy approximation and going in Equation (17.9) from the summation to integration (Appendix 17.A, (2)), we obtain

$$\rho^0_{AB}(\Omega) = \begin{cases} \dfrac{1}{\pi\sqrt{3}}\dfrac{|\Omega|}{t^2}, & |\Omega| \geq |\Delta|, \\ 0, & |\Omega| < |\Delta|. \end{cases} \tag{17.11}$$

The reduced DOS $\bar{\rho}^0_{AB}(x)=\rho^0_{AB}(x)\cdot(\pi\sqrt{3}t)\equiv I_0(x)$ as a function of dimensionless energy $x=\Omega/t$ is shown in Figure 17.3, where $\delta=|\Delta|/t$ is the dimensionless half-width of the gap. At $\Delta=0$, Equation (17.11) is transformed to the DOS of the free single-layer graphene $\rho_g(\Omega)=|\Omega|/\pi\sqrt{3}t^2$ and coincides with Equation (17.15) from [22] without considering the degeneration.

Tables 17.1 and 17.2 give the values of $2\Delta=\varepsilon_p^A-\varepsilon_p^B$, where $\varepsilon_p^{A(B)}$ is the energy of the p state of atom $A(B)$ calculated using the Herman–Skillman [23] and Mann [24] tables of atomic terms (see also Ref. [25]). The same tables also list the results of the first-principle calculations in terms of various variants of the density functional theory [7, 8, 10–12]. As was shown in Refs. [7, 11, 12] and in other works and in their references on earlier

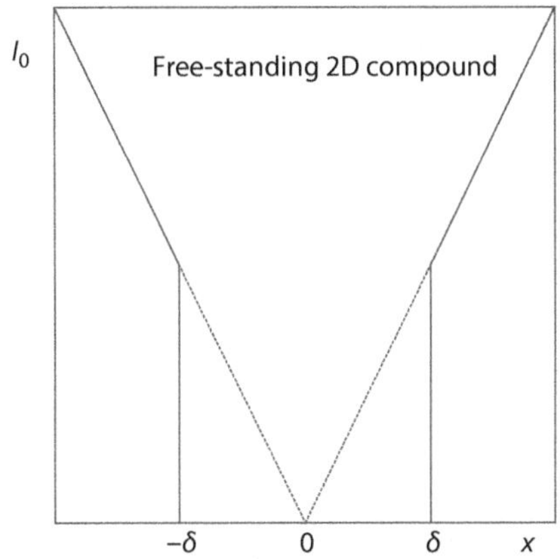

Figure 17.3 Dependence of reduced density of states I_0 from reduced energy $x=\Omega/t$ for a free A_NB_{8-N} monolayer. The dotted line corresponds to the case $\delta=|\Delta|/t=0$ (graphene, silicene, germanene).

Table 17.1 Gap widths $2\Delta=\varepsilon_p^A-\varepsilon_p^B$ (eV) in comparison with the first-principle calculations [7, 8, 11] for free hexagonal Group IV–IV 2D compounds.

Compound	**SiC**	**GeC**	**GeSi**	**SnC**	**SnSi**	**SnGe**
This work, tables from Ref. [23]	2.45	2.65	0.16	3.03	0.58	0.42
This work, tables from Ref. [24]	3.48	3.74	0.26	4.31	0.83	0.57
[7] Structure	2.52* 4.19* F	2.09 3.83 F	0.02 0.00 B	1.18* 6.18* F	0.23 0.68 B	0.23 0.40 B
[8]	3.526	3.160	0.275	–	–	–
[11]	2.547*	2.108	–	–	–	–

*Indirect gap, flat F, and buckled B structures; the upper results of Ref. [7] were obtained in terms of the density functional formalism without the gradient correction, and the lower result of Ref. [7] are obtained taking into account the correction.

Table 17.2 Gap widths $2\Delta = \varepsilon_p^A - \varepsilon_p^B$ (eV) in comparison with the first-principle calculations [7, 8, 10–12] for free hexagonal Group III–V 2D compounds.

Compound	BN	BP	BAs	BSb	AlN	AlP	AlAs	AlSb
This work, tables from Ref. [23]	4.83	1.69	1.27	0.60	6.61	3.47	3.05	2.38
This work, tables from Ref. [24]	5.41	1.11	0.55	-0.29	7.43	3.83	3.27	2.43
[7] Structure	4.61 6.36* F	0.82 1.81 F	0.71 1.24 F	0.39 0.23 F	3.08* 5.57* F	– – –	– – –	1.49* 2.16 B
[8]	6.377	1.912	1.594	–	–	3.453*	2.938*	–
[10] direct	4.48-6.07	0.82-1.36	0.72-1.18	0.29-0.61	–	–	–	–
[11]	4.606	–	–	–	3.037*	–	–	–
[12] indirect Structure	–	–	–	–	4.85 (5.03) F	3.24 (3.93) F	2.49 (3.08) B	2.07 (2.17) B

Compound	GaN	GaP	GaAs	GaSb	InN	InP	InAs	InSb
This work, tables from Ref. [24]	6.57	3.43	3.01	2.34	6.78	3.64	3.22	2.55
This work, tables from Ref. [25]	8.17	3.87	3.31	2.47	8.47	4.17	3.61	2.77
[7] Structure	2.27* 5.00* F	1.92* 3.08* B	1.29* 2.96* B	– – –	0.62* 5.76** F	1.18* 2.88* B	0.86** 2.07** B	0.68** 1.84** B
[8]	–	3.054*	2.475*	–	–	–	–	–
[11]	2.462*	–	–	–	–	–	–	–
[12] indirect Structure	3.23 (4.00) F	2.51 (3.21) B	1.83 (2.39) B	1.43 (1.88) B	1.52 (1.57) F	1.80 (2.32) B	1.41 (1.81) B	1.25 (1.62) B

** The direct gap in point Γ; the upper results of Refs. [7, 12] were obtained in terms of the density functional formalism without the gradient correction, and the lower results of Ref. [7] were obtained taking into account the correction. The other designations are the same as in Table 17.1.

results, some GLC structures did not have a flat structure (F) in a free state, but the structure was certainly transformed. Sublattice atoms were not arranged in the same planes; they were in two planes quite close to one another. Sometimes, this structure is called a buckled structure (B). At the same time, calculations [10] showed that the flat structure was more beneficial in all considered cases, and the possibility of formation of a bucked structure was neglected in Refs. [8, 10]. It should be noted as well that, in Refs. [7, 8, 10–12], the valence band maxima and the conduction band minima in some GLC belong to different points of the Brillouin zone (indirect gaps), which is indicated by "asterisks" in Tables 17.1 and 17.2. In this work, we do not consider the buckled structure for free GLC, and the existence of an indirect gap is not compatible with Equation (17.34) [10]. A comparison of values of $2\Delta = \varepsilon_p^A - \varepsilon_p^B$ results of calculations in Refs. [7, 8, 10–12] (taking into account their scatter due to the use of different variants of the density functional theory) showed a satisfactory agreement, exception for cases of GaN and InN. For BSb, we have $\Delta < 0$, which implies the existence of inequality $\varepsilon_p^A < \varepsilon_p^B$, not the absence of a gap due to overlapping of the valence and the conduction bands. In this case, the gap is $2|\Delta|$. Generally speaking, it is well known that the tight-binding methods, as a rule, overestimate the energy gap width. The DOS of free GLC obtained by numerical calculations were given in Refs. [9, 10, 12, 26]. A comparison showed that the low-energy approximation that we used for π electrons was quite acceptable for describing the GLC DOS near the gap and the band edges.

It is easy to show, using Equation (17.8) that, in free GLC with intrinsic conductivity ($E_F = \varepsilon$), the reciprocal electron and hole masses are $m_{e,h}^{-1} = \pm\hbar^{-2}(\partial^2 R(q)/\partial q^2)_{q=0}$ ($\hbar$ is the reduced Planck constant), respectively, and it follows that

$$\frac{1}{m_{e,h}} = \pm\frac{v_F^2}{|\Delta|}, \quad v_F = \frac{3at}{2\hbar}. \tag{17.12}$$

Here, we introduced the Fermi velocity v_F by analogy with graphene [22], although free nondegenerate semiconducting GLC have no electrons with the Fermi energy. According to the Harrison binding-orbital model, a matrix element of the π interaction of p_z orbitals is $t = \eta_{pp\pi}(\hbar^2/m_0a^2)$, where $\eta_{pp\pi} = 0.63$, and m_0 is the free electron mass [23–25]. Thus, we have $v_F \propto a^{-1}$.

Tables 17.3 and 17.4 give the values of a calculated from the first principles by various authors. It follows from the tables that the ratio of velocities $v_F(AB)/v_F(Gr) = a(Gr)/a(AB)$ always is lesser than 1, since $a(AB) >$ a(Gr) = 1.42 Å. In addition, Tables 17.3 and 17.4 give the values and corresponding ratios Δ/t. It is necessary to note the following: for graphene, $t \approx 2.38$ eV; hence, it follows that $v_F \approx 0.74 \cdot 10^6$ m/s, while the experimental value is $v_F(\mathrm{Gr}) \approx 1.1 \cdot 10^6$ m/s [27]. Because of this, in terms of the tight-binding approximation, it is usually taken that $t \sim 3$ eV. However, in this work, all estimations are carried out using the Harrison theory [23–25].

Tables 17.3 and 17.4 also give ratios $m_e/m_0(= -m_h/m_0)$ calculated by Equation (17.12) in comparison with the calculations performed in Ref. [13]. It can be seen that, in the case of holes, the agreement is quite adequate, but ratio m_e/m_0 obtained in this work is significantly higher than that in Ref. [13].

Table 17.3 Distances between nearest neighboring atoms a (Å), transition energy t (eV), ratios Δ/t for free hexagonal Group IV–IV 2D compounds.

Compound	SiC	GeC	GeSi	SnC	SnSi	SnGe
a	1.77	1.86	2.31	2.05	2.52	2.57
t	1.53	1.39	0.90	1.14	0.76	0.73
Δ/t, tables [24]	0.80	0.95	0.09	1.33	0.38	0.29
Δ/t, tables [25]	1.14	1.35	0.14	1.89	0.55	0.39
m_e/m_0, tables [24]	0.57	0.67	0.06	0.94	0.27	0.20
m_e/m_0, tables [25]	0.81	0.95	0.10	1.34	0.39	0.27
3D ϕ_{AB}, eV	4.95 [38]	–	4.51 [39]	–	–	–

The work function of SiC is given for the 6H polytype; the average of the work functions of Si and Ge was taken as the work function.

17.2.3 Flat Epitaxial Layers

Now, we consider epitaxial layers of GLC, for which we take $\Gamma_a(\omega) = \Gamma_b(\omega) = \Gamma(\omega)$, $\Lambda_a(\omega) = \Lambda_b(\omega) = \Lambda(\omega)$, but $\varepsilon_a \neq \varepsilon_b$. Here, some explanation is necessary. According to the Harrison theory [23–25], matrix elements $V_{a(b)} = \eta^{a(b)}(\hbar^2/m_0 d^2_{a(b)})$ entering in Equations (17.3) and (17.4) for the functions of broadening and shift are determined, first, by the character of adatom–substrate interaction (factor $\eta^{a(b)}$) and, second, by distances d_a and d_b between these adatoms and the substrate surface. The p_z orbitals take a part in the binding of adatoms A and B with substrate, so that $\eta^a = \eta^b$. Thus, the equations taken above suggest a flat F structure of epitaxial GLC for which $d_a = d_b = d$ and $V_a = V_b = V$. From Equation (17.8), we obtain $E_{\pm}(\omega, \mathbf{k}) = \varepsilon + \Lambda(\omega) \pm \sqrt{\Delta^2 + t^2 f^2(\mathbf{k})}$; hence, it follows that the bands of epitaxial GLC differ from the bands of free GLC by the shift along the energy axis by the value $\Lambda(\omega)$, while the value of $2\Delta = \varepsilon_p^A - \varepsilon_p^B$ remains unchanged and independent of energy. The DOS is

$$\rho_{AB}(\tilde{\Omega}) = \frac{I(\tilde{\Omega})}{\pi\sqrt{3}t},$$
$$I(\tilde{\Omega}) = \frac{\Gamma}{2\pi t}\ln\frac{|\xi^4 + b\xi^2 + c|}{c} + \frac{\tilde{\Omega}}{\pi t}\left(\operatorname{arctg}\frac{2\xi^2 + b}{4\Gamma\tilde{\Omega}} - \operatorname{arctg}\frac{b}{4\Gamma\tilde{\Omega}}\right). \tag{17.13}$$

Here, $\tilde{\Omega} = \omega - \varepsilon(\omega)$, $b = -2(\tilde{\Omega}^2 - \Delta^2 - \Gamma^2)$, $c = (\tilde{\Omega}^2 - \Delta^2)^2 + \Gamma^2(\Gamma^2 + 2\Delta^2 + 2\tilde{\Omega}^2)$, $\xi = 3taq_c/2$ is the cutoff energy, and q_c is the cutoff wave vector, for which inequality $q_c << 2\pi/a$ must

Table 17.4 Distances between nearest neighboring atoms a (Å), transition energy t (eV), ratios Δ/t for free hexagonal Group III–V 2D compounds.

Compound	BN	BP	BAs	BSb	AlN	AlP	AlAs	AlSb
a [7, 8]	1.45	1.83	1.93	2.12	1.79	2.28*	2.34*	2.57
a [10, 12]	1.44	1.84	1.93	2.13	1.80	2.27	2.34	2.54
t	2.28	1.43	1.29	1.07	1.50	0.92	0.88	0.73
Δ/t, tables [24]	1.06	0.59	0.49	0.28	2.21	1.88	1.74	1.64
Δ/t, tables [25]	1.18	0.39	0.21	-0.14	2.48	2.07	1.86	1.67
m_e/m_0, tables [24]	0.75	0.42	0.35	0.20	1.56	1.33	1.22	1.15
m_e/m_0, tables [25]	0.83	0.33	0.15	0.10	1.75	1.47	1.30	1.17
m_e/m_0, $\|m_h\|/m_0$, [13]	– –	– –	– –	– –	1.24 2.33	0.59 1.37	0.48 1.20	0.38 1.01
3D ϕ_{AB}, eV [38]	–	–	–	–	–	4.80	4.58	4.41

Compound	GaN	GaP	GaAs	GaSb	InN	InP	InAs	InSb
a [7, 8]	1.85	2.25	2.36	–	2.06	2.46	2.55	2.74
a [10, 12]	1.88	2.26	2.34	2.53	2.10	2.45	2.53	2.70
t	1.40	0.95	0.86	0.75	1.13	0.79	0.74	0.64
Δ/t, tables [24]	2.34	1.81	1.75	1.56	3.00	2.29	2.18	1.99
Δ/t, tables [25]	2.91	2.04	1.92	1.65	3.74	2.63	2.44	2.17
m_e/m_0, tables [24]	1.65	1.27	1.24	1.10	2.12	1.62	1.53	1.40
m_e/m_0, tables [25]	2.06	1.44	1.36	1.16	2.64	1.86	1.72	1.53
m_e/m_0, $\|m_h\|/m_0$, [13]	0.69 1.97	0.41 1.16	0.33 1.06	0.28 0.91	0.43 2.26	0.37 1.39	0.32 1.27	0.28 1.09
3D ϕ_{AB}, eV [39]	–	4.70	4.67	4.25	–	4.80	5.07	4.62

The starred values of a were taken from Ref. [8], others from Ref. [7]; the values of t were calculated using a from the first row of the table.

be fulfilled, according to the low-energy approximation. Following the authors of Ref. [28], by analogy with the Debye model, we took $\pi q_c^2 = (2\pi)^2 / S$ (where $S = 3a^2\sqrt{3}/2$ is the unit cell area); hence, it follows that $q_c = 2\sqrt{2\pi}/a\sqrt{3\sqrt{3}} \approx 2.2/a$ and $\xi = \sqrt{2\pi\sqrt{3}}t \approx 3.3t$. In this case, the DOS of free graphene $\rho_g(\Omega) = 2|\Omega|/\xi^2$ for $||\Omega$ and 0 in other cases [21, 28–30]. At $\Gamma = 0$, density of states (17.13) becomes zero at $|\tilde{\Omega}| < |\Delta|$ and $|\tilde{\Omega}| > \sqrt{\xi^2 + \Delta^2}$ in the range $\sqrt{\xi^2 + \Delta^2} \geq |\tilde{\Omega}| \geq |\Delta|$; we have $\tilde{\rho}_{AB}(\tilde{\Omega}) = |\tilde{\Omega}| / \pi\sqrt{3}t^2$.

Now, we consider the simplest Anderson model of DOS [21] as applied to a metallic substrate; in terms of the model, $r_{sub}(\omega)$ is independent of energy, so that $\Gamma(\omega) = \Gamma = \text{const}$, and $\Lambda(\omega) = 0$. The last equation follows from Equation (17.4) in an approximation of infinitely wide band when integration is performed from $-\infty$ to $+\infty$. In this case, $\tilde{\Omega} = \Omega = \omega - \varepsilon$.

Figure 17.4 shows the dependence of the reduced DOS $\rho^*_{AB}(x) = \rho_{AB}(x)\cdot(\pi\sqrt{3}t) \equiv I(x)$on dimensionless energy $x = \Omega/t$ for various interaction constants $\gamma = \Gamma/t$ and gap parameters $\delta = |\Delta|/t$. By symmetry, $\rho^*_{AB}(x) = \rho^*_{AB}(-x)$, and, because of this, only the region of positive

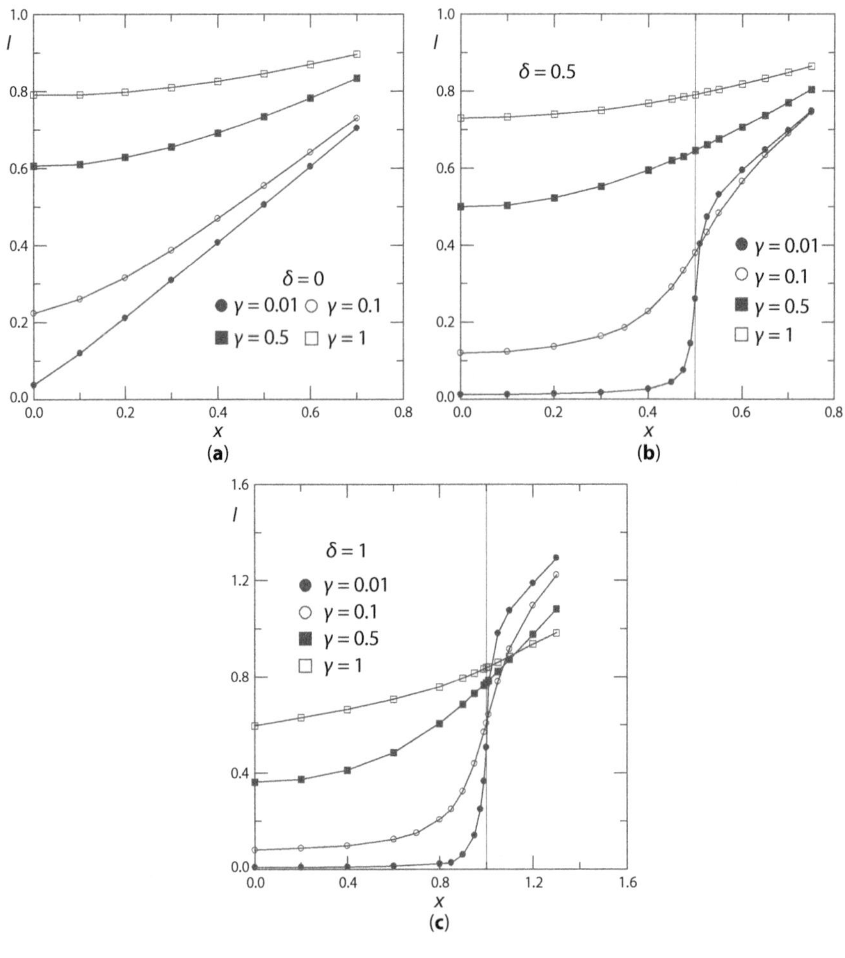

Figure 17.4 Dependence of reduced DOS $I(x)$ from reduced energy $x = \Omega/t$ for a flat structure at various interaction constants $\gamma = \Gamma/t$ and dimensionless parameters of the gap $\delta = |\Delta|/t$ (only the region of positive energies is shown): δ = (a) 0, (b) 0.5, (c) 1.0; γ = 0.01, 0.1, 0.5, 1.0.

energies is shown in Figure 17.4. From Figure 17.4b and c, it follows that epitaxial GLC on a metal do not have a gap as the region of forbidden energies. At low interaction constants γ, this region can be defined as a pseudogap. As γ increases, the influence of the substrate increases, and the pseudogap almost completely disappears at high interaction constants. Figure 17.4a demonstrates that $\rho^*_{AB}(0) \neq 0$ in the Dirac point for graphene, silicone, and germanene, and it is the main effect of interaction homopolar GLC with substrates. Figure 17.4 shows the reduced DOS $I(0)$ as a function of interaction constant γ (Figure 17.4a) and gap half-width δ (Figure 17.4b). It should be noted that the dependences have a substantially nonlinear character. Consider some analytical estimations of dimensionless DOS $I(x)$, the general expression of which is given in Appendix 17.A, (3). In the pseudogap center (at $x = 0$), we have

$$I(0) = \frac{\gamma}{\pi} \ln \frac{\bar{\xi}^2 + \delta^2 + \gamma^2}{\delta^2 + \gamma^2}, \tag{17.14}$$

where $\bar{\xi} = 3.3$. From Equation (17.14), it follows as well that the nonlinearity of dependence $I(0)$ on γ increases as δ decreases, which is demonstrated in Figure 17.5a. Dependence $I(0)$ on δ is similar: the nonlinearity increases with decreasing γ.

It can be shown (Appendix 17.A, (3)) that $I(x) - I(0) \approx x^2/\pi\gamma$ at $\delta = 0$ and $x^2 << \gamma^2 << \bar{\xi}^2$. This result corresponds to the case shown in Figure 17.4a for $\gamma = 0.01$ (quadratic dependence $I(x)$ at $x^2 << 0.01$ is not visible in this scale). In the more general case, $x^2 << \min\{\gamma^2, \delta^2\}$ and $\bar{\xi}^2 >> \max\{\gamma^2, \delta^2\}$, we obtain $I(x) - I(0) \approx (\gamma x^2/\pi)(3\delta^2 + \gamma^2)/(\delta^2 + \gamma^2)^2$. At $\delta^2 >> \gamma^2$, we have $I(x) - I(0) \approx 3\gamma x^2/\pi\delta^2$, which explains the dependences shown in Figure 17.4b and c for cases $\gamma = 0.01$ and 0.1. According to Equation (17.A12), in the case $x^2 = \delta^2$, we have $I(\delta) \approx (\delta/2) + (2\gamma/\pi)\ln\bar{\xi}$ that describes the dependences on δ and γ shown in Figure 17.4b and c. Let $x^2 >> \max\{\gamma^2, \delta^2\}$, but $|x| << \bar{\xi}$. Then, we obtain $I(x) \approx |x| + O(\gamma)$. Thus, as $|x|$ increases, the DOS tends to a linear dependence on energy; in this case, the differences between the values of $I(x)$ corresponding to various parameters γ and δ decrease (Figure 17.4).

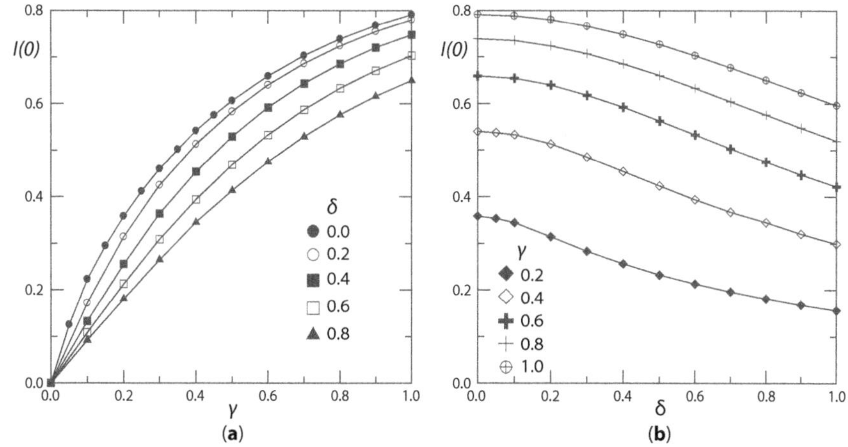

Figure 17.5 Reduced DOS $I(0)$ of the flat structure as a function of (a) interaction constant γ and (b) gap half-width δ.

17.2.4 Buckled Epitaxial Layers

Now, we consider buckled epitaxial layers of compounds $A_N B_{8-N}$ on a metal. The general expressions of the density of states are Equations (17.A13) and (17.A14) given in Appendix A, (4). Let $\Gamma_a = \Gamma$ and $\Gamma_b = \vartheta\Gamma$. Inequality $\vartheta < 1$ ($\vartheta > 1$) means that $V_b < V_a$ ($V_b > V_a$), since adatom B is much (less) far removed from the surface than adatom A. In this case, the DOS is

$$\rho'_{AB}(\Omega) = \frac{1}{\pi\sqrt{3}t} I'(\Omega),$$
$$I'(\Omega) = \frac{\Gamma(1+\vartheta)}{4\pi t}\ln\frac{|\xi^4 + b'\xi^2 + c'|}{c'} + \tag{17.15}$$
$$+\frac{\Omega}{\pi t}\left(\text{arctg}\frac{2\xi^2 + b'}{2\Gamma[(1+\vartheta)\Omega + (1-\vartheta)\Delta]} - \text{arctg}\frac{b'}{2\Gamma[(1+\vartheta)\Omega + (1-\vartheta)\Delta]}\right),$$

where $c' = (\Omega^2 - \Delta^2)^2 + \vartheta^2\Gamma^4 + \Gamma^2[(1+\vartheta^2)\Omega^2 + (1+\vartheta^2)\Delta^2 + 2(1-\vartheta)\Omega\Delta]$ and $b' = 2(\vartheta\Gamma^2 + \Delta^2 - \Omega^2)$. Figure 17.6 shows the typical plots of functions $I'(x)$, where $x = \Omega/t$ as before. Here, it should be noted that the symmetry of density of states is absent for cases $\Delta \neq 0$ and $\vartheta \neq 1$: $I'(\Omega) \neq I'(-\Omega)$. In this case, a specific point $\Omega^* = -\Delta(1-\vartheta)/(1+\vartheta)$ appears in which the second term in Equation (17.14) for $I'(\Omega)$ becomes zero (Equation (17.A15)). Generally speaking, the asymmetry of the DOS is not a quite unusual property. For example, the electron–hole symmetry disappears in graphene as the interactions between the second neighboring atoms are taken into account and in graphite as the interplanar interactions are taken into account (e.g., Ref. [22]).

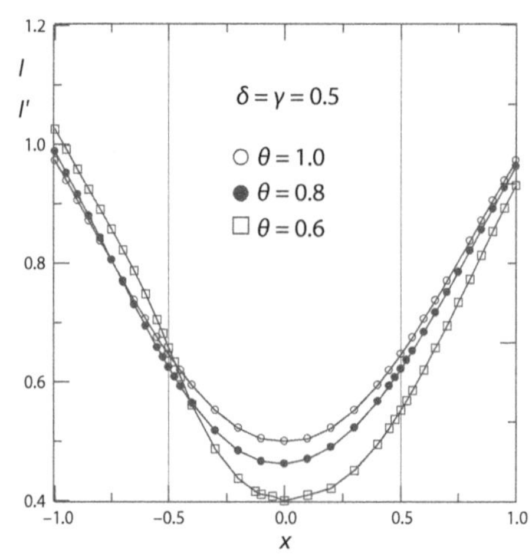

Figure 17.6 DOS of the buckled structure.

Let $1 - \vartheta = \alpha << 1$. Then, it can be shown in a linear approximation with respect to α that the reduced density of states of the buckled structure $I'(0)$ in point $\Omega = 0$ can be represented as

$$I'(0) \approx I(0) - \alpha S(0),$$
$$S(0) = \frac{\gamma}{2\pi}\left(\ln\frac{\bar{\xi}^2 + \delta^2 + \gamma^2}{\delta^2 + \gamma^2} - \frac{2\bar{\xi}^2\gamma^2}{(\delta^2 + \gamma^2)(\bar{\xi}^2 + \delta^2 + \gamma^2)} \right), \quad (17.16)$$

where the reduced DOS of a flat layer $I(0)$ is described by Equation (17.14). Similarly, it is easily shown that $I'(x^*) \approx I(0) - \alpha S(0)$, where $x^* = \Omega^*/t = -\delta(1 - \vartheta)/(1 + \vartheta)$, so that $I'(x^*) \approx I'(0)$. Hence it follows, in particular, that, in the range $x < |x^*|$, the second term in Equation (17.15) for $I'(\Omega)$ has a maximum. However, this maximum is very low and is invisible in the scale of Figure 17.6.

Now, we consider ratios $\eta_{\pm} = I'(\pm\delta)/I(\delta)$, which characterize the degree of asymmetry of the density of states of the buckled structure. Figure 17.6 presents the results of the corresponding calculations. In the case $1 - \vartheta = \alpha << 1$, the reduced densities of states $I'(\pm\delta) \approx I(\delta) - \alpha S(\pm\delta)$, where $I(\delta)$ is given by Equation (17.A12) and $S(\pm\delta)$ is described by Equations (17.A16) and (17.A17). Assuming for simplicity that $\bar{\xi}^2 >> \max\{\gamma^2, \delta^2\}$, we obtain

$$S(\delta) \approx S(-\delta) \approx \frac{\gamma}{\pi}\left(\frac{1}{2}\ln\frac{\bar{\xi}^2}{\gamma(\gamma^2 + 4\delta^2)^{1/2}} - \frac{\gamma^2 + 3\delta^2}{\gamma^2 + 4\delta^2} \right), \quad (17.17)$$

which corresponds to the dependences shown in Figure 17.6 for $\vartheta = 0.8$ and in Figure 17.7 for $\gamma = 0.1$. The asymmetry of the DOS increases as ϑ decreases and γ increases. The nonlinearity of dependences $\eta_{\pm}(\vartheta)$ increases, too. The buckling factor can be estimated as follows. According to Equation (17.3), buckling factor $\vartheta = \Gamma_b/\Gamma_a = (d_a/d_b)^4$. For definiteness, we assume that $d_b - d_a + z_{\perp}$, and we obtain $z_{\perp} = d_b(1 - \vartheta^{1/4})$. According to the data

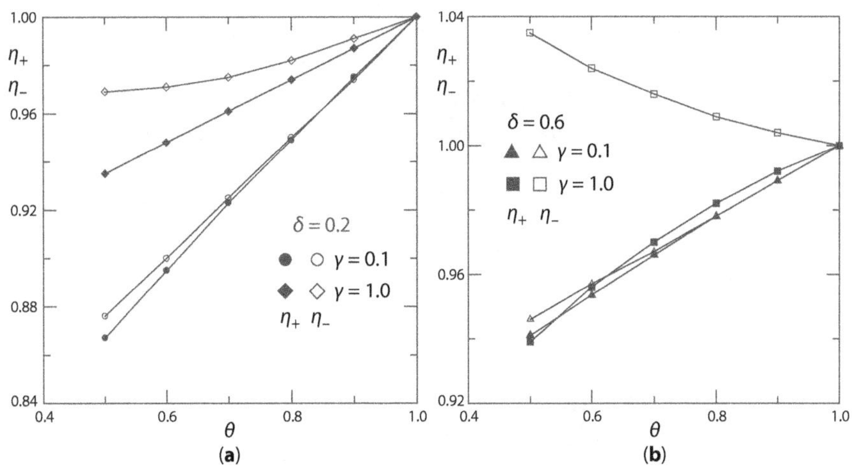

Figure 17.7 Ratios $\eta_{\pm}$ vs. buckling factor ϑ.

from Ref. [13], for III-V epitaxial layers on transition and rare-earth metals, the values $z_{\perp}$ can be both positive and negative, with magnitudes comparable with those of d_a and d_b (Supplemental Material to Ref. [13]). Generally speaking, the layers of GLC can inherit the buckling from the free states (B structures in Tables 17.1 and 17.2) and can form due to interaction with substrate.

17.2.5 Estimations of Charge Transfer and Binding Energy

Let us estimate the charge transition between a flat layer of a GLC and a metallic substrate. From general considerations, it is clear that as Fermi level E_F of a metal coincides with the pseudogap center, the charge transition is absent. Assuming that $\gamma^2 << \delta^2 << \bar{\xi}^2$, we approximate the reduced DOS, according to Equation (17.A9), by the expression

$$I(x) \approx \frac{2\gamma}{\pi}\left(\ln(\bar{\xi}/\delta) + \frac{x^2}{\delta^2}\right). \tag{17.18}$$

At zero temperature, the mean occupation number n_{AB} per atom in the layer is

$$n_{AB} = \int_{-\infty}^{E_F} \rho_{AB}(\omega)\,d\omega = \frac{1}{\pi\sqrt{3}}\int_{-\infty}^{e_F} I(x)\,dx, \tag{17.19}$$

where $e_F = E_F/t$ is the reduced Fermi level, and value $\varepsilon = (\varepsilon_a + \varepsilon_b)$ is taken as energy zero. It follows that, at a low shift of the Fermi level from the center of the band to the region of negative energies, the occupation number decreases by value

$$\nu \approx \frac{2\gamma\,|e_F|}{\pi^2\sqrt{3}}\ln(\bar{\xi}/\delta), \tag{17.20}$$

and, as the Fermi level shifts to the region of positive energies, the occupation number increases by the same value. In the first case, the GLC layer is charged positively and, in the second, negatively. We should emphasize that we consider the transition of π electrons, according to Equation (17.19). Perform some numerical estimates. We assume that the p_z states of a GLC are related to the s states by the σ-bond. Corresponding matrix element is $V_{a(b)} = V_{sp\sigma} = \eta_{sp\sigma}(\hbar^2/m_0 d^2)$, where $\eta_{sp\sigma} = 1.42$ [24, 25] (the estimation of matrix elements of adatom–substrate interaction is discussed in more detail in Ref. [31]). According to the Friedel model (e.g., Ref. [32]), we take for transition metals DOS $\rho_{sub} = N_d/W_m$, where W_m is the conduction band width, and $N_d = 10$. Then, instead of Equation (17.19), we have

$$\nu \approx \frac{2N_d}{\pi\sqrt{3}}\ln(\xi/|\Delta|)\left(\frac{\eta_{sp\sigma}}{\eta_{pp\pi}}\right)^2\left(\frac{a}{d}\right)^4\frac{|E_F|}{W_m}, \tag{17.21}$$

from which we obtain $\nu \sim 20\ |E_F|/W_m$ at $a \sim d$. Note that Equation (17.20) was obtained, assuming that $|E_F|<<t$; hence, it follows that $|E_F|/W_m<<1$ (values of W_m for transition metals are given in Refs. [32, 33]).

In terms of the model taken in this work, the Fermi energy $E_F = \phi_m - |\varepsilon|$, where ϕ_m is the work function of a metallic substrate, $\varepsilon=(\varepsilon_p^A+\varepsilon_p^B)/2$, and $\varepsilon_p^{A(B)}$, as above, is the energy of the p state of atom $A(B)$ counted from the vacuum level. According to the Atomic Term Tables [23–25], ε of carbon ~ 10 eV, while graphene work function ϕ_{Gr} is 4.3–5.1 eV [34–36]. Because of this, the estimations should be performed, assuming that $E_F = \phi_m - \phi_{AB}$, where ϕ_{AB} is the work function of GLC. It was shown in Refs. [12, 13] that most promising substrates for forming hexagonal 2D layers are, in particular, copper and nickel, in which the work functions of faces (100) are $\phi_{Cu} = 4.59$ and $\phi_{Ni} = 5.22$ eV [37], respectively. Thus, values of ϕ_{Gr}, ϕ_{Cu}, and ϕ_{Ni} are close in magnitude, so that inequality $|e_F|<<1$ is fulfilled for epitaxial graphene with an accuracy sufficient for estimations. The same can be said about 2D Group III–V compounds, since all work functions ϕ_{AB} of the hexagonal structures calculated in Ref. [13] have values within 4.25–5.25 eV. The last rows of Tables 17.3 and 17.4 give the estimates of the work functions of intrinsic bulk (3D) semiconducting A_NB_{8-N} compounds, taking $\phi_{AB}=\chi_{AB}+E_g^{AB}/2$, where the first term is the electron affinity and the second term is the forbidden band half-width [38, 39]. The value of ϕ_{AB} shown in the tables not only fall in the range 4.25–5.25 eV, but are also close to the calculated results [13]. According to the theory [20, 21], the change in the work function of a metallic substrate induced by the adsorption of a monolayer of GLC is

$$\Delta\phi_m = -\frac{4\pi e^2 d}{(S/2)}\Delta Z, \tag{17.22}$$

where e is the electron charge, and ΔZ is the change in the charge of an adsorbed atom equal to $\pm\nu$ for the transition of π-electrons from the layer into the substrate $\pm\nu$ and conversely, respectively. We assume once again that $a \sim d$ and obtain $|\Delta\phi_m| \sim 10\ (e^2/a)\nu$. Taking into account that $e^2 = 14.4$ eV A and $a \sim 2$ Å, we obtain $|\Delta\phi_m| \sim 0.7$ eV at $\nu \sim 0.01$. Our estimate of the shift of the Fermi level at a metal surface caused by adsorption $\Delta E_F = -\Delta\phi_m$ is close to that from Ref. [36], that is, 0.47 eV for graphene adsorption on metals. The value of the transfer of ~0.25 electron/atom obtained in Ref. [13] is thought to be overestimated by an order of magnitude. Note that, in the general form, we have

$$|\Delta\phi_m| \propto \frac{e^2a^2}{d^3}\frac{|E_F|}{W_m}. \tag{17.23}$$

However, it should be taken into account what we understand as a monolayer in the case of GLC [21]. The smooth surface model considers an adsorbate monolayer as a 2D structure corresponding to any closely packed plane of a corresponding 3D structure. In this case, the surface concentration of adatoms of GLC is $S/2$ as was taken in Equation (17.21). On the other hand, if the substrate surface is a network of deep potential wells, the monolayer corresponds to the situation when all wells are occupied with adatoms.

Now, we consider the estimation of the energy of adsorption bond E_{ads} of a monolayer of a GLC with a metallic substrate. According to the adsorption theory [21], E_{ads} can be represented as the sum of metallic E_{met} and ionic E_{ion} components. The former is calculated from the standard expression [21, 33]

$$E_{met} = \int_{-\infty}^{E_F} (\omega - E_F)\Delta\rho_{sys}(\omega)d\omega, \tag{17.24}$$

where $\Delta\rho_{sys}$ is the change in the DOS of the layer–substrate system due to adsorption. Without resorting to calculations, we estimate E_{met} using the uncertainty relation $\Delta p_z \cdot \Delta_z \sim \hbar$, where Δp_z and Δz are the momentum and coordinate uncertainties, respectively. Assuming that, as adsorption leads to uncertainty $\Delta z \sim d$ in the electron position in free layer in direction perpendicular to the layer, the gain due to decrease in the kinetic energy is $(\hbar^2/m_0 d^2)$. Now, we take into account that only ν electrons per atom were delocalized. Then, the metallic component of the adsorption energy is

$$E_{met} \sim -\nu \frac{\hbar^2}{m_0 d^2}. \tag{17.25}$$

Since $\hbar^2/m_0 = 7.62$ eV·Å², for $d = 2$ A and $\nu = 0.01$, we obtain $E_{met} \sim 0.2$ eV, which agrees well with the results of Ref. [13]. Here, we neglect the $E_{ion} \sim \nu^2 e^2/4d$ component.

To conclude this section, we make some additional remarks. When writing Equation (17.5), we *a priori* assumed that all adatoms $A(B)$ are in equivalent states, which corresponds to the so-called smooth substrate model in the adsorption theory [21]. If the crystallographic characteristics of a 2D layer and the substrate surface are substantially different, such a situation can occur only in the case when the interaction between atoms A and B are much stronger than their coupling with the substrate. In our model, this corresponds to the case $\gamma \ll 1$. In the analysis, we have not restricted ourselves to only small constants, since we understood that, at $\gamma \sim 1$ and higher, the lattice of the GLC must be changed, at least, transiting to a stressed (strain) state and, as a maximum, aligning (at $\gamma \gg 1$) to a structure similar to the structure of the surface face of the substrate. As was shown in Refs. [12, 13], the possibility exists to choose real substrate A_3B_5 monolayer pairs in which the lattice misfit of contacting structures is relatively small.

We described a metal using a simplest approximation of an infinitely wide band. Let us introduce a "pedestal" model, taking $\rho_{sub}(\omega) = N_d/W_m$ at $|\omega - \varepsilon_m| \le W_m/2$, where ε_m is the energy of the conduction band center, and $\rho_{sub}(\omega) = 0$ at $|\omega| - \varepsilon_m| > W_m/2$. Then, for flat layers, $\Gamma(\omega) = \Gamma$ at $|\omega - \varepsilon_m| \le W_m/2$ and $\Gamma(\omega) = 0$ in other cases, and

$$\Lambda(\omega) = \frac{N_d V^2}{W_m} \ln\left|\frac{W_m - 2\varepsilon_m + 2\omega}{W_m + 2\varepsilon_m - 2\omega}\right|. \tag{17.26}$$

If the conduction band center of a metal ε_m is near the gap center of a GLC (the origin of counting the energy is $(\varepsilon_a + \varepsilon_b) = 0$, when $|\varepsilon_m|/W_m<<1$), then, at low energies, we have $\Lambda(\omega) \approx (4N_d V^2 / W_m^2)(\omega - \varepsilon_m)$. It is clear that this correction changes the DOS only insignificantly, introducing a weak asymmetry at $\varepsilon_m \neq 0$. However, the "pedestal" model makes it possible to simulate the transition from metals of the beginning to metals of the end of the *d* series, shifting the Fermi level from energy $(-W_m + \varepsilon_m)$ to energy $(W_m + \varepsilon_m)$. Rare-earth metal substrates can be described by replacing N_d for $N_f = 14$. Some additional aspects of the considered problems can be found in [40–45].

17.3 Graphene-Like Compounds on Semiconductors

Now we turn to a semiconductor substrate, i.e., to GLC/bulk semiconductor heterostructures. To avoid misunderstandings, it should be emphasized that all the heterostructures considered in this section are currently hypothetical. The only exception is hexagonal boron nitride (*h*-BN): there are theoretical studies of two-layer and three-layer systems graphene/*h*-BN [17] and *h*-BN/graphene/*h*-BN [46]. One more point should also be noted: the A_NB_{8-N} compound, which is unstable in a free-standing two-dimensional state, can become stable in the form of an epitaxial layer, as is the case for metal substrates [12, 13].

17.3.1 Flat Epitaxial Layers

For a semiconductor substrate, the DOS $\rho_{sub}(\omega)$ is chosen as the model used in Ref. [47], but in the approximation of infinitely wide conduction band and valence band [48]

$$\rho_{sub}(\omega) = \begin{cases} A\sqrt{-\omega - E_g/2}, & \omega < -E_g/2; \\ A\sqrt{\omega - E_g/2}, & \omega > E_g/2; \\ 0, & |\omega| \leq E_g/2, \end{cases} \tag{17.27}$$

where A is the factor with the dimension $eV^{-3/2}$ and the zero energy is at the center of the band gap of the substrate. Then, we can write the broadening function $\Gamma_{a(b)}(\omega) = \pi V_{a(b)}^2 \rho_{sub}(\omega)$ and the shift function $\Lambda_{a(b)}(\omega) = AV_{a(b)}^2 \bar{\Lambda}(\omega)$, where

$$\bar{\Lambda}(\omega) = \begin{cases} F_-(\omega), & \omega < -E_g/2, \\ F_-(\omega) - F_+(\omega), & -E_g/2 \leq \omega \leq E_g/2, \\ -F_+(\omega), & \omega > E_g/2 \end{cases} \tag{17.28}$$

and $F_\pm(\omega) = \pi\sqrt{\pm\omega + E_g/2}$. Note that the DOS of epitaxial GLC on semiconductor is still given by Equation (17.13).

The main possible variants of the mutual arrangement of the energy bands in the monolayer of the GLC and the semiconductor substrate in the absence of their interaction $(V = 0, \Lambda(\omega) = 0)$ are shown in Figure 17.8. If these diagrams are considered as analogues

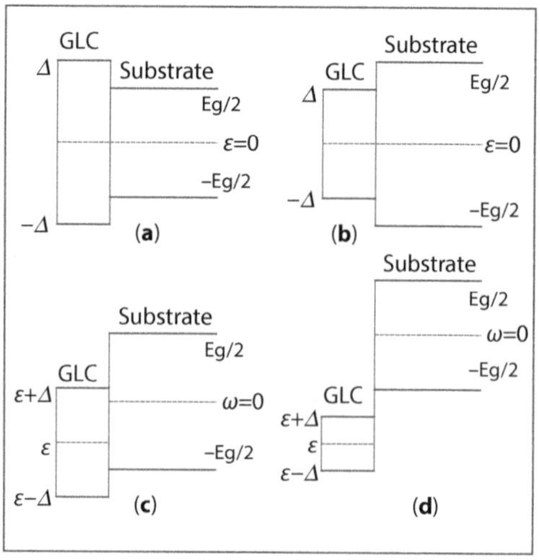

Figure 17.8 Typical cases of the mutual arrangement of energy bands for a layer of the GLC with the energy gap 2Δ and a semiconductor substrate with the band gap E_g in the absence of their interaction with each other. Solid horizontal lines indicate the top of the valence band and the bottom of the conduction band. Dashed lines show the center of the energy gap ε and the center of the band gap, which is taken as zero energy. (a, b) Straddling heterojunction (type I), (c) staggered heterojunction (type II), and (d) broken heterojunction (type III).

of heterojunctions, cases *a* and *b* correspond to straddling transitions of type I; case *c*—to a staggered transition of type II; and case *d*—to a broken transition of type III [39]. Now, we include the interaction V and consider the DOS of epitaxial GLC. The main problem in this case is to elucidate how the interaction affects the diagram shown in Figure 17.8.

Further, we will be interested in the energy regions close to the energy gap of the GLC and the band gap of the substrate. In the latter case, i.e., in the energy region $|\omega| \le E_g/2$, the function $\Gamma(\omega)$ vanishes identically, so that the first term in expression (17.13) disappears. In addition, the second term also disappears if the numerators of the arctangent arguments have the same sign, i.e., satisfy the inequality $b(2\xi^2 + b) > 0$. In this case, the DOS of the epitaxial GLC is equal to zero. The above inequality is satisfied when $|\tilde{\Omega}| < \Delta$ and $\tilde{\Omega}| > R$, where $\tilde{\Omega} = \omega - \varepsilon - \pi A V^2\left(\sqrt{-\omega + E_g/2} - \sqrt{\omega + E_g/2}\right)$ and $R = \sqrt{\xi^2 + \Delta^2}$.

For the further analysis, it is convenient to change over to dimensionless variables by taking the transition energy t equal to unity and setting the following parameters: $x = \omega/t$, $X = \tilde{\Omega}/t = x - e - \lambda$, $e = \varepsilon/t$, $e_g = E_g/t$, $\gamma = \Gamma/t$, $\lambda = \Lambda/t$, $\delta = \Delta/t$, and $r = R/t = \sqrt{2\pi\sqrt{3} + \delta^2}$. The expressions for the functions $\gamma(x)$ and $\lambda(x)$ are given in Appendix 17.B, (1). It is also shown that the coupling constant of the layer between the graphene-like compound and the 6*H*-SiC substrate is given by the formula $C = \pi A V^2/\sqrt{t} \sim v^2$, where $v = V/t$. Thus, the strong and weak couplings between the layer and the substrate correspond to the inequalities $C \gg 1$ and $C \ll 1$, respectively. It should be noted that, in the aforementioned form, expression (17.13) for the DOS $I(\tilde{\Omega})$ in the interval $|x| \le e_g/2$ is transformed into $|X|$, so that condition $\Delta \le |\tilde{\Omega}| \le R$, corresponding to the nonzero epitaxial DOS (17.13), is reduced to the inequality

$$\delta \le |x - e - C(\sqrt{-x + e_g/2} - \sqrt{x + e_g/2})| \le r. \quad (17.29)$$

The reduced DOS $I(x)$ are shown in Figures 17.9 and 17.10. We set $e_g = 2$, which, for the average value of the transition energy $t \sim 1.5$ eV (see Section 17.2), corresponds to the band gap of the 6H-SiC substrate, and considered the cases of $\delta = 1.5$ ($2\delta > e_g$) and $\delta = 0.5$ ($2\delta < e_g$). The coupling constants $C = 0.2$, 1, and 5 simulate the weak, intermediate, and strong layer–substrate coupling regimes, respectively. The values of $e = 0$, -0.6, and -1.4 are chosen so that, in the absence of coupling, they correspond to the diagrams shown in Figure 17.8.

Let us begin the discussion of the obtained results with the region $|x| \leq e_g/2$, where the reduced density of states is $I(x) = |X|$. The graphical method for solving inequality (17.29), which is demonstrated in Figure 17.11, consists in the fact that the points of intersection of the modulus of the function $X(x)$ with straight lines δ (points) and r (points) determine the boundaries of the energy gaps in the region $|x| \leq e_g/2$. By assuming that $C = C' + c'$, where

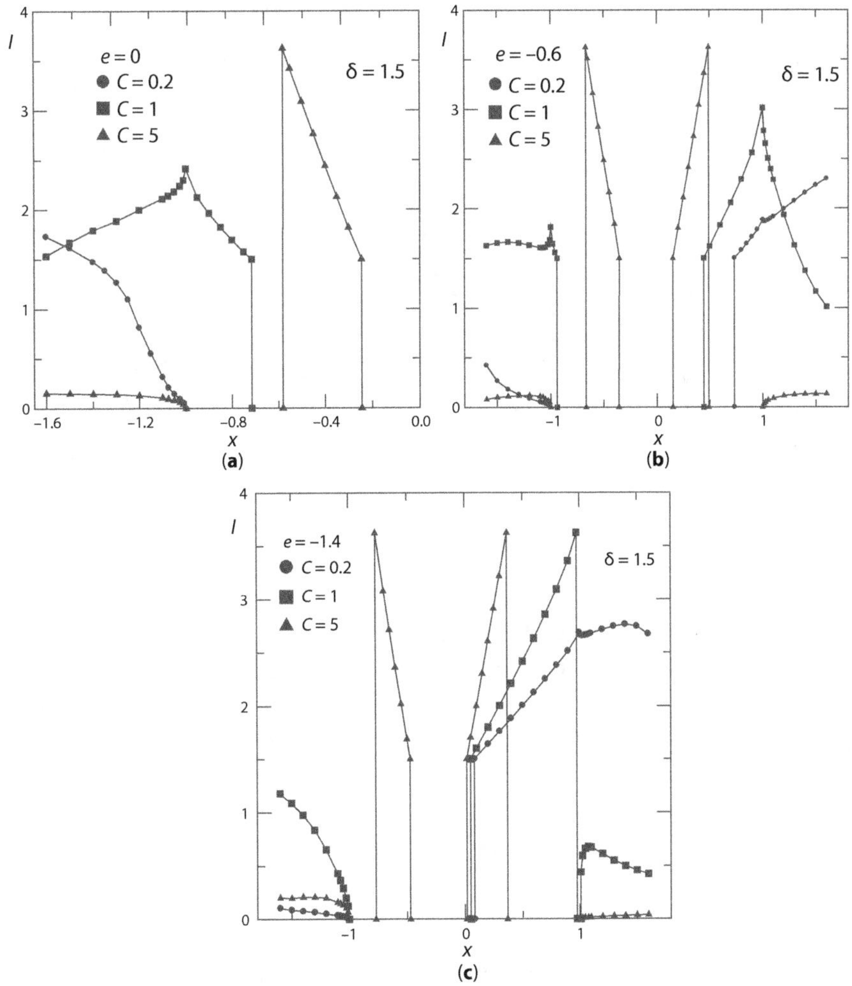

Figure 17.9 Dimensionless DOS $I(x)$ for δ = 1.5 with parameters e = (a) 0, (b) −0.6, and (c) −1.4 and coupling constants C = 0.2 (circles), 1 (squares), and 5 (triangles). For the symmetric case e = 0, only the left half of the figure is presented.

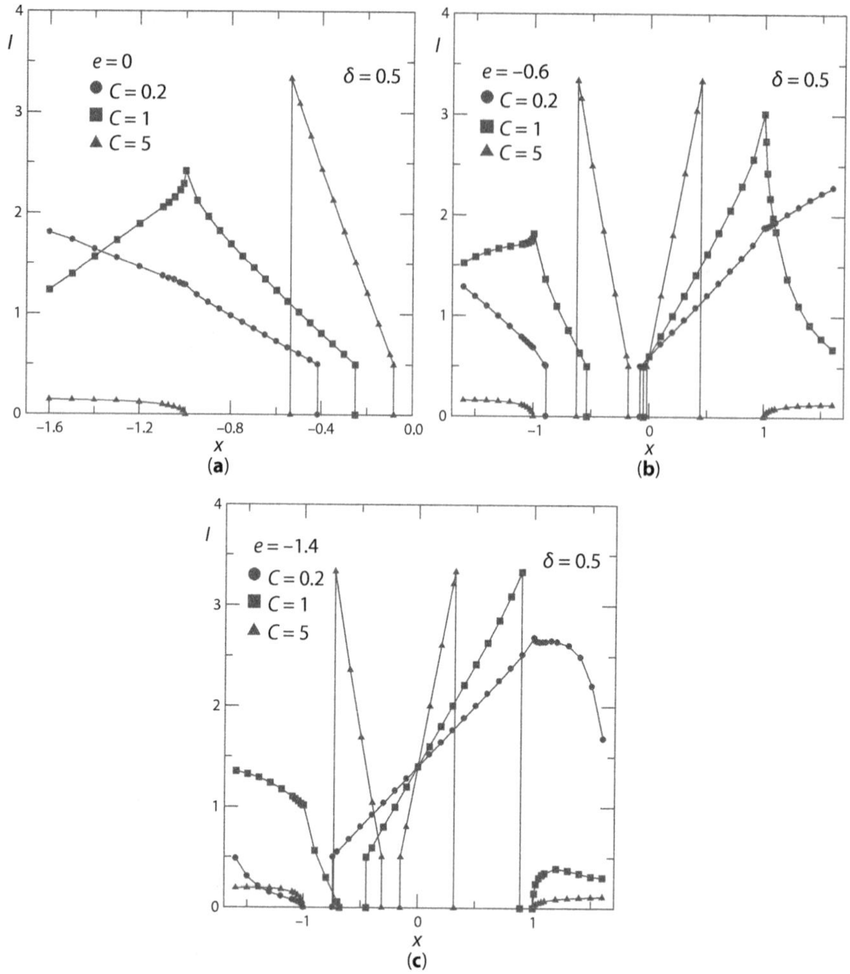

Figure 17.10 Same as in Figure 17.9, but for δ = 0.5.

$C' \gg c'$ and $|x| \le e_g/2$ (here, $\bar{x}_\delta << e_g/2$), we obtain $\bar{x}_\delta \approx c'\sqrt{e_g}$. Thus, the initial narrowing of the region of forbidden states is linear in the coupling constant.

For the case of $e = 0$ (Figure 17.11a), it is easy to show that the nonzero density of states in the region of the band gap of the substrate arises under the condition

$$C \ge C' = \frac{\delta - e_g/2}{\sqrt{e_g}}. \tag{17.30}$$

For the case of $2\delta > e_g$ shown in Figure 17.9a, we can write $C' = 1/2\sqrt{2}$. Therefore, for the coupling constant $C = 0.2 < C'$, the region of the band gap of the substrate remains unchanged. With an increase in the coupling constant C, the slope of the function $|X|$ also increases (see Appendix 17.B, (1)). For $C = 1 > C'$, the DOS of the GLC, by its

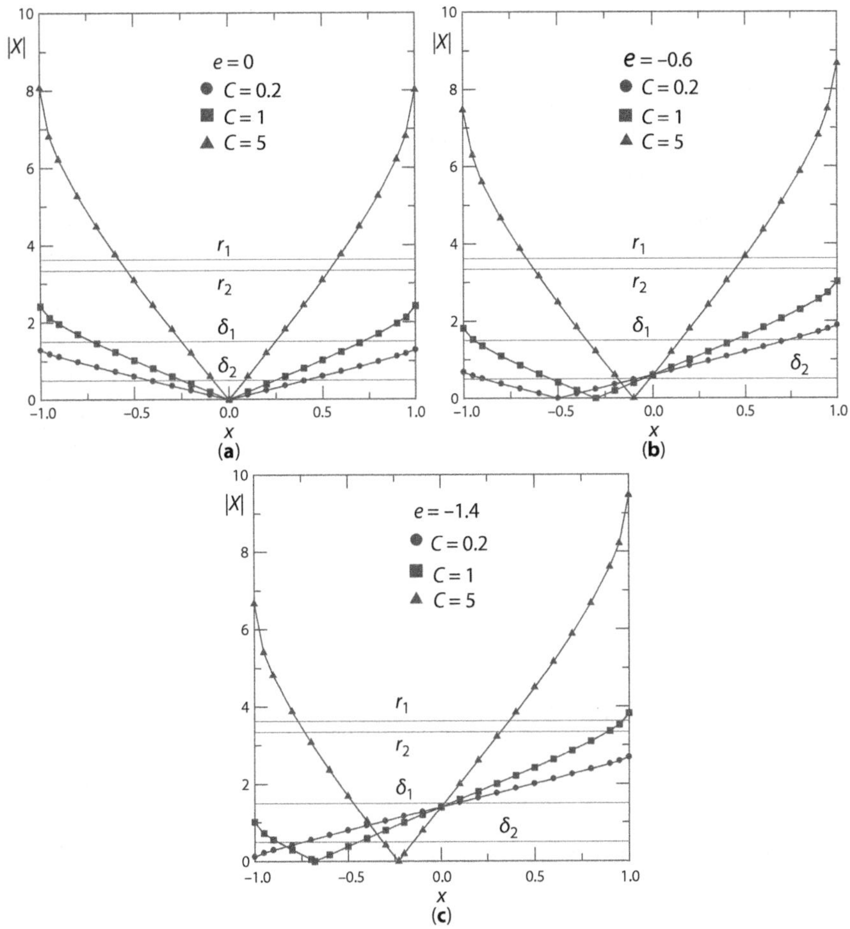

Figure 17.11 To the solution of inequality (17.29). Thin straight lines correspond to the parameters $r_{1,2} = \sqrt{\xi^2 + \delta_{1,2}^2}$, $\delta_1 = 1.5$, and $\delta_2 = 0.5$. The designations are the same as in Figure 17.9.

internal boundaries (the top of the valence band and the bottom of the conduction band), extends into the band gap of the substrate. As a result, the band gap of the system becomes narrower. In the general case, for $C > C'$, the energy gap in the DOS of the epitaxial GLC has the coordinates $(-x_\delta, x_\delta)$. It should be emphasized that, in this case, we can only conditionally speak about the DOS of the GLC and solely in order to simplify the interpretation of the change in the density of states with an increase in the coupling constant. Note also that, for the coupling constant $C = C'$, the internal boundaries of the density of states of the GLC coincide with the boundaries of the band gap of the substrate. Or, in other words, the band gap of the epitaxial layer coincides with the band gap of the substrate.

When the coupling constant C reaches the value

$$C'' = \frac{r - e_g/2}{\sqrt{e_g}}, \tag{17.31}$$

the outer boundaries of the DOS of the GLC (the bottom of the valence band and the top of the conduction band) coincide with the corresponding boundaries of the band gap of the substrate. A further increase in the coupling constant C leads to the situation shown in Figure 17.9a for $C = 5 > C'' = (\sqrt{2\pi\sqrt{3} + 2.25} - 1)/\sqrt{2}$ all DOS of the GLC lies inside the band gap. In the general case, for $C > C''$ in the region of the band gap of the substrate, there are three energy gaps, namely the central energy gap with the coordinates $(-x_\delta, x_\delta)$ and two side energy gaps with the coordinates $(-e_g/2, -x_r)$ and $(x_r, e_g/2)$. By assuming that $x_\delta << 1$, we obtain $x_\delta \approx \delta/(1 + C)$, so that, with an increase in the coupling constant, the DOS of the GLC, which is built in the band gap of the substrate, is shifted toward lower energies.

Let us now turn to the case when, as before, $e = 0$, but $2\delta < e_g$, so that inequality (17.31) is always satisfied (Figure 17.10a). For $C = 0$, the boundaries of the central energy gap $(-x_\delta, x_\delta)$ coincide with the top of the valence band and the bottom of the conduction band of the free-standing GLC, i.e., with the coordinates $(-\delta, \delta)$. In the case of $C \ll 1$, by setting $x_\delta = \delta - \bar{x}_\delta$, where $\bar{x}_\delta << \delta$, we obtain $\bar{x}_\delta = C(\sqrt{\delta + e_g/2} - \sqrt{-\delta + e_g/2})$, i.e., the narrowing of the band gap, which is linear in the coupling constant. A further increase in the coupling constant C results in a shift of the boundaries $(-x_\delta, x_\delta)$ toward the low-energy range (Figure 17.9a for $C = 0.2$ and $C = 1$). In the case of $C = C'' = (\sqrt{2\pi\sqrt{3} + 0.25} - 1)/\sqrt{2}$, the coordinates $(-x_r, x_r)$, which determine the bottom of the valence band and the top of the conduction band of the DOS of the GLC, coincide with the boundaries of the band gap of the substrate $(-e_g/2, -e_g/2)$. For $C = 5 > C''$, the DOS of the GLC lies within the band gap of the substrate, which contains three energy gaps, as in the case of $2\delta > e_g$ (Figure 17.9a). Next, we consider the case of $e \neq 0$ (Figures 17.9 and 17.10b and c). Using the data presented in Figure 17.11b and c, it is easy to show that, instead of conditions (17.30) and (17.31), we now obtain

$$C \geq C'_{\pm} = \frac{\delta \pm e - e_g/2}{\sqrt{e_g}}, C''_{\pm} = \frac{r \pm e - e_g/2}{\sqrt{e_g}}, \tag{17.32}$$

where the signs plus and minus refer to the positive and negative energy regions, respectively. For $e < 0$, we have $C'_+ < C'_-$ and $C''_+ < C''_-$. As a consequence of these inequalities, the DOS of the GLC begins to shift, at lower coupling constants (and, therefore, to a larger extent), from the bottom of the conduction band of the substrate toward the region of its band gap (Figures 17.9 and 17.10b and c). Hence, for the coordinates $(-x_{\delta-}, x_{\delta+})$ and $(-x_{r-}, x_{r+})$, we have $x_{\delta+} < x_{\delta-}$ and $x_{r+} < x_{r-}$. It should be noted that the widths W of the DOS of epitaxial layers of the GLC, which are shifted toward the band gap of the substrate (Figures 17.9 and 17.10 for $C = 5$), are substantially less than the corresponding values of W_0 for the free-standing GLC. Indeed, we have in relative units $W_0 = r - \delta \approx 2.12$ for $\delta = 1.5$, and $W_0 \approx 2.84$ for $\delta = 0.5$.

Figures 17.9 and 17.10 correspond to the energy band diagrams shown in Figure 17.8a–c. In order to illustrate the case shown in Figure 17.8d, it is necessary and sufficient to set $(-e) > r + C\sqrt{e_g} + e_g/2$. Then, we have $|X| > r$, so that the band gap of the substrate is empty.

The specific features of the DOS of the epitaxial GLC in the region of allowed states near the bottom of the conduction band and the top of the valence band of the substrate are considered in Appendix 17.B, (2), with the explanation of the dependences shown in Figures 17.9 and 17.10.

It should also be noted that all the dependences shown in Figures 17.9–17.11 are symmetric with respect to the simultaneous replacements of x with $-x$ and e with $-e$.

17.3.2 Buckled Epitaxial Layers

We now turn to buckled epitaxial layers of the A_NB_{8-N} compounds. By analogy with Section 17.2, we assume that $\Gamma_a = \Gamma$, $\Gamma_b = \vartheta\Gamma$ and $\Lambda_a = \Lambda$, $\Lambda_b = \vartheta\Lambda$, where ϑ is the buckling factor. Since the parameters $\Gamma_{a(b)}$ and $\Lambda_{a(b)}$ are proportional to $V_{a(b)}^2$ and $V_{a(b)} \propto d_{a(b)}^{-2}$, the inequality $\vartheta < 1$ ($\vartheta > 1$) means that the B adatom is more (less) remote from the surface than the A adatom. It should be noted that, in the framework of the adopted model, from the same reasoning it is possible to assume that the two-dimensional layer is flat, whereas the surface of the substrate is "buckled," as is the case, for example, with ionic crystals [39]. Taking into account the degree of buckling of the layer, the DOS of the GLC has the form

$$\rho'_{AB}(\hat{\Omega}) = \frac{1}{\pi\sqrt{3}t} I'(\hat{\Omega}),$$

$$I'(\hat{\Omega}) = \frac{\Gamma(\omega)((1+\vartheta)}{4\pi t} \ln\frac{|\xi^4 + b'\xi^2 + c'|}{c'} +$$

$$+\frac{\hat{\Omega}}{\pi t}\left(\operatorname{arctg}\frac{2\xi^2 + b'}{2\Gamma(\omega)[(1+\vartheta)\hat{\Omega} + (1-\vartheta)\hat{\Delta}]} - \operatorname{arctg}\frac{b'}{2\Gamma(\omega)[(1+\vartheta)\hat{\Omega} + (1-\vartheta)\hat{\Delta}]}\right), \quad (17.33)$$

where $c' = (\hat{\Omega}^2 - \hat{\Delta}^2)^2 + \vartheta^2\Gamma^4(\omega) + \Gamma^2(\omega)[(1+\vartheta^2)\hat{\Omega}^2 + (1+\vartheta^2)\hat{\Delta}^2 + 2(1-\vartheta)\hat{\Omega}\hat{\Delta}], b' = 2(\vartheta\Gamma^2 + \hat{\Delta}^2 - \hat{\Omega}^2)$, $2\hat{\Delta} = |2\Delta + (1-\vartheta)\Lambda(\omega)|$, $\hat{\Omega} = \omega - \varepsilon - ((1+\vartheta)/2)\Lambda(\omega)$. $\Lambda(\omega) = \pi AV^2 \left(\sqrt{-\omega + E_g/2} - \sqrt{\omega + E_g/2}\right)$.

The condition for the existence of a finite DOS for buckled layers of GLC in the region of the band gap of the substrate is similar to inequality $\Delta \le |\tilde{\Omega}| \le R$. By assuming that $1 - \vartheta = 2\alpha \ll 1$ and $\alpha C << \delta_x$, in the dimensionless form we obtain

$$\delta_x^2 \le X^2 + 2\alpha\lambda(X - \lambda) \le r_x^2, \quad (17.34)$$

where, as before, $X = x - e - \lambda$ (flat layer) and $\delta_x = \hat{\Delta}/t$, $r_x = \hat{R}/t$. For $e = 0$, we have the product $\lambda(X - \lambda) < 0$, so that $|X'| < |X|$, where $X' = x - e - ((1 + \vartheta)/2)\,\lambda(x)$ (buckled layer). This result enhances the probability that the right inequality holds true and weakens that for the left inequality. By analogy with the case shown in Figure 17.11a, we come to the following conclusions. In the weak coupling regime $C \ll 1$, when for the flat layer of the GLC, any states in the band gap of the substrate are absent (for example, the case of $C = 0.2$ and $\delta = \delta_1$ in Figure 17.11a), such states are also absent for the buckled layer. With an increase in the coupling constant C, in the regions $|x| \sim 1$ close to the edges of the band gap, the situation can arise where there are solutions $\delta_1 < |X|$ for the flat layer (an example is the case shown in Figure 17.11a), whereas for the buckled sheet, as before, we have $|X'| < \delta_1$. Moreover, the region of forbidden states becomes narrower for the flat layer (Figure 17.10a) and remains unchanged for the buckled layer. In the strong coupling regime $C \gg 1$, when both conditions (17.34) are satisfied for the flat and buckled layers (an example is the case shown in Figure 17.11a for $C = 5$), the width of the bands of allowed states for the flat layer is narrower than that for the buckled layer. It is also easy to imagine the situation where with an increase in the coupling constant C, the right inequality (17.34) still holds true for the buckled layer of the GLC but is no longer valid for the flat layer. This means that, in the first case, there is only the central energy gap in the DOS, whereas in the second case, the density of states additionally includes two side energy gaps adjacent to the boundaries of the allowed bands (Figure 17.10a).

In the cases where the condition $\alpha C << \delta_x$ is not satisfied, the situation is complicated, even though, as before, we have $|X'| < |X|$ (Figure 17.12). First, for the buckled layers of the GLC, the symmetry of the DOS with respect to the change in the sign of x is broken due to the asymmetry of δ_x and r_x. Second, in the regime of strong coupling between the narrow-band-gap layer and the substrate, the dependences of the parameters δ_x and r_x on the buckling factor ϑ are essentially nonlinear. The calculations performed for the case of $C = 5$ and $\delta = 0.5$ demonstrate (Figure 17.13) that, with an increase in the degree of buckling (decrease in ϑ), the bands of allowed states located inside the band gap of the substrate are broadened, while the energy gaps, accordingly, become narrower. For $e = 0$, the differences in the widths of the valence band W_- and the conduction band W_+ are small, whereas for $e = 1.4$, the differences are significant. There are also differences in the band shifts, which correspond to a decrease in the buckling factor ϑ from 1 to 0.5. For $e = 0$, the bottom of the conduction band almost does not shift and remains at the point $r_{\delta+} \approx 0.083$, while the top of the valence band is shifted from $r_{\delta-} \approx 0.091$ to $r_{\delta-} \approx -0.142$. For $e = 1.4$, the bottom of the conduction band is shifted from $r_{\delta+} \approx 0.313$ to $r_{\delta+} \approx 0.394$, whereas the top of the valence band is shifted from $r_{\delta-} \approx 0.150$ to $r_{\delta-} \approx 0.189$. In the first case, the width of the central energy gap increases from 0.174 to 0.225, while in the second case, it increases from 0.163 to 0.205. Thus, the inclusion of the degree of buckling can lead to qualitative changes in the density of states of the epitaxial layer of the GLC of the region of the band gap of the substrate. It should be noted that, in this case, too, the widths of the density of states for the epitaxial layers are significantly less than those for the free-standing layers.

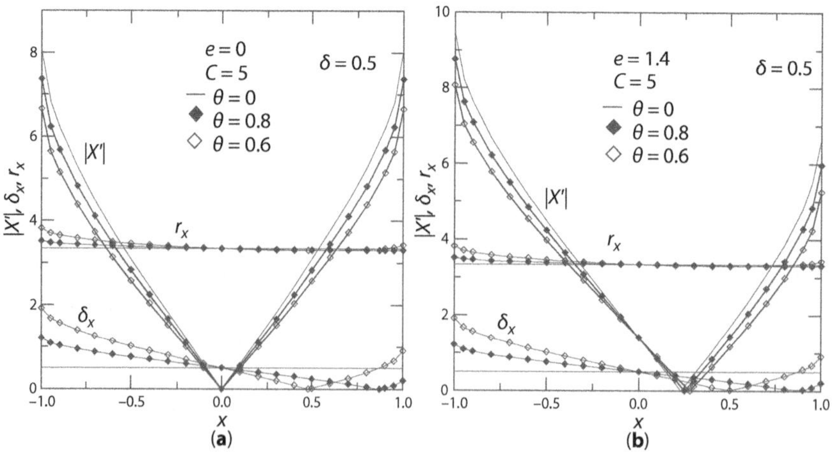

Figure 17.12 To the solution of inequality (17.34) for the parameters $C = 5$; $\delta = 0.5$; $e =$ (a) 0 and (b) 1.4; $\vartheta = 1$ (thin lines corresponding to the smooth structure), 0.8 (closed rhombuses), and 0.6 (open rhombuses); $\delta_x = \hat{\Delta}/t$, and $r_x = \hat{R}/t$.

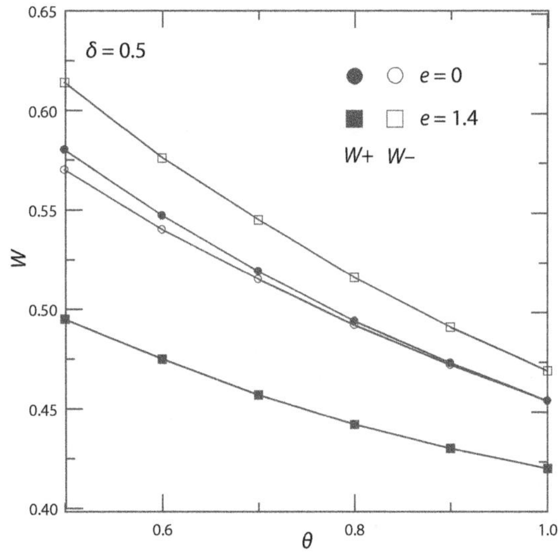

Figure 17.13 Dependences of the conduction band width W_+ and the valence band width W_- of the epitaxial layer on the buckling factor ϑ for $e = 0$ (closed and open circles) and $e = 1.4$ (closed and open squares).

Let us discuss the specific features of the DOS of the buckled sheet of the GLC in the energy regions near the bottom of the conduction band and the top of the valence band of the substrate. Setting $x = \pm(e_g/2) \pm \nu$ ($\nu << e_g$) by analogy with flat layers for the strong coupling regime $C \gg 1$, we obtain $I'(\nu) \propto (1 + \vartheta)$. The weak coupling regime $C \ll 1$ is considered in Appendix 17.B, (3). The analysis demonstrates that the inclusion of the degree of buckling leads only to quantitative differences from the flat case.

Until now, we have thought that the degree of buckling affects only the matrix elements of the layer–substrate coupling $V_{a(b)}$. In the adsorption theory, however, it is considered that the level of the adatom $\varepsilon_{a(b)}$ undergoes a Coulomb shift $V_{im}^{a(b)} = e_0^2 / 4d_{a(b)}$, where e_0 is the electron charge [21]. (This shift is caused by the electron image forces of the adsorbed atom in the substrate. For simplicity, we omit here the factor $(\varepsilon_s - 1)/(\varepsilon_s + 1)$, where ε_s is the static dielectric permittivity of the substrate. We also ignore the possible electrostatic field generated by substrate surface atoms carrying an electric charge.) Hence, it follows that the difference between the shifts is $V_{im}^a - V_{im}^b = V_{im}^a(1-\sqrt[4]{\vartheta})$. Now, the width of the energy gap is determined by the expression $2\hat{\Delta} + V_{im}^a(1-\sqrt[4]{\vartheta})$. In principle, the correction under consideration is relatively small. For $d_a = 2$ Å and $\vartheta = 0.8$, we obtain $V_{im}^a(1-\sqrt[4]{\vartheta}) \approx 0.1$ eV. However, in the case of graphene (and other gapless GLC in the free-standing state), this correction leads to qualitative changes, or, more precisely, to the formation of a substrate-induced energy gap.

17.3.3 Estimations of the Charge Transfer

We assume that the substrate and the GLC in the free-standing state are intrinsic semiconductors. In the case where the middle of the energy gap in the graphene-like compound coincides with the middle of the band gap of the substrate, which, in turn, coincides with the Fermi level E_F, there is no charge transfer between the substrate and the epitaxial layer. It is possible, however, that there is a redistribution of electrons between different regions of allowed states of the epitaxial layer of the GLC. At zero temperature, the average occupation number n_{AB} per atom in the layer has the form Equation (17.19).

Let us consider the occupation numbers of flat layers for the cases shown in Figures 17.9a and 17.10a. In the weak ($C = 0.2$) and intermediate ($C = 1$) coupling regimes, the DOS $I(x)$ differs from zero in the interval from $-\infty$ to $e_V = E_V/t$, where E_V is the energy of the valence band top of the GLC. Integration of expression (17.19) in these limits gives $n_{AB} = n_{val} = 1$, where n_{val} is the contribution to the occupation number from the energy region corresponding to the valence band of the epitaxial layer. In the strong coupling regime ($C = 5$), we have the occupation number $n_{AB} = n_{val} + n_{gap}$, where n_{gap} is the contribution to the occupation number from the nonzero DOS in the region of the lower half of the energy gap. The calculation of the contribution n_{gap} according to formula (17.19) is presented in Appendix 17.B, (4). For the cases shown in Figures 17.9a and 17.10a, we have for the coupling constant $C = 5$ the contribution $n_{gap} \approx 0.15$. However, it is convenient to use the approximation based on a nearly linear dependence of $|X|$ on x (Figures 17.9 and 17.10). Then, for the case where the density of states lies in the range $x^{**} < e_F$, instead of expressions (17.B12) and (17.B13), we can write

$$J \approx \frac{r+\delta}{2}(x^{**} - x^{*}). \qquad (17.35)$$

If the Fermi level lies in the range $x^{*} < e_F < x^{**}$ (partially occupied band), then, in formula (17.35), instead of x^{**}, we should introduce e_F. For the cases shown in Figures 17.9a and 17.10a, formula (17.35) for $C = 5$ gives $n_{gap} \approx 0.16$.

Using estimate (17.35), we calculate the values of n_{gap} for the cases of $e \neq 0$ and $C = 5$ (Figures 17.9 and 17.10b and c), as before, considering that $e_F = 0$. For the completely occupied valence bands of the GLC, at $\delta = 1.5$ we obtain $n_{gap}^{val} \approx 0.15$ for $e = -0.6$ and $n_{gap}^{v} \approx 0.14$ for $e = -1.4$; at $\delta = 0.5$, we have ≈ 0.16 for $e = -0.6$ and ≈ 0.15 for $e = -1.4$. For the cases of $\delta = 0.5$, $e = -0.6$, and $e = -0.14$, the Fermi level crosses the conduction band. The corresponding occupation numbers are as follows: $n_{gap}^{con} \approx 0.006$ and $n_{gap}^{con} \approx 0.053$. Using formula (17.19) and approximation (17.B9) for calculating the contribution from the valence band, we obtain $n_{val} \propto C(\bar{\xi}^2 / X_{-0}^2)$.

Further simplifications are possible if the inequality $\max\{|x_r|,|x_d|\} << e_g/2$ is satisfied. Then, for the strong coupling regime ($C \gg 1$), we obtain $J \sim \pi\sqrt{3}/C\sqrt{2e_g}$, so that $n_{gap} \sim (C\sqrt{2e_g})^{-1}$. For $C = 5$ and $e_g = 2$, we have $n_{gap} \sim 0.1$. This is a rough estimate of the contribution from the completely occupied band of allowed states lying in the gap band to the total occupation number of the epitaxial layer of the GLC. From this estimate, it follows, in particular, that in the strong coupling regime, the specific values of the parameters e and δ have little effect on the contribution n_{gap}.

To this point, we considered the flat layers. In order to change over to the buckled layers, the coupling constant C should be multiplied by the factor $(1 + \vartheta)/2$ (including the case of finding the values of x^{*} and x^{**}).

In this paper, we do not consider the charge transfer between the layer of the GLC and the substrate, because we specified the DOS of the substrate in the form Equation (17.27), i.e., in the approximation of infinitely wide bands, and used the low-energy approximation.

Thus, the main conclusion drawn in this section lies in the fact that a sufficiently strong interaction of GLC with a semiconductor substrate leads to the formation of a complex electronic structure of the system, which cannot be described by a simple superposition of energy diagrams of its components (as is done in Figure 17.8). Therefore, for these systems, we cannot rely on simple schemes (rules) used for the construction of energy band diagrams of the Shockley–Anderson type [39]. Such schemes have been developed for a contact between two "bulk" semiconductors, although even, in this case, in order to describe the interface, sometimes, it is necessary to introduce an intermediate (third) layer. In the case of heterojunctions formed by two-dimensional compounds with bulk semiconductors, adsorption models of multilayer coatings are more adequate to the situation under consideration (see, for example, Ref. [21]). This remark also applies to vertical van der Waals structures [2, 27] consisting of 2D layers.

In this section, we considered the main types of heterojunctions proceeding from the energy band diagrams shown in Figure 17.8 and by varying the layer–substrate coupling constant C. It was shown that, when the coupling constant C reaches the values of and determined by formulas (17.32), instead of one band gap in the system, there can arise three regions of forbidden states, between which, naturally, electronic transitions can occur. This, in principle, opens up possibilities both for the experimental verification and for the design and development of devices with the operating range determined not only by the choice of a "GLC–semiconductor" pair but also by the heteroepitaxial technology, which can be used to significantly change the coupling constant from the values typical of covalent bonds (GLC as a buffer layer) to the energies of van der Waals bonds (quasi-free-standing GLC).

It should be noted that the buckling of layers of GLC with a band gap in the free-standing state does not lead to any qualitative changes in the electronic structure of the system in comparison with flat GLC. However, for the initially gapless 2D structures, the buckling of layers can lead to the formation of a substrate-induced energy gap. We emphasize once again that, in the framework of the model used in this study, the cases of "buckled layer–flat substrate" and "flat layer–buckled substrate" are equivalent to each other.

17.4 Adsorption on Graphene-Like Compounds

17.4.1 Free-Standing GLC

From general considerations [21], Green's function $G_a(\omega)$ for an atom adsorbed on epitaxial graphene can be written as

$$G_a^{-1}(\omega) = \omega - \varepsilon_a - \Lambda_a(\omega) + i\Gamma_a(\omega). \qquad (17.36)$$

where ε_a is the adatom single electron level energy; the adatom quasi-level broadening function is given by

$$\Gamma_a(\omega) = \pi V_{a/sub}^2 \rho_{sub}(\omega), \qquad (17.37)$$

where $V_{a/sub}$ is the matrix element of adatom–substrate coupling, and the quasi-level energy-shift function is given by

$$\Lambda_a(\omega)=\frac{1}{\pi}P\int_{-\infty}^{\infty}\frac{\Gamma_a(\omega')d\omega'}{\omega-\omega'}. \tag{17.38}$$

The adatom density of states corresponding to Green's function (17.36) is written as

$$\rho_a(\omega)=\frac{1}{\pi}\frac{\Gamma_a(\omega)}{[\omega-\varepsilon_a-\Lambda_a(\omega)]^2+\Gamma_a^2(\omega)}. \tag{17.39}$$

Setting ρ_{sub} in Equation (17.37) equal to $\rho_{AB}^0(\Omega)$ from Equation (17.11), we obtain

$$\Lambda_a(\Omega)=\frac{2V_{a/sub}^2}{\xi^2}\Omega\ln\left|\frac{\Omega^2-\Delta^2}{\Omega^2-\Delta^2-\xi^2}\right|. \tag{17.40}$$

At $\Delta = 0$, expression (17.40) transforms to formula (17.39) in Ref. [49] for free-standing graphene. Some analytical properties of the function $\Lambda_a(\Omega)$ are given in Appendix 17.C, (1).

For further analysis, we introduce the coupling constant $\alpha=2V_{a/sub}^2/\xi^2$ and turn to relative units $x = \Omega/\xi$, $\eta_a = \varepsilon_a/\xi$, $\bar{e}=\bar{\varepsilon}/\xi$, $\delta = |\Delta|/\xi$. Then the dimensionless half-width and energy-shift functions are $\gamma_a(x) = \Gamma_a(x)/\xi$ and $\lambda_a(x) = \Lambda_a(x)/\xi$. The functions $\lambda_a(x)$ are shown together with the dimensionless density of states $f_{AB}(x) = \rho_{AB}(x)\cdot\xi$ of free-standing GLC. Of note is the qualitative difference between the functions $\lambda_a(x)$ in the absence and in the presence of the spectral gap. The adatom density of states (17.39) in the presented form can now be written as

$$\bar{\rho}_a(x)\equiv\rho_a(\omega)\cdot\xi=\frac{\alpha f_{AB}(x)}{[x+\eta-\lambda_a(x)]^2+(\pi\alpha f_{AB}(x))^2}, \tag{17.41}$$

where $\eta=(\bar{\varepsilon}-\varepsilon_a)/\xi=\bar{e}-\eta_a$ and $\lambda_a(x) = \alpha x\ln|(x^2-\delta^2)/(x^2-\delta^2-1)|$. The density of states of the atom adsorbed on single-layer graphene was studied in sufficient detail (see, e.g., Refs. [21, 50]); therefore, in the present study, we focus on a GLC with a nonzero gap. It should be emphasized that it was not the low-energy approximation that was used in Refs. [21, 50] but the so-called M-model of the DOS of single-sheet graphene. However, the main results are somehow critically independent of the model, but are associated with the linear energy dependence of the DOS in the Dirac point region and vanishing exactly at this point.

The results of calculations of the functions $\bar{\rho}_a(x)$ are shown for energies corresponding to the continuous GLC spectrum, when $\sqrt{1+\delta^2})\geq|x|\geq\delta$. It follows from the figure that the adatom DOS $\bar{\rho}_a(x)\sim 0.1$ at the typical parameters of the problem, $\eta = 0$, $\delta = 0.25$, $\delta = 0.75$, and the coupling constant $\alpha = 1$. Indeed, according to Equation (17.41), the maximum value of $\bar{\rho}_a(x_r) \sim \alpha\pi^{-2}\bar{f}^{-1}(x\bar{r})$, where x_r^- is the resonant state in the valence-band region (see below). Since $f(x_r^-) \sim 1$ and $\alpha = 1$, $\bar{\rho}_a(x\bar{r}) \sim \pi^{-2} \approx 0.1$. In the case of a strong adatom–substrate bond ($\alpha >> 1$), the DOS $\bar{\rho}_a(x)$ becomes vanishingly small, whereas in the case of the weak bond ($\alpha \ll 1$), the DOS $\bar{\rho}_a(x)$ can reach a significant value. There are no qualitative differences between $\bar{\rho}_a(x)$ for the cases of $\delta = 0.25$ and $\delta = 0.75$.

We now turn to the local states whose dimensionless energies x_l are defined by the roots of the equation

$$x + \eta - \lambda a(x) = 0 \tag{17.42}$$

in the energy region beyond the continuous GLC spectrum ($|x| < \delta$, $|x| > \sqrt{1+\delta^2}$), where the function $f_{AB}(x) \equiv 0$. The graphical method for solving Equation (17.42) in the presence of a gap in the spectrum is shown: the intersection points of the dotted lines $x + \eta$ with the reduced translation function $\lambda_a(x)$ define the levels induced by the adatom. It is easy to suppose that there are two resonance levels x_r^+ and x_r^- in the conduction- and valence-band regions, respectively. There are also three local levels, two of which x_l^+ and x_l^- are, respectively, above the top of the conduction band and below the bottom of the valence band, and a third local level x_{l0} is in the gap region. At $\alpha = 1$ and $\eta = 0$, it is easy to show that $x_{l0} = 0$ and $x_l^+ = \sqrt{\delta^2 + e/(e-1)}$, where e is the natural logarithm base. Then $x_l^+ \approx 1.28$ at $\delta = 0.25$ and $x_l^+ \approx 1.46$ at $\delta = 0.75$.

We can see that the local level x_{l0} appears in the system in the upper gap half at $\eta = -1$. It is easy to suppose (see Equation (17.42)) that there is a similar mirror local level in the lower gap half at $\eta = 1$. Calculation by formula (17.42) for $\alpha = 1$ and $\eta = 1$ yields $x_{l0} \approx -0.20$ and -0.44 at $\delta = 0.25$ and 0.75, respectively. No qualitative differences between the cases $\delta = 0.25$ and 0.75 are observed. The dependence of the roots and for various problem parameters is discussed in Appendix 17.C, (2).

We now turn to calculation of the adatom occupation number n_a. It is convenient to write this number as a sum [21] in which the first term is the occupation number n_{band} representing the contribution of the valence band

$$n_{band} = \int_{-\sqrt{1+\delta^2}}^{-\delta} \bar{\rho}_a(x)\,dx. \tag{17.43}$$

In the cases shown, $n_{band} \approx 0.05$ at $\delta = 0.25$ and $n_{band} \approx 0.02$ at $\delta = 0.75$. A useful qualitative estimate of n_{band} is shown in Appendix 17.C, (3).

The successive terms n_{loc} are defined by local levels. First, it is the contribution of the level x_l^- lying below the bottom of the valence band, and hence always occupied. Second, it is the contribution of the local level x_{l0} in the gap provided that it is below the reduced Fermi level $\varepsilon_F = E_F/\xi$. As is known [21], the occupation number n_{loc} of the local state x_{loc} is given by

$$n_{loc} = \left|1 - \frac{d\lambda_a(x)}{dx}\right|_{x_{loc}}^{-1}. \tag{17.44}$$

The expression for the derivative $d\lambda_a(x)/dx$ is given in Appendix 17.C, (4) (see Equation (17.C3)). In the case of $\alpha = 1$ and $\eta = 0$, the local contributions of the levels $x_l^- \approx -1.28$ ($\delta = 0.25$) and $x_l^- \approx -1.46$ ($\delta = 0.75$) lying below the bottom of the GLC valence band are $n_l^- \approx 0.22$ and $n_l^- \approx 0.21$, respectively. The level x_{l0} makes the contribution $n_{l0} = [1+\alpha \ln((1+\delta^2)/\delta^2)]^{-1}$, from which, at $\alpha = 1$, we obtain $n_{loc} \approx 0.26$ ($\delta = 0.25$) and $n_{l0} \approx 0.49$ ($\delta = 0.75$). Hence, $n_{loc} \approx 0.48$ ($\delta = 0.25$) and $n_{loc} \approx 0.70$ ($\delta = 0.75$), and the total adatom occupation numbers $n_a = n_{band} + n_{loc}$ are $n_a \approx 0.53$ ($\delta = 0.25$) and $n_a \approx 0.72$ ($\delta = 0.75$), respectively. Here we assumed that the Fermi level $\varepsilon_F = 0^+$, i.e., it is shifted by an infinitesimal value to the upper gap half, so that the local level x_{l0} is occupied.

In the case of $\alpha = 1$ and $\eta = 1$, the local contributions of the levels $x_{l0} \approx -0.20$ ($\delta = 0.25$) and $x_{l0} \approx -0.44$ ($\delta = 0.75$) in the lower gap half are $n_{l0}^- \approx 0.20$ and $n_{l0}^- \approx 0.33$, respectively. The general form of the dependences $n_l^-(x_l^-)$ and $n_{l0}(x_{l0})$ is shown, and some particular cases are given in Appendix 17.C, (4). The absence of qualitative differences for the cases of $\delta = 0.25$ and $\delta = 0.75$ should be again noted.

Thus, the contribution of local states $n_{loc} = n_l^- + \vartheta(\varepsilon_F - x_{l0})n_{l0}$, where $\vartheta(z)$ is the Heaviside unit-step function equal to unity at $z > 0$ and zero at $z \le 0$ is prevalent in comparison with the valence-band contribution n_{band} for intermediate and strong adatom–GLC bonds ($\alpha \ge 1$). Such a situation is characteristic, e.g., for hydrogen and halogen-atom adsorption on graphene [50]. As the constant α decreases, the contribution of the band and local states increases and decreases, respectively. In the case of weak adatom–GLC bonds, the contributions n_{band} can become prevalent as in the case of alkali-metal-atom adsorption on graphene [50]. Indeed, to estimate the constant α, simple expressions [21, 50] can be used. For example, for adatoms with outer orbital s_a coupling with the $p_z|$ orbital of GLC, we obtain $\alpha \approx 21.5/d^4t^2$, where the transition energy t is given in eV and the adsorption bond length d is given in Å. Setting $t \approx 3$ eV and accepting $d \approx 1$ Å for atomic hydrogen, we get $\alpha \approx 2$. For adatoms with outer orbital p_a coupling with the p_z orbital of GLC, we obtain $\alpha \approx 52.7/d^4t^2$. Since $d \approx 2$ Å for fluorine adatoms, we have $\alpha \approx 1$. For alkali metals, $d > 2$ Å; hence, $\alpha < 1$.

In principle, to calculate the occupation number n_{band} upon atom adsorption on free-standing GLC, an analytical model similar to the M-model of adsorption on graphene, proposed in Refs. [21, 50], should have been constructed using the GLC parameters given in Section 17.2. However, not only experimental data but also corresponding results calculated by other authors are currently lacking. Therefore, we considered such a problem as premature.

17.4.2 Epitaxial GLC

We begin with the consideration of flat GLC layers formed on metal. As shown in Section 17.2, the metal substrate whose DOS is described by the simplest model $\rho_{sub}(\omega) = \rho_{met} =$ const smoothens all characteristic features of the DOS of free-standing GLC. In this case, the main effect of the metal substrate consists in GLC-gap disappearance as the region of forbidden states (see formula (17.13) and Figure 17.3). According to Section 17.2 and Appendix 17.C, (5), the dimensionless DOS of the 2D layer of epitaxial GLC on metal $\bar{\rho}_{AB}^{met}(x)$ in the normalization of this study is given by formula (17.C6). It is also shown in Section 17.2 that at $\delta^2 << 1$ and $\gamma_m^2 << 1$ at $x^2 << 1$, this DOS can be reduced to the quadratic dependence $\bar{\rho}_{AB}^{met}(x) \approx A + Cx^2$, where constants A and C are defined by expressions (17.C7) and (17.C8) of Appendix 17.C, (5).

Similarly to Equation (17.37), the half-width of the quasi-level of an atom adsorbed on GLC formed on metal is given by

$$\Gamma_a^{AB/met}(\omega) = \pi (V_a^{AB/met})^2 \rho_{AB}^{met}(\omega), \tag{17.45}$$

or, in the reduced form, $\gamma_{am} = \Gamma_a^{AB/met} / \xi$ (here $V_a^{AB/met}$ is the matrix element of adatom interaction with the GLC–metal structure). The energy-shift function of the quasi-level $\Lambda_a^{AB/met}(\omega)$ is given by a formula similar to Equation (17.38). If we set the adatom–GLC coupling constant on metal to be $\beta = (V_a^{AB/met})^2 / \xi^2$. The corresponding reduced adatom DOS is still given by formula (17.43), but with β in place of α, $\bar{\rho}_{AB}^{met}(x)$ in place of $f_{AB}(x)$, and $\lambda_{am}(x)$ in place of $\lambda_a(x)$.

To estimate the charge transfer between the adatom and substrate, for simplicity, we consider that the reduced Fermi level of the GLC/metal system $\varepsilon_F = 0$; therefore, charge transfer between the metal and GLC does not occur. Let then $|\eta| \ll 1$. Let us approximate the adatom DOS by the Lorentzian profile such as

$$\bar{\rho}_a(x) = \frac{\beta A}{(x - x_0)^2 + (\pi \beta A)^2}, \tag{17.46}$$

where we set $\bar{\varepsilon} = 0$, so that $\eta = -\eta_a$, $x_0 \approx \eta_a + \lambda_{am}(\eta_a)$. Calculations show that we have $A \sim C$ at $\delta^2 << 1$ and $\gamma_m^2 << 1$; therefore, in the region $x^2 << 1$, we can set $\bar{\rho}_{AB}^{met}(x) \approx A$. Then the adatom occupation number n_a is approximately given by

$$n_a \approx \frac{1}{\pi} \operatorname{arcctg} \frac{x_0}{\pi \beta A}. \tag{17.47}$$

The results of calculations of the dependences $n_a(\gamma_m)$ and $n_a(\beta)$ are shown, respectively. Of note is not only the qualitative but also the quantitative consistency of the dependences $n_a(\gamma_m)$ and $n_a(\beta)$. This result suggests that both force constants similarly affect the adatom orbital occupation. This problem is considered in Appendix 17.C, (5), in more detail. It is also shown that the dependence of n_a on δ can be considered to be weak.

The decrease in n_a with increasing coupling constants γ_m and β for negative reduced energies of the adatom quasi-level is explained by the positive energy shift function $\lambda_{am}(\eta_a)$: indeed, the adatom quasi-level is depleted shifting to positive energies. At $\eta_a > 0$, the reverse effect takes place: adatom occupation increases with γ_m and β, which is associated with a shift of the adatom quasi-level to negative energies. It also follows from expression (17.47) that the occupation number n_a transforms to $1 - n_a$ with $-n_a$ in place of n_a. Flattening of the dependences $n_a(\gamma_m)$ and $n_a(\beta)$ with increasing γ_m and β is explained in Appendix 17.C, (5). Thus, due to the lack of a gap in the GLC on metal, the local levels of adatoms are lacking, and the contribution to the occupation number n_a is completely defined by states of the continuous spectrum occupying the entire energy space within the used model.

We now turn to adsorption on flat epitaxial GLC layers grown on semiconductors. Possible types of heterojunctions formed by GLC contact with the semiconductor substrate were considered in Section 17.3. In this case, depending on the ratio of the GLC gap 2δ and the substrate band gap E_g, their relative positions in the energy scale and, the main thing, the GLC–semiconductor coupling constant, there is a wide variety of electronic structures that can have a large set of forbidden-state subbands at a certain ratio of parameters. However, the general problem of adsorption on such a structure is formulated simply. Indeed, similarly to Equation (17.45), we can write the quasi-level half-width function in the form $\Gamma_a^{AB/sc}(\omega)=\pi(V_a^{AB/sc})^2\rho_{AB}^{sc}(\omega)$, where $V_a^{AB/sc}$ is the matrix element of the adatom interaction with the GLC/semiconductor structure whose DOS is $\rho_{AB}^{sc}(\omega)$ given in Section 17.3. Substituting $\Gamma_a^{AB/sc}(\omega)$ into expression (17.38), we will find the energy-shift function of the quasi-level $\Lambda_a^{AB/sc}(\omega)$, thus having determined the adatom density of states in the form of Equation (17.39). It is interesting to note that in the case at hand, as in the case of free-standing GLC, both band (resonant) x_r and local x_{loc} states will contribute to the adatom occupation number n_a. We note that the latter are analogues of gap states x_{l0} during adsorption on free-standing GLC.

Note in conclusion that during single-atom adsorption, it, in principle, does not matter with which substrate atom, A or B, it is bound, since this circumstance has an effect on only the coupling constant. Another matter is the problem of the finite concentration of adatoms. In this case, adsorption can change the GLC gap to the free state (by analogy with induced-gap adsorption in graphene), which will certainly manifest itself in epitaxial GLC. In our opinion, the statement of such a problem is also premature.

17.5 Conclusion

Thus, in this work, the analytical expressions of the DOS of freestanding and epitaxial (flat and buckled) layers of GLC on metals and semiconductors have been obtained using the tight-binding method in the low-energy approximation.

For the epitaxial GLC on a metal, the main feature is the absence of a gap as the region of forbidden energies. At low interaction of GLC with a metal, this region can be considered as a pseudogap. As this interaction increases, the pseudogap almost completely disappears. For the epitaxial GLC on a semiconductor, the main conclusion is the possibility of the formation of the complex electronic structure with a number of energy gaps for the systems with strong GLC-semiconductor coupling. Underline once more, that in contrast to 2D GLCs on metal, whose feasibility can be considered as proved theoretically [12, 13], GLCs on semiconductor substrates are still hypothetical structures.

We constructed the general scheme of consideration of the problem of single atom adsorption on freestanding and epitaxial flat 2D layers of $A_N B_{8-N}$ compounds. Numerical calculations and semiquantitative estimations performed for freestanding GLCs and GLC formed on metal allowed a number of conclusions to be drawn on the role of the atomic-level position, GLC band gap, and corresponding coupling constants in the formation of the adatom electronic structure (DOS and occupation numbers). As for GLC epitaxial layers on semiconductors, too many possible situations arise. Therefore, we did not estimate adsorption characteristics in the present study.

As it was mentioned above, it is now generally accepted that the density functional theory is the state of the art in condensed matter physics. We think, however, that the model Hamiltonian approach is still useful for the description of the whole complex of objects and clarifying the corresponding tendencies. Thus, we wish to repeat Anderson's words from his Nobel lecture [51]: "Very often such, a simplified model throws more light on the real workings of nature than any number of "*ab initio*" calculations of individual situations, which even where correct often contain so much detail as to conceal rather than reveal reality"

Appendix 17.A

(1) The Green function defined by Equation (17.7) can be rewritten as $G^{AA(BB)} = \mathrm{Re}G^{AA(BB)} + i\mathrm{Im}\, G^{AA(BB)}$, where

$$\mathrm{Re}\,G^{A(B)} = \frac{\Omega_{b(a)}(\Omega_a\Omega_b - t^2f^2) + \Omega_{a(b)}\Gamma^2_{b(a)}}{|D|^2}, \tag{17.A1}$$

$$\mathrm{Im}\,G^{A(B)} = -\frac{\Gamma_{b(a)}(\Gamma_a\Gamma_b + t^2f^2) + \Gamma_{a(b)}\Omega^2_{b(a)}}{|D|^2}, \tag{17.A2}$$

$$|D|^2 = (\Omega_a\Omega_b - t^2f^2)^2 + \Gamma_a^2\Gamma_b^2 + 2\Gamma_a\Gamma_b t^2 f^2 + \Gamma_a^2\Omega_b^2 + \Gamma_b^2\Omega_a^2. \tag{17.A3}$$

Here, arguments ω and $\mathbf{k}$ entering in the formulas were omitted for the sake of simplicity of the expressions.

(2) As the bond with the substrate is absent, the Green functions (17.7) can be expressed as

$$G^{A(B)}(\Omega, \mathbf{k}) = \frac{\Omega \mp \Delta}{(\Omega - R(\mathbf{k}))(\Omega + R(\mathbf{k}))}, \tag{17.A4}$$

where $\Omega = \omega - \varepsilon$ and $R(\mathbf{k}) = \sqrt{\Delta^2 + t^2 f^2(\mathbf{k})}$. Then

$$G^A(\Omega, \mathbf{k}) + G^B(\Omega, \mathbf{k}) = \frac{1}{\Omega - R(\mathbf{k}) + is} + \frac{1}{\Omega + R(\mathbf{k}) + is}, \quad s = 0^+, \tag{17.A5}$$

and Equation (17.9) takes the form

$$\rho_{AB}(\Omega) = \frac{S}{(2\pi)^2} 2\pi \frac{1}{(3ta/2)^2} \int z\, dz[\delta(\Omega - R(z)) + \delta(\Omega + R(z))]. \tag{17.A6}$$

When deducing Equation (17.A6), we passed from the summation to the integration following the general rule

$$\frac{1}{N}\sum_{\mathbf{k}}(...)\rightarrow\frac{S}{(2\pi)^2}\int(...)d\mathbf{k} \tag{17.A7}$$

($S=3a^2\sqrt{3}/2$ is the unit cell area) and used the low-energy approximation ($z = 3taq/2$, $R(z)=\sqrt{\Delta^2+z^2}$). Performing the integration in Equation (17.A6), we obtain Equation (17.13).

(3) The DOS (17.13) takes the reduced form $\rho^*_{AB}(x)=I(x)$, where

$$I(x)=\frac{\gamma}{2\pi t}\ln\frac{\bar{\xi}^4+\bar{\xi}^2\bar{b}+\bar{c}}{\bar{c}}+\frac{|x|}{\pi t}\left(\text{arctg}\frac{2\bar{\xi}^2+\bar{b}}{4\gamma|x|}-\text{arctg}\frac{\bar{b}}{4\gamma|x|}\right), \tag{17.A8}$$

where $\bar{b}=2(\delta^2+\gamma^2-x^2)$, $\bar{c}=(x^2-\delta^2)^2+\gamma^2(\gamma^2+2\delta^2+2x^2)$, $\bar{\xi}=3.3$. Let $x^2 << \delta^2$. Then

$$\begin{gathered}I(x)\approx I(0)+\frac{\gamma x^2}{\pi}\left(\frac{8\bar{\xi}^2}{\bar{b}_0(2\bar{\xi}^2+\bar{b}_0)}-\frac{\bar{\xi}^2}{\bar{\xi}^4+\bar{\xi}^2\bar{b}_0+\bar{c}_0}+(\delta^2-\gamma^2)\frac{\bar{\xi}^4+\bar{\xi}^2\bar{b}_0}{\bar{c}_0(\bar{\xi}^4+\bar{\xi}^2\bar{b}_0+\bar{c}_0)}\right)\\ I(0)=\frac{\gamma}{\pi}\ln\frac{\bar{\xi}^2+\delta^2+\gamma^2}{\delta^2+\gamma^2}.\end{gathered} \tag{17.A9}$$

Assuming that $\bar{\xi}$ is the largest parameter of the problem, we obtain

$$I(x)\approx I(0)+\frac{\gamma x^2}{\pi}\frac{3\delta^2+\gamma^2}{\delta^2+\gamma^2}. \tag{17.A10}$$

At energies $x^2 = \delta^2$ corresponding to the boundaries of the pseudogap of a graphene-like compound, we have

$$I(\delta)=\frac{\gamma}{2\pi t}\ln\frac{(\bar{\xi}^2+\gamma^2)^2+4\gamma^2\delta^2}{\gamma^2(\gamma^2+4\delta^2)}+\frac{\delta}{\pi t}\left(\text{arctg}\frac{\bar{\xi}^2+\gamma^2}{2\gamma\delta}-\text{arctg}\frac{\gamma}{2\delta}\right) \tag{17.A11}$$

Assuming once again that $\bar{\xi}$ is the largest value, we find that

$$I(\delta)\approx\frac{\gamma}{\pi t}\ln\frac{\bar{\xi}^2}{\gamma(\gamma^2+4\delta^2)^{1/2}}+\frac{\delta}{\pi t}\left(\frac{\pi}{2}-\text{arctg}\frac{\gamma}{2\delta}\right). \tag{17.A12}$$

(4) In the general case, the DOS of an epitaxial GLC is

$$\tilde{\rho}_{AB}(\omega)=\frac{1}{\pi\sqrt{3}t}\tilde{I}(\omega), \tag{17.A13}$$

$$\tilde{I}(\omega)=\frac{\Gamma_a(\omega)+\Gamma_b(\omega)}{4\pi t}\ln\frac{|\xi^4+\tilde{b}\xi^2+\tilde{c}|}{\tilde{c}}+$$
$$+\frac{\Gamma_a(\omega)\Omega_b^2+\Gamma_b(\omega)\Omega_a^2+[\Gamma_a(\omega)+\Gamma_b(\omega)]\Omega_a\Omega_b}{2\pi tC(\omega)}\left(\operatorname{arctg}\frac{2\xi^2+\tilde{b}}{2C(\omega)}-\operatorname{arctg}\frac{\tilde{b}}{2C(\omega)}\right), \tag{17.A14}$$

where $\tilde{b}=2[\Gamma_a(\omega)\Gamma_b(\omega)-\Omega_a\Omega_b]$, $\tilde{c}=\Omega_a^2\Omega_b^2+\Gamma_a^2(\omega)\Gamma_b^2(\omega)+\Gamma_a^2(\omega)\Omega_b^2+\Gamma_b^2(\omega)\Omega_a^2$, and $C(\omega)=\Gamma_a(\omega)\Omega_b+\Gamma_b(\omega)\Omega_a$. At $\Omega=\Omega^*$, coefficients $\tilde{b}$ and $\tilde{c}$ for a metallic substrate are

$$\begin{aligned} b'(\Omega^*)&=2\vartheta\left(\Gamma^2+\frac{4\Delta^2}{(1+\vartheta)^2}\right),\\ c'(\Omega^*)&=\vartheta^2\Gamma^4+2\Gamma^2\Delta^2\left(\frac{(1+\vartheta^2)^2}{(1+\vartheta)^2}-\frac{(1-\vartheta)^2}{1+\vartheta}\right)+\frac{16\vartheta^2\Delta^4}{(1+\vartheta)^4}. \end{aligned} \tag{17.A15}$$

Since $b'(\Omega^*)>0$, the value of $I'(\Omega^*)$ is given by the first term of the second equation from Equation (17.15).

(5) For the buckled layers, the reduced DOS are $I'(\pm\delta)\approx I(\delta)-aS(\pm\delta)$ at $1-\vartheta=\alpha\ll 1$ and $x=\pm\delta$ where $I(\delta)$ is given by Equation (17.A11) and

$$\begin{aligned} S(\delta)=\frac{\gamma}{\pi}&\left(\frac{1}{4}\ln\frac{(\bar{\xi}^2+\gamma^2)^2+4\gamma^2\delta^2}{\gamma^2(\gamma^2+4\delta^2)}-\frac{\gamma^2+\delta^2}{\gamma^2+4\delta^2}+\frac{\gamma^2(\bar{\xi}^2+\gamma^2+\delta^2)}{(\bar{\xi}^2+\gamma^2)^2+4\gamma^2\delta^2}\right)-\\ &-\frac{2\gamma}{\pi}\frac{\bar{\xi}^2(\bar{\xi}^2+2\gamma^2)\delta^2}{(\gamma^2+4\delta^2)[(\bar{\xi}^2+\gamma^2)^2+4\gamma^2\delta^2]}, \end{aligned} \tag{17.A16}$$

$$\begin{aligned} S(-\delta)=\frac{\gamma}{\pi}&\left(\frac{1}{4}\ln\frac{(\bar{\xi}^2+\gamma^2)^2+4\gamma^2\delta^2}{\gamma^2(\gamma^2+4\delta^2)}-\frac{\gamma^2+3\delta^2}{\gamma^2+4\delta^2}+\frac{\gamma^2(\bar{\xi}^2+\gamma^2+3\delta^2)}{(\bar{\xi}^2+\gamma^2)^2+4\gamma^2\delta^2}\right)-\\ &-\frac{2\gamma}{\pi}\frac{\bar{\xi}^2\gamma\delta}{(\bar{\xi}^2+\gamma^2)^2+4\gamma^2\delta^2}. \end{aligned} \tag{17.A17}$$

Appendix 17.B

The presented expressions for the functions $\Gamma(\omega)$ and $\Lambda(\omega)$ take the form

$$\gamma(x)=C\begin{cases}\sqrt{-x-e_g/2}, & x<-e_g/2,\\ \sqrt{x-e_g/2}, & x>e_g/2,\\ 0, & |x|\le e_g/2,\end{cases} \tag{17.B1}$$

$$\lambda(x)=C\begin{cases}\sqrt{-x+e_g/2}, & x<-e_g/2,\\ \sqrt{-x+e_g/2}-\sqrt{x+e_g/2}, & |x|\le e_g/2,\\ -\sqrt{x+e_g/2}, & x>e_g/2,\end{cases} \tag{17.B2}$$

where $C=\pi AV^2/\sqrt{t}$. According to the estimates made in Ref. [47], the 6*H*-SiC substrate has a coefficient $A\approx 0.2$ eV$^{-3/2}$. As was shown in Section 17.2, the transition energy t varies in the range from 0.64 eV for InSb to 2.28 eV for BN. In this case, the coupling constant is estimated as $C\approx(V/t)^2$. Taking the average value of $t\approx 1.5$ eV, we can write $C=(\pi At^{3/2})\nu^2\sim\nu^2$, where $\nu=V/t$.

Considering expressions (17.B1) and (17.B2), it is easy to show that $(dX/dx)_{x\to 0}=1+C(2/e_g)^{3/2}$ and $(dX/dx)_{|x|>e_g/2}=1+C/2(|x|+e_g/2)^{3/2}$.

(2) In the presented form, the DOS (17.13) can be written as

$$I(x)=I_1(x)+I_2(x),\quad I_1(x)=\frac{\gamma}{2\pi}\ln\frac{\bar{\xi}^4+\bar{\xi}^2\bar{b}+\bar{c}}{\bar{c}},$$
$$I_2(x)=\frac{|X|}{\pi}\left(\operatorname{arctg}\frac{2\bar{\xi}^2+\bar{b}}{4\gamma|X|}-\operatorname{arctg}\frac{\bar{b}}{4\gamma|X|}\right), \tag{17.B3}$$

where $\bar{b}=2(-X^2+\delta^2+\gamma^2)$, $\bar{c}=(X^2-\delta^2)^2+\gamma^2(\gamma^2+2\delta^2+2X^2)$, and $X=x-e-\lambda(x)$.

We investigate the behavior of the function $I(x)$ in the regions of the valence band top and the conduction band bottom by setting $x=\pm(e_g/2)\pm\nu$ $(\nu\ll e_g)$, where the signs plus and minus refer to the conduction band and the valence band, respectively. Then, we have $\gamma=C\sqrt{\nu}$ and

$$X_\pm\approx -e\pm(C\sqrt{e_g}+e_g/2)\pm\nu(1+C/2\sqrt{e_g}). \tag{17.B4}$$

As a result, up to the factor ν, we obtain

$$I_{1\pm}(\nu) = \frac{C\sqrt{\nu}}{\pi} \ln \left| \frac{r^2 - X_{\pm 0}^2}{X_{\pm 0}^2 - \delta^2} \right|, \tag{17.B5}$$

where $X_{\pm 0} = X_{\pm}$ for ν = 0 and, as before, $r^2 = \bar{\xi}^2 + \delta^2$. Here and below, we exclude combinations of the parameters corresponding to zeroes and poles of the argument of the logarithm.

Further, it is assumed that $4\gamma |X| \approx 4C\sqrt{\nu}\,|X_{\pm 0}| << 1$. For < δ2 or > r2, up to the factor ν, we obtain

$$I_{2\pm}(\nu) \approx \frac{2C\sqrt{\nu}}{\pi} \frac{\bar{\xi}^2 X_{\pm 0}^2}{(X_{\pm 0}^2 - \delta^2)(X_{\pm 0}^2 - r^2)}; \tag{17.B6}$$

for $\delta^2 < X_{\pm 0}^2 < r^2$, we have

$$I_{2\pm}(\nu) \approx |X_{\pm}| - \frac{2C\sqrt{\nu}}{\pi} \frac{\bar{\xi}^2 X_{\pm 0}^2}{(X_{\pm 0}^2 - \delta^2)(r^2 - X_{\pm 0}^2)}. \tag{17.B7}$$

For $|X| >> \max\{\bar{\xi}, \delta, e_g\}$, i.e., for high energies corresponding to the bands of allowed states, we obtain

$$I(x) \approx I_2(x) \approx \frac{2\gamma(x)}{\pi} \frac{\bar{\xi}^2}{X^2}, \tag{17.B8}$$

where it is taken into account that $I_1(x) \sim I_2(x)(\bar{\xi}/X)^2 << I_2(x)$.

Next, we consider the weak coupling regime by taking as an example the case of $e = 0$, $C = 0.2$, and $X_{\pm 0}^2 \sim (e_g/2)^2 = 1$ (Figures 17.9a, 17.10a, and 17.11a). For the case shown in Figure 17.9a, when $X_{\pm 0}^2 < \delta^2$ (Figure 17.11a), the density of states $I(\nu)$ is described by the sum of expressions (17.B6) and (17.B7), so that we have $I(\nu) \propto \sqrt{\nu}$. For the case shown in Figure 17.10a, when $\delta^2 < X_0^2 < r^2$ δ2 < < r2 (Figure 17.11a), the density of states $I(\nu)$ is the sum of expressions (17.B6) and (17.B8), so that we obtain $I(\nu) \propto A_1 \cdot \sqrt{\nu} - A_2 \cdot \nu$, where $A_{1,2}$ are positive constants.

Let us turn to the strong coupling regime by taking as an example the case of $C = 5$ (Figures 17.9a and 17.10a), when $X_{\pm 0}^2 > r^2$ (Figure 17.11a). By assuming that $C \gg 1$ and $X_{\pm 0}^2 >> r^2$ and summarizing expressions (17.B6) and (17.B7), we obtain

$$I_{\pm}(\nu) \approx \frac{2C\sqrt{\nu}}{\pi} \frac{\bar{\xi}^2}{X_{\pm 0}^2}. \tag{17.B9}$$

For $e \neq 0$, the mirror symmetry with respect to the zero energy is broken. For example, when $e < 0$, we have $X_{+0}^2 > X_{-0}^2$, so that, from relationship (17.B9), we obtain the inequality $I_+(v) < I_-(v)$, which is confirmed by the data presented in Figures 17.9c and 17.10c. It should be noted that, for all values of the parameters at $x = \pm e_g/2$, the dependences $I(x)$ exhibit a root singularity.

(3) Now, we consider the specific features of the density of states for a buckled sheet of the epitaxial GLC (see formula (17.33)) in the ranges of energies near the bottom of the conduction band and the top of the valence band of the substrate. By setting $x = \pm(e_g/2)\pm v$ ($v << e_g$), in analogy with flat epitaxial layers, we obtain $I'_{1\pm}(v) \approx ((1+\vartheta)/2)I_{1\pm}(x)$, where $I_{1\pm}(x)$ is given by relationship (17.B6). In this case, we ignored the dependence of the logarithm (as a slowly varying function) of the buckling factor ϑ.

Next, we analyze the second term in the dimensionless form

$$I'_2(X') = \frac{X'}{\pi}\left(\operatorname{arctg}\frac{2\xi^2 + \bar{b}'}{2\gamma[(1+\vartheta)X' + (1-\vartheta)\delta_x]} - \operatorname{arctg}\frac{\bar{b}'}{2\gamma[(1+\vartheta)X' + (1-\vartheta)\delta_x]}\right) \tag{17.B10}$$

where $X' = x - e - ((1+\vartheta)/2)\lambda$, $\delta_x = |\delta + ((1-\vartheta)/2)\lambda_x|$, $\bar{b}' = 2(\vartheta\gamma^2 + \delta_x^2 - X'^2)$, and $\bar{c}' = (X'^2 - \delta_x^2)^2 + \vartheta^2\gamma^4 + \gamma^2[(1+\vartheta^2)X'^2 + (1+\vartheta^2)\delta_x^2 + 2(1-\vartheta)X'\delta_x]$. It is easy to show that, within the approximation linear in v, for estimates, we can use expressions (17.B7) and (17.B8) by multiplying the coupling constant C by the factor $(1+\vartheta)/2$ and substituting $|X|$ with $|X'|$ in relationship (17.B7).

(4) According to expression (17.19), for the contribution to the occupation number from the density of states lying in the lower half of the energy gap (i.e., in the region $x < e_F = 0$), we have

$$n_{gap} = \frac{1}{\pi\sqrt{3}} J(x^*, x^{**}), \quad J(x^*, x^{**}) = \int_{x^*}^{x^{**}} |X|\,dx, \tag{17.B11}$$

where, as before, $X = x - e - \lambda(x)$ and x^* (x^{**}) is the lower (upper) limit of the density of states. In the general case, when $e \neq 0$ and the function X changes sign in the range $|x| \leq e_g/2$, the range of integration inexpression (17.B11) should be divided into intervals (x_r, x_0) and (x_0, x_δ), where x_0 is determined from the equation $X = 0$. For each such domain (x', x''), we can write

$$J = \left|\frac{x^2}{2} - ex + \frac{2}{3}C[(x + e_g/2)^{3/2} - (-x + e_g/2)^{3/2}]\right|_{x'}^{x''}. \tag{17.B12}$$

It is easy to show that, in the case of $C = 0$ (free-standing layer), when $x_r = -r + e$ and $x_\delta = -\delta + e$, we obtain the integral $J = \pi\sqrt{3}$, so that the occupation number is equal to unity. For the case shown in Figure 17.9a, we have $x_r \approx -0.58$, $x_\delta \approx -0.25$, and, according to expression (17.B12), $n_{gap} \approx 0.15$. For the case shown in Figure 17.10a, we have $x_r \approx -0.54$, $x_\delta \approx -0.08$, and $n_{gap} \approx 0.15$.

All the formulas presented in this section relate to flat epitaxial layers. For buckled layers, the coupling constant C should be multiplied by the factor $(1+\vartheta)/2$.

Appendix 17.C

(1) Using formula (17.40), it is easy to show that at $\Omega \to 0$, the energy-shift function $\Lambda_a(\Omega) \to 0$ as $\Omega \ln(\Delta^2/(\Delta^2 + \xi^2))$ for $\Delta \neq 0$ and as $\Omega \ln(|\Omega|/\xi))$ for $\Delta = 0$. Two other points where $\Lambda_a(\Omega)$ vanishes are $\Omega_0^\pm = \sqrt{\Delta^2 + \xi^2/2}$. At $\Omega_1^\pm = \pm\Delta$ and $\Omega_2^\pm = \pm\sqrt{\Delta^2 + \xi^2}$, the function $\Lambda_a(\Omega)$ logarithmically diverges as $\Lambda_a(\Omega) \propto \Omega_1^\pm \ln(|\Omega^2 - (\Omega_1^\pm)^2|/\xi^2)$ and $\Lambda_a(\Omega) \propto \Omega_2^\pm \ln(\xi^2/|\Omega^2 - (\Omega_2^\pm)^2|)$. At $|\Omega| \to \infty$, we obtain $\Lambda_a(\Omega) \propto \xi^2/\Omega$.

(2) We now turn to the solution of Equation (17.42). Let us consider the roots and corresponding to the states lying, respectively, below the bottom of the valence band and above the top of the conduction band, i.e., at $x^2 > \delta^2 + 1$. Let $|\eta| \gg 1$. Then in the region $|x| >> \sqrt{1+\delta^2}$, we can write $\lambda_a(x) \approx \alpha/x$. As a result, we obtain $x_l^\pm = \eta(1+\sqrt{1+4\alpha^2/\eta^2})$, where the superscripts "plus" and "minus" relate to $\eta > 0$ and $\eta < 0$, respectively.

Now let $|\eta| << 1$. Assuming that the x_l^+ and x_l^- levels are near the outer boundaries of the GLC continuous spectrum, we set $\lambda_a(x) \approx \alpha\sqrt{1+\delta^2} \ln(1/(x^2 - \delta^2 - 1))$. Then $(x_l^\pm)^2 \approx 1 + \delta^2 + \exp(-\sqrt{1+\delta^2}/\alpha)$. As the gap increases, the local levels become closer to the allowed band boundaries. The same is observed with increasing coupling constant α.

We now turn to the local states x_{l0} in the gap region. Let $|\eta| >> 1$. Then it is clear that the x_{l0} level will be near the top of the valence band at $\eta < 0$ and near the bottom of the conduction band at $\eta > 0$. It is easy to show that $x_{l0}^2 \approx 1 + \delta^2 - \exp(-\sqrt{1+\delta^2}/\alpha)$. At $|\eta| << 1$, we have $x + \eta \approx \alpha x \ln(\delta^2/1+\delta^2)$, from which $x_{l0} \approx -\eta/[1 + \alpha \ln((1+\delta^2)/\delta^2)]$.

(3) Taking into account Equation (17.11), we represent the adatom DOS (17.41) for the energy range corresponding to the GLC valence band in the form

$$\bar{\rho}_a(x) \approx \frac{1}{\pi} \frac{2\pi\alpha |x_r^-|}{(x - x_r^-)^2 + (2\pi\alpha x_r^-)^2}. \tag{17.C1}$$

Then according to Equation (17.43), we obtain

$$n_{band} \approx \frac{1}{\pi} \left(\operatorname{arctg} \frac{x_r^- + \sqrt{1+\delta^2}}{2\pi\alpha |x_r^-|} - \operatorname{arctg} \frac{x_r^- + \delta}{2\pi\alpha |x_r^-|} \right). \tag{17.C2}$$

The maximum band contribution n_{band} is on the order of $(2/\pi)\operatorname{arctg}(1/2\pi\alpha)$; at $\alpha \gg 1$, we have $n_{band} \sim (\pi^2\alpha)^{-1}$. The same estimate is also valid at $\alpha \sim 1$ (see text). At $\alpha \ll 1$, we obtain $n_{band} \sim 1$. This result has a simple physical meaning: in the case of a weak bond with the substrate, GLC is a quasi-free structure with a single electron filling the valence band.

(4) Having differentiated expression (17.40), we obtain in the reduced form

$$\frac{d\lambda_a(x)}{dx} = \alpha \ln\left|\frac{\delta^2 - x^2}{\delta^2 + 1 - x^2}\right| - \frac{2\alpha x^2}{(\delta^2 - x^2)(\delta^2 + 1 - x^2)}. \quad (17.C3)$$

It is easy to see that the function $|d\lambda_a(x)/dx| \to \infty$ at band boundaries, i.e., at $x^2 \to \delta^2$ and $x^2 \to \delta^2 + 1$; at $|x| \to \infty$, we obtain $d\lambda_a(x)/dx \to -\alpha/x^2$.

Let us consider the contributions of local states lying below the bottom of the valence band and above the top of the conduction band. At $|\eta|>>1$, according to Appendix 17.C, (2), we have $x_l^{\pm} = \eta(1+\sqrt{1+4\alpha^2/\eta^2})$ and

$$d\lambda_a(x)/dx \sim -\alpha/\eta^2(1+\sqrt{1+4\alpha^2/\eta^2})^2, \quad (17.C4)$$

from which, according to Equation (17.44), we obtain

$$n_l^- \sim [1+\alpha/\eta^2(1+\sqrt{1+4\alpha^2/\eta^2})^2]^{-1}, \quad (17.C5)$$

whence (at reasonable α), we have $n_l^- \sim 1$. We note that the adatom DOS in the valence-band region $\bar{\rho}_a(x) \sim 2\alpha|x_r|/\eta^2 << 1$ in this case, so that $n_{band} \sim 0$. At $|\eta|<<1$, according to Appendix 17.C, (2), we have $(x_l^-)^2 \approx 1+\delta^2+\exp(-\sqrt{1+\delta^2}/\alpha)$, from which, according to Equation (17.44), we obtain $n_l^- \sim 0$.

We now turn to the contributions of local states in the GLC gap. As shown in Appendix 17.C, (2), we have $x_{l0}^2 \approx 1+\delta^2-\exp(-\sqrt{1+\delta^2}/\alpha)$ in the limit $|\eta|>>1$, so that $n_{l0} \sim 0$. In the limit $|\eta|<<1$, we have $n_{l0} \sim [1 + a\ln((1 + \delta^2)/\delta^2)]^{-1}$. Here we supposed that we deal with the intrinsic GLC sample whose Fermi level $\varepsilon_F = 0$.

(5) In Sections 17.2 and 17.3, the transfer integral t is taken as the energy unit, whereas the parameter ξ is used in the present study. According to this replacement, the DOS of epitaxial GLC on the metal substrate is written as

$$\bar{\rho}_{AB}^{met}(x) \equiv \rho_{AB}^{met}(x)\xi = \frac{\gamma_m}{\pi}\ln\frac{|1+b'+c'|}{c'} + \frac{2x}{\pi}\left(\text{arctg}\frac{2+b'}{4\gamma_m x} - \text{arctg}\frac{b'}{4\gamma_m x}\right). \quad (17.C6)$$

Here $b' = -2(x^2-\delta^2-\gamma_m^2)$, $c' = (x^2-\delta^2)^2+\gamma_m^2(\gamma_m^2+2\delta^2+2x^2)$ and $\gamma_m = \Gamma_m/\xi$, where $\Gamma_m = \pi V_m^2 \rho_{met}$ and V_m is the matrix element of the GLC–metal substrate interaction. Under the assumption that $\delta^2 << 1$ and $\gamma_m^2 << 1$, expression (17.C6) at $x^2 << 1$ can be reduced to

$$\bar{\rho}_{AB}^{met}(x) \approx \bar{\rho}_{AB}^{met}(0) + \frac{2\gamma_m}{\pi}\frac{3\delta^2+\gamma_m^2}{\delta^2+\gamma_m^2}x^2, \quad (17.C7)$$

where

$$\bar{\rho}_{AB}^{met}(0) = \frac{2\gamma_m}{\pi} \ln \frac{1}{\delta^2 + \gamma_m^2}. \tag{17.C8}$$

We note that the coefficient at x^2 in the second term of expression (17.C7) is denoted in the text as C, $A = \bar{\rho}_{AB}^{met}(0)$.

We now turn to the analysis of the dependences $n_a(\gamma_m)$ and $n_a(\beta)$ shown. In the low-energy region, $\lambda_{am}(x) \approx -2\beta Ax$. Then, taking into account $A \propto \gamma_m$ (the logarithmic term has a slight effect), we obtain the following: the shifts of the adatom quasi-level $\lambda_{am}(x)$ and its half-width pβA are proportional to the product $\beta\gamma_m$. This is precisely the cause of the similarity of the dependences $n_a(\gamma_m)$ and $n_a(\beta)$. We also note that the coefficient $A = \bar{\rho}_{AB}^{met}(0)$ depends on the gap half-width δ logarithmically, i.e., comparatively weakly.

At the coupling constants γ_m and β, when the quasi-level shift $|\lambda_{am}(\eta_a)|$ begins to significantly exceed its primary energy $|\eta_a|$, the functions $n_a(\gamma_m)$ and $n_a(\beta)$ flatten out. Indeed, at $x^2 << 1$, we obtain that $n_a \rightarrow 0.5 + \pi^{-1}\text{arctg}(2\eta_a/\pi)$.

References

1. Xu, M., Liang, T., Shi, M., Chen, H., Graphene-like two-dimensional materials. *Chem. Rev.*, 113, 3766, 2013.
2. Geim, A.K. and Grigorieva, I.V., Van der Waals heterostructures. *Nature*, London, 499, 419, 2013.
3. Sun, Z. and Chang, H., Graphene and graphene-like two-dimensional materials in photodetection: Mechanisms and methodology. *ASC Nano*, 8, 5, 4133, 2014.
4. Li, P. and Appelbaum, I., Electrons and holes in phosphorene. *Phys. Rev. B: Condens. Matter*, 90, 115439, 2014.
5. Guan, S., Yang, S.A., Zhu, L., Hu, J., Yao, Y., Electronic, dielectric, and plasmonic properties of two-dimensional electride materials X2N (X = Ca, Sr): A first-principles study. *arXiv*, 1502, 0232.
6. Brumme, T., Calandra, M., Mauri, F., First-principles theory of field-effect doping in transition-metal dichalcogenides: Structural properties, electronic structure, Hall coefficient, and electrical conductivity. *arXiv*, 1501, 07223.
7. Sahin, H., Cahangirov, S., Topsakal, M., Bekaroglu, E., Akturk, E., Senger, R.T., Ciraci, S., Monolayer honeycomb structures of group-IV elements and III-V binary compounds: First-principles calculations. *Phys. Rev. B: Condens. Matter*, 80, 155453, 2009.
8. Suzuki, T. and Yokomizo, Y., Silicene: Prediction, synthesis, application. *Physica E*, Amsterdam, 40, 2820, 2010.
9. Wang, S., Studies of physical and chemical properties of two-dimensional hexagonal crystals by first-principles calculation. *J. Phys. Soc. Jpn.*, 79, 064602, 2010.
10. Zhuang, H.L. and Hennig, R.G., Electronic structures of single-layer boron pnictides. *Appl. Phys. Lett.*, 101, 153109, 2012.
11. Mukhopadhyay, G. and Behera, H., Structural and electronic properties of graphene and graphene-like materials. *World J. Eng.*, 10, 39, 2013.

12. Zhuang, H.L., Singh, A.K., Hennig, R.G., Computational discovery of single-layer III-V materials. *Phys. Rev. B: Condens. Matter*, 87, 165415, 2013.
13. Singh, A.K., Zhuang, H.L., Hennig, R.G., *Phys. Rev. B: Condens. Matter*, 89, 245431, 2014.
14. Tong, C.-J., Zhang, H., Zhang, Y.-N., Liu, H., Liu, L.-M., New manifold two-dimensional single-layer structures of zinc-blende compounds. *J. Mater. Chem. A*, 2, 17971, 2014.
15. Feenstra, R.M., Jena, D., Gu, G., Single-particle tunneling in doped graphene-insulator-graphene junctions. *J. Appl. Phys.*, 111, 043711, 2012.
16. Beheshtian, J., Sadeghi, D.A., Neek-Amal, M., Michel, K.H., Peeters, F.M., Induced polarization and electronic properties of carbon-doped boron nitride nanoribbons. *Phys. Rev. B: Condens. Matter*, 86, 195433, 2012.
17. Neek-Amal, M. and Peeters, F.M., Graphene on hexagonal lattice substrate: Stress and pseudo-magnetic field. *Appl. Phys. Lett.*, 104, 041909, 2014.
18. Zoliomi, V., Wallbank, J.R., Fal'ko, V.I., Silicane and germanene: Tight-binding and first-principles studies. *2D Mater.*, 1, 011005, 2014.
19. Padilha, J.E., Fazzio, A., da Silva, A.J.R., van der Waals heterostructure of phosphorene and graphene: Tuning the schottky barrier and doping by electrostatic gating. *Phys. Rev. Lett.*, 114, 066803, 2015.
20. Antonova, I.V., Vertical heterostructures based on graphene and other 2D materials. *Semiconductors*, 50, 66, 2016.
21. Davydov, S. Yu., *Theory of Adsorption: Method of Model Hamiltonians*, St. Petersburg Electrotechnical University "LETI," St. Petersburg, 2013, [in Russian]. twirpx.com/file/1596114/.
22. Castro Neto, A.H., Guinea, F., Peres, N.M.R., Novoselov, R.S., Geim, A.K., The electronic properties of graphene. *Rev. Mod. Phys.*, 81, 109, 2009.
23. Harrison, W.A., *Electronic Structure and the Properties of Solids: The Physics of the Chemical Bond*, vol. 1, Freeman, San Francisco, California, United States, 1980; Mir, Moscow, 1983.
24. Harrison, W.A., Coulomb interactions in semiconductors and insulators. *Phys. Rev. B: Condens. Matter*, 31, 2121, 1985.
25. Davydov, S. YU. and Posrednik, O.V., *Bond-Orbital Method in the Theory of Semiconductors*, St. Petersburg Electrotechnical University "LETI," St. Petersburg, 2007, [in Russian]. twirpx.com/file/1014608/.
26. Mousavi, H., Heat capacity of hexagonal boron nitride sheet in Holstein model. *Semiconductors*, 48, 617, 2015.
27. Zhang, Y., Tan, Y.-W., Stormer, H.L., Kim, P., Experimental observation of the quantum Hall effect and Berry's phase in graphene. *Nat. (London)*, 438, 201, 2005.
28. Peres, N.M.R., Guinea, F., Castro Neto, A.H., Electronic properties of disordered two-dimensional carbon. *Phys. Rev. B: Condens. Matter*, 73, 125411, 2006.
29. Davydov, S. Yu. and Posrednik, O.V., Low-energy approximation in the theory of adsorption on graphene. *Phys. Solid State*, 57, 1695, 2015.
30. Das Sarma, S., Adam, S., Hwang, E.H., Rossi, E., Electronic transport in two-dimensional graphene. *Rev. Mod. Phys.*, 83, 407, 2011.
31. Davydov, S. Yu., Adsorption-induced energy gap in the density of states of single-sheet graphene. *Semiconductors*, 46, 193, 2012.
32. Irkhin, V. Yu. and Irkhin, Yu. P., *Electronic Structure, Correlation Effects and Physical Properties of d- and f-Metals and Their Compounds*, Ural Branch of the Russian Academy of Sciences, Yekaterinburg, 2004; Cambridge International Science, Cambridge, 2007.
33. Einstein, T.L. and Schrieffer, J.R., Indirect interaction between adatoms on a tight-binding solid. *Phys. Rev. B: Solid State*, 7, 3629, 1973.
34. Mattausch, A. and Pankratov, O., *Ab initio* study of graphene on SiC. *Phys. Rev. Lett.*, 99, 076802, 2007.

35. Chan, K.T., Neaton, L.B., Cohen, M.L., First-principles study of metal adatom adsorption on graphene. *Phys. Rev. B: Condens. Matter*, 77, 235430, 2008.
36. Giovannetti, G., Khomyakov, P.A., Brocks, G., Karpan, V.M., van der Brink, J., Kelly, P.J., Doping graphene with metal contacts. *Phys. Rev. Lett.*, 101, 026803, 2008.
37. Grigoriev, I.S. and Meilikhov, E.Z. (Eds.), *Handbook of Physical Quantities*, Energoatomizdat, Moscow, 1991; CRC Press, Boca Raton, Florida, United States, 1996.
38. Davydov, S. Yu., On the electron affinity of silicon carbide polytypes. *Semiconductors*, 41, 696, 2007.
39. Bechstedt, F. and Enderlein, R., *Semiconductor Surfaces and Interfaces: Their Atomic and Electronic Structures*, Akademie, Berlin, 1988; Mir, Moscow, 1990.
40. Davydov, S. Yu., On the specific features of the density of states of epitaxial graphene formed on metal and semiconductor substrates. *Semiconductors*, 47, 95, 2013.
41. Davydov, S. Yu., To the theory of adsorption on epitaxial graphene: Model approach. *Phys. Solid State*, 56, 1483, 2014.
42. Alisultanov, Z.Z., On renormalization of the Fermi velocity in epitaxial graphene. *Tech. Phys. Lett.*, 39, 597, 2013.
43. Davydov, S. Yu., Hexagonal two-dimensional layers of $A_N B_{8-N}$ compounds on metals. *Phys. Solid State*, 56, 849, 2014.
44. Alisultanov, Z.Z., Anomalous increase in the thermopower in a graphene monolayer formed on a tunable graphene bilayer. *JETP Lett.*, 98, 111, 2013.
45. Alisultanov, Z.Z., The thermodynamics of electrons and the thermoelectric transport in epitaxial graphene on the size-quantized films. *Physica E*, 69, 89, 2015.
46. Hashmi, A. and Hong, J., Band gap and effective mass of multilayer BN/graphene/BN: van der Waals density functional approach. *J. Appl. Phys.*, 115, 194304, 2014.
47. Davydov, S. Yu., Appearance conditions for a semiconducting-substrate-induced gap in the density of states in epitaxial graphene. *Tech. Phys.*, 59, 624, 2014.
48. Davydov, S. Yu., Energy gaps in the density of states of a graphene buffer layer on silicon carbide: Consideration for the irregularity of layer-substrate coupling. *Semiconductors*, 48, 46, 2014.
49. Davydov, S. Yu., Energy of substitution of atoms in the epitaxial graphene-buffer layer-SiC substrate system. *Phys. Solid State*, 54, 875, 2012.
50. Davydov, S. Yu. and Sabirova, G.I., Adsorption of hydrogen, alkali metal, and halogen atoms on graphene: Adatom charge calculation. *Tech. Phys. Lett.*, 37, 515, 2011.
51. Anderson, P.W., Local moments and localized states. *Rev. Mod. Phys.*, 50, 191, 1978.

18

Lower Dimensional Materials

B.G. Sidharth

International Institute for Applicable Mathematics and Informatics,
Hyderabad, India, and Udine, Italy
B.M. Birla Science Centre, Adarsh Nagar, Hyderabad, India

Abstract
In the mid-90s, the author explored the behavior of Fermions in one and two dimensions. The finding was that these particles would behave in a neutrino- or even quark-like manner showing handedness, luminal speeds, and so on. Commenting on this, Nobel Laureate Cohen-Tannoudji observed, "it is of course clear that you have done important work on 1D and 2D fermion systems."

It is interesting that 2D crystals can be treated as a case of noncommutative geometry because of their honeycomb-like lattices for graphene. This leads to a noncommutative geometry for the space. Furthermore, because of the two dimensionality of the crystal, the electromagnetic character becomes very different from the usual three-dimensional electromagnetism. For example, we now have the relation. Furthermore, in the limit of a very large sheet of a 2D crystal, we can treat it as a Minkowski space where the Compton length replaces the lattice length. From this, we argue that graphene or even other 2D crystals can be used as a test bed for high energy physics. This is because the situation is similar to a wind tunnel where there are the Reynold numbers. In this case too, there are the counterparts of the Reynold numbers. Furthermore, these 2D crystals have a very strong magnetic field as independently argued by Saito and the author. The author goes further to estimate the strength of the magnetic field. A very surprising deduction is that based on this theory, it is possible to deduce the entire infinite sequence of the fractional quantum Hall effect that was till now an experimental result observed by Von Klitzing. These and other matters are discussed in detail including the rather interesting result that we can observe fermion–boson transmutations, using statistical mechanics. All these results are of an anomalous nature.

Keywords: Nanotubes, graphene, Stanene, handedness, electromagnetism, Pauli matrices, two component Dirac equation, anomalous effects

18.1 2D Crystals

Recently, the author had shown that many puzzling features of graphene (or any 2D crystal), e.g., minimum conductivity or the puzzling fractional Quantum Hall Effect, could be remarkably explained by the non-commutative geometry, which these honeycomb lattices display [7]. Before we embark, it must be noted that 2D materials have very peculiar

Email: iiamisbgs@yahoo.co.in

Tobias Stauber (ed.) Handbook of Graphene: Volume 2, (593–612) © 2019 Scrivener Publishing LLC

characteristics not available in higher dimensions. For example, the electromagnetic field E and the magnetic field B are related by the simple relation $E = B$ without the proportionality, which is available in usual theory.

Further, in the limit of these sheets tending to infinity, the sheet behaves like Minkowski space with the Compton wavelength replacing the lattice length [16]. The author further argued that substances like graphene could be used as a test bed for many high energy physics experiments, rather than use exorbitant particle accelerators. This could be done by "scaling"—for example, the velocity of light c replaces the Fermi velocity.

For graphene as is well known [4], a two-component Dirac equation is obeyed, as for the massless neutrino.

$$\sigma^{\mu}\partial_{\mu}\psi = 0 \tag{18.1}$$

Here, the σ's are the two-component Pauli matrices.

In particular, the following important facts were shown.

A non-commutative geometry applies for graphene. This is due to the homogenous structure of lattices. Earlier the author and Saito had shown that such a non-commutative geometry leads to magnetic fields though from different approaches [10, 13, 14]. Furthermore, the author had deduced the following relation for the magnetic field:

$$Bl^2 = hc/e \tag{18.2}$$

Let us see all this in a little greater detail invoking the connected aspect of non-integrable space [14]. We start with a non-integrable infinitesimal parallel displacement of a four-vector,

$$\delta a^{\sigma} = -\Gamma^{\sigma}_{\mu\nu}\ a^{\mu}\ dx^{\nu} \tag{18.3}$$

The Γ's are the Christoffel symbols. This represents the extra effect in displacements, due to curvature. In a flat space, all the Γ's on the right side would vanish. Considering partial derivatives with respect to the μth coordinate, this would mean that, due to (18.3),

$$\frac{\partial a^{\sigma}}{\partial x^{\mu}} \rightarrow \frac{\partial a^{\sigma}}{\partial x^{\mu}} - \Gamma^{\sigma}_{\mu\nu}\, a^{\nu}, \tag{18.4}$$

The second term on the right side of (18.4) can be written as

$$\Gamma^{\lambda}_{\mu\nu}\, g^{\nu}_{\lambda}\ a^{\sigma} = -\ \Gamma^{\nu}_{\mu\nu}\ a^{\sigma},$$

where we have linearized the metric,

$$g_{\mu\nu} = \eta_{\mu\nu} + h_{\mu\nu},$$

$\eta_{\mu\nu}$ being the Minkowski metric and $h_{\mu\nu}$ a small correction whose square is neglected. From (18.4), we conclude

$$\frac{\partial}{\partial x^{\mu}} \to \frac{\partial}{\partial x^{\mu}} - \Gamma^{\nu}_{\mu\nu} \tag{18.5}$$

We can identify

$$A_{\mu} = \Gamma^{\nu}_{\mu\nu} \tag{18.6}$$

from the above using minimum electromagnetic coupling exactly as in Dirac's monopole theory.

If we use (18.5), we will get the commutator relation,

$$\frac{\partial}{\partial x^{\lambda}}\frac{\partial}{\partial x^{\mu}} - \frac{\partial}{\partial x^{\mu}}\frac{\partial}{\partial x^{\lambda}} \to \frac{\partial}{\partial x^{\lambda}}\Gamma^{\nu}_{\mu\nu} - \frac{\partial}{\partial x^{\mu}}\Gamma^{\nu}_{\lambda\nu} \tag{18.7}$$

Let us now use (18.6) in (18.7): The right side does not vanish due to the electromagnetic field (18.6) and we have a non-commutativity of the momentum components of Quantum Theory. Indeed the left side of (18.7) can be written as

$$\left[p_{\lambda}, p_{\mu}\right] \approx \frac{0(1)}{l^2}, \tag{18.8}$$

l being the Compton wavelength or minimum length. In (18.8) we have utilized the fact that at the extreme scale of the Compton wavelength, the Planck scale being a special case, the momentum is mc. For graphene, this is m times the Fermi velocity.

From (18.6), (18.7), and (18.8), we have

$$Bl^2 \sim \frac{1}{e}\left(= \frac{hc}{e}\right), \tag{18.9}$$

where B is the magnetic field and where we have restored $\hbar$ and c. There is another way of showing Equation (18.9). From the Landau theory of synchroton radiation, the frequency ω is given by

$$\omega = e\,B/mc \tag{18.10}$$

Also the maximum value of ω is given by, as is known

$$\omega = c/l \tag{18.11}$$

Using (18.11) in (18.10), we get (18.9).

18.2 Electromagnetism

In this case, l is the interlattice distance. It must be mentioned that the magnetic field B and the electric field E are spontaneous, in the sense that they are generated by the geometry of the system. In fact, that is the meaning behind the minimum conductivity, a puzzling but well-known feature of graphene (cf. Refs. [16]).

It has further been shown by the author as noted above that as the graphene sheet tends to infinity in all directions, it approaches the Minkowski space including the non-commutativity this time at the Compton wavelength. We can see that both can be scaled into each other exactly. Here the lattice constant $L \sim 2\dot{A}$ replaces the Compton wavelength. To see this, we note that

$$c = 300\ v_F \ and\ m_v \sim 0.05m \tag{18.12}$$

where m_v is the graphene "electron" mass and m is the electron mass. Feeding these figures into the Compton wavelength expression

$$l = \frac{\hbar}{mc}, \tag{18.13}$$

We can easily see that l goes into L and vice versa. Equations (18.12) and (18.13) give the Reynold number type scaling relation.

So the above origin of B and E in graphene due to the geometry of the structures could equally well be applicable to the Minkowski space in general after due scaling.

It is interesting to note that if we go to the case of bilayer graphene, there would be a small mass as in the case of the four-component Dirac equation [5].

$$(\partial_\mu \gamma^\mu - m)\psi = 0 \tag{18.14}$$

where γ's are the Dirac four-component matrices.

So there would be no chirality now, but non-commutativity geometry would still be applicable. So the above relations would still be approximately valid.

It is rather interesting to note that in an earlier era, it was sought to explain the origin of electromagnetism in statistical physics effects via the Bohr–van Leeuwen theorem.

We propose to use (18.9) to derive the otherwise inexplicable Fractional Quantum Hall Effect [15].

$$BL^2 = hc/e \tag{18.15}$$

L^2 defines a quantum of area exactly as in Quantum Gravity approaches. This is the area of individual lattices, in our case.

In these considerations as noted, the Fermi velocity v_F replaces the velocity of light. So we have for the electron mobility and conductivity

$$\mu = v_F/|E| \tag{18.16}$$

$$\sigma = \left(\frac{n}{A}\right) e \cdot \frac{v_F}{|E|}, \ A \sim L^2 \tag{18.17}$$

where A, as in the usual theory, is the area and n is the number of electrons. In our case as noted above A, the area is made up of a number of honeycomb lattice areas, each with area $\sim L^2$, that is,

$$A = mL^2$$

where m is an integer.

We also note that the electric field strength E equals the magnetic field strength B in the case of 2D structures (cf. Ref. [15]). Using these inputs in (18.17), we get

$$\sigma = \frac{n}{m} \cdot \frac{ev_F}{|B|L^2} \tag{18.18}$$

If we now use (18.15) in (18.18) (with v_F replacing c), we get for the conductance

$$\sigma = \frac{n}{m} \cdot \frac{e^2}{h} \tag{18.19}$$

which defines the Fractional Quantum Hall Effect.

Earlier the author had shown that it is this non-commutative space feature in two-dimensional structures that explains also Landau levels or the minimum conductivity that exists in graphene even when there are practically no electrons at the Dirac points. In other words, several supposedly diverse phenomena arise from the non-commutative space of these two-dimensional structures and by extension, in Minkowski spacetime, e.g., the origin of electromagnetism itself.

In 1995, the author had pointed out that in two dimensions and one dimension, electrons will display strange neutrino-like properties. In this case, they obey a two-component equation. This equation is

$$\left(\sigma^{\mu}\partial_{\mu} - \frac{mc}{\hbar}\right)\psi = 0 \tag{18.20}$$

where σ^{μ} denote the 2×2 Pauli matrices. In case $m = 0$, (18.20) gives the neutrino equation. A decade later, graphene was discovered. The neutrino-like equation, which holds for graphene, is

$$v_F\vec{\sigma}.\vec{\Delta}\psi(r) = E\psi(r) \tag{18.21}$$

where $v_F \sim 10^6$ m/s is the Fermi velocity, which replaces c, the velocity of light $\psi(r)$ is a two-component wave function, while $\vec{\sigma}$ and E denote the Pauli matrices and energy.

In any case, Landau had shown several decades ago that such two- and one-dimensional structures would be unstable and as such cannot exist—and this was proved wrong in the case of graphene and nanotubes.

However, there is no Lorentz invariance in the case of Equation (18.21) (except in the case of a hypothetical infinite sheet). Further the two-component wave function $\psi(r)$ in (18.21) comes from the wave functions in two side-by-side honeycomb lattices of graphene. We will see this later. This is rather like spin up and spin down.

18.3 The Graphene Test Bed

Taking the cue from (18.20) and (18.21), we would like to point out that graphene can be a test bed for high energy physics. Firstly Equation (18.21) represents a neutrino like (massless) Fermion. Indeed the massless feature has been experimentally confirmed. These are quasi particles. If we consider bilayer graphene, then even the mass comes in.

Interestingly, graphene behaves like a "chess board," that is, there is a minimum "length" [7]. In such a space, it is known that a non-commutative geometry holds.

In this case, we have

$$[x_i x_j] = \Theta_{ij} l^2 \tag{18.22}$$

where, as can be seen, the coordinates x_i and x_j do not commute. As a result of this, the Maxwell equations get modified with an extra term, as shown in detail elsewhere:

$$\partial^{\mu} F_{\mu\nu} = \frac{4\pi}{c} j_{\nu} + A_{\lambda} \varepsilon F_{\mu\nu} \tag{18.23}$$

where the symbols have their usual meaning. In (18.23), ε is a dimensionless number, which is equal to one for our non-commutative case, namely where (18.22) holds, and is 0 otherwise. With ε = 0, we get back the usual covariant Maxwell equations. Specializing to two dimensions, we get

$$\partial^1 F_{14} = \frac{4\pi}{c} j_4 + A_2 \epsilon F_{14} \tag{18.24}$$

and similar equations for j_1 and j_2. In this case, using the electromagnetic tensor, we get equations like

$$\frac{\partial E_x}{\partial x} = -4\pi \frac{\partial \rho}{\partial t} + \varepsilon A_y E_x \tag{18.25}$$

$$\frac{\partial E_y}{\partial y} = -4\pi \frac{\partial \rho}{\partial t} + \varepsilon A_x \ E_y \tag{18.26}$$

$$-\frac{\partial B_z}{\partial y} = 4\pi \ j_x + \varepsilon \frac{\partial E_x}{\partial t} \tag{18.27}$$

$$\frac{\partial B_z}{\partial y} = 4\pi \ j_x + \varepsilon \frac{\partial E_x}{\partial t} \tag{18.28}$$

As some of these equations are time dependent, we are dealing with non-steady fields, which generate radiation.

This clearly brings out the extra electromagnetic effects. Because of the space geometry (18.22), there appears a magnetic field as was shown by the author and Saito [10]. We have in fact as seen earlier the equation

$$Bl^2 = hc/e \tag{18.29}$$

This clearly can be smoothly carried over to graphene, keeping in mind the somewhat different values for the constants like v_F and l. In fact, we would have in this case

$$Bl^2 = hv_F/e.$$

The energy in the above is given by

$$Energy = \pm \ v_F \left|\vec{p}\right|$$

The positive sign holds for conduction electrons and the negative sign holds for valence particles. These are the analogues of particles and antiparticles in high-energy physics.

The analogy with high energy physics, particularly in the Cini–Toushek ultra relativistic regime, is very strong. There, too, we encounter a massless scenario. In fact, at very high energies, the Dirac equation becomes

$$H \ \psi = \frac{\vec{\alpha} \cdot \vec{p}}{|p|} E(p) \tag{18.30}$$

which resembles the massless version of equation (18.20). In (18.30), we have

$$\alpha^k = \begin{pmatrix} 0 & \sigma^k \\ \sigma^k & 0 \end{pmatrix} \quad \beta = \begin{pmatrix} I & 0 \\ 0 & -I \end{pmatrix} \tag{18.31}$$

$$\gamma^0 = \beta \tag{18.32}$$

This can be readily generalized to the neutrino equation. Importantly we have because of (18.22), as discussed in the literature, the so called Snyder–Sidharth dispersion relation

$$E^2 = p^2 + m^2 + \alpha \frac{l^2}{\hbar^2} p^4 \tag{18.33}$$

For Fermions α in Equation (18.33) is positive showing an extra contribution to the energy.

However, there are differences with the usual Dirac theory—here as pointed out, we do not encounter Lorentz invariance and v_F is not the velocity of light; rather, it is some three hundred times less.

We can see that graphene will be a test bed in some interesting situations. The author had already argued several years ago that for nearly monoenergetic fermions or even bosons, there would be a loss of dimensionality and the collection would behave as if it were in two dimensions. This immediately mimics the two dimensional feature of graphene as seen below.

Our starting point is the well-known formula for the occupation number of a fermion gas [6]

$$\bar{n}_p = \frac{1}{z^{-1} e^b E_p + 1} \tag{18.34}$$

where $z' \equiv \frac{\lambda^3}{v} \equiv \mu z \approx z$ because, here, as can be easily shown, $\mu \approx 1$

$$v = \frac{V}{N}, \lambda = \sqrt{\frac{2\pi\hbar^2}{m/b}}$$

$$b \equiv \left(\frac{1}{kT}\right), \quad and \quad \sum \bar{n}_p = N \tag{18.35}$$

Let us consider in particular a collection of fermions, which is somehow made nearly monoenergetic, that is, given by the distribution,

$$n'_p = \delta(p - p_0)\, \bar{n}_p \tag{18.36}$$

where $\bar{n}_p$ is given by (18.34).

This is not possible in general—here we consider the special situation of a collection of monoenergetic particles in equilibrium.

By the usual formulation, we have

$$N = \frac{V}{\hbar^3} \int d\vec{p} n'_p = \frac{V}{\hbar^3} \int \delta(p - p_0)\, 4\pi p^2 \bar{n}_p dp = \frac{4\pi V}{\hbar^3}\, p_0^2 \frac{1}{z^{-1} e^{\theta} + 1} \tag{18.37}$$

where $\theta \equiv bE_{p_0}$.

It must be noted that in (18.37), there is a loss of dimension in momentum space, due to the δ function in (18.36).

Further, recently the author had pointed out that the neutrinos behaved as if they were a two dimensional collection [15]. Indeed, one could expect such a behavior from the holographic principle. Equally the author (and A.D. Popova) had argued that the universe itself is asymptotically two dimensional.

Furthermore, it has also been argued that not only does the universe mimic a Black Hole, but also that the Black Hole is a two-dimensional object [11]. Indeed the interior of a Black Hole is in any case inaccessible and the two dimensionality follows from the area of the Black Hole, which plays a central role in Black Hole thermodynamics. The author had shown, in his analysis, that the area of the Black Hole is given by

$$A = Nl_p^2 \tag{18.38}$$

For these Quantum Gravity considerations, we have to deal with the Quantum of area [11, 1]. In other words, we have to consider the black hole to be made up of N quanta of area. Thus, we can get an opportunity to test these quantum gravity features in two-dimensional surfaces such as graphene.

In an earlier communication [11], we showed that in the one dimensional case, corresponding to nanotubes, we would have

$$kT = \frac{3}{5}\, kT_F \tag{18.39}$$

where T_F is the Fermi temperature. We can see that for the two dimensional case too kT would be very small. This is because using the well-known formula for two dimensions, we have

$$kT = \frac{e\hbar\pi}{mv_F} \tag{18.40}$$

$$(kT)^3 = \frac{6e\hbar v_F}{\pi} \tag{18.41}$$

Whence we have

$$(kT)^2 = 6\,.\, v_F^2\ \pi^2 m \tag{18.42}$$

Remembering that $\nu_F \sim 10^8$, even for a particle whose mass is that of an electron, kT in (18.42) is very small. By way of a comparison for the Fermi temperature, we get

$$kT_F = \frac{\hbar}{2}(z6\pi)^{\frac{1}{3}}.v_F$$

Another conclusion that could have been anticipated is the following. We get from the above

$$v_F^2 = (\frac{\hbar\pi}{m})^2 \cdot \frac{1}{A} \tag{18.43}$$

where $A \sim l^2$ is the quantum of area. So we get

$$\frac{m^2 v_F^2}{\hbar^2} . l^2 \sim 0\left(1\right) \tag{18.44}$$

as mentioned earlier.

This is perfectly consistent with v_F tending to the velocity of light c and h/mv_F tending to the Compton wavelength. In other words, an infinite graphene sheet would give us back the usual spacetime of Relativity and Quantum Mechanics. In practice, we could expect this for a very large sheet of graphene. In either case, it turns out that whatever be the temperature, it is as if the ensemble behaves like a very low temperature two-dimensional gas. This leads to many possibilities, particularly about magnetism.

As noted we can investigate electromagnetism in this new non-commutative paradigm, which throws up novel features including the Haas Van Alphen type effect. In this case, the magnetization per unit volume, as is known, shows an oscillatory type of behavior.

18.4 Discussion

Fluctuations of the Zero Point Field have been widely studied. Based on this, the author in 1997 predicted a contra model of the universe [12] in which there would be a small cosmological constant, that is, an accelerating universe. In 1998, observations of Perlmutter, Reiss, and Schmidt confirmed this scenario. Today we call this Zero Point Field fluctuations, dark energy. A manifestation of this is a noncommutative spacetime given in (18.22). This lead to the so-called Snyder–Sidharth dispersion relation given in (18.33).

We would like to point out that the extra magnetic effect in equations like (18.25) (and the following) can be attributed to this Zero Point effect of noncommutativity as given in (18.29). Closely related is the Casimir effect, which has been observed even in graphene [2, 3]. This is a Zero Point Field fluctuation effect. The Casimir energy in graphene is given by

$$\frac{Energy}{area} = \frac{\pi^2}{240} \cdot \frac{\hbar c}{a^3} \tag{18.45}$$

The energy itself is given by

$$Energy = \left(\frac{\pi^2}{240}\right) \cdot \frac{\hbar c}{a} \tag{18.46}$$

where we consider the area to be $\sim a^2$.

If following Wheeler [8] we consider ground state oscillators of the Zero Point Field, we can deduce that

$$Energy \sim \hbar c / a$$

resembling (18.46). Similarly if we take the extra term in the dispersion relation (18.33), it is easy to show that this also has the same form. All this is hardly surprising because they are all manifestations of fluctuations in the Quantum Vacuum.

As noted, the Casimir effect in graphene has been observed. What is interesting is that a group of scientists from MIT, Harvard University, Oak Ridge National Laboratory, and other universities have used this Zero Point energy for a compact integrated silicon chip. Clearly the same would be possible for graphene too particularly in the context of Quantum Computers: the "spin" up and down being the qubits [18].

To proceed further, we invoke (18.29) and the well-known result for a coil

$$i = \frac{NBA}{R\Delta t} \tag{18.47}$$

where N is the number of turns, A is the area, and R is the resistance. Use of (18.29) in (18.47) now gives

$$i \approx \frac{NA}{R} \cdot \frac{e}{l^2 \tau} \tag{18.48}$$

Whatever be N, if we think of a coil made up of nanotubes or graphene, remembering that is small and so is the resistance, (18.48) would be observable, like indeed (18.29).

Further observing that nanotubes and graphene can harbor fast-moving Fermions (including neutrons) and of course carbon, we have all the ingredients for manipulating a version of table top fusion possibly using the bosonization of fermions property. In this case, we could use an equation like (18.37) and preceding consideration [11].

We would now have, $kT = \langle E_p \rangle \approx \mathrm{E_p}$ so that $\theta \approx 1$. But we can proceed without giving θ any specific value.

Using the expressions for ν and z given in (18.35) in (18.36), we get

$$\left(z^{-1}e^{\theta} + 1\right) = (4\pi)^{\frac{5}{2}} \frac{z'^{-1}}{p_0} \text{ ; whence}$$

$$z'^{-1}\; A \equiv z'^{-1}\left(\frac{(4\pi)^{5/2}}{p_0} - e^{\theta}\right) = 1, \tag{18.49}$$

where we use the fact that in (18.35), $\mu \approx 1$ as can be easily deduced.

A number of conclusions can be drawn from (18.49). For example, if

$$A \approx 1, \text{ i.e.,}$$

$$p_0 \approx \frac{(4\pi)^{5/2}}{1+e} \tag{18.50}$$

where A is given in (18.49), then $z' \approx 1$. Remembering that in (18.35), λ is of the order of the de Broglie wave length and v is the average volume occupied per particle, this means that the gas gets very densely packed for momenta given by (18.50). In fact, for a Bose gas, as is well known, this is the condition for Bose–Einstein condensation at the level $p = 0$ (cf. Ref. [6]).

In any case, there is an anomalous behavior of the fermions.

18.5 Non-Commutative Maxwell's Equation

Maxwell's equation in covariant format can be written as

$$\partial^{\mu} F_{\mu\nu} = \frac{4\pi}{c} j_{\nu} \tag{18.51}$$

These particular equation is in an abridge form with μ = 1,2,3,4 and ν = 1,2,3,4.

$$\partial_{\mu}\left(\partial_{\mu} A_{\nu} - \partial_{\nu} A_{\mu}\right) = \frac{4\pi}{c} j_{\nu} \tag{18.52}$$

$$\partial_{\mu}\, \partial_{\mu} A_{\nu} - \partial_{\mu}\, \partial_{\nu} A_{\mu} = \frac{4\pi}{c} j_{\nu} \tag{18.53}$$

Let us consider a transformation [17] in which we consider $p_{\mu} \equiv \partial_{\mu}$

$$p_{\mu} p^{\mu} = D_{\mu\mu} - \Gamma^{\mu}_{\lambda\lambda}\, \partial_{\mu} \tag{18.54}$$

$$p_{\mu} p^{\nu} = D_{\mu\nu} - \Gamma^{\mu}_{\lambda\lambda}\, \partial_{\nu} \tag{18.55}$$

which is occurring due to space and time are non-commutative and holds good [17]. As we can see that if RHS last term is zero, then we enter into the continuous space and time. The last term occurred below Compton wavelength where negative energies are dominating. Now we substitute (18.54) and (18.55) in Equation (18.53) and try to analyze the physical behavior.

$$\left(D_{\mu\mu} - \Gamma^{\mu}_{\lambda\lambda}\ \partial_{\mu}\right)A_{\nu} - \left(D_{\mu\nu} - \Gamma^{\mu}_{\lambda\lambda}\ \partial_{\nu}\right)A_{\mu} = \frac{4\pi}{c}\, j_{\nu} \tag{18.56}$$

$$\left(D_{\mu\mu}A_{\nu} - D_{\mu\nu}A_{\mu}\right) = \frac{4\pi}{c}\ j_{\nu} + \left(\Gamma^{\mu}_{\lambda\lambda}\ \partial_{\mu}A_{\nu} - \Gamma^{\mu}_{\lambda\lambda}\ \partial_{\nu}A_{\mu}\right) \tag{18.57}$$

$$\left(D_{\mu\mu}A_{\nu} - D_{\mu\nu}A_{\mu}\right) = \frac{4\pi}{c}\ j_{\nu} + \Gamma^{\mu}_{\lambda\lambda}\left(\partial_{\mu}A_{\nu} - \partial_{\nu}A_{\mu}\right) \tag{18.58}$$

$$\partial^{\mu}\ F_{\mu\nu} = \frac{4\pi}{c}\ j_{\nu} + \Gamma^{\mu}_{\lambda\lambda}F_{\mu\nu} \tag{18.59}$$

$$\gamma^{4}\ F_{\mu\nu} = \frac{4\pi}{c} ij + \partial_{\lambda}\ \varepsilon\ F_{\lambda\nu}$$

or can be written as (18.59) where in (18.59) is a dimensionless number that is equal to 1 for our non-commutative case (18.54) and equal to 0 for the usual commutative case and has been introduced for clarity. In the usual commutative space and time, $\varepsilon = 0$ and we get back the usual covariant Maxwell's equations given in Equation (18.51).

Therefore, due to transformation that represents space and time to be non-commutative, there is an additional term with the Maxwell equation. Now let us try to derive these equations in two dimensional X and Y coordinates. We will see that this is not a trivial case. We get

$$\partial^{1}\ F_{14} = \frac{4\pi}{c}\ j_{4} + \partial_{2}\ \varepsilon\ F_{24} \tag{18.60}$$

$$\partial^{2}\ F_{24} = \frac{4\pi}{c}\ j_{4} + \partial_{1}\ \varepsilon\ F_{14} \tag{18.61}$$

$$\partial^{4}\ F_{42} = \frac{4\pi}{c}\ j_{2} + \partial_{1}\ \varepsilon\ F_{12} \tag{18.62}$$

$$\partial^{1}\ F_{12} = \frac{4\pi}{c}\ j_{2} + \partial_{4}\ \varepsilon\ F_{42} \tag{18.63}$$

$$\partial^2 F_{21} = \frac{4\pi}{c} j_1 + \partial_4 \varepsilon F_{41} \tag{18.64}$$

$$\partial^4 F_{41} = \frac{4\pi}{c} j_1 + \partial_1 \varepsilon F_{21} \tag{18.65}$$

Now from covariant electrodynamics, we know that $F_{\mu\nu}$ is the electromagnetic tensor and is given by [9]

$$F_{\mu\nu} = \mu \downarrow \overset{\nu \rightarrow}{\begin{pmatrix} 0 & -cB_z & cB_y & -E_x \\ cB_z & 0 & -cB_z & -E_y \\ -cB_y & cB_x & 0 & -E_z \\ E_x & E_y & E_z & 0 \end{pmatrix}} \tag{18.66}$$

So this electromagnetic tensor will give Maxwell's equations. We can see that electromagnetic tensor is asymmetric by nature. Now with the different components from (18.60)–(18.64) given by using Equation (18.66), we get

$$\frac{\partial E_x}{\partial x} = \frac{4\pi}{c}\frac{\partial \rho}{\partial t} + \varepsilon \frac{\partial E_y}{\partial x} \tag{18.67}$$

$$\frac{\partial E_y}{\partial y} = \frac{4\pi}{c}\frac{\partial \rho}{\partial t} + \varepsilon \frac{\partial E_x}{\partial x} \tag{18.68}$$

$$-\frac{1}{c}\frac{\partial E_y}{\partial t} = \frac{4\pi}{c^2} j_y + \varepsilon \frac{\partial B_z}{\partial x} \tag{18.69}$$

$$\frac{\partial B_z}{\partial x} = \frac{4\pi}{c^2} j_y - \frac{\varepsilon}{c}\frac{\partial E_y}{\partial t} \tag{18.70}$$

$$\frac{\partial B_z}{\partial y} = \frac{4\pi}{c^2} j_x - \frac{\varepsilon}{c}\frac{\partial E_x}{\partial t} \tag{18.71}$$

$$-\frac{1}{c}\frac{\partial E_x}{\partial t} = \frac{4\pi}{c^2}\, j_x + \varepsilon\,\frac{\partial B_z}{\partial x} \tag{18.72}$$

So from (18.67) to (18.72), we can see that there is an extra term in the Maxwell equations and if ε = 0, then we get the usual Maxwell's equations given below.

$$\frac{\partial E_x}{\partial x} = \frac{4\pi}{c}\frac{\partial \rho}{\partial t} \tag{18.73}$$

$$\frac{\partial E_y}{\partial x} = \frac{4\pi}{c}\frac{\partial \rho}{\partial t} \tag{18.74}$$

$$-\frac{\partial E_y}{\partial t} = \frac{4\pi}{c^2}\, j_y \tag{18.75}$$

$$\frac{\partial B_z}{\partial x} = \frac{4\pi}{c^2}\, j_y \tag{18.76}$$

$$\frac{\partial B_z}{\partial y} = \frac{4\pi}{c^2}\, j_x \tag{18.77}$$

$$-\frac{\partial E_x}{\partial t} = \frac{4\pi}{c}\, j_x \tag{18.78}$$

Therefore, we recover the old Maxwell's equations, which is also shown in (18.73) to (18.78). Moreover when $\varepsilon \neq 0$, equations like (18.67) suggest that there could be electromagnetic radiation as well. Thus, then we can see that introduction of non-commutative space time leads to extra effects in Maxwell's equations (18.67) to (18.72), which indicate an extra electromagnetic field.

In commutative space and time, ε = 0, and the extra magnetic effects will disappear. To get more insight into the extra magnetic effect, we observe an alternate way of expressing the non-commutativity above. This is with a minimum space time extension, which also leads to non-commutative geometry.

We have now a relation such that

$$[x, y] = l^2\theta \tag{18.79}$$

where x and y are coordinates, l is the minimum extension, and θ are the matrices. Because of non-commutative geometry (18.79), it is known that there is the generation of a magnetic field, which was proved independently by Sidharth and Saito [10, 15] with the relation

$$Bl^2 = hc/e \tag{18.80}$$

In other words, the non-commutativity of space–time generates the extra electromagnetic effects.

18.6 Two Dimensional Structures—A Recap

We can now say that due to non-commutative space–time, there is an emergence of magnetic effects. The best example is the recent understanding of space–time in graphene [7], which approximates the above 2D scenario. The authors concluded that electron spin is attributed to the discreteness of space–time, which resembles a chess board, and this type of a discrete partition of space–time will lead to the generation of a magnetic field, which may even affect the intrinsic behavior of the electron.

Here we can see in Figure 18.1 that spacetime is discrete and that during the hopping of electrons, there is change in spin direction, which is represented by the yellow and green dots, which show that the space is not smooth and this gives rise to some sort a of magnetic field that can cause the change of spin in the electron during hopping. So we can see that the above behavior of electron spin can be analogous to the above with appearance of magnetic field in non-commutative space–time.

In this note, we comment on the Hall effect from the following novel point of view, namely the strong parallel between two dimensional Quantum Mechanics and the analysis of graphene. The Hall effect itself was observed in the nineteenth century, in the case of a current (or a moving electron) I_x along the x axis, a magnetic field B_z along the z axis, leading to the so called Hall voltage along the y axis. This is given by

Figure 18.1 Space–nonspace interspersed.

$$V_H = \left(\frac{1}{n|e|}\right) \cdot \frac{I_x}{d} B_z \tag{18.81}$$

The expression $\left(\frac{1}{n|e|}\right)$ is called the Hall resistance and V_H is the Hall voltage, while d is the thickness of the conductor. We would like to point out that the Hall effect has a parallel in relativistic electromagnetic theory.

In two dimensions and one dimension, electrons will display strange neutrino-like properties as had been pointed out by the author, starting in the mid-1990s. The two-component equation that is obeyed is

$$\left(\sigma^\mu \partial_\mu - \frac{mc}{\hbar}\right) \psi = 0 \tag{18.82}$$

where σ^μ denotes the 2×2 Pauli matrices. In case the mass vanishes, Equation (18.82) gives the neutrino equation. This has relevance to graphene that was discovered nearly a decade later. For the electron quasi particles in graphene, we have, as noted,

$$v_F \vec{\sigma} \cdot \vec{\Delta} \psi(r) = E \psi(r) \tag{18.83}$$

$\nu_F \sim 10^6\ m/s$ is the Fermi velocity replacing c, the velocity of light, and $\psi(r)$ being a two-component wave function, with E denoting the energy.

Indeed, going a step further, the author has argued that graphene (or more generally other two dimensional structures) could be a test bed for high energy physics itself, in the sense of the role played by a wind tunnel, given Reynold's numbers, for the actual problem. Surprisingly, we can then resolve puzzles like the minimum conductivity observed in graphene and also get an explanation for the fractional Quantum Hall effect something that has eluded us.

We would now like to observe that the relativity that can be seen in graphene with the Fermi velocity v_F replacing the velocity of light gives a "Lorentz" transform. Furthermore, in the electromagnetic case, this leads to the Lorentz force equal to $\vec{v} \times \vec{B}$, where $\vec{v}$ is the velocity of the moving or conduction electron and $\vec{B}$ is the magnetic field. This Lorentz force can be immediately identified with the Hall effect emf.

More specifically, the Lorentz force is given by

$$Force = \frac{d\vec{p}}{dt} = \frac{e}{c} \vec{u} \times \vec{B} \tag{18.84}$$

So the energy is given by

$$Energy - \frac{1}{c} R \frac{I_x B_z}{d} \tag{18.85}$$

where a factor R has been introduced to indicate a resistance, in case the electron is not a free moving particle. Comparing (18.81) and (18.85), we can immediately see that the Hall Effect is nothing but a manifestation of the Lorentz force in relativistic electrodynamics.

It must also be pointed out that, as shown in references, it is the non-commutative nature of the graphene space that leads to mysteries like the minimum conductance in graphene or the fractional Quantum Hall Effect.

In a recent paper, we had argued that two-dimensional crystals like graphene can be a possible test bed for High Energy Physics experiments as behaviors like Zitterbewegung, Compton scale, non-commutative space–time, etc. are exhibited though at a different scale, much like wind tunnels work with Reynolds scaled down numbers.

For example, the Fermi velocity replaces the velocity of light, which is some 300 times higher.

It was also pointed out that such effects as minimum conductivity and the fractional Quantum Hall Effect get a remarkable derivation due to the non-commutative nature of the space of these structures.

We would like to point out two new and important points in this context. The first is that the magnetic field in this case is stronger than the usual Maxwellian field and in fact is now given by its expression in non-commutative space:

$$Bl^2 = \frac{hc}{e} \tag{18.86}$$

In (18.86), symbols have their usual meaning except that l stands for the minimum length, the lattice length in this case. This was deduced independently by the author and Saito several years ago [10]. The experimentally observed and mysterious minimum conductivity is given by

$$\sigma = 4\frac{e^2}{h}, \tag{18.87}$$

Again symbols have their usual meanings in (18.87). Remarkably, (18.87) can be deduced from the above considerations. What is equally remarkable is that the magnetic field (18.86) and the electric current following from (18.87) arise solely as a result of the non-commutative space geometry of these two dimensional crystalline structures.

Finally, we would like to mention that one of the conclusions of the study was that the two dimensionality of the crystal is all that matters—in other words, it need not be graphene alone. This was borne out by the fact that the same properties seem to apply to other crystals like Stanene and even quasi crystals. We have studied two dimensional crystals and quasi crystals from a fundamental point of view without invoking any special properties of the material. This makes the results applicable over the entire range of such material.

Finally, we comment that as pointed out by the author, a non-commutative space-time imposes a maximal velocity as in special relativity (cf., *New Advances in Physics*, 11(1),2017; to appear in *Zeit fur Natur. A*).

References

1. Baez, J., The quantum of area? *Nature*, 421, 702–703, 2003.
2. Fialkovsky, I.V., Gitman, D.M., Vassilevich, D.V., Casimir interaction between a perfect conductor and graphene described by the Dirac model. *Phys. Rev. B*, 80, 24, 245406, 2009, arXiv:0907.3242.
3. Fialkovsky, I.V., Marachevskiy, V.N., Vassilevich, D.V., Marachevsky; Vassilevich, Finite temperature Casimir effect for grapheme. *Phys. Rev. B*, 84, 35446, 35446, 2011, arXiv:1102.1757.
4. Geim, A., Graphene: Status and prospects. *Science*, 324, 5934, 1530–4, 2009.
5. Greiner, W., Muller, B., Rafelski, J., *Quantum electrodynamics of strong fields*, Springer-Verlag, Berlin, 1985.
6. Huang, K., *Statistical mechanics*, pp. 75ff, Wiley Eastern, New Delhi, 1975.
7. Mecklenburg, M. and Regan, R.C., Spin and the honeycomb lattice: Lessons from graphene. *Phy. Rev. Lett.*, 106, 116803, 2011.
8. Misner, C.W., Thorne, K.S., Wheeler, J.A., *Gravitation*, pp. 819ff, W.H. Freeman, San Francisco, 1973.
9. Panofsky, W.K.H. and Phillips, M., *Classical electricity and magnetism*, pp. 324–339, Addison-Wesley Publishing Company, USA, 1962.
10. Saito, T., Noncommutative spacetime - A short introductory review. *Gravit. Cosmol.*, 6, 22, 130–136, 2000.
11. Sidharth, B.G., Anomalous fermions. *J. Stat. Phys.*, 95, 3/4, 775–784, 1999.
12. Sidharth, B.G., The universe of fluctuations. *Int. J. Mod. Phys. A*, 13, 15, 2599ff, 1998.
13. Sidharth, B.G., Fuzzy, Non-commutative spacetime: A new paradigm for a new century. *Proceedings of Fourth International Symposium on "Frontiers of Fundamental Physics"*, Kluwer Academic/Plenum Publishers, New York, pp. 97–108, 2001.
14. Sidharth, B.G., The elusive monopole. *Nuovo Cimento B*, 118B, 1, 35–40, 2003.
15. Sidharth, B.G., A model for neutrinos. *Int. J. Th. Phys.*, 52, 12, 4412–4415, 2013.
16. Sidharth, B.G., An addendum to the paper Graphene and High Energy Physics. *Int. J. Mod. Phys. E*, 23, 05, May, 2014.
17. Snyder, H.S., The electromagnetic field in quantized space-time. *Phys. Rev.*, 72, 1, 68ff, 1947.
18. Zao, J. *et al.*, Casimir forces on a silicon micromechanical chip. *Nat. Commun.*, 4, 1845, 2013, arXiv: 1207.6163

19

Nature of Graphene, Its Chemical Structure, Composites, Synthesis, Properties, and Applications

Samuel Eshorame Sanni[1]*, Oluranti Agboola[1,2], Rotimi Emmanuel Sadiku[2] and Moses Eterigho Emetere[3,4]

[1]*Department of Chemical Engineering, Covenant University, Ota, Ogun State, Nigeria*
[2]*Department of Chemical, Metallurgical and Materials Engineering, Tshwane University of Technology, Pretoria, South Africa*
[3]*Department of Physics, Covenant University, Ota, Ogun State, Nigeria*
[4]*Department of Mechanical Engineering, University of Johannesburg, Aukland Park, Johannesburg, South Africa*

Abstract
In recent times, graphene, a super-carbon-based nanomaterial, has spurned research interests because it is the most important of the list of available carbon nanomaterials, has distinct properties, has significant impact in human lives, is biodegradable, and finds application in various disciplines, including physics, chemistry, engineering, biomedicine, biotechnology, etc. It is a two-dimensional monolayer material characterized by sp^2 hybridized carbon atoms arranged in a honeycomb lattice/hexagonal array. Its usefulness or functionality is enhanced by the ease with which it undergoes chemical modification to suit a particular application. The new technology behind the synthesis of graphene and its derivatives is benign (green), which complements its wide use. For future applications, its prospective use requires detailed understanding of the technology behind its formation. Its kinetics, anticorrosion properties and photoluminescence mechanism, and how they compare favorably with those of carbon and polymer nanodots (nanoparticles) are also currently being experimented. There are enormous justifications why graphene has gained alarming interest in the fields of science and engineering and some of them include its outstanding carrier mobility, transconductivity, ultimate thickness, and stability. In this chapter, the following subsections shall focus on the nature, chemical structure, properties, synthesis, and applications of graphene and its composites.

Keywords: Anticorrosion properties, chemical modification, graphene, nanocomposites, transconductivity

19.1 Introduction

Graphene is a super carbon allotrope with unique properties that are responsible for its recent attraction as a potential material for the advancement of technology. In this chapter, the nature of graphene, its physicochemical properties, its chemical structure, its synthesis,

**Corresponding author*: adexz3000@yahoo.com

Tobias Stauber (ed.) Handbook of Graphene: Volume 2, (613–636) © 2019 Scrivener Publishing LLC

and applications are discussed. Graphene is a special material comprising of a planar sheet of carbon atoms looking like a honeycomb (Figure 19.1); it is the foundation block for graphitic materials and can be folded into buckyballs/fullerenes (i.e., the third allotrope of carbon), one-dimensional nanotubes, or converted/clustered into graphite. According to Thielemans *et al.* [1], it is the hardest ever measured material with a Young modulus value of 1 TPa and tensile strength of 130 GPa. Its distinct electronic properties (i.e., absence of localized charges, quantum effect, and super-high mobility) are as a result of its pi-electrons, which form its conduction and valence bands; the points where these two bands are superimposed in graphene are called Dirac points. Graphene is known to have high electrical conductivity, which is a result of the presence of fast moving electrons in its crystal lattice when the electrons are exited via thermal application or by current influx into the material. Graphene can be transported with high carrier mobility of approximately 15,000 cm^2/m s at 298 K over very short wavelengths of approximately 4×10^{-6} m. Its high surface area makes it a useful material for making sensors since the entire material plays a significant role in the sensing of signals and other particulates, molecules, atoms, and ions/species. Other areas of application include transistor design, electrochemical cells, capacitors, biosensory devices for detecting enzymes, ferromagnetism, nanoelectronics/nanocomposites of polymeric materials, and materials that are known to exhibit photoelectric effect such as light emitting diodes (LEDs), and also as capacitive sensors in touchscreen devices of android phones. In recent times, graphene was predicted to be a good alternative for silicon-based technologies due to its biodegradability, stability, thickness, high thermal conductivity/ electrical conductivity, as well as its dielectric properties [2]. The structural adjustments/ changes occurring in the material during its refinement give rise to several hybrids that are known to be nonhazardous to the environment and humans, although it has been reported in literature that toxic forms of graphene do exist. Prior to the discovery of the existence of graphene, there were arguments that implied the nonexistence of 2D crystals due to their thermodynamic instabilities; this was backed up by the theory of Landau and Peierls, which

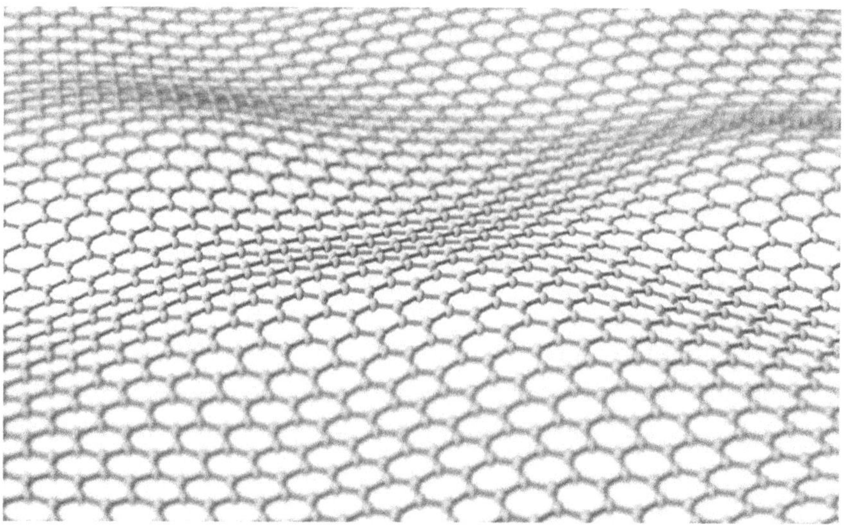

Figure 19.1 Structure of graphene. Adopted from Ref. [3].

states that the thermal fluctuations occurring in low-dimensional crystal lattices may result in displacements that make them appear as interatomic distances at finite temperatures [3]. Their theory was later supported by experimental investigations that showed that the melting point of a material of low thickness drops drastically with a corresponding decrease in its thickness, thus making it unstable at very low thicknesses close to a few tens of atomic layers. This led to the optimal belief that thick monolayers existed only as 3D structures; this also implied that 2D structures could only exist out of 3D structures; however, this idea was later countered when the discovery of graphene, as well as other stand-alone (monolayer boron nitride) 2D crystals, came to light [4]. An investigation on the perpetual layers/entire sheet of graphene was observed, and it was discovered that the inherent layered patterns were not perfectly flat when viewed under a transmission electron microscope (TEM) but appeared wavy with intrinsic tiny/microscopic roughness of several angular displacements and off-plane deformations of approximately 1 nm [5].

19.2 Green Technology/Methods for Synthesizing Graphene

Graphene was first synthesized using the Scotch tape method [6, 7]. Other methods include the following:

i. Production from silicon carbide as base material/molecular beam epitaxy (i.e., the deposition of gaseous graphite anchored on a substrate under high vacuum). Under vacuum or an inert atmosphere, and at a temperature as high as 1500°C, gaseous silicon is released from solid silicon carbide leaving mobile-carbon-rich deposits/atoms, which remain fluidized at that condition but rearrange/realign to establish a stable configuration of graphene. However, the method involves high thermal implications and high costs, which also results in low yield of graphene. Thin-film high-quality graphene can be obtained via this method, which finds application in the making of components of electronic devices [8].
ii. Exfoliation: This can be done in two ways: either by micromechanical exfoliation (MME) of graphite to form single bilayer graphene or by liquid phase exfoliation (LPE) of the graphite, which entails the use of Scotch tape on bulk graphite as starting material while continuously and repeatedly removing the top layer to reveal layers of fine thickness/thin layers of graphene, thus resulting in the formation of reduced graphene oxide (rGO), graphene oxide (GO), and graphene nanoparticles (GNPs). Graphene was first synthesized via the exfoliation of graphene from graphite using adhesive. Exfoliation is a method that employs the use of a solvent to exfoliate/break off or remove graphene scales from its source or parent material. Examples of such methods include the use of a CO_2/H_2O and ethanol system by Gao *et al.* [9] in which a pressurized reactor incorporated with an ultrasonic regenerator takes advantage of the combined effect of the impact of the intensity of the high-pressure force generated from the high-pressure acoustic cavitation and the high penetration of supercritical CO_2 supplied by the supercritical CO_2/H_2O system in a high-pressure batch reactor designed for graphene synthesis; this method has

only been developed on laboratory scale and is yet to be fully commercialized. Graphene concentrations up to 0.2 g/L of 1–2% yield can only be obtained via batch/probe sonication [9]. In sonication, an appropriate solvent is used to extract graphene oxide in layers by forcibly passing the solvent through the oxide layers of graphene. Two types of sonication include probe and batch sonications. In probe sonication, mechanical energy is applied during the sonication process and the higher the mechanical energy, the greater the tendency to break off the graphene layers, increase the sonication strength/intensity, and vice versa, while for batch sonication, the sonication time must be very small since the van der Waals' forces are not strong enough to maintain/withstand high mechanical energies and long exfoliation time as with probe sonication, during the exfoliation process; hence, this will only result in mechanical degradation of the graphene structure if conditions for probe sonication are appropriately applied.

In micromechanical exfoliation (MME), the top layer/exfoliate-graphene removed is adsorbed using an adhesive tape before being anchored on a substrate. If the forces of adhesion of the bottom graphene layer are stronger than the forces of cohesion between successive graphene layers, a layer of graphene is attached to the adsorbent/adhesive material and vice versa. By this method, microsizes of top-quality unit crystals of monolayer graphene can be synthesized. The disadvantage of this method is that it is not scalable. This method finds application in photocopying machines [9, 10]. For LPE, three steps are actually involved, which include (i) in-solvent dispersion, (ii) exfoliation, and (iii) removal of impurities/contaminants by purification, where the graphite/substrate is first suspended/solubilized in a suitable solution, and then subsequent exfoliation of the graphite follows using an ultrasonic device that applies the principle of acoustics, electrochemical method or mixing; by means of a centrifuge/ultracentrifugation, the exfoliated graphene is then separated from the un-exfoliated graphite flakes. Advantages of this process include its scalability, easy reproducibility, a moderate yield of micro- and nanosized graphene can be obtained, as well as low cost of manufacture. Its disadvantages are in the non-uniformity of the nanoparticles/nanocomposites that make up the graphene layers, presence of impurities, low graphene yield, and the graphite-exfoliation process needs to undergo multiple cycles [10]. Its applications include formation of composites, printer ink manufacture, energy storage devices (capacitors and transistors), biocomposites, and surface coatings.

iii. Graphite oxidation. This method involves the oxidation of graphite using strong acids as oxidizing agents/catalysts to form oxidized graphite. Via exfoliation coupled with sonication, continuous mixing, and thermal expansion, graphene-oxide flakes can then be produced from the intermediate product/oxidized graphite [11]. This method is scalable and gives high graphene oxide yield, which can serve as an insulating material. Also, the graphene oxide is dispersible in water and can be modified under different conditions to give other products. The disadvantages of this process are that the reactions of the acids and material lead to the release of toxic/harmful substances, the process requires large volume of water, and it is not cost effective. This method finds application in the production of composites and membranes for filtration purposes, biomedicals/biosensors, and the nuclear sector (production of explosives). Figure 19.2a and b gives the chemical structure of graphite oxide and an image of monolayer graphite oxide as obtained from a scanning tunneling microscope, respectively.

iv. Reduction reaction of graphene oxide: This method employs the action of strong reducing agents such as hydrogen, hydrazine, and sodium borohydride to reduce graphite oxide (Figure 19.2a) to form reduced graphene oxide with heteroatoms embedded within its structural matrix (Figure 19.2a–c). This method is scalable, gives high yield of graphene oxide, and employs a wide variety of reducing agents. Its disadvantage still remains the occurrence of undesirable defects in the final product. Applications include making of coatings, printer ink and composites [12].

v. Thermal expansion of graphene oxide (Figure 19.2b) to form rGO (Figure 19.2c).

vi. Laser desorption/ablation (disintegration of bulk graphite with the aid of rays released in pulses from a laser beam in order to generate structural graphene).

vii. Chemical vapor deposition (CVD) to produce single bilayer graphene. This method involves the three-stage assemblage of graphene sheets on an electrode or the coating of an electrode with graphene sheets, followed by the immersion of the coated graphene electrode in an electrochemical solution of the metallic/graphene precursor and the application of a potential difference to induce the formation of the graphene nanocomposites. The use of group "d" elements/transition metals such as copper/nickel has been discussed [11], whereas evidence on the use of other transition metals, namely gold, copper, platinum, and silver, has also been reported. For the case of silver, a solution of ammoniacal silver, i.e., $Ag(NH_3)_2OH$, is

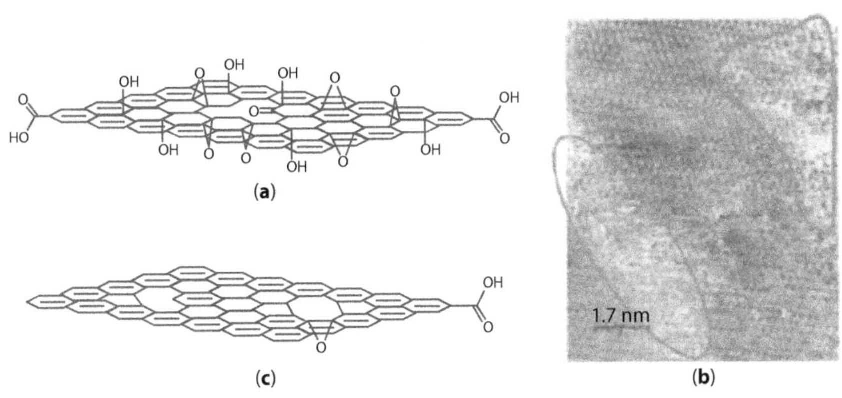

Figure 19.2 (a) Chemical structure of GO. (b) Scanning tunneling microscope (STM) image of a monolayer GO anchored on a thermally treated graphite substrate at high temperature. Oxidized regions are marked green in panel b. (c) Chemical structure of reduced graphene oxide (rGO). Adopted from Refs. [13, 14].

exposed to the cyclic voltammetry of a three-electrode system of indium tin oxide (ITO) electrode, a platinum foil (PF) counter electrode, and a saturated calomel (SC) electrode, which scans in the range of −1.5–0 V at a scan rate of 25 mV; the synthesized nanoparticles had a mean size distribution of 20 nm [15]. According to the method discussed by Davies *et al.* [11], the transition metal is then exposed to a hydrocarbon gas such as CH_4 at temperatures above 1000°C. The stage-wise growth or increase in the formation of graphene is seen on the surface of the substrate by the decomposition of the hot methane gas on the surface of the metal or by the layered separation of carbon as its temperature falls while cooling from a metastable carbon state. The growth of graphene on the surface of the transition metal can also be terminated at lower pressures. This method can also be used to synthesize multilayer graphene. The method finds application in: the formation of surface coats for corrosion prevention and manufacture of transparent conductive materials/electrodes, electronic devices, optoelectronics, and photonics. Rolls of large mass of graphene of greater than 50 cm^2 can be synthesized via this method. The challenges associated with this method include high costs, unfriendly temperatures, and formation of pits or pin holes/polycrystalline defects, while the work of Claussen *et al.* [16] focused on the synthesis of graphene-platinum (G-Pt) nanoparticles carefully decorated on petal-like multilayer graphene nanosheets using chemical/electrochemical vapor deposition. Based on their findings, the multilayer graphene nanoparticles constituted the working electrode for the three-electrode system with the platinum (Pt) gauze acting as the auxiliary/supporting electrode, while silver/silver chloride (i.e., Ag/AgCl) dipped in a mixture of H_2PtCl_6 and NaSO4 served as the reference electrode. The advantage of using this method is the possibility of adjusting the intensity of the current pulses in order to control the density, size, and morphology of the final product (i.e., reduced graphene oxide, rGO) with graphite oxide (GO and H_2PtCl_2) being the parent materials/sources without any need for further treatment, such as the application of heat or addition of reagents to get rGO. Regarding the use of copper, Wu *et al.* [17, 18] succeeded in depositing copper nanoparticles on rGO and understudied the underlying mechanisms responsible for the deposition process using CVD Tafel plots and chronoamperometry. The nucleation/anchoring of copper on rGO was achieved at a potential difference of 0.105 V against Ag/AgCl relative to the use of glassy carbon and pencil-like graphite using an electrolytic solution of $CUSO_4$ in a trio-electrode system comprising of the working rGO electrode. The counter electrode is a platinum mesh while the reference electrode is Ag/AgCl. The rate determining step for the deposition of Cu on the rGO was found to be largely controlled by mass transport with the nucleation of Cu on rGO also reported to have been instantaneous at high concentrations of, say, 50 mM and progressive at lower concentrations of 10 mM depending on the initial concentration of the electrolyte.

19.2.1 Nature of Green Graphene, Toxic Graphene, and Their Hybrids

19.2.1.1 Green and Toxic Graphene

Graphene and its nanocomposite hybrids are highly reactive with free oxygen, are prone to destruction of DNA molecules, and disrupt metabolic activities and apoptosis in the human system. Several interactions of graphene derivatives and the human/biological systems (human erythrocytes, skin cells, and cell lines such as HepG2, A498, etc.) have been identified. The toxicity of graphene oxide, a hybrid of graphene, is largely attributed to the presence of surface charges. Literature also has it that pristine-graphene oxide (P-GO) and carboxylic graphene oxide (i.e., GO-COOH) have good biocompatibility with T-lymphocytes and Albina serum cells in the human blood at concentrations below 25 $\mu g\ mL^{-1}$, while at concentrations above 50 $\mu g\ mL^{-1}$, they become very toxic to the cells [19]. The mechanism behind the high reactivity of graphene oxide is that P-GO interacts directly with protein cells/receptors, thus preventing their ligand binding tendencies, which subsequently results in the occurrence of reactive-oxygen species-dependent apoptosis via the B-cell (B-cl2) lymphoma site. Triethylene glycol reduced graphene (TGRG) oxide is an electrode for energy storage devices including supercapacitors and lithium-ion batteries. As a symmetric supercapacitor with cyclability of 5000 cycles, it can deliver power and energy densities of 60.4 $W\ h\ kg^{-1}$ and 0.15 $kW\ kg^{-1}$, respectively. TGRG has about 80% charge storage potential. The integration of TGRG in lithium-ion batteries improves the reversible capacity of the cell within the limits of 705 $mA\ h\ g^{-1}$ approximately, with good cyclability at constant storage capacity of 37 $mA\ g^{-1}$, which suggests that green graphene is a potential electrode for non-aqueous energy-storing devices. The comparative study of human hepatoma (HepG2) cells treated with graphene and single-walled carbon nanotubes (SWCNTs) showed that the oxidized SWCNTs resulted in oxidative stress in humans, which subsequently altered the cell cycle due to the meddling of the SWCNTs with the formation of protein cells, cytoskeletal systems structure, and intracellular metabolic processes [20]. For the delivery of drugs, proteins, peptides, and genes, nano-based (nanographene-based) compounds have been found to compare favorably with conventional methods especially when they are delivered via intravenous injection and oral administration [21], hence, they enhance drug solubility and bioavailability of suitable enzymes for metabolic activities, and achieve accurate delivery of drugs to target sites by solubilizing the drugs/biomolecules owing to their high specific surface area, their modes of interactions (electrostatic/hydrophilic interactions), as well as their π–πr stacking, which, in turn, reduces the risk of the drug exposure to enzymatic degradation [22].

19.2.1.2 Applications of Nontoxic Graphene and Its Derivatives

- Production of portable water: The membrane of graphene serves as a sieve for the removal of ionic pollutants or contaminants from portable/drinkable water.
- For the production of biosensors or microsensors [23]. Despite the advantages of the applications in terms of selectivity and sensitivity of biosensors such as antibodies, enzymes, and DNA molecules in electrochemical sensors, their limitations (i.e., their high costs, complex immobilization procedures, and low stability) leave room for alternative source materials that are cheaper, easy to reproduce, and highly stable [24]. This has led to the increased interest in the use of graphene nanocomposites/hybrids for detecting electroactive biomolecules at anodic sites.

Reports have it that graphene nanocomposites of gold and platinum have proven to exhibit the aforementioned characteristics especially in the determination of neurotransmitters such as dopamine (DP), uric acid (UA), glucose, and hydrogen peroxide; the sensor mechanism/electrode reduces or oxidizes the analyte at the sensing electrode where there is a difference in electrode potential. The presence of graphene in the nanocomposites helps to avoid excess cell potential.

- It is used in the manufacture of long-lasting batteries [25].
- Graphene can be used for corrosion inhibition of metals [26] and disease detection [27].
- Manufacture of solar cells as discussed in Wang *et al.* [28]. They synthesized graphene electrodes for application in dye-sensitized solar cell whose efficiency was about 0.2% less than the efficiency of a platinum-based counter electrode and at a cost lower than the total cost of producing the platinum. Furthermore, in lieu of the potential application of platinum electrodes in solar cells, two problems associated with the use of platinum electrodes include the availability of platinum, which is too low to guarantee high production of solar cells, and this brings to mind the second problem, which is the cost of synthesizing platinum and hence the need to look into the availability and suitability of other efficient materials/viable alternatives/substitutes such as graphene, which will amount to cost reduction [28]; this has led to the recent proposal that includes the replacement of platinum with graphene because of its probable cost reduction potential in utility bills if used in solar cells.
- Manufacture of electrical circuit boards [29].
- Manufacture of display panels such as liquid crystal display (LCD) and organic light-emitting diode (OLED); LEDs are flexible devices that appear opaque but suddenly become transparent in the presence of an electric field. They consist of a layer of liquid crystals guarded by a polymer and graphene electrodes. The electric field beams the dispersed light from the liquid crystals, which exposes its transparent background with a decal impregnated in the middle of the array of crystals [30]; however, the application of graphene as flexible counter electrode in OLEDs is still widely researched as replacement for indium tin oxide [31] since the latter is known to be brittle and in short supply [32]. The flexibility of graphene leaves room for speculations around its application in the making of touch and curved screens for mobile phones and tablet devices [33].
- Catalysis: Literature has it that the nature of graphene, which comprises a sparse electron density and a large basal plane of graphite, is responsible for its slow electron transfer when it undergoes chemical reactions; hence, it is the cathode/electron receptor of electrochemical cells [34–36]. Also, the industrial inclusion of surfactants in the production of liquid graphene has been reported to be the reason behind the electrochemical synthesis of several target-chemical analytes, which find wide use in analytical chemistry [37–39].
- Energy storage: With a characteristic surface area of 2630 m^2/g, graphene is a proven superconductor/superb capacitor, which has high potential to store and deliver electrical charges at very short time intervals. Its tremendous capacitive storage ability was first reported as a record breaker by Liu *et al.* [40], since no device in the history of conductive materials had ever proven

to have such erroneous conductive potential. Attempts to synthesize and test the ability of graphene-based-superconductors include the works of Chen *et al.* [41] for biosensing applications [34, 42–45], which have led to the consideration of graphene for powering electric cars since these cars require high accelerating powers for their mobility.

Note: The structure shown in Figure 19.3 becomes modified when pits/holes and dangling bonds appear on the pristine graphene. A monolayer graphene has low electrical conductivity as compared to multilayer graphene. However, beliefs of the noncatalytic activity of graphene are tied to the exposure of the basal plane to a target analyte [46].

- An ultrasensitive biosensor made of graphene/gold nanorod or polythionine was developed for the detection of the deadly human papilloma virus in human serum [47]. Huang *et al.* [48] carried out spectrofluorimetric investigation of the presence of glutathione in human blood plasma cells using graphene as adsorbent.

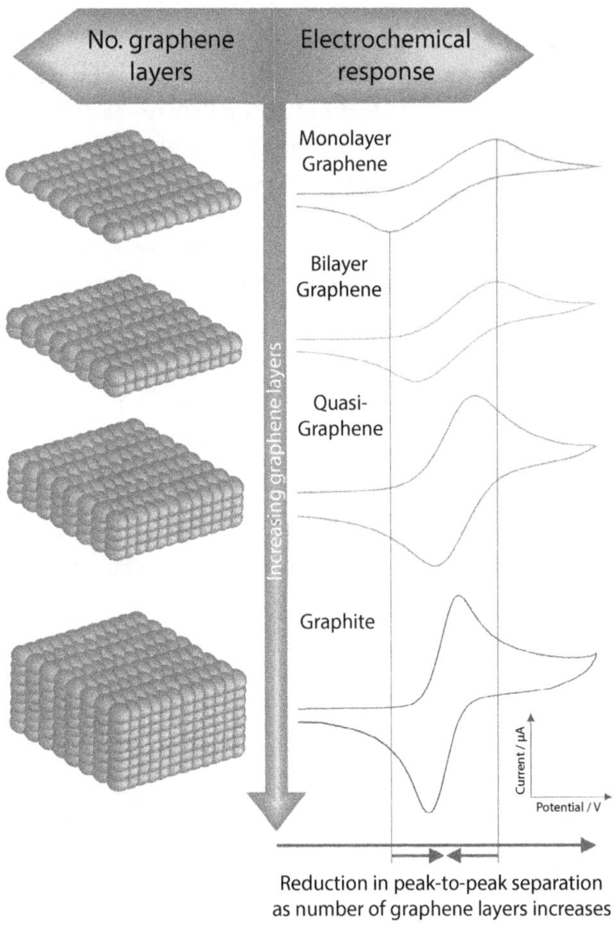

Figure 19.3 N-layered pristine graphene in a redox probe of hexamine-ruthenium (III) chloride with peak–peak separation in voltammetric waves showing an increase in heterogenous electron transfer and increased electrochemical activity. Adopted from Randvirr *et al.* [31].

- In medicine, investigations of the therapeutic potential of graphene to destroy cancer stem cells (CSCs)/tumor initiating cells (TICs), also known as nano-differentiation therapy (NDT), are ongoing. Some derivatives of graphene and their oxides can be dispersed in nano or larger sizes in solvents for industrial applications. Figures 19.4a–d show the vial of 2.3 mg/mL of b-GO dispersed in dimethyl sulfoxide (DMSO) and water [49]. Evidence has it that graphene oxide can successfully inhibit the proliferative enlargement of CSCs across tumor cells [50]. They carried out the inhibitive ability of graphene oxide on six cancer types, which include breast, ovarian, glioblastoma (brain cancer), pancreatic, lung, and prostate cancers with MCF7, SKOV3, U87 MG, MIA-PaCa-2, A549, and PC3 cell lines, respectively; although it was observed that this significant effect of graphene oxide is on single (i.e., stem cells) cancer cells, the reverse is the case for bulk (nonstem) cancer cells of these types and fibroblasts to which it is non-destructive/nontoxic; this is because, unlike for CSCs, graphene is unable to prevent or stop the major signal (i.e., those of Notch, WNT, and STAT) transduction pathways, thus inducing cell differentiation of the CSC cells.

Figures 19.5a and b are illustrations of small and big graphene oxide cells. Graphene hybrids are applied in cells/tissue cells to bring about stem-cell differentiation while also ensuring their proper distribution in the body to accurately and adequately reach the target cells; however, the way to handle cancerous cells needs to be well thought out in order to ensure safety of individuals and maximize the inherent potentials of stem-cell differentiation. For enzyme biosensing and imaging applications (Figures 19.6a–d) where the chemical structure and charge transfer behavior of graphene nanotechnology are of great essence, graphene nanocomposites have been expounded to have future prospects. Graphene composites/nanographenes exist in two forms,

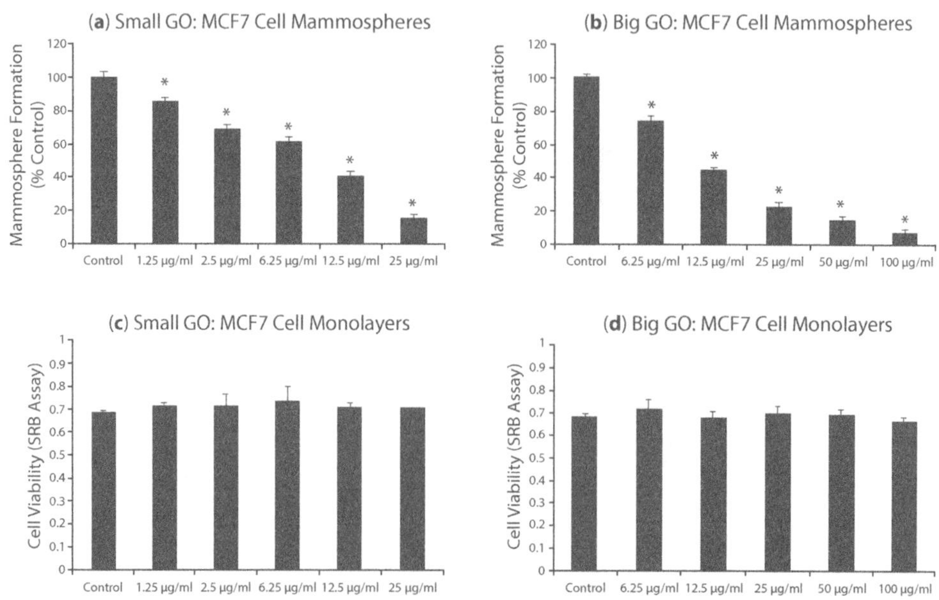

Figure 19.4 (a) Efficacy of small and (b) big graphene oxide in the inhibition of the proliferation of MCF7 CSCs/mammosphere formation. (c) Inefficacy of small and (d) big graphene oxide on the viability of the total MCF7 cell population. Hint: * indicates $p < 0.05$ (based on t-test). Adopted from Hernandez *et al.* [49].

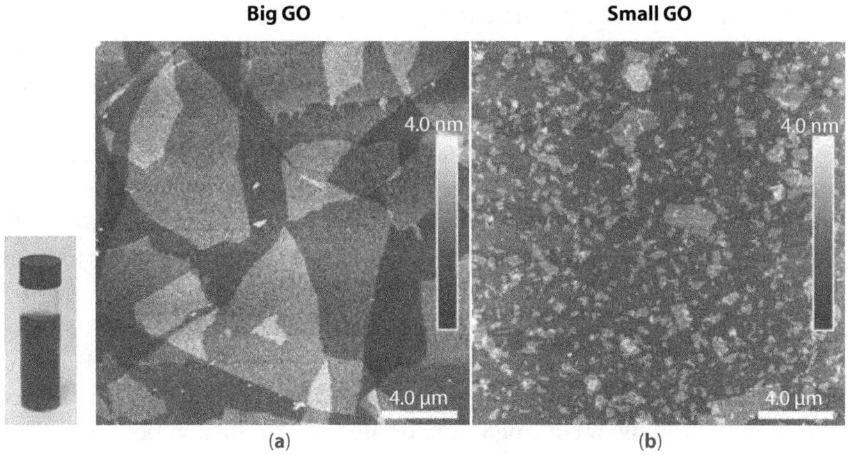

Figure 19.5 Flake size distribution and thickness of atomic force microscopic images of different shades ((a) 40 µm and (b) 40 nm) of graphene oxide anchored on silicon dioxide. Note: Graphene dispersed in water or other solvents does not give a stable aqueous phase. Adopted from Fiorillio *et al.* [50].

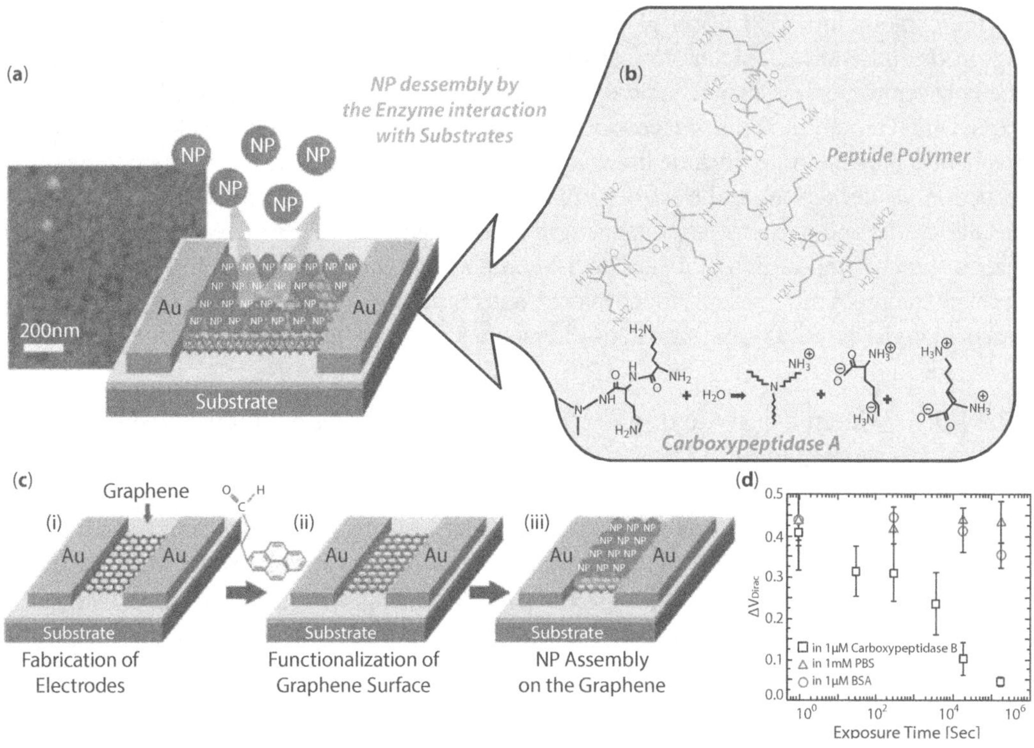

Figure 19.6 Graphene–nanoparticle hybrid sensor for hysteresis-based enzyme detection. (a) Enzyme detection using a graphene–nanoparticle hybrid device. (b) Structure of the polypeptide linker molecule. (c) Stages of fabrication of the hybrid biosensor; construction of the graphene pathway/site between gold electrodes (i). Activation of the surface of graphene by hydrophilic molecules (ii). Assemblage/anchorage of the peptide linker molecules and the gold nanoparticles on the polypeptide layer (iii). (d) Change in VDirac at different exposure times to a 1-µM solution of carboxypeptidase B in PBS, 1-mM PBS solution, and 1-mM solution of BSA in PBS. Adopted from Lee *et al.* [51] and Sun *et al.* [52].

namely the impregnation of nanoparticles in graphene sheets or the encapsulation/embedding of the nanoparticles in graphene sheets to bring about the synergistic/combined effect of both constituents in the composite material for medical applications. Yin *et al.* [23] conducted an extensive review where they highlighted the need to obtain a deeper understanding of the mechanisms responsible for graphene-nanoparticle synthesis so as to be able to monitor/regulate the production process, assembling of the sheets and particles and subsequent binding or entangling of the nanoparticles and graphene, since the morphological orientation depends largely on the ratio of the nanoparticles to the graphene sheets. Thus, a higher ratio of graphene sheets to nanoparticles results in encapsulation/wrapping of the nanoparticles by the graphene sheets; else, the reverse case results where the graphene sheets are rather tangled/embedded within the nanoparticles. Furthermore, there is need to look into ways of accessing the toxicity levels and biodistribution of graphene hybrids in cells as well as targeting better ways of improving the potency or efficacy of treatments in view of limiting or curbing the associated risks in their applications.

19.2.1.3 Graphene Nanocomposites

The doping or anchoring of nanoparticles on graphene to produce graphene nanocomposites is a technology that takes advantage of the synergistic effect of both components for the exploitation/application of their unique properties for several purposes depending on the inherent characteristics of the nanoparticles. Graphene nanocomposites have been proven to possess excellent catalytic properties owing to their highly effective surface areas, which enhance mass transport operations [53]. Two major methods for synthesizing graphene nanocomposites include the *in situ* and *ex situ* methods. The *in situ* method involves reduction, hydrothermal, and electrochemical methods, which are mainly for the synthesis of metallic oxides and noble metals; this method takes advantage of the inherent properties in the matrix of each component that makes up the composite, while the *ex situ* method encompasses the exploitation of the covalent/noncovalent bonds formed between the nanoparticles and the parent material as well as the electrostatic interactions that exist between the separate molecules.

19.2.1.3.1 Chemical Reduction Method for Synthesizing Nanocomposites of Graphene

In this method, the metal precursor and the graphene sheets are mixed in an aqueous solution, and then with the aid of reducing/chemical agents such as ethylene glycol and the borohydride and citrate of sodium, graphene nanocomposites are synthesized from HAuCl4, AgNO3, and K2PtCl4 [41, 54, 55], and according to Zhang *et al.* [56, 57], nanocomposites of gold/silver + graphene/graphene derivatives have a proven degree of biocompatibility with graphene oxide-silver nanoparticle (GO-AgNP) hybrid giving a cell viability of 95% when adenocarcinoma human alveolar basal epithelial cells (A549) were exposed to 1 mg/mL of the hybrid. The reduction reaction is similar to the three-stage conventional methods (reduction, nucleation, and nanoparticle development/growth) for synthesizing graphene nanocomposites. The surface charged negative functional groups of graphene oxide (GO) allow for possible nucleation with metallic salts, which promote the growth of the nanoparticles on GO without altering the excellent electrical properties of graphene. Another beauty of the flexibility of the nanocomposites of GO/rGO is that their densities can be easily adjusted by controlling the attachment of oxygen to the functional groups at sites with high affinity for oxygen.

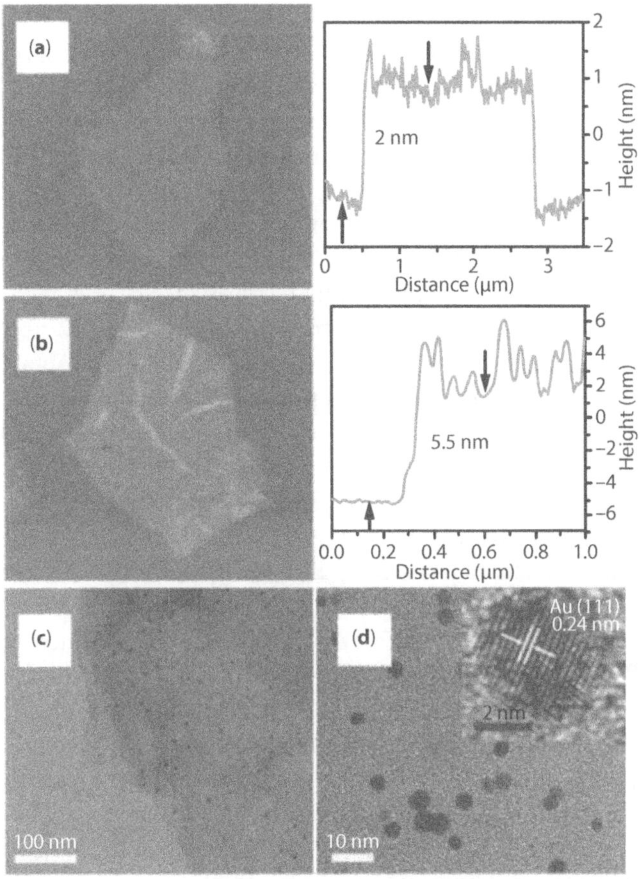

Figure 19.7 Graphene nanocomposites having GO sheets decorated with gold nanoparticles graphene: (a) GO sheet from AFM; (b) 3.5-nm gold nanoparticles decorated GO sheet. To the right side of the figures are curves showing their corresponding thickness of items (a) and (b). (c and d) Images of GO/AuNP sheet as obtained from TEM. Inset of (d) is a high-resolution TEM imaging of a separate gold nanoparticle. Adopted from Zhuo *et al.* [58].

Figures 19.7a–d are illustrations of a nanocomposite of graphene oxide wrapped with gold as obtained from AFM and TEM studies. The surface properties of the GO and rGO are imposed via electrostatic interactions by functional groups such as alcohols, carbonyls, and acids, which also enhance the attachment of free metal ions to the surfaces. Reducing agents help to reduce the attached metal ions in order to encourage the growth of the nanoparticles on the surfaces of the rGO and GO [59], although, despite how highly efficient and easy to adopt this method is, the problem of controlling the structure and size of the metal nanoparticles on the nanocomposites is difficult to handle, thus bringing about a wide size distribution of nanoparticles on the surfaces of the synthesized rGO and GO [58].

19.2.1.3.2 Hydrothermal Method for Synthesizing Nanocomposites of Graphene

This method is used to produce highly crystalline inorganic nanoparticles of oxides of Zn, Ti, Fe^{3+}, and Sn with a narrow size distribution on graphene sheets at high temperatures and pressures without any need for calcination and postannealing steps during the production process; the high temperature and pressure help to induce the growth of the nanoparticles with subsequent reduction of graphene oxide to rGO. In most cases, reducing agents are also added to complete the reduction process in lieu of the fact that long reaction times and high temperatures partly or completely help to reduce the graphene oxide [60, 61]. The aforementioned oxides help to infuse higher electrochemical properties such as improved electrical conductivity, shorter ionic radii, available surface area, and capacitance in their corresponding nanocomposites relative to single graphene owing to size, structure, and crystalline nature of the nanoparticles, which subsequently suppress graphene agglomeration and stacking [17, 18]. The orbiting/cycling performance of one-spot synthesized rGO-SnO_2 nanoparticle composites with initial current discharge of 1662 mA h g^{-1} maintained at 626 mA h g^{-1} when cycled at a current density of 100 mA g^{-1} as obtained from GO was aided by hydrazine as the reducing agent was better relative to that of the isolated tin oxide nanoparticles [62], while the work of Ren *et al.* [63] recorded the use of GO to synthesize high electrically conductive rGO comprising of dense uniformly deposited 7 nm of magnetic nanoparticles on rGO using anhydrous $FeCl_3$, which helped to infuse magnetic properties in the nanocomposite with ethylene glycol/a mixture of diethylene glycol (DG) and ethylene glycol as the reducing agent. According to them, the DG intensity ratio obtained for the synthesized graphene nanocomposite was 2.3:1 with characteristic sp2 domain in the carbon network as compared to 2.45:1, which is the intensity ratio benchmark for pristine graphene with loss of sp2 hybridization in its carbon network. Literature also has it that various chalcogenicides/semi-nanoconductors such as cadmium sulfide [64], zinc sulfide [65], copper I sulfide [66], molybdenum disulfide [67, 68], tin tetrasulfide (Sn_3S_4) [20], and cadmiotellinium [69] with applications in optoelectronics, magnetism, and catalysis were successfully immobilized on graphene via hydrothermal method [70].

Figures 19.8a and b are illustrations of the image of bimetallic-graphene nanocomposite formation process via chemical reduction method.

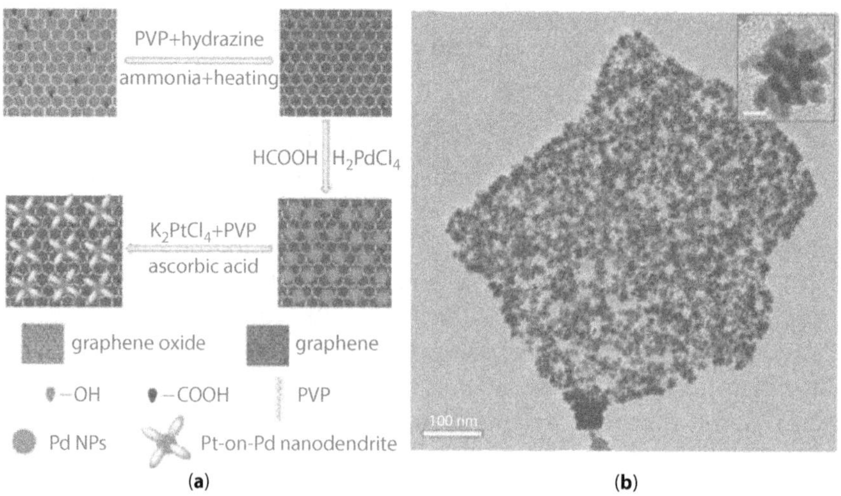

Figure 19.8 Bimetallic nanocomposites of graphene. (a) Stages of production of graphene nanosheet of platinum on palladium bimetallic nanodendrite hybrids. (b) Bimetallic nanocomposite of graphene from TEM. Inset of (b) is the 100-nm magnified view of the bimetallic nanodendrite formation process brought about by anchoring the platinum on palladium to form the bimetallic nanodendrite. Adopted from Guo *et al.* [71].

19.2.1.3.3 The Electrochemical Method

The electrochemical method for synthesizing graphene and its nanocomposites is one of the simplest, quickest, and clean technologies applied till date in the manufacture of graphene and its compounds; it includes the synthesis of graphene nanocomposites of gold [72, 73], silver [15], and platinum particles [16, 74]. The shapes of the nanocomposites formed on the graphene oxide can be controlled by simply manipulating the process conditions, and as highlighted earlier, it encompasses a three-stage process (i.e., the assemblage of the graphene sheets on an electrode, immersion of the graphene coated electrode in solution containing the metallic precursor, and the application of potential difference across the electrodes of the resulting electrochemical cell).

19.2.2 *Ex Situ* Method of Producing Graphene Nanocomposites

By this method, nanoparticles are first synthesized prior to their attachment to the surface of graphene sheets via covalent/noncovalent linking agents, which take advantage of the different interactions (van der Waals, hydrogen bonding, π–π, and electrostatic interactions) taking place in the molecules. Although this method takes longer time and steps for the reaction to be complete, it still offers better advantages in terms of narrower particle size distribution, size and shape adjustment, density, and well-decorated graphene sheets as the final product [75–77].

19.3 Physics and Chemistry of Graphene

19.3.1 Graphene Physics

Previously, the theory of relativistic quantum mechanics, as it relates to nanoparticles bound in semiconductors, also known as the Zitterbewegung behavior [78] and the paradox phenomenon [79], was only known to be associated with vibrations at lower frequencies and high amplitudes than those of free electrons. However, with the recent advancements made in relation to graphene, the atomic layers are indicative of charge carriers moving at Fermi (10^6 m/s) [80] and Femto (10^6 m/s) [81] speeds. In harnessing the novel/potential usefulness of graphene features in device applications, two basic schools of thought are of paramount interest, namely the topological zero mode (TZM) and pseudospin orbital coupling (POC).

19.3.1.1 Topological Zero Modes (TZMs)

Due to the possible interactions of particles and antiparticles in the low energy level of bilayer graphene atoms, it is possible to experience a situation of quantum mechanical tunneling, where electron mobility (transmission/reflection) depends largely on their average kinetic energies, i.e., whether greater or less than the potential step barrier the electrons are incident upon [82]. Furthermore, a third scenario has been theoretically envisaged where particles are held in space by a magnetic force, which is due to the presence of antiparticles; however, the reference to particles by this statement considers only electrons as particles, although it is believed that, in the atomic spectrum, particles include protons, neutrons, and electrons, which then implies that they did not consider particles such as protons and neutrons; hence, the argument here is that, the major possibility of an imposed barrier to the mobility of spatially bound electrons could be as a result of the attractive force exhibited by protons (positively charged particles) in the graphene matrix.

19.3.1.2 Pseudospin Orbital Coupling (POC)

The theories of classical electrodynamics and relativistic quantum mechanics clearly explain the mechanism of the dynamics of the orbital spin of electrons. Based on Dirac's expression, a single spinning particle experiences the Zeeman and coupling effect along its orbit. During orbital coupling, it is thinkable that an electron travelling with a finite speed in an electric field rests/remains in its bound state simply because of the imposition of an effective magnetic field whose strength depends on the angular difference between the electron momentum and the perpendicular magnetic field strength in the same plane. This further implies that a change in electron mobility/procession about the magnetic field may occur along the rotational axis of the orbit, thus defining a change in the first Euler angle, whereas the rotational motion itself is influenced or defined by the third Euler angle; these motions or scattering of electrons resulting from the influence of the field strength on the orbiting electrons may be characteristics of the influence of the attractive positive force of the protons residing in the nucleus of the graphene atom. In addition, the highlighted properties are eye openers or direct implications of their applications in the design of electronic devices/electrical devices such as transistors (i.e., the Rashba and Dresselhaus spin coupling) and semiconductors [83]. Since bilayer graphene has a Hamiltonian feature, which is representative of the Dirac system, the implication therefore, is, in the design of systems; the emphasis here is the pseudospin dynamics of graphene and by the Dirac equation, orbital spin coupling and pseudospin of graphene electrons are very likely in a vacuum characterized by the existence of an electric field. Pseudospin coupling comes to bear when a system is clearly defined by the Hamiltonian behavior [84, 85].

19.3.2 Mobility of Its Electrons, Its Spinning Characteristics and Application

Graphene has high prospects for spintronic applications due to its long spin relaxation times and lengths [86]. This is partly because its carbon configurations are stacks of weakly coupled two-dimensional layers in the third dimension. This quality makes graphene to be potentially adopted for the transport of spin information over relatively long distances in the 100-μm range with limited spin losses. These spin losses brought about the understanding of spin relaxation. Spin relaxation is the process by which a spinning electron loses its initial polarization and behaves random, leading to the disappearance of spin information signal. Spin relaxation mechanism states that, for each carrier momentum scattering event, there is a small probability of spin flip that results in spin information loss. Preliminary studies have shown that spin relaxation increases with the number of layers that are attributed to the improved screening of the external scattering potentials for suspended graphene [87]. Experimentally, there are conflicting views on the spin relaxation mechanism of graphene; however, there have been tremendous insights from theoretical calculations. For example, Emetere and Nikouravan [88] applied the concept of local magnetic moment in spin relaxation to carry out a mathematical experimentation of graphene spins using nuclear magnetic resonance (NMR) to observe the transitions in its different spin states. They solved the spin relaxation problem in graphene by estimating differently, the reduction of the eigen basis blocks of the spin environment-Hamiltonian using the Bloch NMR equation to solve the Schrödinger equations. Three notable observations were made, i.e., identification of the blocking state where the electrons are in a non-equilibrium state; spin flip increase due to the number of electrons with opposite signs at equilibrium; and the bleaching effects on the magnetic moment at quasi-equilibrium state. The validity of the theoretical assumption was strengthened by the experimental verification of electron mobility in mono-, bi-, and trilayer graphenes [89]. It was observed that as the carrier density increases, the mobility decreases for monolayer graphene, while it increases for bilayer and trilayer graphenes. The concept of electron mobility in graphene is believed to have extended effects on its application because of its excellent electrical and thermal properties that hold great promise for integrated-circuit technology [90] and nanoelectronics [91, 92]. Spintronic is particularly interesting for quantum computing and new kinds of fast and efficient memory storage [93]. Dorgan *et al.* [94] showed that the mobility and saturation velocity of electrons in graphene decrease with rising temperature (above 300 K) and rising carrier density. However, it was observed that the SiO_2 substrate had limiting effects on the graphene transport. It is believed that a good substrate should have primarily low spin lifetime steaming from high spin-orbit coupling. This research opened a major research question on the "compatible" substrates for graphene that will enable it to function as a prototype of a transistor-like device for future computer applications due to its excellent tunable electronic properties, as well as its rare behavior known as spintronics. Zhang *et al.* [56] was the first to discover the possibility of spintronics in graphene. In spintronics, the currents are formed from the transport of the electron spins. In graphene devices, an electron's spin can be easily injected either parallel or perpendicular to the planes of the graphene layer. This idea gave credence to interplanar studies of graphene. However, it has been observed that interlayer coupling in graphene alters the behavior of its parallel and perpendicular spin orientations dramatically. Cummings *et al.* [95] showed that the alteration of the interplanar spin orientation leads to the concept of anisotropic spin relaxation from one to several orders of magnitude. Looking at each graphene plane, i.e., for super-thin graphene, it is capable of conveying electrons with coordinated spins over longer distances and preserves the spin for a longer time than any other known material at room temperature. The nature of the spin is believed to have spin-lifetime anisotropy regime [93, 95, 96]. The concept of spin-lifetime anisotropy regime has enormous physics and physical interpretations because of the resulting spin–orbit interaction in the material. For example, Marchenko *et al.* [97] observed that graphene exhibits rotational spins of 180° after traveling about 40 nm when brought in close contact with gold atoms; hence, there are more information on graphene due to the strength of the bonds that keep the carbon atoms together and the peculiar electronic structure of the honeycomb lattice. The application of graphene is gradually expanding due to advancements in research. Graphene is used as atomic scaffold from which other materials are engineered. This is possible because graphene is only 1 atom thick; thus, it is possible to create other materials by interjecting the graphene layers with other compounds. A typical example of its atomic scaffolding properties is the interspersing of alternating boron and magnesium atomic layers of magnesium diboride (MgB_2) superconductors with individual layers of graphene. It has been shown that graphene atomic scaffolding improves the efficiency of the MgB_2 superconductor. Graphenes are good candidates of optoelectronics that can help improve the technology of touchscreens, liquid crystal displays (LCD), and organic light-emitting diodes (OLEDs). Graphenes are used as an ultrafiltration medium to act as a barrier between two substances. This technique is used in water purification among others. The wide application of graphene extends to medicine [98], light processing [99], solar cells, etc.

19.3.3 Chemistry of Graphene and Its Compounds

a. As a reducing agent: Natural graphite acts as a catalyst in the reduction reactions of nitrocompounds, e.g., the reduction of nitrobenzene to aniline using hydrazine as the terminal reducing agent with nitrogen as protective agent [100]. In an extensive study carried out by Larsen *et al.* [101], they unraveled the mechanism of the reaction, where they found that the graphene allotrope acted as an electrical conductor as well as an adsorbent in the course of the reaction.

b. Acidified solid graphene oxide (AS-GO) can act as a catalyst for the condensation reactions of pyrroles with dialkylketones. Its performance was found to compare favorably with zeolites HY, HZSM-5(30), and AL-MCM-41 [102]. Also, AS-GO acts as a catalyst for the reopening of epoxides using methanol. Its catalytic activity was proven to be comparable to H_2SO_4 and gave better conversion as compared to acetic acid when used to carry out the same reaction.

c. Sulfonated graphene oxide (GO) helps to dehydrate furfural to xylose by binding to aryl sulfonic active groups/sites. An appreciable quantity of the product can be obtained within a short reaction time with small catalyst loading as compared to conventional liquid–solid catalysts [103]. Hydrothermal treatment of graphene gives rise to a catalyst, which helps in the reduction of 4-nitrophenol to 2-nitrophenol. The reaction takes advantage of the alkoxy and hydroxyl sites as well as defects at the active sites, while the carboxyl ends impede the strength of the catalyst; also, the rate of reaction is somewhat steady as compared to those of nickel- and cobalt-based catalysts [104].

d. The oxidative coupling of amines to imines can be achieved using chemically treated GO as a catalyst, which takes advantage of the combined action of carboxylic acid groups and unpaired electrons at sites where there is defect in adsorption of molecular oxygen and the amines [105].

e. Graphene oxide promotes oxidative propane dehydrogenation by taking advantage of the hydroxyl groups that are inherent in the active epoxides, thus enhancing the carbon–hydrogen bond activation of C_3H_6 [106].

f. The pi-system in r-GO when used as a catalyst, gives room for proper hydrogen peroxide activation rate and improved adsorption ability of benzene, thus providing the r-GO catalytic control during the catalytic oxidation kinetics of benzene to phenol. High benzene conversion is recorded over the use of titanium silicate [107].

g. Hydrogenated graphene assists in the fenton-degradation of organic dyes. The sp^3 subatomic configuration of the carbon atoms acts as active sites for the reaction [108].

h. Functionalized graphene acts as a preferable catalyst relative to aluminum monohydroxide and silica nanoparticles in the thermal degradation of nitromethane [109, 110].

i. Sulfated and sulfonated graphene are used as catalysts for esterification and hydration reactions, and in the esterification of alkanols and methyl esters, respectively; their catalytic activities were found to be more superior to conventional catalysts such as Amberlyst 15 [110] and Dowex 50W × 2 [111] when used for such reactions.

19.4 Concluding Remarks

The structure of graphene, its nanocomposites, their method of production, as well as their applications have been mentioned. The importance of graphene in various fields of discipline can never be overemphasized as this material and its nanocomposites have peculiar useful properties that comingle to provide additionally improved properties than those of the individual components. Graphene being a unique material helps to destroy cancerous cells as already mentioned; hence, it would not be out of place to term it the "substance of life for cancer victims." Current researches have shown that it is the carbon allotrope of the future with high prospects in biosensing, LED technology, corrosion control, etc. The methods of graphene synthesis as well as the possibilities of producing green and toxic graphene have also been highlighted, which helps to create some form of awareness on the need to apply caution during the production process. It is therefore recommended that future works consider the kinetics of graphene as it undergoes various chemical combinations/reactions as this will help to broaden the current state of knowledge about the substance.

References

1. Thielemans, W., Warbey, C.R., Walsh, D.A., Permselective nanostructured membranes based on cellulose nanowhiskers. *Green Chem.*, 11, 531, 2009.
2. Wei, W., Zhang, X., Huang, Y., Ren, X., Guiding properties of asymmetric hybrid plasmonic waveguides on dielectric substrates. *Nano Res. Lett.*, 9, 1, 2014.
3. Lotya, M., King, P.J., Khan, U., De, S., Coleman, J.N., High-concentration, surfactant-stabilized graphene dispersions. *ACS Nano*, 4, 3155, 2010.
4. Martin, C.A., Sandler, J.K.W., Shaffer, M.S.P., Schwarz, M.K., Bauhofer, W., Schulite, K., Windle, A.H., Formation of percolating networksin multi-wall carbon-nanotube-epoxy composites. *Compos. Sci. Technol.*, 64, 2309, 2004.
5. Carrol, D.L., Czerw, R., Webster, S., Polymer-nanotube composites for transparent, conducting thin films. *Synth. Met.*, 155, 694, 2005.
6. Nosolev, K.S., Geim, A.K., Morozov, S.V., Jiang, D., Zhang, Y., Dubonos, S.V., Grigorieva, I.V., Firsov, A.A., Electric field effect in atomically thin carbon films. *Science*, 306, 666, 2004.
7. Geim, A.K. and Novoselov, K.S., The Rise of Graphene. *Nat. Mater.*, 6, 183, 2007.
8. Stephenson, C.A., Gillett-Kunnath, M., O'Brien, W.A., Kudrawiec, R., Wistey, M.A., Gas source techniques for molecular beam epitaxy of highly mismatched Ge alloys. *Crystals*, 6, 159, 2016.
9. Gao, H., Zhu, K., Hu, G., Xue, C., Large-scale graphene production by ultrasound-assisted exfoliation of natural graphite in supercritical CO_2/H_2O system. *Chem. Eng. J.*, 308, 872, 2017.
10. Microfluidics, Use of microfluidizer technology for graphene exfoliation, in: *Application Note, Idex Material Processing*, pp. 1–2, 2017, www.microfluidicscorp.com.
11. Davies, P., Tzalenchuk, A., Wiper, P., Walton, S., Summary of graphene (and related compounds) chemical and physical properties, nuclear decommissioning authority (nda), Herdus House, Westlakes Science and Technology Park, Moor Row, Cumbria, 2016.
12. Emiru, F.T. and Ayele, D.W., Controlled synthesis, characterization and reduction of graphene oxide: A convenient method for large scale production. *Egypt. J. Basic Appl. Sci.*, 4, 74, 2016.
13. Gomez-Navarro, C., Weitz, R.T., Bittner, A.M., Scolari, M., Mews, A., Burghard, M., Kem, K., Electronic transport properties of individual chemically reduced graphene oxide sheets. *Nano Lett.*, 7, 11, 3449, 2007.

14. Compton, O.C. and Nguyen, S.T., Graphene oxide, highly reduced graphene oxide, and graphene: Versatile building blocks for carbon-based materials. *Small*, 6, 711, 2010.
15. Golsheikh, A.M., Huang, N.M., Lim, H.N., Zakaria, R., Yin, C.R., One-step electrodeposition of silver nano-particle decorated graphene on indium–tin-oxide for enzymeless hydrogen peroxide detection. *Carbon*, 62, 405, 2013.
16. Claussen, J.C., Kumar, A., Jaroch, D.B., Khawaja, M.H., Hibbard, A.B., Porterfield, D.M., Fisher, T.S., Nanostructuring platinum nanoparticles on multilayered graphene petal nanosheets for electrochemical biosensing. *Adv. Funct. Mater.*, 22, 3399, 2012.
17. Wu, Z.S., Wang, D.W., Ren, W., Zhao, J., Zhou, G., Li, F., Cheng, H.M., Anchoring hydrous RuO_2 on graphene sheets for high-performance electrochemical capacitors. *Adv. Funct. Mater.*, 20, 3595, 2010.
18. Wu, Z.S., Zhou, G.M., Yin, L.C., Ren, W., Li, F., Cheng, H.M., Graphene metal oxide composite electrode materials for energy storage. *Nano Energy*, 1, 107, 2011.
19. Singh, Z., Toxicity of graphene and its nanocomposites to human cell lines: The present scenario. *Int. J. Biomed. Clinical Sci.*, 1, 24, 2016.
20. Yuan, J., Gao, H., Ching, C.B., Comparative protein profile of human hepatoma hepg2 cells treated with graphene and single-walled carbon nanotubes: An itraq-coupled 2d lcms/ms proteome analysis. *Toxicol. Lett.*, 207, 213, 2011.
21. Farokhzad, O.C. and Langer, R., Impact of nanotechnology on drug delivery. *ACS Nano*, 3, 16, 2009.
22. Goenka, S., Sant, V., Sant, S.J., Graphene-based nanomaterials for drug delivery and tissue engineering. *J. Controlled Release*, 173, 75, 2014.
23. Yin, P.T., Shah, S., Chhowalla, M., Lee, K., Design, synthesis and characterization of graphene nano-particle hybrid materials for bioapplications. *Chem. Rev.*, 1, 2014.
24. Grieshaber, D., MacKenzie, R., Voros, J., Reimhult, E., Electrochemical biosensors—Sensor principles and architectures. *Sensors*, 8, 1400, 2008.
25. Zhao, X., Hayner, C.M., Kung, M.C., Kung, H.H., In-plane vacancy-enabled high power si-graphene composite electrode for lithium-ion batteries. *Adv. Energy Mater.*, 1, 2011, 1079.
26. Prasai, D., Tuberquia, J.C., Harl, R.R., Jennings, G.K., Bolotin, K.I., Graphene: Corrosion-inhibiting coating. *ACS Nano*, 6, 1102, 2012.
27. Bonanni, A. and Pumera, M., Graphene platform for hairpin-DNA-based impedimetric genosensing. *ACS Nano*, 5, 2356, 2011.
28. Wang, H., Sun, K., Tao, F., Stacchiola, D.J., Hu, Y.H., 3D Honeycomb-like structured graphene and its high efficiency as a counter-electrode catalyst for dye-sensitized solar cells. *Angew. Chem. Int.*, 52, 9210, 2013.
29. Hyun, W.J., Park, O.O., Chin, B.D., Foldable graphene electronic circuits based on paper substrates. *Adv. Mater.*, 25, 4729, 2013.
30. Radivojevic, Z., Beecher, P., Bower, C., Haque, S., Andrew, P., Hasan, T., Bonaccorso, F., Ferrari, A.C., Henson, B., Embodied interaction with complex neuronal data in mixed-reality, in: *Proceedings of the Virtual Reality International Conference*, ACM, Laval, France, vol. 1, ACM New York, New York, USA, 2012.
31. Randvirr, E.P., Brownson, D.A.P., Banks, C.E., A decade of graphene research: Production, applications and outlook. *Mater. Today*, 17, 426, 2014.
32. Wu, J., Agrawal, M., Becerril, H.A., Bao, Z., Liu, Z., Chen, Y., Peumans, P., Organic light-emitting diodes on solution-processed graphene transparent electrodes. *ACS Nano*, 4, 43, 2010.
33. Samsung Press Release (accessed October, 2013).
34. Brownson, D.A.C., Kampouris, D.K., Banks, C.E., Graphene electrochemistry: Fundamental concepts through to prominent applications. *Chem. Soc. Rev.*, 41, 6944, 2012.
35. Brownson, D.A.C. and Banks, C.E., Fabricating graphene supercapacitors: Highlighting the impact of surfactants and moieties. *Chem. Commun.*, 48, 1425, 2012.

36. Lin, W.J., Liao, C.S., Jhang, J.H., Tsai, Y.C., Graphene modified basal and edge plane pyrolytic graphite electrodes for electrocatalytic oxidation of hydrogen peroxide and beta-nicotinamide adenine dinucleotide. *Electrochem. Commun.*, 11, 2153, 2009.
37. Ambrosi, A., Bonanni, A., Sofer, Z., Cross, J.S., Pumera, M., Electrochemistry at chemically modified graphenes. *Chem. Eur. J.*, 17, 38, 10763–10770, 2011.
38. Brownson, D.A.C., Munroe, L.J., Kampouris, D.K., Banks, C.E., Electrochemistry of graphene: Not such a beneficial electrode material. *RSC Adv.*, 1, 6, 978–988, 2011.
39. Brownson, D.A.C., Metters, P.M., Kampouris, D.K., Banks, C.E., Graphene electrochemistry: Surfactants inherent to graphene can dramatically effect electrochemical processes. *Electroanalysis*, 23, 894, 2011.
40. Liu, C., Yu, Z., Neff, D., Zhamu, A., Jang, B.Z., Graphene-based supercapacitor with an ultrahigh energy density. *Nano Lett.*, 10, 4863, 2010.
41. Chen, Y., Li, Y., Sun, D., Tian, D.B., Zhang, J.R., Zhu, J.J., Fabrication of gold nanoparticles on bilayer graphene for glucose electrochemical biosensing. *J. Mater. Chem.*, 21, 7604, 2011.
42. Zhang, F., Tang, J., Shinya, N., Qin, L., Hybrid graphene electrodes for supercapacitors of high energy density. *Chem. Phys. Lett.*, 584, 124, 2013.
43. Cao, J., Wang, Y., Zhou, Y., Ouyang, J., Jia, D., Guo, L., High voltage asymmetric supercapacitor based On MnO_2 and graphene electrodes. *J. Electroanal. Chem.*, 689, 201, 2013.
44. Brownson, D.A.C., Kampouris, D.K., Banks, C.E., An overview of graphene in energy production and storage applications. *J. Power Sources*, 196, 4873, 2011.
45. Song, W., Ji, X., Deng, W., Chen, O., Shen, S., Banks, C.E., Graphene ultracapacitors: Structural impacts. *Phys. Chem. Chem. Phys.*, 15, 4799, 2013.
46. Yuan, W., Zhou, Y., Li, Y., Li., C., Peng, H., Zhang, J., Liu, Z., Dai, L., Shi, G., The edge- and basal-plane-specific electrochemistry of a single-layer graphene sheet. *Sci. Rep.*, 3, 1, 2013.
47. Huang, H., Bai, W., Dong, C., Guo, R., Liu, Z., An ultrasensitive electrochemical DNA biosensor based on graphene/au nanorod/polythionine for human papillomavirus DNA detection. *Biosens. Bioelectron.*, 68, 442, 2015.
48. Huang, K.J., Jing, Q.S., Wei, C.Y., Wu, Y.Y., Spectrofluorimetric determination of glutathione in human plasma by solid-phase extraction using graphene as adsorbent. *Spectrochim. Acta A Mol. Biomol. Spectrosc.*, 79, 1860, 2011.
49. Hernandez, Y., Lotya, M., Rickard, D., Bergin, S.D., Coleman, J.N., Measurement of multicomponent solubility parameters for graphene facilitates solvent discovery. *Langmuir*, 26, 3208, 2010.
50. Fiorillio, M., Verre, A.F., Iliut, M., Peiris-Pages, M., Ozsvari, B., Gandara, R., Cappello, A.R., Sotgia, F., Vijayaraghavan, A., Lisanti, M.P., Graphene oxide selectively targets cancer stem cells across multiple tumour types: Implications for non-toxic cancer treatment via differentiation-based nano-therapy. *Oncotarget*, 1, 2015.
51. Lee, W.C., Lim, C.H.Y.X., Shi, H., Tang, L.A.L., Wang, Y., Lim, C.T., Loh, K.P., Origin of enhanced cell growth and differentiation on graphene and graphene oxide. *ACS Nano*, 5, 7334, 2011.
52. Sun, X.M., Liu, Z., Welsher, K., Robinson, J.T., Goodwin, A., Zaric, S., Dai, H.J., Nano-graphene oxide for cellular imaging and drug delivery. *Nano Res.*, 1, 203, 2008.
53. Chen, W., Rakhi, R.B., Alshareef, H.N., Capacitance enhancement of polyaniline coated curved-graphene supercapacitors in a redox-active electrolyte. *Nanoscale*, 5, 4134, 2013.
54. Tien, H.W., Huang, Y.L., Yang, S.Y., Wang, J.Y., Ma, C.C.M., The production of graphene nanosheets decorated with silver nanoparticles for use in transparent, conductive films. *Carbon*, 49, 1550, 2011.
55. Xu, C., Wang, X., Zhu, J.W., Graphene-metal particle nanocomposites. *J. Phys. Chem. C*, 112, 19841, 2008.

56. Zhang, Q., Chan, K.S., Lin, Z., Spin current generation by adiabatic pumping in monolayer graphene. *Appl. Phys. Lett.*, 98, 032106, 2011.
57. Zhang, X.Y., Yin, J.L., Peng, C., Hu, W.Q., Zhu, Z.Y., Li, W.X., Fan, C.H., Huang, Q., Distribution and biocompatibility studies of graphene oxide in mice after intravenous administration. *Carbon*, 49, 986, 2011.
58. Zhuo, Q.Q., Ma, Y.Y., Gao, J., Zhang, P.P., Xia, Y.J., Tian, Y.M., Sun, X.X., Zhong, J., Sun, X.H., Facile synthesis of graphene/metal nanoparticle composites via self catalysis reduction at room temperature. *Inorg. Chem.*, 52, 3141, 2013.
59. Gao, W., Alemany, L.B., Ci, L.J., Ajayan, P.M., New insights into the structure and reduction of graphite oxide. *Nat. Chem.*, 1, 403, 2009.
60. Bai, S. and Shen, X., Graphene–inorganic nanocomposites. *RSC Adv.*, 2, 64, 2012.
61. Zhou, Y., Bao, Q.L., Tang, L.A.L., Zhong, Y.L., Loh, K.P., Hydrothermal dehydration for the "green" reduction of exfoliated graphene oxide to graphene and demonstration of tunable optical limiting properties. *Chem. Mater.*, 21, 2950, 2009.
62. Park, S.K., Yu, S.H., Pinna, N., Woo, S., Jang, B., Chung, Y.H., Cho, Y.H., Sung, Y.E., Piao, Y., A facile hydrazine-assisted hydrothermal method for the deposition of monodisperse SnO_2 nanoparticles onto graphene for lithium ion batteries. *J. Mater. Chem.*, 22, 2520, 2012.
63. Ren, L.L., Huang, S., Fan, W., Liu, T.X., One-step preparation of hierarchical superparamagnetic iron oxide/graphene composites. *Appl. Surf. Sci.*, 258, 1132, 2011.
64. Yan, S.C., Shi, Y., Zhao, B., Lu, T., Hu, D., Xu, X., Wu, J.S., Chen, J.S., Hydrothermal synthesis of CdS/functionalized graphene sheets nanocomposites. *J. Alloys Compd.*, 570, 65, 2013.
65. Xue, L.P., Shen, C.F., Zheng, M.B., Lu, H.L., Li, N.W., Ji, G.B., Pan, L.J., Cao, J.M., Hydrothermal synthesis of graphene ZnS quantum dot nanocomposites. *Mater. Lett.*, 65, 198, 2011.
66. Su, Y.J., Lu, X.N., Xie, M.M., Geng, H.J., Wei, H., Yang, Z., Zhang, Y.F., A one-pot synthesis of reduced graphene oxide–Cu_2S quantum dot hybrids for optoelectronic devices. *Nanoscale*, 5, 8889, 2013.
67. Chang, K. and Chen, W.X., Single-layer MoS_2/graphene dispersed in amorphous carbon: Towards high electrochemical performances in rechargeable lithium ion batteries. *J. Mater. Chem.*, 21, 17175, 2011.
68. Huang, G.C., Chen, T., Chen, W.X., Wang, Z., Chang, K., Ma, L., Huang, F.H., Chen, D.Y., Lee, J.Y., Graphene-like MoS_2/graphene composites: Cationic surfactant-assisted hydrothermal synthesis and electrochemical reversible storage of lithium. *Small*, 9, 3693, 2013, https://doi.org/10.1002/smll.201300415.
69. Lu, Z.S., Guo, C.X., Yang, H.B., Qiao, Y., Guo, J., Li, C.M., One-step aqueous synthesis of graphene–CdTe quantum dot-composed nanosheet and its enhanced photoresponses. *J. Colloid Interface Sci.*, 353, 588, 2011.
70. Medintz, I.L., Uyeda, H.T., Goldman, E.R., Mattoussi, H., Quantum dot bioconjugates for imaging, labelling and sensing. *Nat. Mater.*, 4, 435, 2005.
71. Guo, S.J., Dong, S.J., Wang, E.K., Three dimensional Pt-on-Pd bimetallic nanodendrites supported on graphene nanosheets: Facile synthesis and used as an advanced nanoelectrocatalyst for methanol oxidation. *ACS Nano*, 4, 547, 2010.
72. Ding, L., Liu, Y.P., Zhai, J.P., Bond, A.M., Zhang, J., Direct electrodeposition of graphene-gold nanocomposite films of ultrasensitive voltammetric determination of mercury (ii). *Electroanalysis*, 26, 121, 2014.
73. Hu, Y.J., Jin, J.A., Wu, P., Zhang, H., Cai, C.X., Graphene-gold toward the oxygen reduction and glucose oxidation. *Electrochim. Acta*, 56, 491, 2010.
74. Zhou, Y.G., Chen, J.J., Wang, F.B., Sheng, Z.H., Xia, X.H., A facile approach to the synthesis of highly electroactive Pt nanoparticles on graphene as an anode catalyst for direct methanol fuel cells. *Chem. Commun.*, 46, 5951, 2010.

75. He, F.A., Fan, J.T., Song, F., Zhang, L.M., Chan, H.L.W., Fabrication of hybrids based on graphene and metal nanoparticles by *in-situ* and self-assembled methods. *Nanoscale*, 3, 1182, 2011.
76. Yang, X., Xu, M.S., Qiu, W.M., Chen, X.Q., Deng, M., Zhang, J.L., Iwai, H., Watanabe, E., Chen, H.Z., Graphene uniformly decorated with gold nanodots: *In-situ* synthesis, enhanced dispersibility and applications. *J. Mater. Chem.*, 21, 8096, 2011.
78. Schliemann, J., Loss, D., Westervelt, R.M., Zitterbewegung of electrons and holes in iii–v semiconductors quantum wells. *Phys. Rev. (for Manes) B*, 73, 085323, 2006.
79. Katsnelson, M.I., Novoselov, K.S., Geim, A.K., Chiral tunneling and the Klein paradox in graphene. *Nat. Phys.*, 2, 620, 2006.
80. Charlier, J.C., Eklund, P.C., Zhu, J., Ferarri, J.C., Electron and phonon properties of graphene: Their relationship with carbon nanotubes, in: *Carbon Nanotubes, Topics: Appl. Physics*, vol. 111, A. Jorio, G. Dresselhaus, M.S. Dresselhaus (Eds.), pp. 673–709, Springer-Verlag Berlin Heidelberg, 2008.
81. Moses, E.E., Sanni, S.E., Agarana, C.M., Virtual observation of the femtosecond spin dynamics mechanism in graphene, in: *Proceedings of the World Congress on Engineering, Vol II WCE*, London, UK, June 29–July 1, 2016.
82. Landau, L.D. and Lifshitz, E.M., *Quantum Mechanics*, 2nd Edition, Pergamon Press Plc, Elsevier-Science, London, 1968.
83. Das, S. and Gupta, N., Effect of ageing on space charge distribution in homogeneous and composite dielectrics. *IEEE Trans. Dielectr. Electr. Insul.*, 22, 541, 2014.
84. Tan, S.G., Jalil, M.B.A., Fujita, T., Pseudospin-orbital coupling for pseudospintronic device in graphene. *J. Magn. Magn. Mater.*, 322, 2390, 2010.
85. Tan., S.G., Jalil, M.B.A., Fujita, T., Monopole and topological electron dynamics in adiabatic spintronic and graphene systems. *Ann. Phys. (N.Y.)*, 325, 8, 1537–1549, 2010b. ISSN 0003-4916.
86. Dugaev, V.K., Sherman, E.Y., Barna´s, J., Spin dephasing and pumping in graphene due to random spin-orbit interaction. *Phys. Rev. B*, 83, 085306, 2011.
87. Bolotin, K.I., Sikes, K.J., Jiang, Z., Klima, M., Fudenberg, G., Hone, J., Kim, P., Stormer, H.L., Ultrahigh electron mobility in suspended graphene. *Solid State Commun.*, 146, 351, 2008.
88. Emetere, M.E. and Nikouravan, B., Femtosecond spin dynamics mechanism in graphenes: The Bloch NMR-Schrodinger probe. *Int. J. Fund. Phys. Sci.*, 4, 105, 2014.
89. Zhu, W., Perebeinos, V., Freitag, M., Avouris, P., Carrier scattering, mobilities and electrostatic potential in mono-, bi- and tri-layer graphenes. *Phys. Rev. B*, 80, 235402, 2009.
90. Ozyilmaz, B., Jarillo-Herrero, P., Efetov, D., Kim, P., Electronic transport in locally gated graphene nanoconstrictions. *Appl. Phys. Lett.*, 91, 192107, 2007.
91. De-Heer, W.A., Berger, C., Wu, X., Sprinkle, M., Hu, Y., Ruan, M., Stroscio, J.A., First, P.N., Haddon, R., Piot, B., Faugeras, C., Potemski, M., Moon, J.S., Epitaxial graphene electronic structure and transport. *J. Phys. D: Appl. Phys.*, 43, 374007, 2010.
92. Jozsa, C. and Van-Wees, B.J., *Handbook of Spin Transport and Magnetism*, E.Y. Tsymbal and Z.I. Zutic (Eds.), pp. 579–598, CRC Press, Boca Raton, FL, 2011.
93. Benítez, A.L., Sierra, J.F., Torres, W.S., Arrighi, A., Bonell, F., Costache, M.V., Valenzuela, S.O., Strongly anisotropic spin relaxation in graphene–transition metal dichalcogenide heterostructures at room temperature. *Nat. Phys.*, 14, 303, 2018.
94. Dorgan, V.E., Bae, M., Pop, E., Mobility and saturation velocity in graphene on SiO_2. *Appl. Phys. Lett.*, 2010, 082112, 2016.
95. Cummings, A.W., Garcia, J.H., Fabian, J., Roche, S., Giant spin lifetime anisotropy in graphene induced by proximity effects. *Phys. Rev. Lett.*, 119, 206601, 2017.

96. Ghiasi, T.S., Ingla-Aynés, J., Kaverzin, A.A., Van-Wees, B.J., Large proximity-induced spin lifetime anisotropy in transition-metal dichalcogenide/graphene heterostructures. *Nano Lett.*, 17, 7528, 2017.
97. Marchenko, D., Varykhalov, A., Sánchez-Barriga, J., Seyller, Th., Rader, O., Rashba splitting of 100 mev in au-intercalated graphene on SiC. *Appl. Phys. Lett.*, 108, 172405, 2016, http://Dx.Doi.Org/10.1063/1.4947286.
98. Li, J., Yoong, S.L., Goh, W.J., Czarny, B., Yang, Z., Poddar, K., Dykas, M.M., Patra, A., Venkatesan, T., Panczyk, T., Lee, C., Pastrorin, G., *In-vitro* controlled release of cisplatin from gold-carbon nanobottles via cleavage linkages. *Int. J. Nanomed.*, 10, 7425, 2015.
99. Li, X., Zhu, M., Du, M., Lv, Z., Zhang, L., Li, Y., Yang, Y., Yang, T., Li, X., Wang, K., Zhu, H., Fang, Y., High detectivity graphene-silicon heterojunction photodetector. *Small*, 12, 595, 2016.
100. Byung, H.H., Dae, H.S., Sung, Y.C., Graphite catalyzed reduction of aromatic and aliphatic nitro compounds with hydrazine hydrate. *Tetrahedron Lett.*, 26, 6233, 1985.
101. Larsen, J.W., Freund, M., Kim, K.Y., Sidovar, M., Stuart, J.L., Mechanism of the carbon catalyzed reduction of nitrobenzene by hydrazine. *Carbon*, 38, 655, 2000.
102. Chauhan, S.M.S. and Mishra, S., Use of Graphite oxide and graphene oxide as catalysts in the synthesis of dipyrromethane and calix [4] pyrrole. *Molecules*, 16, 7256, 2011.
103. Dhakshinamoorthy, A., Alvaro, M., Concepcion, P., Fornes, V., Garcia, H., Graphene oxide as an acid catalyst for the room temperature ring opening of epoxides. *Chem. Commun.*, 48, 5443, 2012.
104. Tang, S.B. and Cao, Z.X., Site-dependent catalytic activity of graphene oxides towards oxidative dehydrogenation of propane. *Phys. Chem. Chem. Phys.*, 14, 16558, 2012.
105. Lam, E., Chong, J.H., Majid, E., Liu, Y., Hrapovic, S., Leung, A.C.W., Luong, J.H.T., Carbocatalytic dehydration of xylose to furfural in water. *Carbon*, 50, 1033, 2012.
106. Kong, X.-K., Chen, Q.W., Lun, Z.-Y., Probing the influence of different oxygenated groups on graphene oxide's catalytic performance. *J. Mater. Chem. A*, 2, 610, 2014.
107. Yang, J.H., Sun, G., Gao, Y.J., Zhao, H.B., Tang, P., Tan, J., Lu, A.H., Ma, D., Direct catalytic oxidation of benzene to phenol over metal-free graphene-based catalyst. *Energy Environ. Sci.*, 6, 793, 2013.
108. Zhao, Y., Chen, W.F., Yuan, C.F., Zhu, Z.Y., Yan, L.F., Hydrogenated graphene as metal-free catalyst for Fenton-like Reaction. *Chin. J. Chem. Phys.*, 25, 335, 2012.
109. Sabourin, J.L., Dabbs, D.M., Yetter, R.A., Dryer, F.L., Aksay, I.A., Functionalized graphene sheet colloids for enhanced fuel/propellant combustion. *ACS Nano*, 3, 3945, 2012.
110. Liu, F., Sun, J., Zhu, L., Meng, X., Qi, C., Xiao, F.-S., Sulfated graphene as an efficient solid catalyst for acid-catalyzed liquid reactions. *J. Mater. Chem.*, 22, 5495, 2012.
111. Wang, L., Wang, D., Zhang, S., Tian, H., Synthesis and characterization of sulfonated graphene as a highly active solid acid catalyst for the ester-exchange reaction. *Catal. Sci. and Technol.*, 3, 1194, 2013.

20

Graphene-Based Nanomaterials in Tissue Engineering and Regenerative Medicine

Sorour Darvishi[1†], Samad Ahadian[2*,†] and Houman Savoji[2,3]

[1]*Laboratoire d'Electrochimie Physique et Analytique, École Polytechnique Fédérale de Lausanne (EPFL), Valais Wallis, Sion, Switzerland*
[2]*Toronto General Research Institute, University Health Network, Toronto, Ontario, Canada*
[3]*Institute of Biomaterials and Biomedical Engineering (IBBME), University of Toronto, Toronto, Ontario, Canada*

Abstract
Tissue engineering and regenerative medicine have rapidly become a novel medical strategy for repair and regeneration of damaged or diseased tissues and organs in the body. The key to advance in this field is the design and development of functional biomaterials that mimic the cellular microenvironment and provide physicochemical cues to enable cell attachment, proliferation, and remodeling. For stem cell-based tissue regeneration, it is crucial to control the chemical and physical properties of materials to stimulate and guide the growth and differentiation of stem cells. Graphene and graphene-derived materials can offer exciting opportunities to regulate different cell behavior toward a desired biological response due to their unique mechanical properties, tunable surface chemistry, and high electrical conductivity. Keeping these points in mind, this chapter describes the applications of graphene and graphene-derived materials in stem cell engineering and tissue regeneration. Moreover, *in vitro* and *in vivo* toxicity issues of graphene are explained. Finally, challenges and future directions of using graphene in tissue engineering and regenerative medicine are described.

Keywords: Graphene, regenerative medicine, stem cells, tissue engineering, toxicity

20.1 Introduction

Tissue engineering is an interdisciplinary research field that strives to design and develop biological substitutes to maintain, restore, or improve the function of tissues or whole organ. The field of tissue engineering has successfully regenerated multiple tissues and organs *in vitro* for therapeutic and pharmaceutical applications [1, 2]. It is important to combine the principles of cell therapy and transplantation, biology, material science, and engineering technologies to develop effective substitutes that can restore and maintain the normal function of diseased or injured tissues and organs [3]. To achieve its goal,

**Corresponding author*: samad_ahadian@yahoo.com
†*These authors contributed equally to this work*

Tobias Stauber (ed.) Handbook of Graphene: Volume 2, (637–658) © 2019 Scrivener Publishing LLC

engineered tissues require high cellular organization and should mimic morphological and physiological features of native tissues *in vivo* [4–6]. Therefore, the development of novel materials to create tissue constructs *in vitro* or regenerate damaged tissues *in vivo* are of great importance. To engineer functional tissues, the scaffold materials have to hold several key parameters, such as guiding cell alignment, growth, modulation, delivering required biomolecules, stimulating mechanical properties of native tissues, and generating proper physical and chemical signal cues. Also, the proposed materials should provide suitable microenvironment for stem cells to grow and differentiate. These factors can be aided by incorporating specific materials that possess unique and tunable physicochemical properties.

Recently, graphene has gained much attention in nanoscience and nanotechnology due to its properties that could have particularly interesting applications in tissue engineering and regenerative medicine [7–9]. Graphene due to its remarkable characteristics, such as high mechanical properties, unique surface chemistry, high electrical conductivity, ease of self-assembling with biomolecules, capability to load various biomolecules through physical or chemical bindings, and biocompatibility has obtained great attention for biomedical applications [10]. In particular, physicochemical properties of graphene-based materials, including topography and surface functionalization, can be precisely controlled and therefore they can provide a well-designed milieu for different stem cell and tissue engineering applications. With rapid development on the synthesis and functionalization methods of high-quality graphene-based materials, they are considered as novel and functional materials with great potential applications in cell therapy and tissue engineering. Specific functional groups of a scaffold material, such as hydroxyl, carboxyl, and amine groups, even at the molecular level, are important to regulate cell behavior and function [11, 12]. Therefore, graphene-based materials with precisely tunable surface chemistry are highly desirable in various cell and tissue engineering applications to elucidate and therefore control complex signaling pathways of cellular behavior [13–15].

In this chapter, we focus on the strategies and current state of the art of graphene-based nanomaterials as attractive class of biocompatible materials for use in tissue engineering and regenerative medicine applications. *In vitro* and *in vivo* studies of graphene biocompatibility are discussed. Major challenges in this area and future potential routes for the research and development of graphene-based nanomaterials are also discussed.

20.2 Biomedical Applications of Graphene

Biomedical application of graphene is a relatively new area with significant potential. The first report of the usage of graphene in biomedical application reported by Liu *et al.* [16] in drug delivery. Several reports have been carried out to explore the application of graphene in wide range of biomedical applications, such as drug/gene delivery, biological sensing, and bio-imaging [17]. The intensive research on the bio-applications of graphene and its derivatives is due to many fascinating properties of graphene, such as exceptional electronic conductivity (mobility of charge carriers, 200,000 $cm^2\ V^{-1}\ s^{-1}$), high specific surface area (2630 m^2/g), thermal conductivity (~5000 W/m/K), mechanical strength (Young's modulus, ~1100 GPa), low cost and scalable production, and facile biological/chemical functionalization [18, 19].

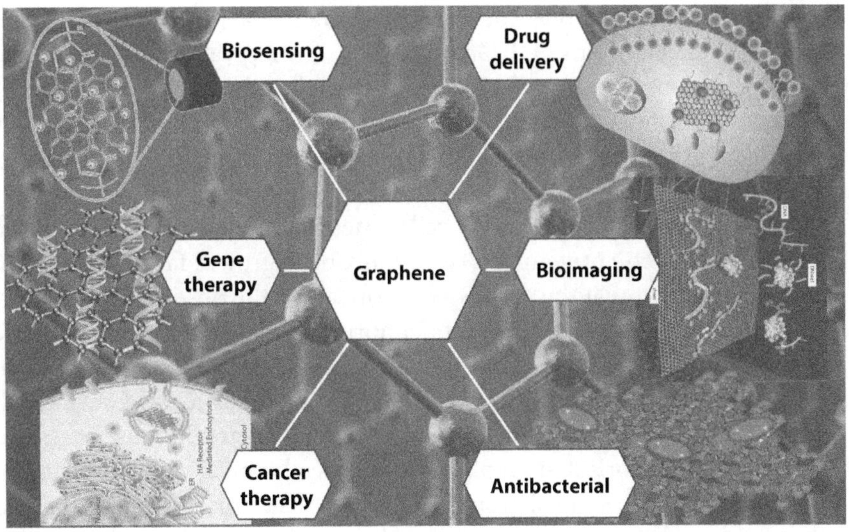

Figure 20.1 Different applications of graphene in the biomedical area. Reprinted with permission from Markovic *et al.*, 2017, *Journal of Colloid and Interface Science* [28], Jung *et al.*, 2014, *RSC Advances* [29], Zhang *et al.*, 2014, *Journal of Materials Chemistry B* [30], Yang *et al.*, 2013, *Materials Today* [31], and Wu *et al.*, 2016, *Nanomaterials* [32].

Various biomedical applications of graphene are schematically shown in Figure 20.1. Graphene-based electrochemical devices have been largely used for detection of different biomolecules. These devices rely on the ballistic electron transport properties of graphene that facilitate the electron transfer between the electrode and underlying sample and thus improve the electrochemical feedback from the sample. In our recent work, reduced GO (rGO) significantly enhanced the glucose biosensing of gelatin methacryloyl (GelMA) hydrogel due to high electrical conductance of rGO [20]. Note that GelMA itself could not be used as an electrochemical glucose biosensor because of low electrical conductivity. The two-dimensional (2D) structure and presence of delocalized π electrons on the graphene surface can be used for effective gene/drug loading via hydrophobic interactions and π–π stacking. Additionally, large surface area of graphene allows for highly dense bio-functionalization via both covalent and noncovalent surface modification. Various studies have proven successful use of graphene and its derivatives as gene/drug carrier *in vivo* [16, 21–26]. Moreover, graphene is able to impede the proliferation of bacteria and can be used in wound dressing materials. The mechanism of antibacterial activity of graphene is not clear yet. However, it is known that when bacteria are exposed to graphene, it is easy for electrons to escape from graphene and move into the cells due to the potential of cell membrane. This extra electrical charge can break the structure of DNA or cell membranes. As a result, the bacteria are not able to grow [27].

20.3 Graphene in Stem Cell Engineering

Stem cells play a pivotal role in the human body because of their ability for continual growth, renewal, and repair. They are critical for numerous groundbreaking therapies in the field of regenerative medicine and tissue engineering. Intensive research has been carried out on stem cells to decipher the myriad of biological and environmental factors that organize their complex molecular and cellular events [33]. Graphene has shown great promise to provide a controlled microenvironment for stem cell culture, proliferation, and differentiation [34]. In this section, we discuss some important applications of graphene in stem cell engineering.

Nayak *et al.* used graphene to control and enhance osteogenic differentiation of human mesenchymal stem cells (hMSCs) [35]. They reported that the graphene could accelerate the stem cell differentiation toward bone cells when cultured even in osteogenic media lacking bone morphogenetic protein-2, a commonly used growth factor for bone stem cell differentiation. To confirm that the graphene is critical for the observed stem cell differentiation, highly oriented pyrolytic graphite and amorphous carbon thin films were set as the control groups. They observed that none of control groups were capable of promoting the cell differentiation. The results pointed out mechanical properties and surface morphology of graphene as the decisive factors for the cell differentiation because micrometer ripples and wrinkles on the fabricated graphene using the chemical vapor deposition (CVD) approach were absent in the control groups. The large-scale surface features caused by the ripples in the CVD graphene were suggested to play a major role in protein adsorption, cell adhesion, and differentiation. Furthermore, the ability of graphene to sustain lateral stress was believed to provide sufficient amount of local cytoskeletal tension for bone stem cell differentiation. In another study, Lee *et al.* [36] investigated the effect of graphene, GO, and polydimethylsiloxane (PDMS) on the adipogenesis of hMSCs. The cells on the graphene films were homogeneously dispersed and showed spindle-like morphology, whereas those cultured on the GO films were more widespread and larger. A direct correlation between the adsorption capacity of substrates for serum proteins and subsequent cell growth was deduced. It was shown that the graphene and GO adsorbed serum proteins up to 8% and 25%, respectively. Serum contains many extracellular matrix (ECM) proteins and glycoproteins, such as albumin and fibronectin (FN) [37]. The more graphene and GO adsorb serum proteins, the higher the density of adhesion molecules is available for cell attachment and growth. The π-electron cloud in graphene was competent to interact with the inner hydrophobic cores of proteins. Due to the presence of oxygenated groups, the hydrophilic GO was able to bind to serum proteins via hydrogen bonding and electrostatic interactions.

Effect of graphene on the growth and differentiation of induced pluripotent stem cells (iPSCs) has been studied [38]. For example, Chen *et al.* [39] showed the differentiation of iPSCs on graphene- and GO-coated glass substrates. Their data demonstrated that the graphene and GO were biocompatible with iPSCs and supported iPSC attachment and proliferation. The immunohistochemical staining results and mRNA expression levels suggested that the graphene appeared to hamper the spontaneous stem cell differentiation especially toward the endodermal lineage while the cells on the glass slide and GO spontaneously lost the pluripotency in a culture environment without leukemia inhibitory factor. The graphene ability to preserve the pluripotency of iPSCs is an asset to design biomimetic materials for stem cell culture and differentiation without using expensive soluble factors.

Certain functional groups (i.e., carboxylic groups) on the GO surface can affect the embryonic stem cell differentiation [40]. However, the mechanism contributing to this cell behavior remains to be investigated. In another study, GO was shown to promote the differentiation of mouse embryonic stem cells to neural cells [41]. A recent study also showed graphene increased the bioelectrical function and development of neural stem cells and thereby offers a controlled way to manipulate the bioelectricity of neural cells [42].

20.4 Applications of Graphene in Tissue Engineering

Interesting properties of graphene-based materials including mechanical and electrical properties can be useful in tissue engineering. Recent literature indicates that graphene and graphene-based materials with the aid of micro- and nanofabrication technologies may lead to the development of functional tissue and organ constructs [43]. In what follows, we outline some applications of graphene in different tissue engineering areas.

20.4.1 Bone Tissue Engineering

Bone tissue regeneration is of high demand to promote the recovery or reconstruction of large bone defects created by tumors or severe trauma where a large portion of bone is removed, is fractured, or has infection and abnormality. The development of this research area requires the use of substrates that enable the attachment, proliferation, and/or differentiation of bone progenitor cells. Several different materials, such as hydrogels and minerals, can initiate complex events that lead to the cell differentiation and osteogenesis [44, 45]. Although these materials can offer suitable surface chemistry for the cell adhesion, growth, and differentiation, their physical and chemical properties may not be tunable [46, 47]. Moreover, they possess poor mechanical properties and may lack specific bioactivities to interact with cells. It is still challenging to create large structures with a well-designed architecture that can modulate different cell behavior and function [45, 47–49]. Therefore, materials with intrinsic characteristics that can sustain cell growth and induce differentiation would have a great potential in bone tissue engineering [50–52].

Graphene-based materials are noncytotoxic and allow the attachment and proliferation of fibroblasts, osteoblasts, and MSCs, as most commonly used cells to regenerate bone tissues [35, 36, 50, 52–56]. Interestingly, human osteoblast-like cells (SAOS-2) and MSCs seeded on CVD graphene presented higher proliferation than those cultured on silicon dioxide (SiO_2) after 48 h of incubation [50]. In another study, self-supporting graphene films induced the osteogenic differentiation of rat bone marrow stem cells [57]. MSCs cultured on reduced GONR (rGONR) and graphene oxide nanoribbon (GONR) grids showed 2.7- and 3.4-time increase in the mineralized deposition than those cultured on PDMS and glass substrates (shown in Figure 20.2A–E) [55]. When glass and Si/SiO_2 were coated with CVD graphene, MSCs presented high expression of osteocalcin (OCN) as compared to uncoated counterparts [35]. OCN is identified as a late bone marker in osteoblasts [58]. These observations are attributed to high Young's modulus and nanotopographies of graphene-based materials.

Three-dimensional (3D) graphene constructs can be synthesized via CVD method using nickel foam as the template. These structures are able to induce spontaneous osteogenic and neuronal differentiation of MSCs [54, 59]. In particular, cells in 3D graphene structures showed an elongated cell morphology having aligned and thin cell nuclei (typical characteristics of osteoprogenitor cells) and expressed osteogenic markers (osteopontin (OPN) and OCN) even without the use of osteogenic medium [59]. Although graphene is able to encourage spontaneous osteogenic differentiation of stem cells, this property is significantly enhanced by the use of soluble factors for osteogenic differentiation [35, 36, 55, 59]. 3D printing has also been used as a novel microscale technology to fabricate 3D graphene scaffolds in bone tissue engineering. For example, 3D-printed graphene flakes-hydroxyapatite

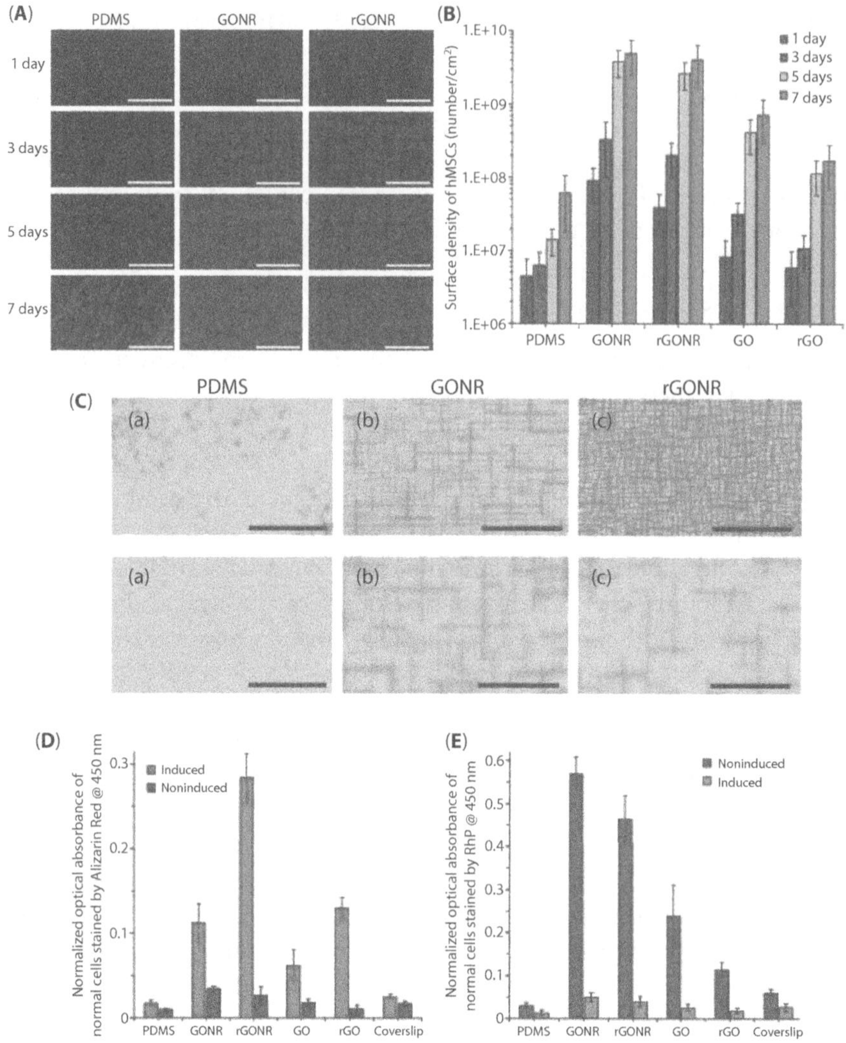

Figure 20.2 (A) Fluorescent pictures of actin filament of hMSCs stained with RhP and DAPI after 1, 3, 5, and 7 days of proliferation on PDMS substrate and GONR and rGONR grids. Scale bars are 10 μm. (B) Surface density of hMSCs on PDMS substrate (as control), GONR and rGONR grids, and GO and rGO sheets (for comparison) after 1, 3, 5, and 7 days of proliferation. (C) Osteogenic differentiation of hMSCs characterized using Alizarin Red staining after 1-week incubation with induction (top row) and without induction (down row) on (a) PDMS substrate and (b) GONR and (c) rGONR grids. Scale bars are 10 μm. (D) Normalized optical absorbance of differentiated cells stained by Alizarin Red and those stained by RhP (E) at the wavelength of 450 nm. Reprinted with permission from Akhavan *et al.*, 2013, *Carbon* [55].

was used as an electrically conductive and mechanically flexible scaffold to induce the osteogenesis in MSCs [60]. It is known that 3D scaffolds have better performance than 2D scaffolds in regulating cell behavior, transport of soluble factors, and tissue development and function [61]. Facile fabrication of 3D graphene-based structures is advantageous to provide complex and biomimetic structures in bone tissue regeneration.

20.4.2 Neural Tissue Engineering

High conductivity, flexibility, and biocompatibility of graphene have been considered as great advantages in fabricating novel graphene electrodes for neural cell recording and stimulating. On this account, the emergence of graphene offers an ideal breakthrough to deal with nervous system models *in vitro* and *in vivo* [62]. In addition, graphene-based materials are chemically stable in biological media and can regulate electrical activity of cells. Transparency of graphene allows *in situ* optical examination of activity and morphological changes of neurons. Therefore, graphene-based materials are suitable for neural cell culture. For example, Heo *et al.* proposed a noble and innovative stimulator based on graphene/polyethylene terephthalate (PET) films and used it to modulate cell-to-cell interactions of neural cells [63]. Li *et al.* assessed how cell neurites are affected by graphene [64]. First, they demonstrated that the graphene was biocompatible with the neural cells. Surprisingly, the viability of neural cells and average neurite length on the graphene substrate were significantly higher compared to conventional polystyrene tissue culture substrate. Further analysis showed that the graphene increased the expression of growth-associate protein-43, leading to the boost of neurite sprouting and outgrowth. The neurite sprouting and outgrowth are indicators of nervous system development. In general, graphene mimics the surrounding matrix of neurons due to nanotopographies on the graphene surface. High electrical conductivity of graphene also enhances the cell–cell communication in neural cells resulting in the neural cell adhesion and neurite outgrowth [65].

Prior to using human neural stem cells (hNSCs) for the brain repair and neural regeneration, it is important to direct the neural differentiation toward neurons rather than glia cells. Park *et al.* showed enhanced differentiation of hNSCs into neurons on graphene-coated glass substrate compared to uncoated glass substrate as the control group [66]. In addition, the hNSC adhesion on the graphene substrate was higher. After 2 weeks of cell differentiation, more cells were remained on the graphene substrate compared to the control group. Moreover, after 3 weeks of differentiation, a significant difference in the morphology of cells cultured on the graphene and glass was observed (Figure 20.3a). The neurite outgrowth was obvious for the hNSCs cultured on the graphene, while the hNSCs on the glass substrate were mostly detached. After 1 month, the hNSCs on the graphene showed typical characteristics of neuronal cells, such as neurite outgrowth and elongated cell shape. The percentage of neurons (TUJ1-positive cells) and glia (GFAP-positive cells) on the graphene and glass are shown in Figure 20.3b and d indicating enhanced neuronal differentiation of hNSCs on the graphene. Taken together, the graphene provided more favorable microenvironments for the hNSC differentiation and promoted the cell adhesion and neurite outgrowth in contrast to the control group. In another study, Akhavan *et al.* fabricated 3D GO foams and used them for the growth of neural cells using electrical stimulation [67]. The low-voltage electrical stimulation increased the differentiation of hNSCs to neurons. Other stimulation techniques, such as pulsed laser and near infrared, can also be coupled with graphene-based scaffolds in neural tissue engineering [68]. Graphene is also a versatile platform to combine with other biomolecules, such as DNA, peptides, and proteins, to make functional scaffolds in neural tissue regeneration as reviewed elsewhere [69].

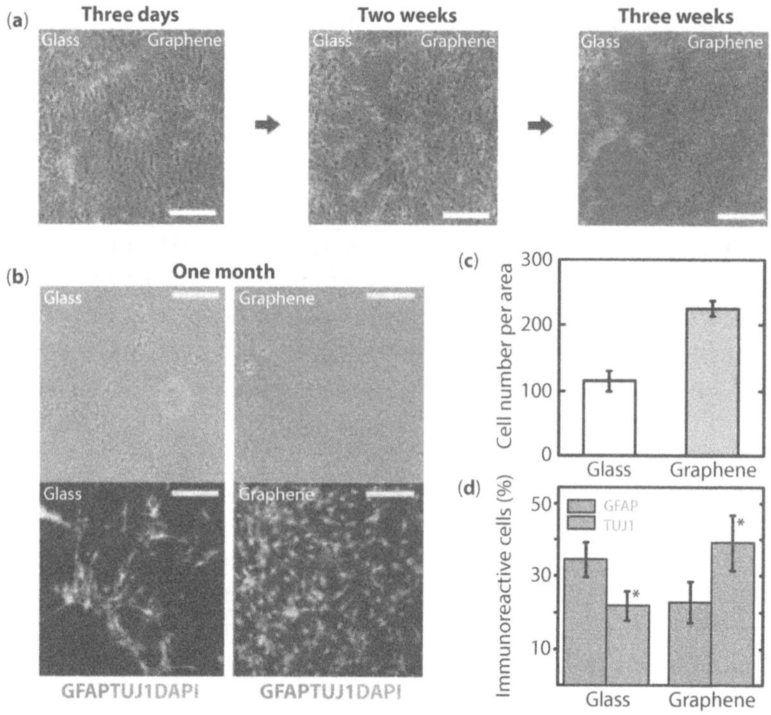

Figure 20.3 Enhanced neural differentiation of hNSCs on graphene films. All scale bars represent 200 μm: (a) Bright-field images of hNSCs differentiated for 3 days (left), 2 weeks (middle), and 3 weeks (right). Note that the hNSCs on the glass were gradually retracted and detached after 2 weeks, while those on the graphene remained stable even after 3 weeks of differentiation. (b) Bright-field (top row) and fluorescence (bottom row) images of hNSCs differentiated on the glass (left) and the graphene (right) after 1-month differentiation. The differentiated hNSCs were immunostained with GFAP (red) for astroglial cells, TUJ1 (green) for neural cells, and DAPI (blue) for cell nuclei. Note that more hNSCs were adhered to the graphene than to the glass. (c) Cell counting per area (0.64 mm^2) on the graphene and the glass regions after 1-month differentiation. Note that much more cells were observed on the graphene in comparison to the glass regions (n = 5, *p<0.001). (d) Percentage of immunoreactive cells for GFAP (red) and TUJ1 (green) on the glass and graphene. Note that glass regions show more GFAP-positive cells (glia) than TUJ1-positive ones (neurons), while graphene regions have more TUJ1-positive ones (neurons) than GFAP-positive ones (glia) (n = 5, *p<0.05). Reprinted with permission from Park, S.Y. *et al.*, 2011, *Advanced Materials* [66].

20.4.3 Cardiac Tissue Engineering

Cardiac tissue injuries as a result of coronary artery diseases or myocardial infarction (MI) are among the most prevalent causes of mortality and morbidity around the globe. The occlusion of coronary arteries causes the loss of blood supply in the myocardium and thereby induces morphological and functional problems in the cardiac tissue. Ischemia also leads to cardiomyocyte death through apoptotic or necrotic processes and scar tissue formation that eventually alter electrophysiological and contractile characteristics of native cardiac tissue [70]. Cardiac tissues possess specific contractile properties that are directly related to their highly organized and anisotropic architecture [71]. Therefore, the proper function of engineered cardiac tissues needs recapitulating the anisotropic structure of native myocardium through a series of biophysical and/or topographical cues. The unique mechanical characteristics of myocardium are due to highly aligned collagen nanofibers with diameters ranging from 10 to 100 nm. Such nanoscale features play a vital role in regulating muscle cell behavior and function. In addition, the native cardiac tissues are electrically conductive (0.005 (transverse) ~ 0.1 (longitudinal) S/m) [72, 73].

Earlier studies showed that stem cells increase the contractile functionality and myocardial perfusion in injured heart through differentiation into cardiomyocytes or vascular cell types. The stem cells release specific cytoprotective factors that enhance the cardiomyocyte survival and to decrease the inflammatory response [74]. A number of microenvironmental factors (i.e., biochemical and biophysical parameters) influence stem cell survival and differentiation to cardiovascular cells. Graphene-based materials have shown great promise to control the cardiac differentiation of stem cells *in vitro* and *in vivo* by enhancing electrical conductivity of scaffolds and providing suitable morphological cues. For example, we recently showed that graphene induced cardiac differentiation in mouse embryoid bodies, and this effect was more pronounced after electrical stimulation of embryoid bodies [9].

It has been thought that MSCs have great potential for cardiac repair using cell therapy approaches. However, available clinical trials have shown low efficiency of MSCs in the cardiac therapy due to limited cell differentiation into cardiomyocytes *in vivo* [75]. Cardiomyogenically differentiated MSCs would significantly increase the efficiency of cell therapy to improve the myocardial contractility and function [76]. Preliminary works showed the ability of graphene to enhance the cardiac differentiation of MSCs by culturing the cells on the graphene substrates [77, 78]. Interestingly, MSCs combined with GO showed an enhanced cellular survival for cardiac repair *in vivo* compared to MSCs because the GO protected the cells from reactive oxygen species (Figure 20.4A–D) [79]. In another study, rGO flakes were incorporated in hMSC spheroids and the results demonstrated the expression of cell-signaling biomolecules and high expression of cardiac-specific biomarkers [79]. The observed behavior is mainly because of high adsorption of FN on the rGO flakes and high electrical conductivity of rGO. Following the implantation of materials in a mouse MI model, the hybrid rGO-hMSC spheroids led to improved cardiac repair and function as compared to rGO alone and pure hMSCs. Moreover, the therapeutic efficacy of implantation procedure was significantly enhanced using the hybrid rGO-MSC spheroids (10-fold more than MSCs alone) [79]. The major advantage of such hybrid system is the delivery of cells at the site of injury in an efficient manner. In addition, different therapeutic molecules can be loaded on the graphene and then delivered in a sustained way.

Graphene-based materials can be combined with other scaffolds to increase their mechanical properties and electrical conductivity. For example, hybrid GO/GelMA hydrogels with tunable electrical and mechanical properties were fabricated for cardiac tissue engineering [80–82]. In the later works, GO was chosen over graphene because it is water-dispersible due to the presence of oxygen-containing functional groups on the GO surface [82]. GelMA is a photopolymerizable hydrogel and allows creating complex scaffold structures. The GO nanosheets made a stable aqueous dispersion in GelMA pre-polymer solution likely by the aid of noncovalent interaction between GelMA and GO [82]. The mechanical stiffness of hybrid GelMA-GO hydrogels was increased by covalent bonding the GO to the hydrogel structure through methacryloyl functional groups on the GO [81]. The covalent bonding of GO to the GelMA increased the GO concentration in the hydrogel (up to 3 mg/ml) without significant increase in the viscosity or aggregation (Figure 20.4E–G) [81]. We recently proposed a novel approach to make aqueous dispersion of graphene by using bovine serum

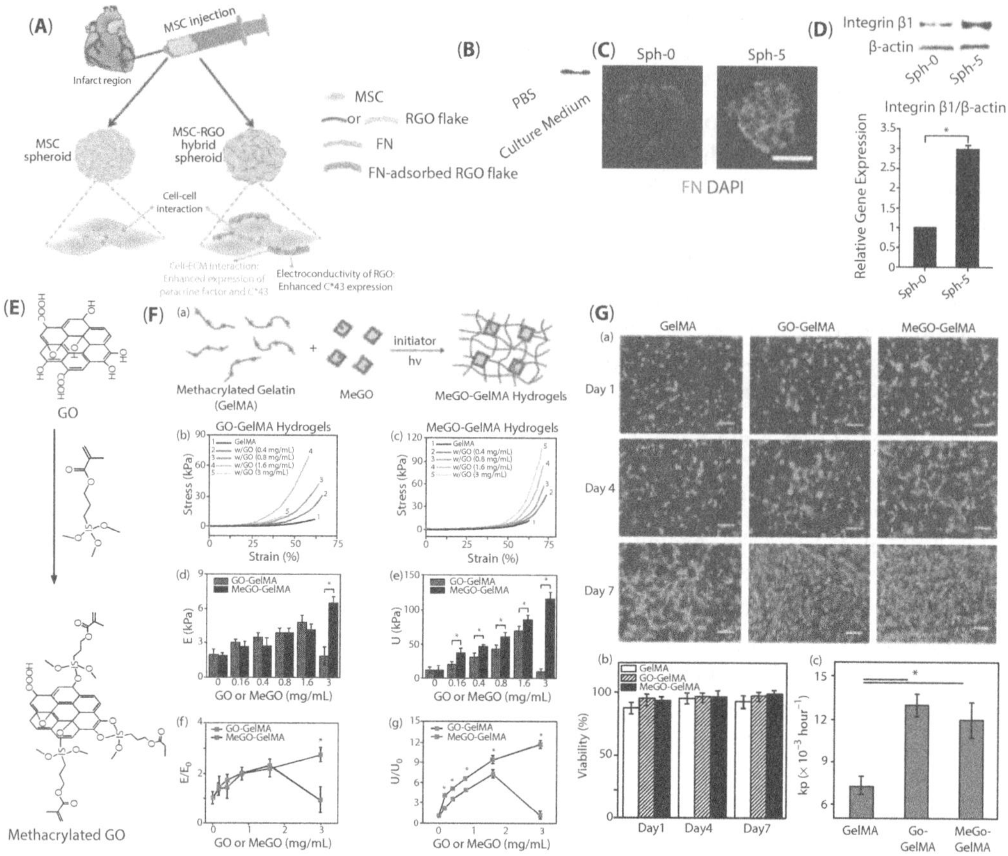

Figure 20.4 (A) Schematic illustration of effect of rGO in MSC spheroids used in the treatment of MI model. The incorporation of rGO flakes in the MSC spheroids enhances cell–ECM and cell–cell communications. (B) Incubation of rGO flakes in a serum-containing culture medium led to FN adsorption on the rGO flakes, whereas the rGO in phosphate-buffered saline shows no FN adsorption, as quantified by Western blot analysis. (C) A homogeneous distribution of FN in the hybrid MSC–rGO spheroids, which confirms enhanced cell–ECM and cell-cell communications, while FN was observed only at the periphery of MSC spheroids, as determined by the cell immunostaining. Scale bar is 100 μm. (D) High expression of integrin β1 (a FN-interacting integrin) in hybrid MSC–rGO spheroids, confirming enhanced cell–FN interaction, as measured by the Western blot analysis, *$p<0.05$. Reprinted with permission from Park, J. *et al.*, 2015, *Advanced Functional Materials* [79]. (E) Functionalization of GO surface with methacrylate using silanization to make methacrylated graphene oxide (MeGO). (F) (a) MeGO-GelMA hydrogel is prepared by radical copolymerization of MeGO and GelMA using UV light. Red dots represent methacrylate groups. Stress–strain curves of GelMA hydrogels with varying amounts of (b) GO or (c) MeGO measured from uniaxial compression. (d) Young's modulus (E) and (e) ultimate stress (U) of GO-GelMA hydrogels and MeGO-GelMA hydrogels. (f) Normalized elastic modulus (E/E_0) and (g) normalized fracture energy (U/U_0) of GelMA hydrogels incorporated with GO or MeGO. The values are normalized with respect to those of pure GelMA hydrogel (E_0, U_0) (*$p<0.05$ at the same concentrations of GO and MeGO). (G) (a) Fluorescent images of fibroblasts encapsulated in GelMA, GO-GelMA, and MeGO-GelMA hydrogels over time. The cells were stained with ethidium homodimer-1 and calcein-AM to visualize dead (red) and live (green) cells. Scale bars are 100 μm. (b) Viability of the encapsulated cells over time. (c) Proliferation rate (kP) of fibroblasts encapsulated in GelMA, GO-GelMA, and MeGO-GelMA hydrogels, *$p<0.05$. The cells in hybrid gels showed higher proliferation rates compared to the cells encapsulated in pure GelMA. Reprinted with permission from Cha, C. *et al.*, 2014, *Small* [81].

albumin [83]. The aqueous graphene dispersion was made in a facile, green, and scalable approach. The combination of graphene with GelMA hydrogel resulted in increasing the mechanical stiffness and electrical conductivity of GelMA. In a following work, we aligned graphene in GelMA hydrogel in either horizontal or vertical directions using dielectrophoresis approach to achieve the hybrid gels with anisotropic mechanical stiffness and electrical conductivity [84]. The hybrid GelMA-graphene hydrogels were biocompatible and suitable for cardiac tissue regeneration using electrical stimulation.

Overall, graphene-based materials have shown great performance in engineering cardiac tissues. They are particularly helpful in stem cell differentiation to cardiomyocytes by providing suitable topographical and electrical cues. The cell–cell communication and electrical current propagation would be easier in graphene-based scaffolds compared to nonconductive scaffolds. Graphene has been used in developing functional hydrogels with high electrical conductivity and tunable mechanical stiffness for cardiac tissue engineering. These hydrogels outperform commonly used hydrogels in cardiac regeneration that are often mechanically weak and nonconductive.

20.4.4 Other Tissue Engineering Applications

Skeletal muscle tissues are contractile tissues composed of bundles of dense and highly organized muscle myofibers. Tumor ablation, traumatic injuries, and external wounds can severely damage the function of muscle tissues in the body. Various regenerative approaches have been used to repair the muscle loss and restore the muscle function. However, skeletal muscle regeneration and engineering are still challenging [85]. The muscle cell differentiation and myotube formation were evaluated on rGO and GO substrates. The myogenic differentiation was significantly increased on the GO compared to the rGO. The observed behavior was attributed to surface roughness and more oxygen-containing functional groups of GO that enhanced the adsorption of serum proteins from culture medium [86]. In another work, we showed that electrical stimulation enhanced the differentiation of muscle cells cultured on graphene substrates due to high electrical conductivity of graphene [11].

Articular cartilage is a nonnervous, nonvascular, and elastic tissue consisting of sparsely distributed chondroblasts in a dense ECM, which is mainly composed of proteoglycan and elastin and collagen fibers [87]. The stem cell differentiation to chondrocytes is conventionally obtained by culturing cells in pellets [88]. However, the diffusion of protein transforming growth factor-β3 (TGF-β3) in the pellets is limited and they provide low cell–ECM interaction. As a result, the chondrogenic differentiation of stem cells using this approach has low efficiency. To tackle these problems, GO sheets were treated with TGF-β3 and FN prior to their encapsulation in pellets for the chondrogenic differentiation of human adipose-derived stem cells (hASCs) [89]. The hybrid hASC-GO pellets significantly enhanced the chondrogenic differentiation of hASCs by supplying the TGF-β3 for the cells and increasing the cell–FN interaction (Figure 20.5). This was an innovative approach in the cartilage regeneration using hASCs. Zhou *et al.* used 3D bioprinting to fabricate GO-gelatin scaffolds for the chondrogenic differentiation of MSCs [90]. The gene expression analysis and immunostaining results confirmed the positive effect of GO on the stem cell differentiation.

Skin is the largest organ in the body that protects the human body against excessive water loss and dangerous pathogens. Several approaches have been developed to regenerate skin tissues *in vitro* that mimic the physiological characteristics of human skin. To this

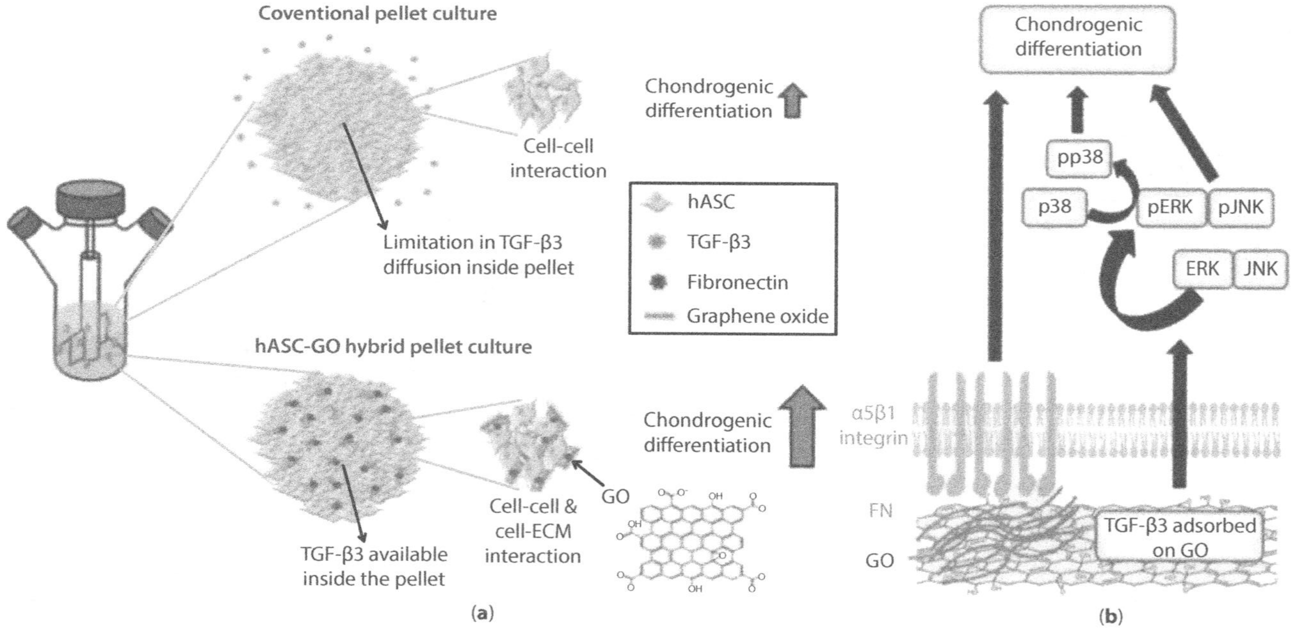

Figure 20.5 (a) A schematic describing enhanced chondrogenic differentiation of hASCs using GO. (b) A schematic diagram of underlying mechanisms describing the cell signaling in the enhanced chondrogenic differentiation in the hybrid hASC-GO pellets caused by TGF-β3 and FN. Reprinted with permission from Yoon, H.H. *et al.*, 2014, *Advanced Functional Materials* [89].

end, natural and synthetic scaffold materials have been used for skin tissue regeneration [91]. To use the advantages of both chitosan (CS) and graphene in wound healing, Lu *et al.* incorporated few-layer graphene on electrospun CS-poly(vinyl alcohol) nanofibers [27]. *In vivo* results showed the positive effect of graphene-containing scaffolds on the wound healing in comparison to control groups in mouse and rabbit models. The authors suggested that free electrons of graphene suppressed the growth of prokaryotic cells and had no influence on the eukaryotic cell multiplication. Therefore, the proliferation of microbes was almost stopped that benefited the wound healing. In another study, GO was introduced in collagen-fibrin (CF) biocomposite films for wound healing applications [92]. The GO increased the mechanical strength of CF films. Higher wound healing rate in rat models was observed for the GO-CF biocomposite films compared with films without GO.

20.5 Biocompatibility of Graphene

Biocompatibility of graphene-based materials is still a hot research subject. Many people have evaluated the potential toxicity of graphene-based materials *in vitro* and *in vivo* [21, 93–95]. While some reports indicated the toxicity of as-made GO or raw graphene without any chemical functionalization, the surface coating of GO and graphene with biopolymers has shown no sign of toxicity to cells *in vitro* and to animal models *in vivo*. Dong *et al.* followed this approach to functionalize graphene nanoribbons with polyethylenimine and used the hybrid material for *in vitro* transfection and *in situ* detection of microRNA [96]. Depan *et al.* exploited this safe use of graphene-based materials in drug delivery. They positively charged folate-conjugated CS and encapsulated doxorubicin (DOX)-loaded GO in it. The DOX–GO–CS–folate was used as a nanocarrier and was able to release drugs in response to pH changes [97]. In addition to biopolymers, biomolecules, such as DNA and proteins, can also be used to make biocompatible graphene and GO. Recently, Hu *et al.* functionalized GO with fetal bovine serum by sonicating the GO in the serum solution. The hybrid GO–protein material showed significantly lower cytotoxicity compared to untreated GO [98]. The serum proteins were immobilized on the surface of GO through hydrophobic interactions. In a similar work, Liu *et al.* used gelatin to functionalize GO [99]. Graphene and GO contain highly localized π-electrons on their surfaces and thereby various aromatic biomolecules are able to bind to them through the π–π stacking. Utilizing this π–π stacking between graphene and single-stranded DNA molecules, Liu *et al.* demonstrated that DNA molecules can be attached to GO during chemical reduction of GO resulting in the DNA-coated rGO with stable dispersion in water [100].

Dextran (DEX)-functionalized GO did not induce any significant toxicity to mouse models after intravenous injection [101]. Polyacrylic acid was also used to functionalize graphene-based materials and the hybrid materials showed biocompatibility as exposed to cells and zebrafish [102]. Duch *et al.* assessed the pulmonary toxicity of graphene after inhalation and discovered that although as-made GO and un-functionalized graphene were highly toxic, pluronic-functionalized graphene exhibited excellent dispersity and the lung toxicity was remarkably decreased without causing any obvious inflammation after the inhalation [103]. As shown in Figure 20.6a–e, they pointed out that when the mice exposed to the graphene aggregates suspended in saline, macrophages surrounded the aggregates with a homogeneous black cytoplasm. In the mice treated with nanoscale dispersed graphene, macrophages surrounded 91±4% of the lung area with a black cytoplasm

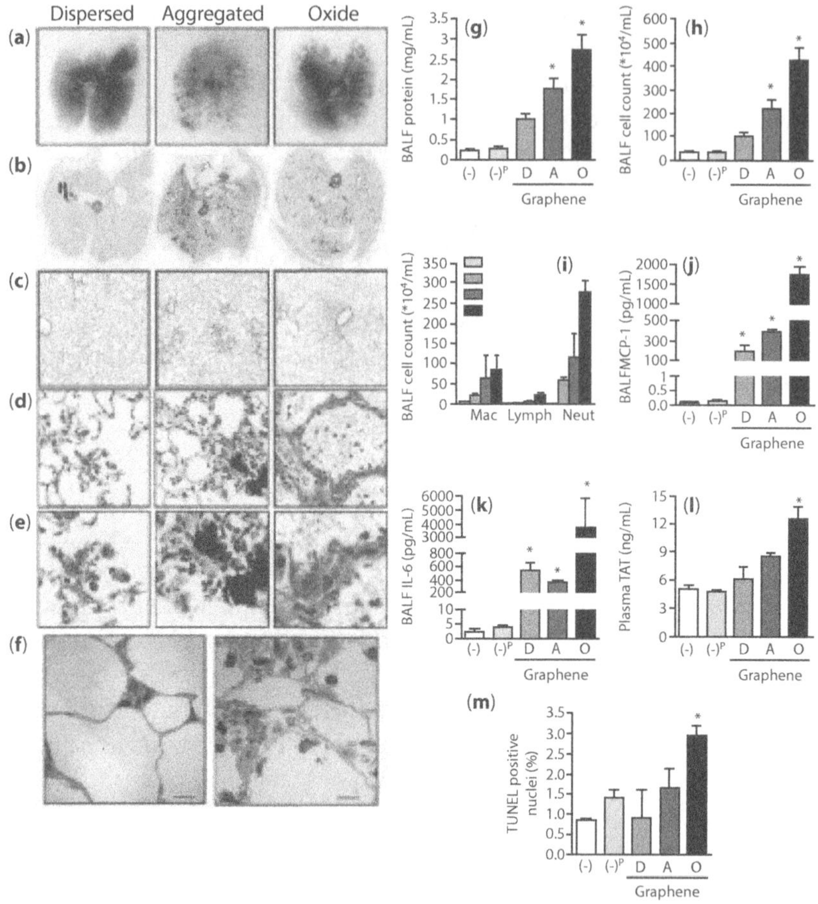

Figure 20.6 Acute lung injury in mouse models induced by GO nanoparticles. Mice were treated intratracheally with three types of graphene-based materials: highly dispersed and purified graphene in 2% pluronic (Dispersed, D), GO (Oxide, O), and graphene aggregates suspended in water (Aggregated, A). The mice were sacrificed 24 h after the treatment to assess the lung injury. (a and b) Photomicrographs of lung fixed in paraffin blocks. The block sectioning was done at approximately the same level to compare the distribution of graphene-based materials, (b–e) Pictures of lung sections at $<1\times$ (c), 50× (d), and 200× (e). (f) Representative pictures of lung sections for the mice treated 24 h earlier with GO (right) and pluronic-dispersed graphene (left); scale bars indicate 10 μm. BALF levels of (g) protein, (h) total cell nuclei, (i) differential cell count macrophages (Mac), Lymphocytes (Lymph), Neutrophils (Neut), and levels of the pro-inflammatory cytokines (j) MCP-1 and (k) IL-6. In the same animals, (l) plasma levels of thrombin–antithrombin (TAT) levels and (m) the percentage of TUNEL positive nuclei in the lung sections were quantified. (m) $n > 3$ for all measures; * indicates $p<0.05$ for comparison with appropriate control. Reprinted with permission from Duch, M.C. *et al.*, 2011, *Nano Letters* [103].

compared with the mice exposed to the graphene aggregates that covered 36±1% of the lung area. This result indicated that distribution of graphene after airway instillation in the lung was largely affected by the size of graphene nanoparticles. However, a severe lung inflammation with alveolar exudates and hyaline membrane formation were found in the mice treated with GO probably due to weak dispersion of these nanomaterials. The GO-treated mice showed serious lung injury as shown in electron micrographs (Figure 20.6f). This injury included bronchoalveolar lavage fluid (BALF) fluid pleiocytosis, a leakage of protein into the alveolar space, and elevated BAL levels of pro-inflammatory cytokines (Figure

20.6g–k). In particular, plasma thrombin antithrombin complexes significantly increased in the GO-treated mice (Figure 20.6l). Also, the number of TUNEL positive nuclei was high for the mice lungs treated with GO (Figure 20.6m). Taken together, published data up to date clearly justify this conclusion that functionalized and well-dispersed graphene-based materials are much less toxic compared to the unfunctionalized and aggregated counterparts [104]. In addition, some physical characteristics of nanomaterials, such as shape, size, and concentration, can largely influence the cytotoxicity of graphene-based materials [105].

Graphene–CS composites were biocompatible to L929 cells as revealed by MTT colorimetric assays [106]. Moreover, many studies indicated low or no cytotoxicity of graphene and GO as exposed to a variety of cell types, such as HeLa cells [107], L929 cells [108], human fibroblasts [105], human hepatoma HepG2 cells [109], and A549 human lung cancer cells [94]. Interestingly, single-layer GO at concentrations lower than 20 μg/mL was able to internalize the cytoplasmic, membrane bound vacuoles of fibroblasts [105] and human lung epithelial cells [110] without considerable cytotoxicity issue after 24 h of treatment. In contrast, when the GO is applied extracellularly, Chang *et al.* found that GO presented minimal toxicity even at a dose higher than 50 μg/mL, but there was no indication of cellular uptake. In another study, Chen *et al.* revealed that GO- and graphene-coated substrates were biocompatible with iPSCs and enabled the cell adhesion and proliferation [39]. L929 cells were able to attach and proliferate on graphene papers and a confluent layer of metabolically active cells was visualized after 48 h of cell culture. The proliferation rate for the cells on the graphene papers was the same as on commercial polystyrene tissue culture plates, indicating the biocompatibility of graphene substrates [111]. Moreover, immunofluorescence images and MTT assays confirmed the cell viability and normal morphology for the cells [35] (Figure 20.7). A general

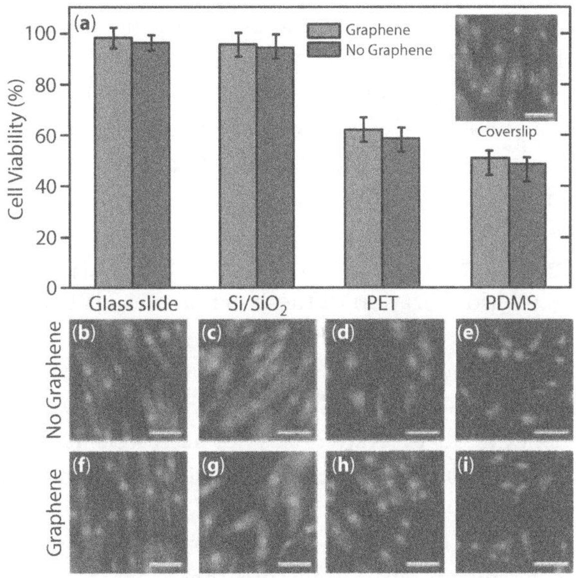

Figure 20.7 Cell morphology and viability of hMSCs cultured on different substrates. Cells were stained with Calcein AM (green) and DAPI (blue). (a) Plot shows the percentage of cell viability as normalized to coverslips used as the control group. (Inset) Morphology of hMSCs cultured on standard coverslips. (b–e) hMSCs cultured on pure glass slide, Si/SiO_2, PET, and PDMS. (f–i) hMSCs cultured on glass slide, Si/SiO_2, PET, and PDMS coated with graphene. Scale bars are 100 μm. Reprinted with permission from Nayak, T.R. *et al.*, 2011, *ACS Nano* [35].

conclusion can be drawn that unless cells are treated with high concentrations of graphene-based materials, it is safe to culture mammalian cells on the materials and use the material advantages in cell proliferation and differentiation.

20.6 Conclusions and Future Directions

Applications of graphene-based materials in tissue engineering and regenerative medicine have grown rapidly in the past few years. Many exciting research works have been done in this research area. The obtained preliminary results are encouraging; however, some remaining challenges need to be addressed particularly prior to the application of graphene in preclinical and clinical studies. Graphene-based materials exhibit excellent electrical conductivity and high mechanical properties. However, the graphene aggregation should be prevented in biological media or in composite materials to take the full advantages of graphene properties. Some novel approaches have been proposed to make homogeneously distributed graphene in biological media. However, these approaches should avoid using toxic chemicals and complicated procedures to make green, high-quality, and scalable graphene production.

Some interesting findings have been reported in using graphene-based materials in stem cell research. However, more investigations are yet required to understand molecular and cellular mechanisms and signaling pathways by which graphene affects stem cell behavior and fate. Such understanding would be greatly helpful in designing highly functional graphene-based materials for stem cell culture, differentiation, and implantation. Stem cell therapy trials should specifically report the fate of graphene *in vivo* and how it might affect other tissues and organs. The use of biodegradable graphene for *in vivo* and clinical applications is highly recommended.

The use of graphene in neural tissue recording and stimulation is interesting. It can be further explored to use graphene electrodes in preclinical and clinical trials. In addition, such electrodes can be integrated in biomedical devices to precisely monitor biological activities of tissues *in vivo*. However, high-quality graphene needs to be used and it should be stable under physiological condition.

Although some concerns still remained regarding cytotoxicity of graphene-based materials, it is known that functionalization of these nanomaterials using biopolymers or biomolecules can largely mitigate their potential toxicity. High-throughput and systematic toxicological experiments should be done to investigate effect of multiple physiochemical characteristics of graphene-based materials on cells and tissues *in vitro* and *in vivo*. Long-term toxicity of graphene and its derivatives has not been well studied yet. Following these studies, a general picture on graphene biocompatibility and safety protocols to avoid the cytotoxicity of graphene-based materials are achieved.

In summary, despite several challenges and issues in using graphene-based materials in tissue engineering and regenerative medicine, these materials have already paved the way for some breakthroughs in regenerative medicine. It is hoped that further research would increase the use of interesting properties of graphene and its derivatives in biomedicine.

References

1. Langer, R. and Vacanti, J., Advances in tissue engineering. *J. Pediatr. Surg.*, 51, 81, 2016.
2. Webber, M.J., Khan, O.F., Sydlik, S.A., Tang, B.C., Langer, R., A perspective on the clinical translation of scaffolds for tissue engineering. *Ann. Biomed. Eng.*, 43, 641, 2015.
3. Atala, A., Engineering tissues, organs and cells. *J. Tissue Eng. Regener. Med.*, 1, 83, 2007.
4. Khademhosseini, A., Vacanti, J.P., Langer, R., Progress in tissue engineering. *Sci. Am.*, 300, 64, 2009.
5. Freed, L.E., Vunjak-Novakovic, G., Biron, R.J., Eagles, D.B., Lesnoy, D.C., Barlow, S.K., Langer, R., Biodegradable polymer scaffolds for tissue engineering. *Nat. Biotechnol.*, 12, 689, 1994.
6. Tamayol, A., Akbari, M., Annabi, N., Paul, A., Khademhosseini, A., Juncker, D., Fiber-based tissue engineering: Progress, challenges, and opportunities. *Biotechnol. Adv.*, 31, 669, 2013.
7. Ahadian, S., Batmanghelich, F., Obregón, R., Rana, D., Ramón-Azcón, J., Sadeghian, R.B., Ramalingam, M., *Carbon-Based Nanobiomaterials, Nanobiomaterials*, pp. 85–104, Wiley-VCH Verlag GmbH & Co. KGaA, 2018.
8. Ahadian, S., Obregón, R., Ramón-Azcón, J., Salazar, G., Shiku, H., Ramalingam, M., Matsue, T., Carbon nanotubes and graphene-based nanomaterials for stem cell differentiation and tissue regeneration. *J. Nanosci. Nanotechnol.*, 16, 8862, 2016.
9. Ahadian, S., Zhou, Y., Yamada, S., Estili, M., Liang, X., Nakajima, K., Shiku, H., Matsue, T., Graphene induces spontaneous cardiac differentiation in embryoid bodies. *Nanoscale*, 8, 7075, 2016.
10. Chung, C., Kim, Y.-K., Shin, D., Ryoo, S.-R., Hong, B.H., Min, D.-H., Biomedical applications of graphene and graphene oxide. *Acc. Chem. Res.*, 46, 2211, 2013.
11. Ahadian, S., Ramón-Azcón, J., Chang, H., Liang, X., Kaji, H., Shiku, H., Nakajima, K., Ramalingam, M., Wu, H., Matsue, T., Electrically regulated differentiation of skeletal muscle cells on ultrathin graphene-based films. *RSC Adv.*, 4, 9534, 2014.
12. Wang, Y., Lee, W.C., Manga, K.K., Ang, P.K., Lu, J., Liu, Y.P., Lim, C.T., Loh, K.P., Fluorinated graphene for promoting neuro-induction of stem cells. *Adv. Mater.*, 24, 4285, 2012.
13. Fisher, O.Z., Khademhosseini, A., Langer, R., Peppas, N.A., Bioinspired materials for controlling stem cell fate. *Acc. Chem. Res.*, 43, 419, 2009.
14. Cha, C., Liechty, W.B., Khademhosseini, A., Peppas, N.A., Designing biomaterials to direct stem cell fate. *ACS Nano*, 6, 9353, 2012.
15. Wheeldon, I., Farhadi, A., Bick, A.G., Jabbari, E., Khademhosseini, A., Nanoscale tissue engineering: Spatial control over cell-materials interactions. *Nanotechnology*, 22, 212001, 2011.
16. Liu, Z., Robinson, J.T., Sun, X., Dai, H., PEGylated nanographene oxide for delivery of water-insoluble cancer drugs. *J. Am. Chem. Soc.*, 130, 10876, 2008.
17. Reina, G., González-Domínguez, J.M., Criado, A., Vázquez, E., Bianco, A., Prato, M., Promises, facts and challenges for graphene in biomedical applications. *Chem. Soc. Rev.*, 46, 4400, 2017.
18. Jiang, H., Chemical preparation of graphene-based nanomaterials and their applications in chemical and biological sensors. *Small*, 7, 2413, 2011.
19. Guo, S. and Dong, S., Graphene nanosheet: Synthesis, molecular engineering, thin film, hybrids, and energy and analytical applications. *Chem. Soc. Rev.*, 40, 2644, 2011.
20. Darvishi, S., Souissi, M., Kharaziha, M., Karimzadeh, F., Sahara, R., Ahadian, S., Gelatin methacryloyl hydrogel for glucose biosensing using Ni nanoparticles-reduced graphene oxide: An experimental and modeling study. *Electrochim. Acta*, 261, 275, 2018.
21. Yang, K., Wan, J., Zhang, S., Zhang, Y., Lee, S.-T., Liu, Z., *In vivo* pharmacokinetics, long-term biodistribution, and toxicology of PEGylated graphene in mice. *ACS Nano*, 5, 516, 2010.

22. Singh, S.K., Singh, M.K., Kulkarni, P.P., Sonkar, V.K., Grácio, J.J., Dash, D., Amine-modified graphene: Thrombo-protective safer alternative to graphene oxide for biomedical applications. *ACS Nano*, 6, 2731, 2012.
23. Yang, Y., Zhang, Y.M., Chen, Y., Zhao, D., Chen, J.T., Liu, Y., Construction of a graphene oxide based noncovalent multiple nanosupramolecular assembly as a scaffold for drug delivery. *Chem. Eur. J.*, 18, 4208, 2012.
24. Yang, K., Zhang, S., Zhang, G., Sun, X., Lee, S.-T., Liu, Z., Graphene in mice: Ultrahigh *in vivo* tumor uptake and efficient photothermal therapy. *Nano Lett.*, 10, 3318, 2010.
25. Hong, H., Yang, K., Zhang, Y., Engle, J.W., Feng, L., Yang, Y., Nayak, T.R., Goel, S., Bean, J., Theuer, C.P., *In vivo* targeting and imaging of tumor vasculature with radiolabeled, antibody-conjugated nanographene. *ACS Nano*, 6, 2361, 2012.
26. Sahu, A., Choi, W.I., Tae, G., A stimuli-sensitive injectable graphene oxide composite hydrogel. *Chem. Commun.*, 48, 5820, 2012.
27. Lu, B., Li, T., Zhao, H., Li, X., Gao, C., Zhang, S., Xie, E., Graphene-based composite materials beneficial to wound healing. *Nanoscale*, 4, 2978, 2012.
28. Marković, Z.M., Matijašević, D.M., Pavlović, V.B., Jovanović, S.P., Holclajtner-Antunović, I.D., Špitalský, Z., Mičušik, M., Dramićanin, M.D., Milivojević, D.D., Nikšić, M.P., Marković, B.M., Antibacterial potential of electrochemically exfoliated graphene sheets. *J. Colloid Interface Sci.*, 500, 30, 2017.
29. Jung, H.S., Lee, M.Y., Kong, W.H., Do, I.H., Hahn, S.K., Nano graphene oxide–hyaluronic acid conjugate for target specific cancer drug delivery. *RSC Adv.*, 4, 14197, 2014.
30. Zhang, L., Xiao, S.J., Zheng, L.L., Li, Y.F., Huang, C.Z., Aptamer-mediated nanocomposites of semiconductor quantum dots and graphene oxide as well as their applications in intracellular imaging and targeted drug delivery. *J. Mater. Chem. B*, 48, 8558, 2014.
31. Yang, Y., Asiri, A.M., Tang, Z., Du, D., Lin, Y., Graphene based materials for biomedical applications. *Mater. Today*, 16, 365, 2013.
32. Wu, B., Zhao, N., Hou, S., Zhang, C., Electrochemical synthesis of polypyrrole, reduced graphene oxide, and gold nanoparticles composite and its application to hydrogen peroxide biosensor. *Nanomaterials*, 6, 220, 2016.
33. Geraili, A., Jafari, P., Hassani, M.S., Araghi, B.H., Mohammadi, M.H., Ghafari, A.M., Tamrin, S.H., Modarres, H.P., Kolahchi, A.R., Ahadian, S., Controlling differentiation of stem cells for developing personalized organ-on-chip platforms. *Adv. Health Mater.*, 2017.
34. Shin, S.R., Li, Y.-C., Jang, H.L., Khoshakhlagh, P., Akbari, M., Nasajpour, A., Zhang, Y.S., Tamayol, A., Khademhosseini, A., Graphene-based materials for tissue engineering. *Adv. Drug Deliv. Rev.*, 105, 255, 2016.
35. Nayak, T.R., Andersen, H., Makam, V.S., Khaw, C., Bae, S., Xu, X., Ee, P.-L.R., Ahn, J.-H., Hong, B.H., Pastorin, G., Graphene for controlled and accelerated osteogenic differentiation of human mesenchymal stem cells. *ACS Nano*, 5, 4670, 2011.
36. Lee, W.C., Lim, C.H.Y., Shi, H., Tang, L.A., Wang, Y., Lim, C.T., Loh, K.P., Origin of enhanced stem cell growth and differentiation on graphene and graphene oxide. *ACS Nano*, 5, 7334, 2011.
37. Oh, S., Brammer, K.S., Li, Y.J., Teng, D., Engler, A.J., Chien, S., Jin, S., Stem cell fate dictated solely by altered nanotube dimension. *Proc. Natl. Acad. Sci. USA*, 106, 2130, 2009.
38. Kenry, W.C., Loh, K.P., Lim, C.T., When stem cells meet graphene: Opportunities and challenges in regenerative medicine. *Biomaterials*, 155, Supplement C, 236, 2018.
39. Chen, G.-Y., Pang, D.-P., Hwang, S.-M., Tuan, H.-Y., Hu, Y.-C., A graphene-based platform for induced pluripotent stem cells culture and differentiation. *Biomaterials*, 33, 418, 2012.
40. Chao, T.-I., Xiang, S., Chen, C.-S., Chin, W.-C., Nelson, A., Wang, C., Lu, J., Carbon nanotubes promote neuron differentiation from human embryonic stem cells. *Biochem. Biophys. Res. Commun.*, 384, 426, 2009.

41. Yang, D., Li, T., Xu, M., Gao, F., Yang, J., Yang, Z., Le, W., Graphene oxide promotes the differentiation of mouse embryonic stem cells to dopamine neurons. *Nanomedicine*, 9, 2445, 2014.
42. Guo, R., Zhang, S., Xiao, M., Qian, F., He, Z., Li, D., Zhang, X., Li, H., Yang, X., Wang, M., Chai, R., Tang, M., Accelerating bioelectric functional development of neural stem cells by graphene coupling: Implications for neural interfacing with conductive materials. *Biomaterials*, 106, Supplement C, 193, 2016.
43. Goenka, S., Sant, V., Sant, S., Graphene-based nanomaterials for drug delivery and tissue engineering. *J. Controlled Release*, 173, Supplement C, 75, 2014.
44. Rosa, V., Della Bona, A., Cavalcanti, B.N., Nör, J.E., Tissue engineering: From research to dental clinics. *Dent. Mater.*, 28, 341, 2012.
45. Lutolf, M.P., Biomaterials: Spotlight on hydrogels. *Nat. Mater.*, 8, 451, 2009.
46. Mizuno, M., Shindo, M., Kobayashi, D., Tsuruga, E., Amemiya, A., Kuboki, Y., Osteogenesis by bone marrow stromal cells maintained on type I collagen matrix gels *in vivo*. *Bone*, 20, 101, 1997.
47. Rosa, V., Zhang, Z., Grande, R., Nör, J., Dental pulp tissue engineering in full-length human root canals. *J. Dent. Res.*, 92, 970, 2013.
48. Lu, Q., Pandya, M., Rufaihah, A.J., Rosa, V., Tong, H.J., Seliktar, D., Toh, W.S., Modulation of dental pulp stem cell odontogenesis in a tunable PEG-fibrinogen hydrogel system. *Stem Cells Int.*, 2015, 2015.
49. Salinas, C.N. and Anseth, K.S., The enhancement of chondrogenic differentiation of human mesenchymal stem cells by enzymatically regulated RGD functionalities. *Biomaterials*, 29, 2370, 2008.
50. Kalbacova, M., Broz, A., Kong, J., Kalbac, M., Graphene substrates promote adherence of human osteoblasts and mesenchymal stromal cells. *Carbon*, 48, 4323, 2010.
51. Mooney, E., Dockery, P., Greiser, U., Murphy, M., Barron, V., Carbon nanotubes and mesenchymal stem cells: Biocompatibility, proliferation and differentiation. *Nano Lett.*, 8, 2137, 2008.
52. Qi, W., Yuan, W., Yan, J., Wang, H., Growth and accelerated differentiation of mesenchymal stem cells on graphene oxide/poly-L-lysine composite films. *J. Mater. Chem. B*, 2, 5461, 2014.
53. La, W.G., Kang, S.W., Yang, H.S., Bhang, S.H., Lee, S.H., Park, J.H., Kim, B.S., The efficacy of bone morphogenetic protein-2 depends on its mode of delivery. *Artif. Organs*, 34, 1150, 2010.
54. Li, N., Zhang, Q., Gao, S., Song, Q., Huang, R., Wang, L., Liu, L., Dai, J., Tang, M., Cheng, G., Three-dimensional graphene foam as a biocompatible and conductive scaffold for neural stem cells. *Sci. Rep.*, 3, 1604, 2013.
55. Akhavan, O., Ghaderi, E., Shahsavar, M., Graphene nanogrids for selective and fast osteogenic differentiation of human mesenchymal stem cells. *Carbon*, 59, 200, 2013.
56. Ryoo, S.-R., Kim, Y.-K., Kim, M.-H., Min, D.-H., Behaviors of NIH-3T3 fibroblasts on graphene/carbon nanotubes: Proliferation, focal adhesion, and gene transfection studies. *ACS Nano*, 4, 6587, 2010.
57. Lu, J., He, Y.S., Cheng, C., Wang, Y., Qiu, L., Li, D., Zou, D., Self-supporting graphene hydrogel film as an experimental platform to evaluate the potential of graphene for bone regeneration. *Adv. Funct. Mater.*, 23, 3494, 2013.
58. Ryoo, H.-M., Hoffmann, H.M., Beumer, T., Frenkel, B., Towler, D.A., Stein, G.S., Stein, J.L., Van Wijnen, A.J., Lian, J.B., Stage-specific expression of Dlx-5 during osteoblast differentiation: Involvement in regulation of osteocalcin gene expression. *Mol. Endocrinol.*, 11, 1681, 1997.
59. Crowder, S.W., Prasai, D., Rath, R., Balikov, D.A., Bae, H., Bolotin, K.I., Sung, H.-J., Three-dimensional graphene foams promote osteogenic differentiation of human mesenchymal stem cells. *Nanoscale*, 5, 4171, 2013.
60. Jakus, A.E. and Shah, R., Multi and mixed 3D-printing of graphene-hydroxyapatite hybrid materials for complex tissue engineering. *J. Biomed. Mater. Res. A*, 105, 274, 2017.

61. Dubey, N., Bentini, R., Islam, I., Cao, T., Castro Neto, A.H., Rosa, V., Graphene: A versatile carbon-based material for bone tissue engineering. *Stem Cells Int.*, 2015.
62. Cohen-Karni, T., Qing, Q., Li, Q., Fang, Y., Lieber, C.M., Graphene and nanowire transistors for cellular interfaces and electrical recording. *Nano Lett.*, 10, 1098, 2010.
63. Heo, C., Yoo, J., Lee, S., Jo, A., Jung, S., Yoo, H., Lee, Y.H., Suh, M., The control of neural cell-to-cell interactions through non-contact electrical field stimulation using graphene electrodes. *Biomaterials*, 32, 19, 2011.
64. Li, N., Zhang, X., Song, Q., Su, R., Zhang, Q., Kong, T., Liu, L., Jin, G., Tang, M., Cheng, G., The promotion of neurite sprouting and outgrowth of mouse hippocampal cells in culture by graphene substrates. *Biomaterials*, 32, 9374, 2011.
65. Hong, S.W., Lee, J.H., Kang, S.H., Hwang, E.Y., Hwang, Y.-S., Lee, M.H., Han, D.-W., Park, J.-C., Enhanced neural cell adhesion and neurite outgrowth on graphene-based biomimetic substrates. *BioMed Res. Int.*, 2014, 2014.
66. Park, S.Y., Park, J., Sim, S.H., Sung, M.G., Kim, K.S., Hong, B.H., Hong, S., Enhanced differentiation of human neural stem cells into neurons on graphene. *Adv. Mater.*, 23, 2011.
67. Akhavan, O., Ghaderi, E., Shirazian, S.A., Rahighi, R., Rolled graphene oxide foams as three-dimensional scaffolds for growth of neural fibers using electrical stimulation of stem cells. *Carbon*, 97, Supplement C, 71, 2016.
68. Akhavan, O., Graphene scaffolds in progressive nanotechnology/stem cell-based tissue engineering of the nervous system. *J. Mater. Chem. B*, 4, 3169, 2016.
69. Li, D., Liu, T., Yu, X., Wu, D., Su, Z., Fabrication of graphene–biomacromolecule hybrid materials for tissue engineering application. *Polym. Chem.*, 8, 4309, 2017.
70. Rowe, W.J., Extraordinary unremitting endurance exercise and permanent injury to normal heart. *The Lancet*, 340, 712, 1992.
71. Mohammadi, M.H., Obregón, R., Ahadian, S., Ramón-Azcón, J., Radisic, M., Engineered muscle tissues for disease modeling and drug screening applications. *Curr. Pharm. Des.*, 23, 2991–3004, 2017.
72. You, J.-O., Rafat, M., Ye, G.J., Auguste, D.T., Nanoengineering the heart: Conductive scaffolds enhance connexin 43 expression. *Nano Lett.*, 11, 3643, 2011.
73. Liau, B., Zhang, D., Bursac, N., Functional cardiac tissue engineering. *Regen. Med.*, 7, 187, 2012.
74. Segers, V.F.M. and Lee, R.T., Stem-cell therapy for cardiac disease. *Nature*, 451, 937, 2008.
75. Psaltis, P.J., Zannettino, A.C., Worthley, S.G., Gronthos, S., Concise review: Mesenchymal stromal cells: Potential for cardiovascular repair. *Stem Cells*, 26, 2201, 2008.
76. Song, H., Hwang, H.J., Chang, W., Song, B.-W., Cha, M.-J., Kim, I.-K., Lim, S., Choi, E.J., Ham, O., Lee, C.Y., Cardiomyocytes from phorbol myristate acetate-activated mesenchymal stem cells restore electromechanical function in infarcted rat hearts. *Proc. Natl. Acad. Sci.*, 108, 296, 2011.
77. Park, J., Park, S., Ryu, S., Bhang, S.H., Kim, J., Yoon, J.K., Park, Y.H., Cho, S.P., Lee, S., Hong, B.H., Graphene-regulated cardiomyogenic differentiation process of mesenchymal stem cells by enhancing the expression of extracellular matrix proteins and cell signaling molecules. *Adv. Health Mater.*, 3, 176, 2014.
78. Bressan, E., Ferroni, L., Gardin, C., Sbricoli, L., Gobbato, L., Ludovichetti, F.S., Tocco, I., Carraro, A., Piattelli, A., Zavan, B., Graphene based scaffolds effects on stem cells commitment. *J. Transl. Med.*, 12, 296, 2014.
79. Park, J., Kim, Y.S., Ryu, S., Kang, W.S., Park, S., Han, J., Jeong, H.C., Hong, B.H., Ahn, Y., Kim, B.S., Graphene potentiates the myocardial repair efficacy of mesenchymal stem cells by stimulating the expression of angiogenic growth factors and gap junction protein. *Adv. Funct. Mater.*, 25, 2590, 2015.

80. Shin, S.R., Aghaei-Ghareh-Bolagh, B., Gao, X., Nikkhah, M., Jung, S.M., Dolatshahi-Pirouz, A., Kim, S.B., Kim, S.M., Dokmeci, M.R., Tang, X.S., Layer-by-Layer assembly of 3D tissue constructs with functionalized graphene. *Adv. Funct. Mater.*, 24, 6136, 2014.
81. Cha, C., Shin, S.R., Gao, X., Annabi, N., Dokmeci, M.R., Tang, X.S., Khademhosseini, A., Controlling mechanical properties of cell-laden hydrogels by covalent incorporation of graphene oxide. *Small*, 10, 514, 2014.
82. Shin, S.R., Aghaei-Ghareh-Bolagh, B., Dang, T.T., Topkaya, S.N., Gao, X., Yang, S.Y., Jung, S.M., Oh, J.H., Dokmeci, M.R., Tang, X.S., Cell-laden microengineered and mechanically tunable hybrid hydrogels of gelatin and graphene oxide. *Adv. Mater.*, 25, 6385, 2013.
83. Ahadian, S., Estili, M., Surya, V.J., Ramón-Azcón, J., Liang, X., Shiku, H., Ramalingam, M., Matsue, T., Sakka, Y., Bae, H., Facile and green production of aqueous graphene dispersions for biomedical applications. *Nanoscale*, 7, 6436, 2015.
84. Ahadian, S., Naito, U., Surya, V.J., Darvishi, S., Estili, M., Liang, X., Nakajima, K., Shiku, H., Kawazoe, Y., Matsue, T., Fabrication of poly(ethylene glycol) hydrogels containing vertically and horizontally aligned graphene using dielectrophoresis: An experimental and modeling study. *Carbon*, 123, Supplement C, 460, 2017.
85. Ostrovidov, S., Hosseini, V., Ahadian, S., Fujie, T., Parthiban, S.P., Ramalingam, M., Bae, H., Kaji, H., Khademhosseini, A., Skeletal muscle tissue engineering: Methods to form skeletal myotubes and their applications. *Tissue Eng. Part B Rev.*, 20, 403, 2014.
86. Ku, S.H. and Park, C.B., Myoblast differentiation on graphene oxide. *Biomaterials*, 34, 2017, 2013.
87. Chung, C. and Burdick, J.A., Engineering cartilage tissue. *Adv. Drug Deliv. Rev.*, 60, 243, 2008.
88. Afizah, H., Yang, Z., Hui, J.H., Ouyang, H.-W., Lee, E.-H., A comparison between the chondrogenic potential of human bone marrow stem cells (BMSCs) and adipose-derived stem cells (ADSCs) taken from the same donors. *Tissue Eng.*, 13, 659, 2007.
89. Yoon, H.H., Bhang, S.H., Kim, T., Yu, T., Hyeon, T., Kim, B.S., Dual roles of graphene oxide in chondrogenic differentiation of adult stem cells: Cell-adhesion substrate and growth factor-delivery carrier. *Adv. Funct. Mater.*, 24, 6455, 2014.
90. Zhou, X., Nowicki, M., Cui, H., Zhu, W., Fang, X., Miao, S., Lee, S.-J., Keidar, M., Zhang, L.G., 3D bioprinted graphene oxide-incorporated matrix for promoting chondrogenic differentiation of human bone marrow mesenchymal stem cells. *Carbon*, 116, 615, 2017.
91. Groeber, F., Holeiter, M., Hampel, M., Hinderer, S., Schenke-Layland, K., Skin tissue engineering—*In vivo* and *in vitro* applications. *Adv. Drug Deliv. Rev.*, 63, 352, 2011.
92. Deepachitra, R., Ramnath, V., Sastry, T., Graphene oxide incorporated collagen–fibrin biofilm as a wound dressing material. *RSC Adv.*, 4, 62717, 2014.
93. Li, Y., Liu, Y., Fu, Y., Wei, T., Le Guyader, L., Gao, G., Liu, R.-S., Chang, Y.-Z., Chen, C., The triggering of apoptosis in macrophages by pristine graphene through the MAPK and TGF-beta signaling pathways. *Biomaterials*, 33, 402, 2012.
94. Chang, Y., Yang, S.-T., Liu, J.-H., Dong, E., Wang, Y., Cao, A., Liu, Y., Wang, H., *In vitro* toxicity evaluation of graphene oxide on A549 cells. *Toxicol. Lett.*, 200, 201, 2011.
95. Yang, K., Wan, J., Zhang, S., Tian, B., Zhang, Y., Liu, Z., The influence of surface chemistry and size of nanoscale graphene oxide on photothermal therapy of cancer using ultra-low laser power. *Biomaterials*, 33, 2206, 2012.
96. Dong, H., Ding, L., Yan, F., Ji, H., Ju, H., The use of polyethylenimine-grafted graphene nanoribbon for cellular delivery of locked nucleic acid modified molecular beacon for recognition of microRNA. *Biomaterials*, 32, 3875, 2011.
97. Depan, D., Shah, J., Misra, R., Controlled release of drug from folate-decorated and graphene mediated drug delivery system: Synthesis, loading efficiency, and drug release response. *Mater. Sci. Eng. C*, 31, 1305, 2011.

98. Hu, W., Peng, C., Lv, M., Li, X., Zhang, Y., Chen, N., Fan, C., Huang, Q., Protein corona-mediated mitigation of cytotoxicity of graphene oxide. *ACS Nano*, 5, 3693, 2011.
99. Liu, K., Zhang, J.-J., Cheng, F.-F., Zheng, T.-T., Wang, C., Zhu, J.-J., Green and facile synthesis of highly biocompatible graphene nanosheets and its application for cellular imaging and drug delivery. *J. Mater. Chem.*, 21, 12034, 2011.
100. Liu, J., Li, Y., Li, Y., Li, J., Deng, Z., Noncovalent DNA decorations of graphene oxide and reduced graphene oxide toward water-soluble metal–carbon hybrid nanostructures via self-assembly. *J. Mater. Chem.*, 20, 900, 2010.
101. Zhang, S., Yang, K., Feng, L., Liu, Z., *In vitro* and *in vivo* behaviors of dextran functionalized graphene. *Carbon*, 49, 4040, 2011.
102. Gollavelli, G. and Ling, Y.-C., Multi-functional graphene as an *in vitro* and *in vivo* imaging probe. *Biomaterials*, 33, 2532, 2012.
103. Duch, M.C., Budinger, G.S., Liang, Y.T., Soberanes, S., Urich, D., Chiarella, S.E., Campochiaro, L.A., Gonzalez, A., Chandel, N.S., Hersam, M.C., Minimizing oxidation and stable nanoscale dispersion improves the biocompatibility of graphene in the lung. *Nano Lett.*, 11, 5201, 2011.
104. Sasidharan, A., Panchakarla, L., Chandran, P., Menon, D., Nair, S., Rao, C., Koyakutty, M., Differential nano-bio interactions and toxicity effects of pristine versus functionalized graphene. *Nanoscale*, 3, 2461, 2011.
105. Liao, K.-H., Lin, Y.-S., Macosko, C.W., Haynes, C.L., Cytotoxicity of graphene oxide and graphene in human erythrocytes and skin fibroblasts. *ACS Appl. Mater. Interfaces*, 3, 2607, 2011.
106. Fan, H., Wang, L., Zhao, K., Li, N., Shi, Z., Ge, Z., Jin, Z., Fabrication, mechanical properties, and biocompatibility of graphene-reinforced chitosan composites. *Biomacromolecules*, 11, 2345, 2010.
107. Lu, C.-H., Zhu, C.-L., Li, J., Liu, J.-J., Chen, X., Yang, H.-H., Using graphene to protect DNA from cleavage during cellular delivery. *Chem. Commun.*, 46, 3116, 2010.
108. Wojtoniszak, M., Chen, X., Kalenczuk, R.J., Wajda, A., Łapczuk, J., Kurzewski, M., Drozdzik, M., Chu, P.K., Borowiak-Palen, E., Synthesis, dispersion, and cytocompatibility of graphene oxide and reduced graphene oxide. *Colloids Surf. B*, 89, 79, 2012.
109. Yuan, J., Gao, H., Sui, J., Duan, H., Chen, W.N., Ching, C.B., Cytotoxicity evaluation of oxidized single-walled carbon nanotubes and graphene oxide on human hepatoma HepG2 cells: An iTRAQ-coupled 2D LC-MS/MS proteome analysis. *Toxicol. Sci.*, 126, 149, 2011.
110. Wang, Y., Liu, J., Liu, L., Sun, D.D., High-quality reduced graphene oxide-nanocrystalline platinum hybrid materials prepared by simultaneous co-reduction of graphene oxide and chloroplatinic acid. *Nanoscale Res. Lett.*, 6, 241, 2011.
111. Chen, H., Müller, M.B., Gilmore, K.J., Wallace, G.G., Li, D., Mechanically strong, electrically conductive, and biocompatible graphene paper. *Adv. Mater.*, 20, 3557, 2008.

Index

CPSIA information can be obtained
at www.ICGtesting.com
Printed in the USA
BVHW011438140619
550976BV00010B/78/P

9 781119 469599